CHEMICAL FIXATION AND SOLIDIFICATION OF HAZARDOUS WASTES

CHEMICAL FIXATION AND SOLIDIFICATION OF HAZARDOUS WASTES

Jesse R. Conner

Chemical Waste Management, Inc.

VAN NOSTRAND REINHOLD
New York

Copyright © 1990 by Van Nostrand Reinhold

Library of Congress Catalog Card Number 89-16522
ISBN 0-442-20511-2

All rights reserved. No part of this work covered by
the copyright hereon may be reproduced or used in any
form by any means—graphic, electronic, or
mechanical, including photocopying, recording, taping,
or information storage and retrieval systems—without
written permission of the publisher.

Printed in the United States of America

Van Nostrand Reinhold
115 Fifth Avenue
New York, New York 10003

Van Nostrand Reinhold International Company Limited
11 New Fetter Lane
London EC4P 4EE, England

Van Nostrand Reinhold
480 La Trobe Street
Melbourne, Victoria 3000, Australia

Nelson Canada
1120 Birchmount Road
Scarborough, Ontario
Canada M1K 5G4

16 15 14 13 12 11 10 9 8 7 6 5 4 3 2 1

Library of Congress Cataloging-in-Publication Data
Conner, Jesse R., 1930-
    Chemical Fixation and solidification of hazardous wastes / Jesse R. Conner.
      p. cm.
    Bibliography: p.
    Includes index.
    ISBN 0-442-20511-2
    1. Hazardous wastes—Purification. 2. Solidification. I. Title.
TD1060.C66 1990
  628.4'2—dc20   89-16522
                                                    CIP

To Nancy, whose support and patience made it possible

CONTENTS

Preface   xi

Chapter 1. Introduction   1
Terminology and Definitions   1
Characteristics of Wastes   3
Regulatory Background   4
Description of CFS Processes   13
Organization of the Book   15

Chapter 2. History and background   18
Origins   18
Use in Nuclear Waste Treatment   19
Other Early Practices   19
The Pre-RCRA CFS Industry   20
The Post-RCRA CFS Industry   21

Chapter 3. Principles of fixation   23
What Is Leaching?   23
Measurement of Leachability   24
Leaching Tests vs. Actual Leaching   25
Leaching Criteria   25
Factors Affecting Leachability   26

Chapter 4. Fixation of metals   58
Metals of Environmental Concern—And Why   60
Solubility of Metals   63
Metal Speciation in Wastes   63
Fixation of Metals in Solution   71
Fixation of Complex Species   71
Fixation of Metals in Solids and Soils   74
Long-Term Stability of Fixed Metals   75
Metal Fixation and Resource Recovery   76
Metal Fixation Mechanisms and Methods   76
Individual Metals   102
Arsenic   104

viii   CONTENTS

    Barium 110
    Cadmium 112
    Chromium 122
    Lead 130
    Mercury 140
    Nickel 148
    Selenium 153
    Silver 154
    Other Metals 156

Chapter 5.   **Fixation of other inorganic constituents** 172
    Fluoride 173
    Cyanides 173
    Sulfides 180

Chapter 6.   **Fixation of organic constituents** 183
    Treatment Techniques 189
    Organic Reaction Principles in CFS 191
    Commercial Processes 198

Chapter 7.   **Choosing the right CFS system** 201
    The Engineered Approach 201
    The Waste 202
    The Waste Source 207
    The Treatment Process 216
    The Disposal Site 223
    Commercial Considerations and Costs 225

Chapter 8.   **Wastes and waste sources** 227
    Waste Classification Systems 227
    Assessment of Waste Quantities and Groupings 246
    Waste Properties as Related to CFS 259
    Waste Streams and Types Suitable for CFS 267
    Characterization of Today's Waste Streams 280

Chapter 9.   **CFS processes: general properties and comparative evaluation** 280
    General Concepts of Solidification 291
    Generic Categories of CFS Processes 304
    Proprietary CFS Processes, Products, and Services 309
    Comparison of Processes 313

Chapter 10.   **Portland cement-based systems** 335
    Chemical and Physical Characteristics of Cement 337
    Generic vs. Proprietary Processes 352
    Physical Properties 355
    Chemical Properties 357
    Laboratory and Pilot Studies 359
    Case Studies, Commercial Projects, and Processes 364

Chapter 11.   **Portland cement/soluble silicate processes** 376
    Chemical and Physical Properties 377
    Laboratory and Pilot Studies 386
    Generic vs. Proprietary Processes 395
    Case Studies, Commercial Projects 403

Chapter 12.	**Lime/flyash and other lime-based processes** 407	
	Lime/Flyash Processes 407	
	Lime/Clay Processes 412	
	Lime-Based Processes: General 414	
Chapter 13.	**Portland cement/flyash processes** 417	
	Chemical and Physical Properties 418	
	Commercial Processes 424	
Chapter 14.	**Kiln dust and flyash-based processes** 427	
	Chemical and Physical Properties 434	
	Commercial Processes 442	
	Case History: Superfund-Type Cleanup 446	
Chapter 15.	**Other CFS and nonchemical processes and systems** 448	
	Miscellaneous Inorganic Processes 448	
	Nuclear Waste CFS: Introduction 450	
	Organic-Based Processes 452	
	Nonchemical Solidification 458	
	Systems Combining Dewatering and CFS 463	
	In Situ Processes and Systems 464	
Chapter 16.	**Delivery systems** 478	
	Fixed Installations 478	
	Mobile Operations 490	
	Regulatory Approval Process 500	
Chapter 17.	**CFS testing and formulation** 515	
	Waste Sampling and Handling 515	
	Sample Containers, Storage, Shipping, and Handling 519	
	Information System: Chain of Custody 519	
	Determination of Waste Characteristics 520	
	Interrogation of Data Base 534	
	CFS Treatability Study: Formulation 541	
	Testing of CFS Products 548	
Chapter 18.	**Information sources, computer applications, and research and development** 570	
	Information Sources 570	
	Computer Applications and Systems 571	
	Research and Development 577	
Appendix 1	**CFS Waste Data Base (CWDB)** 580	
Appendix 2	**Test Methods** 627	
	U.S. EPA Extraction Procedure Toxicity Test (EPT) 627	
	U.S. EPA Multiple Extraction Procedure (MEP) 634	
	U.S. EPA Oily Wastes Extraction Procedure (OWEP) 636	
	U.S. EPA Toxicity Characteristic Leaching Procedure (TCLP) 638	
	State of California Waste Extraction Test (WET) 651	
	American Nuclear Society ANS 16.1 Test Procedure 655	
Glossary 659		
Bibliography 673		
Index 682		

PREFACE

While chemical fixation and solidification technology, CFS for short, has been in commercial existence for more than twenty years, only in the last five years or so has it attracted much attention from environmental scientists and engineers, regulators, waste generators, and the public. Now that the technology is mandated, both implicitly and explicitly, through a growing number of statutes, regulations, and site-specific waste treatment decisions, a severe lack of information has become evident. Few, if any, university courses exist that adequately cover this field, and there is no up-to-date, definitive textbook or reference work on CFS technology, especially on its basic and applied chemistry. Academia is only now beginning to gear up for this challenge. However, industry, academia, and regulatory organizations all suffer from the lack of both experienced people and reliable information. It is the purpose of this book to help correct the latter situation.

More specifically, this book is intended to be of use for several different, but related, purposes and audiences.

- A comprehensive text for advanced courses in environmental chemistry and engineering.
- A reference work with important data for students, laboratory workers, engineers, regulators, and environmentalists.
- A general reference work for engineers, scientists, regulators, chemical suppliers, CFS vendors, buyers of CFS technology, and environmentalists.

Chapter 1

INTRODUCTION

The field of chemical fixation and solidification, CFS for short, has just begun to mature into an accepted environmental technology. Pushed by regulation that essentially mandates its use for many waste streams, it is becoming a standard unit process in liquid and hazardous waste treatment and disposal. Yet, few people working in waste treatment have any knowledge of CFS technology other than that gained from CFS vendors, from superficial reviews in government publications or industry conferences, or from works [1] that are now somewhat dated. While the Environmental Protection Agency (EPA) recently published a useful technical handbook [2], no up-to-date, definitive work has existed in this field. The purpose of this book is to correct that situation in a critical, objective, and authoritative manner.

One reason for this dearth of information is that the technology grew from the efforts of a few vendors offering waste treatment and/or disposal services. They were not inclined to provide information about the mechanisms whereby their processes or products accomplished solidification, fixation, or both. In fact, even the chemical components of the CFS systems were kept secret, at least until patents began to be issued, and this practice is still prevalent. Vendors and users were and are required to show only that the product of their chemistries meets certain regulatory requirements for physical stability and leaching of hazardous constituents. While this has provided some practical sources of data and several proprietary data bases to aid in prediction of long-term stability by extrapolation, the lack of basic information about the chemistry of CFS systems does not provide a sound scientific basis for either prediction of stability or development of better systems. The mechanisms of fixation or immobilization and descriptions of the kinetics that govern leaching behavior are only now beginning to be elucidated.

TERMINOLOGY AND DEFINITIONS

Much of the terminology in the CFS field includes terms borrowed from other technical areas and from the language as a whole. Frequently, these familiar words and phrases are given new and specific meanings. There is no "official" set of definitions as yet, although the American Society for Testing and Materials (ASTM) has been working toward this goal through its D-34 Committee on Waste Disposal, and EPA has unofficially promulgated the definitions given below.

Unfortunately, there has been a tendency to use the words *chemical fixation, stabilization,* and *solidification* interchangeably. Originally, the term stabilization had been used to describe what are essentially pretreatment processes before solidification, or those that physically change liquid or semiliquid wastes, or, in another context, the treatment of sewage sludge with lime to halt biological processes and reduce odor. More recently, the EPA [2] has defined stabilization and solidification as follows.

- *Stabilization* refers to those techniques that reduce the hazard potential of a waste by converting the contaminants into their least soluble, mobile, or toxic form. The physical nature and handling characteristics of the waste are not necessarily changed by stabilization.
- *Solidification* refers to techniques that encapsulate the waste in a monolithic solid of high structural integrity. The encapsulation may be of fine waste particles (microencapsulation) or of a large block or container of wastes (macroencapsulation). Solidification does not necessarily involve a chemical interaction between the wastes and the solidifying reagents, but may mechanically bind the waste into the monolith. Contaminant migration is restricted by vastly decreasing the surface area exposed to leaching and/or by isolating the wastes within an impervious capsule.

Many workers in the field will disagree with this use of terminology, but from a practical standpoint, it is likely to achieve ascendancy purely because of its source. There is a special problem with the use of the word *monolith*, because it conjures up the most familiar literal meaning—a block of stone—which does not fit many of the processes in use today that produce a soil-like solid.

In this book, we will use the words *fixation, chemical fixation,* and *stabilization* interchangeably. The term fixation has fallen in and out of favor over the years, but is still widely used as are its derivatives "to fixate," to "fix," and "fixated." Solidification will mean the conversion of liquids or semiliquids (and in some cases, powders) into solids, but without the requirement of a monolith. The term CFS will be used to describe the combination of both sets of terms, but only when the process involves one or more chemical reactions. This term emphasizes the chemical rather than the engineering aspect of the subject, although the latter outlook is discussed in detail as we look at the applications and uses of CFS technology in the "real world."

The EPA [3] has legally defined the term *treatment*, taking the broadest meaning, to include "any method of modifying the chemical, biological, and/or physical character or composition of a waste." A number of other terms are defined or explained in the Glossary. Some will also be defined as they are used in the CFS field because the meanings are not necessarily identical to those in general usage. Common terms, especially acronyms, with which the reader should quickly become familiar are listed below. Their definitions are given in the Glossary and, often, in the text.

1. Regulatory processes and treatment standards.
   Delisting
   Best demonstrated available technology (BDAT)

Publicly owned treatment works (POTW)
Resource Conservation and Recovery Act (RCRA)
Comprehensive Environmental Response, Compensation and Liability Act (CERCLA) or Superfund
VHS model
2. Facility or treatment types.
Minimum technology unit/requirement (MTU/MTR)
On-site
Off-site
Monofill
Sanitary landfill
CERCLA/RCRA
Secure landfill
In situ
Remedial action
Superfund
Treatment, storage, and disposal unit (TSD)
3. RCRA/CERCLA waste categories.
Listed
Characteristic
Landban
F-waste
California list
First-third
Second-third
Third-third
4. Leaching and other tests.
Extraction procedure toxicity test (EPT)
Toxicity characteristic leaching procedure (TCLP)
Multiple extraction procedure (MEP)

CHARACTERISTICS OF WASTES

In this book, a great deal of emphasis is placed on wastes that may be treated by CFS methods, and their characteristics. Chapter 8 deals with this subject in detail because a good knowledge of the nature of waste streams is vital to understanding CFS technology.

Many waste types are not amenable to CFS treatment, or do not require it. In general, solid, nonhazardous wastes require no treatment. Liquid nonhazardous wastes may require solidification only to prepare them for land disposal. Nonaqueous hazardous wastes, such as spent solvents, are best treated by other means such as recovery or incineration. Aqueous wastes that are designated as hazardous by EPA, especially those that are liquid, are among those most commonly considered for CFS treatment. The recent landbans [4] have brought many new streams into the CFS area, including solid hazardous wastes that had been going untreated to permitted RCRA treatment, storage, and disposal (TSD) facilities for direct landfill in secure cells.

Several new classes of wastes that will be of importance to those working in the CFS field are now appearing, or are likely to appear in the near future as a result of changes in legislation and new regulations coming into force. One of these classes is a large group of residues resulting from provisions of various water pollution control legislation that require pretreatment of industrial discharges to publicly owned treatment works (POTWs). Another class comprises wastes that have been exempted by statute from RCRA, or have turned from nonhazardous to hazardous (by characteristic) as a result of new information or stricter criteria. Examples of such wastes, as stated in [3], are:

> ... mining overburden returned to the mining site; fly ash, bottom ash, and slag wastes; flue gas emission control wastes generated from the burning of fossil fuels; and oil, gas, or geothermal drilling fluids.

Two other major potential uses of CFS technology are in treatment of sewage sludge and for municipal incinerator ash, both of which have been considered nonhazardous in the past. More about these in Chapter 8.

REGULATORY BACKGROUND

Although this book is about an area of technology, not environmental regulation with all of its complexities and uncertainties, some understanding of the development of hazardous waste regulation is vital. Without regulation, CFS technology would scarcely exist, much less flourish, in a world of harsh economic realities. And the characteristics of the technology itself are defined by these regulations.

Much of the impetus for CFS of hazardous wastes has been provided by the Resource Conservation and Recovery Act [5] (RCRA) of 1976, including the subsequent 1984 amendments (HSWA), and the Comprehensive Environmental Response, Compensation and Liability Act [6] (CERCLA), otherwise known as Superfund. The RCRA includes provisions for developing criteria to determine which wastes are hazardous, establishing standards for siting, design, and operation of disposal facilities, encouraging the states to develop their own regulatory programs, and other specifics. The HSWA reauthorized the RCRA and made major changes, including the establishment of more specific criteria and strict deadlines for regulatory action and compliance. Regulations promulgated under both the RCRA and HSWA direct in detail the generation, handling, treatment, and disposal of wastes. The CERCLA, and its reauthorization in 1986—known as SARA—established a massive remedial program for the cleanup of existing sites that threaten the environment.

In addition to the preceeding legislation and the regulatory programs that have sprung from it, other legislation dealing with water pollution control, toxic substances control, and occupational health and safety have had some impact on the CFS field and promise to have even more in the future. These include the following.

The Federal Water Pollution Control Act as amended in 1972 (PL 92-500)
The Clean Water Act of 1977 (PL 95-217)
The Safe Drinking Water Act of 1974 (PL 93-523)
The Toxic Substances Control Act (TSCA) of 1976 (PL 94-469)
The Occupational Safety and Health Act (OSHA) of 1970

The various water laws have effectively prevented the discharge of dilute contaminants into the environment, thus forcing most of them into residuals that eventually must go to the land. Recognizing that this would be the case, Congress has attempted to prevent the reintroduction of the contaminants from another direction by controlling the disposal of the residues. The TSCA covers some special situations, such as PCBs and other concentrated chemical substances, and OSHA furthers this in the workplace. From the standpoint of CFS technology, the recent landbans are designed to prevent environmental degradation when the residuals are disposed of in uncontrolled landfills, or in the event that all the protective measures of secure landfill at TSD facilities fail.

CFS is treated indirectly in most of these laws and in the regulations that they enable. However, the more recent regulations have become more specific, requiring chemical solidification instead of sorption [7] and specifying performance requirements in hazard reduction via CFS treatment. In the promulgation of the various landbans, specific technologies are specified as *best demonstrated available technology* (BDAT). CFS treatment is one of the most important BDATs now, and will continue to be in the future.

Land disposal of hazardous wastes is not the method of choice from an environmental standpoint. In fact, it occupies the lowest position in the EPA's hierarchy of methods. Nevertheless, it is developing an increasingly important place in the overall waste management scheme of the future for several reasons. For example, there are many hazardous wastes that are simply not amenable to techniques such as thermal, chemical, or biological destruction. When these wastes cannot be feasibly reused in any beneficial way, there is no other recourse but land disposal. Even when other techniques are used, they usually generate residues that are themselves hazardous. In the cleanup of abandoned sites under the Superfund program and remediation of other old disposal practices by private entities, on-site or in situ treatment and disposal often remain the safest and least expensive alternatives, and one of the primary techniques here is CFS.

As previously stated, CFS, generally called *stabilization* by the EPA, has been specified as BDAT for certain waste streams in the first-third landbans on the land disposal of hazardous wastes. It had already been accepted as a basis for "delisting" in a number of petitions under the RCRA. CFS, termed *solidification* or *encapsulation*, is specifically cited under the CERCLA (40 CFR 300) [6] as a method to be considered for remedying releases from contaminated soils and sediments [2]. Other than in site-specific remedial action work under the CERCLA/SARA or in private cleanup operations, the two major regulatory systems that impact CFS technology are the landban system and the delisting system. It is worthwhile, even necessary, to explore each of these at this time.

The Landbans

Under the 1984 HSWA, the EPA was directed by Congress to accomplish various tasks that would ultimately result in banning land disposal of hazardous wastes. Some of these directives, the so-called "hammer provisions," were to go into effect automatically; others required EPA interpretation. On May 31, 1985, the EPA published a proposed schedule [8] for the statutory landbans, including the identification of three groups of listed and unlisted hazardous wastes that were to go into effect at different times. These three groups subsequently became known as the "first-third," "second-third," and "third-third," and they were to go into effect on August 8, 1988, June 8, 1989, and May 8, 1990, respectively. They are collectively known as the *landban* and will be so referenced throughout this book. Several of these profoundly affect CFS technology. They are listed here [9] along with their "hammer" dates to provide the novice with some background in the why of CFS technology.

- May 8, 1985. The EPA bans bulk hazardous waste liquids in landfill, whether or not the waste has been treated with absorbents. Solidification of liquid waste to meet this requirement must be accomplished by a "chemical reaction," not by physical absorption alone.
- November 8, 1985. The EPA bans nonhazardous bulk liquids, including water, in landfills unless the operator demonstrates the landfill is the only safe alternative.
- November 8, 1986. Land disposal of dioxins and solvents is banned (with some exceptions). The EPA published the final schedule for reviewing whether to ban all other hazardous wastes from land disposal. The schedule divides wastes into three categories, with the first category containing the most toxic and largest volume wastes. The first one-third of waste is banned as of August 8, 1988. The second one-third is banned as of June 8, 1989. The final category is banned as of May 8, 1990 [4]. The wastes that compose these categories are discussed in more detail in Chapter 8, along with all of the other wastes and categories that are pertinent to CFS treatment.
- July 8, 1987. The so-called "California list" of wastes (including sludges containing differing concentrations of cyanides, arsenic, cadmium, lead, mercury, nickel, selenium, PCBs, thallium, and halogenated organic compounds) were banned from land disposal (except deep wells) unless the EPA determined that certain disposal methods wouldn't affect human health and the environment.

The EPA has met all of these deadlines to date, although regulation of some wastes has been delayed until 1990, and others have been allowed to fall under the "soft-hammer" provisions.* The May and November 1985 bulk liquid bans vastly increased the volume of waste being solidified in the United States, even though the methods were sometimes crude and the resultant solids were not always less hazardous (nor were they required to be). The California list ban [10], which affects only liquid wastes (except for certain halogenated organic compounds), also has undoubtedly increased the use of CFS for certain wastes, although CFS is not yet specified as BDAT in this ban. The dioxins and solvents portion of the November 1986 ban [11] was also met, although the EPA granted variances due to inadequate treatment capacity for certain levels, generators, and sources.

The 1986 ban eliminated CFS technology as a BDAT for organic solvents, but has provided the greatest impetus yet to its use on the first-third wastes. In 1990, it will almost certainly make another major impact when the third-third waste list comes under the ban. This group, which includes the D-wastes (hazardous by char-

---

*If the EPA fails to promulgate regulations for any waste under the various "thirds" of the landban, such waste may be disposed of in landfills only if the facility is an RCRA minimum technology unit (MTR) *and* the generator certifies and demonstrates that "there is no (practically available) treatment that meaningfully reduces toxicity or mobility of the waste" or that treatment has been done to reduce toxicity or mobility.

acteristic) will pick up all the loose ends, including residues from other treatment processes such as incineration. Also, at this time, most time extensions and soft-hammers granted in the other portions of the landbans expire, and *all hazardous wastes* must be treated by the prescribed BDATs and to the required treatment levels. Also, in May 1990, these regulations supercede most previous regulations. In its totality, the landban system (and its ramifications and interactions with other regulations) is extremely complex, almost beyond the comprehension of any individual. However, when examining only the portions that pertain to CFS, and limiting this to areas of general interest,† the system becomes more manageable.

In addition to the landbans, SARA ". . . requires that remedial actions meet all applicable, relevant, and appropriate public health and environmental standards. Therefore, the Superfund program must be consistent with the BDAT approach when disposing of contaminated soils and debris from Superfund sites." [12] Undoubtedly, this will also eventually apply to non-Superfund remedial actions. As other wastes not currently listed as hazardous, or specifically excluded under the RCRA, become regulated they will also become subject to landban provisions and regulations.

The methodology for setting BDATs and the achievement levels required are set forth by the EPA as follows [13]:

> . . . EPA promulgated a technology-based approach to establishing treatment standards under section 3004(m). Section 3004(m) also specifies that treatment standards must "minimize" long- and short-term threats to human health and the environment arising from land disposal of hazardous wastes. . . . the intent is "to require utilization of available technology" and not "process which contemplates technology-forcing standards". . . . EPA has interpreted this legislative history as suggesting that Congress considered the requirement . . . to be met by application of the best *demonstrated* and achievable (i.e., *available*) technology. . . .
> Accordingly, EPA's treatment standards are generally based on the performance of the . . . (BDAT) identified for treatment of hazardous constituents. . . . The treatment standards, according to the statute, can represent levels or methods of treatment, if any, that substantially diminish the toxicity of the waste or substantially reduce the likelihood of migration of hazardous constituents. Wherever possible, the Agency prefers to establish BDAT standards as "levels" of treatment (i.e., performance standards) rather than adopting an approach that would require the use of specific treatment "methods."

In practice, this has meant that the EPA will set a general method of treatment, such as incineration or CFS, but not specific processes. The process used has only to achieve a treatment level (i.e., leachability level) to meet BDAT requirements. The way in which the EPA has set the treatment levels to date is by collecting data from sources that are doing such treatment and using the treatability levels that they can achieve as the standards for the BDAT. Various statistical and other data analysis techniques are used to set the actual levels for each constituent of concern so that analytical and process variability are allowed for. The specific techniques are given in [13].

---

†There are aspects of the landban regulations that apply to residues from treatment at RCRA TSD facilities and that become extremely complex due to the "derived from" rule. These are currently the subject of legal action, and may be simplified in the future.

8 INTRODUCTION

The list of hazardous constituents within the waste codes that are targeted for treatment, referred to as the BDAT constituent list, is given in Table 1-1.

The achievement, or treatability, levels applicable to CFS technology for the wastes in the first-third where CFS is specified as BDAT (either for the original waste itself, or for residue from another treatment in a "treatment train") are given in Table 1-2. It is important to note that the allowable leaching level for any given

TABLE 1-1   BDAT Constituent List

BDAT Reference Number	Constituent	CAS Number
	Volatile Organics	
222	Acetone	67-64-1
1	Acetonitrile	75-05-8
2	Acrolein	107-02-8
3	Acrylonitrile	107-13-1
4	Benzene	71-43-2
5	Bromodichloromethane	75-27-4
6	Bromomethane	74-83-9
223	*n*-Butyl alcohol	71-36-3
7	Carbon tetrachloride	56-23-5
8	Carbon disulfide	75-15-0
9	Chlorobenzene	108-90-7
10	2-Chloro-1,3-butadiene	126-99-8
11	Chlorodibromomethane	124-48-1
12	Chloroethane	75-00-3
13	2-Chloroethyl vinyl ether	110-75-8
14	Chloroform	67-66-3
15	Chloromethane	74-87-3
16	3-Chloropropene	107-05-1
17	1,2-Dibromo-3-chloropropane	96-12-8
18	1,2-Dibromoethane	106-93-4
19	Dibromomethane	74-95-3
20	*trans*-1,4-Dichloro-2-butene	110-57-6
21	Dichlorodifluoromethane	75-71-8
22	1,1-Dichloroethane	75-34-3
23	1,2-Dichloroethane	107-06-2
24	1,1-Dichloroethylene	75-35-4
25	*trans*-1,2-Dichloroethene	156-60-5
26	1,2-Dichloropropane	78-87-5
27	*trans*-1,3-Dichloropropene	10061-02-6
28	*cis*-1,3-Dichloropropene	10061-01-5
29	1,4-Dioxane	123-91-1
224	2-Ethoxyethanol	110-80-5
225	Ethyl acetate	141-78-6
226	Ethyl benzene	100-41-4
30	Ethyl cyanide	107-12-0
227	Ethyl ether	60-29-7
31	Ethyl methacrylate	97-63-2
214	Ethylene oxide	75-21-8
32	Iodomethane	74-88-4
33	Isobutyl alcohol	78-83-1
228	Methanol	67-56-1
34	Methyl ethyl ketone	78-93-3
229	Methyl isobutyl ketone	108-10-1
35	Methyl methacrylate	80-62-6

TABLE 1-1 BDAT Constituent List (*Continued*)

BDAT Reference Number	Constituent	CAS Number
	Volatile Organics	
37	Methacrylonitrile	126-98-7
38	Methylene chloride	75-09-2
230	2-Nitropropane	79-46-9
39	Pyridine	110-86-1
40	1,1,1,2-Tetrachloroethane	630-20-6
41	1,1,2,2-Tetrachloroethane	79-34-6
42	Tetrachloroethane	127-18-4
43	Toluene	108-88-3
44	Tribromomethane	75-25-2
45	1,1,1-Trichloroethane	71-55-6
46	1,1,2-Trichloroethane	79-00-5
47	Trichloroethene	79-01-6
48	Trichloromonofluoromethane	75-69-4
49	1,2,3-Trichloropropane	96-18-4
231	1,1,2-Trichloro-1,2,2-trifluoroethane	76-13-1
50	Vinyl chloride	75-01-4
215	1,2-Xylene	97-47-6
216	1,3-Xylene	108-38-3
217	1,4-Xylene	106-44-5
	Semivolatile Organics	
51	Acenaphthalene	208-96-8
52	Acenaphthene	83-32-9
53	Acetophenone	96-86-2
54	2-Acetylaminofluorene	53-96-3
55	4-Aminobiphenyl	92-67-1
56	Aniline	62-53-3
57	Anthracene	120-12-7
58	Aramite	140-57-8
59	Benz[*a*]anthracene	56-55-3
218	Benzalchloride	98-87-3
60	Benzenethiol	108-98-5
61	Deleted	
62	Benzo[*a*]pyrene	50-32-8
63	Benzo[*b*]fluoranthene	205-99-2
64	Benzo[*gni*]perylene	191-24-2
65	Benzo[*k*]fluoranthene	207-08-9
66	*p*-Benzoquinone	106-51-4
67	Bis(2-chloroethoxy)methane	111-91-1
68	Bis(2-chloroethyl)ether	111-44-4
69	Bis(2-chloroisopropyl)ether	39638-32-9
70	Bis(2-ethylhexyl)phthalate	117-81-7
71	4-Bromophenyl phenyl ether	101-55-3
72	Butyl benzyl phthalate	85-68-7
73	2-*sec*-Butyl-4,6-dinitrophenol	88-85-7
74	*p*-Chloroaniline	106-47-8
75	Chlorobenzilate	510-15-6
76	*p*-Chloro-*m*-cresol	59-50-7
77	2-Chloronaphthalene	91-58-7
78	2-Chlorophenol	95-57-8
79	3-Chloropropionitrile	542-76-7
80	Chrysene	218-01-9

(*continued*)

TABLE 1-1 BDAT Constituent List (*Continued*)

BDAT Reference Number	Constituent	CAS Number
	Semivolatile Organics (*Continued*)	
81	o-Cresol	95-48-7
82	p-Cresol	106-44-5
232	Cyclohexanone	108-94-1
83	Dibenz[a,h]anthracene	53-70-3
84	Dibenzo[a,e]pyrene	192-65-4
85	Dibenzo[a,i]pyrene	189-55-9
86	m-Dichlorobenzene	541-73-1
87	o-Dichlorobenzene	95-50-1
88	p-Dichlorobenzene	106-46-7
89	3,3'-Dichlorobenzidine	91-94-1
90	2,4-Dichlorophenol	120-83-2
91	2,6-Dichlorophenol	87-65-0
92	Diethyl phthalate	84-66-2
93	3,3'-Dimethoxybenzidine	119-90-4
94	p-Dimethylaminoazobenzene	60-11-7
95	3,3'-Dimethylbenzidine	119-93-7
96	2,4-Dimethylphenol	105-67-9
97	Dimethyl phthalate	131-11-3
98	Di-n-butyl phthalate	84-74-2
99	1,4-Dinitrobenzene	100-25-4
100	4,6-Dinitro-o-cresol	534-52-1
101	2,4-Dinitrophenol	51-28-5
102	2,4-Dinitrotoluene	121-14-2
103	2,6-Dinitrotoluene	606-20-2
104	Di-n-octyl phthalate	117-84-0
105	Di-n-propylnitrosamine	621-64-7
106	Diphenylamine	122-39-4
219	Diphenylnitrosamine	86-30-6
107	1,2-Diphenylhydrazine	122-66-7
108	Fluoranthene	206-44-0
109	Fluorene	86-73-7
110	Hexachlorobenzene	118-74-1
111	Hexachlorobutadiene	87-68-3
112	Hexachlorocyclopentadiene	77-47-4
113	Hexachloroethane	67-72-1
114	Hexachlorophene	70-30-4
115	Hexachloropropene	1888-71-7
116	Indeno[1,2,3-cd]pyrene	193-39-5
117	Isosafrole	120-58-1
118	Methapyrilene	91-80-5
119	3-Methylcholanthrene	56-49-5
120	4,4'-Methylenebis (2-chloroaniline)	101-14-4
36	Methyl methanesulfonate	66-27-3
121	Naphthalene	91-20-3
122	1,4-Naphthoquinone	130-15-4
123	1-Naphthylamine	134-32-7
124	2-Naphthylamine	91-59-8
125	p-Nitroaniline	100-01-6
126	Nitrobenzene	98-95-3
127	4-Nitrophenol	100-02-7
128	N-Nitrosodi-n-butylamine	924-16-3
129	N-Nitrosodiethylamine	55-18-5
130	N-Nitrosodimethylamine	62-75-9

TABLE 1-1 BDAT Constituent List (*Continued*)

BDAT Reference Number	Constituent	CAS Number
	Semivolatile Organics (*Continued*)	
131	N-Nitrosomethylethylamine	10595-95-6
132	N-Nitrosomorpholine	59-89-2
133	N-Nitrosopiperidine	100-75-4
134	n-Nitrosopyrrolidine	930-55-2
135	5-Nitro-o-toluidine	99-65-8
136	Pentachlorobenzene	608-93-5
137	Pentachloroethane	76-01-7
138	Pentachloronitrobenzene	82-68-8
139	Pentachlorophenol	87-86-5
140	Phenacetin	62-44-2
141	Phenanthrene	85-01-8
142	Phenol	103-95-2
220	Phthalic annydride	35-44-9
143	2-Picoline	109-06-8
144	Pronamide	23950-58-5
145	Pyrene	129-00-0
146	Resorcinol	108-46-3
147	Safrole	94-59-7
148	1,2,4,5-Tetrachlorobenzene	95-94-3
149	2,3,4,6-Tetrachlorophenol	58-90-2
150	1,2,4-Trichlorobenzene	120-82-1
151	2,4,5-Trichlorophenol	95-95-4
152	2,4,6-Trichlorophenol	88-06-2
153	tris(2,3-dibromopropyl) phosphate	126-72-7
	Metals	
154	Antimony	7440-36-0
155	Arsenic	7440-38-2
156	Barium	7410-39-3
157	Beryllium	7440-41-7
158	Cadmium	7440-43-9
159	Chromium (total)	7440-47-32
221	Chromium (hexavalent)	
160	Copper	7440-50-8
161	Lead	7439-92-1
162	Mercury	7439-97-6
163	Nickel	7440-02-0
164	Selenium	7782-49-2
165	Silver	7440-22-4
166	Thallium	7440-28-0
167	Vanadium	7440-62-2
168	Zinc	7440-66-6
	Inorganics Other Than Metals	
169	Cyanide	57-12-5
170	Fluoride	16964-48-8
171	Sulfide	8496-25-8
	Organochlorine Pesticides	
172	Aldrin	309-00-2
173	alpha-BHC	319-84-6
174	beta-BHC	319-85-7
175	delta-BHC	319-86-8

TABLE 1-2 First-Third Landban BDAT Treatability Levels

Concentration of Constituent (mg/l) in Waste or Leachate

Waste Code	Description[a]	Arsenic	Cadmium	Chromium	Lead	Mercury	Nickel	Selenium	Silver	Cyanide
F006	Electroplating sludge		0.066	5.200	0.510		0.320		0.072	[b]
K001	Wood-preserving WWT sludge				0.510					
K022	Distillation bottom tars			5.200			0.320			
K046	Lead-based initiating compound WWT sludge				0.180					
K048	Petroleum refining DAF float	0.004		1.700			0.048	0.025		
K049	Petroleum refining slop oil emulsion solids	0.004		1.700			0.048	0.025		
K050	Petroleum refining heat exchanger cleaning sludge	0.004		1.700			0.048	0.025		
K051	Petroleum refining API separator sludge	0.004		1.700			0.048	0.025		
K052	Petroleum refining leaded tank bottoms	0.004		1.700			0.048	0.025		
K061	EAF emission control dust		0.140	5.200	0.240		0.320			
K062	Steel finishing spent pickle liquor			0.094	0.370					
K071	Mercury cell brine purification mud					0.025				
K086	Ink manufacturing washes and sludges			0.094	0.370					
K087	Coking decanter tank tar sludge				0.510					
K101	Arsenic-containing pharmaceutical distillation tar residue		0.066	5.200	0.510		0.320			
K102	Arsenic-containing pharmaceutical activated carbon residue		0.066	5.200	0.510		0.320			

[a]All nonwastewaters unless otherwise specified.
[b]Reserved for future action.

constituent is not necessarily the same for all wastes. For example, the lead-leaching level for K061, electric arc furnace dust, is 0.024 mg/l, while in F006, electroplating sludge, it is 0.051 mg/l. This may seem inconsistent, but it must be remembered that the landban BDAT levels are, by statute, to be determined by the best demonstrated available technology, not by health-based standards. In this case, the BDATs were presumably able to achieve a lower level for lead in K061 than in F006, so the standards were set accordingly.

Delisting

Delisting is an amendment to the lists of hazardous wastes, granted by the U.S. EPA when it is shown that a specific waste stream no longer has the hazardous characteristics for which it was originally listed. This may be accomplished by removing the toxic components, or if they are still present, demonstrating that they are "not capable of posing a substantial present or potential threat to human health or the environment" following a specific treatment process. The treatment methods and leaching levels to be achieved are not established by statute or by rule, but are determined by the EPA on a case-by-case basis.

Over the years, the de facto allowable leaching levels have moved from 30 times drinking water standards to 6.3 times, the latter level being established by the use of a groundwater attenuation model called the *vertical and horizontal spread* (VHS) *model* [14]. The delisting approach takes into account the toxicity, persistence, and mobility of constituents, as well as waste quantity and disposal location. In most cases, the present de facto standard is being used.

Delisting is expensive, complex, and time-consuming. This book cannot cover its many aspects and the results of all the completed delistings to date. It is, however, important for CFS technology since many of the delistings granted to date have been based on the use of this technology. Each delisting is site-specific and cannot be used for another waste stream or a different site, even if the wastes are identical. It is an administrative process that involves a great deal of red tape. Since the action is open to the public (except for designated trade secret information), it is an excellent source of information about the physical and chemical, especially leaching, characteristics of CFS-treated wastes. Some of the data given in subsequent chapters comes from studies of delisting petitions and rules.

DESCRIPTION OF CFS PROCESSES

CFS technology can be broken down into three basic classes of systems:

- Solidification only
- Chemical fixation and solidification
- Sorption or gelation

The latter is not really "chemical" treatment at all, but an understanding of its use is important both historically and because it is still allowable as solidification in certain aspects of the regulatory landbans. This subject will be explored more thoroughly in Chapter 15.

Solidification processes utilize chemically reactive formulations that, together with the water and other components in sludges and other aqueous wastes, form stable solids. *Stable*, in this sense, means that the solids are physically stable under *normal or expected*‡ environmental conditions and will not revert to the original liquid, semiliquid, or unstable solid state. The compositions resulting from these processes may or may not be hazardous as defined by waste characteristic tests such as leaching. Chemically solidified wastes normally have load-bearing strengths (compacted) in excess of one ton per square foot and may have other specified physical characteristics. The properties of solidified wastes and the test methods used to determine them are described in detail in Chapters 9 and 17, respectively.

Chemical fixation and solidification systems (CFS) not only solidify the waste by chemical means but also insolublize, immobilize, encapsulate, destroy, sorb, or otherwise interact with selected waste components. The purpose of these systems is to produce solids that are nonhazardous, or less hazardous, than the original waste. The degree of hazard for these kinds of materials and systems is usually defined by leaching tests.

With certain notable exceptions, all present commercial CFS processes are quite simple conceptually [15] and utilize standard mechanical equipment in their operation. Chapter 16 describes the mechanical systems for both mobile/portable and fixed installations. As a matter of fact, an inorganic CFS mechanical system is truly an assembly of mixers, chemical storage and feeding devices, pumps, conveyors, and ancillary equipment. This is shown schematically in Figure 1-1.

The waste to be treated is conveyed by pump, mechanical conveyor, or other means into a surge tank or feed hopper, which in turn feeds the waste into the mixer where it is mixed with the CFS reagents. Depending on the process used, one or more dry and/or liquid components may be added to the waste in the mixer. The mixing process normally takes from one to fifteen minutes, depending on the mechanical system used, the size of the batch, the type of waste, and the amounts and types of reagents being used. After mixing is complete the waste, still in liquid or semisolid form (in many cases), is removed from the mixer by either pumping (in the case of liquid) or screw conveying or dumping (in the case of semisolids). The treated waste is then moved by pump or conveyance device to an area where it can harden and develop its final physical and chemical properties. If the waste has been pretreated, the hazardous components, usually heavy metals, have already been converted into a relatively insoluble form. If it has not been pretreated, the metals are usually immobilized during the mixing process. By the time the waste exits the mixer, it has largely developed its final chemical properties, unless these properties depend upon a physical, monolithic form.

The hardening, or curing, process often takes place in a temporary container or impoundment near the CFS plant, with the solid being subsequently conveyed to the disposal site. Alternatively, the treated waste can be conveyed directly into the disposal site and solidified in place in its final location, regulations permitting. In principle, the latter technique is preferred because it involves fewer steps in handling the waste and thus is less expensive. Also, a waste that is still liquid or semisolid

---

‡Normal or expected refers to the actual conditions known or expected at the disposal site. Such conditions include amount and quality of rainfall, permeation rate into the landfill, temperature variations and freeze-thaw cycling, exposure to ultraviolet light, effects of biological organisms in both aerobic and anaerobic situations, amount and quality of infiltrating groundwater, if any. Testing of wastes for chemical stability is usually done by running leaching tests as specified by the appropriate regulatory agency. Other tests, such as water immersion, freeze-thaw, weatherometer, unconfined compressive strength, and permeability may also be helpful in determining physical stability.

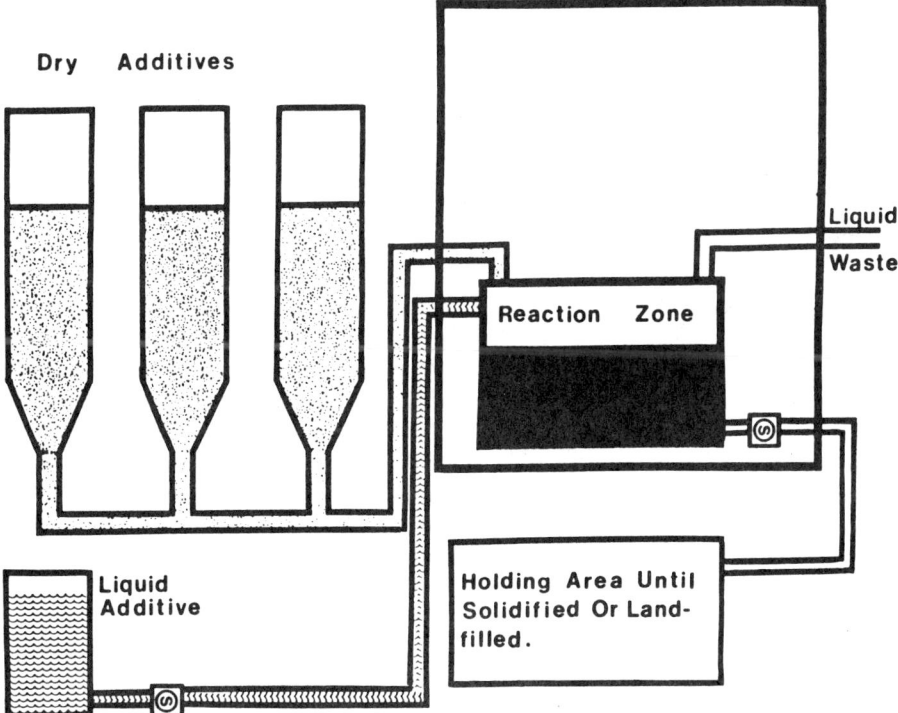

Figure 1-1 Schematic flow diagram of a typical CFS system.

can be poured into place in the landfill with minimum void space, yielding minimal permeability and maximum landfill space utilization in the final product.

In the case where the waste is solidified directly in its final resting place it becomes, essentially, a monolithic form that exhibits the minimal permeability associated with the process [15]-[18]. It is unwise, however, to assume that the chemical properties as measured on such a monolith are necessarily those that ultimately will be associated with the waste over the centuries. Natural processes change the physical character of the material in the landfill just as they do natural soils and rocks [16]. For this reason, chemical stability tests (such as leaching tests described in Chapter 23) are usually conducted in the "worst state" condition, that is, the waste is ground up on the assumption that a variety of factors may cause physical breakdown of the monolith. Conducting tests in this way attempts to assure that the chemical properties that have been developed are the best that can be achieved with the CFS process, and that the leaching rate from the solidified material will probably be considerably less, thereby providing a substantial safety factor.

ORGANIZATION OF THE BOOK

This introduction has given a brief overview of chemical fixation and solidification technology. The following chapters will explore each component of the technology in detail, starting with its technological history in Chapter 2. Because so much of the current interest in CFS involves its use to fix, immobilize, or destroy hazardous constituents in a waste, Chapters 3-6 discuss this area. The placement of this subject in the book is important for another reason: it provides a logical transition from an older, more mature technology—waste water treatment—to CFS. Many of the concepts and much of the technology of fixation find their origins in that field, and environmental engineers and scientists are more familiar with it.

For those more interested in the engineering aspects instead of the chemistry, Chapter 7 provides a rational, practical basis for approaching and solving waste treatment problems that involve, or may involve CFS. Surprisingly, many of the components presented in this chapter are frequently ignored by engineers, regulators, and managers attempting to solve such problems. One component often given only cursory examination is the nature of the waste itself. Consequently, a detailed treatment of wastes, their sources, and their characteristics is presented in Chapter 8. Another component is the final step in any CFS scenario, the ultimate disposal of the treated waste. Disposal is a complex mixture of technology, legislation, regulation, politics, liability, philosophy, and emotion. It can only be touched upon in this book, but some understanding of the issues is vital to any student of CFS technology.

Chapter 9 begins the actual discussion of specific CFS processes and systems, beginning with the general chemistry of solidification, then moving on to the actual system types and the ways in which they are used in practice. Different processes and process types are compared critically and objectively as to their performances, costs, and applicability to different problems. With this as a base, the individual process or system types are discussed in detail in Chapters 10–15.

A CFS system, like any other treatment system, is only as useful as its ability to be applied, or "delivered," in the field. The two types of delivery systems—fixed and mobile/portable—are presented in Chapter 16.

Evaluation of the utility, applicability, and usability of any CFS process or system is accomplished by tests that measure its ability to isolate hazardous constituents of a waste from the environment—to remain physically and chemically "stable" indefinitely. The performance testing of hazardous wastes and the products of their treatment is a large and rapidly growing field in its own right. Volumes have been written on the subject, so it needs no detailed treatment here. Chapter 17 presents the test methods and procedures—past, present, and future—as they apply to process selection and regulatory approval. More importantly, however, this chapter presents a complete and detailed methodology for CFS treatability testing. This is a subject not taught per se in any university, there are no laboratory workbooks, and very few practitioners who have any real-world knowledge of its science and art.

The last chapter is intended for those readers who wish to delve further into the CFS field, or who plan to do research in this area. It includes a brief guide to the information sources and some of the ways in which they may be used, and incorporates a novel use of computer technology in CFS, the expert system. Chapter 18 also provides an introduction to current research in the field and, hopefully, an idea of where we are going in the future.

REFERENCES

1. Pojasek, R. B. (Ed). *Toxic and Hazardous Waste Disposal.* Ann Arbor, MI: Ann Arbor Science Publishers, 1979.
2. Cullinane, M. J., and L. W. Jones. *Stabilization/Solidification of Hazardous Waste.*

Cincinnati: U.S. Environmental Protection Agency (U.S. EPA) Hazardous Waste Engineering Research Laboratory (HWERL), EPA/600/D-86/028, 1986.
3. *Guide to the Disposal of Chemically Stabilized and Solidified Waste.* U.S. EPA SW872. 1982.
4. Federal Register. **51**(102), 19305–19308 (May 28, 1986).
5. Resource Conservation and Recovery Act. PL 94-580, 1976.
6. Comprehensive Environmental Response, Compensation and Liability Act of 1980. PL 96-510, 1980.
7. U.S. EPA. *Prohibition on the Disposal of Bulk Liquid Waste in Landfills—Statutory Interpretive Guidance* (Draft). Washington, DC: Mar. 19, 1985.
8. U.S. EPA Federal Register. **50**(105), 23250–23258 (May 31, 1985).
9. *Chemical Week,* 24–30 (Apr. 17, 1985).
10. *The California List Ban Made Easy—Almost.* Chicago: Chemical Waste Management, 1986.
11. *The November Eighth Solvent Ban Made Easy—Almost.* Chicago: Chemical Waste Management, 1986.
12. U.S. EPA. *Proceedings of the Fourteenth Annual Research Symposium: Land Disposal, Remedial Action, Incineration and Treatment of Hazardous Waste.* Cincinnati: HWERL, 1988.
13. U.S. EPA. *Best Demonstrated Available Technology (BDAT) Background Document for K061.* Washington, DC: 1988.
14. U.S. EPA. Federal Register. **50**(38), 7896–7900 (Feb. 26, 1985).
15. Conner, J. R. *The Modern Engineered Approach to Chemical Fixation and Solidification Technology.* Solidtek, Inc., Morrow, GA, 1981.
16. Conner, J. R. Ultimate disposal of liquid wastes by chemical fixation. In *Proceedings of the 29th Annual Purdue Industrial Waste Conference.* May 1974.
17. Darcel, F. *Recent Studies on Leach Testing—A Review.* Unpublished Draft, Ontario Ministry of Environment, June 1984.
18. *Simulated Field Leaching Test for Evaluation of Surface Runoff for Land Disposal of Waste Materials.* I.U. Conversions Systems, Horsham, PA, 1977.

Chapter 2

HISTORY AND BACKGROUND

To really understand the technology of CFS, how and why it has developed as it has, its history is important. As with most technologies used in the environmental field, CFS has its roots in other processes and equipment developed for different purposes. Its early uses in waste treatment also reflect a very different situation, both technologically and environmentally. These origins and earlier uses have had a strong influence on today's technology, on the way in which it is applied, and on the attitudes of regulators, engineers, and potential users toward it.

ORIGINS

With a few exceptions, the history of the development of CFS systems for general use on waste residues dates only from about 1970. However, the roots of most present-day commercial CFS systems go back to four primary areas of technology that were practiced long before 1970. These are:

- Radioactive waste solidification and disposal
- Mine backfilling
- Soil stabilization and grouting
- Production of stabilized base courses for road construction

Of these, only radioactive waste treatment is a CFS process in the present sense. The other three applications had other utilitarian purposes, although they frequently used wastes such as flyash in the process. Portland cement or flyash or both were used in mine backfilling, sodium silicate plus setting agents, cement and organic polymerizing systems for grouting and soil stabilization, and lime/flyash for road-base construction.

There are many isolated instances where waste residue generators, especially waste disposal site operators, used cement, flyash, lime, soil, and various combinations of these materials to solidify liquids for disposal in landfills where some stability was required in the fill material. Nearly all of this early work involved a need for solidification only, and rarely, if ever, were leaching or other performance tests conducted or required.

USE IN NUCLEAR WASTE TREATMENT

The genesis of most modern-day CFS systems comes from the radioactive waste (rad-waste) solidification field that began in the 1950s. Early on, the nuclear industry recognized the need for solidification of rad-waste in drums and other containers before these wastes could be shipped or buried at government controlled disposal sites in the United States. Much of the liquid waste containing low-level radioactivity was simply absorbed into various mineral sorbents, such as vermiculite, or solidified by making a concrete mixture with very large quantities of Portland cement. In Europe, radioactive wastes were typically solidified in concrete and buried at sea in drums.

The large amount of cement used in solidification of these wastes, especially when the water content is high, results in low volume efficiency—the percentage of the final container occupied by the waste, not the cement. This makes the container more difficult to handle because of its high bulk density, and greatly increases the disposal cost, already very high. Since the major cost in disposal of containerized radioactive wastes is in the cost of the container, its transportation, and disposal, the obvious way to cut costs was by getting more waste and less CFS agent into the container. Also, the solidification process with cement at that time was not well controlled and was somewhat unpredictable, particularly when constituents that retard the setting of Portland cement were present in the waste streams, as they often were. Because of these problems, urea-formaldehyde and asphalt systems came into use to provide more consistency, lower weight, and better space efficiency. By the late 1950s, it was realized that the addition of sodium silicate to the Portland cement process often provided better results overall than did other processes [1]. Later, organic polymer processes were also used for rad-waste solidification [2]. The nuclear industry also experimented with, and used, deep underground disposal of intermediate-level wastes using cement/flyash/clay compositions, pumping the fluid mixtures into fractured shale zones where they solidified and became immobilized [3].

OTHER EARLY PRACTICES

For decades, mines in Canada and the United States have been using Portland cement alone, with flyash, or in combination with solid mine waste to produce cemented hydraulic backfill, which helped eliminate the need for structural pillars or walls within an ore body. Portland cement was also used by various waste generators and operators of waste disposal sites, but little documentation was ever made of these practices, probably because there was no evaluation of the chemical characteristics of the solidified material. At that time there were essentially no technical requirements for work of this sort, other than scattered local and state regulations about the water content of materials that could be placed into a sanitary landfill. In fact, in those days most liquid wastes were co-disposed with municipal refuse in sanitary landfills or open dumps, if they were not ocean dumped or held in lagoons. The concept was that the refuse would act as an absorbent for the liquid. Another method used was to drill or dig holes in completed cells of sanitary landfills and pour liquid wastes into the holes until the area became saturated with liquid or plugged with solids that prevented further absorption. This "solidification" practice was widespread in England and in many areas of the United States.

In the 1950s relatively little industrial liquid waste residue was produced because water and air pollution control were not widely practiced by industry. Most of the solids that later were to form the residues ended up in streams, oceans, lakes, and in the air. However, some concentrated wastes were produced directly from processes such as steel pickling, and it is with these wastes that lime/flyash processes first came to be used [4]. Lime neutralized the acid content and flyash, soil, or Portland cement was then added to produce a solid that could be easily landfilled. Very little control was exercised in such processes and the treatment was usually done on a batch basis with no set formulation and with inconsistent mixing and handling. Efficiency was of little concern, since the lime and flyash were usually obtained as wastes themselves, and there was no requirement for testing the solid for leachability. The only purpose of the treatment was to change the physical form and neutralize strong acid so that a landfill would accept the waste. Later, operators of these systems learned that it was possible to incorporate other types of wastes, such as the oily sludges from petroleum refining, into lime/flyash systems and still produce a solid with reasonable physical properties. While the waste mixtures being solidified became more complex, and the solid produced more potentially hazardous, there was still no requirement for leachability testing and, in fact, no standard leaching tests were available anyway.

THE PRE-RCRA CFS INDUSTRY

Other than internal studies at various nuclear installations and government nuclear regulatory agencies [5], there was no published work done on leachability, environmental degradation, or any other performance characteristics of solidified waste until the early 1970s. At that time, International Utilities Conversion Systems (now Conversion Systems, Inc.) (CSI) and Chemfix, Inc. (now Chemfix Technologies, Inc.) began applying scientific principles to solidification of various waste materials. CSI used a lime/flyash process that incorporated sulfates [6] from the waste to contribute to strength development in the final solid. Chemfix used a (later) patented process [7] based on the combination of soluble silicates and silicate setting agents, usually sodium silicate solution and Portland cement. CSI's activities were aimed primarily at the sludges expected to be produced by lime/limestone scrubbing of flue gases to remove sulfur oxides. These air pollution control processes produced calcium sulfite/sulfate sludges that were compatible with CSI's process and patents [6,8,9], as were sludges produced from neutralization of sulfuric acid pickle liquor and from the battery industry. Operation of the CSI process was designed to be at fixed installations where large quantities of wastes are produced or at areas such as construction sites where the solid produced could be used as construction base for roads and parking lots. The latter applications used a portable unit that could be set up for the duration of the construction project.

Chemfix, on the other hand, aimed at a wider variety of sludges, especially those emanating from manufacturing, metal finishing, and metal-producing operations, and that contained large concentrations of the polyvalent metals that are considered toxic. Recognizing that most such wastes were stored in lagoons at the point of generation, Chemfix predicated its business on a mobile service whereby treatment units could be brought to the location, set up, the waste lagoon emptied and treated in a matter of days, weeks, or months, and the unit moved to the next site. Later, the process was expanded to include sludges containing organics, and a considerable amount of work was done in the chemical industry, on biological sludges, and various mixed wastes. In the mid-1970s, Crossford Pollution Control (later Stablex) in England began commercial utilization there of its cement/flyash process [10] for solidifying industrial inorganic waste streams at central treatment sites.

From a business standpoint, which also directly affected the technological development of CFS processes, this period was not the most propitious for anyone entering the waste residuals treatment field. There were few laws or regulations in the United States or elsewhere concerning the disposal of waste residues. Most of the off-site (i.e., not at the generation site) business went to various landfills inland, and to ocean dumping near the shores, because these were the cheapest methods available. The abovementioned companies spent millions of dollars developing and marketing their products prior to the coming of the RCRA in 1976. While a considerable volume of waste was treated—Chemfix alone solidified more than 100 million gallons of sludges between 1970 and 1976—none of these operations were financially successful in the industrial waste area. In fact, the hazardous waste *treatment* industry—until the passage of the RCRA—was characterized by undercapitalization, overcapacity, and financial losses.

THE POST-RCRA CFS INDUSTRY

Beginning about 1975, in anticipation of the coming of the RCRA, CFS systems began to receive attention from governmental agencies, waste generators, and engineering firms, all of which had largely ignored the technology and the industry until that time. A series of projects was initiated in the EPA's Office of Solid Waste Management (OSW) to evaluate existing processes, to explore other possible systems, and to develop the necessary testing and evaluation criteria. Until recently, the EPA's effort had been almost entirely concentrated in the latter area as they struggled to promulgate test methodologies in accordance with the dictates of the RCRA. This culminated in the May 19, 1980, promulgation of regulations for a federal hazardous waste management system under Subtitle C of the RCRA [11], and the subsequent regulations outlined in Chapter 1. Only in the last few years has the EPA begun to fund or conduct research and development in the technology of CFS at the EPA laboratories, contractors, and academic "centers of excellence."

After the passage of the RCRA, solid waste disposal companies began to take an interest in hazardous waste treatment technology, and a number of vendors appeared on the scene offering processes, chemicals, and services. While considerable technical development has been done by the vendors, especially by those in the industry prior to the RCRA, little detail has been published, and most of this is found in the patent literature. Vendors, understandably, have been reluctant to share their knowledge with others who paid little attention to the CFS field earlier, and this feeling still persists today. Now that CFS technology has become an accepted, if little understood, segment of the hazardous waste management field, it has begun to mature into a science that is attracting investigators from industry, government, and academia. Unfortunately, much of the work now taking place is simply a repeat of that already done by the vendors. One purpose of this book is to minimize this waste of resources and focus attention on areas that have not been explored. The history of CFS from 1980 on is really part of the present scene, and so it is discussed in succeeding chapters.

REFERENCES

1. Commissariat A L'Energie Atomique French Patent 1,246,848 (Oct. 17, 1960).
2. Bell, N. E., *et al*. Solidification of low-volume power plant sludges. Electric Power Research Institute Project 1260-20, **CS-2171** (1981).
3. DeLaguna, W. Radioactive waste disposal by hydraulic fracturing. *Industrial Water Engineering,* 32–37 (Oct. 1970).
4. Brink, R. H. Use of waste sulfate on Transpo'72 parking lot. Federal Highway Administration, Washington, DC, 1972.
5. Winsche, W. E. U.S. Patent 3,152,984 (Oct. 13, 1964).
6. Smith, C. L., and W. C. Webster. U.S. Patent 3,720,609 (Mar. 13, 1973).
7. Conner, J. R. U.S. Patent 3,837,872 (Sept. 24, 1974).
8. Smith, C. L., and W. C. Webster. U.S. Patent Reissue 29,783 (Sept. 26, 1978).
9. Webster, W. C., R. G. Hilton, and R. F. Cotts. U.S. Patent 4,028,130 (June 7, 1977).
10. Chappell, C. L. U.S. Patent 4,116,705 (Sept. 26, 1978).
11. Federal Register. **45**(98), 33063–33285 (May 19, 1980).

Chapter 3

PRINCIPLES OF FIXATION

Although it is usually spoken of and often used in conjunction with solidification, fixation (as defined in Chapter 1 and the Glossary) is a separate subset of CFS technology. It can be practiced where there is no need to change a waste's physical properties, for example, in the detoxification of contaminated soil. And in another sense, it has been used for many years in waste water treatment. In fact, much of what we do understand about fixation is attributable to that field, and where specific fixation processes or additives are used, they have usually been derived from water and waste water treatment chemistry.

The mechanisms of fixation are different for the three primary groups of pollutants: metals, other inorganics, and organics. Chapters 4–6 discuss these groups separately. In this chapter, we explore fixation in general as an introduction to the subject and to provide a context for the later, more detailed discussions. Since fixation is generally defined as a treatment product's resistance to leaching in the environment, it is essential to understand the leaching process and the tests by which it is measured. The reader is referred to Chapter 17 for a discussion of leaching tests; those requiring a more in-depth treatment of leach testing should consult the references to Chapter 17, Appendix 2, and the Bibliography.

WHAT IS LEACHING?

If ground- or surface water contacts or passes through a material, each constituent dissolves at some finite rate. Even in the most impermeable solidified waste—or clay, concrete, brick, or glass for that matter—water will eventually permeate if there is a driving force. Wherever water penetrates, some of the waste dissolves—there is no such thing as a completely insoluble material. Therefore, when a waste, treated or not, is exposed to water a *rate* of dissolution can be measured. We call this process *leaching*, the water with which we start the *leachant*, and the contaminated water that has passed through the waste the *leachate*. The capacity of the waste material to leach is called its *leachability*.

Leaching is a rate phenomenon and our interest, environmentally, is in the *rate* at which hazardous or other undesirable constituents are removed from the waste and passed into the environment via the leachate. This rate is usually measured and expressed, however, in terms of *concentration* of the constituent in the leachate. This is because concentration determines the constituent's effects on living organisms, especially humans.* Concentration is the primary basis for water quality standards, and water quality standards, especially drinking water standards [1], are normally the basis for leaching standards. Thus, we speak of the leaching rate of a constituent, but usually measure it as concentration in the leachate.

When evaluating a material for leachability, we usually compare the concentration of the hazardous constituent in the leachate to that in the original waste. This tells us what proportion of the constituent dissolved out during the test, which becomes a measure of the leachability of the material. If the test conditions can be expressed in terms of a time-related number, such as equivalent years of field exposure to rainfall, then we can state leachability in true rate terms; for example, pounds of constituent per year or percent of the original content per year. Multiplying the hazardous constituent's concentration in the original waste by the total quantity of the waste gives us the total amount of the constituent, and thus the hazard potential posed by that quantity of waste. Interestingly, as the hazardous constituents leach, the hazard potential gradually diminishes. Thus, if the leaching rate is controlled so as not to exceed the allowable environmental standards in the ground- or surface water, leaching should really be a beneficial process in the long term. This assumes, of course, that the leaching rate remains constant at an acceptable level, or decreases with time. As we shall see later, a number of scenarios are possible, depending on the disposal conditions.

MEASUREMENT OF LEACHABILITY

For obvious reasons, actual tests cannot be run on the solidified waste in the specific disposal site for thousands of years—the only absolutely certain way—before choosing a CFS technology. As with the testing of any other product requiring long-term stability, we must make judgments and reach decisions based on information that is available or can be obtained rather quickly. In this case, the information consists of actual field data—about 20 years worth at present—laboratory test results, some understanding of the chemistry of the system, and models that tie everything else together. A large body of information has been built up in this fashion, primarily for the metal constituents specified by the EPA [2]—arsenic, barium, cadmium, chromium, lead, mercury, selenium and silver.

It is not possible here to explore the details of leaching test methods and results, but a good summary is given by Darcel [3]. The environmental acceptability of a hazardous waste for land disposal in the United States is based on the U.S. EPA Extraction Procedure Toxicity (EPT) test [2] or, more recently, the Toxicity Characteristic Leaching Procedure [4] (TCLP). While the latter test has not yet been promulgated as a final rule by the EPA, its use in the recent landbans has made it the defacto standard, replacing the EPT for many purposes. In Canada (Ontario Province), the standard leachate extraction procedure is very similar to the EPT. In addition, certain states, such as California, have their own tests that must be used

---

*There are also cumulative effects determined on total exposure over a long period of time.

in addition to the EPT or TCLP. The EPA has also used two other test methods in delistings: the Multiple Extraction Procedure (MEP) [5] and the Oily Waste Extraction Procedure (OWEP) [6].

All of these tests used for regulatory purposes are batch procedures in which the waste is contacted with a leachant for a specific period of time, agitating the mixture to achieve continuous mixing. Chemical equilibrium is often obtained [7], especially when the solidified waste is crushed before extraction. After extraction and separation of the fluid from the solids, the leachate is analyzed for specific constituents. It should be noted that most of these tests use a leachant-to-waste ratio of 20:1, so that the maximum concentration of constituent that can be attained in the leachate is 5 percent of that in the original solid. The leachant in most cases is dilute acetic acid, buffered in some procedures. The total amount of acid added varies with the test and/or with the alkalinity of the waste. The final pH of the leachate at the end of the test is controlled by the alkalinity of the waste in most cases where the leachant is deionized water or dilute acid. As we will see, final pH is one of the prime controlling factors in metal leaching. The TCLP test is designed to simulate the leaching potential of a waste in a so-called "mismanagement" scenario, where it is disposed in a landfill designed for municipal refuse. Such landfills are known to generate organic acids during decomposition of organic matter in the refuse. The purpose of acetic acid in the leachant is to simulate those acids.

LEACHING TESTS VS. ACTUAL LEACHING

Neither the TCLP nor its predecessor, the EPT, actually simulate any real-world set of conditions. Arguably, however, they do perhaps create a worst-case environment for leaching. Attempts to correlate laboratory leaching tests with field data have not been successful [8]. Natural minerals do not exhibit leach rates even approaching those measured for waste forms in the laboratory. If they did, the world's land masses would be very different from what they actually are now. The same processes that inhibit leaching of natural substances may apply also to buried waste forms in the real world.

Nevertheless, these procedures are so widely used and required as specification tests that arguments over their validity are largely academic today. In subsequent chapters, when leaching or leachability are discussed, the reader may assume that the TCLP or EPT was used unless otherwise stated. It is not unusual, however, to use several tests or even a whole test battery for specific disposal situations.

LEACHING CRITERIA

The other aspect of leachability testing is the criteria that are applied in judging whether the concentrations of constituents in the leachate are acceptable, that is, the impact on the environment will be negligible. The basis for these criteria in the United States has been the National Interim Primary Drinking Water Standards [9]. For the characterization of a nonlisted waste (see Chapter 8 for definitions of the various categories of wastes) as nonhazardous, the concentration of any listed constituent may not exceed 100 times the drinking water standard [2]. This standard covers only the eight metals previously listed, plus six pesticides/herbicides. Acceptable levels for other toxic constituents were not established by the EPA at that time (1980) and have only been proposed at this time in connection with the proposed TCLP procedure. However, many other toxic constituents were covered under the listing system (see Chapter 8) and regulatory limits have been, or will be established under the landbans (see Chapter 1).

Table 3-1 lists the regulatory limits for the various test protocols and regulatory systems. It is immediately obvious that the allowable concentrations of constituents in the leachate are not the same for all tests, not even for the same test used for different wastes under different regulatory schemes. This is partly deliberate—for example, to allow for the specific environmental impact of a particular disposal scenario. However, it is often the result of changing attitudes and stricter controls that have not yet been retrospectively applied to the older rules and regulations, thus creating inequities in the regulatory scheme as a whole. Chapter 8 discusses this subject in more detail, but it is impossible to thoroughly explore the complexities of even the U.S. regulatory system in this book. Again, the reader is referred to the references in Chapters 1, 8, and 17, and to the Bibliography.

FACTORS AFFECTING LEACHABILITY

There are two sets of factors, or variables, that affect the leachability of a treated waste: (1) those that originate with the material itself, and (2) those that are a function of the leaching test. The combination of the two sets determines the leachability of the material.

Test Method Factors

The major variables are normally specified for given test protocol, but latitude in the specification and controllability of the parameter can, and do, cause significant problems with reproducibility. The following discussion is aimed primarily at the EPT and TCLP, but most of the comments would apply to other batch testing protocols and, to a lesser extent, to column and dynamic leach testing. While this discussion would seem to be more fittingly placed in Chapter 17 on test methods, it is essential here to an understanding of leachability as we currently measure it, and therefore to the subject of fixation.

Surface Area of the Waste. Ideally, the waste should be tested in the same physical condition expected in the landfill. In practice, this is not possible and, in fact, many test protocols deliberately increase the surface area by grinding or crushing the waste to achieve a worst-case condition. CFS processes may produce the waste form as a monolith, as a soil-like material, and even in granular or powder form. From the standpoint of fixation, the ideal CFS process should contain the constituents of concern even if the physical structure is destroyed. In practice, most processes rely on a combination of chemistry and micro- or macroencapsulation. The EPT test allows the use of either of two compromises. The waste can be crushed to moderate particle size (9.5 mm) to represent the effects of handling and other environmental conditions. Alternately, a monolithic form can be subjected to an impact procedure (the structural integrity option) and tested in whatever condition results from that procedure. Presumably, any specimen surviving the impact as a monolith would

TABLE 3-1 Comparison of Regulatory Limits for Various Test Procedures

Contaminant	HWNO	EPT	TCLP	CCWE[a]	1988 First-Third Landban TCLP[a]	1988 First-Third Landban Total[a]	California TTLC	California STLC	Canadian CGSB
Acetone			0.59	0.590		0.370			
Acrylonitrile	D018		5.000						
Aldrin									0.07
Antimony							140	15.00	
Arsenic	D004	5.000	5.000		0.004		500	5.00	5.00
Barium	D005	100.000	100.000		(1)		500		100.00
Benzene	D019		0.070			0.710			
Beryllium							75	0.75	
bis(2-Chloroethyl) ether	D020		0.050			5.600			
Boron				5.000					500.00
Butyl alcohol						0.370			
Cadmium	D006	1.000	1.000		0.066		100	1.00	0.50
Carbaryl									7.00
Carbon disulfide	D021		14.400	4.810					
Carbon tetrachloride	D022		0.070	0.960					
Chlordane	D023		0.030				250		0.70
Chlorobenzene	D024		1.400	0.050		6.000			
Chlorodibenzofurans (all tetra)				0.001		0.001			
Chlorodibenzofurans (all hexa)				0.001		0.001			
Chlorodibenzofurans (all penta)				0.001		0.001			
Chlorodibenzo-p-dioxin (2,3,7,8 tetra)				0.001		0.001			
Chlorodibenzo-p-dioxin (all hexa)				0.001		0.001			
Chlorodibenzo-p-dioxin (all penta)				0.001		0.001			
Chlorodibenzo-p-dioxin (all tetra)				0.001		0.001			
Chloroform	D025		0.070			6.000			

(continued)

TABLE 3-1 Comparison of Regulatory Limits for Various Test Procedures (*Continued*)

						1988 First-Third Landban		California		Canadian
Contaminant	HWNO	EPT	TCLP	CCWE[a]	TCLP[a]	Total[a]	TTLC	STLC	CGSB
Chromium (+3)	D007	5.000	5.000		0.094		2500	560.00	5.00
Chromium (+6)		5.000	5.000				500	5.00	5.00
Cobalt							8000	80.00	
Copper							2500	25.00	
o-Cresol	D026		10.000	0.750		2.200			
m-Cresol	D027		10.000	0.750					
p-Cresol	D028		10.000	0.750					
Cyanide						0.900			20.00
Cyclohexanone						1.800			
2,4-D	D016	10.000	1.400			0.490			10.00
DDT									3.00
Diazinon									0.02
1,2-Dichlorobenzene	D029		4.300	0.125		0.490			
1,4-Dichlorobenzene	D030		10.800						
1,2-Dichloroethane	D031		0.400			6.000			
1,1-Dichloroethylene	D032		0.100						
2,4-Dinitrotoluene	D033		0.130						
Endrin	D012	0.020	0.003				20		0.02
Ethyl acetate				0.750		0.370			
Ethyl benzene				0.053		0.031			
Ethyl ether				0.750					
Fluoride							18000	180.00	150.00
Heptachlor	D034		0.001				470		0.03
Hexachlorobenzene	D035		0.130			28.000			
Hexachlorobutadiene	D036		0.720			5.600			
Hexachloroethane	D037		4.300			28.000			
Isobutanol	D038		36.000	5.000					
Lead	D008	5.000	5.000		0.180		1300	5.00	5.00
Lindane	D013	0.400	0.060				400		0.40
Mercury	D009	0.200	0.200		0.025		20	0.20	0.10
Methanol				0.750		0.370			
Methoxychlor	D014	10.000	1.400						10.00
Methyl parathion									0.70
Methylene chloride	D039		8.600	0.960		0.037			

FACTORS AFFECTING LEACHABILITY

Methyl ethyl ketone	D040		7.200	0.750	0.370	
Methyl isobutyl ketone				0.330	0.370	
Molybdenum						3500	350.00
Nickel						2000	20.00
Nitrate and nitrite					0.048			1000.00
Nitriloacetic acid								5.00
Nitrite								100.00
Nitrobenzene	D041		0.130	0.125	0.490	
1,2-Nitrophenol					13.000	
Parathion								3.50
PCBs (all)						5000		0.30
Pentachlorophenol	D042		3.600	0.010	37.000	
Phenol	D043		14.400		2.400	
Pyridine	D044		5.000	0.330		
Selenium	D010	1.000	1.000			100	1.00	1.00
Silver	D011	5.000	5.000			500	5.00	5.00
1,1,1,2-Tetrachloroethane	D045		10.000			
1,1,2,2-Tetrachloroethane	D046		1.300		5.600	
Tetrachloroethylene	D047		0.100	0.050	6.000	
2,3,4,5-Tetrachlorophenol				0.050		
2,3,4,6-Tetrachlorophenol						
Thallium	D048		1.500		0.031	700	7.00
Toluene	D049		14.400	0.330		
Toxaphene	D015	0.500	0.070			500		0.50
1,1,1-Trichloroethane	D050		30.000	0.410	0.044	
1,1,2-Trichloroethane	D051		1.200			
Trichloroethylene	D052		0.070	0.091	0.031	
2,4,5-Trichlorophenol	D053		5.800	0.050		
2,4,6-Trichlorophenol	D054		0.300	0.050		
1,1,2-Trichloro-1,2,2-trifluoroethane				1.050		
Trichlorofluoromethane				0.960		
Trihalomethanes								35.00
2,4,5-TP (Silvex)	D017	1.000	0.140					1.00
Uranium								2.00
Vanadium						2400	24.00
Vinyl chloride	D055		0.050		0.015	
Xylene							
Zinc				0.150		5000	250.00

[a]Non-waste waters, lowest applicable values.

remain that way in the disposal environment, an assumption that is not justified by supporting data.

There are environmental conditions other than direct mechanical force that can destroy the monolithic structure: freeze–thaw cycling, wet–dry cycling with or without abrasion, and long-term soaking in groundwater. The use of tests to simulate these conditions is extremely controversial. Waste forms in real disposal conditions are rarely exposed to freeze–thaw conditions, at least not for any length of time. Wet–dry cycling may occur in landfills, although landfills containing hazardous wastes, treated or not, are increasingly constructed and operated in such a fashion as to minimize this phenomenon. Certainly, wastes in landfills are not subjected to abrasion after the initial filling and compaction.

Under the EPT procedure, neither preparation method is sufficiently controlled to allow good replication of results, let alone comparison of various CFS products that fracture in different ways. As a result, the sample preparation is dependent on the technician doing the preparation and on the equipment used. For example, the crushing procedure specifies the maximum particle size, but not the minimum size nor the size distribution. Anyone who has actually conducted sample preparation will realize that, within the specifications of the test, the EPT sample can be deliberately or accidentally varied to give widely variable results if the results depend on physical properties. Unfortunately, the newer TCLP procedure does little to correct this fault in the methodology, although it does eliminate the variable of the structural integrity option. The variability of test results between laboratories and waste forms is therefore not decreased much by the TCLP.

Extraction Vessel. None of the batch tests adequately spell out the geometry of the extraction vessel, except when the zero headspace extractor (ZHE) is used for leaching of organic constituents in the TCLP. This geometry affects the degree of abrasion to which the waste form is subjected during the tests and, therefore, the ultimate particle size of the waste being tested. The EPA is currently investigating the use of a stainless-steel screen inside the vessel to make the agitation more uniform. However, this also may affect the ultimate particle size distribution by acting as a grater on larger waste pieces.

Agitation Technique and Equipment. All batch methods use some form of agitation to permit more rapid attainment of equilibrium between the waste specimen and the leachant. The EPT does not specify either equipment or technique, but simply states that they must impart sufficient agitation to prevent stratification and that they must ensure that all sample surfaces are continuously in contact with the extraction fluid. The TCLP requires a rotary extraction device and gives suitable specific examples. Most workers now use a rotary device of the National Bureau of Standards (NBS) design [10].

There is no analogy for such agitation techniques in a modern waste landfill, where the leachant is normally stationary or flowing very slowly around the waste particles, and the leaching process diffusion limited [8]. The major objection to this sort of agitation technique is that it changes the particle size distribution of the waste, thus increasing the effective surface area exposed to the leachant. However, with CFS processes that produce soil-like products, the physical state of the waste is usually of relatively little importance, as we shall see in Chapter 4.

Nature of the Leachant. In principle, the leachant should be that which is actually in contact with the waste in the landfill. However, this condition is impossible to reproduce in a practical leaching test. Leaching water pH, oxidation-reduction potential (E_h), even composition all change with time and are usually not known with any degree of accuracy in any case. The trend in present testing protocols is to use an aggressive leaching solution with moderately low pH—thus the EPT and TCLP leachants.

The leachant must also be usable in everyday laboratory situations, such as ordinary laboratory reagents, easy and reproducible preparation of solutions, atmospheric pressure, in air. The choice of mildly acidic solutions using carbonic acid (CO_2-saturated water) or acetic acid meets these requirements and also has some basis in natural systems and from other, older and more established technologies. Acetate buffer systems have been used in estimating availability of trace metals in agriculture [8]. Also, natural precipitation and some soil waters contain CO_2. Thus, the choice of these systems is a compromise, even though they do not exactly reproduce any real condition.

Ratio of Leachant to Waste. It is obvious that no ratio can be selected that would represent real conditions in any given landfill at all times, let alone in a variety of landfill scenarios. Wastes, especially after CFS treatment, contain large amounts of soluble, nontoxic components that generate common ion and total ionic strength effects that can reduce the solubility of certain constituents of concern when the ratio is low. Investigators have generally determined that higher ratios are more appropriate [11].

Number of Elutions Used. In some test protocols, such as the MEP [5], the waste is extracted with successive batches (elutions) of leachant. This procedure is designed to approach the natural, continuous replacement of fresh leachant that occurs in many landfills, and it has been used to simulate column leaching test procedures that are more difficult to set up and to reproduce. The MEP has been used by the EPA to simulate long-term leaching of up to 1000 years [5], and by other investigators in sequential batch methods [12]. It has been generally assumed that the initial elutions contain the maximum concentrations of constituents, because they are exposed to the highest concentrations present on the fresh waste surfaces. However, as we will see in Chapter 4, in the case of metals this is not necessarily true [12]. Since nearly all of the regulatory tests practiced at this time use a single elution, this should be a cause of concern and a reason for further investigation of test methods.

Time of Contact. Early in the development of standard leaching test procedures, the effect of time was studied in some detail [13]. The goal was to achieve or approach equilibrium at reasonable test times. It was found that 24 hours was a good compromise in most situations, and most protocols have used this time as a standard. Recently, the EPA modified this period for the TCLP, reducing test time to 18 hours for practical scheduling reasons in a laboratory. It is believed that this change does not substantially affect the results, compared to other uncontrolled variables.

Temperature. Solubility of constituents is a function of temperature, and leaching test results are, at least partially, functions of the solubility of the species being investigated. Temperature in the disposal site varies as a function of time, depth, location, and chemical reactions occurring in the landfill. Obviously, all of these conditions cannot be used; in fact, they are not even known in most cases. Therefore, the laboratory standard temperature of 20°C–25°C (68°F–77°F) is normally used. The TCLP standard, for example, is 22°C ± 3°C (72°F ± 37°F).

pH Adjustment. Probably the greatest operating variable in the EPT has been the pH adjustment procedure. Here, pH is required to be adjusted at various intervals to maintain pH 5.0 ± 0.2. The rate of addition of 0.5 N acetic acid to do this is not specified, and is the source of much of the variability, since solubility of metal species, especially of the most common hydroxides, is quite sensitive to pH in certain ranges that vary with each metal. The TCLP methodology improves on this by eliminating the periodic adjustment procedure and by using one of two solutions depending on the alkalinity of the waste (see Table 16-12).

The level and control of pH are extremely important factors in evaluating leachability, especially for metals. The final pH of the leachant should always be recorded, since it is this value that often determines whether a waste meets regulatory standards for metals. Unfortunately, measurement of final TCLP leachate pH is not required in the test methodology, and so is often neglected by testing personnel.

Separation of Extract. This seemingly straightforward test element is quite important. Many metals can exist in waste forms and in the waste–leachant mixture in colloidal state. The filtration procedures are designed to remove these colloids, but don't always do so. Any filtrate that is cloudy should be considered suspect. Also, filtration takes time, and the longer the time, the more contact with the leachant. While not always controllable, filtration time should always be recorded.

Analysis. Analytical procedures are spelled out in detail in SW-846 [14]. Laboratories experienced in water analysis but new to analysis of hazardous waste residuals and leachates often have difficulty in both accuracy and reproducibility in the leachate. Matrix effects and interferences are the rule rather than the exception, and are often very strong in the high ionic strength leachates encountered in this field. In the past, regulatory levels were typically in the milligram/liter range and were expressed only to 0.1 mg/l accuracy. Present levels are falling rapidly, sometimes requiring accuracy to several micrograms/liter (parts per billion) or below. This has placed greater demands on both analyst and equipment.

Waste Form Factors and Fixation Mechanisms

A number of factors, characteristics, mechanisms, or containment systems affect the degree of immobilization, or fixation, of constituents in the waste. Usually, several are in operation at the same time. A list of the major factors follows.

- pH control
- Redox potential control
- Chemical reaction
    Carbonate precipitation
    Sulfide precipitation
    Silicate precipitation

Ion-specific precipitation
　　Complexation
- Adsorption
- Chemisorption
- Passivation
- Ion exchange
- Diadochy
- Reprecipitation
- Encapsulation
　　Microencapsulation
　　Macroencapsulation
　　Embeddment
- Alteration of waste properties
　　Particle size

Figure 3.1 [15] demonstrates schematically how they work. The leaching mechanisms that control, and are controlled by these containment systems are listed below.

- Solubilization
- Transport through the solid
　　Convective transport
　　Molecular diffusion
　　　Solid state
　　　Pore solution
- Transport through the leached layer
- Transport through the solid–liquid boundary layer
- Bulk diffusion in the liquid leachant
- Chemical reactions in the leachant
- Biological attack

First, we will discuss containment mechanisms in some detail. Much of our knowledge of this subject applies primarily to metals, since the main thrust of regulation of wastes most susceptible to CFS technology has been toward the priority pollutant metals. However, many of the fixation mechanisms also apply to other inorganic species and to organics.

Equilibrium Constants. Before discussing specific mechanisms, it is worthwhile reviewing the equilibrium constant concept, since so many of the reactions that affect leaching occur between relatively simple molecules and ions, and are reversible. While this may seem elementary, it is frequently misunderstood. For any reaction

$$aA + bB \rightleftharpoons cC + dD$$

the equilibrium constant is expressed as

$$K_e = \frac{[C]^c \times [D]^d}{[A]^a \times [B]^b} \qquad (3\text{-}1)$$

where $[A]$, $[B]$, $[C]$, and $[D]$ are the activities of the respective species. In the case of an ionizable species, such as most metal compounds in aqueous media

Adsorption

Example: Interaction of single layer of clay mineral with polar groups

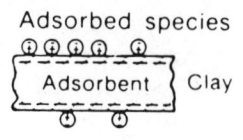

Chemsorption

Example: Organic compounds on clays

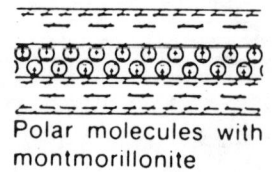

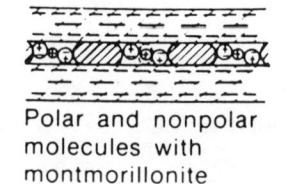

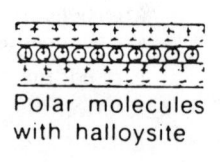

Polar molecules with montmorillonite

Polar and nonpolar molecules with montmorillonite

Polar molecules with halloysite

Passivation

Example: Reaction of gypsum with barium solution

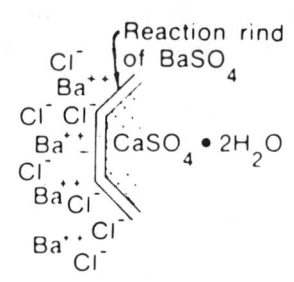

Diadochy

Example: Substitution in the calcite crystal lattice

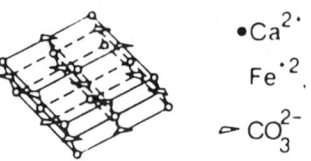

Reprecipitation

Example: Production of sideronatrite from sodium sulfate

$$Fe_2(SO_4)_3 + 2Na_2SO_4 + 8H_2O$$
$$\longrightarrow 2Na_2Fe(SO_4)_2 \cdot 3H_2O + H_2O + H_2SO_4$$

Comparative Solubilities at 25°C

$Na_2SO_4 \cdot 7H_2O$ 524 g/ℓ

$Na_2Fe(SO_4)_2 \cdot 3H_2O$ 0.16 g/ℓ

Figure 3-1 Hazardous constituent containment systems. (From Malone and Larson [15].)

$$AB \rightleftharpoons A + B$$

the dissociation constant is expressed as

$$K_d = \frac{[A] \times [B]}{[AB]} \tag{3-2}$$

The reactions of interest in leaching occur as heterogeneous reactions, that is, they involve both a solid and a liquid. In the preceding equation, the solid is AB and its concentration is a constant. Multiplying both sides of Eq. (3-2) by $[AB]$ results in a different value for K_d, which we call K_{sp}, the solubility product

$$K_{sp} = [A] \times [B] \qquad (3\text{-}3)$$

From this comes the important conclusion that, for dissociation of solids such as metal compounds, the concentration of the solid metal compound in the waste does not affect the concentration of metal ion in the leaching solution. This remains true as long as any solid metal species exists in contact with the leaching solution. Furthermore, since K_{sp} is a constant for a given system at a given temperature, changing the concentration of either A or B automatically changes the concentration of the other, giving rise to the common-ion effect. This is quite important in metal fixation for systems at equilibrium, since the metal concentration in the leachate can be controlled by the addition or removal of its associated anion from another species. It is rarely possible to actually calculate the expected effect with any degree of accuracy in complex waste–reagent systems; however, the degree of departure of measured values from the theoretical concentration indicates the extent of other factors operating in the system.

K_d may be determined experimentally, in which case the presence of the solid is automatically included in the value determined. Alternately, K_d may be calculated from the equation

$$\Delta G_r^0 = -RT \ln K_d \qquad (3\text{-}4)$$

where:

ΔG_r^0 = the standard free-energy change of a reaction
R = universal gas constant
T = absolute temperature (degrees Kelvin)

pH Control. It is widely accepted that cement and pozzolan-based waste forms rely heavily on pH control for metal containment. In fact, some investigators believe that this is the only important factor in metal fixation. However, more recent work has shown that this is not so, at least when very low leachabilities are required. Using a sequential batch extraction technique, Bishop [12] demonstrated that even after the alkali was leached from cement-based waste forms, lead and chromium leaching was much lower than would be expected from metal hydroxide solubilities. In this case, the metals were probably bound into the silica matrix itself. In other CFS systems, the same mechanism may not be controlling. Nevertheless, pH control is a necessary, if not sufficient condition for metal fixation in most systems. It has also been shown to affect the leachability of some other inorganic and organic species.

Normally, high pH is desirable because metal hydroxides have minimum solubility in the range of pH 7.5–11. However, some metals such as chromium that exhibit amphoteric behavior have higher solubility at both low and high pH. Unfortunately, all metals do not reach minimum solubility at the same pH, so that the optimum pH of the system must be a compromise. Solubility curves for various metal hydroxides, derived from several sources [16, 17], are shown in Figures 3-2

36   PRINCIPLES OF FIXATION

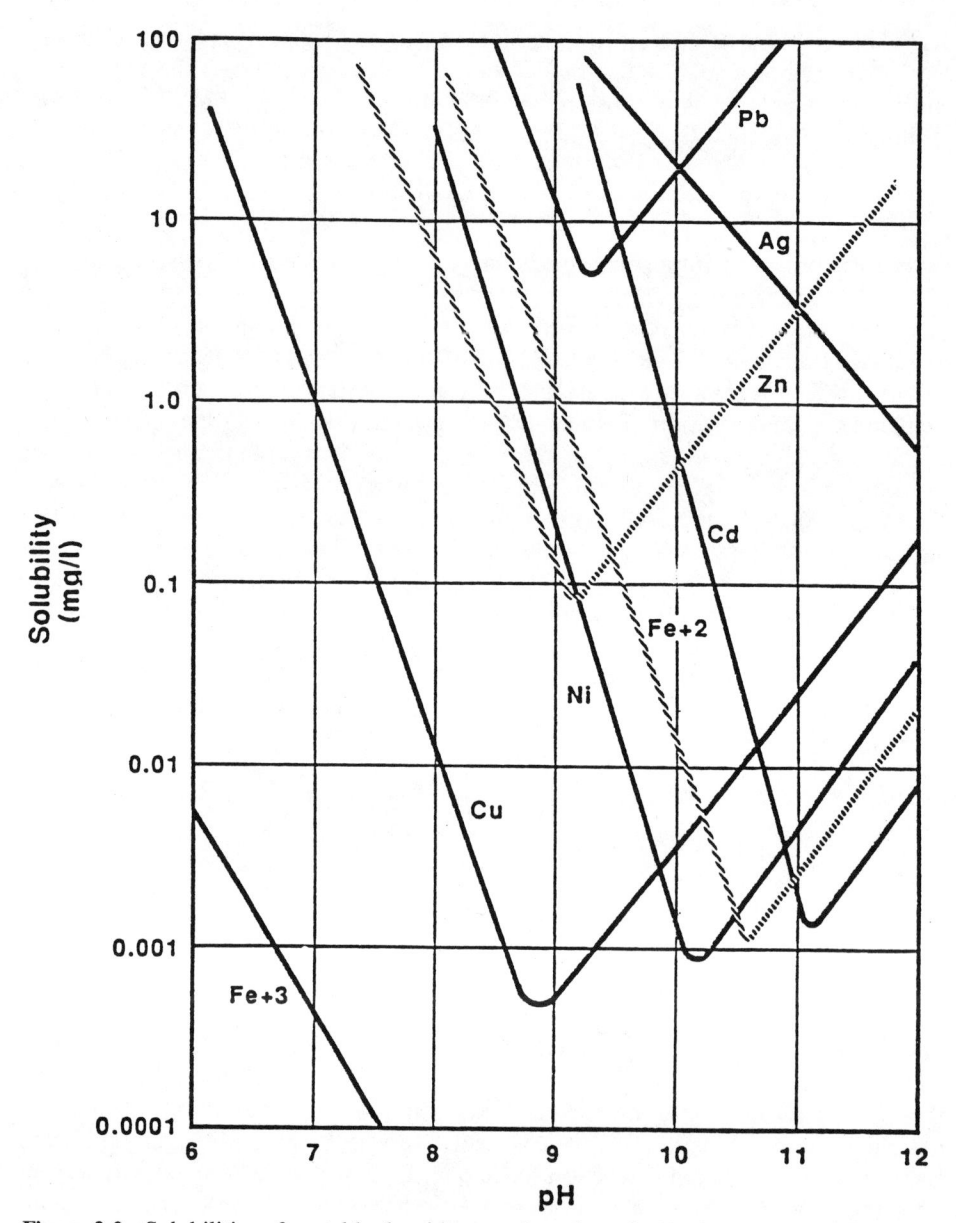

Figure 3-2 Solubilities of metal hydroxides as a function of pH. (From EPA [17].)

and 3-3. It is obvious from examination of these graphs that care must be exercised in using solubility data to predict leachability of any constituent. Such data are obtained experimentally or calculated from stability constants for individual species alone, or for relatively simple combinations. Depending on the assumptions used in the calculations, or the experimental conditions used, widely different results are obtained. Waste forms, moreover, contain many other soluble species that may change the actual solubility of any metal due to effects such as common ion, complexation, total ionic strength, and redox potential. Certain metals, notably chro-

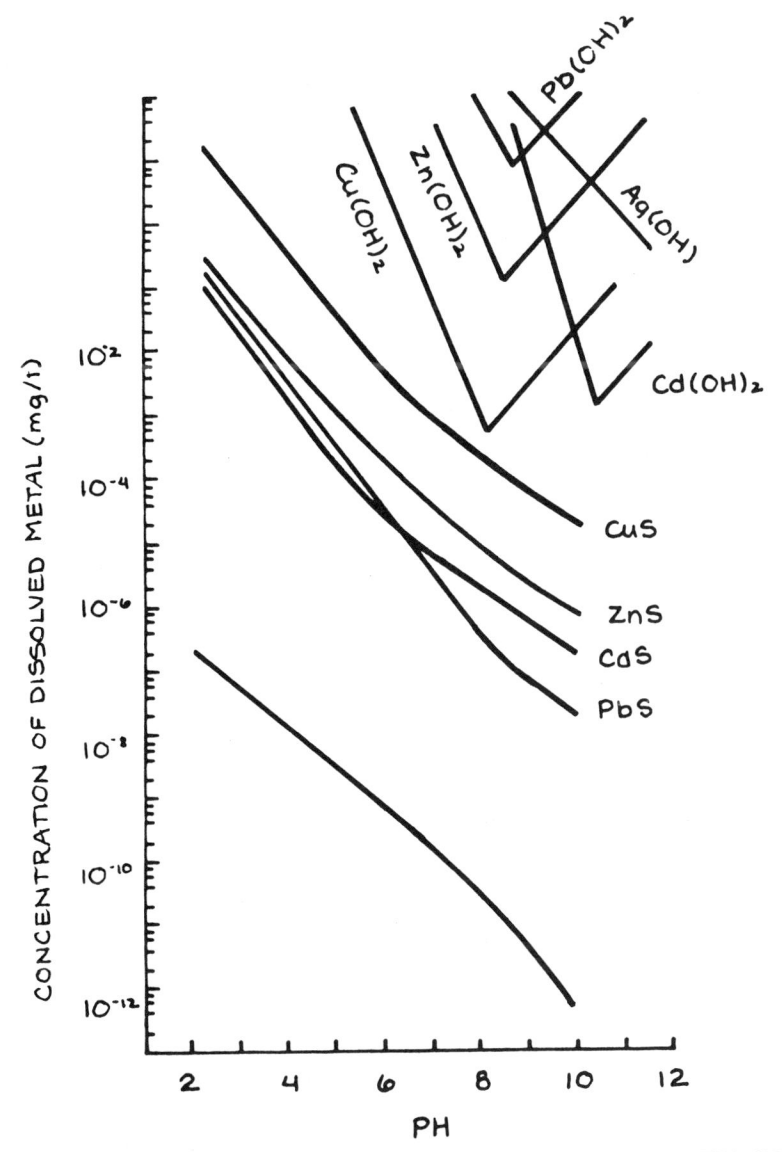

Figure 3-3 Solubilities of metal hydroxides and sulfides. (From EPA [17].)

mium, exhibit completely different solubilities in their various valence states, and may exist in cationic or anionic form, or both, in the same waste.

Several good examples of the differences between calculated hydroxide solubility–pH curves and those determined empirically for real CFS systems are given by Cote [16]. Figures 3-4–3-6 show these data schematically. While the general shape is the same, the calculated hydroxide curves show a much greater dependence on pH than the measured data. Minimum solubility for hydroxides is fairly constant through a smaller pH range in each case, about one pH unit for lead, two units for chromium, and three for cadmium. Empirical data show similar solubilities through pH ranges of four, eight, and greater than three units, respectively. The calculated

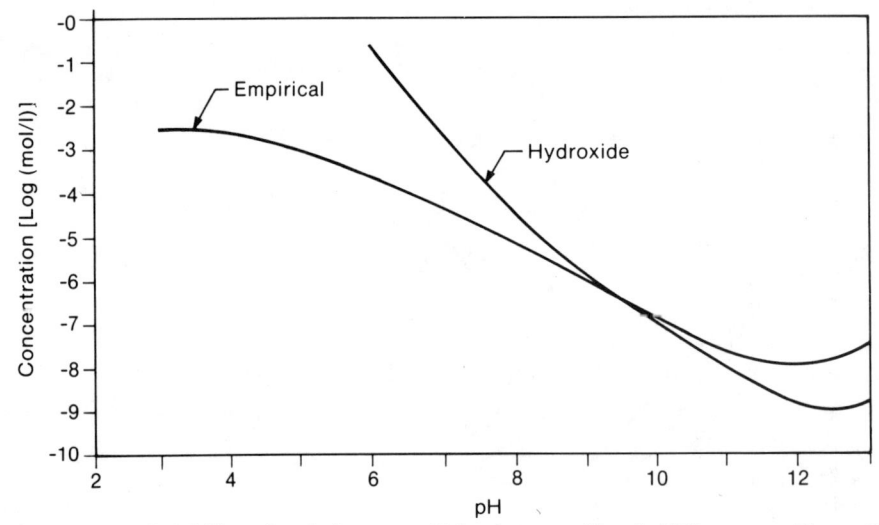

Figure 3-4 Solubility of cadmium vs. pH in a cement/flyash CFS system. (From Cote [16].)

minimum solubilities for hydroxides, however, are lower by factors of 5 to 10 than the empirically obtained solubilities. All of the empirical data were obtained from a real CFS system—cement/flyash—on a synthetic waste, which is less complex than most real wastes. Obviously, factors other than hydroxide precipitation are in operation even in such a simplified system.

Redox Potential. Chromium is only one example of the influence of valence state on solubility. Arsenic is another. The presence of strong oxidants or reductants can change the valence state of a number of the metals, affecting their chemical speciation and therefore their mobilities by orders of magnitude. For some metals, such

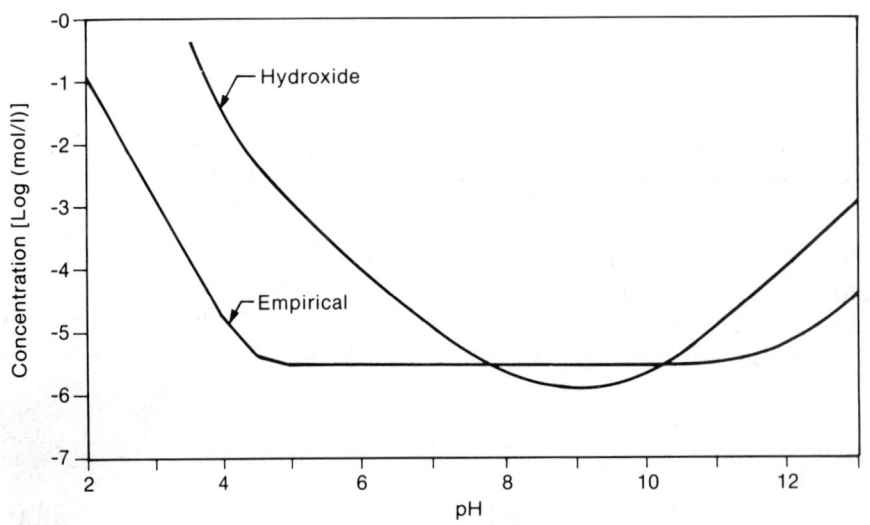

Figure 3-5 Solubility of chromium vs. pH in a cement/flyash CFS system. (From Cote [16].)

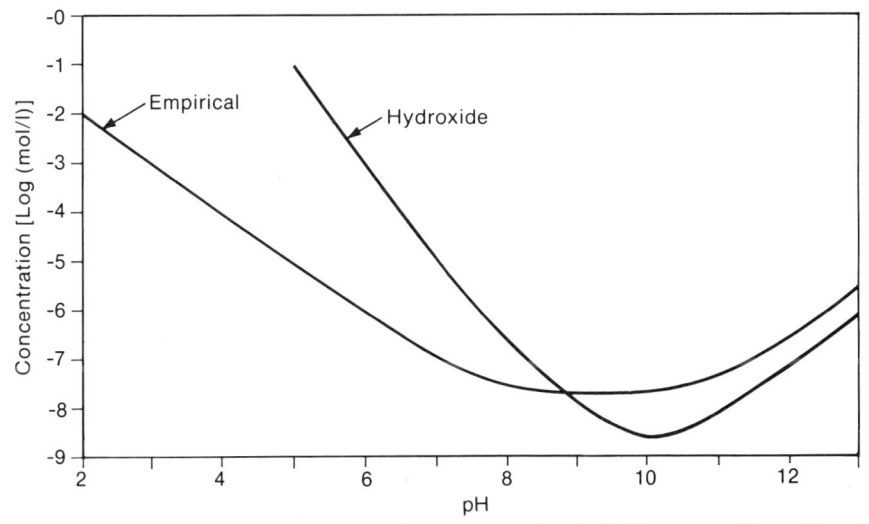

Figure 3-6 Solubility of lead vs. pH in a cement/flyash CFS system. (From Cote [16].)

as arsenic, both valence and speciation as either cation or anion can change easily with redox potential [18]. Of the metals of interest in nonnuclear CFS work, seven have more than one possible valence state in aqueous systems (As, Cr, Fe, Hg, Mn, Ni, and Se). Also, nitrogen and sulfur have multiple valence states that affect speciation of the metals in a given system. And Ag, Cu†, Cd, and Zn† can be strongly influenced by redox processes even though they have only one valence state in water [19].

The redox potential, E_h, is the oxidation-reduction potential referred to the hydrogen scale, expressed in millivolts. It is usually measured against another reference electrode for convenience, then converted to E_h. Redox potential is also referred to as oxidation-reduction potential (ORP). ORP measurement establishes the ratio of oxidants and reductants existing within the waste.

The effects of redox processes are commonly expressed as E_h–pH diagrams, such as that given for cadmium by Dragun [19] in Figure 3-7. The scale ranges from oxidizing (positive) to reducing (negative) potentials, and the resulting diagram shows the domains, or stability fields, in which different species can exist. Such diagrams are useful primarily for conceptual purposes in CFS systems, due to the complexity of the systems and the fact that the boundary lines are regions of transition rather than sharp delineations. Also, the diagram presupposes that the anionic species shown are in fact available in the system.

In the case of cadmium, the E_h–pH diagram shows that the metal forms CdS, a relatively insoluble species, only under reducing conditions but through a wide pH range. In more oxidizing conditions, Cd can form either the soluble cation at pH <8, or the very low solubility carbonate or hydroxide forms at pH >8. Knowledge of the redox potential of this system could allow estimation of the probable speciation of the metal and its leachability.

Unfortunately, relatively little attention has been paid to this factor by workers

---

†Cu and Zn are not currently regulated under the RCRA, but have been included in some state regulations.

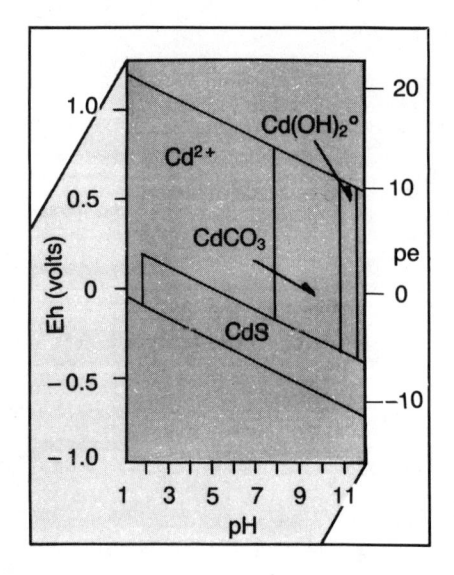

Figure 3-7 E_h–pH diagram for cadmium. (From Dragun [19].)

in the field. It is an area that needs much more elucidation, especially as stricter regulatory standards require fixation of metals such as arsenic to low leachability levels, even for disposal in secure landfills.

Chemical Reaction. We have already introduced one chemical reaction—hydroxide precipitation—under the subject of pH control. While this is undoubtedly the most important for metal fixation, others are also significant. Metals can be precipitated as carbonates, sulfides, silicates, sulfates, and other simple species, and also as complexes. These species will be discussed in detail in Chapter 4. Many nonmetal inorganic species can also be precipitated in various forms or destroyed chemically, but little work has been done in this area because there are only a few species listed as priority pollutants under the RCRA at present. Examples are amenable (free or soluble) cyanide, soluble sulfide, and fluoride. The former two species can be either destroyed or precipitated as low-solubility metal complexes. Fluoride can be removed as the calcium salt, which is naturally present in most CFS systems.

It has been stated that, in the case of fixation of metals by soils, the reactions are preceded by surface adsorption followed by dehydration and rearrangement of the solid phase [19]. Undoubtedly, many fixation reactions occur in several stages, involving two or more of the mechanisms described in this chapter. One reaction, chelation, which is known to take place in soils, may also be incorporated into CFS systems, or in some cases, may occur naturally. Chelation is usually thought of as contributing to leaching by the formation of soluble anionic complexes from metal cations. However, certain high-molecular-weight organic ligands, such as those of fulvic and humic acids, may contribute to fixation by adsorption onto solid particles.

Some organics can be changed chemically by commercial CFS systems alone or with minor additives. Examples are precipitation of phenolics as calcium salts, and various reactions, such as hydrolysis, that occur under strongly alkaline conditions. Recent work at Louisiana State [20] and elsewhere have opened up interesting possibilities in this area. Generally, the EPA and environmentalists feel that organics should be destroyed by thermal or other methods, since it is possible to do so, at least to fairly low levels. However, there is a growing need for CFS systems that can immobilize trace quantities of toxic organics in contaminated soils and residues from other processes such as incineration. Chapter 6 will explore the work to date and future possibilities for organic CFS in the context of the present and forthcoming landban requirements.

Adsorption. In a solid, molecules are held together by a variety of types and magnitudes of cohesive forces. Ionic, valence bonds are very strong. Covalent bonds are weaker, and van der Waals forces are weaker yet. At solid surfaces, cohesive forces are unbalanced, giving rise to the phenomenon of adsorption. Adsorption occurs when a molecule or ion becomes attached to a surface, usually as a monolayer (Fig. 3-1). Therefore, the adsorbing ability of a material or system is directly related to its surface area, which in turn is a function of both particle size and shape, and porosity. The phenomenon is reversible in general, but the ease and completeness of desorption vary with the nature of the solid surface, and depend on factors such as concentration and pH. Adsorption is also highly specific as to species; this aspect will be discussed further in Chapters 4–6.

Adsorption is commonly described in terms of the Langmuir adsorption isotherm [21]:

$$C_2^s = \frac{C_m^s C_2}{C_2 + a} \quad (3\text{-}5)$$

where

C_2^s = surface concentration of the adsorbed species, moles/cm^2

C_m^s = surface concentration of the adsorbed species, moles/cm^2 at monolayer adsorption

C_2 = concentration of the adsorbed species in the liquid phase at equilibrium, moles/l

a = constant $\left\{ 55.5 \exp\left[\dfrac{\Delta G^0}{RT}\right] \right\}$, moles/l, at temperature T where

ΔG^0 is the free energy of adsorption at infinite dilution.

Equation (3-5) is valid only under the following conditions:

1. Adsorbent is homogeneous
2. Both solute and solvent (water phase) have equal molar surface areas
3. Both phases exhibit ideal behavior
4. The adsorption film is monomolecular

Plots of log absorption of the adsorbed species versus log concentration in the liquid phase give the isotherm for a given system. The shape of the isotherm can give information about the actual processes taking place in the system, indicating one or more of the mechanisms by which adsorption takes place. Rosen [21] describes these for surfactants.

1. Ion exchange: Replacement of counterions adsorbed onto the surface from the solution by similarly charged ions [22].

2. Ion pairing: Adsorption of surfactant ions from solution onto oppositely charged sites unoccupied by counterions [22].
3. Hydrogen bonding: Adsorption by hydrogen bond formation between substrate and adsorbate.
4. Adsorption by polarization of pi electrons: Occurs when adsorbate contains electron-rich aromatic nuclei and adsorbent has strongly positive sites [23].
5. Adsorption by dispersion forces: Occurs via London–van der Waals dispersion forces. This mechanism is important because it acts not only alone but as a supplementary mechanism in all other types. Its degree increases with molecular weight of the adsorbate.
6. Hydrophobic bonding: Caused by aggregation of molecules via mutual attraction of their hydrophobic groups, which results in a stronger hydrophobicity of the aggregate and its subsequent adsorption onto the solid surface [24].

Many simple systems show Langmuir-type behavior, even when these conditions are not met. Whether or not many CFS systems would do so, the concepts and mechanistic models resulting from adsorption theory are valuable in exploring actual and possible immobilization of constituents in CFS systems. Another adsorption isotherm by Freundlich is an empirical relationship that, in its linear form, is more often used in the analysis of experimental data. This form is given in Eq. (3-6) [19]:

$$\log \frac{X}{m} = \log K + \frac{1}{n} \log C \qquad (3\text{-}6)$$

where

X = amount of species adsorbed per milligram of adsorbent, m
C = equilibrium concentration of the species
K = equilibrium constant indicative of the strength of adsorption
$1/n$ = constant describing the degree of nonlinearity

For many organic compounds, $1/n$ approaches unity and Eq. (3-6) becomes

$$\frac{X}{m} = K_s C \qquad (3\text{-}7)$$

where K_s is the the adsorption or distribution constant. Readers interested in exploring this aspect of "fixation" are referred to the references for this chapter and others given in the Bibliography. Specific applications of adsorption theory will be discussed in Chapters 4-6 when individual constituents are considered.

Adsorption is most effective at low concentrations of the species in solution where it is well described by Eqs. (3-6) and (3-7). Most research treats the phenomenon as one that reaches equilibrium rapidly. However, this is not necessarily true, especially with the desorption process. Therefore, the effectiveness of adsorption should be evaluated to take time into account.

Two major sorptive ingredients that can be used in CFS are activated carbon and clays. Adsorption on activated carbon is well documented in the literature [25]–[27], and will be discussed in more detail in Chapter 6. It takes place through forces of predominantly physical nature—van der Waals forces—and desorption is relatively easy. Most use of activated carbon has been for adsorption of organics, although the material will also adsorb metal ions and other inorganics [28]. From a cost efficiency standpoint, activated carbon is not normally a method of choice for the latter purpose due to the availability of cheaper, more effective substrates, such as natural and modified clays.

The use of clays and other substrates for adsorption has been described by Chan *et al.* [29] and Theis *et al.* [30]. Griffin *et al.* [31] found that adsorption of the cationic metals [lead, cadmium, zinc, copper, and chromium(III)] increased with increasing pH, while the opposite was true with anionic metals [chromium(VI), arsenic, and selenium]. Adsorption of cations by certain clays is caused by the negative surface charge that results from the clay structure itself. The resulting adsorption is not as reversible as that of simple surface adsorption [32], and the mechanism is more that of chemisorption than of Langmuir-type adsorption.

In surface layers of metal oxides, the metal ions have a reduced coordination number. In the presence of water, they develop an hydroxylated surface [33] with a net negative charge at high pH. This causes the adsorption of cations at any pH higher than that of zero point charge. The adsorption occurs before overall precipitation takes place, saturating the surface sites. This is a rapid but reversible reaction. Since metal oxides are present in most CFS systems in the waste or in the matrix, their adsorption properties are of interest.

Clays, because of their normally hydrophillic surfaces, are usually not good adsorbents for non-water-soluble organics. However, considerable recent work has been done on the modification of clays to create so-called "organoclays." Boyd *et al.* [34]–[36] have described various materials of this sort and their usefulness in fixation of organics. This work and others will be discussed further in Chapter 6.

Chemisorption. Close-range chemical or physical forces can cause the retention of one species by another, usually at the surface of the adsorbent. The binding forces are greater than those of pure adsorption, but do not represent a true chemical bond. There is no sharp boundary between adsorption and chemisorption, and some of the mechanisms described by Rosen [21] really can be categorized as the latter. Schematically, one possible difference between chemisorption and adsorption is illustrated in Figure 3-1. Functionally, the major difference is that chemisorbed species do not desorb easily. Further discussion of chemisorption will take place in subsequent chapters in conjunction with adsorption and chemical reactions of the individual species.

Passivation. This phenomenon is important in CFS treatment. Metal ions dissolving from a solid surface may precipitate on the surface after contacting an anion in solution that forms a less soluble species. If the precipitate forms a tight, impermeable layer, it may block or inhibit further reaction at the surface (Fig. 3-1). Such phenomena are highly specific to the constituents and the system as a whole, and may operate only temporarily. One such system is that of the formation of a silica gel on the surface of metal salts exposed to a soluble silicate solution. The gel layer prevents the passage of further metal ion into solution, thus slowing dissolution for a time. However, it also acts as a semipermeable membrane, allowing the passage of water that builds osmotic pressure on the solid side. Eventually, the membrane ruptures, spilling concentrated metal solution into the liquid phase. Depending on the relative concentrations of the constituents and their stoichiometry, and on pH, the metal may precipitate as a low-solubility silicate, as a hydroxide, or may remain in solution. In some cases, this process may repeat itself periodically, possibly giving rise to fluctuations in leaching rate with time.

Ion Exchange. A number of natural materials, such as soils and zeolites, as well as various synthetics, have the ability to exchange ions that they contain for others in solution. The most important are cation exchangers, and their ability to do this is expressed as cation exchange capacity (CEC), the magnitude of which is determined by the charge of the hydrated cation and its size in solution. Removal efficiency of an ion from solution is directly proportional to its charge and inversely proportional to its size. Organic synthetic ion exchangers contain acidic groups such as –COOH, –SO$_3$H and –OH. Natural and synthetic inorganic ion exchangers may exchange H$^+$ or a monovalent cation such as Na$^+$ for polyvalent metals in solution.

The charge on a clay surface may arise from either of two sources [19]: (1) ionization of the hydroxyl groups attached to silicon atoms at the surface, or (2) isomorphous substitution. The latter phenomenon is the result of substitution of Al^{+3} and Fe^{+3} for Si^{-4}, or Mg^{+2} for Al^{+3}. Isomorphous substitution results in a fairly uniform charge distribution over the surface of the clay particle in the "2:1"-class clays, such as montmorillonite, beidellite, and vermiculite. Two-to-one clays are composed of layers consisting of one alumina sheet between two sheets of silica. The other class of clays, "1:1" (one layer each of silica and alumina), have a lower surface charge resulting from hydroxyl group ionization.

The deliberate addition of ion exchange materials to CFS formulations has had limited use, but common reagents (clays and other soils, cements, and flyashes) often have limited ion exchange properties. Synthetic resins are usually too expensive for most CFS use, primarily because the resin will also exchange ions in solution other than the toxic metals, for example, calcium. This has another negative effect: it may inhibit solidification reactions, such as cement setting, by removing calcium before it can take part in the setting reactions (see Chapter 10).

Anion exchange also can take place. This type of reaction is favored by low pH and is much greater in 1:1-type clays. Dragun [19] describes anion exchange in soil clays. Specific ion exchange reactions will be discussed in Chapters 4 and 5 in connection with discussions of the individual species.

Diodochy. When one element substitutes for another of similar size and charge in a crystalline lattice, it is called *diodochy* [37]. This is shown schematically in Figure 3-1. Examples are zinc or manganese for calcium in calcite, or arsenic for phosphorus in apatite. Bishop [12] postulates that chromium and lead are bound into the silica matrix in Portland cement waste forms, possibly as "silicates." Whether these elements actually replace silicon in the lattice is questionable, but it appears that they coexist with silica, perhaps as hydrated metal oxides, at the molecular level rather than merely being microencapsulated. The latter mechanism is in keeping with theories about the actual structure of heavy metal silicates [38], [39]. It is similar to diodochy except that the substitution is by metal oxides embedded in the silica lattice rather than ion for ion replacement. Coprecipitation, as with iron in water treatment [40], might also be very similar.

Reprecipitation. This may be one of the most important containment mechanisms for metals in sludges, where they are originally speciated as hydroxides and then subjected to acid leaching. It is generally assumed that low-solubility species, such as hydroxides, do not respeciate. If this were the case, they should leach at rates approximated by the respective hydroxide solubilities at any given pH, especially under the conditions prevalent in a leaching test such as the TCLP. Even allowing for other fixation mechanisms such as those presented previously in this chapter, the shape of the leaching versus pH curve should be the same.

In fact, this is not the case in cement- and pozzolan-based systems. One explanation is that as the acid leachant dissolves alkali from the solid matrix and thus creates a localized acidic condition, metal hydroxide dissolves. Subsequently, the metal is reprecipitated by one of the mechanisms that we have just discussed. Another possibility is that the metal is reprecipitated before the leaching takes place during pozzolanic reactions or the formation of cement paste in cement-based systems. In both systems, similar reaction steps occur, the first of which is the formation of highly alkaline conditions as calcium dissolves from the cement or lime. This results in the dissolution of certain of the metal hydroxides (Fig. 3-2) at the same time that silica is being solublized as the silicate. The heavy metals then compete with the alkali earth metals, such as calcium, for the silicate, both precipitating the heavy metals and interfering with the normal setting reaction. The latter effect is known in cement chemistry [41].

Encapsulation. In the early days of CFS development, the word *encapsulation* was used to describe many of the processes. If encapsulation is the only operative mechanism, then the degree of protection to the environment is determined entirely by the matrix and its continuing ability to isolate the waste from the environment. Isolation, in turn, depends primarily on the permeability of the matrix. Encapsulation without any other fixation mechanism rarely occurs in commercial CFS operations, but it has been used to a limited degree in the case of organic polymer and bitumen CFS systems. Actually, there are three levels of encapsulation: micro-, macro-, and embeddment.

Microencapsulation. Most waste forms resulting from the CFS treatment of sludges and other suspended solids-containing wastes hold the waste particulate matter in a microencapsulated form. The sizes of the microencapsulated particle may range from very small occlusions to larger agglomerates visible under low-power light microscopy, or even to the naked eye. Individual particles or agglomerates retain their identities even after thorough mixing with high-energy devices. Recent work by Eaton *et al.* [42] using scanning electron microscopy has demonstrated the presence of microscopically concentrated regions of raw waste dispersed throughout the solidified matrix. This is to be expected in view of the original nature of the wastes. As the system ages, interaction between the waste and the matrix may eventually result in a homogeneous mass even at the microscopic scale, but this transformation might only occur on a geologic time scale.

Macroencapsulation. The term *macroencapsulation* has been used to describe processes in which a solid or cemented waste form is coated with, or contained in, an impervious layer to limit access of leachant to the waste itself. It is also sometimes used to describe the encapsulation of large waste agglomerates in a solidification matrix—a gross version of microencapsulation. The degree of protection attained depends completely on the matrix, coating, or container. The coating method has not been used in the hazardous waste field, except for nuclear wastes. However, the gross encapsulation of waste in a solidification matrix is quite common in actual commercial-scale CFS projects.

Embeddment. It is sometimes necessary or desirable to incorporate discrete waste masses of discernible size into a solid matrix before disposal. Examples are contaminated trash and debris from remedial actions; contaminated personal protective equipment from laboratories, from environmental investigations, or at hazardous waste management facilities; medical laboratory solid wastes (syringes, etc.); and highly radioactive objects. In these cases, it may be unsafe or impractical to reduce the size of the mass, but it is necessary to limit the mobility of its hazardous constituents. The CFS formulation used for such a matrix may require special properties, such as high cohesive strength, low permeability to water, and proven long-term stability. One matrix used for embeddment in many such situations is concrete, reinforced if necessary.

Alteration of Waste Properties. Frequently, the highest degree of containment can be obtained by pretreating the waste in some manner, altering its properties before CFS. Examples include reduction of hexavalent chromium to the less soluble trivalent state, destruction of cyanide, and breaking of soluble nickel complexes. In some cases, particle-size reduction to allow better reaction of the waste with the reagents may be useful. In others, dewatering may provide a better final waste form at lower cost. Often, removal of immiscible organic phases is necessary to achieve acceptable leaching and meet recent landbans based on total constituent analysis. Acids are usually neutralized, at least partially, prior to CFS. In some cases, it may be necessary to lower the pH of caustic wastes.

The ultimate in waste property alteration is to immobilize the constituents of concern *before* the CFS treatment is applied. To some degree, this is already done with waste water treatment sludges, where solubility must be reduced to a low level so that the water phase is suitable for discharge to a Publicly Owned Treatment Works (POTW) (sewage treatment plant) or to a stream under a National Pollutant Discharge Elimination System (NPDES) permit (the U.S. permit system for point source discharge to any waterway). However, these residues are typically metal hydroxides with the pH sensitivity already discussed briefly in this chapter. Pretreatment can, however, speciate the metals as silicates or other forms less susceptible to acid attack. Organics can be removed or destroyed, and anions can be destroyed or insolublized. Subsequent CFS treatment will then function to create a stable solid and to provide an extra measure of protection against leaching. Obviously, this approach is expensive, but it is already being used to some degree in central hazardous waste management facilities. One example is the proposed facility to be built and operated by Ontario Waste Management Corporation (a Crown corporation) in the Province of Ontario, Canada [43]. This plant will use a very complete and complex pretreatment unit to prepare waste streams for CFS. The subject of pretreatment will be brought up again in Chapter 8 from a different viewpoint. Knowledge and understanding of the waste itself is vital to proper selection and use of CFS technology.

Leaching Mechanisms

A list of the main mechanisms that are known to cause leaching of constituents from a waste form was presented earlier in this chapter. Figure 3-8 illustrates most

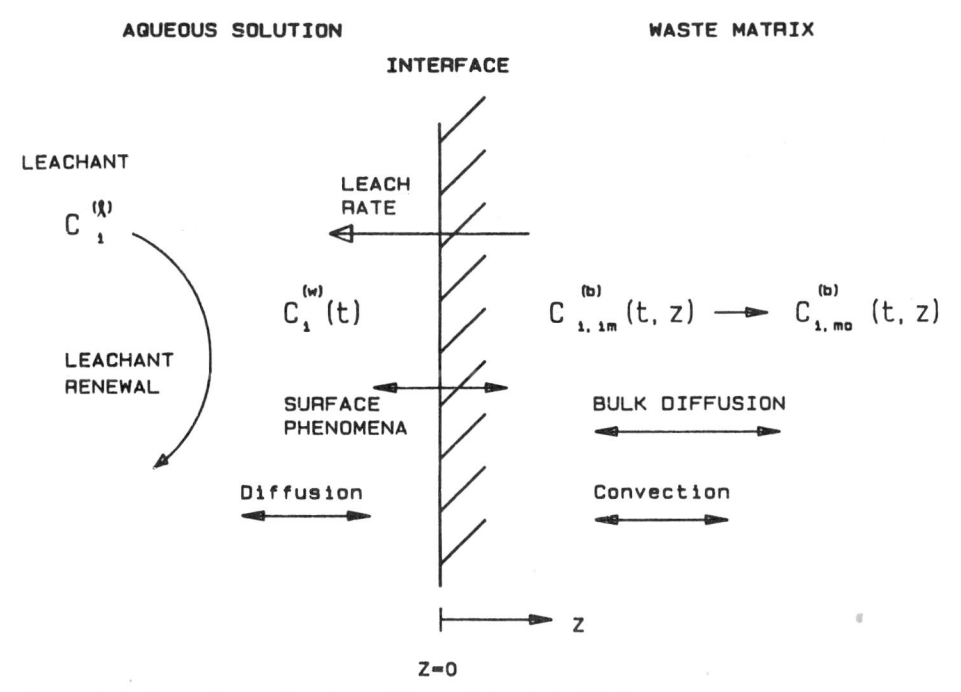

```
                    AQUEOUS SOLUTION                    WASTE MATRIX
                                    INTERFACE

       LEACHANT
                            LEACH
          (ℓ)               RATE
        C
         i

                        (w)                    (b)              (b)
                      C  (t)                  C      (t, z)  →  C      (t, z)
       LEACHANT        i                       i,im              i,mo
       RENEWAL
                                  SURFACE              BULK DIFFUSION
                                  PHENOMENA            ←—————→

                              Diffusion                Convection
                              ←—————→                  ←—————→

                                              ├——→  z
                                      z=0
```

Figure 3-8 Leaching mechanisms. (From Cote et al. [44].)

of them schematically. Cote et al. [44] present the solid-state mechanisms in another way by plotting the various concentration gradients versus position in the solid matrix (Fig. 3-9). A solidified waste is a porous solid at least partially saturated with water. It consists of one or more solid phases, entrapped air in the form of air voids, and a liquid phase called the pore solution—all in chemical equilibrium (or close to it). When the solid is exposed to leaching conditions, equilibrium is disturbed. The resulting difference in chemical potential between the solid and the leaching solution causes a mass flux between the solid surface and the leachant. This in turn causes concentration gradients (Fig. 3-9) that induce bulk diffusion in the solid. Chemical reaction can also occur between constituents of the solid and of the leachant. Various submechanisms are in action in the solid, the various interfacial surfaces, and the bulk leachant.

Referring again to the list of containment systems, we can now discuss the various mechanisms individually.

Solubilization. For any constituent to leach, it must first dissolve in the pore water of the solid matrix or in the leachant permeating the solid. Some species dissolve more slowly than others, with the rate of solubilization being controlled both by basic solubility considerations and by the concentration in the solution near the surface. The latter, in turn, is controlled by boundary layer transport phenomena. The thinner the boundary layer, the faster the rate of dissolution, until the bulk leachant is saturated.

For most CFS matrices in contact with neutral groundwater, the leaching rate will be controlled by molecular diffusion of the solubilized species. Under acidic conditions, the rate is initially limited by the supply of H^+. After a time, molecular or boundary layer diffusion of the waste constituents again becomes rate limiting, since H^+ diffuses much faster than any other species [16].

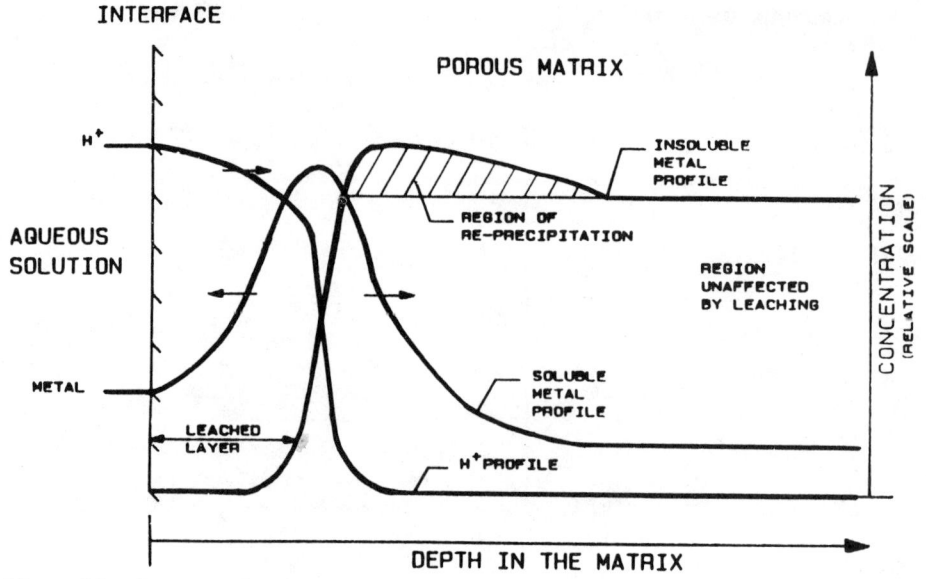

Figure 3-9 Concentration gradients during leaching. (From Cote *et al.* [44].)

Even in the case of a finely divided solid and a rapidly agitated leaching solution as, for example, in the EPT or TCLP test methodologies, the leachant requires 18 to 24 hours to approach equilibrium constituent concentrations for metals. Under real leaching conditions, equilibrium is attained much more slowly. If the leachant flows slowly around or through the waste form, near equilibrium conditions may develop. Under faster flow conditions, transport of constituents may be solubility limited. In the former case, molecular diffusion through the pore water will probably be the limiting factor, since it is likely to be more rapid than true solid-state diffusion. This scenario is the more likely one in a properly managed hazardous waste landfill, for example, one meeting RCRA minimum technology requirements (MTR). This is true, however, only after the landfill cell is closed so that water inflow is minimal. When the cell is "open," that is, being filled, excessive rainfall can create rapid flow conditions through the waste, and the leachate will not become saturated.

Transport Through the Solid. Within the solid waste form, transport could occur either by convection or by diffusion. Most waste forms have relatively low permeability, so convective transport is probably not important in either a leaching test or in actual field conditions. Therefore, diffusion is the only effective process operating.

Molecular diffusion. Diffusion occurs by the random motion of individual molecules or ions. Assuming that the solid is in chemical equilibrium when leaching begins, diffusion is driven by the difference in chemical potential (constituent concentration) between the solid and the fluid leachant. The chemical gradient thus created causes constituents to migrate from the solid to the leachant. The flux of the constituent at a position within the solid is described by Fick's first law:

$$J = -D\frac{dC}{dz} \qquad (3\text{-}8)$$

where

    C = concentration (mass/unit volume of solution)
    D = diffusion coefficient (area of solution/time)
    J = flux (mass/area of solution x time)
    z = distance (L)

The "area of solution" for a porous solid is not known, so the flux is normally expressed on an "area of porous solid" basis. Cote [16] has treated this more rigorously, including a term for connected porosity, and also a modification for tortuosity. The latter term expresses the fact that diffusion in a porous solid "takes place in a tortuous path of fluid between and around solid particles." The product of these modifications, taking into account simplifying assumptions, gives us the following equation for diffusion in a porous solid:

$$\frac{dC}{dt} = D_s \frac{d^2C}{dz^2} \qquad (3\text{-}9)$$

where

    t = time
    D_s = molecular coefficient of diffusion corrected for tortuosity of the matrix (L^2/T)
    T = tortuosity, defined as the ratio of the length of the actual simuous path over a depth interval

Since this equation assumes that the constituent concentration in the leachant does not increase, it most closely approximates a dynamic leaching scenario. Also, for the remainder of this discussion it will be assumed that diffusion is through the pore solution rather than the solid matrix itself. True solid-state diffusion is slow compared to diffusion through the solution contained in the interconnected pores, so this assumption is justified.

A widely accepted model for the leaching of CFS-treated wastes was proposed by Godbee *et al.* [45]:

$$\sum \frac{-a_n}{A_0} \frac{V}{S} = \left[2 \frac{D_e}{\pi} \right]^{0.5} t_n^{0.5} \qquad (3\text{-}10)$$

where

    a_n = contaminant loss during leaching period n, mg
    A_0 = initial amount of a contaminant present in the specimen, mg
    V = volume of the specimen, cm^3
    S = surface area of the specimen, cm^2
    t_n = time to the end of the leaching period p, s
    D_e = effective diffusion coefficient, cm^2/s

Assumptions for this model are that the contaminant concentration in the initial solid is uniform, leaching is controlled by diffusion through the solid waste form (but not solid-state diffusion as the term is used here), and that there is always zero surface concentration at the solid–liquid interface. As we shall see later, the latter assumption may not be justified in real leaching situations.

Godbee's model can be combined with the formula used in the American Nuclear Society's (ANS) leaching protocol [46], which is a series of seven sequential batch leachings. The results of the ANS leaching test are presented as the leachability index, LX:

$$LX = \frac{1}{7} \sum_{1}^{7} \log \left[\frac{B}{D_e} \right]_n \qquad (3\text{-}11)$$

When combined with Godbee's model, we get

$$\sum \frac{a_n}{A_0} = (1.128)(10^{-0.5L})(t_n^{0.5}) \left[\frac{S}{V} \right] \qquad (3\text{-}12)$$

Most of the variables that effect metal leaching from CFS-treated waste are accounted for by Eq. (3-12). These include constituent speciation, particle size, and the initial concentration of the constituent. However, boundary layer effects and chemical reactions occurring during the leaching process can significantly affect the practical utility of such mathematical models. These effects are discussed later in this chapter.

A mathematical model such as Eq. (3-12) can be used to determine leaching rates that, when coupled with data about the disposal site and groundwater, can be used to predict in situ leaching rates over extended time periods [12]. A number of other models have been developed and used for such predictions, including the EPA VHS (vertical and horizontal spread) model described in more detail in Chapter 17. All mathematical models, however, should be used with caution because of the simplifications used in their development, and because of other factors not considered nor even recognized at the present stage of development.

Diffusion Through the Leached Layer. As more soluble constituents are leached from a relatively insoluble solid matrix, a layer deficient in the leached constituents develops [12], [16]. Under low pH conditions, both H^+ and the leachable constituents must diffuse through this layer in opposite directions. As we saw earlier, the leaching rate in the leached layer should eventually be limited by diffusion of constituents, since H^+ diffuses much faster than other species. However, this layer may not be rate-limiting in the overall leaching process. As constituents leach, the layer may become more porous compared to the unleached solid, so that molecular diffusion in the pore water and boundary layer phenomena become the limiting factors.

Transport Through the Solid–Liquid Boundary Layer. Throughout the discussion of constituent transport, we have ignored the boundary layer at the solid–liquid interface. In many leaching scenarios, however, this layer may create the limiting condition for leaching, as it does in electochemical phenomena. The speed with which a system comes to equilibrium is frequently limited by the thickness and nature of the boundary layer. In attempting to develop fast leaching assessment tests for routine quality control, for example, this limitation is thought to be controlling (P. L. Cote, private communication).

The solid–liquid boundary layer is frequently described as the "electrical double layer." At any interface between two phases, there is an unequal distribution of electrical charges, with a net negative charge on one side and a corresponding net positive charge on the other. Because overall electrical neutrality must be maintained, the net charges must equal each other. This localized unequal distribution gives rise to a potential across the interface. The phenomenon was first treated mathematically by Helmholtz [47], later modified to the Gouy-Chapman model, and finally by Stern [48], whose model is still used today. Stern's model is shown in Fig. 3-10. He divided the solution side of the double layer into two sublayers: (1) an inner region —now called the "Stern layer"—of strongly adsorbed counterions about one hydrated ion radius thick at fixed sites, corresponding to Langmuir-type adsorption, and (2) a diffuse region of counterions distributed according to electrical forces and thermal motion. The electrical potential drops rapidly through the Stern layer, possibly even changing the sign resulting from the original charged surface, and then more gradually through the diffuse layer. The shear plane between the surface and the surrounding solution is somewhere in the diffuse layer, and the potential difference between the shear plane and the solution is known as the zeta, or electrokinetic, potential. Zeta potential is a measurable value that is widely used in colloid and surface chemistry.

The effective thickness of the diffuse portion of the double layer, called the Debye length, is given by Eq. (3-13):

$$\frac{1}{k} = \left[\frac{e_r e_0 RT}{4\pi F^2 \ \Sigma C_i Z_i^2} \right]^{1/2} \tag{3-13}$$

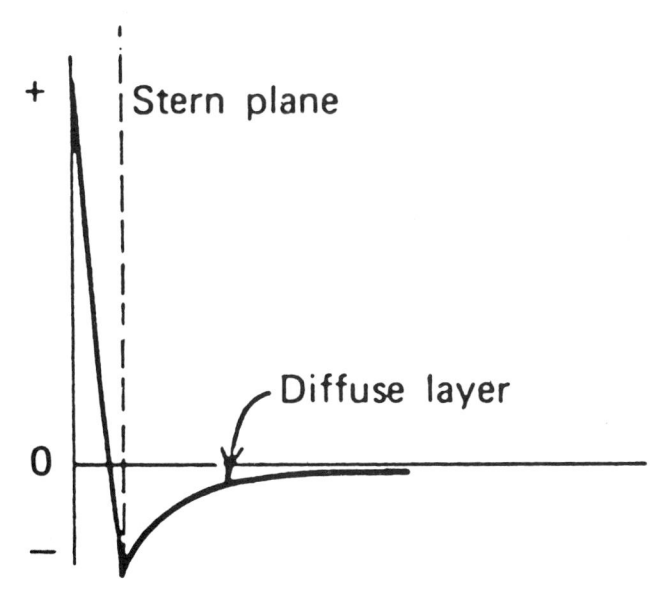

Figure 3-10 Stern model of the electrical double layer. (From Rosen [21].)

where

$e_r = (e/e_0)$ = the relative static permitivity or dielectric constant of the solution e = static permitivity of the solution and e_0 = the permitivity of a vacuum)
R = the gas constant
T = the absolute temperature
F = the Faraday constant
C_1 = molar concentration of any ion in the solution phase
Z_1 = valence of any ion in the solution phase

The Debye length is important to our discussion of leaching, since any species moving from the solid into solution must traverse this layer. From Eq. (3-13), it is apparent that the Debye length is inversely proportional to the valence of ions in the solution, and directly proportional to the square roots of the absolute temperature and dielectric constant. In pure water, with a high dielectric strength, electrical effects extend relatively far into solution; in the presence of an electrolyte, the layer is compressed.

Grutsch [40] has described a model that incorporates a number of these concepts, shown here in Figure 3-11. While we may not be able to determine from such models what the leaching rate will be in a given system, the concepts are useful in understanding the factors that affect it. They are also important in the fixation mechanisms that we have outlined in this chapter and will discuss further later on.

Bulk Transport in the Liquid Leachant. The effects of bulk transport in the leachant—or in the environment or the groundwater—depend primarily on the hydraulic regime. In batch tests, the system approaches equilibrium and the leaching rate decreases as the chemical potential between the solid and the solution phases (the driving force) decreases to near zero. This may also occur in a landfill that is saturated, but with no movement of groundwater. It results in the highest possible concentration of constituents in the leachate, but the overall leaching rate may be lower than in a dynamic system. Fast-moving groundwater (or a column leaching test in the laboratory), on the other hand, will transfer constituents from the solid as rapidly as the solid and boundary layer transport mechanisms will allow. With a highly soluble constituent in a porous matrix, leaching rate will be very rapid.

The various possibilities may be visualized as Cote [16] has done. He described "four master variables . . . [which] . . . develop long term leaching scenarios: hydraulic regime of the groundwater, chemical characteristics of the groundwater, permeability of the waste form and chemical speciation of the contaminant." From this, he has developed eight leaching scenarios and predicted the leaching rate as a function of time for each scenario. These are shown schematically in Figure 3-12 and described briefly below.

Master Variables: Hydraulic regime, groundwater chemistry, permeability, mobility (chemical speciation of contaminant)
Scenario A: Static groundwater. Leaching decreases rapidly with time; groundwater chemistry, permeability, and mobility are not important.
Scenario B: Groundwater flowing around waste, soluble contaminant. Leaching is

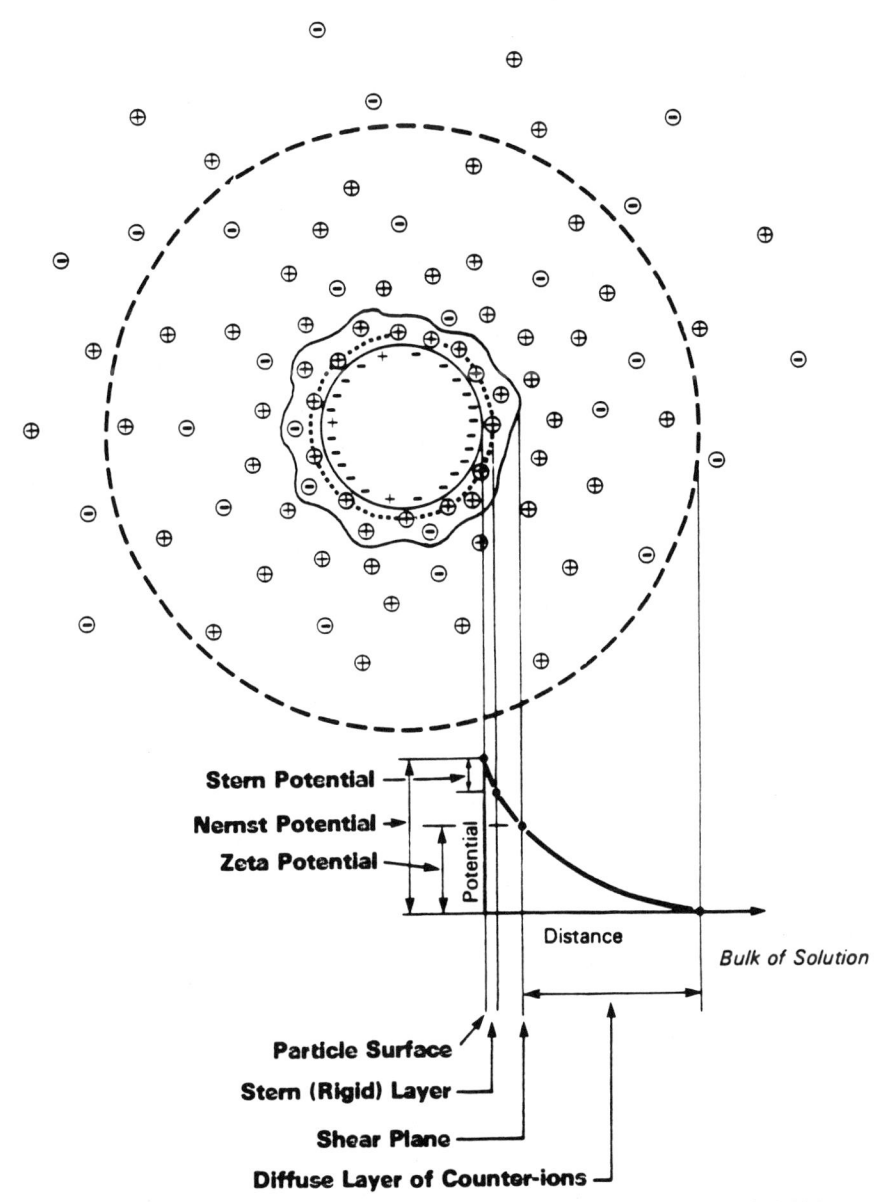

Figure 3-11 Representation of the electric double layer. (From Grutsch [40].)

diffusion-limited and decreases from high to moderate with time; independent of groundwater chemistry, permeability.

Scenario C: Groundwater flowing around waste, insoluble contaminant under acidic leaching conditions. Leaching initially controlled at moderate level by available acid. Diffusion at waste–groundwater interface eventually becomes rate-limiting and reduces rate.

Scenario D: Groundwater flowing around waste, insoluble contaminant under neutral leaching conditions. Leaching controlled by diffusion in waste matrix.

54   PRINCIPLES OF FIXATION

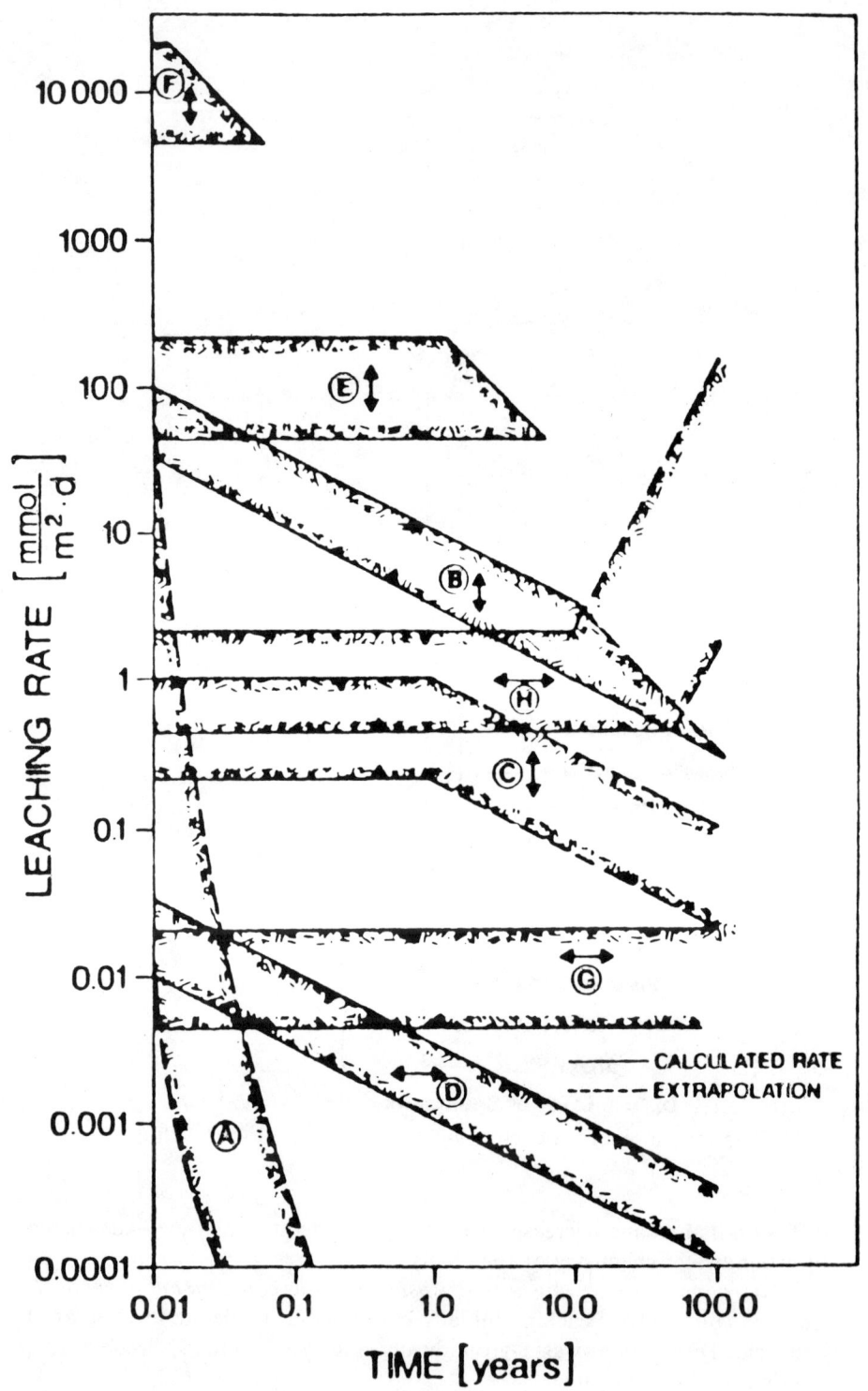

Figure 3-12 Inference of leaching rates following different disposal scenarios. (From Cote [16].)

Chemical equilibrium of contaminants maintained at all times. Low initial leach rate decreasing with time.

Scenario E: Groundwater flowing through the waste, soluble contaminant, low permeability. Leaching controlled by convective transport, stops when matrix is emptied of contaminants at an early stage. Leach rate is high and constant.

Scenario F: Groundwater flowing through the waste, soluble contaminant, high matrix permeability. Similar to Scenario E, except more rapid leaching.

Scenario G: Groundwater flowing through waste, insoluble contaminant, low matrix permeability. Chemical equilibrium of contaminants ensures constant leach rate at low level.

Scenario H: Groundwater flowing through waste, insoluble contaminant, high matrix permeability. Similar to Scenario G, but higher (moderate) leach rate.

They give an interesting, relative view of the overall behavior of various waste forms under different field conditions.

Chemical Reactions. The various scenarios just described and the leaching mechanisms that we have discussed to this point assume that no chemical reactions (other than those involved in dissolution of the constituents in the solid) occur. While this might be true in laboratory leaching tests, since they were so designed, it is not the case in the real environment. Rain, surface-, and groundwaters all contain constituents that may increase or decrease the leaching rate. Redox potential, pH, anions such as carbonate, sulfide, and silicate, organic chelating agents, and adsorptive particulates, all can affect leaching. Precipitates may passivate waste form particles, slowing or completely blocking transport through the solid or the boundary layer by clogging up the pores. Alkalinity in the waste form may neutralize acid leachants, preventing them from solubilizing acid-labile constituents.

On the other hand, chelating agents can increase solubility by preventing precipitation and converting ionic species into soluble organic complexes, preventing saturation in the leachant of the ionic species. Oxidizing conditions can increase the solubilization rate of some species, while a reducing environment may do the same for others. Some of these reactions will be discussed in the following chapters when we look at the chemistry of individual constituent species.

Biological Attack. In the long term, potential biological attack of CFS waste forms is a matter of concern to regulatory agencies. Most CFS systems produce conditions both of high pH and high total alkalinity. These conditions are not conducive to the activity or even survival of most microorganisms, as demonstrated by the long-standing practice of "stabilizing" sewage sludges by addition of lime. Organic CFS systems, such as urea-formaldehyde (UF), on the other hand, have been shown to be biodegradable [10].

For most waste forms, the primary concern is that of indirect attack through the known biological mechanisms that produce acids that in turn can attack certain waste constituents or even the matrix itself. Plant roots produce carbonic acid when alive, and organic acids when decaying. Plant roots, however, are not normally in contact with hazardous wastes, treated or otherwise, in properly managed disposal scenarios.

A more reasonable concern is that of the biologically enhanced [49] oxidation of insoluble metal sulfides to soluble sulfates, releasing the metal and also generating sulfuric acid that can attack the matrix. This reaction is partly responsible for the acid mine drainage from iron sulfides. It requires, however, oxidizing conditions that are not normally present in hazardous waste landfills after completion of the cover system. In uncontrolled sites, these conditions could occur near the surface, especially with low-alkalinity systems. In any case, no such process has ever been observed with inorganic waste forms produced by CFS treatment.

REFERENCES

1. Safety of Public Water Systems (Safe Drinking Water Act). Public Law 93-523. Washington, DC: 93rd Congress, Dec. 16, 1974.
2. U.S. EPA Extraction Procedure. Toxicity Test Federal Register, 40CFR Part 261.24, Appendix II (May 19, 1980).
3. Darcel, F. *Recent Studies on Leach Testing—A Review.* Ontario MOE, 1984.
4. U.S. EPA. *Solid Waste Leaching Procedure Manual.* SW-924. Cincinnati, 1985.
5. U.S. EPA. Federal Register **47**(225), 52687 (Nov. 22, 1982).
6. U.S. EPA. Federal Register. **49**(206), 42591 (Oct. 23, 1984).
7. Cote, P. L., and D. Isabel. Presented at the 3rd Hazardous and Industrial Waste Management and Testing Symposium. Philadelphia: ASTM, pp. 48-60, 1984.
8. U.S. EPA. *Guide to the Disposal of Chemically Stabilized and Solidified Waste.* SW 872. Washington, DC, 1982.
9. U.S. EPA. Federal Register. 45CFR: 57332 (Au2g. 27, 1980).
10. U.S. EPA. Federal Register. **51**(114), 21686 (June 13, 1986).
11. Lowenbach, W. A. *Compilation and Evaluation of Leaching Test Methods.* EPA-600/2-78-095. Cincinnati, 1978.
12. Bishop, P. L. Leaching of inorganic hazardous constituents from stabilized/solidified hazardous wastes. *Hazardous Waste Hazardous Mater.* **5**(2): 129-143 (1988).
13. Bause, D. E., and K. T. McGregor. *Comparison of Four Leachate-generation Procedures for Solid Waste Characterization in Environmental Assessment Programs.* EPA-600/7-80-118. Cincinnati, 1980.
14. U.S. EPA.. *Test Methods for Evaluating Solid Waste.* SW-846. Washington, DC: Office of Solid Waste and Emergency Response, 1986.
15. Malone, P., and R. Larson. Symposium Poster Presentation. Presented at the 4th Annual Hazardous Waste Symposium, Atlanta, GA, May 1987.
16. Cote, P. *Contaminant Leaching from Cement-Based Waste Forms Under Acidic Conditions.* Ph.D. Thesis, McMaster Univ., Hamilton, Ont., Canada, 1986.
17. U.S. EPA. Federal Register. **52**(155): 29999 (Aug. 12, 1987).
18. Moore, J. N., W. H. Ficklin, and C. Johns. Partitioning of arsenic and metals in reducing sulfidic sediments. *Environ. Sci. Technol.* **22**:432-437 (1988).
19. Dragun, J. The fate of hazardous materials in soil. *Hazardous Mater. Contr.:* 41-65 (May/June 1988).
20. Tittlebaum, M., F. Cartledge, and H. Eaton. Applicability of solidification to organic wastes. Louisiana State Univ., Baton Rouge, 1988.
21. Rosen, M. J. *Surfactants and Interfacial Phenomena.* New York: Wiley-Interscience, 1978.
22. Law, J. P., and G. W. Kunze. *Soil Sci. Soc. Am. Proc.* **30**: 321 (1966).
23. Snyder, L. R. *J. Phys. Chem.* **72**: 489 (1968).
24. Giles, C. H., A. P. D'Silva, and I. A. Easton. *J. Colloid Interface Sci.* **47**: 766 (1974).
25. Chereminisoff, P. N., and F. Ellerbusch. *Carbon Adsorption Handbook.* Ann Arbor, MI. Ann Arbor Science Publishers, 1980.
26. Perrich, J. R., Ed. *Activated Carbon Adsorption for Wastewater Treatment.* Boca Raton, FL: CRC Press, 1981.

27. Suffet, I. H., and M. J. McGuire. *Activated Carbon Adsorption of Organics from the Aqueous Phase.* Ann Arbor, MI: Ann Arbor Science Publishers, 1981.
28. Huang, C. P., P. K. Wirth, and D. W. Blankenship. Removal of Cd(II) and Hg(II) by activated carbon. In *Proc. 2nd Natl. Conf. on Environmental Engineering.* Atlanta: ASCE, pp. 382–390, 1981.
29. Chan, P. C., J. W. Liskowitz, A. Perna, and P. Trattner. *Evaluation of Sorbents for Industrial Sludge Leachate Treatment.* Cincinnati: U.S. EPA, EPA-600/2-80-052, 1980.
30. Theis, T. L., J. L. Wirth, R. O. Richter, and J. J. Marley. Sorptive characteristics of heavy metals in fly-ash-soil environments. In *Proc. 31st Industrial Waste Conference.* West Lafayette, IN: Purdue University, 1976.
31. Griffin, R. A., R. R. Frost, and N. F. Shimp. In *Effect of pH on Removal of Heavy Metals from Leachates by Clay Minerals.* Urbana, IL: Illinois State Geological Survey.
32. Kinniburgh, D. G., and M. L. Jackson. In *Effect of pH on Removal of Heavy Metals from Leachates by Clay Minerals.* Urbana, IL: Illinois State Geological Survey.
33. Schindler, P. W. In *Adsorption of Inorganics at a Solid-Liquid Interface.* Ann Arbor, MI: Ann Arbor Science Publishers, 1981.
34. Boyd, S. A., M. M. Mortland, and C. T. Chiou. Sorption characteristics of organic compounds on hexadecyltrimethylammonium-smectite. *Soil Sci. Soc. Am. J.* **52**:652–657 (1988).
35. Boyd, S. A., S. Shaobai, J. Lee, and M. M. Mortland. Pentachlorophenol sorption by organo-clays. *Clays Clay Miner.* **36**(2): 125–130 (1988).
36. Boyd, S. A., J. F. Lee, and M. M. Mortland. Attenuating organic contaminant mobility by soil modification. *Nature* **333**(6171): 345–347 (1988).
37. Immobilization and leachability of hazardous wastes. *Environ. Sci. Technol.* **16**(4): 219a–223a (1982).
38. Vail, J. G. *Soluble Silicates,* Vol. 1. New York: Van Nostrand Reinhold, 1952.
39. Iler, R. K. *The Chemistry of Silica.* New York: Wiley, 1979.
40. Grutsch, J. F. *The Chemistry and Chemicals of Coagulation and Flocculation.* In *American Petroleum Institute Manual on Disposal of Refinery Wastes,* Chicago, 1983.
41. Cullinane, M. J., R. M. Bricka, and N. R. Francingues, Jr. An assessment of materials that interfere with stabilization/solidification processes. In *Proc. 13th Annual Research Symposium* (Cincinnati), pp. 64–71, 1987.
42. Eaton, H. C., M. B. Walsh, M. E. Tittlebaum, F. K. Cartledge, and D. Chalasani. Microscopic characterization of the solidification/stabilization of organic hazardous wastes. *Energy-Sorces and Technology Conference and Exhibition.* ASME Paper No. 85-Pet-4 (1985).
43. Ontario Waste Management Corp. *Facilities Development Process,* Phase I Report. Toronto, Ont., Canada: (1982).
44. Cote, P. L., T. R. Bridle, and A. Benedek. An approach for evaluating long term leachability from measurement of intrinsic waste properties. In *Proc. 3rd International Symposium on Industrial and Hazardous Waste* (Alexandria, Egypt, 1985).
45. Godbee, H., et al. Application of mass transport theory to the leaching of radionuclides from solid waste. *Nucl. Chem. Waste Manage.* **1**:29 (1980).
46. American Nuclear Society. *Measurement of the Leachability of Solidified Low-Level Radioactive Wastes.* (1981).
47. Von Helmholtz, H. *Wiederherstellungschir. Ann. Phys.* **7**: 337 (1879).
48. Stern, O. *Z. Electrochem.* **30**: 508 (1924).
49. U.S. EPA. *Inorganic Sulfur Oxidation by Iron Oxidizing Bacteria.* Washington, DC: U.S. Government Printing Office, June 1971.

Chapter 4

FIXATION OF METALS

More is known about metal fixation than about the fixation, destruction, and immobilization of any other hazardous constituent group encountered in CFS technology. Metals are the only really hazardous constituents that cannot be destroyed or altered by chemical or thermal methods, and so must be converted into the most insoluble form possible to prevent their reentry into the environment. While this can be done with all metals, the difficulty and cost of such treatment varies greatly with the form of the metal in the waste—its speciation—as well as with the amount present.

The discussion of metal fixation can be approached systematically from several different viewpoints: by metal, by species in the raw waste, by fixation mechanism or process, or by waste type. The following lists contain the various parameters under the first three headings; waste types are discussed separately in Chapter 8. Column *a* in the first list contains metals that are regulated by federal or state agencies in the United States.* These are the metals of primary concern in CFS technology, and are commonly referred to as the "toxic metals" or "RCRA metals." Column *b* lists additional metals of possible concern in certain situations.

a. Regulated Metals
Antimony
Arsenic†
Barium†
Beryllium
Cadmium†
Chromium†
Cobalt
Copper
Lead†
Mercury†

b. Other Metals
Aluminum
Magnesium
Manganese
Potassium
Sodium
Tin

---

*This discussion excludes metals and other constituents regulated by the Nuclear Regulatory Commission. Radioactive nuclides are normally not present in industrial and domestic wastes. They have different chemistries, are regulated at different levels and by different test methods. Chapter 15 gives a brief description of nuclear waste CFS.
†RCRA metals [1].

Molybdenum
Nickel†
Selenium†
Silver†
Thallium
Vanadium
Zinc

Next we list the types of metal compounds that may be found in waste streams. The asterisk denotes the more common species. This listing does not include anions containing the toxic metals themselves, or individual organic complexes.

Acetates
Bromides
Carbonates*
Chlorides*
Citrates
Cyanides*
Ferricyanides
Ferrocyanides
Fluorides
Hydroxides*
Iodides
Nitrates*
Oxalates
Oxides*
Phosphates*
Silicates*
Sulfates*
Sulfides*
Organic complexes*

The various categories of fixation mechanisms or processes are given in the following list. This is similar to the listing of general fixation mechanisms given in the last chapter, but here it is specific to *metal* fixation. These mechanisms and processes are those that come under the general technical category of fixation, or that can be included integrally or as pretreatments in CFS systems. In their forms discussed here, they are also in the realm of economic and operational feasibility. More exotic techniques that may be practical in areas such as waste water treatment and metal recovery are usually not applicable to residue treatment.

- pH control
- Redox potential control
- Precipitation
    Hydroxide/oxide
    Sulfide
    Silicate
    Carbonate
    Phosphate

Coprecipitation
  Inorganic complexation
  Organic complexation
- Bonding to an insoluble substrate
- Sorption and chemisorption
- Passivation
- Ion exchange
- Diadochy
- Encapsulation
  Microencapsulation
  Macroencapsulation
  Embeddment
- Extraction
- Miscellaneous

The reactions of various species and the available mechanisms for a given metal will be examined later in this chapter when we discuss each of the metals individually. Before doing so, however, there are general principles and practices that are common to many or all metals.

METALS OF ENVIRONMENTAL CONCERN—AND WHY

It seems obvious today why certain metals are of environmental concern. They are hazardous in one way or another to humans and/or to other forms of life. The hazard to humans may be in the form of acute or chronic toxicity, or the metal may act in more subtle ways, causing cancer and other secondary-effect diseases or damage to fetuses. Aquatic organisms in fresh water or marine environments are often extremely sensitive to very small concentrations of metals. Plant growth may be adversely affected by metals in soils and irrigation water, or may concentrate them in leaves, stems, or roots where they can subsequently affect the food chain. Mammals may be affected both through the food chain and by drinking contaminated water. On the other hand, trace amounts of many metals, including the "toxic metals," are vital to the human diet and that of lower forms of life. Metals exist naturally in the environment, originating from the leaching of soils by rain, ground-, and surface water. So, the question is one of degree: How much of a constituent is harmful, to what organisms, and in what media? In the context of the CFS field, the primary medium is usually groundwater, and the organism of most concern is man.

The regulated metals listed previously have their regulatory basis in drinking water standards, more specifically those promulgated under the Safe Drinking Water Act of 1974 [2] and the revisions of 1986 [3]. Allowable leaching limits were developed from models that attempt to determine attenuation factors for leachate migrating from a landfill, and thus are multiples of the drinking water limits. Other factors may also be taken into account in setting standards for specific waste disposal scenarios. Most recently, in the "first-third landban" rule [4], the allowable leaching levels for various metals in a number of "listed" wastes were based on what was achievable with best demonstrated available technology (BDAT). These levels are generally lower than those set previously, and are used within the context of disposal in a minimum technology unit (MTU), which is a state-of-the-art secure landfill.

The various standards were previously listed in Table 3-1. The setting of these leaching standards is the subject of intense, ongoing debate between government, industry, and environmentalists. As new toxicology information is developed, and more sensitive analytical procedures become available, the allowable leachate concentrations have fallen to levels that were barely measurable a few years ago. This has impacted the CFS business to a very large degree, making more sophisticated chemistry essential.

One consequence has been the necessity of looking at the leaching properties of CFS reagents themselves. Most solidification reagents are relatively low purity products manufactured for other purposes—Portland cement, lime, calcined clays, soluble silicates—or wastes, such as flyash and cement or lime kiln dust. Many of these products contain substantial amounts of metals, such as chromium, copper, lead, nickel, and zinc. They also contain lesser quantities of antimony, arsenic, barium, cadmium, mercury, selenium, silver, and thallium. These metals originate mostly in the natural raw materials used in reagent manufacture, that is, clay, limestone, coal, sand, and miscellaneous minerals. However, they may become concentrated in the reagent or by-product.

Typical total and leachable concentrations of the metals in various reagents are given in Tables 4-1 and 4-2. The leaching tests were run on the reagents in the

TABLE 4-1 Leachability of Common CFS Reagents, with Test Type

| Reagent Name | Test Type | Concentration of Constituent in Leachate (mg/l) |||||
		Arsenic	Barium	Cadmium	Chromium	Copper
Kiln dust #1	Total		92.700	3.140	31.900	44.800
	TCLP-2		2.740	<0.010	0.050	0.160
Kiln dust #2	Total	10.600	104.000	10.100	59.200	58.700
	TCLP-2	<0.010	2.740	<0.010	0.100	0.110
Kiln dust #3	Total	1.460	9.020	2.200	13.100	22.800
	TCLP-2	<0.010	0.210	<0.010	0.070	0.120
Kiln dust #4	Total	22.900	157.000	10.700	54.900	55.600
	TCLP-2	<0.010	0.610	<0.010	0.090	0.130
Kiln dust #5	Total	10.600	63.200	4.030	24.700	28.900
	TCLP-2	<0.010	0.580	<0.010	0.080	0.100
Kiln dust #6	Total	4.450	93.800	1.880	26.500	33.100
	TCLP-2	<0.010	0.770	<0.010	0.200	0.120
Portland cement, Type I	Total	31.600	104.000	1.110	33.700	47.800
	TCLP-2	<0.010	1.010	<0.010	0.260	0.180
Portland cement, Type I	Total	26.300	109.000	1.380	36.700	
	TCLP-2	<0.420	0.840	<0.010		
Portland cement, Type II	Total	21.400	215.000	2.050	35.800	64.200
	TCLP-2	<0.010	1.770	<0.010	0.240	0.210
First-third landban BDAT—F006		(1)	(1)	0.066	5.200	
First-third landban BDAT—K061				0.140	5.200	
RCRA characteristic levels			100.000	1.000	5.000	
Delisting levels			6.300	0.063	0.315	
			1.000	0.010	0.050	

TABLE 4-2 Leachability of Common CFS Reagents, without Test Type

Reagent Name	Concentration of Constituent in Leachate (mg/l)					
	Lead	Mercury	Nickel	Selenium	Silver	Zinc
Kiln dust #1	156.000		12.600	8.670	4.130	65.600
	0.290		0.020	0.030	0.020	0.040
Kiln dust #2	432.000	<0.034	24.100	1.640	14.200	145.000
	0.390	<0.001	0.020	0.030	0.030	<0.010
Kiln dust #3	94.400	<0.034	13.000	<0.240	6.700	9.560
	0.430	<0.001	0.020	<0.010	0.030	<0.010
Kiln dust #4	227.000	<0.034	28.000	<0.240	10.500	171.000
	0.490	<0.001	0.030	<0.010	0.040	0.040
Kiln dust #5	155.000	0.197	15.800	<0.250	7.090	34.100
	0.290	<0.001	0.020	<0.010	0.010	0.030
Kiln dust #6	107.000	<0.034	13.100	<0.240	5.170	41.400
	0.390	<0.001	0.020	<0.010	0.030	0.030
Portland cement, Type I	57.500	<0.033	24.900	<0.25	3.490	43.800
	0.260	<0.001	0.020	<0.010	0.010	0.070
Portland cement, Type I	70.100	<0.025		<0.230	4.140	
	0.250	<0.001		<0.230	<0.010	
Portland cement, Type I	85.900	<0.032	9.210	<0.280	7.780	120.000
	0.340	<0.001	0.020	<0.010	0.020	0.020
First-third landban BDAT—F006	0.510		0.320	(1)	0.072	0.086
First-third landban BDAT—K061	0.240		0.320			
RCRA characteristic levels	5.000	0.200		1.000	5.000	
Delisting levels	0.315	0.013		0.063	0.315	
	0.050	0.002		0.010	0.050	

nonsolidified form, that is, in powder or granular state. If the reagents are mixed with water and allowed to set and cure, leaching levels may be lower than the values shown in the tables, depending on the leaching test used and the sample preparation. Nevertheless, these data point up two important facts: (1) reagent composition must be taken into consideration in the overall leachability of CFS-treated wastes, and (2) regulators should consider both content and leachability of materials all around us that are "generally recognized as safe"‡ when setting allowable leaching levels for wastes. The latter point seems to have been almost universally ignored in recent years by regulators and environmentalists alike. Earlier, these materials were excluded from regulation under the RCRA because they were considered nonhazardous or represented a low level of risk.

At that time, the definition of a waste hazardous by characteristic [1] was set at metal leaching levels much higher than those promulgated under the recent landbans [4], in some cases by an order of magnitude. While the new levels may be achievable, they may not make sense in view of the real environment in which we live.

---

‡This term, abbreviated as "GRAS," is commonly used in the food-and-drug area to recognize that such materials do not require toxicological or other safety testing before use.

SOLUBILITY OF METALS

As we saw in Chapter 3, the theoretical solubilities of metals or those measured in pure water or other simple systems are often much different than actual solubilities in complex waste–leachant systems. Nevertheless, as a starting point in choosing possible chemical reactions for fixation purposes, it is useful to know the basic water/acid/base solubilities of various species. Tables 4-3–4-5 are compilations of information about metal compound solubilities taken from various literature sources [5]-[8].

Basic solubility concepts were presented in Chapter 2. We saw there examples of the differences between measured solubilities in water and hydroxide values from the literature for various CFS-treated wastes. It is difficult, perhaps impossible, to separate out the effects that cause these differences. They can be the result of any combination of the containment or leaching mechanisms discussed earlier, or simply due to the leaching test itself. One way to view this subject is to compare published value ranges for individual species with those obtained in leaching tests on treated wastes. This is done in Table 4-6 for the more common species.

In considering Table 4-6, it is important to understand that such comparisons assume a certain speciation for the metal, which in most cases has not been confirmed. Nevertheless, the comparison serves a useful function, that is, to point out the limitations of fixation technology in real applications, or in some cases, to show how the complex interactions of several mechanisms can achieve results better than expected from theory.

METAL SPECIATION IN WASTES

The aspect of metal fixation that seems most confusing to workers in the field involves the speciation of the metal before CFS treatment. Most wastes that are treated by CFS are not solutions, but are sludges, filter cakes, and other residues from waste water treatment systems. The metals have been precipitated with lime or other alkali, or with other agents, such as sulfide, to produce metal hydroxides, sulfides, or other compounds that have low solubility under the conditions of precipitation, usually in the pH range of 6 to 8 and a dilute water medium. In this state, there is little immediate reaction between the metal species and the CFS reagents. If the metal species in the waste is more soluble than the species that would be formed by anions or ligands introduced by the CFS system, there may be gradual respeciation at the particle surface, but total respeciation would be expected to occur only over a long period of time. Furthermore, the soluble anion or ligand may not be available as such for long if it can react rapidly with other components of the CFS system. The ultimate result, then, is a mixture of metal hydroxides, sulfides, etc., dispersed in a cementitious matrix. The leachability of such a product will be determined by the solubility of the metal compounds and by the other factors previously discussed.

In the design of CFS systems to fix metals, it is essential to know the speciation of the metals in the raw waste. Unfortunately, this is easier said than done. Analytical methods used in the environmental field are designed to determine metal as the element. Anionic species are seldom determined, except those of environmental concern such as cyanide, phosphate, and fluoride. When they are analyzed, it is as the

FIXATION OF METALS

TABLE 4-3 Solubility of Metal Species: Al to Cr^{+6}

Species	Al	As^{+3}	As^{+5}	Sb^{+3}	Sb^{+5}
Acetate—$(C_2H_3O_2)$					
Arsenate—(AsO_4)	a	—	—	A	
Arsenite—(AsO_2)		—	—	A	
Bromide—(Br)	W	W		d	
Carbonate—(CO_3)		d			
Chloride—(Cl)	700000	I	W	9100000	d
Chromate—(CrO_4)					
Citrate—$(C_6H_5O_7)$	5000				
Cyanide—(CN)					
Ferricyanide—$(Fe(CN)_6)$					
Ferrocyanide—$(Fe(CN)_6)$	w				
Fluoride—(F)	6700			4440000	W
Hydroxide—(OH)	A/L				
Iodide—(I)	W	60000		d	d
Nitrate—(NO_3)	1300000				
Nitrite—(NO_2)					
Oxalate—(C_2O_4)	A				
Oxide—(O)	a/l	12000	658000	w/l(KOH)	w/l(KOH)
Phosphate—(PO_4)	A/L				
Silicate—(SiO_3)	I				
Sulfate—(SO_4)	364000			A	
Sulfide—(S)	d	L 0.3	L 3	A 1.2–20	0.7
Sulfite—(SO_6)					
Tartrate—$(C_4H_4O_6)$	w			W	
Thiocyanate—(SCN)					

Note: All table notes apply to Tables 4-3–4-5.
Note: I = insoluble in both water and acids; W = water soluble; w = slightly water soluble; A = insoluble in water, soluble in acids; a = insoluble in water, slightly soluble in acids; L = insoluble in water, soluble in alkalies; l = insoluble in water, slightly soluble in alkalies; d = decomposes; * = salt occurs in two modifications. All numbers are in mg of salt per liter of water in pH 7 water, unless otherwise stated.

METAL SPECIATION IN WASTES 65

Ba	Be	B	Cd	Ca	Cr^{+3}	Cr^{+6}
760000			W	374000	W	
w/A		W	A	130		
555				w		
450000		d	950000–990000	1430000	W/I*	
A	3600		A/L(NH4OH)	A	W	
15–20			0.06–0.028	13		
360000	420000	d	1200000–1286000	420000	I	
A			A	166000	—	
1.4–10						
570			A	2200		
800000			17100	d	A	
1700			A	W		
W			A	1040000		
1600	W	1050000	40000–43500	a	I	
	(slow)			8.8–20		
39000	w		A	A	A	
			2.6	1700	2×10^{-8}(pH8)	
			0.002(pH11)			
204000	W	d	847000–86000	676000	L	
90000	1660000		1090000–1500000	1520000	208000	
67500				845000		
90	635000		33	A	W	
				6		
34800	0.5	3300	A	A	a	1670000
		(slow)	9.6	1300		
A	W		A	A	w	
				300		
1700	I		A			
a	(4H$_2$O)		766000–1130000	A	W/I*	
1.5	391000			2000		
79000	I		A	200	A(HNO$_3$)	
d			3×10^{-9}–1.3	d		
200				A		
				40		
260			A	340		
1700000				1500000		

Source: J. A. Dean, *Lange's Handbook of Chemistry,* 12th ed., McGraw-Hill, New York, 1979; Seidell, A. *Solubilities of Inorganic and Metal Organic Compounds,* 3d ed., Van Nostrand Reinhold, New York; G. A. Lewandowski and M. F. Abd-El-Bary, *Water and Sewage Works.* Jan. 1981; Kirk-Othmer, *Encyclopedia of Chemical Technology,* 3rd Ed., Wiley, New York, 1979.

TABLE 4-4 Solubility of Metal Species: Co to Ni

Species	Co	Cu^{+1}	Cu^{+2}	Fe^{+2}	Fe^{+3}
Acetate—(C$_2$H$_3$O$_2$)	W		80000	W	W
Arsenate—(AsO$_4$)	A/L(NH$_4$OH)		A/L(NH$_4$OH)	A	A(HCl)
Arsenite—(AsO$_2$)	A		A		
Bromide—(Br)	1120000	w	1260000	1170000	W
Carbonate—(CO$_3$)	A 1800	0.8	0.755–1.9	A 720	
Chloride—(Cl)	529000	A(HCl) 240	730000	625000	740000
Chromate—(CrO$_4$)	A	A(HNO$_3$)			A
Citrate—(C$_6$H$_5$O$_7$)	3000		A 170		W
Cyanide—(CN)	A 42	A(HCl) 2.6	A/L	a	
Ferricyanide—(Fe(CN)$_6$)	I		I	I	
Ferrocyanide—(Fe(CN)$_6$)	I		I	I	A(HCl)
Fluoride—(F)	13600	A(HCl)	A 750	w/A	A 910
Hydroxide—(OH)	A 1.8		A 1.9 3×10^{-5}(pH9)	A 3.75–60	A 0.12
Iodide—(I)	2030000		A(HCl)	W	W
Nitrate—(NO$_3$)	1550000		1380000	1340000	1380000
Nitrite—(NO$_2$)					
Oxalate—(C$_2$O$_4$)	A 20		L(NH$_4$OH) 20	A 440	W
Oxide—(O)	A	A(HCl)	A	A	A(HCl)
Phosphate—(PO$_4$)	A/L(NH$_4$OH)		A	A	A(HCl)
Silicate—(SiO$_3$)	A		A	I	
Sulfate—(SO$_4$)	W	d	143000	450000	w
Sulfide—(S)	A	A(HNO$_3$) 1.62×10^{-10}	A(HNO$_3$) 2×10^{-14}–0.12	A 3.9–6	A(d)
Sulfite—(SO$_3$)		a(HCl)		w	
Tartrate—(C$_4$H$_4$O$_6$)	w		w	w	W
Thiocyanate—(SCN)	78000	L(NH$_4$OH) 4.4	d	W	W

Pb	Mg	Mn	Hg^{+1}	Hg^{+2}	Mo	Ni
456000	534000	380000	A 10000	250000		160000
A(HNO$_3$)	A	w	A	w		A
A(HNO$_3$)	W	A	A	A		A
A 840	1010000	1470000	A	5600	L(d)	1130000–1310000
A/L 0.05–109	A 31–100	A 65	A(HNO$_3$)	0.002		A 90
9900	546000	740000	a 2.7	66000	A/L	608000–640000
A 0.08–0.11	1370000		w	w		A
W	200000	w	w			W
w	W		a	93000		a/L(NH$_4$OH) 60
w	W			A		I
a	W	A		I		I
510–640	A(HNO$_3$) 130	A 10600	d	d		25600–40000
A/L 155 0.001(pH10)	A 9.7–11.5	A 20		A 5×10^{-9}(pH8)	a(HCl) 2000	13 6.1
L 630	1400000	W	A	60	a	1240000–1430000
560000	1200000 W	1950000	W d	W		1500000
A/L 1.5	A 420	A 350	10	A(HCl) 110		A 3
A(HNO$_3$) 17	A	A	A	A 47–50	A	A
A/L 0.08	A 200	w	A	A		A
A	A	I				
29–40	337000	629000	A(HNO$_3$) 600	d		293000
A 2×10^{-9}–120		A 6 5	I	I 2×10^{-21}–0.01	I	A(HNO$_3$)
	660000					A(HCl)
A/L 25	w	w	I			A
A(HNO$_3$) 4400	W	W	A	A(HCl) 630		

68 FIXATION OF METALS

TABLE 4-5 Solubility of Metal Species: Se^{+4} to Zn

Species	Se^{+4}	Se^{+6}	Ag	Sr	Tl	Sn^{+2}	Sn^{+4}	V	Zn
Acetate—($C_2H_3O_2$)			10400–11000	411000	W	d	W		300000–400000
Arsenate—(AsO_4)			A/L (NH_4OH)	w					$A(HNO_3)$
Arsenite—(AsO_2)			A				A		A
Bromide—(Br)	d	d	a 0.14	1000000	500	850000	W	W	4460000
Carbonate—(CO_3)			A 25–33	A 10	53000				A/L 10–200
Chloride—(Cl)	d	d	a/L (NH_4OH) 1.17–1.9	529000	3300	840000	W	W	3950000–4320000
Chromate—(CrO_4)			20–36	A(HCl) 900	a 300	A	W		A
Citrate—($C_6H_5O_7$)			w	A					w
Cyanide—(CN)			a 0.023	W	168000				A/L 50–580
Ferricyanide—(Fe(CN)$_6$)			I	W		A			A
Ferrocyanide—Fe(CN)$_6$			I	W		a			L
Fluoride—(F)	Reacts		1720000–1800000	110	780000	300000	W	$I^{(+3)}$ $W^{(+4, +5)}$	16000

Anion							
Hydroxide—(OH)		2.1(pH11)	17700	259000–350000	A 1.5	w	w/A/L 2
Iodide—(I)			1 0.0026	1780000	A(HCl) 9800	d	4320000–5700000
Nitrate—(NO$_3$)			2160000	695000	d	d	930000–1460000
Nitrite—(NO$_2$)				715000			
			4100–4200	403000			
Oxalate—(C$_2$O$_4$)			40	15800	A		A/L 20
Oxide—(O)	380000	W	20–22	6900	A/L	A	A/L 1.3–4.2
Phosphate—(PO$_4$)			A	5000	A		A/L(NH$_4$OH) 26000
Silicate—(SiO$_3$)			A				A
Sulfate—(SO$_4$)			8000–8300	48700	189000	W	419000–538000
				130			
Sulfide—(S)		I	A 2×10^{-12}–0.14	200	A	A	A 1×10^{-4}–7
Sulfite—(SO$_3$)			A				A 1600
Tartrate—(C$_4$H$_4$O$_6$)			2000	A(HCl) 1100	133000	W	w
Thiocyanate—(SCN)			L (NH$_4$OH) 0.13	3200			1400

TABLE 4-6 Comparison of Published Metal Solubility Values with Actual Leaching Results on CFS-Treated Wastes

Metal	Published Value[a] (mg/l)	Leaching Test Result[b] (mg/l)
Antimony	1.2	<0.05
Arsenic	0.3	<0.01
Barium	1.4	<0.02
Cadmium	3.0×10^{-9}	<0.01
Copper	2.0×10^{-14}	<0.002
Chromium	0.2	<0.01
Lead	0.001	0.003
Mercury	2.0×10^{-21}	<0.001
Nickel	3.0	<0.005
Selenium	—	<0.05
Silver	2.0×10^{-12}	<0.003
Zinc	1.0×10^{-4}	<0.005

[a] Tables 4-3 through 4-5.
[b] For wastes with significant metal content.

anion alone, not as the associated metal (or other) compound. Analysis for organics is normally limited to those priority pollutants of possible concern in the waste stream. Organic complexing agents are seldom determined. A frequent source of confusion to the engineer is the practice of reporting certain parameters *as* a particular compound, for example, total alkalinity *as* CaO. This means that the parameter is calculated from the measurement method as if it were the stated species, while the actual species may not even be present in the waste. It is a useful method of expression to the chemist, but can be misleading to, and misused by, the novice.

Why, then, don't we determine the actual species in the waste? The answer is cost. In these complex systems, really complete analysis is very time-consuming and expensive, requiring specialized equipment in many instances. In one case in the author's experience, the determination of the speciation of lead in a flyash required a combination of scanning electron microscopy/energy dispersive analysis of X-rays (SEM/EDX) and microscopic Fourier transform infrared spectroscopy (FT-IR) on individual particles. This work cost several thousand dollars and required a very sophisticated laboratory. Yet, possession of this prior, certain knowledge of the species would have saved money and time in the overall project. Unfortunately, few laboratories are equipped and staffed to conduct such analyses on CFS systems.

An alternative is the indirect determination of speciation by other, less expensive means. Examples are the use of techniques such as thermal analysis, titration, and sequential chemical extraction [9]. While limited in scope and degree of definition, such methods are faster and much less expensive, and may give useful insights.

A third alternative is a thorough knowledge of the source of the waste stream. Most industrial streams come from a single process or are combined from several processes. The possible metal speciation can usually be narrowed down by knowledge of the chemicals used or generated in the process, or modified by the treatment process used, if any. For example, let us take the case of EPA waste F006, a waste water treatment sludge from electroplating operations. The possible metals from most plating operations (there are specialized processes that are different) are cadmium, chromium, copper, lead, nickel, silver, and zinc. The formulation of plating baths is well known. Waste water treatment nearly always consists of the following chemical steps: chromium reduction (if required); hydroxide precipitation by neutralization with lime, sodium hydroxide, or soda ash; polishing treatment with sulfides (if required). From this, we can determine with considerable accuracy the composition of the waste sludge simply by more careful querying of the generator, which requires skill and experience on the part of the questioner.

Obviously, wastes such as those found in lagoons at superfund sites are not amenable to this approach. In the past, generators were reluctant to furnish information on the composition of their wastes for proprietary reasons and out of fear of regulatory agencies. However, more stringent disclosure requirements coupled with a more enlightened attitude on the part of generators has alleviated this situation to a large degree. The primary reason for the devotion of such a large proportion of this book (Chapter 8 and Appendix 1) to the characteristics and properties of waste streams is to aid the reader in intelligently choosing applicable CFS technology with a minimum of tedious and expensive testing.

FIXATION OF METALS IN SOLUTION

The problem of speciation is minimal when the metal species to be fixed is in solution. Unless it is complexed in soluble, stable form, the metal can usually be precipitated from solution as a species that exhibits minimum solubility under the expected disposal conditions (or in meeting the required leaching test). This is especially important for the heavy metals of environmental interest that exhibit amphoterism, such as arsenic, cadmium, chromium, lead, and zinc. A neutral hydroxide that is both an acid and a base is said to be *amphoteric,* and the term is also applied to oxides and sulfides. The hydroxides of these metals exhibit minimum solubility through a narrow pH range, usually in the area of pH 7.5 to 10, with solubility increasing rapidly as pH increases or decreases on both ends of the range (see Figs. 3-2–3-6). Because most CFS systems are quite alkaline, usually above pH 11 (at least initially), the solubility of the metal hydroxide in the CFS-treated waste may actually be higher than in the original, untreated sludge.

Wastes whose metals are in solution have an important advantage from the fixation standpoint. The metal can be precipitated as the species of choice: silicate, sulfide, carbonate, etc. The solubility of the species produced will, of course, depend to some extent on other species present in solution, but the system can be optimized in any case. There are two major factors that place limits on the ability of a CFS system to immobilize dissolved metals: valence and complexation. The latter will be discussed in the next section. Valence state may limit the insolubilization of certain metals, notably chromium. Chromium exists in two valence states, Cr^{+3} and Cr^{+6}. The only practical way to fix hexavalent chromium when it is present in substantial amounts is to reduce it to the trivalent state. Chromium reduction will be discussed in detail later in this chapter.

FIXATION OF COMPLEX SPECIES

Metals exist in solution in forms other than the simple ions or molecules that we have discussed. Actually, metal ions in solution are usually solvated, that is, they are associated with water molecules in a definite arrangement. The maximum number of water molecules associated with the ion depends on the metal's coordination number. When the water molecules are replaced by other ions or molecules, the result is termed a metal complex. The chemical bonds involved are covalent rather than ionic, that is, electrons are shared by each of the bonded atoms. A common example of such a complex often encountered in environmental chemistry is the cupriammonium complex:

$$\begin{bmatrix} H_2N & NH_2 \\ & Cu & \\ H_2N & NH_2 \end{bmatrix}^{++}$$

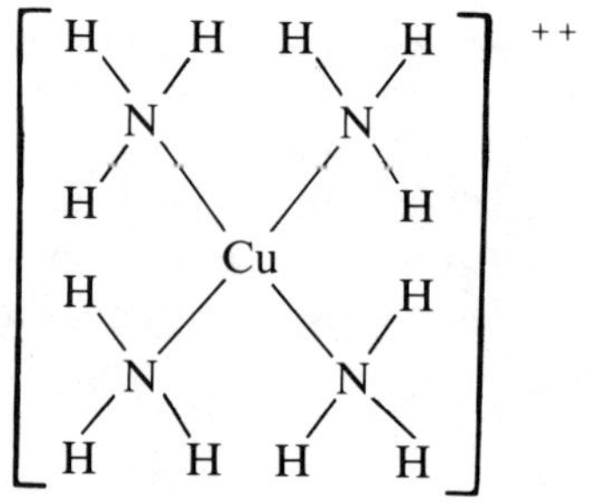

A metal complex may be either inorganic, such as the copper complex ion shown here, or organic. It may be an ion like the cupriammonium complex or a neutral molecule. Finally, it may be either soluble or insoluble. In this section, our concern is with soluble complexes; the insoluble type provides a means of immobilizing some metals and other species, and will be discussed under the individual metals and in Chapter 5. Many complexes, such as the cupriammonium complex, are relatively easy to break up by simple means, such as pH adjustment. Others, like the ferricyanide complex, are more stable:

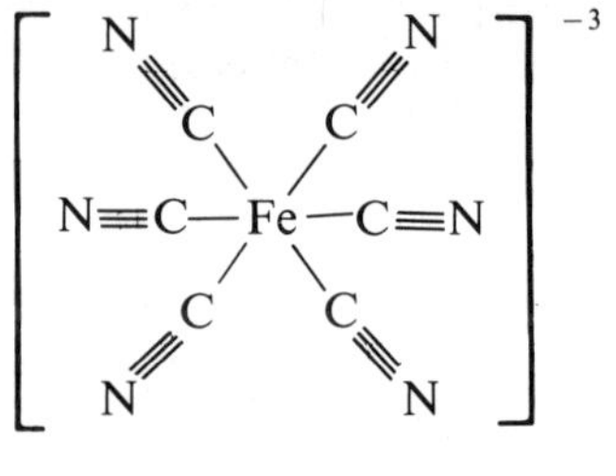

$$[Fe(CN)_6]^{-3}$$

Even more stable is a special type of complex formation known as a *chelate*. In chelates, the metal is bound chemically to at least two sites on the the ligand. This results in a ring formation, inherently more stable than the simple structures formed by ammonia and cyanide. Most chelating agents are organic, but some are inorganic ligands such as phosphate:

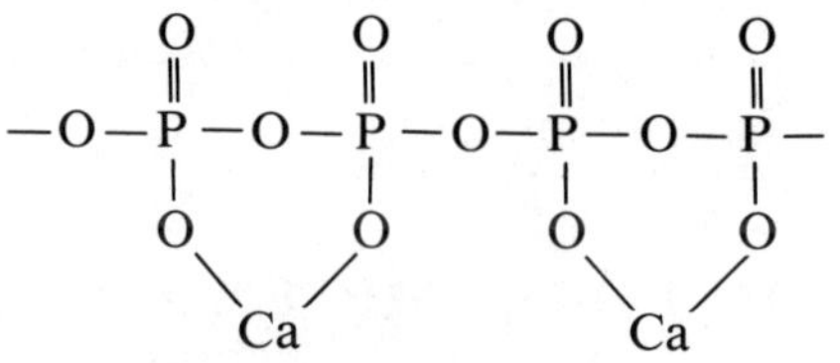

$$-O-\underset{O}{\overset{O}{\underset{\|}{P}}}-O-\underset{O}{\overset{O}{\underset{\|}{P}}}-O-\underset{O}{\overset{O}{\underset{\|}{P}}}-O-\underset{O}{\overset{O}{\underset{\|}{P}}}-$$
$$\quad\quad\quad\searsow\swarrow \quad\quad \searrow\swarrow$$
$$\quad\quad\quad\quad Ca \quad\quad\quad\quad Ca$$

Citrate, gluconate, glycine, EDTA (ethylene diamine tetraacetic acid), and nitrilotriacetate are all chelating agents that form stable, water-soluble metal complexes. A relatively simple structure, calcium gluconate, illustrates the ring structure:

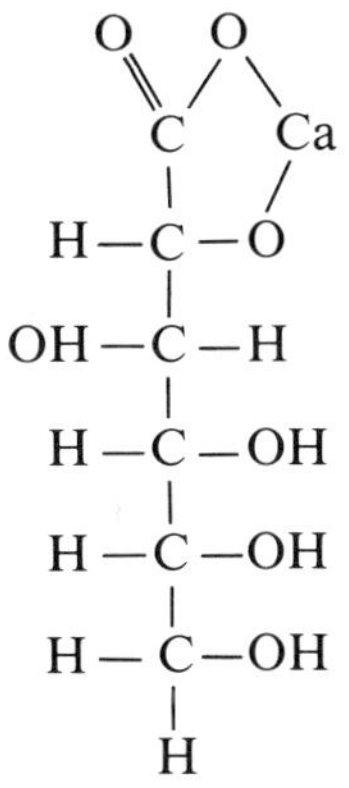

$$\begin{array}{c}
\text{O} \quad \text{O} \\
\parallel \quad / \\
\text{C} \quad \text{Ca} \\
| \quad / \\
\text{H–C–O} \\
| \\
\text{OH–C–H} \\
| \\
\text{H–C–OH} \\
| \\
\text{H–C–OH} \\
| \\
\text{H–C–OH} \\
| \\
\text{H}
\end{array}$$

EDTA forms a more complex, multiple ring structure with metals:

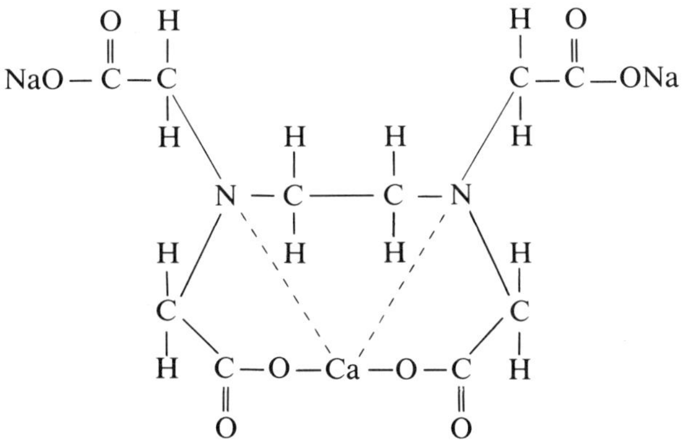

[Structure of EDTA-Ca chelate complex showing two NaO-C(=O)-C(H)(H)- groups connected via N-C-C-N backbone with Ca coordinated through carboxylate oxygens]

EDTA is the prototypical chelating agent, but many others, including natural organic acids, are also found in wastes. When a ligand forms a water-soluble complex, it is called a sequestering agent. In this form, the metallic ion is inactivated and no longer participates in its usual chemical reactions. However, since it remains in solution, it is readily leached from wastes, treated or not, and shows up in the subsequent analysis.

Since the sequestered metal species is not chemically reactive, it may not be precipitated by the usual methods, including pH adjustment. In fact, most CFS reagents at ambient temperature are ineffective in precipitating chelated metals. Strong oxidizing agents, often at elevated temperature, are required to destroy the complex and release the metal ion within a reasonable time frame. Like other chemical species, complexing agents vary in their ability to solubilize metals. And, just as with other compounds, the chelate structures exist in solution in an equilibrium mixture with chelating agent (L) and metal ion (M):

$$M^{+n} + L^{-n} \rightleftharpoons ML$$

The measure of effectiveness of the chelating agent is the stability constant, K, which is expressed as:

$$K_{ML} = \frac{[ML]}{[M^{+n}][L^{-n}]} \qquad (4\text{-}1)$$

The larger the positive value of K, the greater is the tendency of the chelate to form.

Chelating agents are used commercially not only to keep a metal in solution, as in electroplating, but also to dissolve insoluble metal species in operations such as boiler tube cleaning in the power industry [10]. In this case, an "exchange" reaction takes place:

$$ML + B \rightleftharpoons MB + L$$

where

B = precipitating anion
MB = insoluble species

This reaction can be described in terms of the stability constant, K_{ML}, and the solubility product for MB [11]:

$$K_{\text{exchange}} = \frac{1}{K_{ML} \times K_{MB}} \qquad (4\text{-}2)$$

If K_{exchange} is less than one, chelation will take precedence over precipitation, and the insoluble phase will dissolve. This assumes the absence of interfering factors such as very low pH or competition of other cations that may be complexed.

In addition to strong oxidizing conditions, some chelates may be broken up by very high pH, by sodium sulfide, and by proprietary organosulfur precipitants [12]. However, all of these methods involve more complex or expensive treatment than is common in the CFS field: multiple reaction steps, elevated temperatures, corrosive or toxic reagents. Often the best method for treatment of metal complexes is pretreatment by the generator before mixing with other waste streams that may complicate the process. For example, a small amount of complexed nickel from an electroless plating bath may necessitate the treatment of a much larger volume of waste water treatment sludge. The cost would be much less if the plating bath were managed separately.

Many of the metals of environmental interest form very stable chelates: cadmium, cobalt, copper, lead, mercury, nickel, and zinc. Further discussion of their treatment will take place under the individual metal sections later in this chapter.

FIXATION OF METALS IN SOLIDS AND SOILS

The treatment of contaminated soils and other solids is a special case. Such wastes are becoming increasingly more important in the CFS field as remedial programs become more active. Also, the greater use of incineration is generating residues that are themselves hazardous, usually because of toxic metal content. These wastes are generated in ways very different from the typical waste water treatment sludge, and so their compositions, especially with regard to metal speciation and distribution, are different. Natural soils contain clay, rock, silt, sand, and many natural organic substances. When they become contaminated, it is usually by infiltration of metal species (and other hazardous components) in solution. Such contaminants may originate in accidental spills, from deliberate dumping, or from leaching of older landfills. The interaction of metals with soils is very complex. Adsorption and ion exchange by clay minerals, reaction with insolubilizing anions present in the soil, and complexation by humic substances in the organic fraction of the soil all occur. Complexation may lead either to increased or decreased solubility of the metal, depending on the ligands present.

Residues from incineration and other high-temperature processes have a totally different set of characteristics. Slags and bottom ashes are often nonhazardous by EPT characteristic, probably due primarily to their glassy natures. These materials, when generated by burning nonhazardous waste, are commonly recycled as road building and other structural base. Flyashes, on the other hand, are frequently hazardous. While individual particles may have a vitreous character, they are small with a resultant large surface area that increases the potential leachability on a mass basis. In addition, metals such as lead, zinc, and cadmium, which are relatively volatile at high-temperature, vaporize and subsequently deposit on the surfaces of the particles. The speciation of these metals in flyashes and their distributions throughout the particles are not well understood. Some studies have found enrichment of the metals at the particle surface [13], but this would likely vary with different sources, temperatures, and compositions. For example, ash from waste incinerators might well have metals speciated as chlorides because of the presence of organochlorine compounds and plastics in the waste feed, while metals in electric arc furnace dust (K061) would more likely be speciated as oxides. However, the distinction may not be so clear-cut, especially in residues from waste incineration.

Solids such as these usually do not require solidification, although ashes may be processed to suppress dust and provide greater stability in a landfill. Also, they require the *addition* of water in treating for fixation purposes, unlike other waste streams where water removal is practiced to reduce volume. In the case of soils, soluble metal salts are often distributed throughout large, hard pieces of clay and porous rock. This may have occurred over a period of many years, and outward diffusion of the species so that it can react with fixation agents can be very slow, yet rapid enough to leach at unacceptable rates. This is especially true where the species must be oxidized, reduced, or otherwise treated in a multistep process such as Cr^{+6} reduction. Here, excess reducing agent is destroyed in the final process steps and is not available to reduce chromium diffusing from the particle interior at a later time. In some such situations, the only solution may be to grind the waste to a very fine powder so that the reactions may be carried to completion within a reasonable time. In others, long-lived agents may be incorporated in the CFS matrix, remaining there to react with the metal as it diffuses out of the soil particles. There are even situations in which the agent may be generated within the matrix as it is used. These techniques will be discussed later under individual metal sections in this chapter, and under various CFS system types in other chapters.

LONG-TERM STABILITY OF FIXED METALS

Much concern has been expressed by environmentalists and regulators about the long-term stability of fixed species. Certain leaching test procedures purport to be equivalent to tens or hundreds of years of natural leaching action in the environment

[14]. However, no actual long-term data are available because the technology has only been practiced for about 20 years at this time. Nevertheless, except for recovery, there is no alternative to CFS technology for management of hazardous metals. Therefore, it is especially important to understand the fundamental chemistry of these systems so we can make intelligent estimates of long-term effects. As previously pointed out, slow leaching of metals at a controlled rate is not detrimental to human health and the environment.

The real concern is the sudden release of contaminants due to breakdown of the matrix. This could occur, for example, when the fixation mechanism is pH control and the buffering action of the alkali is finally used up in an acid environment. Knowledge of the structures and species formed in a given CFS system can allay these fears by allowing us to simulate the sudden release conditions that could occur and measure the resulting effects. As we shall see, work of this sort is being done and the results are encouraging.

METAL FIXATION AND RESOURCE RECOVERY

A constant refrain is heard from environmentalists: Why not recover and reuse the valuable metals that are present in wastes? The common answers are that it is not economical to do so, or that the technology is not available, or that the recovered products would cause economic chaos in the marketplace. There is, however, another more compelling reason in many cases: that the recovery processes themselves generate residues that still contain metals, often in more leachable form. *Practical* technology simply is not available to remove most metals in most wastes to below the levels of environmental concern. There are exceptions, and these are currently being pursued by generators and hazardous wastes management firms. However, most metal-containing wastes now generated will necessarily be managed by CFS technology in the foreseeable future.

METAL FIXATION MECHANISMS AND METHODS

A list of the methods, processes, and mechanisms that are used in CFS technology appeared earlier in this chapter. Each of these will be discussed separately before exploring the reactions and properties of the individual metals. The CFS technologies discussed here are limited to those that are useful in a practical sense today, or have near-term potential applicability in the future. Other feasible approaches to management of metal wastes, such as resource recovery, are outside the scope of this book.

pH Control

This is the most common, simplest method of fixation of metal-bearing waste streams. The principles were introduced in Chapter 3 (Figs. 3-2–3-6) and earlier in this chapter (Tables 4-3–4-5), and will be discussed further in the subsection ''Hydroxide Precipitation'' later in this chapter. Any alkali might be used for the purpose of pH control, but the most common are lime (CaO or $Ca(OH)_2$), soda ash (sodium carbonate, Na_2CO_3), sodium hydroxide (NaOH) and, less commonly, magnesium hydroxide (Mg(OH)$_2$). Each of these reagents is available in several different forms. They are also used extensively in waste water treatment, sometimes as combination neutralization/fixation agents.

In addition to the common alkalies, most of the solidification reagents are alkaline. Some can totally take the place of the traditional alkalies in CFS work, acting both as pH control agents and as cements or pozzolans. These include Portland cement, cement and lime kiln dusts, flyash (especially Type C), and sodium silicate. Others, such as limestone and some clays, act as buffers because of their ability to neutralize acids in leachants. Limestone is well known in this regard. Of the clays, sodium montmorillonite, especially, was found to be an effective, fast-reacting, acid neutralization agent [15].

Lime. The most common alkali used in CFS technology, other than the lime-containing cements and pozzolans, is lime. The properties of various types of lime are given in Table 4-7.

Lime products may be either high calcium or dolomitic. The latter contains substantial amounts of MgO or Mg(OH)$_2$ in place of the calcium. Both quicklime (anhydrous) and hydrated lime are used in CFS work, but the hydrated form is more common because it is easier and safer to store and handle. Hydration of CaO has

TABLE 4-7 Physical and Chemical Data Pertaining to Lime Products

QUICKLIMES

	High Calcium	Dolomitic
Primary constituents	CaO	CaO and MgO
Specific gravity	3.2–3.4	3.25–3.45
Bulk density (pebble lime), lb/ft^3	55–60	55–60
Specific heat at 100°F (38°C), BTU/lb	0.19	0.21
Angle of repose (approximate average for pebble)	50–55[a]	50–55[a]

HYDRATES

	High Calcium	Normal Dolomitic	Pressure Dolomitic
Primary constituents	Ca(OH)$_2$	Ca(OH)$_2$ + MgO	Ca(OH)$_2$ + Mg(OH)$_2$
Specific gravity	2.3–2.4	2.7–2.9	2.4–2.6
Bulk density, lb/ft^3	25–35[b]	30–40[b]	30–40[b]
Specific heat at 100°F (38°C), BTU/lb	0.29	0.29	0.29
Angle of repose (most common)	70[a]	70[a]	70[a]

LIMESTONES

	High Calcium	Dolomitic
Primary constituents	CaCO$_3$	CaCO$_3$ and MgCO$_3$
Specific gravity	2.65–2.75	2.75–2.90
Bulk density (¾-in. stone), lb/ft^3	87–95	87–95
Specific heat at 100°F (38°C), BTU/lb	0.21	0.21

[a] Approximate average values, since considerable variance may occur.
[b] In some instances these values may be extended. The Scott method is used for determining the bulk density values.

a heat of reaction of 15,300 cal/g mol with water and much higher with acids, so it can generate high temperatures, and even explosive steam with aqueous wastes. The solubility of Ca(OH)$_2$ is 1200 mg/l at 70°F (21°C), 670 mg/l at 175°F (79°C). The pH of Ca(OH)$_2$ solutions ranges from 11.27 at 64 gm/l to 12.53 at 1160 gm/l.

Soda Ash. Soda ash (Na$_2$CO$_3$) is much less frequently used than lime in CFS work, but is sometimes present in raw wastes where it was used for neutralization of waste waters or other process wastes. Properties of sodium carbonate are listed in Table 4-8:

Another form of sodium carbonate, trona or sodium sesquicarbonate, has potential use in CFS systems because of its low cost. It is a natural product, mined in Wyoming. One problem with the use of sodium alkalies is that the pozzolanic or cementitious reactions that serve to moderate high pH in those systems [16] are counteracted by the presence of sodium ion. This phenomenon will be discussed further in Chapters 10–14.

Other Alkalies. Sodium hydroxide and magnesium hydroxide have been seldom, if ever, used in commercial CFS systems. A major drawback to their use is cost, especially since lower cost alkalies are available. However, Mg(OH)$_2$ has interesting possibilities. It does not immediately raise the pH due to its insolubility in water, yet it is available to react with and counteract any acid contacting the treated waste. Thus, it may be useful where more precise pH control is required, as for example when fixing amphoteric metals. Magnesium oxide has also been found to be useful as an additive to Portland cement solidification in the nuclear waste area. This will be discussed in Chapter 17.

Redox Potential Control

Oxidation and reduction techniques are frequently employed in CFS technology to convert metals to more desirable valence states for precipitation. The examples of chromium and arsenic have already been presented. With these metals, the reduced

TABLE 4-8 Properties of Soda Ash

Molecular weight	105.989	
Density	2.533 g/cc at 25°C	
Specific heat	26.41 cal/g/°C	
Heat of hydration (25°C)	Monohydrate	30.0 cal/g
	Heptahydrate	156.4 cal/g
	Decahydrate	208.8 cal/g
Heat of solution[a]	Na$_2$CO$_3$	55.4 cal/g
	Na$_2$CO$_3$·H$_2$O	25.4 cal/g
	Na$_2$CO$_3$·7H$_2$O	−101.0 cal/g
	Na$_2$CO$_3$·10H$_2$O	−153.4 cal/g
Bulk density	Light ash: 35 lb/ft^3 loose	
	46 lb/ft^3 packed	
	Dense ash: 68 lb/ft^3 loose	
	78 lb/ft^3 packed	
Stability in air	Slowly absorbs moisture and CO$_2$ to form trona	

[a] At 25°C, one mole dissolved in 200 moles of water.

form produces the more insoluble species. Other metals can be reduced to lower valence states, or even to the elemental form [17], at least in waste water treatment.

Reduction. The major reducing agents or systems are listed in Table 4-9 and described below. Specific examples that have been used in CFS treatment are discussed in the sections on individual metals.

Several of the more commonly used reducing agents from Table 4-9 are compared in Table 4-10 [18]. The values given are in pounds of chemicals used or pounds of sludge created in treating 100 pounds of chromic acid (CrO_3). These are theoretical consumptions. In practice, larger, often much larger, quantities must be used to produce quantitative reduction of metals. Deviation from stoichiometry will vary with the chemistry of the waste, especially where other reducible species are present, and with the concentration of the metal being reduced.

Ferrous sulfate. This is probably the most widely used metal reducing agent in the CFS field, primarily on wastes containing Cr^{+6}. It is safe to use, inexpensive (as reducing agents go), and often produces additional benefits by coprecipitating the toxic metals. Its main drawbacks are very large volume increase and the requirement of low pH for acceptable treatment times. The latter problem is especially acute in the case of wastes with high alkalinity, where very large quantities of acid may be required for pH adjustment. Reduction with ferrous sulfate is a three-step process: (1) pH adjustment to <3, (2) addition of ferrous sulfate and mixing until reaction is complete, and (3) raising pH to >7 to precipitate metal as the hydroxide or as a coprecipitate with ferric and ferrous iron. The last step allows oxidation of ferrous iron to the Fe^{+3} valence state by oxygen in solution, and any residual reagent should be destroyed. However, Eary and Rai [19] have reported rapid and stoichiometric reduction of Cr^{+6} in solution up to pH 10. This will be discussed further in the section on chromium in this chapter.

TABLE 4-9 Reducing Agents

Agent	pH	References	Comments
$FeSO_4 \cdot 7H_2O$	<3	[18, 19]	Reduction occurs over a wide pH range, but at lower rates
$Na_2S_2O_5/NaHSO_3$	2–3	[18]	Reduction rate depends on pH. The reactions of both reagents are the same
$Na_2S_2O_4$	>7	[18]	Operates at high pH
SO_2	<4.5	[18]	No additional acid required. Problems in handling SO_2
FeS	4.5–7	[20]	Slow and/or requires elevated temperatures
$Fe^{+2} + S^{-2}$	7–10	[21]	Rapid, stoichiometric reduction
$NaBH_4$	>9	[17]	Expensive reagent. Reductions possible to elemental state
Reductive resins Hydrazine		[22]	Expensive; used with precious metals
Electrochemical reduction		[23]	May not be suitable for sludges and filter cakes

80   FIXATION OF METALS

TABLE 4-10   Characteristics of Reducing Agents

	\ Reducing Agent			
	$Na_2S_2O_5$	$Na_2S_2O_4$	SO_2	$FeSO_4 \cdot 7H_2O$
Reducing agent	147	261	96	834
H_2SO_4 (66 Be°)	80	—	—	316
$Ca(OH)_2$	111	—	111	444
Na_2CO_3	159	—	159	631
NaOH	120	120	120	480
Sludge from				
$Ca(OH)_2$	307	—	307	1346
NaOH	103	103	103	423
Na_2CO_3	103	—	103	423

Note: Pounds of chemicals used and sludge produced per 100 pounds of chromic acid in waste.

Sodium metabisulfite/bisulfite.   These two reagents react in the same way, since metabisulfite is the anhydrous form of bisulfite. The former is often used because it does not cake in storage. These reagents are effective reducing agents for Cr^{+6} and require much less acid and alkali than does ferrous sulfate, and therefore generate less sludge. While the reagent is considerably more expensive than ferrous sulfate, which can often be obtained as a waste product, much less is used. The overall cost of bisulfite reduction is generally less than that of ferrous sulfate. The principal objection to the use of the bisulfites in CFS systems is their tendency to generate sulfur oxides in volume on contact with acids. In water treatment systems, with very dilute waste and low addition rates of chemicals, this may not be a problem. In CFS, it is a very serious problem, so much so that such use would require an effective air pollution control system. The solids in waste residuals seem to increase the evolution of SO_2, perhaps by a surface catalysis reaction. This is especially evident when treating Cr^{+6}-contaminated soils, a frequent use of reduction techniques.

The use of sulfur dioxide is potentially the least expensive reduction method, especially for Cr^{+6}, but shipping, handling, storage, and air pollution considerations may make it impractical for most CFS systems. Also, its use would not be feasible with very high solids, nonfluid wastes unless water is added.

Sodium hydrosulfite.   This reagent is effective at the alkaline pH of many waste residuals; hence, it normally does not require pH adjustment before addition. Also, it remains effective after addition of the usually highly alkaline solidification reagents. This allows the addition of both types of reagent in rapid sequence, or even together. The major drawback to the use of sodium hydrosulfite is cost: it is several times as expensive as metabisulfate and more of it is required, although no acid is used. Nevertheless, where the Cr^{+6} level in the waste is not too high, the additional cost may be justified by simplification of the process.

Sulfides.   The use of sulfides to precipitate metals as low solubility species is well known and is discussed throughout this book. Sulfides, such as Na_2S, are also known reducing agents. However, Higgins *et al.* [21] found that sodium sulfide alone was not effective as a reducing agent for Cr^{+6} in the range of pH 7 to 10. This is consistent with the author's findings in numerous unpublished treatability studies. While such reduction is known to take place at pH values less than 3, the use of alkali sulfides at low pH is unsafe in practical CFS usage. On the other hand, freshly precipitated FeS—the Sulfex™ process—has been reported to be effective for chromium reduction [24]. Higgins *et al.* found that sodium sulfide, in combination with ferrous sulfate, aided in Cr^{+6} reduction at high pH. Conflicting experimental results, combined with the other problems with sulfides (see discussion later in this chapter) may make their use for this purpose unjustified, since there are a number of alternatives.

Sodium borohydride. This powerful reducing agent is widely used as a reducing agent in industrial applications [17] and has more recently been applied to waste water treatment [25]. It is generally used for the latter purpose as an aqueous caustic solution. $NaBH_4$ has been used to reduce mercury and lead to the elemental states in waste water treatment applications. However, attempts by the author and others to use it in CFS processing have been unsuccessful, and it apparently has not been used commercially in the CFS field. It is claimed by manufacturers to be competitive to other methods of metal removal, in spite of its high reagent cost.

Reductive resins. Resins such as those produced under the trade name Amborane, are used for precious metal recovery [22]. Such resins have the potential for removing other metals such as silver, arsenic, mercury, and antimony in waste streams, since they are selective. The reduced metal, in elemental form, is held within the porous matrix of the resin bead, which could then be encapsulated in a CFS solidification matrix. There do not appear to be any commercial uses of resins to date in CFS work, possibly due to cost.

Hydrazine. Hydrazine (N_2H_4) is used as a reducing agent to remove oxygen and chromates from waste waters, and may be useful in CFS work. However, it has not been used commercially for the latter purpose to date. Actually, hydrazine can function both as an oxidizing agent and as a reducing agent. In basic solution, it is a powerful reducing agent; in acid solution it is less so. Figure 4-1 shows the metals reduced by hydrazine, either to lower valence state or to elemental form. Hydrazine will reduce metals such as nickel, cobalt, iron, and chromium to the elemental form,

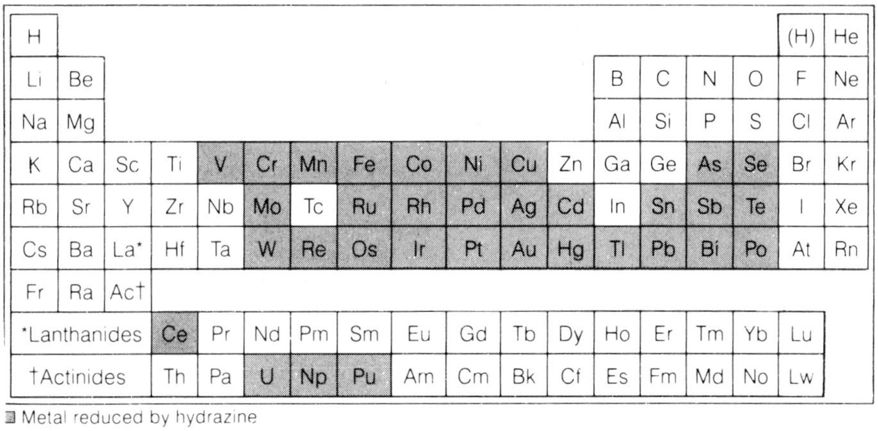

H																(H)	He
Li	Be											B	C	N	O	F	Ne
Na	Mg											Al	Si	P	S	Cl	Ar
K	Ca	Sc	Ti	V	Cr	Mn	Fe	Co	Ni	Cu	Zn	Ga	Ge	As	Se	Br	Kr
Rb	Sr	Y	Zr	Nb	Mo	Tc	Ru	Rh	Pd	Ag	Cd	In	Sn	Sb	Te	I	Xe
Cs	Ba	La*	Hf	Ta	W	Re	Os	Ir	Pt	Au	Hg	Tl	Pb	Bi	Po	At	Rn
Fr	Ra	Act†															
*Lanthanides		Ce	Pr	Nd	Pm	Sm	Eu	Gd	Tb	Dy	Ho	Er	Tm	Yb	Lu		
†Actinides		Th	Pa	U	Np	Pu	Am	Cm	Bk	Cf	Es	Fm	Md	No	Lw		

▨ Metal reduced by hydrazine

Figure 4-1 Metals reduced by hydrazine. (From Olin Chemical Co.)

and is used for this purpose in electroless plating. It also enters into a variety of organic reactions that could be useful in CFS work in specific instances.

Metals, including iron oxide (as rust on containers) catalyze decomposition of hydrazine, so it must be handled with great care, especially in high concentrations [26]. Anhydrous hydrazine is volatile, toxic, corrosive, and forms explosive mixtures with air, so it is not used in this form for water treatment. Even its aqueous solutions must be handled and stored properly, but this is done routinely in water treatment and may be feasible in CFS work. One advantage of the use of hydrazine is that it leaves no solid residue from its reactions. However, because of the various problems associated with its use, other reducing agents should be evaluated first for CFS work.

Oxidation. Oxidation as a step in CFS systems is less common than reduction. Certain metals, such as iron, form less soluble species at higher valence states; these species are highly specific to the metal. One use of oxidation is to destroy stable chelates and other metal complexes so that they can be precipitated as a less soluble species. There are many oxidizing agents that could, in principle, be used, but most are impractical for one reason or another, usually cost. Major problems in the use of oxidants are nonspecificity and competing constituents. If a waste contains both nickel chelate and Cr^{+3} ion, oxidizing the chelate to release nickel may also oxidize the chromium to Cr^{+6}, which must subsequently be reduced again. This is costly not only in the extra processing required, but in the use of an oxidizing agent where it is not needed or wanted. Organics such as oil and grease will compete for the oxidizing agent, making oxidation a very costly procedure if a selective agent cannot be found. Oxidizing agents that have been used in CFS work include oxygen, ozone, hypochlorites, permanganates, and persulfates. The most common usage has been for nonmetals, that is, cyanide, hydrogen sulfide, and phenol.

Precipitation

By far the most important fixation mechanism for metals in CFS systems is chemical precipitation as low-solubility species. All of the inorganic CFS processes and methods that are discussed in Chapters 9–15 precipitate dissolved metals as hydroxides, silicates, or sulfides, and less frequently as carbonates, phosphates, or various complexes. However, few of these systems are really as simple as they appear. Usually, a combination of mechanisms is active and the products of treatment are frequently not simple compounds. Equally important, the wastes being treated by CFS technology often are already speciated as relatively insoluble compounds in sludges, filter cakes, and soils. In spite of these complexities, this section will attempt to delineate the different precipitation types to allow their discussion in a systematic way.

Hydroxide Precipitation. As we have seen, hydroxide precipitation occurs when the pH of a solution of dissolved metal ions is raised to some optimum level for a specific metal. The optimum pH is different for each metal and, often, for different valence states of a single metal. It may also vary for a specific metal ion with the presence of other species in solution, with the redox potential, and with aging of the hydroxide. As the pH is increased, the metal hydrolyzes to form several different species whose total concentration increase. In the presence of excess alkalinity, the metal may start to resolubilize. Aging effects have been explained by Sturm and

Morgan [27], as the slow conversion of an active, crystalline matrix to a more inactive amorphous form. However, crystalline phases were not detected by X-ray diffraction [28] in another study.

Cote [16] has described the dissolution of metal hydroxides in actual waste forms, and we have discussed some of the mechanisms in Chapter 3, where solubility–pH curves are also presented for both calculated and experimental data. As has been pointed out previously, these curves are useful in understanding the effects of pH on hydroxide precipitation, but they should not be used to predict actual solubility in a real system.

Sulfide Precipitation. There are three classes of sulfide precipitation reagents that have been investigated and used in CFS work. They are shown in Table 4-11.

Soluble inorganic sulfides. Other than precipitation as hydroxides, sulfide precipitation of metals has probably been the most widely used method to remove metals from waste water [23]. This method has also been used in CFS treatment, especially for achieving regulatory limits for wastes containing mercury. Most metal sulfides are less soluble than the hydroxides at alkaline pH (arsenic is an exception). Table 4-12 shows the solubility products and solubilities of some of the metals in water.

It is evident from Table 4-12 why sulfide precipitation is effective. The low solubilities required for highly toxic metals such as mercury are often achievable only by speciation as sulfides, since the metal sulfides have solubilities several orders of magnitude lower than the hydroxides throughout the pH range. Also, their solubilities are not as sensitive to changes in pH [27]. This is illustrated in the comparative solubility curves shown in Figure 3-3. However, metal sulfides can resolubilize in an oxidizing environment [29], and there is currently disagreement over the acceptability of metal sulfide sludges in uncontrolled landfills. In some cases, the soluble sulfides will precipitate chromium directly without pretreatment to reduce it to the trivalent state, and sulfides are claimed to precipitate complexed metals [23]. It is necessary to maintain pH 8 or above to prevent evolution of H_2S. While excess sulfide ion is necessary for the precipitation reaction, the excess must be kept to a minimum so that free sulfide removal treatment is not required before the waste can be landfilled.

Precipitation is normally conducted with Na_2S or $NaHS$, not with H_2S, since

TABLE 4-11 Sulfide Precipitation Reagents

Soluble inorganic sulfides	
Sodium sulfide	Na_2S
Hydrogen sulfide	H_2S
Sodium hydrosulfide (sodium sulfhydrate)	NaHS
Calcium sulfide (low solubility)	CaS
"Insoluble" inorganic sulfides	
Ferrous sulfide	FeS
Sulfur	S
Organosulfur compounds	
Dithiocarbamates	[R-NH-CS-S]$^-$
Thiourea	H_2N-CS-NH_2
Thioamides	R-CS-NH_2
Xanthates	[RO-CS-S]$^-$

TABLE 4-12 Solubilities of Metal Sulfides

Species	Solubility (mol/l)	K_{sp}
Cadmium sulfide	6.0×10^{-15}	3.6×10^{-29}
Copper sulfide	9.2×10^{-23}	8.5×10^{-45}
Iron sulfide	6.1×10^{-10}	3.7×10^{-19}
Lead sulfide	1.8×10^{-14}	3.4×10^{-28}
Mercuric sulfide	4.5×10^{-25}	2.0×10^{-49}
Nickel sulfide	1.2×10^{-12}	1.4×10^{-24}
Zinc sulfide	3.5×10^{-12}	1.2×10^{-23}

Source. Adapted from K. F. Cherry, *Plating Waste Treatment*, Ann Arbor Science Publishers, Ann Arbor, MI, 1982.

the reaction must be carried out in an alkaline environment in any case and H₂S is very dangerous to handle in CFS treatment scenarios. The soluble sulfide is usually added as a solution. The stoichiometry is difficult to calculate in CFS systems because the total contents and proportions of sulfide reactive metals are seldom known with any degree of accuracy. Therefore, the addition of sulfide is determined empirically by making up a number of mixtures and determining the minimum addition level that achieves the target leaching goal. The sulfide is added *before* any of the solidification reagents, since the latter contain calcium, magnesium, iron, and other metals that will compete for the soluble sulfide. However, calcium sulfide, which has limited solubility, may act as a sulfide buffer to maintain a small excess available sulfide level in the system, similar to the action of the alkalies in hydroxide precipitation. In fact, one commercial process has been described using CaS for treatment of copper-bearing waste water.

"Insoluble" inorganic sulfides. One answer to the problems encountered with the use of soluble sulfides is the use of very low-solubility species such as FeS or elemental sulfur. In the case of FeS, its solubility is about 0.0001 mg/l in water. The former technology has been commercialized as the Sulfex process by the Permutit Company [24, 30, 31]. The advantage of this approach is that very little excess sulfide is present in the system at any time, so the problem of H₂S odor and toxicity is eliminated.

However, since the solubility of FeS is considerably higher than that of the RCRA toxic metals, the latter will precipitate the available S^{-2} ion, causing more FeS to dissolve. This process continues until all of the metals have been precipitated, or until all of the FeS has been used up. It would also tend to respeciate the hydroxides as sulfides, given sufficient time.

Table 4-13 compares the solubilities of hydroxides and sulfides of a number of metals of interest. The advantage of sulfide precipitation, and the driving force for respeciation, are evident from these data. Cadmium, lead, nickel, and mercury hydroxides all have water solubilities above the recent landban levels (see Chapter 1 and the individual metals later in this chapter), and the leaching levels in tests such as the TCLP are even higher.

The data given in the table apply to an environment in the alkaline pH range. Under acid conditions and/or oxidizing environments, the sulfides can resolubilize. However, with the use of the highly buffered, alkaline CFS systems that constitute

TABLE 4-13 Comparison of Hydroxide and Sulfide Solubilities

	Approximate Solubility (mg/l)		
Metal	Hydroxide	Sulfide	Difference Factor
Iron	5×10^1	1×10^{-4}	5×10^3
Cadmium	3×10^0	1×10^{-8}	3×10^8
Chromium	1×10^{-3}	None	None
Copper	2×10^{-2}	2×10^{-13}	1×10^{11}
Lead	2×10	6×10^{-9}	3×10^8
Mercury	6×10^{-4}	1×10^{-21}	6×10^{17}
Nickel	7×10^{-1}	6×10^{-8}	1×10^7
Silver	2×10	4×10^{-12}	5×10^{12}
Zinc	3×10^2	1×10^{-6}	3×10^8

$$\text{Difference factor} = \frac{\text{hydroxide solubility}}{\text{sulfide solubility}}$$

Source: Adapted from Permutit Company, *Sulfex*™ *Heavy Metals Waste Treatment Process,* Tech. Bull. XIII, No. 6, Paramus, NJ.

by far the largest category, and disposal under controlled landfill conditions, the likelihood of resolubilization is minimal.

One interesting exception in sulfide precipitation is that of chromium, which does not precipitate as the sulfide, but as the hydroxide. Therefore, chromium leaching will be controlled by hydroxide precipitation, and hence by pH, at least in theory. As we shall see later in this chapter, chromium's expected solubility at higher pH takes place only at very high pH levels, and is generally not a problem at the present regulatory levels. It is also claimed by Permutit [32] that FeS has the ability to reduce Cr^{+6} to the trivalent state by the reaction:

$$CrO_4^{-2} + FeS + 4H_2O \longrightarrow Fe(OH)_3 + Cr(OH)_3 + S^0 + 2OH^- \quad (4\text{-}3)$$

It is also claimed that the Sulfex process is effective with complexed metals, a claim also made for soluble sulfides. Feigenbaum [31] has given the test results shown in Table 4-14. Obviously, FeS is more effective on some species than on others. This is not too surprising, given the considerable differences in the complexes.

TABLE 4-14 Sulfex Treatment of Complexed Metals[a]

		Complex		
Metal	Precipitant	Rochelle Salt	Na EDTA	Rochelle Salt/EDTA
Cadmium	Hydroxide	0.67	4.8	4.4
	Sulfide	<0.005	2.9	2.9
Chromium	Hydroxide	0.20	0.05	0.47
	Sulfide	0.20	0.05	0.40
Copper	Hydroxide	0.18	1.8	1.2
	Sulfide	0.01	0.01	0.01
Nickel	Hydroxide	1.2	1.3	1.8
	Sulfide	0.18	1.4	0.88
Zinc	Hydroxide	0.02	2.2	1.5
	Sulfide	0.02	1.5	1.4

[a] Metal concentration in solution (mg/l).

Iron sulfide works best as a freshly prepared slurry, not as pulverized iron sulfide. The slurry is prepared by reaction of ferrous sulfate, a soluble sulfide, and lime. The slurry is then added to the waste. Knocke *et al.* [30] have found that the sulfides produced by the FeS technique have superior dewatering characteristics compared to the analogous hydroxides.

It is important to remember that this technology is not known to have been used on concentrated waste residuals, only on waste water to date. As in so much of the chemistry of metal fixation, results with waste water may not be translatable to the CFS field. It is unlikely, in any case, that anything approaching the sulfide insolubilization levels indicated by Table 4-13 could be achieved in CFS work.

Organosulfur compounds. In principle, organosulfur compounds may have certain advantages over the inorganics. Lo *et al.* [33] found that hexadecyl mercaptan was effective in reducing leachability (TCLP) of mercury in a spiked electroplating waste from 2.12 mg/l in the normal Chemfix® process (Portland cement plus sodium silicate) to 0.0006 mg/l at a mercaptan addition level of 8000 mg/l. The reagent was not effective in reducing leachability of any other RCRA metals. Several investigators [34, 35] have found thiourea to be useful in reducing mercury leaching from caustic-chlorine production wastes and other residues to very low levels, 0.0001 mg/l, in water. Nakaaki *et al.* [36] lowered the mercury leachability of seabed sludge with a fixing agent containing polydithiocarbamates and an iron or copper salt.

Notably, all of these applications of organosulfur compounds have been on mercury. This may be due to the extremely low regulatory leaching levels for mercury, or to the basic insolubility of mercury sulfides, or both. Little has been reported on the use of these reagents for other metals in sludges and other waste residuals. It is likely that most investigators are keeping the results of their studies proprietary. Glod and Ader have reported (private communication) the use of a proprietary system to fix mercury in caustic-chlorine production wastes to very low levels. Environmental Technology Inc. [37] has announced a line of "organosulfur" reagents for use in waste water treatment, and there are a number of other reagents that may be of this class offered on a proprietary basis. Another approach to the use of "sulfides" is in the insoluble xanthates, which react with the metal ion, but fix it in the form of an organic polymer insoluble in water. This will be discussed further under the subsection "Bonding to Insoluble Substrate."

Silicate Precipitation. The overall subject of soluble silicates in CFS systems is discussed in Chapter 11. Here, we deal only with the fixation reactions of this system.

Soluble silicate chemistry. The reactions of polyvalent metal salts in solution with soluble silicates have been studied extensively over many years [38]. Nevertheless, the "insoluble" precipitates that result from such interactions are not usually well characterized, especially in the complex systems representative of most wastes. Metal silicates are nonstoichiometric compounds in which the metal is coordinated to silanol groups, SiOH, in an amorphous silica matrix.

The reactions of soluble silicates in solution is best summarized by Vail [39], who states: "The precipitates formed by the reaction of the salts of heavy metals with alkaline silicates in dilute solution are not the result of the neat stoichiometric reactions describing the formation of crystalline silicates, but are the product of an interplay of forces which yield hydrous mixtures of varying composition and water content." These reaction products are usually noncrystalline and therefore very difficult to characterize structurally. They are most often described as hydrated metal ions associated with silica or silica gel. Iler [40] mentions "that many ions are held irreversibly on silica surfaces by forces still poorly understood in addition to ionic attraction."

The composition and form of the metal "silicates" formed from metal ions and soluble silicates are functions of the conditions under which they are formed: temperature, concentration, addition rate, metal ion speciation, or presence of other species. For example, in dilute solutions, the species may remain in colloidal suspension, while at higher concentration or in the presence of other destabilizing conditions, it will precipitate [39]. Metals are often present as soluble complexes or as negatively charged anions that should not bind to the silica surface. The pH is very important, for, as Iler [40] has pointed out, "silica suspended in a solution of most polyvalent metal salts begins to adsorb metal ions when the pH is raised to within 1-2 pH units below the pH at which the polyvalent metal hydroxide is precipitated." ("Silica" here can refer to either colloidal silica or polymeric silicates.) Petit and Rouxhet [41] found the same phenomenon using infrared spectroscopy and X-ray diffraction. If the pH is not raised, the metal is not precipitated; if the pH of the system is already above the adsorption point, the metal may precipitate as the hydroxide instead. Different metals have different precipitability by soluble silicate solutions; according to Vail [39], the order, beginning with the most precipitable, is copper, zinc, manganese, cadmium, lead, nickel, silver, magnesium, calcium.

The preceding discussion deals only with the complexities of reactions in solution, and only with those between the metal species and the soluble silicate. As we have seen, with real wastes the components may be present also as suspended solids or in immiscible liquid phases. Metals may already be speciated as relatively insoluble hydroxides or other solid phase compounds. Nearly always, cement or other solidification reactant is added to the system, and these materials themselves often interact with the metal species, as well as with the soluble silicate. If too little soluble silicate is added to the system, the reaction products that form will depend on which competing reactions are most successful, and also on the order of addition of the reactants. If an excess is added, unwanted leachable metal ion silica complexes may form (J. S. Falcone, personal communication). The rate of addition, degree of agitation, and temperature will also affect the nature of the reaction products.

Frequently, the metals have been precipitated as low solubility species, such as hydroxides. In this state, there is little reaction between the metal species and the soluble silicate. If the metal hydroxide is more soluble than the silicate, there may be gradual respeciation of the hydroxide at the particle surface, but total respeciation would not be expected to occur except over a very long period of time. Furthermore, the soluble silicate will not be available as such for long, because it will react rapidly with other components of the CFS system, such as calcium. The ultimate result is a mixture of metal hydroxides, sulfides, etc. dispersed in a cementitious matrix. In this situation, the primary function of the soluble silicate may be to reduce permeability [42]. It does this by forming precipitates in the matrix that block pores, reducing the effective pore volume and slowing the movement of any mobile species through the matrix into the environment. This effect is probably also responsible for the reduced leachability of species that would not be expected to react with soluble silicates: monovalent cations, anions, and organics. It also accounts for the fact that monoliths produced by soluble silicate systems often show higher levels of leachability when subjected to mechanical degradation, either naturally or during leaching test procedures.

The preceding problem of speciation is avoided when the metal species to be fixed is in solution. Unless it is complexed in soluble, stable form, the metal can usually be precipitated from solution as a silicate that exhibits low solubility through a wide pH range [43], approximately from pH 2 to pH 11 or higher. This is especially important for the heavy metals of environmental interest that exhibit amphoterism, such as chromium, zinc, and lead. The hydroxides of these metals exhibit minimum solubility through a narrow pH range, usually in the area of pH 7.5 to 10. Because most CFS systems are quite alkaline, usually above pH 11 (at least initially), the solubility of the metal species in the CFS-treated waste may actually be higher than in the original, untreated sludge. This problem tends to be exacerbated with increased surface area. While the soluble silicate, if available, will respeciate the metal hydroxide as it dissolves, it usually will not be available because it will have been used up in other reactions. There are two silicate-speciation solutions to this problem: (1) create a system that continuously produces soluble silicate as it is needed, analogous to the insoluble sulfide approach described earlier, or (2) pretreat the waste by reducing pH to dissolve the metal hydroxide, then respeciate as silicate.

In situ silicate formation. Insoluble silicates and silica dissolve under both very low and very high pH conditions. Calcium silicates dissolve in the alkaline environment created when cement is mixed with water, releasing $Ca(OH)_2$ (private communication). Highly alkaline conditions that result from the use of cement and lime kiln dusts do not cause the metal leaching that would be expected of metal hydroxides under such conditions, possibly because the dissolving metals are reprecipitated as silicates. The only source of soluble silicates for this reaction would be dissolving silicates from the reagents. Bishop [44] postulated that the observed low leaching of metals in Portland cement CFS systems was due to the association of the metals with silica, possible as silicates. Perhaps most interesting is a patent application (EnviroGuard, Inc., private communication) that claims the continuous production of soluble silicates from biogenetic, amorphous silica, and alkali. This discovery will be explored further in Chapter 11.

The second method of creating metal silicates from metal hydroxides is by deliberate dissolution of the metal species—hydroxide, carbonate, sulfide, etc.—in acid followed by reprecipitation by silicates or polymeric dissolved silica. One version of this approach is the SoliRoc™ process [45] (Chapter 11).

Other silicate processes. An unusual amount of activity in silicate fixation and solidification processes took place in the mid- to late 1970s, and continued into the 1980s. Relatively little notice was taken of this activity, however, because nearly all of it was either proprietary or was published only in the Japanese patent application literature. Recently, a number of U.S. patents have issued in this area. Perhaps the best known process, Chemfix, uses a combination of Portland cement and sodium silicate (see also Chapter 11). The process is said to be effective in preventing certain metals—cadmium, chromium, copper, zinc, and nickel—from leaching in the TCLP test [16, 46]. However, it was not effective on arsenic, antimony, and mercury, actually promoting the release of the latter elements.

The various soluble silicate processes and techniques will be discussed in detail in Chapter 11, since most relate to complete CFS systems, not just fixation of metals.

Carbonate Precipitation. In certain cases, metal carbonates are less soluble than their corresponding hydroxides. This can be seen in Tables 4-3 and 4-4 for barium, cadmium, and lead, and has also been mentioned by Sturm and Morgan [27]. In cement chemistry, the natural formation of carbonates from carbon dioxide from the air is termed "carbonation." Cote [16] has shown that carbonate ion concentration in a system depends on both CO_2 partial pressure and pH. The carbonate species, CO_3, dominates at pH values larger than 10.3. The carbonation process at alkaline pH takes place according to Eq. (4-4):

$$Me(OH)_2(s) + H_2CO_3 \longrightarrow MeCO_3(s) + 2H_2O \qquad (4-4)$$

The pH at which carbonation occurs depends on the solubility products of the carbonate and hydroxide species, and the CO_2 concentration.

Patterson et al. [47] found that the formation of hydroxide precipitates controlled the solubility of zinc and nickel over a range of pH values, but cadmium and lead solubilities were controlled by carbonate precipitates in a narrower range. This is in keeping with the results of treatability tests conducted by the author on lead-bearing wastes, using soluble and "insoluble" carbonates. However, precipitation by addition of carbonates has not been widely used in either waste water or CFS treatment. One problem is that the carbonates are decomposed at low pH, such as that encountered in the TCLP test, if the pH of the leaching solution in contact with the solid actually drops too low. If CO_2 is evolved, the reaction is irreversible even if the final pH of the leachant is high, and the final speciation of the metal will be as the hydroxide. This may explain the widely varying results reported (informally) by CFS investigators testing carbonates, since the efficacy of carbonate precipitation would seem to be unusually sensitive to test variables; the sample being tested could either lose carbonate as CO_2, or gain it in the same way from the atmosphere (air) in the test container. As we have seen in Chapter 3, the test variability for the TCLP and other regulatory protocols is large.

Carbonate precipitation may explain the highly variable results obtained with commonly used reagents, such as kiln dusts, that contain carbonates from the ingredients used in manufacturing both lime and cement. Some observations along these lines will be discussed in the individual metals sections in this chapter, and in the later chapters on specific CFS systems. However, more investigation is needed in this area of fixation technology.

Phosphate Precipitation. There appears to have been very little work done on the use of phosphates for metal fixation, although ferric chloride is commonly used for phosphate removal in waste water treatment. Tables 4-3 to 4-5 indicate that there are possibilities here for those metals that form low-solubility phosphate species, but the CFS literature contains no examples of the use of phosphates in these systems.

Phosphate chemistry is very complex and varied. Compounds containing monomeric PO_4^{-3} are called orthophosphates or simply phosphates. The latter broad term is also used to describe all compounds in which phosphorus atoms are surrounded by a tetrahedron of four oxygen atoms. Since the oxygen atoms can also be shared among tetrahedra, chains, branched chains, and rings can be formed, in one-, two-, and three-dimensional networks [26]. The linear formula can be expressed as $M_{n+2}P_nO_{3n+1}$. For orthophosphates, $n = 1$. Linear chains are known as *polyphosphates* for $n = 2, 3, \ldots$ (pyro- or dipoly-, tripoly-, etc.); cyclic rings as *metaphosphates;* and branched or cage structures as *ultraphosphates*. When n becomes very large, the composition becomes indistinguishable from that of the cyclic metaphosphates $(MPO_3)_n$.

The simple phosphate salts of the toxic metals have low water solubility, although they are soluble in acids (Tables 4-3 to 4-5). However, as we saw earlier in discussing metal complexes, phosphates with $n > 1$ have the potential to sequester the metals as water-soluble species. This property is the basis for the detergent and water-treatment applications of phosphates. Therefore, the presence of phosphates in the waste or the CFS system may be harmful or beneficial to fixation, depending on the phosphate species.

Coprecipitation. The removal of toxic metals from waste water with systems that coprecipitate and/or flocculate them with iron salts is well known and widely used [48]-[50]. More recently, it has been used to reduce the solubility of various toxic metals in hazardous waste treatment [51]. The ratio of Fe^{+2} to Fe^{+3} is important, with ratios of 1:1 to 1:2 reported as yielding optimum results. Sano *et al.* [52] attribute the removal of zinc and cadmium to the formation of ferrite crystals that could subsequently be removed magnetically. The crystals capture the other metals in the lattice, or adsorb them on the surfaces. This mechanism has not been confirmed by others, who ascribe the results obtained to coprecipitation followed by flocculation.

Sols of hydrous metal oxides are stabilized by the presence of excess ferric ion, but acquire a negative charge, destabilize, and flocculate under alkaline conditions. As the system becomes alkaline, the ferrous ion is also easily oxidized to ferric, and precipitates as the hydroxide. These reactions remove other metal ions from solution, reducing their concentrations to levels below those obtained with simple hydroxide precipitation. Whatever the mechanism, the net effect is reduction of leachability in many instances. In addition to the iron species, coprecipitation with calcium carbonate and other species has been reported [53]. Specific uses for fixation of individual metals will be discussed later in this chapter.

Inorganic Complexation. We have previously discussed complexation as a problem in fixation, due to the presence of very stable, nonreactive, soluble metal species in solution. However, other complexes can have the opposite effect, producing insoluble metal species that may have lower solubilities than the simple metal compounds. This approach to fixation in the CFS field has not been deliberately used commercially, but it is likely that the fixation of metals in commercial systems is at least partially due to this phenomenon. Many metals exist in ores and rocks as complex crystalline structures. The curing of cement-based solidification systems may eventually result in such complexes. Another example is the formation of ferrite crystals mentioned in the previous section, if indeed these actually form in CFS systems. Both ferricyanide and ferrocyanide form low-solubility species with a number of metals (Tables 4-3–4-5).

This is an area that needs exploration, because it may have important benefits in CFS work. We have already observed that wastes and waste forms resulting from

CFS often show changed leachability with aging, and this may very well be due to changes in crystal structure and other morphological phenomena. If mechanisms for the formation of low-solubility, complex metal species can be elucidated, it may be possible to create favorable conditions for them to operate. Since inorganic complexes are often crystalline, they can be studied by X-ray and electron diffraction techniques—a definite advantage over the amorphous species that we usually encounter.

Organic Complexation. Ordinarily, CFS investigators think in terms of fixation of metals with inorganic species. However, many organic compounds also form low-solubility species with certain metals. Tables 4-3 and 4-4 show low solubility for the tartrates of cadmium, lead, mercury(+1), and nickel. Unfortunately, little information is available on the actual numbers for metal organic compound in aqueous solution, or on the effects of pH and other ions. Manahan and Smith [54] report that humic acids formed in the decay of vegetable matter immobilize metal ions in sediments and soil. Humic substances—humic acids and fulvic acids—are known to accumulate heavy metals [55], possibly adsorbing on inorganic solids that remove them from solution. Saar and Weber [56] state that the "many oxygen-containing functional groups, particularly –COOH and –OH . . . enable fulvic acid to behave as a polyelectrolyte." The complexed metals, even if soluble, are less toxic to aquatic organisms.

One area in which some work has been done is in the use of organosulfur compounds previously discussed in the section on sulfide precipitation. These agents are expensive and have been used primarily in the water treatment field, but they demonstrate that the general subject of organic complex precipitation warrants exploration. Takatomi and Tukada [57] claimed leachability reduction of 10- to 100-fold for Hg, Cd, Cr^{+6}, As, Pb, and Cu in an incinerator ash using additions of amino acids to cement-based CFS systems, specifically L-lysine. Lo et al. [33] found that the addition of polyethylene imine at the 4 percent level helped to immobilize As^{+5} in cement-soluble silicate CFS systems; however, it promoted the release of several other metals, including mercury.

A different aspect of this approach, bonding of metals to insoluble substrates, is discussed next.

Bonding to Insoluble Substrates

Rather than precipitate metals from solution in the usual sense—by formation of low-solubility species from ions in solution—another approach is to react the metal with an active area or functional group on the surface of an insoluble substrate. Nelson [58] used a treated leather waste to remove heavy metal ions, particularly lead and cadmium, from nitrate and acetate solutions. He achieved levels of 0.3 to 0.7 ppm from solutions containing 78 to 200 ppm. The leather was treated with oxalic acid and ammonia to remove the chromium present from the tanning operation.

Another process [59] uses modified casein to remove cadmium, chromium, copper, mercury, nickel, and zinc from waste water. Chromium reportedly can be removed as directly as Cr^{+6}, without prior reduction, to levels below 1.0 ppm from streams containing up to 500 ppm Cr^{+6}. The casein is modified by treating with formaldehyde to form a cross-linked, insoluble product. The process works on the principle of ion exchange at pH >4. Casein is less expensive than conventional ionexchange resins.

The most widely publicized insoluble substrate for heavy metal fixation has been insoluble starch xanthate (ISX) [60, 61]. ISX is produced by treatment of starch with cross-linking agents, then xanthating it with CS_2 in the presence of an alkali metal base, such as sodium hydroxide. The resulting product is a particulate solid with the structure

$$\left[\text{starch} - O - \overset{\overset{S}{\|}}{C} - S - \right]$$

Since 1980, ISX has been a commercially available product with wide usage in the treatment of waste waters from metal finishing and similar sources [62]. Removal efficiencies for this material are shown in Table 4-15 [63].

In addition to the removal ability previously shown, ISX is also reported to act as a reducing agent for Cr^{+6}. Chromium is reduced to Cr^{+3} and removed by neutralization to precipitate the hydroxide [64]. It should be noted that ISX operates at low pH (the Table 4-15 tests were done at pH 3.7) and that the process is less effective as pH rises. Tests with selenium [65] indicated that the metal after immobilization with ISX could not be extracted with either high- or low-pH water, but the chloride ion does cause some release of selenium. It would therefore appear that this agent could be used as an additive in CFS systems in a two-stage process: addition of ISX followed by solidifying agents.

Bricka and Hill [66] incorporated xanthated metal sludge into a CFS system in a three-step process. Both cellulose and starch xanthates were added to synthetic waste waters to produce a sludge that was subsequently dewatered and solidified with Portland cement. The results were compared with sludge precipitated as hydroxides, and with the unsolidified xanthate sludge, and are presented in condensed form in Table 4-16.

The conclusions to be drawn from Table 4-16 are: (1) solidification with Portland cement provides good fixation of cadmium, chromium, and nickel in either

TABLE 4-15 Heavy Metal Removal with ISX

Metal	Initial Concentration (mg/l)	ISX Addition (% of solution)	Residual Concentration (mg/l)
Ag	53.94	0.032	0.016
Au	30.00	0.050	<0.010
Cd	56.20	0.064	0.012
Co	29.48	0.064	0.090
Cr^{+3}	26.00	0.064	0.024
Cu	31.77	0.032	0.008
Fe^{+2}	27.92	0.032	0.015
Hg^{+2}	100.00	0.064	0.001
Mn	27.47	0.064	0.015
Ni	29.35	0.064	0.160
Pb	103.60	0.064	0.035
Zn	32.69	0.032	0.294

Source: Adapted from *Insoluble Starch Xanthate*, Pollution Technology Systems, Inc.

TABLE 4-16 Leachability of Xanthates and Hydroxides Solidified and Unsolidified

Sample Tested	EPT Leachate Concentration (mg/l)			
	Cd	Cr	Hg	Ni
Solidified				
Starch xanthate	<0.0001	0.0643	<0.0008	0.1168
Cellulose xanthate	<0.0001	0.0750	0.002	0.0038
Hydroxide	<0.0001	0.0138	0.565	0.0025
Unsolidified				
Starch xanthate	0.3030	12.66	<0.0008	50.95
Cellulose xanthate	26.05	3.385	0.0133	248.3
Hydroxide	57.93	242.0	0.8392	148.8

Source: Adapted from R. M. Bricka and D. O. Hill, Metal immobilization by solidification of hydroxide and xanthate sludges. In *Proc. 4th International Hazardous Waste Symposium on Environmental Aspects of Stabilization/Solidification of Hazardous and Radioactive Wastes,* Atlanta, GA, May 3-6, 1987.

xanthate or hydroxide systems; (2) acceptable fixation of mercury is achieved only by xanthates, even after solidification; (3) starch xanthate fixes mercury better than cellulose xanthate; (4) xanthates alone are not adequate for fixation of cadmium, chromium, and nickel—solidification is also required. The latter finding probably demonstrates the necessity for pH control in the alkaline range for adequate fixation of cadmium, chromium, and nickel, a result that is not surprising. This study also underscores a basic fundamental of metal fixation to today's standards—that each metal must be considered separately in choosing a CFS system. There are no "magic" solutions that work for all wastes.

The results of studies such as that just shown must be used with caution in choosing a CFS system in the real world. This work was done with synthetic solutions where the metals were not already speciated as partially soluble compounds. Also, the xanthate fixation was done in dilute solution, presumably at an optimum pH for the system, before the products were concentrated and solidified. Based on the author's experience, it is doubtful whether such low leaching levels could be achieved in the treatment of real concentrated wastes, especially sludges. However, the possibilities for use of this process, at least for mercury fixation, are interesting if the cost is not too high. This is becoming increasingly important in view of the new, much lower landban leaching requirements.

Sorption

In this section, we use the term *sorption* to cover adsorption at surfaces, as discussed in Chapter 3, absorption into the interior of solid substrates, and chemisorption. Often, it is impossible to distinguish between these various effects in real, complex systems, and in the practical sense, it doesn't matter anyway. As Swallow *et al.* [50] point out:

> The capacity of amorphous hydrous ferric for metal ions makes the meaning of the terms "surface" and "adsorption" somewhat arbitrary. It becomes necessary to provide a more satisfactory image of the system than that of an interface separating two phases. One is clearly not concerned with the interface between the visible precipitate and the bulk solution, but rather the interface between some microstructure and its immediately surrounding water. Amorphous hydrous ferric oxide has been described as "amorphous, randomly cross-linked aggregates

containing large and indefinite amounts of water" [67]. This description calls to mind the structure of a swollen ion-exchange resin which is permeable to hydrated ions. The ions are free to diffuse throughout the structure and are not restricted to external "surface" sites.

While this description was applied only to one sorption system, it probably describes the real situation with many sorbents in CFS work.

The other important consideration to be kept in mind while reading this section is that most data on sorption of metals, and other species, result from experimental work with water or waste water treatment. In these systems, the sorbent is contacted with the contaminants for a specific time period, then removed. Furthermore, the contact is often of a dynamic nature, with the water flowing through the sorbent, and equilibrium is not attained. In CFS work, the sorbent remains in the waste mixture, and eventually achieves some sort of equilibrium. As the solidification part of the system cures, pH and other changes occur that can alter the immobility of the sorbed species, for better or worse.

Sorbents can be broken down into a number of classes or groupings. For the purposes of this discussion, the following groups are pertinent and convenient:

- Metal Oxides: Iron, manganese, aluminum, etc.
- Clays: Bentonite, montmorillonite, attapulgite, illite, kaolinite; natural and modified
- Natural Materials: Peat moss, natural zeolites, sawdust, vermiculite, sand, etc.
- Synthetic Materials: Zeolites, flyash, cullite, activated alumina, organic polymers, etc.
- Activated carbon

Chan et al. [68] evaluated ten potential sorbents, natural and synthetic, for their ability to remove a variety of pollutants from landfill leachate. They found that no single sorbent is effective in removing all pollutants, and some were not attenuated by any of the materials tested (bottom and flyashes, vermiculite, illite, sand, activated carbon, kaolinite, natural zeolite, activated alumina, and cullite). As expected, activated carbon was most effective for organics, with illite also being successful. Specific sorbents were effective on fluoride and cyanide, but not on chloride. Activated alumina was moderately effective in removing nickel, but none of the other sorbents achieved any success with this metal. None of the sorbents tested were effective for lead.

Other investigators have found that flyash, kaolin, and sawdust [69], and a calcined mixture of bentonite and flyash [70] were effective in immobilizing a variety of metals. Benson [69] passed solutions of metal salts through columns containing the fixing agents and measured the attenuation of contaminants in the effluent. The fixing capacities for the various metals and fixants at pH 4.0 are shown in Table 4-17. The efficacy of sawdust may be due to organic complexation reaction, while precipitation is believed to be the predominant factor with flyash. Ion exchange may be most important with kaolin, since the cation exchange capacity measured is close to that predicted for this clay.

Peat moss was found to effectively remove metals from a metal finishing effluent [71] containing chromium, copper, nickel, zinc, and cyanide. Removal efficiency varied with the metal, pH, and redox potential (degree of aeration). A Japanese

TABLE 4-17 Metal Fixing Capacities of Natural Materials

Metal	Fixing Capacity (mg/g metal to fixant)		
	Kaolin	Flyash	Sawdust
Chromium	0.69	0.91	1.28
Copper	0.26	0.64	0.63
Cadmium	0.05	0.22	0.11
Lead	0.28	1.60	1.42
Zinc	0.13	0.51	0.39

Source: Adapted from R. E. Benson, Jr., Natural fixing materials for the containment of heavy metals in landfills. In *Proc. Mid-Atlantic Industrial Waste Conference*, pp. 212–215, 1980.

patent disclosure [72] claims that an activated coral sand has a large adsorptivity for metals.

Clays. A large amount of work has been done on the sorption of metals by clays. Much of this information comes from the soil sciences and agricultural fields. Clays vary widely in their compositions, structures, and other characteristics, so wide variation in sorptive capacity would be expected. Grim states the present concept of clay minerals [73]:

> ... clays are essentially composed of extremely small crystalline particles of one or more members of a small group of minerals which have come to be known as clay minerals. The clay minerals are essentially hydrous aluminum silicates, with magnesium or iron proxying wholly or in part for the aluminum in some minerals and with alkalies or alkaline earths present as essential constituents in some of them. Some clays are composed of a single clay mineral, but in many there is a mixture of them. In addition to the clay minerals, some clay materials contain varying amounts of so-called nonclay minerals, of which quartz, calcite, feldspar and pyrite are important examples. Also many clay materials contain organic matter and water-soluble salts. ...

Grim also sets forth the clay classification system shown below, which bases distinctions on the shape of the clay minerals and the expandability of the structure.

I. Amorphous
   A. Allophane group
II. Crystalline
   A. Two-layer type (sheet structures composed of units of one layer of silica tetrahedrons and one layer of alumina octohedrons)
      1. Equidimensional—kaolinite, nacrite, etc.
      2. Elongate—halloysite
   B. Three-layer types (sheet structures composed of two layers of silica tetrahedrons and one central dioctahedral or trioctahedral layer)
      1. Expanding lattice
         a) Equidimensional
            (1) Montmorillonite, sauconite, etc.
            (2) Vermiculite
         b) Elongate—nontronite, saponite, hectorite

    2. Nonexpanding lattice
       a) Illite group
  C. Regular mixed-layer types (ordered stacking of alternate layers of different types)
    1. Chlorite group
  D. Chain-structure types (hornblend-like chains of silica tetrahedrons linked together by octahedral groups of oxygens and hydroxyl containing Al and Mg atoms)
    1. Attapulgite
    2. Sepiolite
    3. Palygorskite

For the purposes of this book, to this classification should be added the bentonite clay found in the Wyoming area of the United States, and other similar clays sold commercially as bentonites. This material is a highly colloidal, plastic clay that has the unique characteristic of swelling to several times its original volume when placed in water.

The structural differences between the two-layer and three-layer clays are seen in Figures 4-2 and 4-3, and we have already briefly mentioned these differences in Chapter 3. Figure 4-4 gives a schematic representation of the attapulgite structure. Because this is a three-dimensional chain structure, it does not swell or expand like certain montmorillonites or vermiculite. The many subvariations in crystal structure contribute strongly to the different properties of the clays in their interactions with metal ions and other constituents in hazardous wastes. Some aspects of this were discussed in Chapter 3, and others will be covered in subsequent chapters. The prop-

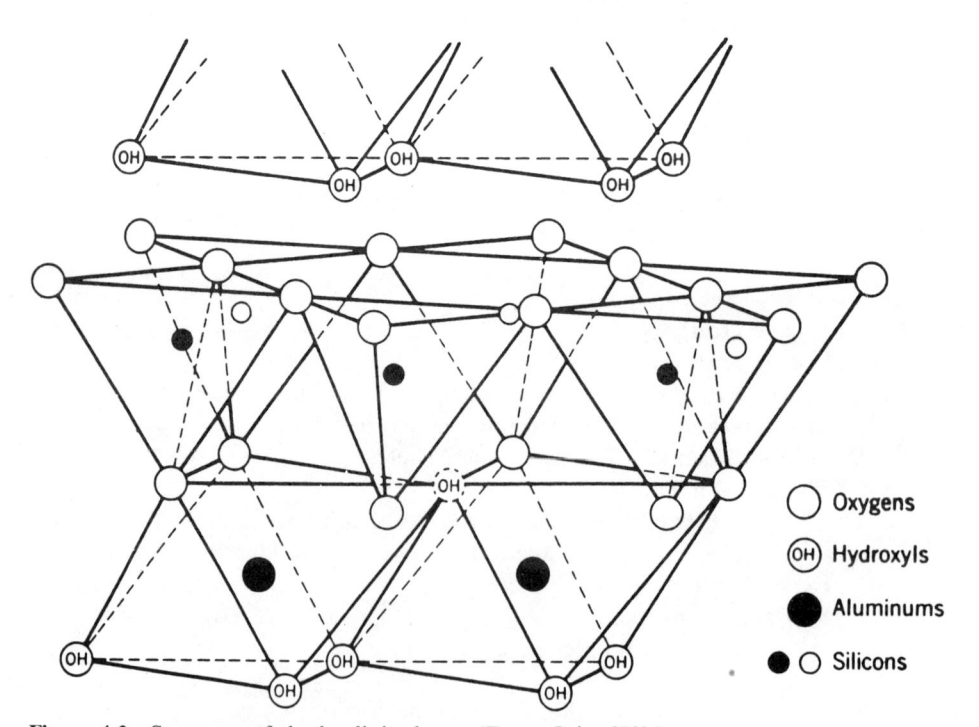

Figure 4-2 Structure of the kaolinite layer. (From Grim [73].)

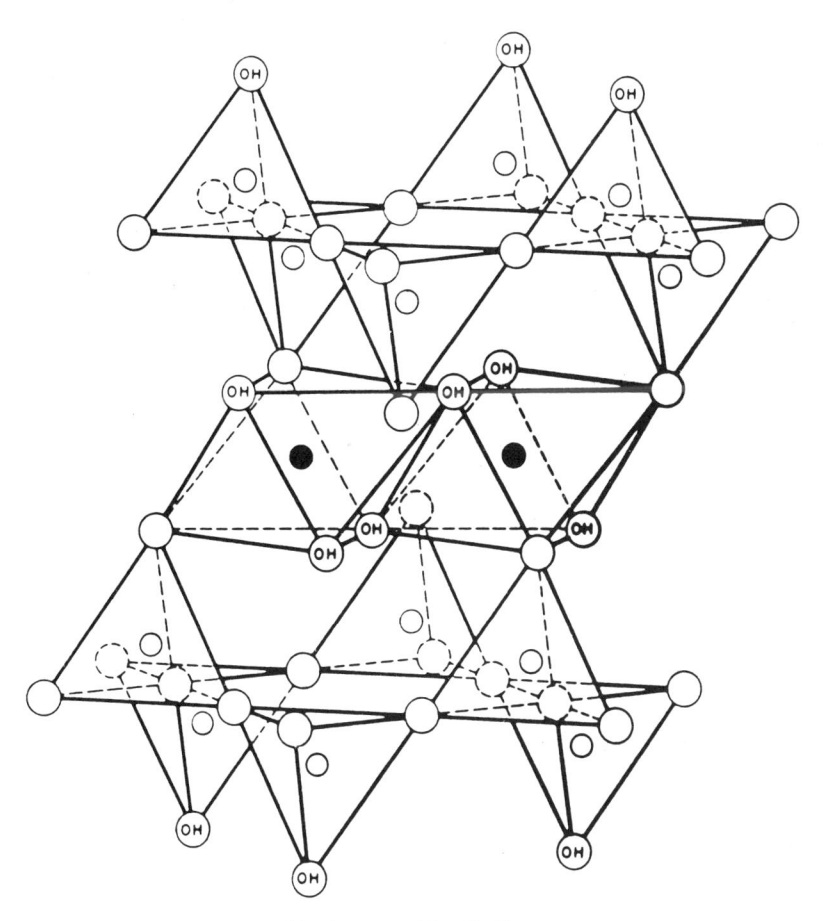

Exchangeable Cations
n H$_2$O

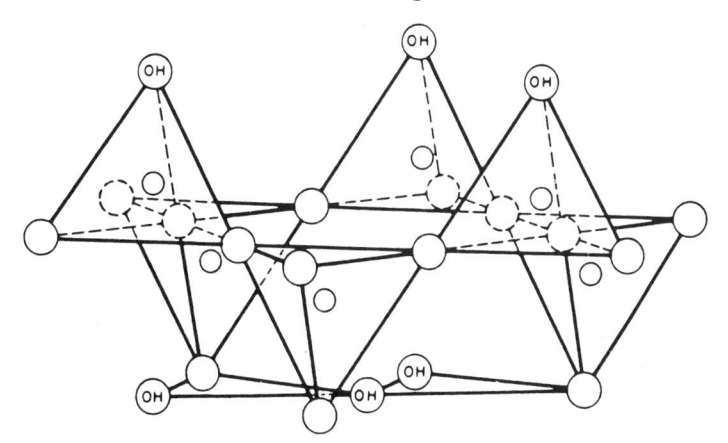

Figure 4-3 Structure of smectite. (From Grim [73].)

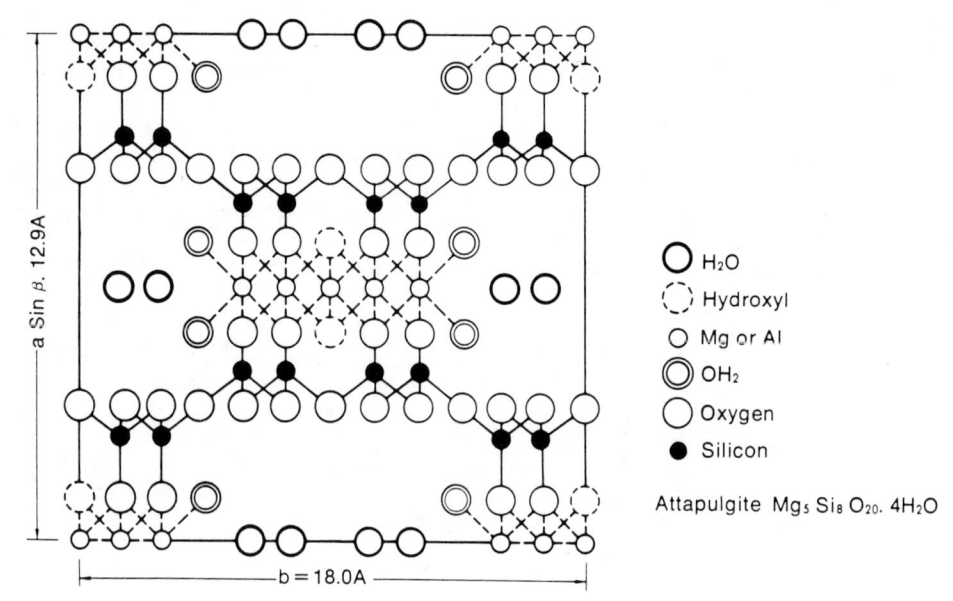

Figure 4-4 Schematic presentation of the structure of attapulgite. (From Grim [73].)

erties of clays are important not only for fixation of metals and other species, but also in certain solidification processes that will also be discussed in later chapters.

Griffin et al. [74] found that kaolinite and montmorillonite are good sorbents for Cd, Zn, Cu, Cr^{+3}, and Pb at high pH, especially above pH 7–8. As anions, As^{+5}, Cr^{+6}, and Se^{+4} show maximum sorption at pH 5 or below, and are not sorbed under alkaline conditions. These findings are very important for CFS technology, since nearly all CFS systems operate at high pH. Many other investigations have shown the sorptivity of clays for metals in water and waste water, and in soil chemistry. These will be discussed later as we explore the reactions of the individual metals and other species of interest. However, the usefulness of clays for metal fixation has not been adequately explored in the author's opinion. The structures, functional groups, and possibilities for diodochic substitution in the lattice seem to hold numerous possibilities for immobilization of metals by a variety of mechanisms. Since clays are basically very inexpensive materials, modification for specific purposes may be feasible for metals, as it appears it will be for organics.

Active Oxides. We have already discussed the role of metal oxides/hydroxides in coprecipitation [49, 50]. Actually, it is difficult to separate the effects of coprecipitation from those of sorption in systems where the sorbent is formed in situ. However, in situ formation is the most common method in investigations of metal removal from water and waste water, since it is the normal mode in water treatment operations. Another example of sorption of metals on active oxides is given by Posselt and Anderson [75]. They used hydrous manganese dioxide formed in situ by reduction of permanganate ion (Mn^{+7}). Metal ion sorption was rapid, with equilibrium attained within minutes. Dzombak and Morel [76] found considerable data that show that metals adsorb strongly to iron and aluminum oxides. Lagvankar et al. [77] described an interesting process using iron filings that are activated by the waste stream itself. Activation involves the formation of an active surface oxide layer as the iron oxidizes in an aerated solution. This is done in a reactor, and requires hours to reach equilibrium, but good removal efficiencies are attained on electroplating, metal finishing, and utility waste waters.

Synthetic Sorbents. Most synthetic metal removal agents and processes are ion-exchange resins and zeolites. These will be discussed in the next section. Various proprietary "adsorbents" are marketed for metals removal from waste water, such as the ALM-series from Nippon Soda Co. [78]. ALM-525 is said to achieve levels of <0.5 ppb for Cd, <0.1 ppb for Hg, <1.0 ppb for Pb, and <10.0 ppb for Cu and Cr. How well these agents would work in CFS systems is not known. Generally, the primary problem with the use of proprietary additives is cost. Unless they are selective, much of their capacity may be used in sorbing metals of no interest environmentally (in CFS), thus requiring very large and costly additions that may even interfere with the solidification reactions by removing calcium. Thus, it is necessary to test each additive in the overall system in which it is to be used.

Activated Carbon. The most common use of activated carbon in the environmental areas of technology is for adsorption of organics. However, activated carbons of various types can be effective sorbents for heavy metals [79], at least for waste water treatment. Powdered activated carbon (PAC) was much more effective than the granular form (GAC), although the equilibrium capacities of the two may be nearly the same [80]. Huang et al. [81] found that removal of Cd was more effective in the high-pH region, but the opposite was true of Hg. The latter was partially removed by volatilization after reduction to Hg^0 by the carbon. Kinetically, PAC achieved equilibrium within 30 minutes, while GAC required about ten times as long.

The usefulness of activated carbon for metal fixation in CFS work is unknown at this time. Since so many techniques are available for immobilization of metals, carbon would not seem to be an attractive approach because of its cost and its relative nonspecificity. Also, carbon sorption is known to be a highly reversible phenomenon, probably more so than many of the other sorptive systems already discussed. Nevertheless, it may be a useful tool in special situations. And for organics and some other species of interest in CFS, it may be quite attractive. The latter possibilities will be discussed in Chapter 6.

Ion Exchange

It is likely that all fixation (and solidification) methods for metals involve some degree of ion exchange activity. Many materials exhibit this property in some degree. Some of the most common are organic ion-exchange resins, natural or synthetic zeolites, silica gels, soils, and metal oxides. Although ion exchangers can contribute to fixation of metals, they can also interfere with the setting and curing reactions of cementitious and pozzolanic CFS systems by removing calcium ion from the system (see Chapter 10). While organic resins and other synthetic materials will usually be too expensive for use in CFS systems, with the possible exception of the nuclear wastes area, the natural materials such as clays may be quite useful. Many of the reagents used for solidification also contribute ion exchange properties to the system.

It is not necessary here to discuss the theory and practice of ion exchange.

Kennedy [81] offers a good review of the theory and a system for selecting the conditions to sorb metals from specific systems. Many books and other writings are available on the subject, and the manufacturers of ion exchange materials offer much practical advice. However, it must be remembered that, like sorption, most of the data apply to water or waste water treatment, not CFS. And, like sorption, it is fundamentally a reversible process. In waste water treatment, the ion exchange material is removed from the water when it is exhausted or has fulfilled its function, thereby eliminating any chance for the contaminant to get back into its source vehicle. In CFS, the contaminated material remains in the solid product. If conditions favoring the reverse reaction occur, the contaminants would reenter the mobile phase of the solid and be available for leaching.

Since many of the solidification reagents and fixants used in CFS probably function for metal immobilization by more than one mechanism, it is better to discuss their utility later under the individual metals species. There, we will be more interested in results than in mechanisms.

Miscellaneous Mechanisms, Materials, and Methods

A number of mechanisms, processes, techniques, and reagents have been suggested, and some are being used, for fixation of individual metals or metal groups. These include cementation, hydrophobizing, biological, and electrochemical methods, as well as proprietary reagents that have proved effective in certain situations.

Cementation. Certain metals can be precipitated from solution by electrochemical reduction in contact with a metal (usually in elemental form) of higher oxidation potential. An example of this technique is the use of iron shot, particulate iron, or granular ferrosilicon alloys to recover copper from concentrated solutions. The method is simple and inexpensive since it requires no external applied current, but has limited applicability—primarily to concentrated, acidic solutions. It may be useful with the latter type of waste to immobilize most of the metal, using the more common fixation methods for polishing.

Hydrophobizing. The author has found that the use of waterproofing or hydrophobizing agents is sometimes beneficial to inhibit leaching of metals and other species from solidified wastes, and there are several references to this method in the literature. Portland cement is made waterproof by the addition of a small amount of stearate (calcium, aluminum, etc.) to the clinker during final grinding [82]. Bolsing [83] claims:

> A method for rendering harmless an oily waste material comprising mixing an alkaline earth oxide with a surface active agent which delays reaction between the alkaline earth metal oxide and water. . . . The alkaline earth oxide is preferably calcium oxide and . . . it is also mixed with a hydrophobizing agent. . . .

While the purpose of Bolsing's method is not for metal fixation per se, the principle is the same. It uses fatty acids or paraffin oil, such as stearic acid or palmitic acid.

One problem with this approach is that the metal is not actually fixed. The hydrophobic agent simply limits access of the leachant to the waste surface and/or decreases permeability of the waste form. The presence of large amounts of oil in a waste accomplishes something of the same purpose. The EPA has recognized this by developing the "Oily Waste Extraction Procedure" (OWEP) [84] to eliminate the effect of the hydrophobic oil, on the principle that the oil will eventually biodegrade, resulting in a higher leaching rate than that measured in the EPT or TCLP tests. In the OWEP, a succession of powerful solvents is used to extract the organics before the regular EPT is run. If this method were to be applied to hydrophobized waste forms, it is likely that the hydrophobic agent would be removed. The legitimacy of such a test depends entirely on whether the hydrophobizing agent is really biodegradable. If it is not, the test should not be applied.

Another method of rendering a waste impermeable through the incorporation of a hydrophobic agent was patented by Conner [85]. This technique, which is really a complete CFS process, uses emulsified asphalt (a stabilized emulsion of asphalt in water). The waste is mixed with an asphalt emulsion that has a particle charge opposite to that of the emulsion, such that the emulsion will "break" and coalesce into a hydrophobic mass. This process will be discussed in more detail in Chapter 15.

Biological Methods. It is known that a wide range of microorganisms can concentrate various dissolved metals by incorporation into cellular structures, removing them from aqueous solution. The organisms, or certain fractions of them, can be immobilized onto biosorbent beads that can be used in various types of contact reactors for waste water treatment. It is conceivable that these immobilized organisms could be incorporated into CFS systems to fix metals. To the author's knowledge, this has never been done, and there are problems associated with the concept, foremost among which is the alkaline nature of most CFS systems. In general, biological systems are inactivated or destroyed at high pH; lime is used to "stabilize" sewage sludge by killing many of the organisms. However, given the rapid growth in the science of genetic engineering, it seems that altered or newly designed organisms active at high pH may soon be feasible. This is an area to watch closely in the future.

Electrochemical Methods. Electrolysis, electrophoresis, and other electrochemical phenomena have been used, although sparingly compared to chemical techniques, in waste water treatment and metal recovery. They have not been applied to CFS to date, and it is difficult to visualize how they might be. However, the subject is broached here to perhaps suggest another line of investigation for workers in this field. Dewalens' paper on codeposition of copper arsenide [86] gives an example of such a technique. One possible approach would be the pretreatment of a waste stream before solidification. Another would be the imposition of an electrical potential across a treated waste, causing electrochemical reactions to occur slowly in the waste form. Since the electrolyte concentration of most wastes, before or after CFS treatment, is high, current flow through the form should be significant, even at low applied voltage. In any case, this is another possible approach to metal fixation.

Miscellaneous Methods and Reagents. In the last several years a plethora of proprietary reagents and processes has appeared on the CFS scene. Most are derived from water and waste water treatment, and often are ineffective or too expensive for CFS use. Some, however, have already proved useful, and as CFS technology grows and matures, proprietary products designed specifically for this field can be expected to appear. Some of the specific products now on the market are described as follows.

Polynucleolyte P-s, P-CR, P-CX, Ensol. Ensotech Inc., North Hollywood, California. This family of reagents for waste water treatment is said to remove metals to low levels, including Cr^{+6} and copper and nickel complexes. The composition of the reagents is not disclosed, but Material Safety Data Sheets show a variety of clays, polymer, and caustic compounds.

Toxsorb HM and HC™. Silicate Technology Corp., Scottsdale, Arizona. Toxsorb HM™ is a CFS product claimed to reduce metal leaching to <0.1 mg/l for copper, zinc, and nickel at competitive costs.

FERSONA PROCESS. Industrial Resources Inc., Chicago, Illinois. A method to insolublize sodium by reaction with ferric ion and sulfuric acid to produce basic sodium hydroxyferric sulfate/sulfite compounds. While not directly related to the fixation of toxic metals, the process is interesting because it takes the approach of formation of low-solubility complexes. In view of the regulatory trend toward ever lower leaching limits, fixation as simple species may not be sufficient for some metals, such as arsenic and lead.

X-154. Iso-Clear Systems Corp., Yorkville, Illinois. X-154 is a cellulose xanthate that is claimed to reduce copper below detection limits, and cadmium, chromium, zinc, and selenium to below 0.01 ppm in waste water. See also the discussion of xanthates earlier in this chapter.

OMNI-FIX. Ajax Floor Products Co., Great Meadows, New Jersey. Based on "modified, natural and synthetic aluminum silicate/calcium aluminum zeolites," this product line fixes metals primarily by cation exchange. The manufacturer also markets solidification compounds.

The preceding listing is exemplary, not complete, and does not include the general CFS processes that are discussed in later chapters. It is expected that the list of proprietary fixation agents will grow rapidly in the near future as the new regulatory landbans go into effect.

INDIVIDUAL METALS

In the subsequent sections of this chapter, we will discuss the fixation of individual metals. Before doing so, however, it is useful to put leachability values into the regulatory context, since this is the primary driving force for fixation technology. Table 4-18 lists the metals and their present (1989) compliance levels under various regulations. The applicable regulation is determined by location (state), choice of disposal site, and the waste type involved. This is one of the most confusing aspects of the CFS field. It was discussed briefly in Chapter 1 and, hopefully, the reader has some understanding of how the present system came about and now operates.

Another confusing facet of fixation is comparison of vendor statements about the efficacy of their processes and products. Frequently, the vendor will publish numbers that clearly do not meet one or another current compliance standard. This is usually due to vendor ignorance of the regulations and, too often, of the CFS field as a whole. However, the recent changes in allowable leaching levels due to the

TABLE 4-18 Comparison of Regulatory Limits for Various Test Procedures[b]

Contaminant	HWNO	EPT	TCLP	California List Total	1988 First-Third Landban TCLP[a]	1988 First-Third Landban Total[a]	California TTLC	California STLC	Canadian CGSB
Antimony							500	15.00	
Arsenic	D004	5.000	5.000	500	0.004		500	5.00	5.00
Barium	D005	100.000	100.000		(1)				100.00
Beryllium							75	0.75	
Boron									500.00
Cadmium	D006	1.000	1.000	100	0.066		100	1.00	0.50
Chromium(+3)	D007	5.000	5.000	500	0.094		2500	560.00	5.00
Chromium(+6)		5.000	5.000				500	5.00	5.00
Cobalt							8000	80.00	
Copper							2500	25.00	
Lead	D008	5.000	5.000	500	0.180		1000	5.00	5.00
Lead, organic							1300		
Mercury	D009	0.200	0.200	20	0.025		20	0.20	0.10
Molybdenum							3500	350.00	
Nickel				134	0.048		2000	20.00	
Selenium	D010	1.000	1.000	100	0.250		100	1.00	1.00
Silver	D011	5.000	5.000		0.072		500	5.00	5.00
Thallium				130			700	7.00	
Uranium									2.00
Vanadium							2400	24.00	
Zinc							5000	250.00	

[a] Non-waste waters, lowest applicable values.
[b] Expressed as concentration of constituent in leachate or waste, mg/l.

1988 landbans have caused much of the published literature to become instantaneously outdated and useless. Worse, this will continue to be the case for at least several years to come, as the second- and third-third landbans take effect and other regulations are revised (hopefully) to make the federal system more logical, and scientifically and environmentally justifiable.

The following sections are structured in the same way for each of the RCRA and landban metals. The other metals—antimony, copper, zinc—are lumped together in the final section. These latter metals are of general environmental interest, and are regulated in some states such as California, but they have been of less interest in the CFS field, and so relatively little data are available for them. In addition to the metal-specific data in the following sections, information will be presented in Chapters 10–15 about the properties and abilities of general-purpose CFS systems of both the generic and proprietary nature. Much of these data include leaching test results on individual metals. Also, nonstandard test results for certain metals were given in Chapter 3, as well as earlier in this chapter. To exhaustively explore the literature on a given metal, the reader should review these portions of the book and look up some of the references.

ARSENIC

Arsenic is actually classified chemically as a nonmetal or metalloid, although it is grouped with the metals for most environmental purposes. Its principal valence states are +3, +5, and −3. It combines with many metals to form arsenides. Arsenic is widely distributed around the earth in more than 150 arsenic bearing minerals [8], but the most common commercial source is as by-product from the refining of copper, lead, cobalt, and gold ores.

Most arsenic compounds are highly toxic, causing dermatitis, acute and chronic poisoning, and possibly cancer. As little as 0.1 grams of arsenic trioxide, As_2O_3, the most important commercial form, can be fatal when ingested. It is also highly toxic to other forms of animal life, and damage to plants has been observed [87]. The other arsenic species are also poisonous, although generally less so.

The commercial uses of arsenic compounds in 1974, as given by Kirk-Othmer [8] are given in Table 4-19.

Arsenic Trioxide

The major use for As_2O_3 is in the production of agricultural pesticides, including calcium arsenate, arsenic acid, lead arsenate, sodium arsenate, various arsenites, and organic arsenicals. It is also used in the glass industry.

TABLE 4-19   Uses of Arsenic Compounds

Use	Metric Tons	Percentage
Pesticides	34,000	65
Wood preservation	9,500	18
Glass	5,000	10
Alloys and electronics	1,500	3
Miscellaneous	2,000	4

Cacodylic Acid and Sodium Cacodylate

Also called dimethylarsinic acid, the compound was widely used for weed killers and defoliants. The formula for cacodylic acid is $(CH_3)_2AsO_2H$.

Metal Arsenates

Calcium, copper, lead, sodium, zinc, and manganese arsenates are used as pesticides, herbicides, and fungicides. Their formulas are:

$Ca_3(AsO_4)_4$
$PbHAsO_4$—acid lead arsenate
$Pb_4(PbOH)·(AsO_4)_3·H_2O$—basic lead arsenate
$Cu(CuOH)AsO_4$
$5ZnO·2AsO_5·4H_2O$
Na_2HAsO_4
$MnHAsO_4$

The use of these compounds has declined significantly in recent years because of their buildup in agricultural soils.

Metal Arsenites

Sodium arsenite, $NaAsO_2$, is used as a weed killer and corrosion inhibitor. Copper acetoarsenite, $(CH_3CO_2)_2Cu·3Cu(AsO_2)O_3$, or Paris green, is an insecticide that has been replaced by lead arsenate and organic pesticides for crop plants, but is finding use as a mosquito larvicide.

Arsenic Sulfides

Arsenic trisulfide, As_2S_3, is used in the manufacture of glass, semiconductors and photoconductors, as a pigment, in pyrotechnics, and for depilating hides. It is also found as a waste product from phosphoric acid manufacture, since it is nearly insoluble in acids. However, it is quite soluble in alkalies, giving dithioarsenites. Arsenic sulfide, As_4S_4, is used in many of the same uses of the trisulfide. Arsenic pentasulfide, As_2S_5, is used in pigments.

Species of Interest in CFS

The compounds of arsenic previously described are those commonly found in waste materials. Systems containing the simple As^{-3}, As^{+3}, and As^{+5} ions in solution are unknown. The oxide is amphoteric, and thus is soluble in both acids and bases. Arsenites are often present as complexes, such as Scheele's green, $CuHAsO_3$, and copper acetoarsenite described previously. Arsenates derived from the arsenic acids are oxidizing agents.

Arsenic forms the three sulfides just described. The trisulfide is soluble only in bases. It is often encountered in waste from phosphoric acid manufacture where it is precipitated from strong acid solution. This property is an unusual one for metal sulfides, which are usually decomposed by acids, but are stable and have low solubility in bases. A number of other thiocompounds of arsenic are known

H_3AsS_3—thioarsenious acid
H_3AsS_4—thioarsenic acid

as well as the salts of these acids that form in basic solution.

The organic arsenicals are of special interest in CFS because they are frequently found in wastes, especially in remediation work at old lagoons and other contaminated disposal sites. Arsenic combines readily with carbon to form a wide variety of compounds, many of which are manufactured and used commercially, but may also be formed as waste products in manufacturing and during waste treatment processes. They may be derived from both trivalent and pentavalent arsenic, with the former being the more toxic. The organoarsines are derived from arsine, AsH_3, by replacement of one to three hydrogens by alkyl, cycloalkyl, aryl, or heterocyclic groups. The tertiary arsines are of most importance, and they are widely used as ligands in coordination chemistry. The arsinic acids, $R_2AsO(OH)$, are important commercially, especially the dimethyl compound (cacodylic acid). Arsonic acids and salts are used in the poultry industry and in weed control. Arsenicals are also used as medicinals.

Arsenic chemistry is complex, involving a variety of valence states, anionic and cationic species, and both inorganic and organic compounds. And these are all commonly found in arsenic wastes, often at the same time. To further complicate the issue, the valence state changes easily and reversibly with redox potential. It is not unusual to find cycling between one state and another in the same waste system depending on, for example, the degree of aeration resulting from different depths in a waste lagoon. Moore *et al.* [29] demonstrated this variability in sediments in a reservoir, and it may be equally true in waste lagoons, soils, and even waste piles. Turner [88] found that As^{+5} concentration in coal ash disposal systems varied systematically as a function of pH, but As^{+3} remained nearly constant through a wide pH range (3–12). Oxidation of the trivalent form to the pentavalent form with dissolved oxygen is slow at neutral pH, but much faster at either extreme. Both valence states can form soluble complexes with Fe^{+3}. Overall, Turner states that "arsenic is subject to a variety of reactions in heterogeneous aqueous systems, including oxidation-reduction, sorption, complexation, and precipitation."

Johnson and Lancione [89] identified three forms of arsenic-bearing wastes that are of primary concern because of their total quantity and arsenic concentration.§ These are as follows.

1. Residues from the manufacture of organic arsenical herbicides (see previous discussion for a description of the specific compounds), most of which are stored in lagoons or waste piles at or near the generating facility. The typical composition is about 2 percent organic arsenicals (about 1 percent expressed as As) in sulfate and chloride salts.
2. Filter cake from the production of food-grade phosphoric acid. These wastes are currently being shipped to RCRA TSD facilities. They contain about 2.5 percent As_2S_3, 25 percent filter aid, and the remainder water (private communication).
3. Flue dusts from nonferrous smelting operations, currently being handled by a

---

§ It is important to note, however, that arsenic is present in many, perhaps most, waste streams in widely varying concentrations. Most of these do not leach at RCRA levels or even at drinking water levels, but some may be of concern.

variety of storage, disposal, and recycling routes. The arsenic form is As_2O_3, up to 90 percent, although the concentration is much lower in the typical dry dust.

Using nonstandard leaching tests and proprietary fixation technologies, they determined that the As_2S_3 and As_2O_3 wastes could be effectively fixed, but that CFS technology was not available to adequately fix the organic species. When the study was repeated using the EPT, the same conclusion was reached.

CFS Treatment Processes

A number of CFS processes specifically for arsenic fixation have been described in the literature. These are listed as follows.

Process	Reference
Portland cement	[90]
Portland cement, sand, and Ca or Mg salt	[91]
Sulfate, ferrous/ferric ions, $Ca(OH)_2$	[92]
H_2O_2 and ferric ions	[93]
$Fe(OH)_3$ and pH >6.5	[94]
$Fe_2(SO_4)_3$ and CaO or $CaCO_3$	[95]
Portland cement, sodium silicate, and EDTA	[33]
Portland cement, sodium silicate, and polyethylene imine	[33]
$CaCO_3$ at pH 6	[96]
Red mud	[97]

In addition to these specific references, information on arsenic fixation in the literature is available in a number of general leaching studies [98]–[105], as well as in reference works such as those by Pojasek [51], Cote [16], and Conner [38].

Table 4-20 gives some typical leaching results for a number of arsenic wastes, containing a variety of arsenic compounds at different levels, and using different fixation processes. It is evident that low leaching levels can be attained at moderate inorganic arsenic concentrations, well below the EPT level of 5 mg/l and approaching or even meeting the drinking water standards (0.05 mg/l). With high concentrations of arsenic, the EPT requirements can often be met, again only with the inorganic species, although the author has found that difficulty may be encountered at very high levels. This latter problem seems to be especially true with As_2S_3, and careful control of pH and E_h, or use of special reagents may be required to prevent resolubilization of the sulfide due to the high pH of the solidification system. Moore et al. [29] provide a useful basis for determining how this may need to be accomplished. Shively et al. [99] found that arsenite did not precipitate with the addition of alkali at pH 8.5. The Portland cement system that they used was not very effective in fixing arsenic in this form. In sequential extraction tests, arsenite was the only heavy metal to be measured in the leachate from the first extractions. Cote and Hamilton [106] found that arsenic as $NaAsO_2$ at 7200 mg/l leached cumulatively at about 2 to 9 percent (fraction leached) in a dynamic leaching test, depending on the CFS system used. This was higher by a factor of 10 to 100 than the other metals tested

TABLE 4-20 Summary of Arsenic Fixation Data

Reference[a]	Waste Description	Waste Code	CFS System	Mix Ratio	Treatment Scale	Leaching Test	Arsenic Content (mg/kg or mg/l) Total	Arsenic Content (mg/kg or mg/l) Leachate
4:98	Metal finishing	F006-9, D002-3	Lime/sulfide	—	Commercial	EPT	<1	<0.500
4:98	Metal finishing	F006-9, D002-3	Lime/sulfide	—	Commercial	TCLP		<0.100
4:98	Metal finishing	F006-9, D002-3	Lime/sulfide	—	Commercial	EPT	1	<0.050
4:98	Metal finishing	F006-9, D002-3	Lime/sulfide	—	Commercial	TCLP		<0.010
4:98	Oil waste		Flyash	1.20	Commercial	TCLP		<0.100
4:98	Paint sludge		Flyash	0.35	Commercial	EPT		<0.100
4:98	Paint sludge		Flyash	0.35	Commercial	TCLP		<0.100
4:98	Biological sludge		Flyash	1.10	Commercial	TCLP		<0.100
4:100	TSD incinerator ash	Various	Cement kiln dust	0.50	Lab	TCLP	16	<0.010
4:100	TSD incinerator filter cake	Various	Cement kiln dust	0.50	Lab	TCLP	5	0.020
4:100	TSD landfill leachate	Various	Proprietary	0.50	Lab	TCLP	9	<0.010
4:104	Arc furnace dust	K061	Proprietary		Lab	EPT	<20	<0.020
4:105	Mixed wastes	Various	Portland cement	1.00	Commercial	EPT	<1	0.025
4:90	Arsenic sludge				Lab	Static water	20,000	0.760
4:92	Organic arsenical sludge		Sulfate, ferrous/ferric ion, lime	1.00	Lab	Stirred water	6,850	16.000
4:93	Arsenic trioxide		Hydrogen peroxide, ferric ion					0.100
4:33	Mixed		Cement, sodium silicate, EDTA	0.20			40	0.016

ARSENIC

4:114	Electroplating sludge	F006	Lime, sulfide		Commercial	EPT	5	<0.050
						TCLP		0.016
4:119	Arc furnace dust	K061	Portland cement	0.05	Lab	TCLP	36	<0.010
4:119	Arc furnace dust	K061	Kiln dust	0.05	Lab	TCLP	36	<0.010
4:119	Arc furnace dust	K061	Lime/flyash	0.10	Lab	TCLP	36	<0.010
	CCA wood-preserving waste		Unstabilized		Lab	EPT	11,500	1.800
	CCA wood-preserving waste		Cement/silicate	0.40	Lab	EPT	11,500	2.300
	CCA wood-preserving waste		Potassium silicate		Lab	EPT	11,500	<0.010
	CCA wood-preserving waste		Proprietary	0.18	Lab	EPT	11,500	3.000
	CCA wood-preserving waste		Portland cement	0.10	Lab	EPT	11,500	13.8
	CCA wood-preserving waste		Portland cement	0.20	Lab	EPT	11,500	4.3
	Arc furnace dust	K061	Potassium silicate		Lab	EPT		<0.400
4:143	Synthetic sludge		Raw waste			TCLP		1.570
4:143	Synthetic sludge		Cement/silicate + hexadecyl mercaptan	0.2	Lab	TCLP		2.280
4:159	Mixed waste		Cement/flyash		Lab	EPT	635	0.022
4:159	Mixed waste		Cement/flyash		Lab	MEP[b]	635	0.172
4:160	Waste treatment plant sludge		Cement/flyash		Lab	ELT,[c] 1 da	0.8	<0.005
4:160	Waste treatment plant sludge		Cement/flyash		Lab	ELT,[c] 28 d	0.8	0.020

[a] References are listed at the end of Chapter 4.
[b] Maximum concentration.
[c] Sequential batch test, CO_2-saturated water.

(Cd, Cr, and Pb). A study by Cote and Isabelo [107] describes the leaching of arsenic from waste forms in a static leaching test, which measures the diffusivity of arsenic as measured by the leachability index (LX).

One of the techniques that seems promising is the use of iron salts, in both valence states, to fix arsenic as iron complexes or coprecipitates [92, 93]. Sittig [108] mentions several such processes used in the waste water treatment field, along with the use of charcoal filtration. Sandesara [92] describes the reaction of inorganic arsenicals, such as $As(ONa)_3$ with $FeSO_4$ and lime as

$$4As\,(ONa)_3 + 5FeSO_4 + 4Ca\,(OH)_2 + 5H_2SO_4 \longrightarrow$$
$$FeAs_2 + 2FeAsO_4 + 6Na_2SO_4 + 2Fe\,(OH)_3 + 4CaSO_4 + 6H_2O \quad (4\text{-}5)$$

The problem with organic arsenicals is more serious. It may be necessary to alter or destroy the organic species to adequately fix arsenic in this form, particularly if the concentration is high. In many ways, this situation is similar to the complexation problem we find with some nickel-bearing wastes—the organic "complexes" are soluble and stable, and considerable chemical energy may be required to "break" them. Sandesara [92] claims that cacodylic acid reacts with lime and that monosodium methane arsenate reacts with sulfuric acid, both forming "insoluble" calcium salts. Unfortunately, these salts are not sufficiently insoluble to meet present-day leaching standards. This deficiency is shown by the relatively high leaching levels of about 16 mg/l achieved by his process (see Table 4-20).

The challenge of arsenic leaching is likely to get more critical if current regulatory trends continue. The EPA has established a BDAT leaching level of 0.004 mg/l in several of the landban waste codes (K048–K052). This is below the current commercial laboratory detection limit of about 0.01 mg/l for routine analysis, and more than an order of magnitude below the drinking water limit. It is unattainable with current technology for many arsenic-containing wastes, and certainly seems unnecessary from an environmental protection standpoint.

BARIUM

Barium is classed as an alkaline earth metal (Group IIA of the periodic table), along with calcium, whose chemical behavior it resembles. It is widely distributed in the earth's crust, in igneous rocks, sandstone and shale [8], almost always in the form of barite ($BaSO_4$), which is the only form mined commercially. Other barium compounds are manufactured from barite, usually starting with high-temperature reduction to produce barium sulfide. Its valence state is +2.

The insoluble compounds, such as the sulfate, are considered nontoxic, but the soluble compounds—carbonate, chloride, hydroxide, nitrate, acetate, and sulfide—are highly toxic when ingested.

Production of barium compounds in the United States in 1972 is estimated in Table 4-21.

Barium Borate

Barium borate (BaB_2O_4) is used as a paint pigment. Its solubility in water is 3000 mg/l.

TABLE 4-21 Production of Barium Compounds in the United States

Compound	Production (tons)
Barite (BaSO$_4$) ore	96,000
Sulfide	73,000
Carbonate	42,000
Hydroxide	12,000
Chloride	9,000
Sulfate	5,000
Others	4,500

Source: After Kirk-Othmer, *Encyclopedia of Chemical Technology,* 3d ed., Wiley, New York, 1979.

Barium Carbonate

Barium carbonate was formerly mined, but is now produced by precipitation from a sulfide solution. It is the most widely used barium chemical with the exception of barite mineral. The largest single use is in brick manufacture to immobilize soluble salts in the interior of the brick. Barium carbonate is used in oil-well drilling as an additive to drilling mud to insolublize gypsum and thus stabilize the mud. Along with barite, this is a major source of barium in wastes that may require CFS treatment. Current regulations exempt drilling muds from the RCRA, along with other so-called mining wastes, but this may change in the future, as it already has in some states. Other uses that may generate wastes for CFS are in the manufacture of photographic paper and television picture tubes, in the ceramic industry, and in the manufacture of other compounds.

Barium Chloride

Barium chloride is manufactured from barium sulfide and hydrochloric acid. It is used in the manufacture of barium pigments, and in metal case-hardening and heat-treating baths.

Barium Hydroxide

Much of the production of barium hydroxide is by one company, using barium sulfide as the starting material. It is used as an additive in plastics and paper, as a dispersant for pigments, and in self-extinguishing polyurethane foams. It was also used in the United States in beet sugar refining, which results in a waste mixture of barium carbonate, barium sulfate, and barium silicate. Some of this waste still exists in large waste piles, and is a potential source of barium pollution.

Barium Nitrate

A suspension of barium carbonate and nitric acid is used in the manufacture of barium nitrate. Its primary use is in pyrotechnics and explosives. It is also used as a source of barium oxide for enamels.

Barium Sulfate

The most widely used barium compound, barium sulfate is not toxic due to its low solubility, 1.4 mg/l, as barium in water. It is not a subject for CFS treatment, since the EPT regulatory level is 100 mg/l. Drilling muds consume 90 percent of all barite produced, and it is also used in the manufacture of other barium chemicals, in paints, plastics, glass, and ceramics. The precipitated form, lithopone, is a widely used white pigment, and it is also used in photographic paper.

Barium Sulfide

Barium sulfide is used in the manufacture of other barium chemicals, not by itself. However, it may show up in waste streams from such manufacture. It is soluble in water.

Species of Interest in CFS

Barium rarely leaches from wastes above EPT levels, and so is seldom a subject for fixation technology. The exceptions occur in old waste piles from sugar refining and in certain industrial dusts where the total concentration is very high. Fixation can usually be easily accomplished by treatment with a sulfate compound such as sodium or calcium sulfate.

Table 4-22 lists a number of leaching studies that included barium in the analysis. Obviously, barium is present in many waste streams, and is also easily fixed to well below the RCRA levels. It seems unlikely that regulatory levels would be lowered sufficiently to create any barium leaching problem for CFS technology.

CADMIUM

Cadmium is a Group IIB element, with only one valence state for all practical purposes (+2). It is found primarily as sulfide minerals in zinc, lead, and copper ores, and is recovered as a by-product from the refining of those ores. Unlike zinc, the cadmium ion is not very amphoteric, and its hydroxide, $Cd(OH)_2$, has quite low solubility in an alkaline medium. However, it forms stable ammonium, cyanide, and halide complexes [8].

Cadmium and its compounds are highly toxic, but most cases of cadmium poisoning have occurred by inhalation of fumes or dusts. Fumes are formed at the temperatures of many industrial processes, such as welding or brazing, and may be volatilized from incinerators where no pollution control devices are in operation. Chronic poisoning due to ingestion has been recorded; for example, the outbreak of Itai Itai disease in Japan has been traced to cadmium in the environment. Cadmium may also be a carcinogen, although there is little evidence to support this from human studies [108]. Some shellfish tend to concentrate cadmium, as does the use of sewage sludge in land application, thus affecting both the soil and the food chain [109]. However, there appears to be no cumulative effect in humans [87]. Cadmium is toxic toward fish and other aquatic organisms.

Cadmium consumption in the United States in 1979 was estimated at 6000–7000 metric tons. Elemental cadmium is used primarily in electroplating to protect ferrous metal parts against corrosion. It is usually plated from a cyanide bath, so cadmium plating wastes are a primary potential source of cadmium and cyanide pollution. It is also used as the negative electrode in rechargeable nickel–cadmium and silver–cadmium batteries, in pigment manufacture, as heat and light stabilizers in plastics, in alloys, electronic applications, and catalysts.

Major cadmium compounds and their uses are as follows.

- Cadmium arsenides, antimonides, and phosphides: Used in electronic and semiconductor applications.
- Cadmium borates: Used as phosphors, and for electrodeposition of cadmium in high-strength steels (fluoborate).
- Cadmium carbonate: Used as catalysts, in the production of other cadmium compounds, and in phosphors.
- Cadmium complexes: A number of organic ligand complexes are known: acetic acid, dimethylglyoxime, EDTA, glycolic acid, methylamine, oxalic acid, pyridine, sulfamic acid, tartaric acid, and thiourea. The major complex in use, however, is inorganic—$Cd(CN)_4^{-2}$, which is used in electroplating. Other plating electrolytes include the sulfate, sulfamate, chloride, fluoborate, and pyrophosphate.
- Cadmium halides: The chloride is used in photography, dying and cloth printing, vacuum tubes, cadmium yellow pigment, special mirrors, and in lubricants. Cadmium fluoride is used in phosphors.
- Cadmium hydroxide: An important application is the active anode material in storage batteries; it is also used in the manufacture of other cadmium compounds.
- Cadmium nitrate: Other cadmium salts are made from the nitrate. It is also used in photographic emulsions and in ceramics.
- Cadmium oxide: Widely used in electroplating baths dissolved in excess sodium cyanide, the oxide is also used in phosphors, semiconductors, glass, battery electrodes, catalysts, ceramic glazes, and in the manufacture of silver alloys.
- Cadmium selenide and telluride: These compounds are used in electronic applications and, along with sulfide, as a series of pigments and colorants.
- Cadmium sulfate: This salt is used in electrodeposition of cadmium, copper and nickel, and in phosphors.
- Cadmium sulfide: The sulfide is used widely for pigments, especially in the plastics industry, where approximately 2600 metric tons were consumed in 1976. Cadmium colorants are also used in paints, soaps, rubber, paper, glass, inks, ceramic glazes, and textiles.
- Cadmium tungstate: Phosphors and catalysts are the primary consumers of this compound.
- Dialkyl cadmium compounds: These organic compounds are used as polymerization catalysts in organic and polymer synthesis, and as heat and light stabilizers in plastics.
- Organocadmium soaps: Cadmium salts of organic acids, in combination with barium soaps, are widely used as heat and light stabilizers in plastics. Approximately 22,000 metric tons of such compounds were consumed for non-food-contact applications in the United States in 1976.

Species of Interest in CFS

Electroplating generates an estimated 1.5 million pounds of cadmium wastes per year [87] in the following forms:

114 FIXATION OF METALS

TABLE 4-22 Summary of Barium Fixation Data

Reference[a]	Waste Description	Waste Code	CFS System	Mix Ratio	Treatment Scale	Leaching Test	Barium Content (mg/kg or mg/l) Total	Barium Content (mg/kg or mg/l) Leachate
4:98	Metal/cyanide	F006-8	Lime	—	Commercial	EPT		0.130
4:98	Metal finishing	F006-9, D002-3	Lime/sulfide	—	Commercial	EPT	30	<0.020
4:98	Metal finishing	F006-9, D002-3	Lime/sulfide	—	Commercial	TCLP		0.180
4:98	Metal finishing	F006-9, D002-3	Lime/sulfide	—	Commercial	EPT	<10	<0.020
4:98	Metal finishing	F006-9, D002-3	Lime/sulfide	—	Commercial	TCLP		<0.100
4:98	Oil waste		Flyash	1.20	Commercial	TCLP		0.510
4:98	Paint sludge		Flyash	0.35	Commercial	EPT		1.300
4:98	Paint sludge		Flyash	0.35	Commercial	TCLP		0.540
4:98	Biological sludge		Flyash	1.10	Commercial	TCLP		0.690
4:98	Metal hydroxide		Flyash	0.25	Commercial	EPT		0.370
4:98	Metal hydroxide		Flyash	0.25	Commercial	TCLP		0.330
4:100	TSD incinerator ash	Various	Cement kiln dust	0.50	Lab	TCLP	3	0.690
4:100	TSD incinerator filter cake	Various	Cement kiln dust	0.50	Lab	TCLP	739	0.880
4:100	TSD landfill leachate	Various	Cement kiln dust	0.50	Lab	TCLP	29	0.460
4:100	TSD incinerator ash	Various	Cement kiln dust	0.50	Lab	TCLP	2270	0.570
4:101	Electroplating sludge	F006	Unstabilized raw waste			TCLP	19	0.280
4:101	Electroplating sludge	F006	Cement kiln dust	0.20	Lab	TCLP		0.190
4:101	Electroplating sludge	F006	Cement kiln dust	0.50	Lab	TCLP		0.110

4:102	Electroplating sludge	F006	Unstabilized raw waste		Lab	TCLP	1.510
4:102	Electroplating sludge	F006	Cement kiln dust	0.50	Pilot	TCLP	
4:102	Electroplating sludge	F006	Cement kiln dust	0.50	Commercial	TCLP	0.350
4:104	Arc furnace dust	K061	Proprietary		Lab	EPT	0.700
4:105	Mixed wastes	Various	Proprietary		Commercial	EPT	<500	1.900
4:114	Electroplating sludge	F006	Lime, sulfide		Commercial	EPT	63	1.000
						TCLP	40	0.030
								0.280
4:119	Arc furnace dust	K061	Portland cement	0.05	Lab	TCLP	238	0.580
4:119	Arc furnace dust	K061	Kiln dust	0.05	Lab	TCLP	238	0.507
4:119	Arc furnace dust	K061	Lime/flyash	0.10	Lab	TCLP	238	0.464
4:119	Arc furnace dust	K061	Potassium silicate		Lab	EPT		2.800
4:143	Synthetic sludge		Raw waste			TCLP		0.800
4:143	Synthetic sludge		Cement/silicate + hexadecyl mercaptan	0.2	Lab	TCLP		0.900
4:159	Mixed waste		Cement/flyash		Lab	EPT	1730	0.073
4:159	Mixed waste		Cement/flyash		Lab	MEP[b]	1730	0.158

[a] References are listed at the end of Chapter 4.
[b] Maximum concentration.

- Aqueous slurry of 5 percent cadmium cyanide and 5 percent NaCN in NaOH
- Liquid containing 1.5 percent cadmium cyanide and 8.5 percent NaCN
- Solid containing 3 percent CdO and 16 percent cadmium metals with alkali carbonates
- 300–500 ppm cadmium chromate mixture with alkaline salts and organic detergents

Electroplating is by far the largest source of cadmium-bearing wastes. A significant portion of these wastes is from small plating shops [110] (job shops), although large amounts are also generated in captive shops where they are often mixed with other waste streams from the plant. Most producers treat the waste waters with lime, producing calcium hydroxide sludge, which now usually goes to RCRA TSD facilities.

Cadmium waste from battery manufacturing includes battery-plate impregnation-bath sludge and liquid from the bath or the rinsing operation. It is estimated that less than two tons of such waste is generated each year in the United States.

In the solvent-based paint industry, cadmium sludges are generated in both the washup system and in the solvent recovery stills. This waste probably goes to incineration now, and does not enter the CFS area of treatment, although the cadmium will end up in the incineration residues, which require CFS treatment under the new landbans.

Other sources of wastes that may require CFS treatment are cadmium compound manufacturers, alloy producers, smelters, and parts manufacturers that use cadmium metal or compounds.

CFS Treatment Processes

Freshly precipitated cadmium hydroxide reportedly leaves approximately 1 mg/l in solution at pH 8, but only 0.1 mg/l at pH 10 [108]. When coprecipitated with iron at pH 8.5, Cd concentration in the solution is not detected. In view of this, and the nonamphoteric nature of Cd, the primary treatment method has been precipitation with alkali, usually lime. However, in the presence of complexing agents, such as cyanide, the cadmium ion will not precipitate [111]. It is then necessary to destroy the cyanide complex, usually by alkaline chlorination (Chapter 5), which precipitates the Cd as the hydroxide. Oxidation of cyanide by hydrogen peroxide is also reported to break the complex and precipitate CdO (see Chapter 5).

If the carbonate ion is present in a waste containing cadmium, the carbonate will form preferentially at pH 7–8. This is shown for a four-ligand system—CO_3^{-2}, Cl^-, SO_4^{-2}, and OH^- in Figure 4-5 [112]. Between pH 8 and 11, about 99 percent of the cadmium will be speciated as the solid carbonate. Above pH 9, the relative proportion of cadmium speciated as hydroxide begins to rise, until the system is primarily hydroxide above pH 11 (Figure 3-7). At very low E_h (reducing conditions), the presence of the sulfide ion will speciate cadmium primarily as CdS across a broad pH range of about 2 to 12 or above.

A number of CFS processes for the fixation of cadmium have been reported. These are listed as follows:

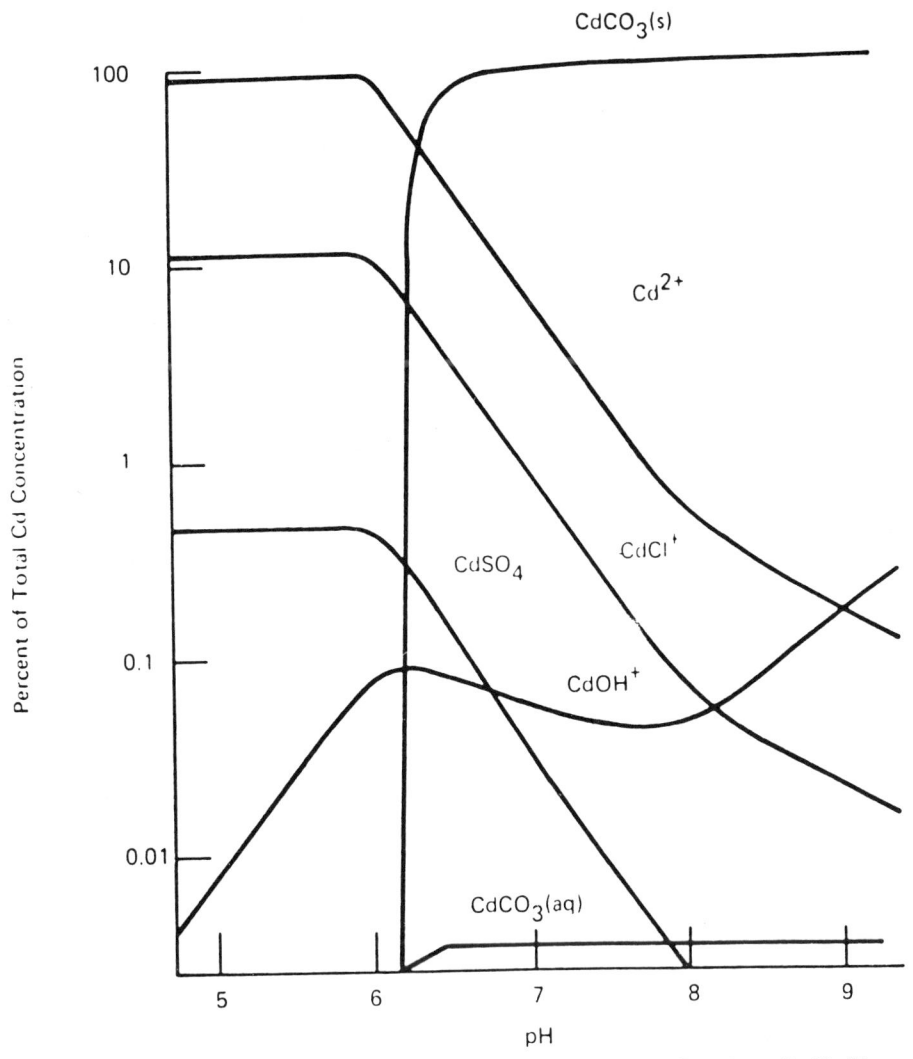

Figure 4-5 Speciation of cadmium in a four-ligand system as a function of pH. (From Morel *et al.* [112].)

Process	Reference
Portland cement + aluminum hydroxide	[113]
Ferrous/ferric ions + NaOH	[114]
Flyash/lime	[106]
Flyash/cement	[106]
Cement/silicate	[106, 9]
Cement/bentonite	[106]
Lime	[99]
Sulfide	[99]
Lime/flyash	[99]
Portland cement	[115]

Process	Reference
Portland cement	[116]
Cement/silicate	[117]
Lime/sulfide	[118]
Portland cement	[119]
Kiln dust	[119]
Lime/flyash	[119]

Additional information on cadmium fixation can be obtained from a number of leaching studies [98–105], as well as in several reference works [16, 38, 52]. Table 4-23 gives typical leaching results for many cadmium-containing wastes treated with a variety of CFS processes.

There is no difficulty in reaching very low cadmium leaching levels—less than the current landban level of 0.066 mg/l—with generic CFS systems, so long as the alkalinity of the system is maintained throughout the leaching test. However, when the final pH of the leachant drops below about 7, cadmium leaches out at much higher levels. This is evident in Table 4-23 where low mix ratios were used or where the reagent is not very alkaline, as with flyash alone. This presents a potential environmental problem, however, wherever the natural environment is acidic even when using highly alkaline systems, since the alkalinity will eventually be removed. Bishop [44] has demonstrated this problem for cadmium by using a sequential-batch acid leaching test (see Chapter 23) that eventually neutralized the alkalinity of a Portland cement system. After fifteen extractions, about 75 percent of the cadmium leached out, versus only 25 percent and 15 percent for lead and chromium, respectively. The cadmium cumulative leaching curve followed the alkalinity leaching curve very closely, while lead and chromium followed the silicon leaching curve. Even more dramatically, the beginning of cadmium leaching coincides exactly with the sudden drop of final leachant pH from 11 to about 5. It is evident that the primary fixation mechanisms are very different for unlike metal species, with lead and chromium being bound into the silica matrix itself, while cadmium is not.

The dependency of cadmium leaching on pH has been verified by a number of investigators. Brunner and Baccini [120] found the same effect with treated (Portland cement) and untreated municipal incinerator ash. As the pH dropped below 9, cadmium began to leach significantly, rising very rapidly below pH 7. It is obvious that the EPT and TCLP tests do not counter the alkalinity of most CFS treated waste forms. Various studies [117–119] yield almost identical, and very low leachability (0.014–0.073) in different wastes and CFS processes when the cadmium levels are moderate (<1000 mg/l) and the mix ratios are high. At low mix ratios (0.05) one study [119] found high cadmium leaching in the range of 0.5 to 3.6 mg/l. Only when very high cadmium levels (about 10,000 mg/l) were present did cadmium leaching rise somewhat for highly alkaline waste forms [98]. Butler et al. [117] attribute the behavior of cadmium to the early formation of $Cd(OH)_2$, which then provides nucleation sites for the precipitation of calcium silicate hydrate (in cement systems, see Chapter 10), resulting in cadmium being in the form of the hydroxide with an impervious coating.

In some systems, cadmium may be sorbed or fixed by cation exchanged mechanisms. Some "fixing" capacities for cadmium are as follows:

CADMIUM 119

TABLE 4-23 Summary of Cadmium Fixation Data

Reference[a]	Waste Description	Waste Code	CFS System	Mix Ratio	Treatment Scale	Leaching Test	Cadmium Content (mg/kg or mg/l) Total	Leachate
4:98	Metal/cyanide	F006-8	Lime	—	Commercial	EPT	7200	0.260
4:98	Metal finishing	F006-9, D002-3	Lime/sulfide	—	Commercial	EPT	20	0.008
4:98	Metal finishing	F006-9, D002-3	Lime/sulfide	—	Commercial	TCLP		<0.020
4:98	Metal finishing	F006-9, D002-3	Lime/sulfide	—	Commercial	EPT	<5	0.007
4:98	Metal finishing	F006-9, D002-3	Lime/sulfide	—	Commercial	TCLP		<0.020
4:98	Oil waste		Flyash	1.20	Commercial	TCLP		<0.100
4:98	Paint sludge		Flyash	0.35	Commercial	EPT		<0.100
4:98	Paint sludge		Flyash	0.35	Commercial	TCLP		<0.100
4:98	Biological sludge		Flyash	1.10	Commercial	TCLP		<0.100
4:98	Metal hydroxide		Flyash	0.25	Commercial	EPT		34.000
4:98	Metal hydroxide		Flyash	0.25	Commercial	TCLP		4.100
4:100	TSD incinerator ash	Various	Cement kiln dust	0.50	Lab	TCLP	3	0.030
4:100	TSD incinerator filter cake	Various	Cement kiln dust	0.50	Lab	TCLP	685	0.910
4:100	TSD landfill leachate	Various	Cement kiln dust	0.50	Lab	TCLP	2	0.010
4:100	TSD incinerator ash	Various	Cement kiln dust	0.50	Lab	TCLP	17	0.190
4:101	Electroplating sludge	F006	Unstabilized raw waste		Lab	TCLP	5	0.010
4:101	Electroplating sludge	F006	Cement kiln dust	0.20	Lab	TCLP		<0.010
4:101	Electroplating sludge	F006	Cement kiln dust	0.50	Lab	TCLP		<0.010
4:101	Electroplating sludge	F006	Cement-based	0.10	Lab	TCLP		0.010
4:101	Electroplating sludge	F006	Cement-based	0.15	Lab	TCLP		0.010
4:101	Electroplating sludge	F006	Cement-based	0.20	Lab	TCLP		0.080
4:102	Electroplating sludge	F006	Unstabilized raw waste		Lab	TCLP		<0.010
4:102	Electroplating sludge	F006	Cement kiln dust	0.50	Pilot	TCLP		<0.010
4:102	Electroplating sludge	F006	Cement kiln dust	0.50	Commercial	TCLP		<0.010
4:103	Arc furnace dust	K061	Unstabilized raw waste		Lab	EPT	600	1.700
4:103	Arc furnace dust	K061	Unstabilized raw waste		Lab	TCLP		1.090

(*continued*)

120 FIXATION OF METALS

TABLE 4-23 Summary of Cadmium Fixation Data *(Continued)*

Reference[a]	Waste Description	Waste Code	CFS System	Mix Ratio	Treatment Scale	Leaching Test	Cadmium Content (mg/kg or mg/l) Total	Leachate
4:103	Arc furnace dust	K061	Cement-based	0.20	Lab	TCLP		<0.010
4:103	Arc furnace dust	K061	Cement/silicate	0.20	Lab	TCLP		<0.010
4:103	Arc furnace dust	K061	Proprietary	0.20	Lab	TCLP		<0.010
4:103	Arc furnace dust	K061	Lime kiln dust	0.20	Lab	TCLP		<0.010
4:104	Arc furnace dust	K061	Proprietary		Lab	EPT		0.030
4:105	Mixed wastes	Various	Proprietary		Commercial	EPT	200	0.500
4:114	Electroplating sludge	F006	Lime, sulfide		Commercial	EPT	115	0.031
						TCLP	53	<0.020
4:117	Cadmium hydroxide sludge		Portland cement	0.45	Lab	TCLP	15340	0.026
4:117	Cadmium hydroxide sludge		Cement/silicate	0.45	Lab	TCLP	15340	0.014
4:117	Cadmium nitrate sludge		Portland cement	0.41	Lab	TCLP	18980	0.043
4:119	Arc furnace dust	K061	Portland cement	0.05	Lab	TCLP	481	3.290
4:119	Arc furnace dust	K061	Kiln dust	0.05	Lab	TCLP	481	1.743
4:119	Arc furnace dust	K061	Lime/flyash	0.10	Lab	TCLP	481	0.052
4:119	Electroplating waste	F006	Unstabilized		Lab	EPT	6700	12.000
			Cement-based		Lab			0.030
4:136	Flyash A		Unstabilized		Lab	EPT		0.860
4:136	Flyash A		Gelled silicate	0.26	Lab	EPT		0.190
4:136	Flyash A		Lime	0.10	Lab	EPT		0.270
4:136	Flyash B		Unstabilized		Lab	EPT		64.000

4:136	Flyash B		Gelled silicate	0.26	Lab	EPT	24.000
4:136	Flyash B		Lime	0.10	Lab	EPT	26.000
4:136	Plating waste		Unstabilized		Lab	EPT	0.33
4:136	Plating waste		Gelled silicate	0.16	Lab	EPT	0.050
4:136	Plating waste		Cement/silicate	0.24	Lab	EPT	0.050
4:136	Mixed municipal/ industrial waste		Unstabilized		Lab	EPT	0.200
4:136	Mixed municipal/ industrial waste		Gelled silicate	0.16	Lab	EPT	0.060
4:136	Mixed municipal/ industrial waste		Cement/silicate	0.24	Lab	EPT	0.100
	Arc furnace dust	K061	Potassium silicate		Lab	EPT	0.050
4:143	Synthetic sludge		Raw waste			TCLP	34.800
4:143	Synthetic sludge		Cement/silicate + hexadecyl mercaptan	0.2	Lab	TCLP	0.074
4:159	Mixed waste		Cement/flyash		Lab	EPT	79.1	<0.050
4:159	Mixed waste		Cement/flyash		Lab	MEP[b]	79.1	<0.050
4:160	Zinc phosphate waste		Cement/flyash		Lab	ELT,[c] 1 da	1.9	<0.001
4:160	Zinc phosphate waste		Cement/flyash		Lab	ELT,[c] 28 d	1.9	<0.001
4:160	Zinc phosphate waste		None		Lab	ASTM,[c] 2 d	1.9	0.040
4:160	Zinc phosphate waste		None		Lab	ASTM,[c] 10	1.9	0.007
4:160	Zinc phosphate waste		Cement/flyash		Lab	ASTM,[c] 2 d	1.9	0.001
4:160	Zinc phosphate waste		Cement/flyash		Lab	ASTM,[c] 10	1.9	<0.001

[a] References are listed at the end of Chapter 4.
[b] Maximum concentration.
[c] Sequential batch test, CO_2-saturated water.

Kaolin clay   0.05 mg/g or 54 meq/kg
Flyash   0.22 mg/g or 93 meq/kg
Sawdust   0.11 mg/g or 113 meq/kg

In nonacidic leaching media, such as seawater, sequential batch tests leached much less cadmium [116], only about 1 percent after sixteen extractions. However, in the same medium and test, lead leaching was undetectable. The system tested had very high metal content, about 5 percent, in the form of the hydroxide. It was observed that the cadmium sludge seemed to create a more porous microstructure with a large amount of ettringite, an expansive compound found in Portland cement pastes that causes destructive cracking of the waste form. A lead sludge tested in the same way did not have any observable physical effect on the waste form microstructure. This is in keeping with the results of the other studies previously mentioned.

When waste forms containing cadmium are subjected to a diffusion-controlled leaching test scenario with neutral water as the leachant, cadmium leaching is very low. Cote et al. [106] found only 0.008 percent of the cadmium to be leached from a cement/flyash waste form after 11 months in such a system. In this case, cadmium leaching was less than that of chromium or lead.

It is obvious that the leaching scenario used is critical when evaluating cadmium leachability from the products of any CFS process, with pH and redox potential being factors of prime importance. Meaningful estimation of leaching potential must include tests using the actual environment conditions expected at the disposal site, especially the groundwater composition. However, testing with actual or simulated groundwater should include control of redox potential as well as the compositional parameters. While this is true to some extent of all the toxic metals, cadmium appears to be more pH-sensitive than most. In acidic environments, it may be necessary to fix cadmium with additives that can immobilize the metal at pH values of about 4 or above, where nearly all ambient environments fall.

CHROMIUM

Chromium belongs to Group VIB of the periodic table. It has three valence states, +2, +3, and +6, but the latter two are the most common. Cr^{+6} is acidic, forming chromates $(CrO_4)^{-2}$ and dichromates $(Cr_2O_7)^{-2}$, while the other valence states are basic. Chromium is produced by roasting the chromite ore with lime and potash, producing sodium chromate, which is then further processed into various other compounds. The largest single use of the metal is in alloying, where the ore is processed by pyrometallurgical means to ferrochrome and other materials for direct use in steelmaking and in other alloys.

The primary environmental problems associated with chromium are with Cr^6 compounds. While the trivalent state is also poisonous [87], these compounds have no established toxicity [8] and, due to their low solubility, create little risk. Acute systemic poisoning due to Cr^6 compounds is rare; the acute effects of chromates are mainly on the skin and mucous membranes. Lung cancer in humans has been observed in the manufacturing industries, and certain chromium compounds have been found to be carcinogenic in animal experiments. Cr^6 compounds are very toxic to aquatic plant and animal life, as evidenced by their widespread use as algaecides.

Total chrome ore consumption in the United States in 1975 was 881,000 metric tons; the demand for sodium dichromate was about 150,000 tons. The two primary industrial chromium compounds are sodium chromate and sodium dichromate, which are used directly and also to produce secondary compounds [8]. The distribution of uses of chromium chemicals is shown in Table 4-24.

The major chromium compounds and their uses [8] are in the following areas.

Metal Finishing and Corrosion Control

The use of chromic acid in chromium electroplating is the most important application of chromium compounds in this area. The plating baths all have about the same composition, being primarily CrO_3, with minor additions of sulfate, NaOH, complexing agents, and surfactants. Another use is anodizing of aluminum in a chromic acid solution. Dichromates or chromic acid are used as sealers to improve corrosion resistance of various coatings on steel. Chromates are used to inhibit corrosion in recirculating water systems, and in other similar applications.

The use of chromium in metal finishing is the major generator of chromium wastes in the United States. About 40,000 tons of chromium compounds, calculated as chromic acid, were consumed in 1972, and it is estimated that about 80 percent, or 32,000 tons, would ultimately end up as waste [87]. Improvements in waste reduction and recycling have occurred since then, but a substantial portion of the chromium still ends up as waste sludge and filter cake, mostly as a result of waste water treatment operation on process rinse water. Most of this chromium has already been reduced to the trivalent state in the water treatment process, but some Cr^{+6} still finds its way into the concentrated waste encountered in CFS treatment. Since the advent of pretreatment requirements before discharge into municipal sewage systems, most of the waste chromium is disposed of in RCRA facilities. Until recently, most of this sludge and filter cake was untreated except for solidification to bind free water, but the August 8, 1988 first-third landban has made the application of CFS technology necessary in many, if not most cases.

Pigments

Chromium-containing pigments are based on either lead chromate, chromium oxide greens, or corrosion-inhibiting pigments using a difficultly soluble chromate [8]. Of the lead chromate group, different colors are achieved with various crystalline forms, additives, basic lead chromate, $PbCrO_4 \cdot PbO$, and lead silicochromate. Chromium oxide greens are anhydrous or hydrated oxide or Guignet's green, a hydrous oxide. They are very stable colors. Corrosion-inhibiting pigments include zinc chro-

TABLE 4-24 Uses of Chromium Chemicals

Use	Percent
Metal finishing and corrosion control	37
Pigments and allied products	26
Leather tanning and textiles	20
Wood preservation	5
Drilling muds	4
Other uses (catalysts, intermediates, etc.)	8

mate, zinc tetroxychromate, basic lead silicochromate, strontium chromate, and barium potassium chromate. They are used in paints and other coatings.

Chromium-containing wastes are generated both from pigment manufacture and from cleanup operations in solvent-based paint manufacture. Perhaps 100 tons per year of chromium compounds enter the wastes streams from the latter source. Probably less chromium is lost in pigment manufacture itself.

Leather Tanning and Textiles

Although the chemical purchased in the tanning industry is sodium dichromate, it is reduced to chromium sulfate before being impregnated into the hide, where it forms stable complexes with proteins. Therefore, the waste consists of Cr^{+3} compounds that are easily insolubilized as the hydroxide. In textiles, the chromate compounds are used both as oxidizing agents and as dye mordants (fixants). Most textile mills operate waste water treatment plants that convert any chromate to the trivalent form in the sludge or filter cake. Chromate oxidizing agents are being replaced by other agents as a result of environmental considerations [87].

Wood Preservation

There has been a large recent increase in the use of chromium compounds for this purpose, in the form of chromated copper arsenate (CCA). In 1976, more than 5000 metric tons of sodium dichromate was used for this purpose (including use in fire-retardants) in the United States. Much of the waste from this process is in the form of contaminated soil from around the treatment operations, most of it from older operations that have since been discontinued.

Drilling Muds

These formulations contain both soluble and difficultly soluble chromates for corrosion control, and chromium lignosulfonates. After use, most of the drilling mud is disposed of as a waste. It is, however, excluded from the RCRA system as a mining waste, and so is seldom treated except in certain states that have more stringent regulations. One CFS vendor is known to be treating waste drilling mud in California as of this writing (see Chapter 11).

Other Uses

Chromium compounds are sold for a wide variety of other uses. Some of these are as follows.

Catalysts. About 1500 metric tons of sodium dichromate are sold annually for this purpose. Chromic acid, ammonium dichromate, and chromic nitrate are also sold for catalyst manufacture. Most of this material, when discarded, probably goes to RCRA TSD facilities without further treatment at this time. The third-third land-ban, however, may require CFS treatment of these wastes by 1990.

Photosensitive Compounds and Products. Combinations of chromium compounds and colloids have been used since the early days of photography. Chromates are used in gravure (printing) platemaking operations, and wastes originate in the washing of the plates as well as in the periodic dumping of the tanks.

Other uses of chromium compounds are in batteries, magnetic tapes, chemicals, ceramics, pyrotechnics, electronics, and fungicides.

Species of Interest in CFS

The major chromium compounds that may be encountered in CFS work are composed of Cr^{+3} and Cr^{+6} valence states. Divalent compounds are used primarily in laboratories. The chromate compounds discussed previously are usually reduced to the trivalent state and precipitated with bases in waste water treatment. The resultant $Cr(OH)_3$ is the form in which the vast majority of chromium is encountered in CFS treatment. Because of its low solubility, fixation of chromium as this species has not been a problem in the past, even though chromium is nearly ubiquitous in waste streams. Judging from the most recent landban leaching levels (0.094 for several waste codes) it may be a problem in the future. Also, in special situations, such as the EPA's delisting process, the required maximum leaching level of 0.31 mg/l may be difficult to meet with some waste streams.

CFS Treatment Processes

In view of the previously described status of chromium fixation work, most technical emphasis on CFS treatment of chromium-containing wastes has been, and likely will be, on more efficient and cost effective reduction of Cr^{+6}, or in the direct fixation of this species. The hexavalent form is encountered in a number of wastes, either by mixing of untreated chromate solutions with waste water treatment sludge or because it is intrinsic to the waste, as in some incinerator residues and in wood-treating wastes. It is also a major problem in the remediation of sites where soil has been contaminated with chromate or chromic acid solutions, a frequent situation with old plating operations. The classical way to manage the environmental problem with Cr^{+6} has been a two-step process: reduce the chromium to the trivalent state and then precipitate it as the hydroxide. In waste water this is an efficient process, but in the treatment of concentrated residues containing high Cr^{+6} levels, it often is not. We discussed this technology, and its problems in CFS work, earlier in this chapter.

In addition to the standard reduction methods listed in Table 4-9, a number of other approaches to the fixation of Cr^{+6}-containing wastes have been proposed, and some even used in pilot and commercial operations. However, most were developed for waste water treatment, not CFS work. These methods can be divided into two types: reduction processes and direct fixation processes. They are listed below. Leach testing results for a number of these processes, as well as for chromium in general, are given in Table 4-25.

Reduction Processes	*Reference*
Metal sulfide	[121]
Iron filings, brass or aluminum chips	[87]
Jet cement	
Jet cement + $Fe(NO_3)_2$	[122]
Cement + $Al_2(SO_4)_3$ + $CaSO_3$	[123]

126 FIXATION OF METALS

TABLE 4-25 Summary of Chromium Fixation Data

Reference[a]	Waste Description	Waste Code	CFS System	Mix Ratio	Treatment Scale	Leaching Test	Chromium Content (mg/kg or mg/l) Total	Leachate
4:98	Metal/cyanide	F006-8	Lime	—	Commercial	EPT	11200	0.380
4:98	Metal finishing	F006-9, D002-3	Lime/sulfide	—	Commercial	EPT	16300	<0.010
4:98	Metal finishing	F006-9, D002-3	Lime/sulfide	—	Commercial	TCLP		<0.050
4:98	Metal finishing	F006-9, D002-3	Lime/sulfide	—	Commercial	EPT	10000	<0.010
4:98	Metal finishing	F006-9, D002-3	Lime/sulfide	—	Commercial	TCLP		<0.050
4:98	Oil waste		Flyash	1.20	Commercial	TCLP		0.320
4:98	Paint sludge		Flyash	0.35	Commercial	EPT		<0.100
4:98	Paint sludge		Flyash	0.35	Commercial	TCLP		<0.100
4:98	Biological sludge		Flyash	1.10	Commercial	TCLP		<0.100
4:98	Metal hydroxide		Flyash	0.25	Commercial	EPT		0.430
4:98	Metal hydroxide		Flyash	0.25	Commercial	TCLP		0.260
4:100	TSD incinerator ash	Various	Cement kiln dust	0.50	Lab	TCLP	97	0.120
4:100	TSD incinerator filter cake	Various	Cement kiln dust		Lab	TCLP	420	0.450
4:100	TSD landfill leachate	Various	Cement kiln dust		Lab	TCLP	12	0.720
4:100	TSD incinerator ash	Various	Cement kiln dust	0.50	Lab	TCLP	233	0.120
4:101	Electroplating sludge	F006	Unstabilized raw waste		Lab	TCLP	644	0.010
4:101	Electroplating sludge	F006	Cement kiln dust	0.20	Lab	TCLP		0.050
4:101	Electroplating sludge	F006	Cement kiln dust	0.50	Lab	TCLP		0.240
4:101	Electroplating sludge	F006	Cement-based	0.10	Lab	TCLP		0.010
4:101	Electroplating sludge	F006	Cement-based	0.15	Lab	TCLP		0.010
4:101	Electroplating sludge	F006	Cement-based	0.20	Lab	TCLP		0.090
4:102	Electroplating sludge	F006	Unstabilized raw waste		Lab	TCLP		33.300
4:102	Electroplating sludge	F006	Cement kiln dust	0.50	Pilot	TCLP		2.710
4:102	Electroplating sludge	F006	Cement kiln dust	0.50	Commercial	TCLP		1.230
4:103	Arc furnace dust	K061	Unstabilized raw waste		Lab	EPT	1100	0.900
4:103	Arc furnace dust	K061	Unstabilized raw waste		Lab	TCLP		0.060
4:103	Arc furnace dust	K061	Cement-based	0.20	Lab	TCLP		0.610
4:103	Arc furnace dust	K061	Cement/silicate	0.20	Lab	TCLP		0.730
4:103	Arc furnace dust	K061	Proprietary	0.20	Lab	TCLP		0.040
4:103	Arc furnace dust	K061	Lime kiln dust	0.20	Lab	TCLP		0.500

4:104	Arc furnace dust	K061	Proprietary		Lab	EPT	900	0.160
4:104	Electroplating sludge	F006	Lime, sulfide		Commercial	EPT	16300	<0.010
						TCLP		<0.050
4:119	Arc furnace dust	K061	Portland cement	0.05	Lab	TCLP	1370	0.043
4:119	Arc furnace dust	K061	Kiln dust	0.05	Lab	TCLP	1370	0.032
4:119	Arc furnace dust	K061	Lime/flyash	0.10	Lab	TCLP	1370	0.073
4:122	Electroplating waste	F006	Jet cement/Fe(NO$_3$)$_2$	0.20	Lab	Water	3000 (Cr^{+6})	0.010
4:123	Incinerator ash		Portland cement + Al$_2$(SO$_4$)$_3$ + CaSO$_3$	0.83	Lab	Water	25000 (Cr^{+6})	0.200
4:125	Sodium chromate waste		Cement + additives	0.50	Lab	Water	1050 (Cr^{+6})	<0.010
4:132	Electroplating waste	F006	Cement + zeolite + FeSO$_4$ + CaCO$_3$/CaCl$_2$	1.05	Lab	Water		0.120
4:136	Flyash A		Unstabilized	0.26	Lab	EPT		0.140
4:136	Flyash A		Gelled silicate	0.10	Lab	EPT		0.090
4:136	Flyash A		Lime		Lab	EPT		0.080
4:136	Flyash B		Unstabilized	0.26	Lab	EPT		0.130
4:136	Flyash B		Gelled silicate	0.10	Lab	EPT		0.060
4:136	Flyash B		Lime		Lab	EPT		0.260
4:136	Plating waste		Unstabilized	0.16	Lab	EPT	56	2.500
4:136	Plating waste		Gelled silicate	0.24	Lab	EPT		0.420
4:136	Plating waste		Cement/silicate		Lab	EPT		1.400
4:136	Mixed municipal/industrial waste		Unstabilized		Lab	EPT	353	0.130
4:136	Mixed municipal/industrial waste		Gelled silicate	0.16	Lab	EPT	353	0.150
4:136	Mixed municipal/industrial waste		Cement/silicate	0.24	Lab	EPT	353	0.190
	CCA wood preserving waste		Unstabilized		Lab	EPT	16000	90.000
	CCA wood preserving waste		Cement/silicate	0.40	Lab	EPT	16000	16.000

(*continued*)

128 FIXATION OF METALS

TABLE 4-25 Summary of Chromium Fixation Data (*Continued*)

Reference[a]	Waste Description	Waste Code	CFS System	Mix Ratio	Treatment Scale	Leaching Test	Chromium Content (mg/kg or mg/l) Total	Chromium Content (mg/kg or mg/l) Leachate
	CCA wood preserving waste		Potassium silicate		Lab	EPT	16000	0.500
	CCA wood preserving waste		Proprietary	0.18	Lab	EPT	16000	150.000
	CCA wood preserving waste		Portland cement	0.10	Lab	EPT	16000	13.390
	CCA wood preserving waste		Portland cement	0.20	Lab	EPT	16000	4.110
	Arc furnace dust	K061	Potassium silicate		Lab	EPT		0.120
4:143	Synthetic sludge		Raw waste			TCLP		7.100
4:143	Synthetic sludge		Cement/silicate + hexadecyl mercaptan	0.2	Lab	TCLP		<0.050
4:158	FeCl₃ Etching waste		Lime	1.5	Lab	EPT		0.004
4:158	FeCl₃ Etching waste		Lime	0.7	Lab	EPT		0.026
4:159	Mixed waste		Cement/flyash		Lab	EPT	1100	1.09
4:159	Mixed waste		Cement/flyash		Lab	MEP[b]	1100	0.08
4:160	Chromium waste		Cement/flyash		Lab	ELT,[c] 1 d	1750	0.090
4:160	Chromium waste		Cement/flyash		Lab	ELT,[c] 28 d	1750	0.005
4:160	Zinc phosphate waste		Cement/flyash		Lab	ELT,[c] 1 d	26	0.300
4:160	Zinc phosphate waste		Cement/flyash		Lab	ELT,[c] 28 d	26	<0.002
4:160	Waste treatment plant sludge		Cement/flyash		Lab	ELT,[c] 1 d	136	0.070
4:160	Waste treatment plant sludge		Cement/flyash		Lab	ELT,[c] 28 d	136	0.002
4:160	Zinc phosphate waste		None		Lab	ASTM,[c] 2 d	26	0.080
4:160	Zinc phosphate waste		None		Lab	ASTM,[c] 10	26	0.003
4:160	Zinc phosphate waste		Cement/flyash		Lab	ASTM,[c] 2 d	26	0.060
4:160	Zinc phosphate waste		Cement/flyash		Lab	ASTM,[c] 10	26	0.002

[a] References are listed at the end of Chapter 4.
[b] Maximum concentration.
[c] Sequential batch test, CO₂-saturated water.

Reduction Processes	Reference
Cement + FeSO$_4$	[124]
Portland cement + CaO + NaSO$_3$ + Ca(H$_2$PO$_4$)$_2$ + sodium lignosulfonate + sodium laurylbenzenesulfonate	[125]
Portland cement + sulfur	[126]
Sugar	Private communication
Cellulose	Private communication
Sodium sulfide/ferrous sulfate	[127]
Electrochemical reduction	[128]

Direct Fixation Processes	Reference
Ion exchange	[108]
Barium chromate precipitation	[129]
Lead chromate precipitation	[130]
Activated carbon	[87]
Portland cement + BaCl$_2$	[131]
Portland cement + zeolite + FeSO$_4$ + CaCl$_2$/CaCO$_3$	[132]
Portland cement + red mud	[133]
Portland cement + metallurgical slag	[134]

When we discussed cadmium fixation, the paper by Bishop [44] was discussed. In this study on a cement-based system, he found that even after all of the alkalinity had been leached from the test specimen, trivalent chromium and lead were still largely retained (85 percent for chromium) in the silicate matrix. This gives rise to the theory that chromium has been respeciated (it was originally present as the hydroxide) as the "silicate." In actuality, it is probably not a simple silicate, almost certainly not a stoichiometric form. As we will see in Chapter 11, most metal silicates are probably complex metal—hydrated oxide/silicate structures of indeterminate composition. However, some investigators believe that the metal hydroxide is simply encapsulated in a silica matrix that prevents its removal except by destroying the matrix. This seems unlikely, since diffusion would presumably allow dissolution of the acidsoluble hydroxide particle at a faster rate than is evident in Bishop's work. Furthermore, Bishop found the chromium (and lead) dissolution rate to follow almost exactly the silicon dissolution rate, which provides strong presumptive evidence of the dissolution of an actual "silicate." In any case, chromium is obviously strongly bound in the CFS product matrix, at least in cement-based systems. Since the formation of silicates is also the basis for pozzolanic systems (the other major CFS system type), it is likely that chromium is also respeciated there. Small wonder then that chromium, in the trivalent state, is so easily immobilized.

Another interesting study is presented in a patent [122] on the use of "jet" cement and ferrous nitrate to directly fix Cr^{+6} ion. The presence of the Fe^{+2} ion would seem to indicate that the Cr^{+6} is simply being reduced to the trivalent state. However, the jet cement alone reduced Cr^{+6} leaching by an order of magnitude—from about 12 mg/l to about 1.2 mg/l—while ferrous nitrate additions to regular Portland cement were not effective. Some characteristic of this type of cement, therefore, would seem to be controlling the reaction, so it is useful to compare the compositions of the two cements, as shown in Table 4-26.

The obvious differences between the two cements, other than in fineness, is the presence of a large amount of the calcium aluminate/calcium fluoride complex ($C_{11}A_7 \cdot CaF_2$) in the jet cement, more than replacing the calcium aluminate (C_3A) in the Portland cement. Aluminum oxides/hydroxides are known coprecipitants for various metals [123], and they work even better in some cases when combined with the ferrous ion. The action of the halide, F, in this process is not known. Jet cement is a rapid hardening cement, but this action alone probably does not account for the difference.

Another combination fixation system is that of ferrous sulfate and sodium sulfide. Carpenter [127] showed that this combination is effective in waste water treatment at the ratios of 2:1.5:1 for $S^{-2}:Fe^{+2}:Cr^{+6}$, respectively. This system definitely operates by chromium reduction, but the presence of the two reductants is synergistic, and S^{-2} compounds are also quite insoluble in their own right.

LEAD

Lead is a member of Group IVA of the periodic table. It has two valence states, +2 and +4, with the +2 state being the most common. Pb^{+4} compounds are regarded as being covalent, while Pb^{+2} compounds are primarily ionic. Lead is amphoteric and forms anionic plumbites and plumbates as well as both cations. While lead reacts rapidly with oxygenated water, small quantities of carbonates and silicates will passivate the surface and greatly reduce further corrosion; thus, its use in many applications for corrosion resistance. The world consumption of lead in 1978 was 4.4 million metric tons, of which 22 percent was used in the United States. It occurs as the minerals galena (PbS, 87 percent Pb), anglesite ($PbSO_4$, 68 percent Pb), and cerrusite ($PbCO_3$, 77 percent Pb), the latter two of which result from the natural weathering of galena. Other minerals containing lead are listed in Table 4-27 [161]. The solubilities of six lead minerals are shown diagrammatically in Figure 4-6.

Lead and its compounds are cumulative poisons. Its use in vessels for making wines and grape syrups in Roman times is said to have caused gout, mental retardation, and personality changes in the Roman aristocracy and its offspring [162]. Lead poisoning may be acute or chronic, but the effects are varied and severe. It may occur by ingestion or inhalation. Because of its long industrial and consumer use

TABLE 4-26   Cement Compositions

	Mineral Composition (%)	
Component[a]	Jet Cement	Portland Cement
$C_{11}A_7CaF_2$	20.6	0.0
C_3S	50.7	52.7
C_2S	1.7	23.9
C_3A	—	8.2
C_4AF	4.7	9.7
Al_2O_3/SO_3	1.1	3.2
Fineness, Blaine (cm²/g)	5300	3200

[a] This is the standard mineral nomenclature for cements (see Chapter 10 for more detail): C = CaO; A = Al_2O_3; S = SiO_2; F = Fe_2O_3.

TABLE 4-27 Minerals Containing Lead

Mineral	Structure
Anglesite	$PbSO_4$
Bromopyromorphite	$Pb_5(PO_4)_3Br$
Cerrussite	$PbCO_3$
Chloropyromorphite	$Pb_5(PO_4)_3Cl$
Dumonite	$Pb_2(UO_2)_3(PO_4)_2(OH)_4 \cdot 3H_2O$
Fluoropyromorphite	$Pb_5(PO_4)_3F$
Galena	PbS
Hydroxypyromorphite	$Pb_5(PO_4)_3OH$
Parsonite	$Pb(UO_2)(PO_4)_2$
Plumapatite	$Pb_2Pb_3(PO_4)_3 \cdot H_2O$
Plumbogummite	$PbAl_3(PO_4)_2(OH)_5 \cdot H_2O$
Wulfenite	$PbMoO_4$

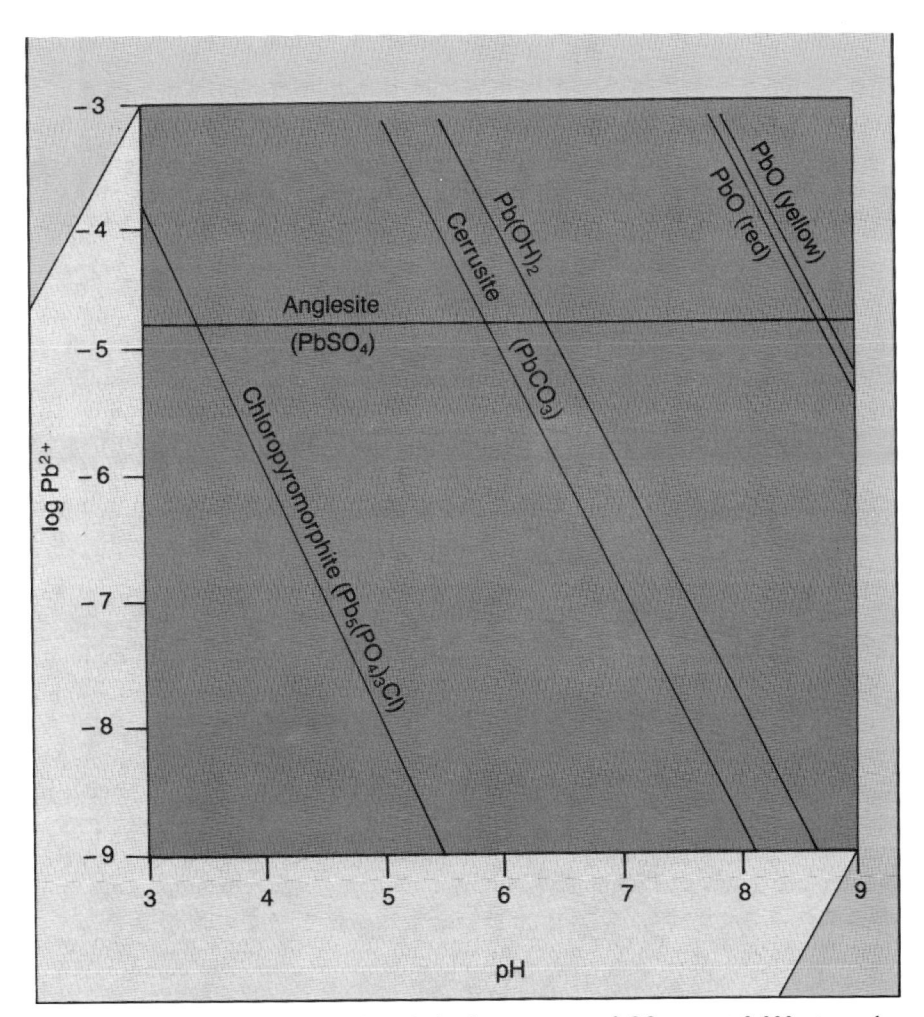

Figure 4-6 Solubility of lead minerals in the presence of CO_2 gas at 0.003 atmospheres. (From Dragun [161].)

history, more is known about lead poisoning than about that of any other metal. Allowable lead levels have been established for drinking water, and in the workplace for airborne lead and its compounds under OSHA regulations.

The usual valence of lead is $+2$, and Pb^{+4} exists as the free cation only in small concentrations. Unlike the bivalent ions of other Group IVA metals such as tin, the plumbous ion does not have reducing properties. In general, the chemistry of inorganic lead compounds is similar to that of the alkaline earth elements [8]. Many inorganic lead compounds possess two or more crystalline forms whose properties are different. Pb^{+2} forms a soluble nitrate, chlorate and acetate, a slightly soluble chloride, and a low-solubility sulfate, carbonate, chromate, phosphate, molybdate, and sulfide. It also forms anhydrous and hydrous basic lead salts such as $4PbO \cdot PbSO_4$, and complexed mixed salts such as $2PbCO_3 \cdot Pb(OH)_2$ (white lead). PbO combines easily with SiO_2 to form a low-melting glass. Pb^{+4} is formed when the metal is subjected to strong oxidizing conditions, such as in wet oxidation by chlorine water, and in the reversible reaction of the lead storage battery:

$$Pb + PbO_2 + 2\ H_2SO_4 \longrightarrow 2\ PbSO_4 + 2\ H_2O \qquad (4\text{-}6)$$

Most lead compounds are produced from pig lead, usually from the monoxide.

The primary uses of lead and lead compounds, in order of importance, are in storage batteries, manufacture of tetraethyllead, pigments, ammunition, solders, plumbing, cable covering, bearings, and caulking [8]. Major lead compounds (Pb^{+2} unless otherwise stated) are as follows.

Halides

Lead fluoride is used in fuses, as a catalyst, in glass coatings and glasses, in television tubes and other electronics, and in electroless plating activators. The chloride is the starting material for the production of many organolead compounds; it is also used in electronics, as a catalyst, a flame retardant, and a flux for the galvanizing of steel. Lead bromide is used in photoduplication and in many of the applications of lead chloride. In addition to the uses of the bromide, lead iodide is used in brake linings, batteries, and lubricants.

Oxides

In addition to the divalent and tetravalent oxides, lead forms a mixed oxide, Pb_3O_4. The largest single use of lead monoxide is in storage battery electrodes. The second largest use is in the ceramics industry for glasses, glazes, and vitreous enamels, where the oxide is converted to silicates. The paint and rubber industries are the third largest consumer of lead oxides. The dioxide is used as an oxidizing agent in pyrotechnics, chemicals, and polymers. The tetroxide, Pb_3O_4, is used as an anticorrosion pigment, in positive storage battery plates, in glass, ceramics, and in catalysts. Lead hydroxide is used in nickel–cadmium batteries, catalysts, lubricants, and in waste water filters to remove chromates (see the section on chromium).

Sulfide and Telluride

Lead sulfide, or galena, is the chief ore of lead. It is used in electronic applications, in catalysts, pigments, and lubricants. Lead telluride is used in electronics.

Sulfates

The sulfates of lead include normal, acid, and a number of basic compounds. PbSO$_4$ forms in lead storage batteries during the discharge cycle, and has also been used in stabilization of clay soils. The basic lead sulfates, PbO·PbSO$_4$, 2PbO·PbSO$_4$, 3PbO·PbSO$_4$·H$_2$O, and 4PbO·PbSO$_4$, are used in plastics stabilizers, pigments, rubber fillers, and in textile dyeing and printing.

Lead Nitrate

This compound is used in ore processing (floatation agent), pyrotechnics and explosives, stabilizers, coatings, catalysts, and in electronics.

Acetates

The trihydrate is used primarily in the preparation of other compounds. Other uses include antioxidants, mildew treatments, and as mordant. The tetraacetate, Pb(C$_2$H$_3$O$_2$)$_4$, is used in organic synthesis.

Carbonates

Lead carbonate has a wide range of applications: catalyst, bonding agent, coating, stabilizer; in lubricants, stabilizers, storage batteries, and electronics. Basic lead carbonate, or white lead, is the old basic white hiding pigment in paints; it has been largely replaced by titanium dioxide due to its toxicity. It is still used as a catalyst, in ceramic glazes, as a specialized pigment, stabilizer, and lubricant.

Silicates

Three lead silicates—mono (1·5PbO·SiO$_2$), bi (PbO·0·03Al$_2$O$_3$·1.95 SiO$_2$), and tri (or tribasic) (3PbO·SiO$_2$)—are used primarily in glasses, ceramics, and glazes.

Other Lead Compounds

Phosphite: Polymer stabilizer
Azide: Detonating agent
Antimonate: Pigment, colorant
Phthalates: Polymer stabilizers
Borate: Glazes and enamels
Zirconate: Electronics
Arsenicals: See the section on arsenic

Organolead Compounds

Until recently, widely used in the form of tetraethyllead as an antiknock additive in gasoline, organolead compounds are the largest group of organonmetallics. These compounds are highly toxic, especially since they can be introduced into the body by absorption through the skin and are easily inhaled as a vapor. The only nongasoline use has been in seed disinfectants and for fungi control. These uses also are being questioned for environmental reasons.

Species and Uses of Interest in CFS

Lead is nearly ubiquitous in the environment. It occurs in soils, rocks, and minerals, and so is found in substantial amounts in CFS reagents such as cement, kiln dusts, and lime. It is also widely distributed from its use in gasoline, from lead-bearing pesticides, and from industrial emissions, where it is discharged into the atmosphere and settles in soils. The range of lead in soil is 2 to 200 ppm [161], with an average of about 10 ppm. Because of various fixation reactions, however, lead in groundwater is usually in the low parts-per-billion (ppb) range. As we saw earlier, lead forms numerous phosphate minerals in soil, and since the average soil phosphate content is about 600 ppm, phosphate may be the major controlling factor over lead in solution in groundwater. Fulvic acids from humic substances also interact with lead in the soil [163].

A number of other nonwaste sources of lead are environmental problems. For example, older buildings are covered, inside and out, with lead-bearing paints. Lead plumbing and solders are now believed to be a problem. Most of these sources, however, are not encountered in CFS work, although the leachability of lead in CFS reagents is becoming an increasing problem as allowable leaching levels continue to get lower.

Lead is a common constituent of plating wastes [48] and is often the biggest fixation problem with these materials. Mining and smelting wastes are another source of lead pollution, but mining wastes are not currently covered under RCRA. Another major source of lead that poses a real challenge to CFS technology is electric arc furnace dust (K061). This waste contains high lead levels, typically 0.5 to 5 percent, and often in very leachable forms. Lead is also a problem in other air pollution control residues and in bottom ashes from some sources. In certain cases, the lead is known to be speciated as the soluble chloride; in others, as oxides and sulfates.

Lead from process wastes other than electroplating is especially prevalent in the petroleum industry, where it originates from refineries, tank bottoms, and other sources. These wastes may contain both inorganic and organic lead compounds. Actually, most waste residuals from industry contain at least small amounts of lead which, in many cases, cause no leaching problems. In general, however, it can be said that lead is the most common and widespread problem encountered in metal fixation technology. Its pervasive presence and the many species in which it appears make each waste type, and even each new source, an individual challenge. And the newest regulatory leaching levels have exacerbated the problem.

CFS Treatment Processes

The literature contains a great deal of data on the precipitation or fixation of lead. As with the other metals, most of this comes from the water treatment area, but an increasing body of information is specific to fixation of lead in CFS systems. The standard methods of lead removal from waste water are to precipitate it as the hydroxide, carbonate, or basic carbonate, all of which are relatively insoluble at alkaline pH. Rohrer [164] reports the limiting solubilities of these species as:

$Pb(OH)_2$	133	mg/l as Pb
$PbCO_3$	0.85	mg/l as Pb
$Pb_3(PO_4)_2$	0.11	mg/l as Pb
$Pb_3(CO_3)_2(OH)_2$	1.3	mg/l as Pb

However, he found that a treatment consisting of trisodium phosphate (TSP) plus caustic soda to be the most effective, reducing lead concentration below 0.02 mg/l in waste water. TSP alone gave concentrations at about the limiting solubility level.

Lead in waste water has been precipitated out with lime and ferrous sulfate [165] and with dolomite (CaCO$_3$·MgCO$_3$) to yield lead carbonate [166]. Organic lead compounds in waste waters can be oxidized to yield inorganic lead that is precipitated at pH 8 to 9.5 in the presence of carbonates [48]. Sulfides have also been used for this purpose, as has ferrous sulfate at pH 10.4 to 10.8.

As we have seen, waste water data are often not very useful in CFS work. The complex matrices and high ionic strengths create a very different environment for precipitation, and equilibrium concepts have limited applicability. We do know that pH control is important. Minimum lead leaching in nearly all CFS systems occurs when the pH is maintained between about 8 and 10 in the leachate. Jirka *et al.* [167] found that the technique used to adjust pH in the EPT test was critical in determining lead leaching by that method. In their tests, if the pH *ever* fell below 4.8, lead leaching is significantly elevated. In general, in any case, the best balance of pH and fixant must be determined individually for each waste stream.

Table 4-28 gives typical leaching results for many lead-containing wastes treated with various CFS processes. A summary of the various processes that have been reported for fixation of lead in residues is given in the following list.

Process	*Reference*
Portland cement	[44], [101], [106], [115]–[117], [119], [120]
Portland cement + aluminum sulfate	[137]
Cement/soluble silicate	[103], [136], [138], [139]
Potassium silicate	Private Communication, 1987
Cement/soluble silicate + sodium sulfide	[138], [139]
Cement/soluble silicate + ammonium phosphate	[138], [139]
Lime	[98]
Sulfide	[98]
Lime/flyash	[119]
Kiln dust	[100]–[103], [119]
Proprietary	[104], [105]

Some of these processes are worthy of special note. As with most metals, there is no difficulty in reaching low lead leaching levels—less than the current RCRA level of 5.0 mg/l—with generic CFS systems, so long as the alkalinity of the system maintains the proper pH level throughout the leaching test. However, when the final pH of the leachant drops below about 8, lead leaches out at much higher levels. This presents a potential environmental problem, however, wherever the natural environment is acidic even when using highly alkaline systems, since the alkalinity will eventually be removed.

When we addressed chromium fixation, the paper by Bishop [44] was discussed. In this study on a cement-based system, he found that even after all of the alkalinity had been leached from the test specimen, trivalent chromium and lead were still largely retained (75 percent for lead) in the silicate matrix. This gives rise to the theory that lead has been respeciated (it was originally present as the hydrox-

136 FIXATION OF METALS

TABLE 4-28 Summary of Lead Fixation Data

Reference[a]	Waste Description	Waste Code	CFS System	Mix Ratio	Treatment Scale	Leaching Test	Lead Content (mg/kg or mg/l) Total	Lead Content (mg/kg or mg/l) Leachate
4:98	Metal/cyanide	F006-8	Lime	—	Commercial	EPT	970	<0.100
4:98	Metal finishing	F006-9, D002-3	Lime/sulfide	—	Commercial	EPT	375	0.080
4:98	Metal finishing	F006-9, D002-3	Lime/sulfide	—	Commercial	TCLP		<0.10
4:98	Metal finishing	F006-9, D002-3	Lime/sulfide	—	Commercial	EPT	42	0.080
4:98	Metal finishing	F006-9, D002-3	Lime/sulfide	—	Commercial	TCLP		<0.100
4:98	Oil waste		Flyash	1.20	Commercial	TCLP		<0.100
4:98	Paint sludge		Flyash	0.35	Commercial	EPT		<0.100
4:98	Paint sludge		Flyash	0.35	Commercial	TCLP		<0.100
4:98	Biological sludge		Flyash	1.10	Commercial	TCLP		<0.100
4:98	Metal hydroxide		Flyash	0.25	Commercial	EPT		0.150
4:98	Metal hydroxide		Flyash	0.25	Commercial	TCLP		0.150
4:100	TSD incinerator ash	Various	Cement kiln dust	0.50	Lab	TCLP	910	2.770
4:100	TSD incinerator filter cake	Various	Cement kiln dust		Lab	TCLP	2740	0.780
4:100	TSD landfill leachate	Various	Cement kiln dust	0.50	Lab	TCLP	96	0.690
4:100	TSD incinerator ash	Various	Cement kiln dust	0.50	Lab	TCLP	928	1.060
4:101	Electroplating sludge	F006	Unstabilized raw waste			TCLP	24500	50.200
4:101	Electroplating sludge	F006	Cement kiln dust	0.20	Lab	TCLP		1.070
4:101	Electroplating sludge	F006	Cement kiln dust	0.50	Lab	TCLP		0.400
4:101	Electroplating sludge	F006	Cement-based	0.10	Lab	TCLP		20.000
4:101	Electroplating sludge	F006	Cement-based	0.15	Lab	TCLP		0.220
4:101	Electroplating sludge	F006	Cement-based	0.20	Lab	TCLP		0.100
4:102	Electroplating sludge	F006	Unstabilized raw waste		Lab	TCLP		0.210
4:102	Electroplating sludge	F006	Cement kiln dust	0.50	Pilot	TCLP		<0.020
4:102	Electroplating sludge	F006	Cement kiln dust	0.50	Commercial	TCLP		<0.020
4:103	Arc furnace dust	K061	Unstabilized raw waste		Lab	EPT	38000	139.000
4:103	Arc furnace dust	K061	Unstabilized raw waste		Lab	TCLP		14.300
4:103	Arc furnace dust	K061	Cement-based	0.20	Lab	TCLP		0.190
4:103	Arc furnace dust	K061	Cement/silicate	0.20	Lab	TCLP		0.170
4:103	Arc furnace dust	K061	Proprietary	0.20	Lab	TCLP		0.220
4:103	Arc furnace dust	K061	Lime kiln dust	0.20	Lab	TCLP		3.400

LEAD   137

4:104	Arc furnace dust	K061	Proprietary		Lab	EPT	12000	0.100
4:105	Mixed wastes	Various	Proprietary		Commercial	EPT	840	<0.100
4:114	Electroplating sludge	F006	Lime, sulfide		Commercial	EPT	2800	0.180
4:117	Lead hydroxide sludge		Portland cement	0.45	Lab	TCLP	17200	70
4:117	Lead hydroxide sludge		Portland cement	0.41	Lab	TCLP	51600	490
4:119	Arc furnace dust	K061	Portland cement	0.05	Lab	TCLP	20300	1.157
4:119	Arc furnace dust	K061	Kiln dust	0.05	Lab	TCLP	20300	0.787
4:119	Arc furnace dust	K061	Lime/flyash	0.10	Lab	TCLP	20300	0.095
4:136	Flyash A		Unstabilized		Lab	EPT		3.200
4:136	Flyash A		Gelled silicate	0.26	Lab	EPT		1.200
4:136	Flyash A		Lime	0.10	Lab	EPT		0.440
4:136	Flyash B		Unstabilized		Lab	EPT		170.000
4:136	Flyash B		Gelled silicate	0.26	Lab	EPT		74.000
4:136	Flyash B		Lime	0.10	Lab	EPT		350.000
4:136	Plating waste		Unstabilized		Lab	EPT	1048	8.600
4:136	Plating waste		Gelled silicate	0.16	Lab	EPT		2.700
4:136	Plating waste		Cement/silicate	0.24	Lab	EPT		12.000
4:136	Mixed municipal/industrial waste		Unstabilized		Lab	EPT	823	4.500
4:136	Mixed municipal/industrial waste		Gelled silicate	0.16	Lab	EPT	823	2.500
4:136	Mixed municipal/industrial waste		Cement/silicate	0.24	Lab	EPT	823	3.200
4:137	Arc furnace dust	K061	Potassium silicate		Lab	EPT		0.26
4:137	Pigment sludge		Cement + Al$_2$(SO$_4$)$_3$		Lab	Water		<0.010
4:138	Ceramic slip waste		Cement/silicate	0.30	Lab	EPT	10000	0.18
4:138	Ceramic slip waste		Cement/silicate	0.12	Lab	EPT	10000	2.7
4:138	Ceramic slip waste		Cement/silicate + Na$_2$S	0.22	Lab	EPT	10000	0.18
4:138	Ceramic slip waste		Cement/silicate + Na$_2$S	0.14	Lab	EPT	10000	5.6
4:138	Ceramic slip waste		Cement/silicate + (NH$_4$)2HPO$_4$	0.22	Lab	EPT	10000	0.73

(continued)

TABLE 4-28 Summary of Lead Fixation Data (Continued)

Reference[a]	Waste Description	Waste Code	CFS System	Mix Ratio	Treatment Scale	Leaching Test	Lead Content (mg/kg or mg/l) Total	Leachate
4:138	Ceramic slip waste		Cement/silicate + $(NH_4)2HPO_4$	0.14	Lab	EPT	10000	0.21
4:139	Ceramic glaze waste		Cement/silicate	0.30	Lab	EPT	450000	43
4:139	Ceramic glaze waste		Cement/silicate	0.12	Lab	EPT	450000	87.8
4:139	Ceramic glaze waste		Cement/silicate + $(NH_4)2HPO_4$	0.22	Lab	EPT	450000	38.1
4:139	Ceramic glaze waste		Cement/silicate + $(NH_4)2HPO_4$	0.14	Lab	EPT	450000	81.7
4:139	Ceramic glaze waste		Cement/silicate + Na_2S	0.22	Lab	EPT	450000	41.8
4:139	Ceramic glaze waste		Cement/silicate + Na_2S	0.14	Lab	EPT	450000	1290
4:143	Synthetic sludge		Raw waste			TCLP		3.970
4:143	Synthetic sludge		Cement/silicate + hexadecyl mercaptan	0.2	Lab	TCLP		0.300
4:159	Mixed waste		Cement/flyash		Lab	EPT	3320	0.003
4:159	Mixed waste		Cement/flyash		Lab	MEP[b]	3320	0.029
4:160	Zinc phosphate waste		Cement/flyash		Lab	ELT,[c] 1 d	23.4	<0.005
4:160	Zinc phosphate waste		Cement/flyash		Lab	ELT,[c] 28 d	23.4	<0.005
4:160	Waste treatment plant sludge		Cement/flyash		Lab	ELT,[c] 1 d	79	0.010
4:160	Waste treatment plant sludge		Cement/flyash		Lab	ELT,[c] 28 d	79	0.006
4:160	Zinc phosphate waste		None		Lab	ASTM,[c] 2 d	23.4	0.050
4:160	Zinc phosphate waste		None		Lab	ASTM,[c] 10	23.4	0.007
4:160	Zinc phosphate waste		Cement/flyash		Lab	ASTM,[c] 2 d	23.4	0.030
4:160	Zinc phosphate waste		Cement/flyash		Lab	ASTM,[c] 10	23.4	<0.005

[a] References are listed at the end of Chapter 4.
[b] Maximum concentration.
[c] Sequential batch test, CO_2-saturated water.

ide) as the "silicate." Bhatty [115] has postulated that the lead is fixed by an addition reaction, forming a metallic calcium silicate hydrate. In actuality, it is probably not a simple silicate, almost certainly not a stoichiometric form. Some investigators believe that the metal hydroxide is simply encapsulated in a silica matrix that prevents its removal except by destroying the matrix. This seems unlikely, since diffusion would presumably allow dissolution of the acid-soluble hydroxide particle at a faster rate than is evident in Bishop's work. Furthermore, Bishop found the lead dissolution rate to follow almost exactly the silicon dissolution rate, which provides strong presumptive evidence of the dissolution of an actual "silicate." In any case, lead is obviously strongly bound in the CFS product matrix, at least in cement-based systems. Since the formation of silicates is also the basis for pozzolanic systems, it is likely that lead is also respeciated there.

The dependency of lead leaching on pH has been verified by a number of investigators. Brunner and Baccini [168] found the same effect with treated (Portland cement) and untreated municipal incinerator ash. As the pH dropped below 8, lead began to leach significantly, with the leaching rate rising very rapidly as the pH dropped. Lead leaching also increased very rapidly above pH 12, demonstrating the amphoteric nature of the lead species in this system. This effect is also shown in the work of Mundell and Hill [169] in the in situ treatment of lead-contaminated sands at an old foundry site. Lime additions of 10 to 15 percent reduced lead leaching in the EPT test by a factor of 185, but excess lime (above 15 percent) resulted in high leaching rates. The investigators compared actual and theoretical lead carbonate solubility versus pH for cerrusite, and also showed that the minimum theoretical solubility point occurred at about pH 8.5.

It is important to note that the lead compounds in incinerator and other ashes, formed through thermal processes, may be quite different from those produced chemically in wastes such as electroplating sludges. In addition, the spatial distribution of the species through the solid may be nonuniform, and very different from one waste type to another.

It is obvious that the EPT and TCLP tests do not counter the alkalinity of most CFS-treated waste forms. Various studies (see Table 4-28) claim lead leachability as low as <0.02 in different wastes and CFS processes when the lead levels are moderate—<20,000 mg/l—and the mix ratios are high. At low mix ratios (0.05) one study [119] found high lead leaching in the range of 0.79 to 1.16 mg/l. Butler et al. [117] attribute the high leachability of lead in their tests to the solubilization and reprecipitation of lead as moderately soluble salts on the surfaces of the cement minerals.

When total lead content is high and/or the speciation is such that adequate insolubilization cannot be achieved by the simple generic processes, a number of additives can be used to enhance fixation. Some of these were just listed. In one set of treatability studies [138], [139], soluble silicates, sulfide, and phosphate were found to be effective at lead levels of about 1 percent (10,000 mg/l), but not at the 45 percent level. It is important to note from these data (see Table 4-28) that the relative proportions of the additives and the solidification reagents must be balanced to achieve optimum results; more is not necessarily better. This is true with all CFS systems, but becomes especially critical with high metal content and in the perspective of the new, lower allowable leaching levels under the landbans.

In nonacidic leaching media, such as seawater, sequential batch tests [116] demonstrated that lead leaching was undetectable after 16 extractions. The system tested had very high metal content, about 5 percent, in the form of the hydroxide. The investigators observed that lead did not have any observable physical effect on the waste form microstructure. This is in keeping with the results of the other studies previously mentioned.

When waste forms containing lead are subjected to a diffusion-controlled leaching test scenario with neutral water as the leachant, lead leaching is very low. Cote et al. [106] found only 0.177 percent of the lead to be leached from a cement/flyash waste form after an 11-month test duration in such a system. In this case, lead leaching was somewhat higher than that of chromium or cadmium, but much lower than that of arsenic. Other CFS systems—lime/flyash, bentonite/cement, and cement/soluble silicate—leached lead at somewhat higher rates than cement/flyash, but still below 1 percent. It is obvious that the leaching scenario used is critical when evaluating lead leachability from the products of any CFS process, with pH, and perhaps redox potential, being factors of prime importance. Meaningful estimation of leaching potential must include tests using the actual environmental conditions expected at the disposal site, especially the groundwater composition. Testing with actual or simulated groundwater should include control of redox potential as well as the compositional parameters.

MERCURY

Mercury and its prime ore, cinnabar or vermillion (HgS), have been known since antiquity. They were used for cosmetic and medical preparations and as a pigment. Mercury is estimated to occur in natural rocks at the level of 10 to 1000 ppb, and in petroleum in the range of 2000 to 20,000 ppb. Today, mercury is used in the chemical industries, in the manufacture of paints and paper, in pesticides, and in the electrical industries. The metal is produced by retorting the sulfide ores to liberate the metal as a vapor, then condensing it to the liquid metal. Other methods include the leaching of ores and concentrates with sodium sulfide and sodium hydroxide, or with sodium hypochlorite, but these techniques are not used now [8]. World production of mercury in 1978 was about 6000 metric tons. The standard unit of trade is the "flask," which weighs 76 pounds.

Mercury salts exist in two valence states, +1 and +2. The mercurous compounds often exist as double salts, for example, Hg_2Cl_2. Many mercury compounds are volatile; they are also labile and easily decomposed by light, heat, and reducing agents, even weak reducing agents such as amines, aldehydes, and ketones. Because of their covalent nature and ability to form a variety of organic complexes, mercury compounds have unusually wide solubility. Small amounts of amines, ammonia, and so forth, can exercise a large solubilizing effect [8].

Although the toxic nature of mercury and its compounds was known for centuries, the relatively small scale of use and its familiarity prevented recognition of the environmental hazards. The metal, its vapors, and most of its compounds are protoplasmic poisons that can be fatal to all forms of life, including man. Poisoning may occur through oral, dermal, and inhalation routes. The most toxic forms of mercury are the alkyl organic compounds. In 1960, it was reported that more than 100 people had died or suffered severe neurological damage from eating fish and shell fish contaminated with methyl mercury and mercuric chloride from plastics manufacturing near Minamata, Japan [87]. Later, it was demonstrated that inorganic mercury could be biologically methylated in the aquatic environment and that marine organisms could concentrate the mercury up to 3000 times. In the following years, other incidents in Japan, Sweden, Canada, and the United States resulted in regulatory action controlling the discharges and uses of mercury compounds.

Commercial Uses

Consumption of mercury in the United States by use is given in Table 4-29. Ottinger *et al.* [87] divide mercury's uses into recyclable and dissipative. Recyclable uses include its important use as a cathode in caustic-chlorine production by electrolysis, and in various electrical and electronic applications, the two largest consumers of mercury in the United States. The others are basically dissipative uses.

Caustic-Chlorine Production. Even though it is classified as a recyclable use, there is a large loss of mercury from this process into the environment, and it has been by far the largest single source of mercury pollution. This waste is classified by the EPA as:

K071— Brine purification muds from mercury cell process in chlorine production, where separately prepurified brine is not used.

K106—Wastewater treatment sludge from the mercury cell process in chlorine production.

Both of these listed wastes are in the first-third landban, but K106 final BDAT standards were not set by the EPA and have been deferred until later. K071 BDAT standards require solubilizing of the mercury in the waste so that it can subsequently be respeciated as the very low-solubility sulfide. This is the only case to date where such respeciation has been formally required prior to fixation, and may constitute

TABLE 4-29 Uses of Mercury in the United States (Metric Tons)

Use	1959	1968	1978
Agriculture	110	118	21[a]
Amalgamation	9	9	<0.5[a]
Catalysts	33	66	29[a]
Dental preparations	95	106	18
Electrical applications	426	677	619
Caustic-chlorine	201	602	385
General laboratory	38	69	14
Industrial and control	351	275	120
Paint			
Antifouling	34	14	
Mildew proofing	87	351	309
Paper and pulp	150	14	
Pharmaceuticals	59	15	15[a]
Other	298	285	216
Totals	1892	2600	1681

[a] 1975 figures.

Source: Adapted from Kirk-Othmer, *Encyclopedia of Chemical Technology*, 3d ed., Wiley, New York, 1979.

a precedent. In all other cases, CFS or fixation techniques have been left to the treater, with only the performance level being a requirement.

The brine sludge contains small amounts of mercury as well as mercuric ions, mostly in the form of the tetrachloro complex, $HgCl_2^{-2}$. This waste is being reduced by the use of other chlorine production methods that do not require mercury. Nevertheless, there is still a significant amount of brine sludge being generated, and probably a large number of lagoons and other areas contaminated by mercury from this source. These will require cleanup in the future, probably using CFS technology. Waste from caustic-chlorine production is definitely the most important of all sources to CFS technology.

Agriculture. As is obvious from Table 4-29, this use has declined drastically, due to the 1972 banning of all mercury compounds in agriculture in the United States. At present, only mercuric and mercurous chloride are permitted for use on turf as fungicides [8]. However, in view of the past consumption, it is likely that significant amounts of mercury may be encountered in the cleanup of old pits, ponds and lagoons, contaminated soils, and underwater sediments under CERCLA and in private remediation work.

Electrical Applications. Mercury and its compounds may be found in wastes from certain manufacturing operations, for example, battery manufacture, but these are not important impacts on CFS technology.

Catalysts. Mercury is used as a catalyst for the production of vinyl chloride monomers and urethane foams. Discharges from these plants are sources of mercury-containing wastes, and may require CFS treatment to clean up old disposal areas.

Industrial and Control. Most of these uses involve elemental mercury, and the environmental effects are primarily of an occupational health nature, or result from the disposal of the products (thermometers, batteries, etc.) into landfills. There is little impact from these sources on CFS technology.

Paint. This use of mercury and its compounds has decreased drastically in recent years.

Paper and Pulp. The use of mercury compounds as slimicides in the pulp and paper industry has also decreased greatly since 1970.

Other. Mercury from mine tailings may be an important issue to CFS work in the future, but such wastes are exempted from the RCRA at this time.

Major Compounds

Mercuric Chloride. This salt is very soluble in water and in polar organic solvents. It is used as a catalyst and an intermediate in organic synthesis. $HgCl_2$ is also a component of fungicides and antiseptics, and is used in photography and in batteries.

Mercuric Oxide. Mercuric oxide exists in both red and yellow forms. The red form is more dense, and is used in the Ruben-Mallory dry cell, where it is mixed with graphite and serves as a depolarizer.

Mercuric Nitrate. This compound is used as the starting material for many other inorganic mercury salts, and is also used in organic synthesis.

Mercuric Sulfate. Mercuric sulfate is used as a catalyst.

Mercuric Sulfide. The red hexagonal form of mercuric sulfide is found in nature as cinnabar, while the black sulfide is a tetrahedral form. Both are used as pigments due to color and insolubility.

Organomercury Compounds. Alkyl compounds of the general formula RHgX are no longer manufactured in most of the world because of extreme and long-lasting toxicity. They were previously used as seed disinfectants. Phenylmercuric acetate, or PMA, finds its primary use in latex paint as a preservative and antifungicide. Other organomercury compounds are also used for this purpose.

CFS Treatment Processes

Nearly all of the literature references refer to the removal of mercury from waste water. Some exceptions are listed below, including mercury leaching results for CFS processes.

Process	*Reference*
Portland cement	[115]
Cement/soluble silicate	[143]
Cement + thiourea	[140]
Cement + Fe^{+2} + thiourea	[141]
Cement + metal cyanides	[149]
Cement + vermiculite	[151]
Cement + coal	[150]
Coprecipitation with Fe^{+2}/Fe^{+3}	[156]
Fe/Cu salt + polydithiocarbamate	[142]
Calcium polysulfide	[153]
Inorganic sulfur compounds	[155]
Sulfides	
Thiosulfates	
Dithionates	
Hypophosphorous acid reduction	[144]
Formaldehyde reduction	[145]
Sodium borohydride reduction	[146]
Zinc or iron reduction	[48], [154]
Ion exchange	[48], [147]
Activated carbon	[48], [148]
Iron cementation	[152]
Lime	[98]

Process	Reference
Sulfide	[8], [48], [98], [156]
Bacteria	[157]
Absorption	[87]

Table 4-30 summarizes fixation processes for mercury in both CFS systems and in waste water treatment. These data are presented because waste water fixation methods have been found to work, and be necessary, for fixation of mercury to sufficiently low levels in CFS systems. The requirement for mercury fixation under the RCRA, especially in the first-third BDAT standards, is lower than that for any other metal except arsenic (0.025 mg/l in K071). In the nonlisted waste categories, the mercury leaching standard is also the lowest of the metals (0.2 mg/l).

Precipitation with Sulfur Compounds. Most of the information in the preceding list comes from processes for treatment of brine, brine-sludge, or waste water from caustic-chlorine production. Mercury in wastes does not form low-solubility hydroxides that would allow pH control to be effective in its fixation, and the oxide solubility is too high. The most common way to remove mercury from waste water or fix it in CFS systems is by the use of various sulfur compounds, typically sodium or calcium sulfide. The solubility of the sulfide, HgS, at 18°C is 0.0107 mg/l. However, the use of excess sulfide increases the solubility of HgS, as shown in Table 4-31 [8]. In view of Table 4-31, it is advisable to use no more than a 20 percent excess of the alkali sulfide over the calculated requirement. In doing a treatability study, a range of alkali sulfide additions should be used, and the results graphed to obtain the minimum requirement. Also, HgS typically precipitates in a very fine particle size, making filtration of the leachate difficult and risking the carryover of solid HgS particles into the filtrate. Obviously, this has been a problem in waste water treatment, and has been resolved in many cases by using flocculating agents such as ferric chloride, ferric sulfate, or starch.

Other sulfur compounds that may be used are in the preceding list. Several patents describe the use of thiourea and cement, with or without other additives, to reduce mercury leaching to low levels—0.001 mg/l and below. Another uses polydithiocarbamates along with iron or copper salts. Unfortunately, all of these investigators used the Japanese leaching test, with water as the leachant, to assess the success of the process. Their results are not directly translatable for use in the United States. One study in the United States [143] achieved excellent TCLP leaching results (0.0006 mg/l) with hexadecyl mercaptan as an additive to a cement/soluble silicate process. The leaching of other metals was not affected by this additive; the sulfhydryl group apparently is highly specific to mercury, at least in this system.

Kaczur et al. describe a process [155] for insolubilizing mercury in situ, using inorganic sulfur compounds. The additives, in this case, were sprinkled on the surface of the waste and eluted into and through the waste by flushing with water. The additives dissolve in the eluting water and subsequently react with mercury in the waste or landfilled material, "insolubilizing" it. Typical results from this process showed soluble mercury in the effluent reduced by factors of 55 to 90 percent. Ottinger et al. [87] reference various processes for precipitating mercury from solutions by Na_2S and H_2S. Sittig [48] also references a number of reported techniques using sodium, potassium, and lithium sulfides. Other examples of such processes are the

TABLE 4-30 Summary of Mercury Fixation Data

| | | | | | | Mercury Content (mg/kg or mg/l) | |
						Total	Leachate
Reference[a]	Waste Description	Waste Code	CFS System	Mix Ratio	Treatment Scale	Leaching Test		
4:98	Metal/cyanide	F006-8	Lime	—	Commercial	EPT		<0.002
4:98	Metal finishing	F006-9, D002-3	Lime/sulfide	—	Commercial	EPT	<1	<0.002
4:98	Metal finishing	F006-9, D002-3	Lime/sulfide	—	Commercial	TCLP		<0.002
4:98	Metal finishing	F006-9, D002-3	Lime/sulfide	—	Commercial	EPT	<1	<0.002
4:98	Metal finishing	F006-9, D002-3	Lime/sulfide	—	Commercial	TCLP		<0.002
4:98	Oil waste		Flyash	1.20	Commercial	TCLP		<0.100
4:98	Paint sludge		Flyash	0.35	Commercial	EPT		<0.100
4:98	Paint sludge		Flyash	0.35	Commercial	TCLP		<0.100
4:98	Biological sludge		Flyash	1.10	Commercial	TCLP		<0.100
4:100	TSD incinerator ash	Various	Cement kiln dust	0.50	Lab	TCLP	<0.007	<0.001
4:100	TSD incinerator filter cake	Various	Cement kiln dust	0.50	Lab	TCLP	3	0.003
4:100	TSD landfill leachate	Various	Cement kiln dust	0.50	Lab	TCLP	<0.007	<0.001
4:100	TSD incinerator ash	Various	Cement kiln dust	0.50	Lab	TCLP	0.048	<0.001
4:104	Arc furnace dust	K061	Proprietary		Lab	EPT	<1	<0.002
4:105	Mixed wastes	Various	Proprietary		Commercial	EPT	<1	0.005
4:114	Electroplating sludge	F006	Lime, sulfide		Commercial	EPT	<1	<0.002
						TCLP		<0.001
4:119	Arc furnace dust	K061	Portland cement	0.05	Lab	TCLP	3.8	0.001
4:119	Arc furnace dust	K061	Kiln dust	0.05	Lab	TCLP	3.8	<0.001
4:119	Arc furnace dust	K061	Lime/flyash	0.10	Lab	TCLP	3.8	0.002
4:119	Arc furnace dust	K061	Potassium silicate		Lab	EPT		<0.001

(*continued*)

TABLE 4-30 Summary of Mercury Fixation Data (Continued)

Reference[a]	Waste Description	Waste Code	CFS System	Mix Ratio	Treatment Scale	Leaching Test	Mercury Content (mg/kg or mg/l) Total	Mercury Content (mg/kg or mg/l) Leachate
4:140	Cl-caustic brine sludge	K071	Cement + thiourea	0.65	Lab	Water	15	<0.010
4:140	Cl-caustic brine sludge	K071	Cement + NaS	0.65	Lab	Water	15	0.017
4:141	Acetylene sludge		Cement + Fe^{+2} + thiourea	2.5	Lab	Water	381	<0.001
4:142	Seabed sediment		Fe^{+2}/Cu^{+2} + polydithiocarbamate	0.01	Lab	Water	10	<0.001
4:143	Synthetic sludge		Raw waste			TCLP		2.070
4:143	Synthetic sludge		Cement/silicate + hexadecyl mercaptan	0.2	Lab	TCLP		<0.001
4:159	Mixed waste		Cement/flyash		Lab	EPT	0.377	0.003
4:159	Mixed waste		Cement/flyash		Lab	MEP[b]	0.377	0.027
4:160	Zinc phosphate waste		Cement/flyash		Lab	ELT,[c] 1 d	2.6	<0.002
4:160	Zinc phosphate waste		Cement/flyash		Lab	ELT,[c] 28 d	2.6	<0.002
4:160	Zinc phosphate waste		None		Lab	ASTM,[c] 2 d	2.6	<0.002
4:160	Zinc phosphate waste		None		Lab	ASTM,[c] 10	2.6	<0.002
4:160	Zinc phosphate waste		Cement/flyash		Lab	ASTM,[c] 2 d	2.6	0.003
4:160	Zinc phosphate waste		Cement/flyash		Lab	ASTM,[c] 10	2.6	0.002

[a] References are listed at the end of Chapter 4.
[b] Maximum concentration.
[c] Sequential batch test, CO$_2$-saturated water.

TABLE 4-31   Solubility of HgS in Excess Na$_2$S

Na$_2$S in g/100g of Solution	HgS in Solution (mg/l)
0.95	2100
1.50	5700
2.31	14500
3.58	29100
4.37	41200
6.07	72700
9.64	155900

Source: Adapted from Kirk-Othmer, *Encyclopedia of Chemical Technology,* 3d ed., Wiley, New York, 1979.

uses of sulfur-based "getter" systems, including elemental sulfur and thiourea-type organics [170].

Chemical Reduction. Reductants are used to reduce mercury to the elemental state in water, after which it can be removed and reclaimed. This technique probably has limited use in CFS processes, since the mercury would remain in the waste and could be leached out by acidic leachants under oxidizing conditions. Under reducing conditions, organomercury compounds may be formed. It is also claimed that bacteria [157] become resistant to mercuric salts by metabolizing the salts to elemental mercury, the end result of which is reduction. Processes [48] where the reduction is accompanied by amalgamation—for example, with zinc or iron—may produce a species more resistant to leaching, but no data are available on leachability of the amalgams in a CFS matrix.

Ion Exchange and Sorption. A number of references show the use of ion exchange for waste water treatment. In both ion exchange and absorption, the metal species remains in the matrix and the process is reversible, so these processes are probably less effective in CFS treatment than they are with waste water. Exceptions may be when the mercury is fixed with thiol-type [145] or other reactive resins. Also, Jones [150] reports the removal of mercury from water with powdered coal. Although described as an absorbent, the sulfur content of the coal may comprise the real fixing agent. Chen and Majewski [151] claim that vermiculite, usually considered an absorbent, mixed with cement is effective in fixing mercury.

Other Precipitants. A Japanese patent [149] claims the fixation of mercury with cement containing copper cyanide. According to the authors, both the mercury and the cyanide were "strongly fixed" in the solidified product. Nippon Electric developed a process for fixing mercury as magnetic ferrite by coprecipitation with Fe^{+2} ion. The reaction is described as shown in Eq. (4-7).

$$x\text{M}^{++} + \text{Fe}^{+2}_{(3-x)} + 6(\text{OH})^- \longrightarrow \text{M}_x\text{Fe}_{(3-x)}(\text{OH})_6 \qquad (4\text{-}7)$$

Oxidation with air converts the product in Eq. (4-7) to a black ferrite, M$_x$Fe$_{(3-x)}$O$_4$. This latter compound can be separated by magnetic means. Mercury in the effluent is claimed to be 0.005 mg/l, compared with 6.0 mg/l in the influent. Since the ferrite has low solubility, this may be a usable procedure in the CFS area. In fact, it may be what is actually happening in other processes using reagents that contain iron.

Bhatty [115] found that while mercury in a dilute tricalcium silicate (the major component of cement) slurry was strongly fixed at short hydration times, it was released at longer times, suggesting that cement-based systems alone may not fix mercury in a stable form. Other metals (see prior sections in this chapter) did not display this effect. No explanation was given for possible mechanisms.

With the exception of a few very specific waste streams such as K061, it can probably be said that mercury leaching is not a major problem in the CFS field. Most wastes show leaching levels below routine detection limits. For K061 and other residuals where mercury leaching is a problem, sulfides or other sulfur compound additives provide the most fruitful approach in every case, although the best additive may vary from waste source to waste source. This is especially true now, in view of the allowable mercury leaching levels set in recent landbans and those likely to be set in the near future.

NICKEL

Nickel comprises about 3 percent of the earth's composition, but only about 0.009 percent of the earth's crust. Two types of ores are mined commercially: sulfide and laterite. The former is currently the most important, and the most common form is pendlandite, $(Ni,Fe)_9S_{16}$, which is found along with chalcopyrite, $CuFeS_2$, and pyrrhotite, Fe_7S_8. Lateric ores were formed as a result of weathering, and nickel silicate in solid solution in other minerals, or nickel in solid solution in hydrated iron oxides [8]. Nickel is produced from sulfide ores by pyrometallurgical processes, followed by electrolytic or carbonyl refining where pure nickel is required.

Nickel is a Group VIII transition element, along with iron and cobalt. It forms compounds in which the nickel atom has the oxidation states of -1, 0, $+1$, $+2$, $+3$, and $+4$. The Ni^{+2} valence state, however, represents the majority of all nickel compounds. The most common structural form for nickel is the octahedral configuration, in which nickel has the coordination number 6 and the compounds are the familiar green color. Less common forms exist where the coordination number is 4. Nickel is not amphoteric, but the ease with which it forms coordination complexes at high pH gives it the appearance of being so.

Nickel and its compounds are inhalation hazards, and some aqueous solutions of nickel compounds are skin irritants and may cause allergic dermatitis. Nickel and most of its salts are not considered to cause systemic poisoning. Some nickel compounds are carcinogenic via the inhalation route, and are regarded as carcinogenic via ingestion on an experimental basis [178].

Major Compounds

Nickel Oxide. Nickel oxide (NiO) has the characteristic "nickel green" color, but there is also a black crystalline form that is an incomplete calcination product with slightly more than the stoichiometric oxygen content. The oxides are used in alloying, in catalysts, ceramic frits, ferrites, and pigments, and in the manufacture of other nickel chemicals.

Nickel Sulfate. The common commercial form is the hexahydrate, $NiSO_4 \cdot 6H_2O$. Its principal use is in formulating the electrolyte in nickel electroplating; it is also used in electrorefining and electroless nickel plating. Other uses include catalyst intermediates and the manufacture of other nickel chemicals.

Nickel Nitrate. Also a hexahydrate, the nitrate decomposes on heating to form nickel oxide. It is used in the manufacture of nickel catalysts and in nickel–cadmium batteries.

Nickel Halides. The chloride hexahydrate is used with nickel sulfate in the conventional Watts electroplating bath. It is also used in catalysts, to absorb ammonia in gas masks, and as an intermediate in the formation of coordination compounds.

Nickel Carbonate. There are two carbonates, $NiCO_3$ and basic nickel carbonate, $2NiCO_3 \cdot 3Ni(OH)_2 \cdot 4H_2O$. Carbonates are used in catalysts, in the manufacture of colored glass in pigments, and in electroplating solutions.

Nickel Hydroxide. Nickel hydroxide [$Ni(OH)_2$] has low water solubility, which accounts for the adequate fixation of nickel in most wastes by simple pH control. It is used as a catalyst intermediate and in nickel–cadmium batteries.

Nickel Fluoroborate. Nickel fluoroborate [$Ni(BF_4)_2 \cdot 6H_2O$] is used in specialty high-speed nickel plating.

Nickel Cyanide. The simple cyanide is used in organic chemical manufacture. The complex cyanides, formed when $Ni(CN)_2 \cdot 4H_2O$ is dissolved in bases, are water soluble and quite stable. $K_2[Ni(CN)_4]$ does not precipitate with H_2S from aqueous solution.

Nickel Sulfamate. Nickel sulfamate [$Ni(SO_3NH_2)_2 \cdot 4H_2O$] is used as an electrolyte in electroforming of nickel.

Nickel Sulfide. Nickel sulfide (NiS) occurs naturally as the mineral millerite, and has low water solubility. It is found in nickel wastes from waste water treatment polishing operations.

Other Inorganic Nickel Compounds. A number of other nickel compounds are used in industry and may be present in waste streams. These include: nickel arsenate (hydrogenation catalysts), nickel phosphate (steel pretreatment before coating), a number of double salts with NH_4^+ (dye mordants and metal finishing), and nickel amine complexes.

Organonickel Compounds. Organic compounds of interest in CFS chemistry are primarily salts such as the acetate, formate, oxalate, stearate, and nickel chelates. Nickel acetate is found in metal finishing operations such as aluminum anodizing and electroplating. The fatty acid salts are used in the dying of synthetic fibers. Various soluble chelates are formed, and these are the source of most of the problems encountered with fixation of nickel. They are quite stable and very difficult to "break" so that the nickel can be precipitated as an insoluble species.

150   FIXATION OF METALS

Commercial Uses and Sources of Wastes

The largest use of nickel in the United States is in alloying: 44 percent in stainless steels and 33 percent in nonferrous and high-temperature alloys. The waste from this use consists of pickling baths and rinse waters, and emission control dusts such as electric arc furnace dust, K061. In most cases, however, the concentrations and species of nickel in the wastes cause no particular problems for fixation technology. They are normally fixed adequately (at present regulatory levels) by pH control in standard CFS systems.

The other major use is in electroplating (17 percent), with the remaining 6 percent used in catalysts, ceramics, magnets, and chemicals. Total U.S. consumption in 1978 was 240,000 metric tons, and this figure had remained constant for a decade [8]. The largest use for nickel chemicals is in catalysts, especially hydrogenation catalysts. However, there is little waste from this source that affects the CFS industry. The second largest use, electroplating, is by far the largest generator of nickel-bearing wastes requiring CFS treatment. The nickel comes not only from the chemicals used to formulate the baths, but from the nickel anodes used in most plating operations. Additives used in certain baths may form stable, soluble nickel complexes that do not precipitate with the usual CFS reagents and additives.

Nickel compounds are used in a number of other applications where nickel-containing wastes are produced. Most of these, however, pose no fixation problems. Other uses in ceramics, plastics additives, dyes and pigments, and agricultural chemicals are not important contributors to waste streams requiring CFS treatment at present, although the eventual full landban standards to be in effect by May 1990 may change this situation.

CFS Treatment Processes

Table 4-32 gives nickel leaching results for a number of different CFS systems, usually where other metals were of prime interest and nickel analysis was simply recorded. Analysis of these data shows that, where proper pH control was exercised, there was no difficulty in fixing nickel to current BDAT levels or below. There is really very little in the CFS literature about nickel fixation, for two primary reasons: (1) nickel has not been a "RCRA" metal, and thus has not been regulated under RCRA programs, except in case-specific instances by states and under CERCLA or other remedial action programs, and (2) nickel leaching has not been a problem in any case except where nickel complexes are involved. This situation may change somewhat under the landbans, since nickel fixation is now required in the TCLP test procedure and in some of the landban wastes, for example F006, K022, K048-52, K061, K101, and K102, to date. However, test work indicates that there will be little problem with nickel fixation (again, with the exception of complexed nickel) unless the BDAT levels change drastically.

Nearly all work on the fixation of complexed nickel has been done in the waste water treatment field, where the purpose is to precipitate the nickel from solution. Some of the processes and agents used to do this are listed in Table 4-33. The complexes may be citrates, EDTA complexes, cyanides, or others formed from proprietary additives. Many come from electroless nickel baths, where this problem is most severe. The simpler complexes may be easier to "break" with any of the reagents listed in Table 4-33, but the more sophisticated plating formulations may resist all but the most energetic, for example, sodium borohydride or hypochlorite

NICKEL 151

TABLE 4-32 Summary of Nickel Fixation Data

Reference[a]	Waste Description	Waste Code	CFS System	Mix Ratio	Treatment Scale	Leaching Test	Nickel Content (mg/kg or mg/l) Total	Nickel Content (mg/kg or mg/l) Leachate
4:98	Metal finishing	F006-9, D002-3	Lime/sulfide	—	Commercial	EPT	1700	0.100
4:98	Metal finishing	F006-9, D002-3	Lime/sulfide	—	Commercial	EPT	1600	0.060
4:98	Oil waste		Flyash	1.20	Commercial	TCLP		<0.100
4:98	Paint sludge		Flyash	0.35	Commercial	EPT		<0.100
4:98	Paint sludge		Flyash	0.35	Commercial	TCLP		<0.100
4:98	Biological sludge		Flyash	1.10	Commercial	TCLP		<0.100
4:100	TSD incinerator ash	Various	Cement kiln dust	0.50	Lab	TCLP	398	1.190
4:100	TSD incinerator filter cake	Various	Cement kiln dust	0.50	Lab	TCLP	214	1.980
4:100	TSD landfill leachate	Various	Cement kiln dust	0.50	Lab	TCLP	7	0.100
4:100	TSD incinerator ash	Various	Cement kiln dust	0.50	Lab	TCLP	326	1.440
4:101	Electroplating sludge	F006	Unstabilized raw waste		Lab	TCLP	5730	16.100
4:101	Electroplating sludge	F006	Cement kiln dust	0.20	Lab	TCLP		0.350
4:101	Electroplating sludge	F006	Cement kiln dust	0.50	Lab	TCLP		0.030
4:101	Electroplating sludge	F006	Cement-based	0.10	Lab	TCLP		3.000
4:101	Electroplating sludge	F006	Cement-based	0.15	Lab	TCLP		0.110
4:101	Electroplating sludge	F006	Cement-based	0.20	Lab	TCLP		0.090
4:102	Electroplating sludge	F006	Unstabilized raw waste		Lab	TCLP		133.000
4:102	Electroplating sludge	F006	Cement kiln dust	0.50	Pilot	TCLP		0.090
4:102	Electroplating sludge	F006	Cement kiln dust	0.50	Commercial	TCLP		0.030
4:104	Arc furnace dust	K061	Proprietary		Lab	EPT	<5	<0.100
4:105	Mixed wastes	Various	Proprietary		Commercial	EPT	1060	0.120
4:114	Electroplating sludge	F006	Lime, sulfide		Commercial	EPT	4700	0.090
						TCLP		
4:119	Arc furnace dust	K061	Portland cement	0.05	Lab	TCLP	243	0.019
4:119	Arc furnace dust	K061	Kiln dust	0.05	Lab	TCLP	243	<0.012
4:119	Arc furnace dust	K061	Lime/flyash	0.10	Lab	TCLP	243	<0.012
4:135	Electroplating waste	F006	Unstabilized		Lab	EPT	240	1.100
			Cement-based		Lab	EPT		0.130
4:136	Flyash A		Unstabilized		Lab	EPT		1.600
4:136	Flyash A		Gelled silicate	0.26	Lab	EPT		0.280

(continued)

TABLE 4-32 Summary of Nickel Fixation Data (*Continued*)

Reference[a]	Waste Description	Waste Code	CFS System	Mix Ratio	Treatment Scale	Leaching Test	Nickel Content (mg/kg or mg/l) Total	Leachate
4:136	Flyash A		Lime	0.10	Lab	EPT		0.840
4:136	Flyash B		Unstabilized		Lab	EPT		0.400
4:136	Flyash B		Gelled silicate	0.26	Lab	EPT		0.230
4:136	Flyash B		Lime	0.10	Lab	EPT		0.280
4:136	Plating waste		Unstabilized		Lab	EPT	23.4	44.000
4:136	Plating waste		Gelled silicate	0.16	Lab	EPT		1.400
4:136	Plating waste		Cement/silicate	0.24	Lab	EPT		1.900
4:136	Mixed municipal/ industrial waste		Unstabilized		Lab	EPT	564	4.700
4:136	Mixed municipal/ industrial waste		Gelled silicate	0.16	Lab	EPT	564	1.700
4:136	Mixed municipal/ industrial waste		Cement/silicate	0.24	Lab	EPT	564	2.200
4:143	Synthetic sludge		Raw waste		Lab	TCLP		31.900
4:143	Synthetic sludge		Cement/silicate + hexadecyl mercaptan	0.2	Lab	TCLP		0.190
4:158	FeCl$_3$ etching waste		Lime	1.5	Lab	EPT		0.024
4:158	FeCl$_3$ etching waste		Lime	0.7	Lab	EPT		0.030
4:159	Mixed waste		Cement/flyash		Lab	EPT	645	0.168
4:159	Mixed waste		Cement/flyash		Lab	MEP[b]	645	0.114
4:160	Nickel waste		Cement/flyash		Lab	ELT,[c] 1 d	2900	<0.005
4:160	Nickel waste		Cement/flyash		Lab	ELT,[c] 28 d	2900	<0.005
4:160	Chromium waste		Cement/flyash		Lab	ELT,[c] 1 d	40	<0.005
4:160	Chromium waste		Cement/flyash		Lab	ELT,[c] 28 d	40	<0.005
4:160	Zinc phosphate waste		Cement/flyash		Lab	ELT,[c] 1 d	1040	<0.005
4:160	Zinc phosphate waste		Cement/flyash		Lab	ELT,[c] 28 d	1040	0.030
4:160	Zinc phosphate waste		None		Lab	ASTM,[c] 2 d	1040	50
4:160	Zinc phosphate waste		None		Lab	ASTM,[c] 10	1040	26
4:160	Zinc phosphate waste		Cement/flyash		Lab	ASTM,[c] 2 d	1040	<0.005
4:160	Zinc phosphate waste		Cement/flyash		Lab	ASTM,[c] 10	1040	0.030

[a] References are listed at the end of Chapter 4.
[b] Maximum concentration.
[c] Sequential batch test, CO_2-saturated water.

TABLE 4-33 Complexed Nickel Fixation Techniques

Process or Agent	Mechanism
Iron filings	Reduction
Formaldehyde	Reduction
Potassium permanganate	Oxidation
Potassium chlorate	Oxidation
Potassium persulfate	Oxidation
Calcium hypochlorite	Oxidation
Hydrogen peroxide	Oxidation
Ferric chloride	Oxidation, coprecipitation
Iron sulfide [171]	Sulfide precipitation
Proprietary reagent [172]	
Dimethylthiocarbamate	Sulfide precipitation
Insoluble starch xanthate [173]	Sulfide precipitation
Sodium borohydride [174]	Reduction

at elevated temperature. The former is expensive, and the latter can cause secondary problems in concentrated waste residuals. For example, if chromium is present, it will be oxidized to Cr^{+6} and must then be reduced before precipitation. With either oxidizing or reducing agents, species other than nickel may compete for the reagent. For these reasons, it is highly advisable for the generator to destroy the nickel complexes when treating the waste water. In cases where this is not feasible, or in remedial projects, there is a real need for inexpensive, specific agents or processes for the destruction of complexed nickel.

SELENIUM

Selenium is widely dispersed in igneous rocks, in volcanic sulfur deposits, in hydrothermal deposits, and in copper ores. Nearly all of the metal is obtained as a byproduct of precious metal recovery from electrolytic copper refinery slimes. It is next to sulfur in group VIA, and between arsenic and bromine in period 4 of the periodic table. It forms compounds, both inorganic and organic, similar to those of sulfur. Its valence states are -2, 0, $+4$, and $+6$. The most important compounds of selenium are halides, oxides, oxyacids [H_2SeO_3 (selenious) and H_2SeO_4 (selenic)], and selenides. Selenium reacts with active metals to form ionic compounds containing the selenide ion, Se^{-2}, but forms covalent compounds with most other substances [8]. Selenides resemble sulfides in composition and properties.

Selenium poisoning of humans and animals comes primarily from ingestion of foods containing toxic quantities of selenium, bioaccumulated through the food chain. The source of the selenium is primarily natural, particularly in the Midwest region of the United States [48]. Industrial sources have seldom presented a threat, but selenium compounds are believed to cause damage to body organs when inhaled or ingested. Selenium is an essential nutrient in animals and, possibly, humans.

Uses and Species of Interest in CFS

The most visible use of selenium is in electronics and photoconductors, as photoelectric cells, rectifiers, and in xerography. Selenium and its compounds are also used as an alloying element, in chromium plating, glass, ceramics, pigments, rubber, lubricants, photography, pharmaceuticals, and organic chemical synthesis. Most of these uses do not impact CFS technology, but selenium compounds are present in small quantities in some industrial waste streams.

CFS Treatment Processes

There is no evidence of any specific treatment technologies for selenium having been developed. Table 4-34 shows why this is so. Rarely is selenium found in most industrial wastes in appreciable amounts, and selenium leaching even from the untreated waste is never (in the author's experience) above the RCRA level of 1.0 mg/l. In CFS-treated wastes, leaching is usually below detection limits. In waste waters, there is some interest in selenium removal, but usually from natural waters. The few reported analyses of selenium in waste waters are below the current lowest landban BDAT level of 0.025 mg/l. Selenium usually is present in water as the selenite (SeO_3^{-2} or selenate (SeO_4^{-2}) ion, which can be removed by certain specific electron exchange resins by electron transfer between hydroquinone functional groups and the ions [175].

SILVER

There are 55 silver minerals, mostly associated with lead, copper, zinc, and gold. The majority of world silver production comes from by-products of the other metals. It is a noble metal with only one normal valence state. Most of its compounds are relatively insoluble, especially the halides, cyanide, sulfide, and thiocyanate.

Soluble silver compounds, such as silver nitrate, may cause acute adverse health effects, but low-solubility compounds and complexes are considered to be safe [8]. Toxic effects from chronic exposure appear to be limited to the skin. The environmental effects of silver seem to be primarily in the aquatic area, where the magnitude of the effect is determined by silver ion concentration, not by total silver content. Microorganisms convert silver complexes such as the thiosulfate to silver sulfide and metallic silver.

Because of its value, most silver in wastes and waste water streams is recovered. This is a mature, well-established technology in the photographic, electronic, and plating industries.

Uses and Species of Interest in CFS

The primary uses of silver in the United States, in order of consumption, are:

Jewelry, flatware, coins, and electroplated objects
Photographic materials
Electrical and electronic products
Brazing alloys and solders
Catalysts
Dental and medical supplies
Mirrors
Bearings

TABLE 4-34 Summary of Selenium Fixation Data

Reference[a]	Waste Description	Waste Code	CFS System	Mix Ratio	Treatment Scale	Leaching Test	Selenium Content Total (mg/kg)	Selenium Content Leachate (mg/l)
4:98	Metal finishing	F006-9, D002-3	Lime/sulfide	—	Commercial	EPT	<10	<0.010
4:98	Metal finishing	F006-9, D002-3	Lime/sulfide	—	Commercial	TCLP		<0.010
4:98	Metal finishing	F006-9, D002-3	Lime/sulfide	—	Commercial	EPT	<10	<0.010
4:98	Metal finishing	F006-9, D002-3	Lime/sulfide	—	Commercial	TCLP		<0.010
4:98	Oil waste		Flyash	1.20	Commercial	EPT		<0.100
4:98	Paint sludge		Flyash	0.35	Commercial	TCLP		<0.100
4:98	Paint sludge		Flyash	0.35	Commercial	TCLP		<0.100
4:98	Biological sludge		Flyash	1.10	Commercial	TCLP		<0.100
4:100	TSD incinerator ash	Various	Cement kiln dust	0.50	Lab	TCLP	<0.05	<0.010
4:101	Electroplating sludge	F006	Unstabilized raw waste		Lab	TCLP		0.450
4:101	Electroplating sludge	F006	Cement kiln dust	0.20	Lab	TCLP		0.110
4:101	Electroplating sludge	F006	Cement kiln dust	0.50	Lab	TCLP		0.170
4:102	Electroplating sludge	F006	Unstabilized raw waste		Lab	TCLP		<0.010
4:102	Electroplating sludge	F006	Cement kiln dust	0.50	Pilot	TCLP		0.150
4:102	Electroplating sludge	F006	Cement kiln dust	0.50	Commercial	TCLP		0.030
4:104	Arc furnace dust	K061	Proprietary		Lab	EPT	200	<0.050
4:105	Mixed wastes	Various	Proprietary		Commercial	EPT	<1	<0.010
4:114	Electroplating sludge	F006	Lime, sulfide		Commercial	EPT	<10	<0.010
						TCLP		<0.040
4:119	Arc furnace dust	K061	Portland cement	0.05	Lab	TCLP	<5.0	<0.010
4:119	Arc furnace dust	K061	Kiln dust	0.05	Lab	TCLP	<5.0	<0.010
4:119	Arc furnace dust	K061	Lime/flyash	0.10	Lab	TCLP	<5.0	<0.010
4:143	Synthetic sludge		Potassium silicate		Lab	EPT		<0.400
4:143	Synthetic sludge		Raw waste			TCLP		0.020
			Cement/silicate + hexadecyl mercaptan	0.2	Lab	TCLP		0.066
4:159	Mixed waste		Cement/flyash		Lab	EPT	1275	0.096
4:159	Mixed waste		Cement/flyash		Lab	MEP[b]	1275	0.36
4:160	Copper waste		Cement/flyash		Lab	ELT,[c] 1 d	0.36	0.020
4:160	Copper waste		Cement/flyash		Lab	ELT,[c] 28 d	0.36	<0.001

[a] References are listed at the end of Chapter 4.
[b] Maximum concentration.
[c] Sequential batch test, CO$_2$-saturated water.

The most common silver compounds are the nitrate, used as a starting material for all other silver compounds, as well as for numerous direct applications, and the halides used in photography. Perhaps the most common waste product is a combination of complexes produced by dissolving silver halides in excess sodium or ammonium thiosulfate during the processing of photographic film and paper. However, the silver from this source is mostly recovered. The other major source of silver in wastes is in sludges from electroplating waste water treatment.

CFS Treatment Processes

Because of the value of silver, and the ease with which most dissolved species can be precipitated by chloride ion, fixation of silver in CFS systems is rarely, if ever, a problem. This is easily seen by examining Table 4-35. The solubility of silver chloride in water is about 1.4 mg/l, and even less with a slight excess of chloride ion, although large chloride excesses increase the solubility by formation of soluble complexes. The RCRA leaching level is 5.0 mg/l, but the landban BDAT level for F006 is 0.072 mg/l. Nevertheless, silver leaching is not a problem now and is not expected to be.

In cases where a stable, soluble complex is present in large amounts, several techniques are available. One is precipitation by magnesium sulfate and lime [176], which likely precipitates a mixed sulfate-oxide. Another is alkaline chlorination [177], which breaks the complex and precipitates silver chloride. Sulfides and hydrosulfites can also be used to treat silver complexes.

OTHER METALS

In addition to the RCRA metals, a number of other metals or metalloids are of environmental interest and are specifically regulated under some program. These are listed in Table 4-36. With the exception of uranium, all are regulated in California as to either total concentration in the waste (TTLC) or leachability (STLC). It is possible that additional metals may be regulated by the EPA in the future, since copper and zinc were proposed for leachability limitations in the Proposed Rule under the first-third landban, but later dropped. State and local jurisdictions may limit leaching of others. The metals in Table 4-36 are the most likely to see some sort of control, and so they will be briefly discussed here.

Copper

The EPA has determined that copper is an aquatic toxin [4], and has considered adding copper compounds to Appendix VIII [1] (Hazardous Constituents) for that reason. However, it has done so only for copper cyanide because of the cyanide radical. California sets the maximum copper content in a waste at 2500 mg/l and the allowable leachable level at 25 mg/l for the waste to qualify as nonhazardous. In general, it is not difficult to fix copper to levels well below this when using the EPT, TCLP, or other similar tests for evaluation. The California WET, however, can pose problems for copper fixation if the content is very high and/or the copper is complexed, as it often is. Such a situation is rare, but occasionally appears in remedial action work.

TABLE 4-35 Summary of Silver Fixation Data

Reference[a]	Waste Description	Waste Code	CFS System	Mix Ratio	Treatment Scale	Leaching Test	Silver Content (mg/kg or mg/l) Total	Silver Content (mg/kg or mg/l) Leachate
4:98	Metal finishing	F006-9, D002-3	Lime/sulfide	—	Commercial	EPT	<2	0.010
4:98	Metal finishing	F006-9, D002-3	Lime/sulfide	—	Commercial	TCLP		<0.020
4:98	Metal finishing	F006-9, D002-3	Lime/sulfide	—	Commercial	EPT	<2	0.090
4:98	Metal finishing	F006-9, D002-3	Lime/sulfide	—	Commercial	TCLP		<0.020
4:98	Oil waste		Flyash	1.20	Commercial	TCLP		<0.100
4:98	Paint sludge		Flyash	0.35	Commercial	EPT		<0.100
4:98	Paint sludge		Flyash	0.35	Commercial	TCLP		<0.100
4:98	Biological sludge		Flyash	1.10	Commercial	TCLP		<0.100
4:100	TSD incinerator ash	Various	Cement kiln dust	0.50	Lab	TCLP	18	0.070
4:100	TSD incinerator filter cake	Various	Cement kiln dust	0.50	Lab	TCLP	10	0.050
4:100	TSD landfill leachate	Various	Cement kiln dust	0.50	Lab	TCLP	3	0.070
4:100	TSD incinerator ash	Various	Cement kiln dust	0.50	Lab	TCLP	13.8	0.050
4:101	Electroplating sludge	F006	Unstabilized raw waste		Lab	TCLP	19	<0.010
4:101	Electroplating sludge	F006	Cement kiln dust	0.20	Lab	TCLP		0.020
4:101	Electroplating sludge	F006	Cement kiln dust	0.50	Lab	TCLP		0.020
4:101	Electroplating sludge	F006	Cement-based	0.10	Lab	TCLP		0.010
4:101	Electroplating sludge	F006	Cement-based	0.15	Lab	TCLP		0.010
4:101	Electroplating sludge	F006	Cement-based	0.20	Lab	TCLP		0.050

(continued)

TABLE 4-35 Summary of Silver Fixation Data (*Continued*)

Reference[a]	Waste Description	Waste Code	CFS System	Mix Ratio	Treatment Scale	Leaching Test	Silver Content (mg/kg or mg/l) Total	Leachate
4:102	Electroplating sludge	F006	Unstabilized raw waste		Lab	TCLP		<0.010
4:102	Electroplating sludge	F006	Cement kiln dust	0.50	Pilot	TCLP		<0.010
4:102	Electroplating sludge	F006	Cement kiln dust	0.50	Commercial	TCLP		<0.010
4:104	Arc furnace dust	K061	Proprietary		Lab	EPT	10	<0.050
4:105	Mixed wastes	Various	Proprietary		Commercial	EPT	3	0.090
4:114	Electroplating sludge	F006	Lime, sulfide		Commercial	EPT	14	0.060
						TCLP		<0.020
4:119	Arc furnace dust	K061	Portland cement	0.05	Lab	TCLP	59	<0.003
4:119	Arc furnace dust	K061	Kiln dust	0.05	Lab	TCLP	59	<0.003
4:119	Arc furnace dust	K061	Lime/flyash	0.10	Lab	TCLP	59	<0.003
4:119	Arc furnace dust	K061	Potassium silicate		Lab	EPT		<0.030
4:143	Synthetic sludge		Raw waste			TCLP		0.010
4:143	Synthetic sludge		Cement/silicate + hexadecyl mercaptan	0.2	Lab	TCLP		0.020
4:159	Mixed waste		Cement/flyash		Lab	EPT	18.1	<0.050
4:159	Mixed waste		Cement/flyash		Lab	MEP[b]	18.1	<0.050

[a] References are listed at the end of Chapter 4.
[b] Maximum concentration.

TABLE 4-36 Other Regulated Metals

Metal	Regulatory System
Antimony	California TTLC and STLC
Berylium	California TTLC and STLC
Boron	Canadian CGSB
Cobalt	California TTLC and STLC
Copper	California TTLC and STLC
Lead, organic	California TTLC
Molybdenum	California TTLC and STLC
Thallium	California TTLC and STLC, California list
Uranium	Canadian CGSB
Vanadium	California TTLC and STLC
Zinc	California TTLC and STLC

Copper is relatively abundant in minerals and soils. The metal is required for all forms of aerobic life and most anaerobic forms [8]. No chronic copper poisoning has been reported in humans, although ingestion of very large amounts can cause temporary systemic effects. The Cu^{+2} ion is toxic to marine invertebrates, but chelates with EDTA and NTA protect fish from its effects.

Relatively little data on copper leaching are available for CFS treatment systems, except where it has been analyzed for in studies of other metals. It is present in most industrial wastes, especially in electroplating sludges. The results of CFS treatment of this kind of waste are shown in Table 4-37. Because of its aquatic toxicity, however, considerable information has been published on removal of copper compounds from waste water [48]. The primary method is precipitation with lime and coagulants, which routinely reduces the level in water to less than one milligram/liter, often to fractional parts per million. The primary problem with copper fixation would be with stable complexes, which copper forms with many ligands, but even this should not be a problem in CFS technology unless unreasonably low allowable leaching levels were set.

Zinc

The EPA has also determined that zinc is an aquatic toxin [4], and has considered adding zinc compounds to Appendix VIII of 40CFR, but again has not done so. California has set the TTLC and STLC levels for zinc even higher than those for copper, 5000 mg/l and 250 mg/l, respectively. There should be little problem in meeting these levels for zinc, even in the WET leaching procedure, simply by pH control.

Zinc is found in all natural waters and soils, and is an important nutrient in plant and animal life [8]. Normal soils contain 10–30 ppm. Zinc is not toxic to humans—it is used as a dietary supplement—and animals have a high tolerance for it. Marine organisms are more sensitive, but little data are available on its aquatic toxicity. It is necessary for plant growth, but very high levels can cause toxic effects.

Like copper, most information on zinc fixation comes from the waste water treatment industry. Again, pH control with lime is the most common treatment method. Residual zinc concentrations in the low to fractional ppm range are typical. Complexation with cyanide is common in electroplating wastes, but data on CFS treatment of electroplating sludges and other wastes indicate no problems with fixa-

160 FIXATION OF METALS

TABLE 4-37 Summary of Copper Fixation Data

Reference[a]	Waste Description	Waste Code	CFS System	Mix Ratio	Treatment Scale	Leaching Test	Copper Content (mg/kg or mg/l) Total	Leachate
4:98	Metal finishing	F006-9, D002-3	Lime/sulfide	—	Commercial	EPT	330	0.050
4:98	Metal finishing	F006-9, D002-3	Lime/sulfide	—	Commercial	EPT	432	0.010
4:98	Oil waste		Flyash	1.20	Commercial	TCLP		<0.100
4:98	Paint sludge		Flyash	0.35	Commercial	EPT		<0.100
4:98	Paint sludge		Flyash	0.35	Commercial	TCLP		<0.100
4:98	Biological sludge		Flyash	1.10	Commercial	TCLP		<0.100
4:100	TSD incinerator ash	Various	Cement kiln dust	0.50	Lab	TCLP	17000	5.520
4:100	TSD incinerator filter cake	Various	Cement kiln dust	0.50	Lab	TCLP	719	0.024
4:100	TSD landfill leachate	Various	Cement kiln dust	0.50	Lab	TCLP	21	0.290
4:100	TSD incinerator ash	Various	Cement kiln dust	0.50	Lab	TCLP	569	0.350
4:101	Electroplating sludge	F006	Unstabilized raw waste		Lab	TCLP	27400	16.900
4:101	Electroplating sludge	F006	Cement kiln dust	0.20	Lab	TCLP		0.780
4:101	Electroplating sludge	F006	Cement kiln dust	0.50	Lab	TCLP		0.320
4:101	Electroplating sludge	F006	Cement-based	0.10	Lab	TCLP	73.000	
4:101	Electroplating sludge	F006	Cement-based	0.15	Lab	TCLP		0.160
4:101	Electroplating sludge	F006	Cement-based	0.20	Lab	TCLP		0.450
4:102	Electroplating sludge	F006	Unstabilized raw waste		Lab	TCLP		34.900
4:102	Electroplating sludge	F006	Cement kiln dust	0.50	Lab	TCLP		0.840
4:102	Electroplating sludge	F006	Cement kiln dust	0.50	Pilot	TCLP		0.470
4:114	Electroplating sludge	F006	Lime, sulfide		Commercial	EPT	775	0.030
4:119	Arc furnace dust	K061	Portland cement	0.05	Lab	TCLP	2240	0.024
4:119	Arc furnace dust	K061	Kiln dust	0.05	Lab	TCLP	2240	0.012
4:119	Arc furnace dust	K061	Lime/flyash	0.10	Lab	TCLP	2240	0.008
4:135	Electroplating waste	F006	Unstabilized		Lab	EPT	11000	20.000
4:135	Electroplating waste	F006	Cement-based		Lab	EPT		0.350

4:136	Flyash A	Unstabilized		Lab	EPT	2.000
4:136	Flyash A	Gelled silicate	0.26	Lab	EPT	0.130
4:136	Flyash A	Lime	0.10	Lab	EPT	0.480
4:136	Flyash B	Unstabilized		Lab	EPT	37.000
4:136	Flyash B	Gelled silicate	0.26	Lab	EPT	14.000
4:136	Flyash B	Lime	0.10	Lab	EPT	20.000
4:136	Plating waste	Unstabilized		Lab	EPT	2.600
4:136	Plating waste	Gelled silicate	0.16	Lab	EPT	0.140
4:136	Plating waste	Cement/silicate	0.24	Lab	EPT	0.480
	CCA wood preserving waste	Unstabilized		Lab	EPT	13.000
	CCA wood preserving waste	Cement/silicate	0.40	Lab	EPT	0.200
	CCA wood preserving waste	Potassium silicate		Lab	EPT	<0.050
	CCA wood preserving waste	Proprietary	0.18	Lab	EPT	2.000
	CCA wood preserving waste	Portland cement	0.10	Lab	EPT	9.910
	CCA wood preserving waste	Portland cement	0.20	Lab	EPT	7.410
4:143	Synthetic sludge	Raw waste		Lab	TCLP	23.500
4:143	Synthetic sludge	Cement/silicate + hexadecyl mercaptan	0.2	Lab	TCLP	0.110
4:159	Mixed waste	Cement/flyash		Lab	EPT	0.173
4:159	Mixed waste	Cement/flyash		Lab	MEP[b]	0.112
4:160	Nickel waste	Cement/flyash		Lab	ELT,[c] 1 d	<0.002
4:160	Nickel waste	Cement/flyash		Lab	ELT,[c] 28 d	<0.002
4:160	Copper waste	Cement/flyash		Lab	ELT,[c] 1 d	0.020
4:160	Copper waste	Cement/flyash		Lab	ELT,[c] 28 d	<0.002

[a] References are listed at the end of Chapter 4.
[b] Maximum concentration.
[c] Sequential batch test, CO_2-saturated water.

162   FIXATION OF METALS

TABLE 4-38   Summary of Zinc Fixation Data

Reference[a]	Waste Description	Waste Code	CFS System	Mix Ratio	Treatment Scale	Leaching Test	Zinc Content (mg/kg or mg/l) Total	Leachate
4:98	Metal finishing	F006-9, D002-3	Lime/sulfide	—	Commercial	EPT	375	0.010
4:98	Metal finishing	F006-9, D002-3	Lime/sulfide	—	Commercial	EPT	68	0.010
4:98	Oil waste		Flyash	1.20	Commercial	TCLP		<0.100
4:98	Paint sludge		Flyash	0.35	Commercial	EPT		<0.100
4:98	Paint sludge		Flyash	0.35	Commercial	TCLP		<0.100
4:98	Biological sludge		Flyash	1.10	Commercial	TCLP		<0.100
4:100	TSD incinerator ash	Various	Cement kiln dust	0.50	Lab	TCLP	2580	46.400
4:100	TSD incinerator filter cake	Various	Cement kiln dust	0.50	Lab	TCLP	4860	22.200
4:100	TSD landfill leachate	Various	Cement kiln dust	0.50	Lab	TCLP	36	0.020
4:100	TSD incinerator ash	Various	Cement kiln dust	0.50	Lab	TCLP	900	7.490
4:101	Electroplating sludge	F006	Unstabilized raw waste		Lab	TCLP	322	1.290
4:101	Electroplating sludge	F006	Cement kiln dust	0.20	Lab	TCLP		0.030
4:101	Electroplating sludge	F006	Cement kiln dust	0.50	Lab	TCLP		0.020
4:101	Electroplating sludge	F006	Cement-based	0.10	Lab	TCLP		1.000
4:101	Electroplating sludge	F006	Cement-based	0.15	Lab	TCLP		0.010
4:101	Electroplating sludge	F006	Cement-based	0.20	Lab	TCLP		0.110
4:102	Electroplating sludge	F006	Unstabilized raw waste		Lab	TCLP		38.300
4:102	Electroplating sludge	F006	Cement kiln dust	0.50	Pilot	TCLP		<0.010
4:102	Electroplating sludge	F006	Cement kiln dust	0.50	Commercial	TCLP		<0.010
4:103	Arc furnace dust	K061	Unstabilized raw waste		Lab	TCLP		1460.000
4:103	Arc furnace dust	K061	Cement-based	0.20	Lab	TCLP		0.040
4:103	Arc furnace dust	K061	Cement/silicate	0.20	Lab	TCLP		0.030
4:103	Arc furnace dust	K061	Proprietary	0.20	Lab	TCLP		0.150
4:114	Electroplating sludge	F006	Lime, sulfide		Commercial	EPT	2100	0.100
4:119	Arc furnace dust	K061	Portland cement	0.05	Lab	TCLP	244000	22.633
4:119	Arc furnace dust	K061	Kiln dust	0.05	Lab	TCLP	244000	3.180
4:119	Arc furnace dust	K061	Lime/flyash	0.10	Lab	TCLP	244000	0.390
4:135	Electroplating waste	F006	Unstabilized		Lab	EPT	33000	220.000
			Cement-based		Lab	EPT		0.320

OTHER METALS

4:136	Flyash A	Unstabilized		Lab	EPT	36.000
4:136	Flyash A	Gelled silicate	0.26	Lab	EPT	11.000
4:136	Flyash A	Lime	0.10	Lab	EPT	7.000
4:136	Flyash B	Unstabilized		Lab	EPT	8600.000
4:136	Flyash B	Gelled silicate	0.26	Lab	EPT	4200.000
4:136	Flyash B	Lime	0.10	Lab	EPT	4100.000
4:136	Plating waste	Unstabilized		Lab	EPT	20.000
4:136	Plating waste	Gelled silicate	0.16	Lab	EPT	6.200
4:136	Plating waste	Cement/silicate	0.24	Lab	EPT	4.100
4:136	Mixed municipal/industrial waste	Unstabilized		Lab	EPT	30.000
4:136	Mixed municipal/industrial waste	Gelled silicate	0.16	Lab	EPT	8.300
4:136	Mixed municipal/industrial waste	Cement/silicate	0.24	Lab	EPT	22.000
4:143	Synthetic sludge	Raw waste		Lab	TCLP	158.000
4:143	Synthetic sludge	Cement/silicate + hexadecyl mercaptan	0.2	Lab	TCLP	0.054
4:159	Mixed waste	Cement/flyash		Lab	EPT	4300	<0.050
4:159	Mixed waste	Cement/flyash		Lab	MEP[b]	4300	<0.050
4:160	Chromium waste	Cement/flyash		Lab	ELT,[c] 1 d	100	<0.005
4:160	Chromium waste	Cement/flyash		Lab	ELT,[c] 28 d	100	<0.005
4:160	Zinc phosphate waste	Cement/flyash		Lab	ELT,[c] 1 d	27625	<0.005
4:160	Zinc phosphate waste	Cement/flyash		Lab	ELT,[c] 28 d	27625	0.200
4:160	Waste treatment plant sludge	Cement/flyash		Lab	ELT,[c] 1 d	10735	<0.005
4:160	Waste treatment plant sludge	Cement/flyash		Lab	ELT,[c] 28 d	10735	0.200
4:160	Zinc phosphate waste	None		Lab	ASTM,[c] 2 d	27625	226
4:160	Zinc phosphate waste	None		Lab	ASTM,[c] 10	27625	134
4:160	Zinc phosphate waste	Cement/flyash		Lab	ASTM,[c] 2 d	27625	<0.005
4:160	Zinc phosphate waste	Cement/flyash		Lab	ASTM,[c] 10	27625	0.700

[a] References are listed at the end of Chapter 4.
[b] Maximum concentration.
[c] Sequential batch test, CO_2-saturated water.

TABLE 4-39   Summary of Antimony Fixation Data

Reference[a]	Waste Description	Waste Code	CFS System	Mix Ratio	Treatment Scale	Leaching Test	Antimony Content (mg/kg or mg/l) Total	Antimony Content (mg/kg or mg/l) Leachate
4:114	Electroplating sludge	F006	Lime, sulfide		Commercial	EPT	<10	0.010
4:119	Arc furnace dust	K061	Portland cement	0.05	Lab	TCLP	294	<0.050
4:119	Arc furnace dust	K061	Kiln dust	0.05	Lab	TCLP	294	<0.050
4:119	Arc furnace dust	K061	Lime/flyash	0.10	Lab	TCLP	294	<0.050
4:143	Synthetic sludge		Raw waste			TCLP		0.500
4:143	Synthetic sludge		Cement/silicate + hexadecyl mercaptan	0.2	Lab	TCLP		1.310

[a] References are listed at the end of Chapter 4.

tion of zinc in these residuals. This is illustrated in Table 4-38. Bhatty [115] showed excellent fixation of zinc by the tricalcium silicate component of Portland cement.

Antimony

Most antimony compounds are toxic via oral and inhalation routes [178], but because of their association with lead and arsenic and industry, their toxicity as antimony is often difficult to assess. Certain antimony compounds are used for the treatment of parasitic diseases.

It may exhibit a valence of $+3$, $+5$, or -3, and is classified as a nonmetal or metalloid; the trivalent state exhibits metallic characteristics. It forms a number of inorganic and organic compounds, many of complex structure.

Antimony compounds in wastes are considered hazardous only by California regulations, with a reasonably high allowable leaching level. Very little information is available on fixation methods, although leachate analyses often include antimony. Lo [33] found that activated carbon was somewhat effective in binding low concentrations of antimony and thallium in a cement-soluble silicate process. Other fixation data for antimony compounds are given in Table 4-39.

Thallium

A relatively abundant metal, thallium is found in potash minerals and a number of others. It is a member of Group IIIA of the periodic table with boron, aluminum, gallium, and indium [8]. Unlike the others, it exists in both monovalent and trivalent forms, with the former being generally the most stable. It is used in alloys and in electrical applications.

Toxicity of thallium compounds varies with their solubilities and valence states. Thallous sulfate, acetate, and carbonate are especially toxic. Fatal poisoning has been known to occur, but cases of industrial poisoning are rare [178]. It does not occur naturally in body tissues and is not an essential element to mammals, but does accumulate in humans. It is known to have deleterious effects when ingested during pregnancy.

Virtually nothing appears in the literature on CFS treatment or fixation in general. However, thallium is regulated in wastes in California, and appears on the EPA California list.

REFERENCES

1. U.S. EPA Toxicity Test Procedure. Federal Register, 40CFR Part 261.24, Appendix II (May 19, 1980).
2. Safety of Public Water Systems (State Drinking Water Act). Public Law 93-523. Washington, DC: 93rd Congress, Dec. 16, 1974.
3. Drinking water bill sails through Congress. *Chem. Eng. News* (June 2, 1986).
4. U.S. EPA. Federal Register. 53(159) (Aug. 17, 1988).
5. Dean, J. A. *Lange's Handbook of Chemistry*, 12th ed. New York: McGraw-Hill, 1979.
6. Seidell, A. *Solubilities of Inorganic and Metal Organic Compounds,* 3d ed. New York: Van Nostrand Reinhold.
7. Lewandowski, G. A., and M. F. Abd-El-Bary. *Water Sewage Works* (Jan. 1981).
8. Kirk-Othmer. *Encyclopedia of Chemical Technology,* 3d ed. New York: Wiley, 1979.

9. Bridle, T. R., P. L. Cote, T. W. Constable, and J. L. Fraser. Evaluation of heavy metal leachability from solid wastes. *Water Sci. Tech.* **19:** 1029–36 (1987).
10. Brennan and Mace. Waste treatment of chemical cleaning wastes in the power industry. In *Proc. Purdue Industrial Waste Conference.* West Lafayette, IN: Purdue University, pp. 899–907, 1977.
11. Bell, W. E. Chelation chemistry. *Mater. Prot.* **79** (Feb. 1965).
12. *Chem. Eng.* **17** (Feb. 15, 1988).
13. Clapp, T. L., J. F. Magee, R. C. Ahlert, and D. S. Kosson. Municipal solid waste composition and behavior in incinerator ashes. *Environ. Prog.* **7**(11): 22–30 (1988).
14. U.S. EPA. Federal Register **47**(225), 52687 (Nov. 22, 1982).
15. Thibodeaux, L. J. *Directions in Hazardous Waste Research.* Louisiana State University, Baton Rouge (Jan. 1987).
16. Cote, P. *Contaminant Leaching from Cement-Based Waste Forms Under Acidic Conditions.* Ph.D. Thesis, McMaster University, Hamilton, Ont., Canada, 1986.
17. Jula, T. F. *Inorganic Reductions with Sodium Borohydride.* Beverly, MA: Ventron Corp., 1974.
18. *Treatment of Chromium Waste Liquors.* New York: Allied Chemical Co., 1976.
19. Eary, L. E., and D. Rai. Chromate removal from aqueous wastes by chromium reduction with ferrous ion. *Environ. Sci. Technol.* **22**(8): 972–977 (1988).
20. Lancy, L. E. U.S. Patent 3,294,680 (1966).
21. Higgins, T. E., M. Asce, and S. G. TerMaath. Alkaline ferrous and sulfide reduction of chromium. In *Proc. ASCE Environmental Engineering Division Specialty Conference.* Atlanta, GA, 374–381, 1981.
22. Amborane™ 345/355 Reductive Resin(s). Philadelphia: Rohm and Haas Co., 1980.
23. Cherry, K. F. *Plating Waste Treatment.* Ann Arbor, MI: Ann Arbor Science Publishers, 1982.
24. Scott, M. C. Sulfide process removes metals, produces disposable sludge. *Ind. Wastes* 34–38 (July/Aug. 1979).
25. Cook, M. M., and J. A. Lander. Sodium borohydride controls heavy metal discharge. *Pollut. Eng.* 36–38 (Dec. 1981).
26. Kirk-Othmer. *Encyclopedia of Chemical Technology,* 3d ed. New York: Wiley, 1979.
27. Sturm, W., and J. J. Morgan. *Aquatic Chemistry,* 2d ed. New York: Wiley, 1973.
28. Malone, P. G., R. B. Mercer, and D. W. Thompson. The effectiveness of fixation techniques in preventing the loss of contaminants from electroplating wastes. Presented at the First Annual EPA/AES Conference on Advanced Pollution Control for the Metal Finishing Industry. (Jan. 17, 1978).
29. Moore, J. N., W. H. Ficklin, and C. Johns. Partitioning of arsenic and metals in reducing sulfidic sediments. *Environ. Sci. Technol.* **22:** 432–437 (1988).
30. Knocke, W. R., L. P. Croy, and R. T. Kelley. An evaluation of the solids handling characteristics of sludge produced by two metals' precipitation techniques. Virginia Polytechnic Institute and State University, Blacksburg, 1977.
31. Feigenbaum, H. N. The Sulfex™ process for removal of heavy metals in a textile waste stream. In *Proc. Textile Wastewater Treatment and Air Pollution Control Conference,* Jan. 1977.
32. Permutit Company. *Sulfex™ Heavy Metals Waste Treatment Process,* Tech. Bull. XIII, No. 6. Paramus, NJ.
33. Lo, P. C., R. S. Reimers, M. C. Metcalf, J. Shamas, L. F. Roberts, and T. C. Akers. *Chemical stabilization—More than a Fixation Process.* Chemfix Technologies, Kenner, LA, 1986.
34. Yagi, T., and S. Matsunaga. Japan Kokai 76 123,775 (Oct. 28, 1976).
35. Yagi, T. Japan Kokai 75 105,541 (Aug. 20, 1975).
36. Nakaaki, O., Y. Horie, M. Idohara, and J. Shiraogi. Japan Kokai 75,99,962. (Aug. 8, 1975).

37. *Chemical Engineering:* p. 17 (Feb. 15, 1988).
38. Conner, J. R. Ultimate disposal of liquid wastes by chemical fixation. In *Proc. 29th Annual Purdue Industrial Waste Conference.* West Lafayette, IN: Purdue University (1974).
39. Vail, J. G. *Soluble Silicates.* New York: Reinhold, 1952.
40. Iler, R. K. *The Chemistry of Silica.* New York: Wiley, 1979.
41. Petit, L., and P. G. Rouxhet. Incorporation r'ions metalliques dans la silice. In *Proc. 7th International Congress on the Chemistry of Cement.* (1980).
42. Falcone, J. S. *Soluble Silicates.* New York: Reinhold, 1982.
43. Gowman, L. P. Chemical stability of metal silicates vs. metal hydroxides in ground water conditions. In *Proc. 2nd National Conference on Complete Water Reuse.* (1975).
44. Bishop, P. L. Leaching of inorganic hazardous constituents from stabilized/solidified hazardous wastes. *Hazardous Wastes Hazardous Mater.* **5**(2): 129–143 (1988).
45. Rousseaux, J. M., and A. B. Craig, Jr. *Stabilization of Heavy Metal Wastes by the Soliroc Process.* Brussels, Belgium: Cemstobel, S.A., 1980.
46. Falcone, J. S., Jr., R. W. Spenser, R. H. Reifsnyder, and E. P. Katsanis. Chemical interactions of soluble silicates in the management of hazardous wastes. *ASTM Special Technical Testing Publication 851.* Philadelphia: ASTM, 1984.
47. Patterson, J. W., H. E. Allen, and J. J. Scala. Carbonate precipitation from heavy metals pollutants. *J. Water Pollut. Control Fed.* **12**: 2397–2410 (1977).
48. Sittig, M. *Pollutant Removal Handbook.* London: Noyes Data Corp., 1973.
49. LeGendre, G. R., and D. D. Runnells. Removal of dissolved molybdenum from wastewaters by precipitates of ferric iron. *Environ. Sci. Technol.* **9**(8): 744–749 (1975).
50. Swallow, K. C., D. N. Hume, and F. M. M. Morel. Sorption of copper and lead by hydrous ferric oxide. *Environ. Sci. Technol.* **14**(11): 1326–1331 (1980).
51. Pojasek, R. B. *Toxic and Hazardous Waste Disposal,* Vols. 1–4. Ann Arbor, MI: Ann Arbor Science Publishers, 1980.
52. Sano, M., I. Sugano, and T. Okuda. Japan Kokai 75,133,654 (Oct. 23, 1975).
53. *Environ. Sci. Technol.* **16**(12): 641a (1982).
54. Manahan, S. E., and M. J. Smith. The importance of chelating agents. *Water Sewage Works:* 102–106 (1973).
55. Humic substances. *Environ. Sci. Technol.* **16**(1): 20a–23a (1982).
56. Saar, R. A., and J. H. Weber. Fulvic acid: Modifier of metal-ion chemistry. *Environ. Sci. Technol.* **16**(9): 510a–517a (1982).
57. Takatomi, H., and S. Tokuda. Japan Kokai 8,001,865 (Jan. 9, 1980).
58. Nelson, D. A. Removal of heavy metal ions from aqueous solution with treated leather. *American Chemical Society:* Houston, TX, 1980.
59. *Chem. Eng.* 83–84 (May 21, 1979).
60. Wing, R. E. Corn starch compound recovers metals from water. *Ind. Wastes* 26–27 (Jan./Feb. 1975).
61. Wing, R. E. U.S. Patent 3,294,680 (Sept. 27, 1977).
62. Wing, R. E., and D. W. McFeeters. Insoluble starch xanthate (ISX) dissolved heavy metal removal case history reports.
63. *Insoluble Starch Xanthate.* Pollution Technology Systems, Inc.
64. Heavy metal removal?—Try starch xanthate. *Prod. Finish.* 72–74 (Sept. 1978).
65. Navickis, L. L., R. E. Wing, and E. B. Bagley. ISX reduces selenium count in process wastewaters. *Ind. Wastes* 26–30 (Jan./Feb. 1979).
66. Bricka, R. M., and D. O. Hill. Metal immobilization by solidification of hydroxide and xanthate sludges. In *Proc. 4th International Hazardous Waste Symposium on Environmental Aspects of Stabilization/Solidification of Hazardous and Radioactive Wastes* (May 3–6, 1987).
67. Walton, H. F. *Principles and Methods of Chemical Analysis.* Englewood Cliffs, NJ: Prentice-Hall, 1964.

68. Chan, P. C., J. W. Liskowitz, A. Perna, and R. Trattner. *Evaluation of Sorbents for Industrial Sludge Leachate Treatment.* Cincinnati: U.S. EPA, EPA-600/2-80-052, 1980.
69. Benson, R. E., Jr. Natural fixing materials for the containment of heavy metals in landfills. In *Proc. Mid-Atlantic Industrial Waste Conference.* 212–215, 1980.
70. Japan Kokai 70 40,797 (Oct. 20, 1980).
71. *HazTECH News:* 100 (June 18, 1987).
72. Japan Kokai 62 144,747 (June 27, 1987).
73. Grim, R. E. *Clay Mineralogy.* New York: McGraw-Hill, 1968.
74. Griffin, R. A., R. R. Frost, and N. F. Shimp. *Effect of pH on Removal of Heavy Metals from Leachates by Clay Minerals.* Urbana, IL: Illinois State Geological Survey.
75. Posselt, H. S., and F. J. Anderson. Cation sorption on colloidal hydrous manganese dioxide. *Environ. Sci. Technol.* **2**: 1087–1093 (1968).
76. Dzombak, D. A., and F. M. M. Morel. Development of a data base for modelling adsorption of inorganics on iron and aluminum oxides. *Environ. Prog.* **6**(2): 133–137 (1987).
77. Lagvankar, A., K. Mayenkar, and P. A. Pherson. Innovative process for removing heavy metals from wastewater. In *Proc. 59th Annual Meeting Central States Water Pollution Control Association* (May 14–16, 1986).
78. *Chem. Eng.* 49–50 (Sept. 29, 1975).
79. Huang, C. P., P. K. Wirth, and D. W. Blankenship. Removal of Cd(II) and Hg(II) by activated carbon. In *Proc. 2nd National Conference on Environmental Engineering.* ASCE, Atlanta, GA, pp. 382–390, 1981.
80. Rubin, A. J., and D. L. Mercer. Adsorption of free and complexed metals from solution by activated carbon. Columbus, OH: Ohio State University, 1979.
81. Kennedy, D. C. Predict sorption of metals on ion-exchange resins. *Chem. Eng.* 106–118 (1980).
82. Lea, F. M. *Chemistry of Cement and Concrete.* New York: St. Martin's Press, 1970.
83. Bolsing, F. U.S. Patent 4,018,679 (1977).
84. U.S. EPA. Oily Waste Extraction Procedure. Federal Register. **49**(206): 42591 (Oct. 23, 1984).
85. Conner, J. R. U.S. Patent 4,623,469 (1986).
86. Dewalens, J., L. Heerman, and L. Van Simaeys. The codeposition of copper and arsenic from H_2SO_4-$CuSO_4$-As_2O_3 solutions. *J. Electrochem. Soc.* **122**(4): 477–482 (1975).
87. Ottinger, R. S., J. L. Blumenthal, D. F. Dal Porto, G. I. Gruber, M. J. Santy, and C. C. Shih. *Recommended Methods of Reduction, Neutralization, Recovery or Disposal of Hazardous Waste.* Washington, DC: EPA-670/2-73-053-f, 1973.
88. Turner, R. R. Oxidation state of arsenic in coal ash leachate. *Environ. Sci. Technol.* **15**(9): 1062–1064 (1981).
89. Johnson, J. C., and R. L. Lancione. Laboratory assessment of fixation and encapsulation processes for arsenic-laden wastes. In *Proc. 4th Annual Research Symposium of the Southwest Research Institute.* pp. 326–341, 1978.
90. Thompson, S. R. U.S. Patent 3,980,558 (1976).
91. Young, D. A. U.S. Patent 4,142,912 (1979).
92. Sandesara, M. D. U.S. Patent 4,118,243 (1978).
93. *Chem. Eng.* 23 (May 9, 1988).
94. DeCarlo, E. H., and D. M. Thomas. Removal of arsenic from geothermal fluids by adsorptive bubble flotation with colloidal ferric hydroxide. *Environ. Sci. Technol.* **19**(6): 538–543 (1985).
95. Bloom, P. A., J. H. Maysilles, and H. Dolezai. Hydrometallurgical treatment of arsenic-containing lead-smelter flue dust. Bureau of Mines, Salt Lake City, UT: PB82-262775, 1982.
96. Vasil'ev, B. T., V. S. Sokolov, and M. N. Shchemer. Precipitation of arsenic with calcium carbonate in the cinder hydroremoval system. *Khim. Prom-st. Miner. Udobr. Sernoi Kisloty* **2**: 7–8 (1980).

97. Japan Kokai 80 132,633 (Oct. 15, 1980).
98. Shively, W. E., and M. A. Crawford. EP toxicity and TCLP extractions of industrial and solidified hazardous waste. Boston, MA: CH2M Hill, 1986.
99. Shively, W., P. Bishop, D. Gress, and T. Brown. Leaching tests of heavy metals stabilized with Portland cement. *J. WPCF* **58**(3): 234–241 (1986).
100. Chemical Waste Management Inc., Internal Report on Residue Management, 1988.
101. Chemical Waste Management Inc., Internal Report on F006 Wastes, 1987.
102. Chemical Waste Management Inc., Internal Report on Mixing, 1988.
103. Chemical Waste Management Inc., Internal Report on K061 Waste, 1988.
104. Federal Register. **53**(121): 23661–23671 (June 23, 1988).
105. Federal Register. **51**(199): 36707–36730 (Oct. 15, 1986).
106. Cote, P. L. and D. P. Hamilton. Evaluation of pollutant release from solidified aqueous wastes using a dynamic leaching test. In *Proc. Hazardous Wastes and Environmental Emergencies Conference.* 302–308, 1984.
107. Cote, P. L., and D. Isabel. Application of a static leaching test to solidified hazardous waste. Presented at the ASTM International Symposium on Industrial and Hazardous Solid Wastes (Philadelphia, 1983).
108. Sittig, M. *Pollutant Removal Handbook.* Park Ridge, NJ: Noyes Data Corp., 1973.
109. Sanjour, W. *Cadmium and Environmental Policy.* Washington, DC: U.S. EPA, 1974.
110. Powers, P. W. *How to Dispose of Toxic Substances and Industrial Wastes.* Park Ridge, NJ: Noyes Data Corp., 1976.
111. Weiner, R. F. Acute problems in effluent treatment. *Plating* **54**: 1354–1356 (1967).
112. Morel *et al.* In *Proc. of 2nd Symposium on Hazardous and Industrial Waste Testing:* ASTM: Lake Buena Vista, FL, Jan. 1982.
113. Nakanishi, K. Japan Kokai 76 120,974 (Oct. 22, 1976).
114. Sano, M., and T. Okuda. Japan Kokai 75 133,654 (Oct. 23, 1975).
115. Bhatty, M. S. Y. Fixation of metallic ions in Portland cement. *Superfund '87:* 140–145 (1987).
116. Campbell, K. M., T. El-Korchi, D. Gress, and P. Bishop. Stabilization of cadmium and lead in Portland cement paste using a synthetic seawater leachant. *Environ. Prog.* **6**(2): 99–103 (1987).
117. Butler, L. G., F. K. Cartledge, D. Chalasani, H. C. Eaton, F. Frey, M. E. Tittlebaum, and S. L. Yang. Immobilization mechanisms in solidification/stabilization using cement/silicate fixing agents. Louisiana State Univ., Baton Rouge, pp. 42–61, 1988.
118. Driscoll, K., and B. Kaplan. Assessment of alternative technologies for treating spent electroplating solutions and sludges. In *Proc. 13th Annual Research Symposium at Cincinnati, Ohio.* Cincinnati, pp. 431–441, 1987.
119. U.S. EPA. *Best Demonstrated Available Technology (BDAT) Background Document for K061,* Vol. I. Washington, DC: EPA/530-SW-88-031D, 1988.
120. Brunner, P. H., and P. Baccini. The generation of hazardous waste by MSW-incineration calls for new concepts in thermal waste treatment.: Swiss Federal Institute for Water Resources, Dubendorf, Switzerland, 1988.
121. Lancy, L. E. U.S. Patent 3,294,680 (Dec. 27, 1966).
122. Uchikawa, H., and M. Shimoda. U.S. Patent 4,132,558 (Jan. 2, 1979).
123. Mizumoto, Y., H. Fujita, S. Mitsuoka, and S. Masegama. Japan Kokai 77 42,469 (Apr. 2, 1977).
124. Oshikata, T., and K. Ogawa. Japan Kokai 78 15,264 (Feb. 10, 1978).
125. Onata, H. Japan Kokai 77 62,186 (May 23, 1977).
126. Unimura, T. Japan Kokai 76 148,667 (Dec. 21, 1976).
127. Carpenter, C. J. Ferrous sulfate/sodium sulfide chromium reduction metals precipitation. In *Proc. 5th National Conference on Hazardous Waste and Hazardous Metals.* Las Vegas, NV, pp. 52–56 (1988).
128. *Andco Chromate and Heavy Metal Removal System.* Buffalo, NY: Andco Environmental, 1979.

129. Richards, R. U.S. Patent 3,371,034 (Feb. 27, 1968).
130. Nieuwenhuis, G. J. U.S. Patent 3,493,328 (Feb. 3, 1970).
131. Chichibu Cement Co. Japan Kokai 80 165,195 (Dec. 23, 1980).
132. Mitsubishi Heavy Industries. Japan Kokai 80 109,260 (Aug. 22, 1980).
133. Shimizu, S., and T. Ishii. Japan Kokai 79 62,168 (May 18, 1979).
134. Ono, M., F. Kitamure, and T. Fumio. Japan Kokai 75 131,860 (Oct. 18, 1975).
135. *Hazardous Waste Management Solutions From STC.* Scottsdale, AZ: Silicate Technology Corp., 1988.
136. Spencer, R. W., R. H. Reifsnyder, and J. C. Falcone. *Applications of Soluble Silicates and Derivative Materials in the Management of Hazardous Wastes.* Lafayette Hill, PA: The PQ Corp., 1986.
137. Kotani, K. Japan Kokai 79 147,172 (Nov. 17, 1979).
138. Conner, J. R. Unpublished treatability study.
139. Conner, J. R. Unpublished treatability study.
140. Yagi, T. Japan Kokai 75 105,541 (Aug. 20, 1975).
141. Yagi, T., and S. Matsunaga. Japan Kokai 76 123,775 (Oct. 28, 1976).
142. Oda, N., Y. Horie, M. Idohara, and J. Shiraogi. Japan Kokai 75 99,962 (Aug. 8, 1975).
143. Chemfix Technologies Inc., Kenner, LA, 1987.
144. Tenneco process recovers mercury from wastewater. *Chem. Week:* 35 (Aug. 24, 1977).
145. Neipert, M. P., and C. D. Bon. U.S. Patent 2,885,282 (May 5, 1959).
146. *Chem. Eng.* **71** (Feb. 27, 1971).
147. Grain, G. E., and R. H. Judice. U.S. Patent 3,213,006 (Oct. 19, 1965).
148. MacMillan, J. B. U.S. Patent 3,502,434 (Mar. 24, 1970).
149. Mihara, T., K. Endo, and T. Ando. Japan Kokai 78 64,950 (June 9, 1978).
150. Jones, C. T. Canadian Patent 1,034,686 (July 11, 1978).
151. Chen, K. S., and H. W. Majewski. U.S. Patent 4,113,504 (Sept. 12, 1978).
152. Gould, J. P., M. Y. Masingale, and M. Miller. Recovery of silver and mercury from COD samples by iron cementation. *J. Water Pollut. Control Fed.* **56**, Pt. 1: 280–286 (1984).
153. Webb, W. M. *Environ. Pollut. Control.* p. 222, 1980.
154. New Jersey Zinc Co. *The Use of Zinc to Remove Mercury from Plant Waste Water,* Tech. Information Bull. 65-557, Bethlehem, PA, 1971.
155. Kaczur, J. J., J. C. Tyler, Jr., and J. J. Simmons. U.S. Patent 4,354,942 (Oct. 19, 1982).
156. Mercury cleanup routes—II. *Chem. Eng.:* 36 (Feb. 3, 1975).
157. How bacteria detoxify mercury made clearer. *Chem. Eng. News:* 25 (Oct. 7, 1985).
158. Oberkrom, S. L., and T. R. Marrero. Detoxification process for a ferric chloride etching waste. Hazard. Waste & Hazard. Mater. 2, No. 1: 107-112 (1985).
159. McCoy & Assoc. EPA confirms that Stablex solidification process immobilizes heavy metals. Hazard. Waste Consult. 2, No. 2: (1984).
160. Pojasek, R. B. ed. Chapter 7. Toxic and Hazardous Waste Disposal. Ann Arbor, MI: Ann Arbor Science Pub. Inc., 1980.
161. Dragun, J. The fate of hazardous materials in soil. *Hazardous Mater. Control:* 41–65 (May/June 1988).
162. *Environ. Sci. Technol.* **17**(5): 197A (1983).
163. Saar, R. A., and J. A. Weber. Lead (II)-fulvic acid complexes. *Environ. Sci. Technol.* **14**(7): 877–880 (1980).
164. Rohrer, K. L. Chemical precipitants for lead-bearing wastewaters. *Ind. Water Eng.:* 13–18 (1976).
165. *J. Ins. Munic. Eng.* **84** (Nov. 1957).
166. Volnesenskic, S. A., A. V. Evallanova, and R. V. Suvorova. *Water Pollut. Abstr.* **13**: 135 (1940).
167. Jirka, A., M. Shannon, J. Morris, and P. Parikh. Factors affecting EP toxicity metals

results. In *Proc. 3d Annual Symposium Solid Waste Testing and Quality Assurance* **1**: 5-47–5-52 (1987).
168. Brunner, P. H., and P. Baccini. The generation of hazardous waste by MSW-incineration calls for new concepts in thermal waste treatment.:U.S. EPA, Cincinnati, OH, 343–350 (1988).
169. Mundell, J. A., and K. R. Hill. In-place precipitation immobilization: Technical and economic assessment at the A. Y. McDonald foundry site, Dubuque, Iowa. In *Proc. Hazardous Wastes and Environmental Emergencies.* Houston, TX, pp. 177–181, 1984.
170. Advanced Technology Center Inc. *Mercury Pollution Control in Stream and Lake Sediments.* Washington, DC: U.S. Government Printing Office, 1972.
171. Parker, K. The waste treatment of spent electroless nickel baths. *The First AES Electroless Plating Symposium.* Cincinnati, OH, Mar. 1982.
172. Schering Chemical AG. Germany.
173. Wing, R. E. Process for heavy metal removal from plating wastes. USDA Northern Regional Research Center, Peoria, IL, 1980.
174. Zickgraf, J. R. Electroless nickel waste treatment. *Finisher's Management:* 15 (Apr. 1983).
175. *Chem. Eng. News:* 17 (May 13, 1985).
176. Pool, S. C. U.S. Patent 2,507,175 (May 9, 1950).
177. Hendrickson, T. N., and T. J. Dagon. U.S. Patent 3,594,157 (July 20, 1971).
178. Sax, N. I. *Dangerous Properties of Industrial Materials.* New York: Van Nostrand Reinhold Co., 1979.

Chapter 5

FIXATION OF OTHER INORGANIC CONSTITUENTS

A number of anions and other inorganics are of importance environmentally, and of interest in CFS technology. Toxic, metallic cations and anions have been discussed in Chapter 4 generally and under the individual metals. Other species that impact our discussion of fixation are listed in Table 5-1. There are, of course, many other species of environmental concern in other media—water and air. Those listed here, however, are more specific to CFS technology for the following reasons.

1. They are quite toxic, as in the case of cyanide and sulfide (and to a lesser extent, fluoride).
2. Their high solubilities or negative aesthetic properties cause secondary drinking water pollution.
3. They are commonly present in CFS-treated wastes.

TABLE 5-1 Inorganic Species

Species	Regulatory MCLs[a]
Cyanide (CN$^-$)	0.2
Sulfide (S^{-2})	0.1[b]
(H$_2$S)	0.05[c]
Fluoride (F$^-$)	1.4[d]
Chloride (Cl$^-$)	250.[c]
Nitrate (NO$_3^-$)	10.[d]
Nitrite (NO$_2^-$)	
Sulfate (SO$_4^{-2}$)	250.[c]
Sulfite (SO$_3^{-2}$)	
Ammonia or Ammonium (NH$_3$ or NH$_4^-$)	
Iron (Fe)	0.3[c]
Manganese (Mn)	0.05[c]
Zinc (Zn)	5.[c]
Copper (Cu)	1.[c]
Sodium (Na$^+$)	
Potassium (K$^+$)	

[a] Maximum contaminant limits.
[b] National pollution discharge elimination system (NPDES).
[c] Secondary drinking water standards.
[d] Primary drinking water standards.

The primary drinking water standards are based on human health effects; the secondary standards are based on aesthetic or recreational factors. The national pollution discharge elimination system (NPDES) standards involve human, animal, and plant effects, especially effects on aquatic life.

The polyvalent metals—iron, copper, zinc, and manganese—are fixed to low leaching levels in CFS-treated wastes and are of little concern. Due to the presence of excess calcium ion in inorganic CFS processes, the solubility of sulfites is quite low, in the order of 20 mg/l in neutral or alkaline media. Similarly, sulfate solubility is about 1500 ml/l maximum, and generally less in the actual leachate, which is in the range of secondary drinking water standards. The primary problem arises with the monovalent cations and anions. These are quite soluble in practically all their forms, and leach at high levels that are controlled only by total species content and diffusion (or other mechanism of removal) rate from the waste. Currently, the only real control method for these species is to limit the transport rate into the environment by permeability control or isolation. Several sorbents have been described for nitrate [1] and ammonia [2],* and perhaps there are fixation mechanisms for the others, but practical solutions have not yet been specifically applied in this area.

This leads us to a discussion of the three species on which considerable fixation/destruction work has been done and applied in commercial systems. First, let us examine the easiest, the fluoride ion.

FLUORIDE

Fluoride is easily insolubilized as calcium fluoride (CaF), whose solubility is only 20 mg/l in water, and that is only slightly soluble in acids. Since calcium is virtually omnipresent in commercial CFS systems, leaching of fluoride is little problem. If the principle of 100 times attenuation in a landfill were applied, the allowable leaching limit for fluoride would be 140 mg/l, well above the equilibrium solubility in a CFS system. The only U.S. regulation (California) that covers leaching of fluoride from solidified wastes allows 150 mg/l in the leachate.

CYANIDES

The fixation or destruction of cyanides is of importance in CFS. Most often, the latter method of control has been used, and likely will continue to be used in the future for high cyanide level wastes; the EPA is expected to propose destruction techniques as mandatory under the landbans. Nevertheless, small amounts of cyanide in waste streams such as F006—electroplating waste water treatment sludge—are common and do not justify expensive destructive techniques such as those described later in this chapter. Before discussing the fixation processes or techniques that may be used, it is useful to explore the sources, characteristics, and current destruction methods used for cyanide wastes.

From an environmental standpoint, cyanide occurs in three types, each having somewhat different characteristics [3].

---

*At the high pH of commercial CFS systems, ammonia and most ammonium compounds are decomposed and driven off as ammonia gas. In essence, the CFS process itself acts as a stripping system to remove ammonia to fairly low levels.

1. "Simple" or free cyanide, such as sodium and potassium cyanides.
2. Loosely complexed cyanides, such as potassium nickel cyanide and sodium cadmium cyanide. These are soluble and toxic.
3. Strongly complexed cyanides, such as ferricyanides and ferrocyanides.

The three groups are a rough classification of the ability of cyanide salts and complexes to yield the toxic cyanide ion or hydrogen cyanide to the environment. Simple cyanides are directly toxic, whereas strongly complexed cyanides are resistant to breakdown and are low in toxicity.

The cyanides in between these extremes cause analytical problems because they form a continuum of stability. The cyanides that are actually measured in any test are quite dependent on the test method, and this produces some controversy in various regulatory frameworks. The U.S. EPA Method 9010 [4] incorporates the *total cyanide* test that tends to measure all three groups. Method 9010 also includes the "cyanide amenable to chlorination test" that measures the first group and mostly the second group. Because of the problems of reproducibility of the *amenable cyanide* test, some workers in the field prefer to use the Standard Methods Simple Cyanide Test [5] or a variation of it.

Cyanides are produced and used industrially in four main groups: hydrogen cyanide, alkali and alkaline earth metal cyanides, heavy metal cyanides and cyanide complexes, and organic cyanides.

Hydrogen Cyanide

Prussic acid, or hydrogen cyanide (HCN), has a boiling point slightly above room temperature, and while supplied industrially as a liquid shipped in steel cylinders and tank cars, it is usually encountered environmentally as a gas. It is highly poisonous. HCN can enter the body by inhalation, skin absorption, or orally; it has the classic odor of bitter almonds, but approximately one out of five people are unable to detect the odor, making it all the more hazardous for them. The toxicity of HCN is about the same as hydrogen sulfide, but it has a very high "knock-down" power that makes it more dangerous. Production of HCN in 1964 was approximately 220,000 tons per year, more than half of which was used in the production of acrylonitrile. Other major uses are for nitriles, acrylates for plastics, and intermediates, as in the production of sodium cyanide. The remainder of HCN production goes into the production of ferrocyanides, chelating agents, optical laundry bleaches, and pharmaceuticals. The HCN used industrially is not normally encountered directly in the CFS field; however, it is encountered as a waste generated in various industrial processes and when other cyanide compounds are acidified.

Alkali and Alkaline Earth Metal Cyanides

The important alkali metal compounds are sodium and potassium cyanides, produced in 1963 at the rate of 100,000 tons per year, and 5000 tons per year, respectively. Major uses for sodium cyanide are in gold and silver metallurgical extraction, heat treating of metals, ore floatation, and the synthesis of organic chemicals (dyes, pharmaceuticals, vitamins, plastics, ferrocyanides, etc.). Electroplating (copper, zinc, cadmium, gold, and silver) is also a fairly large user. Potassium cyanide is used in electroplating, along with sodium cyanide in the nitriding of steel, in refining platinum, and in metal coloring processes. Oral and dermal toxicity are similar to that of HCN. In contact with air, these compounds can absorb moisture and evolve HCN.

The only commercially important alkaline earth metal cyanide is calcium cyanide, which finds a large use in ore refining (extraction and floatation). It is also used as a fumigant, rodenticide, and insecticide, has a large use in the production of ferrocyanides or prussiates, and is used in the case hardening of steel. Like the other soluble cyanide compounds, it is extremely toxic and can decompose in moist air to become HCN gas.

Heavy Metal and Complex Cyanides

Commercial metal cyanides include nickel cyanide, silver cyanide, and zinc cyanide; the latter two are used in the respective plating baths and nickel cyanide is used as a brightener in other plating baths. These and other heavy metals form complex cyanides such as the ferrocyanide and ferricyanide complexes with iron. In general, these compounds are less toxic than the sodium, potassium, and calcium cyanides, largely due to their low solubility and high stability. (However, at the same levels complex metal cyanides are more toxic than sodium cyanide to certain aquatic life.) While the complex cyanides are less toxic, they are also more resistant to removal by cyanide treatment processes, and they are sensitive to decomposition by ultraviolet light, raising the possibility of the generation of HCN in streams containing ferrocyanides that have been discharged by industry. In the CFS field, this would not seem to be a problem, since the products are not normally exposed to such photolysis effects. Cyanates, a product of partial oxidation of cyanides, are about 1000 times less toxic than cyanides, and in the past have been discharged directly into streams. One of the problems involved with cyanates is the possible reconversion to cyanide under anaerobic conditions in sewer systems. The ferrocyanides and ferricyanides encountered frequently in industry are stable under normal handling conditions and are not particularly toxic. The U.S. PHS drinking water standard for cyanide is 0.2 ppm.

Organic Cyanides

The organic cyanides are called nitriles. They are not nearly as toxic as the inorganic cyanides and are used as intermediates, plasticizers, and additives in various plastics and chemical products. For the purposes of this book, they fall into the category of organic pollutants.

Waste Sources

To compare cyanide waste treatment processes it is necessary to understand the different sources and the nature of the cyanide wastes from those sources, since they vary considerably and this variation will affect the choice of treatment process. One of the primary distinctions between different sources is in whether the waste is concentrated or dilute with respect to cyanide. Certain treatment processes apply only to dilute cyanide wastes, others only to concentrated wastes, and very few to both.

176 FIXATION OF OTHER INORGANIC CONSTITUTENTS

This is one of the most sharply defined examples of the differences in treatment technology between waste water and concentrated chemical wastes. Following are the primary industrial sources of cyanide waste in the United States.

Electroplating. When one thinks of cyanide waste one usually thinks of the electroplating industry, and it is widely assumed that this industry generates most of the waste cyanide in the United States. However, this is not necessarily true or at least no published figures support this claim. The supposition is probably due to the fact that there are a large number of small plating shops throughout the country, that many of these shops use cyanide plating processes, and that the raw rinsewaters will therefore generally contain cyanide. It has been stated that the plating industry generates about 21 million pounds of "cyanide waste" per year, but the form of this waste is not given. Even if one assumes 21 million dry pounds of cyanide radical, this is still only a fraction of the total production of sodium and potassium cyanides each year.

In any case, metal finishing is a major source of cyanide pollution. This pollution comes primarily from copper, zinc, cadmium, silver, and gold plating processes that use large quantities of sodium and potassium cyanides as well as smaller quantities of the metal cyanides. A considerable volume of dilute cyanide rinsewater is produced, and smaller quantities of much more concentrated plating baths are disposed of from time to time. Rinsewaters generally contain 10 to 770 ppm cyanide, while the plating baths themselves are up in the percent range. Much of the literature on the treatment of cyanide waste has been developed around the plating industry. Most of the producers of cyanide treatment equipment and processes seem to concentrate their marketing activities in that industry. However, there is some doubt about the future generation of cyanide waste in the electroplating industry, due to the increasing utilization of cyanide recovery systems and the replacement of cyanide plating baths by noncyanide processes.

Mining. The most extensive use of sodium and calcium cyanides in mining is in the ore floatation processes for copper, lead, and zinc, and in silver and gold extraction processes. Since the tailings from mining are generally not treated other than by settling ponds, most of the cyanide is discharged with the effluent water, although there is the possibility for some biological decomposition. It is apparently possible to replace the cyanide floatation processes by other methods. Mining wastes are exempt from the RCRA as of this writing.

Primary Metals. In the steel industry, considerable amounts of cyanide are generated from coking operations and from carryover into blast furnace dust and sludges. Cyanides are generated in the aluminum industry during the production of aluminum from cryolite. Cyanide wastes are also generated from cold finishing operations in the steel industry. Large quantities of cyanides can be discharged by primary metal producers; for example, one plant in Maryland discharged approximately 2 million pounds of cyanide per year, or about 10 percent of the output of the whole electroplating industry.

Machinery (and Metal Heat-Treating Operations). The machinery industry generates cyanides from plating processes and therefore would come under the same general comments as the electroplating industry. Metalworking and finishing operations are large users of sodium, potassium, and calcium cyanides in heat-treating and nitriding operations. The quenching process in these operations produces cyanide-contaminated oils and waters that are eventually disposed of. In addition, heat-treating baths are periodically dumped, in which case solid cyanide salts ranging from 10 to 70 percent of the bath by weight must be disposed of. Quench water from metal-hardening operations, such as nitriding, contains ferrocyanide as well as cyanide; ferrocyanides may subsequently form HCN under the influence of ultraviolet light in surface waters.

Photographic Processing. The photographic industry discharges dilute solutions containing ferrocyanides (but never free cyanides) from bleaching processes. This is not a major source, and the present practice of regeneration of bleaches is further reducing the disposal of concentrated baths. Other cyanides are used in photographic laboratories, but generally in very small quantities.

Treatment Processes

There are a large number of processes that have been developed for cyanide treatment and destruction, and quite a few that have been utilized commercially. Here, we will discuss in detail only the processes that deal with the destruction or other treatment of concentrated cyanide solutions. For purposes of comparison, listed below are the currently known processes for treating cyanide in both concentrated and dilute wastes.

Electrolytic oxidation
Chemical oxidation
Chemical reaction
Acidification followed by HCN destruction ("Cyan-Cat")
Evaporative recovery
Incineration or catalytic oxidation
Evaporative recovery
Irradiation (gamma)
Reverse osmosis or electrodialysis (recovery)
Carbon absorption
Carbon bed catalytic destruction
Foam separation
Waste-plus-waste
Freeze-out (recovery)
"Kastone" process
Nascent oxygen
Solvent extraction (recovery)
Aeration
Polymerization
Starch process

"Cyan-Cat" Process [6]. The cyan-cat process consists basically of acidifying the concentrated cyanide wastes with a strong acid, such as sulfuric acid, thereby decomposing both simple and complex cyanides to form other metal salts and HCN. The HCN is stripped from the solution in a conventional desorption tower and is then catalytically oxidized in a thermal reactor (incinerated) to transform it into carbon dioxide, nitrogen, and water vapor. The water discharge from the system (the solution flowing from the desorption tower) still contains small amounts of cyanide that are then removed by conventional dilute cyanide treatment processes, most frequently alkaline chlorination. The removal of cyanide from solution by acidification is not new; it has been practiced for many years in industry on dilute streams where the liberated and desorbed hydrogen cyanide is simply diluted with air and discharged to the atmosphere. Others have treated concentrated cyanide wastes this way, desorbing the HCN and passing it to a conventional furnace or incinerator where it is burned in air. One acidification process involves neutralizing the alkaline cyanide solution to only pH 7, recovering the metal cyanides precipitated at that point, and then either reusing them or heating them in air to produce metal oxides. Another variation on this method is to liberate HCN with acid and then scrub the liberated HCN in a sodium hydroxide–sodium cyanide scrubber to produce more sodium cyanide [7].

There have been at least three cyan-cat installations in Europe, one of them at a central site. Aside from cost, the only apparent deterrent to the use of such a process at anything except a large and fairly sophisticated waste-producing operation (or at a well-controlled central site) is the inherent danger of any process that acidifies cyanide solutions and produces HCN gas. Little auxiliary fuel is used because the exothermic combustion of HCN provides most of the heat necessary to maintain the catalyst at a temperature of 350° to 400°C (662°F–752°F). This is an advantage over conventional incineration. Capital costs are high, probably significantly higher than any of the other available concentrated cyanide destruction processes. Amortization of this cost at a reasonable level would require good utilization, again limiting the use to large cyanide producers and to central sites. The requirement for an alkaline chlorination treatment unit for the stripper effluent would add capital cost unless the plant already has (or needs) such a process for the treatment of other dilute wastes; in any case, the added operating costs for the alkaline-chlorination unit must also be included.

Electrolytic Oxidation. During the 1950s and 1960s, a great deal of work was done on processes for anodic oxidation in electrochemical cells. Destruction of cyanide takes place in a two-step process (as with most other oxidation techniques), first to cyanate and then to carbon dioxide and nitrogen; however, the electrolytic conversion is very rapid at high cyanide concentrations. Depending upon the solution being treated, metal may be collected at the cathode in metallic form and precipitated at the anode as the hydroxide. Therefore, electrolytic units, properly used, have the potential for removing metals as well as the cyanides. The effluent from electrochemical treatment typically contains up to 10 ppm of free cyanide and therefore must be posttreated, usually by alkaline chlorination. This has been one of the problems that has prevented its economic use in dilute solutions until fairly recently. Also, care must be taken in very concentrated solutions to prevent overheating, with the resulting release of HCN. This is sometimes done by dilution to a somewhat lower concentration. One solution to these problems has been offered by Shockcor [8]. The space between anode and cathode is filled with carbon cylinders or spheres which, acting as a semiconductor bed, produce a multiplicity of anodes and cathodes that eliminate the problems of low conductivity in dilute solutions. A fairly typical unit removes 2500 pounds of cyanide per year from water with a concentration of about 100 ppm, producing a dischargeable effluent.

Another approach to the use of electrochemical methods is the production of sodium hypochlorite in an electrochemical process separate from the cyanide destruction process. The hypochlorite is subsequently used for cyanide destruction. The claimed advantage of this approach is that the hypochlorite can be produced for half the cost of commercial hypochlorite, and is therefore competitive with chlorine systems, which are cheaper than hypochlorite but are more troublesome to use and require more expensive equipment.

CFS Treatment Processes

The preceding processes, while interesting, do not have a direct bearing on CFS technology; that is, they cannot be used as part of a CFS process, although the residues from them may require further "polishing" treatment by chemical destruction or fixation techniques. The following processes have potential use in the CFS field.

Chemical Oxidation. Numerous processes have been evolved that use oxidants such as hydrogen peroxide [9], potassium permanganate [10], ozone, nascent oxygen, and so forth. Probably the best example of these processes is the alkaline chlorination [11] used primarily for dilute waste streams. Aside from the high cost of chemicals, alkaline chlorination is not used on concentrated cyanides, because the first step in the oxidation of cyanide by chlorine is to form cyanogen chloride, which is just as toxic a gas as HCN. In dilute solutions, cyanogen chloride is further oxidized in solution to cyanate; however, in strong cyanide solutions, cyanogen chloride concentration will exceed the solubility limit and may be evolved as a gas, creating a very dangerous situation. While other oxidation processes do not necessarily have the same problem, the chemical costs for chlorine oxidation are very high per pound of cyanide destroyed. However, this technique is good at reducing relatively low cyanide concentrations (10 to several hundred ppm) to a low fractional parts-per-million range suitable for discharge.

Another process for treating dilute wastes is hydrogen peroxide oxidation, for example, by the Kastone process [12], which uses a combination of hydrogen peroxide and formaldehyde (or, in the case of copper-bearing rinses, hydrogen peroxide alone) plus a patented catalytic additive such as a complex magnesium salt (Epsom salt). Ozone has also been used in a number of cases. Although it has the advantage of lower chemical costs, it has a high fixed cost.

A primary problem with all of these processes is that while the reaction of cyanide to cyanate is very rapid, oxidation of cyanate to carbon dioxide and nitrogen is quite slow. In the past, it has been possible in many instances to discharge cyanate because of its lower toxicity; however, it is unlikely that this will be continued to be allowed in the future. In the case of the Kastone process, organic glycolic compounds are formed that may require biological treatment before discharge. An additional difficulty with most of the oxidation processes is that complex metal cyanides are much more difficult to chemically oxidize—so much so, that it is essentially impossible to operate any kind of continuous process and the batch treatment operations take a very long time (hours to weeks) for complete destruction. Finally, any oxidation method is likely to be costly and inefficient if the cyanide is in a waste that also contains other oxidizable species, especially organics, that compete for the oxidizing agent.

Nonoxidative Chemical Reaction. The nonoxidative chemical reaction approach has not been very much used, but there are a variety of cyanide reactions that convert the toxic compounds into nontoxic or less toxic compounds, or perhaps even into recoverable chemicals of commercial value. One of these has been used to some extent in the waste-plus-waste approach. This is the conversion of cyanide to ferrocyanide at high pH using ferrous sulfate waste [13]. This is not a very good approach for dilute waters to be discharged, because ferrocyanide may be converted back into cyanide under the influence of ultraviolet radiation in streams, lakes, or rivers. However, it is useful for the detoxification of concentrated solutions and in CFS work.

Other processes that have been proposed include organic polymerization, such as reaction with aldehydes at controlled higher temperatures to form nitriles, which in turn can be hydrolyzed to amino acids, which are biodegradable and in any case will not revert to active cyanide upon acidification [14]. A pharmaceutical industry process reportedly uses a starch conversion syrup to produce a nontoxic reaction product [15]. Additional processes for fixing cyanide are listed below. These methods are generally capable of reducing cyanide leaching to the 0.02 to 1.0 mg/l range.

Process or Additive	*Reference*
Portland cement + active Al(OH)$_3$	[16]
Portland cement + anion exchange resin	[17]
Portland cement + calcium polysulfide	[18]
Portland cement + CaO + iron salt	[19]
Cement + lime + FeSO$_4$	[20]
Cement + CaCl$_2$ + FeCl$_2$ + AlCl$_3$	[21]
Cement + surfactant + iron salt	[22]

SULFIDES

Sulfides occupy a peculiar position in CFS technology, since they are the only group of compounds commonly used as fixation additives that are, at the same time, themselves the subject of environmental concern in the long-term leaching of treated wastes. The use of inorganic and organic sulfides to fix metals has been discussed in Chapters 3 and 4. Actually, the problem is not with the fixation of sulfides in wastes—this is easy to achieve by the simple addition of ferrous iron—but with the stability of the metal sulfides that are formed. Free sulfides can also be oxidized with hydrogen peroxide [23] or other oxidants to the sulfate. Free, that is, soluble, sulfide is determined by EPA Method 9030 [4].

Hydrogen sulfide (H_2S) is extremely toxic, even more so than hydrogen cyanide, but its characteristic odor at very low concentrations gives warning. At higher concentrations, 30-100 ppm in air, it has a deceptively sweet smell, and above this range, it deadens the senses. Hydrogen sulfide is frequently encountered in CFS work, from anaerobic sludges and other sources, but usually at very low levels that are easily detected. The greater danger comes from the accidental acidification of wastes containing free sulfides, such as the alkali or alkaline earth metal sulfides, liberating large amounts of H_2S in enclosed areas. The alkali metal sulfides are caustic alkalies and are corrosive to animal tissues. They are classed as systemic poisons primarily because they can decompose with gastric acids to liberate H_2S [24].

Analytically, sulfide is determined by EPA Method 9030 [4] for dissolved sulfides. If insoluble sulfides are to be determined, a prior distillation step (the same as that used for cyanide in EPA Method 9010) is required.

Metallic Sulfides

The metal sulfides often have unusual stoichiometries. They frequently have nonstoichiometric phases rather than classical compounds, are often polymorphic, and some are semimetallic in behavior [24]. Metal sulfides tend to be more covalent than the corresponding oxides, and they typically have different structures. Some transition metal sulfides—FeS, CoS, and NiS—have what is known as the nickel arsenide structure where the metal–metal distances are too small for normal ionic bonding and are more representative of metal alloys. Iron sulfide is a good example of a nonstoichiometric sulfide where an Fe/S ratio of unity is rare; formulas of Fe_6S_7 and $Fe_{11}S_{12}$ have often been assigned to it. Another class of sulfides is the disulfides typified by FeS_2, pyrite. These compounds contain discrete S_2 units.

Most of the RCRA metals form very low-solubility sulfides, the most insoluble of which is mercury sulfide, HgS. This fact is used extensively in waste water treatment to "polish," or remove the last traces of the metal, in water after hydroxide precipitation has been used to remove the bulk of the metals. In fact, to achieve the very low allowable discharge levels of mercury, sulfide precipitation is indispensible. It has been used to some degree in CFS work also, again especially for mercury. Arsenic impurities are removed from food-grade phosphoric acid by sulfide precipitation, forming arsenic trisulfide, which has the unusual property of low solubility in acids, but high solubility in alkalies. Sulfides of the other RCRA metals have the opposite, and more common, characteristic of increased solubility under acidic conditions.

Recently, a controversy has developed over the use of sulfides in CFS treatment to insolubilize metals. Sulfides are often useful in this application not only to form less soluble species, but to react with certain soluble metal complexes (see Chapters 3 and 4). However, concerns have arisen that the metal sulfides so formed may not be stable under specific disposal conditions, such as a highly oxidizing state. Examples are given of acid mine drainage, where the sulfide in iron pyrite is oxidized to sulfate, forming acidic conditions and releasing the iron in soluble form. Release of metals is the concern, not the formation of H_2S. Moore *et al.* [25] have shown that this can occur in the case of arsenic, copper, and zinc in underwater sediments at the redox interface. Unfortunately, there appear to be no published studies of similar phenomena in solidified waste systems, or in solid waste disposal of any sort.

In general, one expects that the redox conditions throughout most of the depth of a landfill tend toward the reducing state, which would favor the formation, not the decomposition of sulfides. This is shown directly by the formation of methane in such landfills where organic matter is present. The conditions in buried, solidified waste are very different from those in sedimentary layers in water bodies. Furthermore, there have been no reports of the oxidation of metal sulfides in solidified wastes, either in the field or in laboratory leaching tests. Since most of the latter tests are run under at least mildly oxidizing conditions, that is, nondeaereated water-based leachants tumbled vigorously in air, one would expect some evidence of sulfide dissolution if it were occurring. But, in fact, metal sulfide species apparently remain intact under oxidizing conditions as long as the leaching environment does not become too acidic—the same as with metal hydroxides and most other species.

This is, however, an important issue that needs further study. The use of sulfides in CFS fixation is increasing as regulatory leaching limits for metals continue to drop to lower values. And even more important, many sludges from waste water treatment contain sulfides from the polishing operations described previously. If metal sulfide speciation were not allowed, the impact on the technology of waste water treatment could be profound.

REFERENCES

1. Barney, G. S., and L. E. Brownell. U.S. Patent 4,028,265 (June 7, 1977).
2. Japan Kokai 80 104,687 (Aug. 11, 1980).
3. Taylor, J. *Guidance Memorandum*. Georgia Department of Natural Resources, Atlanta, Apr. 20, 1984.
4. U.S. EPA *Solid Waste Testing Manual SW-846*. Washington, DC, 1986.
5. American Public Health Association. *Standard Methods for the Examination of Water and Wastewater* (15th ed.). Washington, DC: American Public Health Association, 1980.
6. Jola, M. Destruction of cyanides by the Cyan-Cat process. *Painting Surf. Finish.*: 42–44 (Sept. 1976).
7. U.S. Patent 3,592,586.
8. Shockcor, J. H. U.S. Patent 3,692,661 (Sept. 19, 1972).
9. FMC Corp. *A Guidebook to Hydrogen Peroxide for Industrial Wastes*. Philadelphia, 1975.
10. *Potassium Permanganate to Treat Cyanide Effluents*. Spain: Asturquimica, S.A., 1975.
11. Zievers, J. F., R. W. Crain, and F. G. Barclay. Waste treatment in metal finishing: U.S. and European practice. *Plating* **55**: 1171–1179 (1968).
12. New process detoxifies cyanide waste. ES&T **5**(6): 496–497 (1971).
13. Ottinger, R. S., J. L. Blumenthal, D. F. Dal Porto, G. I. Gruber, M. J. Santy, and C. C. Shih. *Recommended Methods of Reduction, Neutralization, Recovery or Disposal of Hazardous Waste*. Washington, DC: EPA-670/2-73-053-f, 1973.
14. U.S. Patent 3,505,217.
15. U.S. Patent 3,697,421.
16. Nakanishi, K. Japan Kokai 76 120,976 (Oct. 22, 1976).
17. Miyaharam, S., K. Tayama, and M. Komatsu. Japan Kokai 80 18,209 (Feb. 8, 1980).
18. Uemura, T., and E. Hirotsu. Japan Kokai 78 14,953 (Feb. 10, 1978).
19. Oyama, S., T. Uemura, and T. Ashida. *Mizu Shori Gijutsu* **19**(4): 349–355 (1978).
20. Shimototki, T., T. Ando, T. Uemura, and T. Ashida. Japan Kokai 78 07,728 (Jan. 24, 1978).
21. Uemura, T., and E. Hirotsu. Japan Kokai 78 130,852 (Nov. 15, 1978).
22. Saigu, Y., and S. Maida. Japan Kokai 78 14,954 (Feb. 10, 1978).
23. Strunk, W. G. Hydrogen peroxide treats diverse wastewaters. *Ind. Wastes*. (Jan./Feb. 1979).
24. Kirk-Othmer. *Encyclopedia of Chemical Technology*, 3d ed. New York: Wiley, 1979.
25. Moore, J. N., W. H. Ficklin, and C. Johns. Partitioning of arsenic and metals in reducing sulfidic sediments. *Environ. Sci. Technol.* **22**: 432–437 (1988).

Chapter 6

FIXATION OF ORGANIC CONSTITUENTS

There are five distinct types of organic-containing wastes that might be encountered in CFS treatment.

1. Oil- and solvent-based wastes, such as used solvent, distillation bottoms, and refinery wastes, that are hazardous according to the RCRA, Appendix VIII of Federal Register 45, No. 98 [1], the waste listings, the CCW and CCWE tables, the California list, and the landbans.
2. Aqueous wastes containing large amounts—1 to 20 percent or more—of water-soluble or -insoluble, emulsified organics that are hazardous according to the preceding regulations.
3. Aqueous wastes containing large amounts—1 to 20 percent or more—of water-soluble or -insoluble, emulsified organics that are not hazardous, or that are hazardous only by the characteristic of ignitability or are marginal, like oil.
4. Aqueous wastes containing small amounts of nonhazardous organics—less than 1 percent and usually in the 10–1000 mg/l range—that are of interest in CFS only when they affect cementitious and other reactions of the CFS system.
5. Aqueous wastes containing small amounts of hazardous organics—less than 1 percent and usually in the 10–1000 mg/l range [2].

The first type is of interest in CFS only in very specialized applications, where solidification is required temporarily for safety in transportation or storage, or in spill-control work. These wastes will normally be incinerated if they are hazardous. Many of these wastes, and those in the second group, have been effectively removed from the CFS area by the landban regulations. Table 6-1 gives the allowable maximum total concentrations of hazardous organics that can be landfilled in any circumstance, even after treatment, unless the treatment destroys the organic. The third group of organic-containing wastes comprises oily refinery wastes and other industrial residues where the only question is the containment of the organic in the solid matrix over time. A considerable amount of CFS treatment work has been done commercially on this type of waste [3], and will likely continue in the future. The fourth waste type listed previously affects only the solidification reactions, and will be discussed later when we look at CFS process types.

TABLE 6-1 Comparison of Organic Regulatory Limits for Various Test Procedures (Non-waste waters, lowest applicable values)

Contaminant	HWNO	EPT	TCLP	CCWE	California List Total	1988 First-Third Landban TCLP	1988 First-Third Landban Total	California TTLC	California STLC	Canadian CGSB
Acenaphthalene							3.400			
Acetone			0.59	0.590			0.370			
Acetophenone							19.000			
Acrylonitrile	D018		5.000							
Aldrin					1000			140		0.07
Aniline							5.600			
Anthracene							6.200			
Asbestos								10000		
Benzene	D019		0.070				0.710			
Benzo[a]anthracene							1.400			
Benzo[a]pyrene							0.084			
BHC (alpha)					1000					
BHC (beta)					1000					
BHC (delta)					1000					
BHC (gamma)					1000					
bis(2-Chloroethyoxy)ethane					1000					
bis(2-Chloroethyl)ether	D020		0.050		1000		5.600			
bis(2-Chloroisopropyl)ether					1000					
bis(2-Ethylhexyl)phthalate							0.490			
Bromodichloromethane					1000					
Bromoethane					1000					
Butyl alcohol				5.000						
Carbaryl							0.370			7.00
Carbon disulfide	D021		14.400	4.810	1000					
Carbon tetrachloride	D022		0.070	0.960	1000					
Chlordane	D023		0.030		1000			250		0.70
Chloroaniline (1,4-)					1000					
Chlorobenzene	D024		1.400	0.050	1000		6.000			
Chlorodibenzofurans (all tetra)				0.001			0.001			

FIXATION OF ORGANIC CONSTITUTENTS 185

Chlorodibenzofurans (all hexa)			0.001	1000	0.001
Chlorodibenzofurans (all penta)			0.001	1000	0.001
Chlorodibenzo-*p*-dioxin (2,3,7,8 tetra)			0.001	1000	0.001
Chlorodibenzo-*p*-dioxin (all hexa)			0.001	1000	0.001
Chlorodibenzo-*p*-dioxin (all penta)			0.001	1000	0.001
Chlorodibenzo-*p*-dioxin (all tetra)			0.001	1000	0.001
Chlorodibromomethane				1000	
2-Chloro-1,3-butadiene				1000	
Chloroethane				1000	6.000
2-Chloroethyl vinyl ether				1000	
Chloroform	D025	0.070		1000	6.000
Chloromethane				1000	
2-Chloronaphthalene				1000	
2-Chlorophenol				1000	
3-Chloropropene				1000	
3-Chloropropionitrile				1000	
Chrysene					2.200
o-Cresol	D026	10.000	0.750		2.200
m-Cresol	D027	10.000	0.750		
p-Cresol	D028	10.000	0.750		0.900
Cyanide				1000	1.800
Cyclohexanone			0.750		0.490
2,4-D	D016	1.400		1000	
DDD				1000	
DDE				1000	
DDT				1000	
Diazinon					20.00
1,2-Dibromo-3-chloropropane				1000	10.00
Dibromoethane				1000	3.00
					0.02

(*continued*)

TABLE 6-1 Comparison of Organic Regulatory Limits for Various Test Procedures (Continued)

Contaminant	HWNO	EPT	TCLP	CCWE	California List Total	1988 First-Third Landban TCLP	1988 First-Third Landban Total	California TTLC	California STLC	Canadian CGSB
Dibromomethane					1000					
Di-n-butyl phthalate							4.200			
1,2-Dichlorobenzene	D029		4.300	0.125	1000		0.490			
1,3-Dichlorobenzene					1000					
1,4-Dichlorobenzene	D030		10.800		1000					
3,3'-Dichlorobenzidine					1000					
1,4-Dichloro-2-butene (trans-)					1000					
1,2-Dichlorodifluoromethane					1000					
1,1-Dichloroethane					1000		6.000			
1,2-Dichloroethane	D031		0.400		1000		6.000			
Dichloroethene (trans-1,2-)					1000					
1,1-Dichloroethylene	D032		0.100		1000					
2,4-Dichlorophenol					1000					
2,6-Dichlorophenol					1000					
2,4-Dichlorophenoxyacetic acid					1000		1.000	10000		
1,2-Dichloropropane					1000					
Dichloropropene (cis-1,3-)					1000					
Dichloropropene (trans-1,3-)					1000					
Dieldrin					1000			800		
2,4-Dinitrophenol							5.600			
2,4-Dinitrotoluene	D033		0.130						1	
Dioxin							13.000			
Diphenylamine + diphenylnitrosamine							0.100			
Disulfoton					1000					
Endosulfan I					1000					
Endosulfan II					1000					
Endrin	D012	0.020	0.003		1000			20		0.02
Endrin aldehyde					1000					
Ethyl acetate				0.750			0.370			
Ethyl benzene				0.053			0.031			
Ethyl ether				0.750						

FIXATION OF ORGANIC CONSTITUTENTS

Fluoranthene				3.400		0.03
Heptachlor	D034	0.001			470	
Heptachlor epoxide						
Hexachlorobenzene	D035	0.130		28.000		
Hexachlorobutadiene	D036	0.720		5.600		
Hexachlorocyclopentadiene				5.600		
Hexachloropentadiene			1000			
Hexachloroethane	D037	4.300	1000	28.000		
Hexachlorophene			1000			
Hexachloropropene			1000	19.000		
Indeno(1,2,3-cd)pyrene				3.400		
Iodomethane			1000			
Isodrin			1000			
Isobutanol	D038	36.000	5.000			
Kepone			1000		2100	
Lindane	D013	0.060	0.400		400	0.40
Methanol						10.00
Methoxychlor	D014	1.400	0.750	0.370		0.70
Methyl parathion			1000			
4,4'-Methylene-bis(2-chloroaniline)			1000			
Methylene chloride	D039	8.600	0.960	0.037		
Methyl ethyl ketone	D040	7.200	0.750	0.370		
Methyl isobutyl ketone			0.330	0.370		
Mirex					2100	
Naphthalene				0.490		
Nitriloacetic acid				14.000		
1,2-Nitroaniline	D041	0.130	0.125	0.490		5.00
Nitrobenzene				13.000		
1,2-Nitrophenol						
p-Chloro-m-cresol			1000			
Parathion			50		5000	3.50
PCBs (all)			1000			0.30
PCBs (individual)			1000			
Pentachlorobenzene			1000	28.000		
Pentachloroethane			1000	5.600		
Pentachloronitrobenzene			1000			

(continued)

TABLE 6-1 Comparison of Organic Regulatory Limits for Various Test Procedures *(Continued)*

Contaminant	HWNO	EPT	TCLP	CCWE	California List Total	1988 First-Third Landban TCLP	1988 First-Third Landban Total	California TTLC	California STLC	Canadian CGSB
Pentachlorophenol	D042		3.600	0.010	1000		37.000			
Phenanthrene							3.400			
Phenol	D043		14.400				2.400			
Phthalic acid							6.000			
Pronamide					1000					
Pyrene							2.000			
Pyridine	D044		5.000	0.330			13.600			
1,2,4,5-Tetrachlorobenzene					1000					
1,1,1,2-Tetrachloroethane	D045		10.000		1000		5.600			
1,1,2,2-Tetrachloroethane	D046		1.300		1000		6.000			
Tetrachloroethene					1000		6.000			
Tetrachloroethylene	D047		0.100	0.050						
2,3,4,5-Tetrachlorophenol				0.050						
2,3,4,6-Tetrachlorophenol	D048		1.500		1000					
Toluene	D049		14.400	0.330			0.031			
Toxaphene	D015	0.500	0.070		1000		19.000	500		0.50
1,2,4-Trichlorobenzene					1000		0.044			
1,1,1-Trichloroethane	D050		30.000	0.410	1000					
1,1,2-Trichloroethane	D051		1.200		1000		0.031			
Trichloroethylene	D052		0.070	0.091	1000					
2,4,5-Trichlorophenol	D053		5.800	0.050	1000					
2,4,6-Trichlorophenol	D054		0.300	0.050	1000					
2,4,5-Trichlorophe-noxypropionic acid					1000			1000		
1,1,2-Trichloro-1,2,2-trifluoroethane			1.050							
Trichlorofluoromethane				0.960	1000					
1,2,3-Trichloropropane					1000					
Trihalomethanes										35.00
2,4,5-TP (Silvex)	D017	1.000	0.140							1.00
tris(2,3-Dibromo-propyl)phosphate										
Vinyl chloride	D055		0.050		1000					
Xylene				0.150			0.015			

The last waste group just listed is the one of real interest in the exploration of fixation technology. Aqueous, inorganic waste streams containing parts-per-million levels of hazardous organics are quite common, and will actually become even more so as a result of landban regulations that require destruction of organics, but that leave some residue that may leach above the allowable levels shown in Table 6-1.

TREATMENT TECHNIQUES

Hazardous organic wastes lend themselves primarily to destructive treatment by processes such as incineration, biodegradation, chemical oxidation, and dechlorination. When these methods are feasible, all the problems associated with long-term effects are eliminated; thus, they are high on everyone's list of preferred treatment methods. However, many industrial wastes and contaminated materials contain small amounts of toxic organics at levels that make organic destruction processes not only very expensive, but sometimes ineffective. Examination of Table 6-1 will reveal that hazardous organic total content for specific compounds is limited to 1000 mg/kg under the California list landban and much lower, often in the parts-per-billion (ppb) range, where applied to date under the first-third landban. Furthermore, under the CCWE tables (Table 6-1) and the forthcoming TCLP rule [4] the allowable leaching levels for a number of organics are even lower, sometimes at a few parts per billion. Destruction processes cannot always destroy these constituents to the point that leaching is below the allowed limits.

For these reasons, it is important to develop "fixation"* techniques for low-level organics. We have already discussed reactions in which organic ligands fix metal species, but most toxic organics are not nearly as reactive. There is indication that "reactions" do occur between some organics and the inorganic, cementitious CFS systems in use today. However, it has been difficult to determine whether seemingly positive results are due to sorption effects, dilution by reagent additions and the leachant, volatilization of volatile and semivolatile constituents, or true chemical reactions. Weitzman et al. [5] determined that most of the volatile organic carbon (VOC) in stabilized wastes is lost during the mixing and curing processes. Since considerable temperature rise is associated with certain CFS systems, it is possible that semivolatile organics are lost as well. The EPT and TCLP tests cause a 20:1 dilution of all constituents, and CFS reagents also dilute the waste by factors of 10 to 100 percent (weight basis), resulting in automatic dilution factors in the range of 22 to 40. Finally, the large body of knowledge on destruction/removal of organics in waste water treatment is not very applicable. Much of this involves biodegradation, which probably does not occur to significant degree in CFS systems due to their alkalinity. Many of the other methods use sorption techniques, especially with activated carbon, which are at least partially reversible in CFS systems where the sorbent remains in the product.

In spite of the dearth of information about true immobilization of organics in CFS systems, some data are available. These are shown in Table 6-2 for several CFS process types and wastes. It will be noted that the leaching levels achieved are not always very good. The work cited did not meet CCWE criteria for xylene or

---

*The term *fixation* is really a misnomer in the sense that it is used for the immobilization of metals, but it is commonly used in practice.

TABLE 6-2 Summary of Organic Fixation Data

Waste Description	CFS System	Mix Ratio	Treatment Scale	Leaching Test	PCBs Total	PCBs Leachate	1-Chlorohexa-decane Total	1-Chlorohexa-decane Leachate
Synthetic waste	Cement/ soluble silicate		Lab	TCLP	380	165.000	1300	700.000
Synthetic waste	Cement/ soluble silicate + FeCl$_2$		Lab	TCLP	380	460.000	1300	640.000
Synthetic waste	Cement/ soluble silicate + Na$_2$S		Lab	TCLP	380	800.000	1300	980.000
Leaded gasoline	Proprietary sorbent	1.00	Lab	TCLP				
Unleaded gasoline	Proprietary sorbent	1.00	Lab	TCLP				
Diesel fuel	Proprietary sorbent	1.00	Lab	TCLP				
Kepone-contaminated sediment	None		Lab	Water				
Kepone-contaminated sediment	Cement/ soluble silicate		Lab	Water				
Kepone-contaminated sediment	Organic		Lab	Water				
Kepone-contaminated sediment	Sulfur-based		Lab	Water				
Waste caulking compound latex	Cement		Lab	EPT				
Lagoon sludge	Kiln dust		Lab	EPT	68	<0.001		
Regulatory Level—TCLP Max.								
Regulatory Level—CCWE								

[a] References listed at end of Chapter 6.

ethylbenzene, or the TCLP criterion for benzene. Several processes met the TCLP criterion for toluene, and the kiln dust method achieved a very low level, <0.001 mg/l, for PCBs. The latter result is quite common; most CFS processes leach PCBs at less than 0.01 even with constituent levels in the waste of 50 to 100 mg/kg. This is believed to be due largely to the low solubility of PCBs in water, but even so, some other immobilization mechanism appears to be operative. In general, organics with low water solubility are immobilized fairly well in commercial CFS processes, while those with higher solubility are not.

In addition to the methods for which specific leaching data are available, a number of others are described by vendors or investigators in more general terms. Processes that have been mentioned are listed in Table 6-3.

Organic Content (mg/kg or mg/l)											
Benzene U109		Toluene U220		Xylene U239		Kepone	bis(2-Ethylhexyl) phthalate		Ethylbenzene	
Total	Leachate	Total	Leachate	Total	Leachate	Leachate	Total	Leachate	Total	Leachate
100000	9.200	50000	12.400	50000	3.250					
100000	12.200	50000	3.250	50000	13.400					
	0.100		0.180		0.420					
						0.180				
						0.454				
						0.040				
						0.310				
							100000	<1.000		
		27000	820.000						3700	68.000
	0.070		14.400							
					0.150					0.053

ORGANIC REACTION PRINCIPLES IN CFS

The number of organic reactions that might occur in hazardous waste treatment is almost infinite. In practice, however, inorganic CFS systems operating at ambient temperatures and pressures in nonexotic aqueous environments can produce only relatively few reaction schemes. Aside from adsorption, volatilization, and biodegradation, the most likely reactions fall into four categories: hydrolysis, oxidation, reduction, salt formation. Some general and specific reactions within these categories are shown in Table 6-4. R indicates any organic grouping and X means a halide. The reaction products are not all shown in some cases where they are not pertinent to the discussion, and the stoichiometry is not necessarily as stated.

FIXATION OF ORGANIC CONSTITUENTS

TABLE 6-3   Organic Immobilization Processes

Method, Process, or Reagent	Constituent	Reference
Activated carbon	General, phenol	[6][7]
Potassium permanganate	Phenol, odor, color, general	[8]
Cement/soluble silicate	Substituted phenols	[9]
Lime/flyash	Phenol	[10]
Cement/flyash	Phenol	[10]
Cement/bentonite	Phenol	[10]
Cement	Substituted phenols, ethylene glycol	[11]
Hydrogen peroxide + Fe^{+2}	Phenol	[12]
Organoclays	Substituted phenols, benzene, trichloroethene, other chlorinated organics	[13]–[17]
Catalytic destruction	Chlorinated organics	[15]
Potassium ferrate oxidation	Nitrilotriacetic acid	[18]
$Fe(OH)_3$ precipitation	Chlorendic (HET) acid, humic acid, PCBs, other organics	[19]
Cement	Phenol, trichloroethylene	[20]
Kiln dust/flyash	Miscellaneous organics	[21]
CaO/flyash	Miscellaneous organics	[21]

Hydrolysis

Hydrolysis refers to the reaction of a compound with water. This usually results in the exchange of the hydroxyl group (-OH) for another functional group at a reaction center. Hydrolysis may be catalyzed by acidic or basic species (OH^-, H^+ or H_3O^+) and may involve intermediates. The rate law for hydrolysis of chemical RX is usually expressed as [22]:

TABLE 6-4   Some Possible Organic Reactions in CFS Systems

Reactants	Products[a]	Reference
Hydrolysis		
$RX + H_2O$	$ROH + HX$	[22]
Organoaminos	Organics + NH_3	[23]
Oxidation		
Phenol + $14H_2O_2$ + Fe^{+2}	$6 CO_2 + 17 H_2O$	[24]
$R-CH_3$	$R-COOH$	[8]
$R-CH_2OH$	$R-COOH$	[8]
$RCHOH-CHOHR'$	$R-COOH + R'-COOH$	[8]
$R-CHO$	$R-COOH$	[8]
R_2CH_2	R_2CO	[8]
$R_2CH(OH)$	R_2CO	[8]
R_3CH	$R_3C(OH)$	[8]
$R_3CH + HCR'_3$	$R_3-C-C-R'_3$	[8]
$R_2N-H + H-NR'_2$	$R_2N-NR'_2$	[8]
$RCH=CHR'$	$RCHOH-CHOHR'$	[8]
$2R-SH$	$R-S-S-R$	[8]
$R-S-S-R'$	$R'SO_3H + RSO_3H$	[8]
Reduction		
$Fe + 2H_2O + 2RCl$	$2ROH + Fe^{+2} + 2Cl^- + H_2$	[25]
Salt Formation		
Oxalic acid	Calcium oxalate	[26]

[a] All reaction products are not necessarily listed; stoichiometry is not complete.

$$K_h = \frac{K_B K_W}{H^+} + K_A [H^+] + K_N \qquad (6\text{-}1)$$

where K_B, K_A, and K_N are the second-order rate constants for acid and base and neutral processes, respectively, and K_h is the pseudo-first-order rate constant observed for hydrolysis at a specific pH and temperature.

It is evident from Eq. (6-1) how pH affects the overall reaction rate, causing one of the three rate constants to become dominant. At a given pH, the rate process is pseudo-first-order and the half-life $t_{1/2}$ of the substrate is independent of its concentration:

$$t_{1/2} = \frac{0.693}{K_h} \qquad (6\text{-}2)$$

Equation (6-1) can be expressed graphically as three equations (see Dragun [22]).

The pH versus rate curves for some important alkylhalides are shown in Figure 6-1. Note that the time scale for half-lives is logarithmic, and scans nine orders of magnitude between chloroform and butyl chloride at low pH. For purposes of comparison, let us assume that a reasonable reaction time (after treatment) for CFS products in most hazardous waste disposal scenarios might be from one month to one year, since the long-term effects are of most concern. In that case, it would be necessary to set the pH of the system at the maximum allowable under the RCRA definition of noncorrosive (pH 12.5) to hydrolyze all of these compounds. This is feasible if metal leaching is not of concern; however, if lead, for example, is present, we would want the pH between about 9 and 11 for minimum leaching. In general, pH values between 8 and 11 are best, so methyl fluoride and chloroform destruction

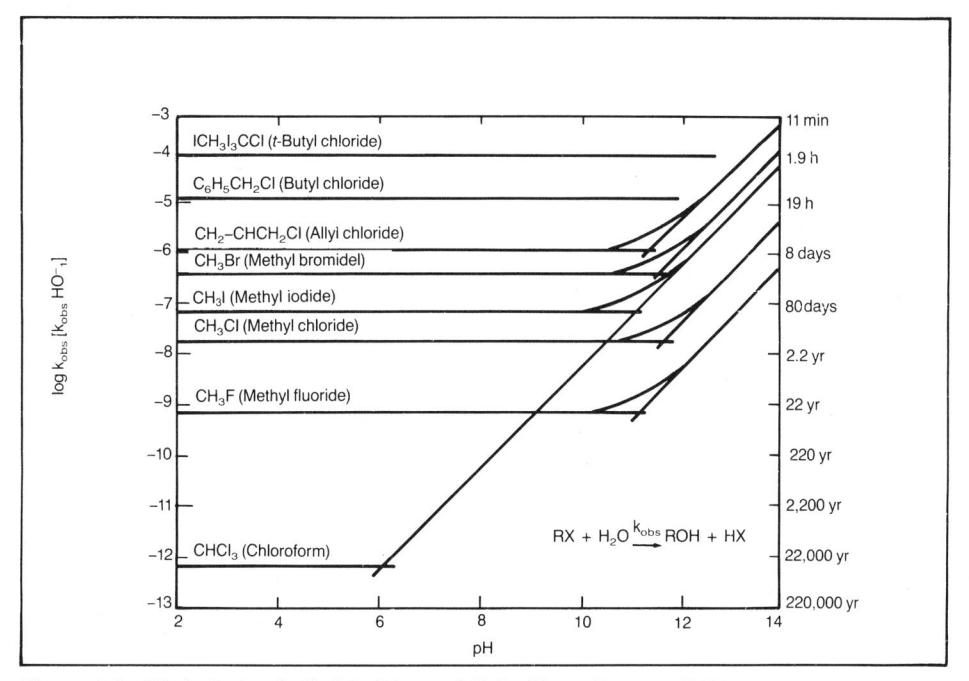

Figure 6-1 Hydrolysis of alkyl halides at 25°C. (From Dragun [22].)

would be too slow. Dragun [22] provides a summary of hydrolyzable chemical structures that may be expected to persist in water for significantly more than a year at pH 7 and 25°C (77°F) (Table 6-5). Those less resistant to hydrolysis are listed below. Several typical hydrolysis reactions [27] are shown in Figure 6.2.

Alkyl and benzyl halides
Polymethanes
Substituted epoxides
Aliphatic acid esters
Chlorinated acetamides
Certain carbamates
Phosphoric acid halides
Phosphoric acid and thiophosphoric acid esters and acylating and alkylating agents

The preceding discussion illustrates a basic reality of CFS technology—that necessary and practical trade-offs are not only economical, but technical. Mabey and Mill [28] have published a useful summary of estimated hydrolysis rates for organic chemicals in pure water systems. Such data are useful for general comparative work, but suffer from the same limitations as do solubility tables, that is, other constituents and characteristics of a real waste/CFS system are not taken into account. For example, the effects of catalytic agents are not considered. In real systems, metal ions such as copper and calcium may act as catalysts for certain chemical structures, and adsorption on surfaces such as clay and activated carbon may also accelerate reactions [15]. These effects apply not only to hydrolysis, but to all of the other reaction types presented here, as well as to sorption, biological degradation, and even volatilization.

TABLE 6-5 Chemical Persistent to Hydrolysis[a]

Category	General Structure	Persistent Chemicals[b]
Esters	$R_1C(O)OR_2$	All Al esters of Al, Ar, or allylic acids
Amides	$R_1C(O)NR_2R_3$	All amides where R_1–R_3 are Al or Ar; only amides with halogenated alkyl R_1 hydrolyze rapidly
Nitriles	RCN	All aliphatic or aromatic nitriles
Acyl chlorides	RC(O)Cl	No acyl chlorides
Carbamates	$R_1OC(O)NR_2R_3$	All carbamates having only Al or Ar on N and O
Alkyl halides	RX	All AlF and polychloro- or poly-bromo-methanes
Phosphorous acid esters	$R_1P(O)R_2R_3$	All esters where R_1 is Al or Ar and R_2 and R_3 are AlO and ArO (phosphonates); no esters where R_1–R_3 are AlO or ArO (phosphates); only esters where R_1 and R_2 are Al_2N and R_3 is F (phosphonohalidates)
Epoxides, lactones, sultones		Only hindered, bicyclic epoxides; no simple lactones or sultones

[a] Having a half-life >1 yr at 25°C and at pH 7.
[b] Al = aliphatic, Ar = aromatic.

Source: Dragun, J. The fate of hazardous materials in soil. *Hazardous Mater. Control:* 24-43 (Sept./Oct. 1988).

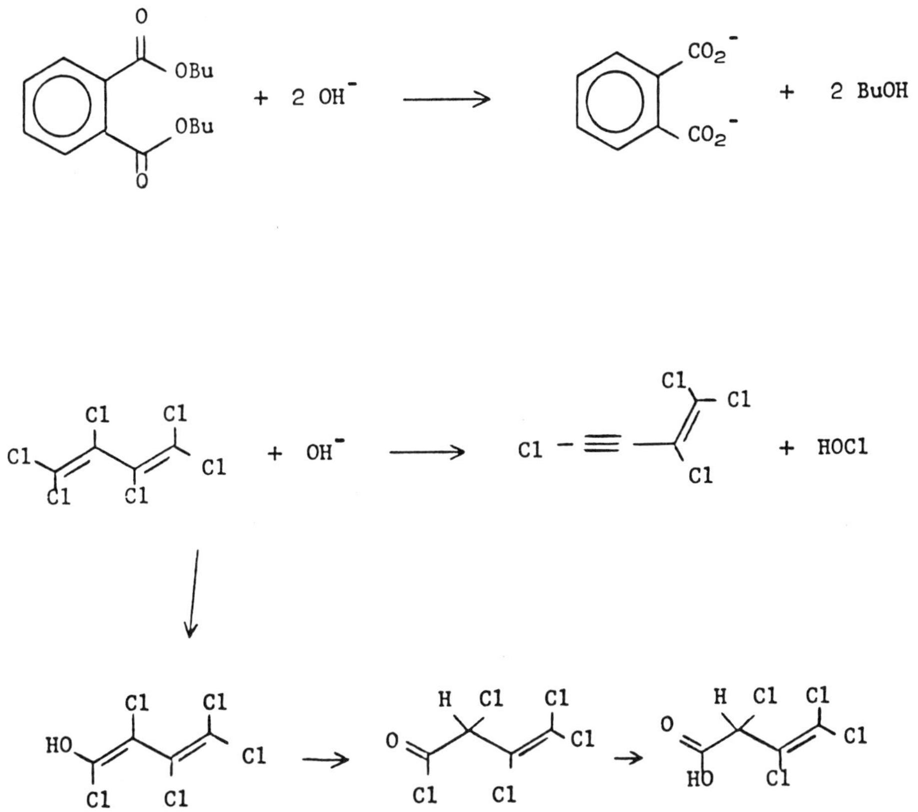

Figure 6-2 Typical reaction chemistry of selected organic waste constituents under basic conditions. (From Tittlebaum et al. [27].)

Oxidation

A number of oxidation reactions are shown in Table 6-4. Oxidation, along with hydrolysis, is probably the most common reaction for organics in CFS systems. We have discussed oxidation of metals and inorganic species; oxidation of organics occurs via two pathways [22]. In one, an electrophilic agent attacks an organic molecule and removes an electron pair; in the other, only one electron is removed, forming a free radical. The former is called *heterolytic*; the latter, *homophilic*. Free radical reactions require much less energy than do oxidation of a polar compound or cleavage of a covalent bond. Free radical reactions are summarized nicely by Dragun [22] in Figure 6-3. They involve three steps: initiation or formation of a free radical; propagation, which forms other free radicals; and termination, which destroys the free radicals. Termination can take place in three ways: simple coupling, disproportionation, and abstraction.

Based on the assumption that the concentration of each oxidant is constant during the process, the oxidation rate law simplifies to Eq. (6-3):

$$t_{1/2} = \frac{0.693}{\text{sum } K_{\text{ox } n}} \qquad (6\text{-}3)$$

where $K_{\text{ox } n}$ = specific second-order rate constant for each oxidant.

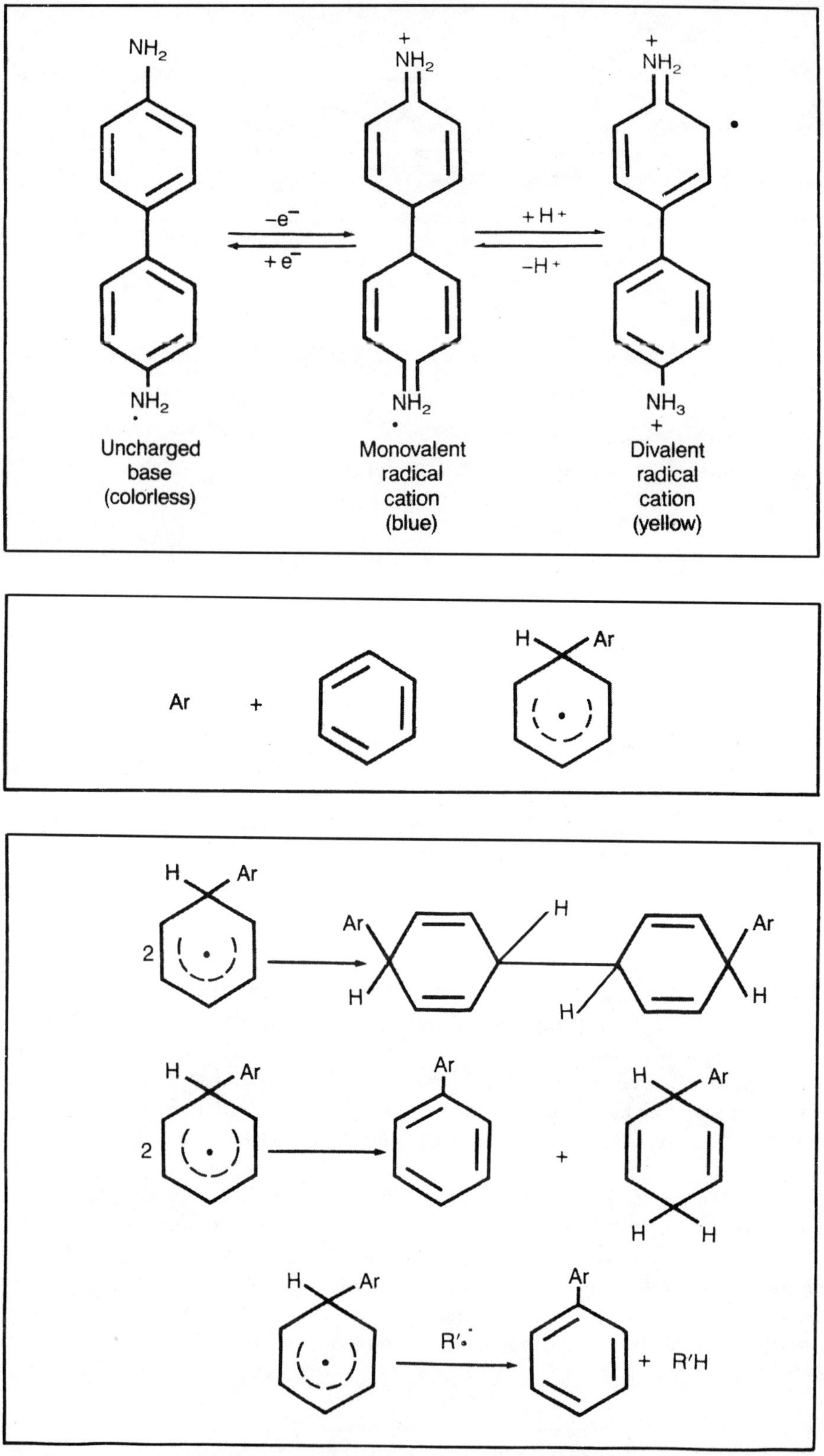

Figure 6-3 Free radical reactions (oxidation). (a) Initiating step. (b) Propagation step. (c) Termination step: Top—simple coupling; middle—disproportionation; bottom—abstraction. (From Dragun [2].)

Organic oxidation reactions have been catalyzed for many years in the chemical industry by crystalline aluminosilicates at elevated temperatures and pressures. Recently, it has been recognized that this catalysis also occurs at ambient temperature and pressure with clays and soils, not only in oxidation, but in reduction, hydrolysis, and neutralization reactions. Iron, aluminum, and trace metals within the layered silicate minerals have been identified as the specific catalysts, but not all clays exhibit this property. Some general predictive guidance can be obtained from Dragun and Helling [29], but the same caveats as given before in the discussion of hydrolysis also apply here. Based on this guidance, the following generalities can be stated.

1. Many substituted aromatics undergo free radical oxidation, for example, benzene, benzidine, ethyl benzene, naphthalene, phenol and others among the organic priority pollutants.
2. Chlorinated aromatics and polynuclear organics are unlikely to be oxidized.

Water content may be one important factor in soils, and possibly also in waste forms. Partially saturated systems are more likely to undergo oxidation than fully saturated ones.

The preceding comments apply to "natural" oxidation by reagents normally used in CFS systems, or characteristics of the waste itself. The deliberate addition of strong oxidants such as potassium permanganate and hydrogen peroxide are a different matter. Some of these reactions and reaction products were shown in Tables 6-3 and 6-4. These agents have been used in the CFS field for oxidation of phenols and other organics, as well as for inorganics and breaking of metal complexes, and are described in the literature [8], [12]. An important consideration in all organic reaction schemes is the product of the reaction, which may also be hazardous. This is especially true for oxidation processes. Also, the use of oxidants or other agents in CFS systems may destroy the contaminant of interest but create hazardous species from other organics in the waste that were previously nonhazardous or less hazardous. Furthermore, the use of strong oxidants in wastes that contain chromium may result in the formation of Cr^{+6}, necessitating a subsequent reduction step. A thorough treatability study, including a final-product priority pollutant scan, is necessary when employing any organic reaction scheme.

Reduction

A metal is said to be reduced if it is changed to a lower valence state, but the definition for organics is different. Reduction is defined as an increase in its hydrogen content or decrease in its oxygen content. An example of the former is the reaction:

$$Cl_2C=CCl_2 + H^+ \longrightarrow ClCH \quad CCl_2 + Cl^-$$

and of the latter:

$$R-\overset{\overset{\displaystyle O}{\diagdown}}{CH} - \overset{\diagup}{CH_2} + H_2 \longrightarrow R-CH_2=CH_2 + H_2O$$

There is also reduction defined in terms of electron transfer: an organic chemical is said to be reduced if it experiences a net gain of electrons. Reduction can occur in clay systems, for example, with montmorillonite [22] acting as the reducing agent. Reductive alteration of organic contaminants in wastes is probably the least studied of the reaction areas, and subsequently is least understood. Absent the use of strong reducing agents, the system redox potential gives an indication of whether reduction is possible. If the oxygen supply is deficient, species such as nitrates and sulfates are known to be reduced. One would expect reduction of organic compounds to also occur if the system potential is less than that of the organic species. One reduction reaction described in the literature is given in Table 6-4 [25].

Salt Formation

The CFS literature is practically devoid of references to this type of organic immobilization, yet many organic compounds react easily with metals and other cationic species to form less soluble salts. This may be the cause of the changes in morphologies and electron diffraction patterns observed by Chou *et al.* [30] in cement matrices. In this study, they observed these changes with ethylene glycol and *p*-bromophenol, and could not correlate the data with any expected products of hydration of cement itself. Examination of any handbook of chemistry will reveal numerous salts of organic compounds that are less soluble than their corresponding acid forms. For example, the solubility of oxalic acid is 95,000 mg/l, compared with 6 mg/l for the calcium salt. It is likely that a number of organic acids of environmental concern could be effectively immobilized in the calcium-rich environment of commercial CFS systems.

In addition to salt formation, a number of other direct reactions are possible between organic contaminants and organic or inorganic reagents under ambient conditions. It would seem that this is a fertile area for research in the immobilization of trace organics.

COMMERCIAL PROCESSES

In Table 6-2, a few examples of organic "fixation" processes were listed. In the last few years, a number of CFS processes, mostly proprietary, have been offered that claim to fix, destroy, or immobilize organic priority pollutants and other organics of environmental concern. We have already discussed a few of the generic types. Unfortunately, virtually no credible data are available. Low leaching levels are claimed without any indication of the initial levels in the waste, or the effects of dilution or volatilization of the constituents. For this reason, no listing of processes or references is presented in this chapter, with the exception of the generic processes already described.

Several investigators have conducted research in this area, and this work is continuing. The interested reader is referred especially to the work at Louisiana State University (LSU) [27], the University of New Hampshire (UNH) [20], and Michigan State University (MSU) [15-17] for further reading. The LSU work has been going on for some years now, and is centered around electron microscopy, both transmission and scanning, and electron diffraction to identify morphological and diffraction pattern characteristics and changes and correlate them with other observations and tests. Some interesting SEM photographs are presented in the original literature. The investigators at UNH have found some correlation between curing time and leachability for several organic constituents.

The work at MSU is based on organoclays, which are also the active constituents in some proprietary products on the market, and work is also going on at other research centers on these materials [13]. These clays are natural products that have been altered by various organic chemicals, usually quaternary organoammonium compounds. This alteration changes the clays' surfaces from hydrophilic to hydrophobic, increasing the clay's sorptive capacity for organic constituents in wastes. Many of the modified clays are available commercially. While most of this work was done with waste water treatment concepts in mind, the materials may be useful in the CFS area if their cost is reasonable. These products have the possibility of immobilizing organics so that they will not desorb under the conditions of CFS system use and disposal practices; some appear to catalyze reactions that might detoxify the organics. Other surface active reagents, such as activated carbon and ion exchange resins, may also be useful in CFS work, but their high cost has generally been a major deterrent.

REFERENCES

1. Federal Register. **45**(98): 33119-33133 (May 19, 1980).
2. Conner, J. R. *CONTEC Data Base,* 1985.
3. Delchad, S. Chemical treatment: An inexpensive alternate to handling oily sludge. In *Proc. 5th National Conference on Hazardous Wastes and Hazardous Materials.* pp. 85-88, Apr. 19-21, 1988.
4. Federal Register. **51**(114): 21685 (June 13, 1986).
5. Weitzman, L., L. E. Hamel, and S. R. Cadmus. *Volatile Emissions from Stabilized Waste in Hazardous Waste Landfills,* Contract 68-02-3993. Research Triangle Park, NC: U.S. EPA, Aug. 28, 1987.
6. Mahajan, O. P., C. Moreno-Castilla, and P. L. Walker, Jr. Surface treated activated carbon for removal of phenol from water. Washington, DC: NTIS, 1980.
7. O'Brien, R. P., D. M. Jordan, and W. R. Musser. Trace organic removal from groundwaters with granular activated carbon. *Proc. Am. Chem. Soc.* Atlanta, GA, Mar. 29-Apr. 3, 1981.
8. Spicher, R. G., and R. T. Skrinde. Effects of potassium permanganate on pure organic compounds. *J. Am. Water Works Assoc.* **57**(4): 472-484, (1965).
9. Cartledge, F. K., D. Chalasani, H. C. Eaton, A. Roy, and M. E. Tittlebaum. Mechanisms of interaction of phenols with cementitious stabilization matrices. Louisiana State Univ., Baton Rouge.
10. Cote, P. L., T. R. Bridle, and D. P. Hamilton. Evaluation of pollutant release from solidified aqueous wastes using a dynamic leaching test. *Hazardous Wastes Environ. Emergencies.* Houston, TX (Mar. 12-14, 1984).
11. Sheffield, A., M. E. Tittlebaum, F. K. Cartledge, and H. C. Eaton. *Dynamic Leachability of Hazardous Organic/Cement Mixtures.* Baton Rouge: Louisiana State Univ., 1987.
12. FMC Corporation, Technical Brochure, Philadelphia, PA.
13. Sheriff, T. S., C. J. Sollars, D. Montgomery, and R. Perry. *The Use of Activated Charcoal and Tetra-alkylammonium-Substituted Clays in Cement-based Stabilization/Solidification of Phenols and Chlorinated Phenols.* London: Imperial College, 1987.
14. Boyd, S. A., J. Lee, and M. M. Mortland. Attentuating organic contaminant mobility by soil modification. *Nature* **333**(6171): 345-347 (1988).
15. Boyd, S. A., M. M. Mortland, and T. J. Pinnavaia. Use of modified clays for adsorption and catalytic destruction of contaminants. In *Proc. 13th Annual Research Symposium at Cincinnati,* July 1987.
16. Boyd, S. A., M. M. Mortland, and C. T. Chiou. Sorption characteristics of organic

compounds on hexadecyltrimethylammoniumsmectite. *Soil Sci. Soc. Am. J.* **52:** 652–657 (1988).
17. Boyd, S. A., S. Shaobai, J. Lee, and M. M. Mortland. Pentachlorophenol sorption by organo-clays. *Clays Clay Miner.* **36**(2): 125–130 (1988).
18. Carr, J. D., P. B. Lelter, and A. T. Ericson. Ferrate (VI) oxidation of nitrilotriacetic acid. *Environ. Sci. Technol.* **15**(2): 184–187 (1981).
19. *HazTech News,* p. 100, June 18, 1987.
20. Kolvites, B., and P. Bishop. *Column Leach Testing of Phenol and Trichloroethylene Stabilized/Solidified With Portland Cement.* Durham, NH: Univ. of New Hampshire, 1987.
21. *The Hazardous Waste Consult.* 1-14-1-18, (Nov./Dec. 1987).
22. Dragun, J. The fate of hazardous materials in soil. *Hazardous Mater. Control:* 24–43 (Sept./Oct. 1988).
23. Reimers, R. S., C. P. Lo, and P. P. Meehan. *Chemically Stabilizing Organic, Metallic and Pathogenic Materials.* Kenner, LA: Chemifix Technologies, Inc., 1986.
24. FMC Corp., Bulletin, Philadelphia, PA, 1980.
25. *Water Sewage Works:* 40–42 (Jan. 1979).
26. Dean, J. A., Ed. *Lange's Handbook of Chemistry,* 13th ed. New York: McGraw-Hill, 1985.
27. Tittlebaum, M., F. Cartledge, and H. Eaton. *Applicability of Solidification to Organic Wastes.* Baton Rouge, LA: Louisiana State Univ., 1987.
28. Mabey, W., and T. J. Mill. *J. Phys. Chem. Ref. Data* **17**(2): 383 (1978).
29. Dragun, J., and C. S. Helling. *Soil Sci.* **139**(2): 100–111 (1985).
30. Chou, A. C., H. C. Eaton, F. W. Cartledge, and M. E. Tittlebaum. A transmission electron microscope study of solidified/stabilized organics. *Hazardous Waste Hazardous Mater.* **5**(2): 145–153 (1988).

Chapter 7

CHOOSING THE RIGHT CFS SYSTEM

We have discussed the fixation of metals, other inorganics and organics, and seen fixation as an independent technology, not necessarily tied to solidification. From a practical standpoint, however, the two usually go together. Very often, the solidification technology *is* the fixation technology, as for example with most Portland cement systems that we discuss in Chapters 9 and 10. We have defined these terms (in Chapter 1 and the Glossary) and discussed the regulatory and historical background of CFS technology. Now it is time to put everything together into a system that provides a sound, practical scientific and engineering basis for selection of the appropriate CFS solution for a given problem. The author, in 1981, called this "the engineered approach" [1, 2], because it approaches the project from an engineering point of view, that is, definition, solution, and implementation. This chapter discusses that approach, preparing the way for more detailed coverage of waste types (Chapter 8), CFS processes (Chapters 9–15), and delivery systems (Chapter 16).

THE ENGINEERED APPROACH

The engineered approach, as we mean it when applied to CFS technology, has five basic steps:

1. Definition of the problem and collection of data on the important design parameters and project considerations.
2. Analysis of the data and comparison of them with past experience. This can now be aided by using a computerized "expert system," along with the experience and knowledge of a specialized staff.
3. CFS process testing. In the laboratory, possible test formulations established in step 2 are applied to actual waste samples using efficient, well-established test protocols. This fine-tunes the process for a particular waste stream and verifies that formulations are correct.
4. Regulatory interaction and approval.
5. Implementation of the solution.

In theory, it would seem that much of steps 2 and 3 could be bypassed by pretesting various wastes, charting them, and then finding the required process on the chart. Attempts to apply this technique in practice, however, have been unsuccessful. Waste streams vary so widely in chemical composition and other characteristics that it is not possible to pretest all of the combinations.

On the other hand, some CFS companies and waste generators have built up an enormous body of knowledge during the testing of some tens of thousands of individual waste samples, representing more than thousands of distinct waste streams, over a period of some 20 years. Furthermore, these experimental data have been scaled up into hundreds of commercial projects over this same period. It became evident that this mass of data required some sort of automated data processing for practical use. This finally has become possible through the development of a branch of artificial intelligence technology called *expert systems* (ES). Put simply, ES is a computerized technique for simulating the mental processes (conscious and unconscious) of a human expert while combining it with the vast information storage, retrieval, and processing abilities of the computer. ES has two basic parts: a data or knowledge base, and an "inference engine," or set of rules that experts apply to the information. The latter part is accomplished by "heuristic" programming—the use of intuitive, informal rules and experiential knowledge rather than the orderly algorithmic progression of most of today's computer programs.

Only within the last five years has it become practical to apply ES technology to hazardous waste treatability. The author completed the first stage of a rudimentary expert system, called WISE™ (Waste Immobilization System Expert) [3] in 1985. WISE incorporated part of a commercial CFS company's data base together with a number of conscious rules, and has undergone testing of its general design and refinement of the data base. It could predict, with about 75 percent accuracy, an approximate optimum formulation for any type of waste and disposal scenario that has previously been tested.

To build such a system requires considerable information about the waste and a very extensive data base. However, its value lies beyond simply selecting a CFS process, because the ES is essentially a decision matrix, logic tree, or screening flow chart. Regulations, engineering alternatives, and other options can easily be incorporated to form a rather complete decision-making system that is especially valuable for students, permit writers, regulators, legislators, and others without long experience in the hazardous waste field. The WISE system is described in more detail in Chapter 18, for those interested in pursuing this approach.

THE WASTE

The waste properties to consider in choosing a CFS system are as follows.

1. Toxicity
   (a) leachable (soluble)
   (b) nonleachable (insoluble)
2. Medium (water, solvent, etc.)
3. Physical properties
   (a) solids content

(b) particle size, distribution, and shape
(c) rheological characteristics
4. Chemical properties
    (a) inhibitors
    (b) accelerators
    (c) bioactivity
5. Working hazards
    (a) odors
    (b) vapors
    (c) flammability
    (d) toxicity
    (e) reactivity (pyrophoric, etc.)

A complete listing of pertinent waste characteristics is given below.

Waste name
Generating process
Industry
EPA hazardous waste number(s)
EPA handling methods number
  Storage
  Treatment
  Disposal
NFPA hazard identification
  Flammability
  Health
  Reactivity
  Special
Toxicity rating
  Inhalation
  Dermal
  Oral
Annual generation rate in gallons
Quantity stored in gallons
Type of storage
Medium (aqueous, oil, solvent, etc.)
Physical state (solution, sludge, powder, etc.)
Phases/layering (none, bilayered, multilayered)
Total solids (%)
Suspended solids (%)
Type of solids (organic, inorganic, mixed)
Specific gravity
Bulk density (for solids)
Grain size distribution
Viscosity
Flash point
pH
Alkalinity/acidity (%)

Odor
Color
Waste analysis
   Metals, total
   Metals, leached
   Inorganics, total
   Inorganics, leached
   Organics, total
   Organics, leached

These characteristics are all measurable or determinable if the proper waste information system is used. The methods of measurement are described in more detail in Chapter 17. Characteristics of wastes and waste classification systems are discussed more completely in Chapter 8.

Toxicity

Of primary importance is the toxicity (or potential toxicity) of the waste. It determines not only the CFS process to be applied but also the required regulatory interaction and method of disposal. For example, the presence of more than 50 ppm of PCB in a waste removes it from the jurisdiction of the RCRA and places it under the TSCA. The treatment and disposal requirements of the TSCA are totally different. Toxic components of the waste can be described as either leachable or nonleachable with reference to the particular leaching test and leachant being applied. While most concern has been about the leachable portion of the toxic components in the waste, the nonsoluble components also represent a potential hazard if the actual leaching conditions in the environment are not the same as those tested in the laboratory, if the conditions change, or if there is a change in the disposal operation.

    A common example is the problem with metal hydroxide sludges. This sludge usually comes from a water treatment plant where the pH of the system has been adjusted to achieve minimum solubility of the metals, precipitating them and allowing discharge of the clear water phase of the waste stream. The sludge, then, as it is fed into the system is essentially nonleachable with distilled water or a neutral leaching medium. However, under acid (or sometimes alkaline) conditions, these metals may be resolubilized and become available to the environment through leaching. If the actual groundwater conditions are slightly acid, then the process should be chosen so that the metals will not be leached under these real conditions. The chosen disposal site should also properly protect the environment from extremely toxic constituents that might be leached out either normally or under changing groundwater conditions.

Medium

The medium in which waste constituents are dispersed is either water, an organic of some sort, or a combination of the two (unless the waste is a dry solid). Today, most CFS processes are designed to work on water-based waste streams, or on emulsions in which the liquid organic (such as oil) is reasonably well dispersed in the water medium. Pure solvent systems and inverse emulsions generally do not solidify well with inorganic systems, although some organic systems can handle these wastes. In certain systems, such as those using bitumens, the water is evaporated before or during incorporation of the waste into the matrix.

Physical Properties

Physical properties of a waste significantly affect the treatment process used and the proper chemical formulation for treatment. The primary physical properties of interest are type of medium, phase separation, solids content and type, specific gravity, particle size distribution and shape, and viscosity. Inorganic systems generally work well on aqueous wastes with a high solids content because these wastes normally require smaller amounts of chemicals to be added, provided that the treatment system itself does not react with the solids. Several systems, such as cement-soluble silicate, are also designed to solidify low solids wastes. In aqueous media, only two phases are normally present: water with dissolved solids, and suspended or settled solids. Occasionally, a third, floating phase may be present. If this is a liquid organic, particular attention must be paid to the choice of process and/or pretreatment. If it is a low specific gravity floc or organic solid, it may be treated separately or, more often, homogenized into the other phases.

Particle size, distribution, and shape in sludges have an effect on the mechanical properties of the solid produced by the CFS process, and on the handling characteristics of the waste solution or slurry. They also can affect the solidification reactions for certain particulates and process types. From the handling standpoint, for example, certain waste from paper mills appears nearly solid at only a few percent solids (by weight) in water due to the fibrous nature of the particles in suspension. Conversely, flyash slurries can be pumped at solids concentrations of 50 percent or higher because of the smooth, spherical shape of the flyash particles.

Specific gravity and viscosity due to factors other than particle size and shape also affect the choice of CFS system because in all cases the system chemicals must be mixed into the waste (or vice versa), and the equipment and energy to use for mixing depends on these physical characteristics. Materials handling and energy economy make pumping and low-energy mixing the desirable way to process and transport wastes whenever possible. If the physical properties of the waste require other means of conveyance or high-energy mulling or blending, the mechanical requirements of the system will sometimes dictate the choice of chemical system.

With regard to viscosity, most wastes handled in CFS systems exhibit non-Newtonian rheological properties; that is, viscosity is not independent of the rate of shear. As some unfortunate truckers have experienced, a load of apparently solid filter cake, under shear stress set up by the vibrations of the truck, can turn into a fluid that will either pour out of the back of the truck or be deposited on the windshield and hood when the driver comes to a quick stop. This situation of viscosity decreasing with increasing rate of shear occurs with plastic and pseudoplastic materials. Because of this phenomenon, many solid-appearing sludges can be pumped quite easily once sufficient shear force has been applied to decrease the viscosity. On the other hand, certain materials, called *dilatant,* demonstrate increasing viscosity with increasing rate of shear. Such a sludge can literally turn solid inside a pump.

Finally, certain materials are thixotropic; that is, their viscosity at any particular rate of shear depends on the amount of previous shearing the waste has undergone, that is, this waste has a memory. Thixotropy occurs as a result of the reversible gel–sol–gel transition, and is caused by the building up of a definite structure within the waste. (The word "thixotropy," however, is often misused to describe all the non-Newtonian flow characteristics.) A waste's rheological properties can make a nonsolid appear to be a solid, leading to disastrous results in a landfill. Equally important, these properties strongly affect the physical and mechanical design of a CFS system. The measurement of viscosity is described in Chapter 17.

Chemical Properties

Obviously, the chemical nature of a waste affects the choice of a CFS process for reasons other than fixation of specific constituents. The possible interactions between waste ingredients and system components are myriad, complex, and often not very well understood. Basically three types of chemical interactions between a waste and a CFS system can affect solidification: (1) inhibition, (2) acceleration, and (3) biological activity.

Good examples of both inhibition and acceleration occur in the Portland cement processes, and are readily understood because a great deal is known about the chemistry of Portland cement (Chapter 10). Examples of inhibitors in Portland cement setting and hardening include boric acid and borate salts, various organic compounds, silt, coal, lignite, and a number of inorganic salts and metal compounds. Accelerators for Portland cement setting and curing include lime, calcium chloride, and sodium silicate. Probably all of the processes will be affected, either positively or negatively, by some reactive waste constituents. All inhibitors, however, are not necessarily detrimental, nor are all accelerators desirable. It depends on how the system is to be used and upon the desired rate of setting and curing.

Biological activity in waste can cause problems in many CFS systems and must be considered separately because it is so unpredictable. For example, a biologically active organic sludge sometimes cycles between aerobic and anaerobic states, as evidenced by the evolution of hydrogen sulfide and by the obvious (by olfactory detection) presence of various organic sulfides. With certain systems, the waste will harden when in one state, but not in the other (or hardening will be greatly retarded). Usually the aerobic state is the most desirable, but not always.

Working Hazards

In applying a CFS system, the user must be aware of any health hazards associated with handling the waste and operating the system. Common hazards that occur are:

- Inhalation toxicity due to vapors
- Dermal toxicity
- Corrosiveness (both to humans and to equipment)
- Pyrophoric activity
- Exothermic reactions (especially occurring during the mixing of the chemicals with the waste)
- Infection from biological sludges (especially sewage sludge)
- Unpleasantness associated with foul-smelling waste

These problems can all be managed properly by using such personal protective equipment as goggles, masks, gloves, boots, disposable protective clothing, and inhalators, and by proper equipment design. A waste should first be thoroughly tested in a laboratory to determine the possibility of these hazards, especially those that may occur during the mixing or reaction of the waste with the chemicals or additives used in the CFS system. Information on health effects, reactive hazards, and compatibility of different wastes with each other and with the CFS reagents can be obtained from sources such as Sax [4] and Bretherick [5]. Obviously, to use this information it is necessary to know the composition of the waste; this again emphasizes the importance of *complete* waste characterization as the vital first step in choosing a CFS system.

THE WASTE SOURCE

A common mistake in evaluating CFS processes is to consider only the interaction of the waste, the process, and the disposal site, and to ignore problems associated with the generation or holding facility from which the waste will come. While these factors most often affect cost, sometimes they can dictate the choice of the CFS system, the disposal site, and even determine whether or not CFS technology is the proper approach. Waste generation factors to consider include:

- The holding facility: its size, age, condition, physical geometry, accessibility, availability of utilities, and climate
- Transport to CFS system: distance, topography, obstacles (railroad tracks, roads, pipelines, etc.), necessary delivery rate, type of transport system (pump, conveyor, pipeline, etc.)
- Process mode: continuous, semicontinuous, batch
- Effect of plant stoppage on CFS system
- Effect of CFS stoppage on the plant operation
- Waste removal system: dredge, pump, drag line, clam shell

This list is not all-inclusive, but it outlines areas to consider and also helps indicate the overall complexity of a total CFS system. Several of these areas deserve more detailed consideration.

Holding/Generating Facility

While more and more waste is being treated as generated, most CFS projects have been done, and are still being done, on already-generated waste contained in a wide variety of facilities. Operating plants usually hold or store wastes in lagoons, tanks, drums, pits, or piles. Remedial action jobs often involve dredging of contaminated sediments from waterways, lakes, and harbors, or excavation of contaminated soils. In many cases, the waste will be treated in situ, that is, without removing it from its present location. When the waste source is the newly generated stream, then the prime consideration is matching the treatment plant capacity to that of the source and making the necessary allowances for stoppage. When the source is a holding facility or remedial site, removal, homogenization, and transport become critical. In fact, in many cases this is the most difficult part of the project.

It is nearly impossible to describe in words the situations in which waste storage sometimes occurs. Even photographs, such as those in Figures 7-1 through 7-8, hardly do justice to the actual sites. Expecting a neat, well-defined lagoon such as

Figure 7-1 Photograph of a well-managed waste holding lagoon. (Photo by Fotographics of Elyria.)

that shown in Figure 7-1, those inexperienced in remedial work are often stunned when they first see an old waste disposal area (Fig. 7-2) or a waste pile full of debris and without any real delineation of the battery limits of the site (Fig. 7-3). Many disposal sites contain multiple pits and piles, with narrow or unstable berms that will not support equipment or even allow decent access. An example is shown in Figure 7-4. Not only is access difficult but each of these pits contained a different type of waste, some incompatible with others. Inadvertent breaching of a dividing wall or exterior berm might result in an environmental disaster, especially if the site is next to a waterway, which is often the case. And operating in poor weather conditions, a frequent necessity, greatly adds to the difficulty (Fig. 7-5).

Waste contained in facilities such as the clarifier shown in Figure 7-6 is often ideal, because the system is already designed to deliver sludge in pumpable or conveyable form. Tanks (Fig. 7-7), however, may be misleading because of stratification and hard settling inside. Access for homogenization equipment may be impossible, necessitating removal in phases. Even well set up and managed lagoons like the one shown in Figure 7-8 require agitation to homogenize the waste and provide a reasonably uniform feed to the CFS plant. When the facility is large, special equipment is required to do this without destroying the integrity of the lagoon, particularly the delicate synthetic liner.

Waste Removal

In remedial CFS work, homogenization of pumpable waste is the key to efficient processing. CFS is a chemical process, and requires a reasonably uniform feed to produce a uniform and predictable product. While it is not always possible to do

Figure 7-2 A poorly managed waste holding lagoon.

Figure 7-3 A poorly delineated waste pile.

Figure 7-4 A waste lagoon system with poor access and treatment areas.

Figure 7-5 Operation of a CFS remediation project under poor weather conditions.

Figure 7-6 An ideal waste holding facility: clarifiers and thickeners.

so, the objective should be to generate that uniform feed. This may be done by circulating the lagoon contents through pumps, as shown in Figure 7-9. When the lagoons are small and not too deep, and the sludge does not resettle too rapidly, this is an effective, economical method. It can be useful in depths up to 10–15 feet, as shown in Figure 7-10, if the waste is not too dense and viscous.

In removal of contaminated sediment from waterways or waste water settling

Figure 7-7 Waste contained in a tank.

Figure 7-8 A large, well-managed holding lagoon. (Photo by George W. Sommer.)

Figure 7-9 Homogenization of lagoon waste by pumping. (Print by Associated Photographers, Inc.)

Figure 7-10 Homogenization of lagoon waste by pumping in a deep lagoon. (Photo by George W. Sommer.)

ponds, small dredges with adjustable cutting heads such as that shown in Figure 7-11 are practical if there is not too much debris present. These machines are capable of removing the sediment with minimal impact on the overlying water layer. Sometimes, in remediation work where the lagoon or settling pond is to be eliminated, the best approach is to dredge and treat the sediment and then pump out the water for treatment and/or discharge. Large projects in lakes, bays, and open ocean usually require much larger dredges. A problem arises here in matching the output of the dredge to the CFS plant, since dredging rates are usually much higher than a cost effective process plant can handle. The solution is to create a holding lagoon to feed the CFS plant.

In addition to pumps, backhoes, draglines, clam shells, and other mechanical devices are used for large lagoons and high specific gravity wastes, including materials that are not pumpable. One such system used extensively by CFS contractors is shown in Figure 7-12. It consists of a double drum winch, cable, and floating pontoon raft with cutters suspended below the surface. The cutting head operates in both directions and can be adjusted up and down to the desired depth. The anchor points on the lagoon berms are moved as necessary to cover the lagoon completely. Many variations on this theme have been used, adapting to meet the specific situation in the field. In fact, imaginative adaptation, tempered with experience, is the key to effective waste removal and handling.

Sometimes it is better to treat different waste phases separately, using CFS

Figure 7-11 Homogenization and removal of waste from a holding facility by use of a floating dredge.

formulations and ingredients appropriate to each phase. This may be the case when the phases are not miscible, or separate too rapidly after mixing. One typical example is a lagoon with a floating oily layer, overlying a contaminated water layer, which in turn covers a bottom sludge layer.

In the case of contaminated soils and other solids, the wastes may be homogenized if the depth is less than about 5 feet, or if the waste pile can be spread out to this depth or less. Homogenization is done with machines like the one shown in

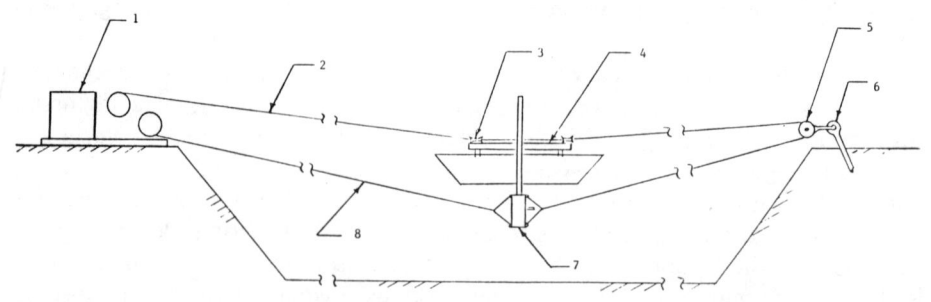

Figure 7-12 Schematic of a winch-operated dragline system for homogenization of waste in a lagoon. 1. Double drum winch. 2. Cable (top drum). 3. Cable guides. 4. Pontoon raft. 5. Snatch block (deadman). 6. Earth anchor. 7. Cutter/agitation head. 8. cable (bottom drum).

Figure 7-13 Auger-type sludge homogenization machine.

Figure 7-13, which function like a huge Rototiller. This type of device is also used in shallow sludge lagoons where it is often the most effective method. In most cases, however, solid wastes are treated as they are excavated, varying the formulation if required as the project proceeds. This is possible because the site can be grid sampled prior to treatment, and the processing system set up to match the changing, but known, waste composition.

Waste Transport

Once the waste is out of the lagoon, the removal system may also be used to transport it to the process plant. This is usually the case with pumpable materials. Nonpumpable wastes may be conveyed to the mixer via conveyor belt, or by batch loading with front loaders, clam shell cranes, or similar equipment. Either method can feed either continuous or batch processing modes of operation. If the waste must be transported a long way, this is usually done by truck. If this is the case, another dump pad, bin, or pit must be provided, along with a loading mechanism to feed the mixer. For this reason, it is desirable to locate the treatment plant close to the waste storage or holding area wherever possible. In any case, standard materials handling equipment and procedures can be used for waste handling, so a wide variety of options are available.

Process Mode

Processing can be either continuous or batch, with many possible variations of each. Actual delivery systems will be described in Chapter 16. The choice of mode depends on many factors: waste type, throughput rate, total project size, CFS process type,

THE TREATMENT PROCESS

The factors to consider in choosing a CFS process are given below.

- Chemical and operational cost
- Type of system—fixed or mobile/portable
- Capital write-off period (in the case of fixed systems)
- Required redundancy—for backup
- Space requirements
- Labor and personnel considerations
- Occupational health considerations (dust, noise, toxicity)
- Throughput rate
- Waste feed system
- Mixing system
- Chemical storage requirements
- Chemical feed system
- Utility and power requirements
- Setting and curing times
- Handling of solid after solidification
- Volume increase

Most of these factors will be discussed in subsequent chapters, and need no elaboration here. Others are covered later in this chapter. Cost, of course, is a primary consideration. It affects and is affected by the type of treatment system used, for example: whether it is fixed or portable; the necessary write-off period on purchased, fixed equipment; redundancy requirements that assure the plant or process line is not affected by the waste disposal operation; space and labor required to house and operate the unit; and occupational health considerations.

The CFS Process

The various commercial and developmental CFS systems are actually of three basic system types. Within the general systems are specific subcategories and specific processes.

Nonchemical Systems. The basic approaches to nonchemical solidification include dewatering, mixing with absorbents or absorption into solids, and vegetative stabilization. This approach is discussed here to provide a context in which to consider alternative methods of solidification, even though nonchemical solidification is no longer acceptable in the United States for bulk hazardous waste.

Dewatering can take many forms. Water may be removed using thermal drying equipment including incinerators (although this is an expensive technique). More commonly, dewatering is done by filtration or centrifugation. Sludges especially amenable to filtration are those that are not gelatinous in nature and that can be dewatered to less than about 75 percent water, although the water content of "solid" residues varies anywhere from near zero to 80 percent or more. While the waste treatment industry has no commonly accepted standard definition of "solid," a minimum of 1.0 ton per square foot unconfined compressive strength is, for many purposes, a usable specification.

Fluegas cleaning sludges, some dredge spoils, and certain industrial sludges, especially after being conditioned, can sometimes be dewatered by filtering to a suitable state for disposal as a solid. Most organic or organic-containing sludges, such as sewage sludge, cannot be adequately dewatered this way.

Settling basins can also be used for dewatering, either with subsequent removal of waste to a disposal site, or with basins operated in situ at the disposal site itself with recycling or discharge of the supernatent water. Certain wastes such as the dense particulate slurries from steel industry rolling mills settle and compact naturally and sufficiently so that the settled residue can be placed in landfill without further treatment. In a variation, by alternating layers of sand and sludge in a specially designed landfill, the weight of the overlying layers squeezes the water from the sludge into the permeable sand layers. Then it is removed by a suitable drain or pumping system. This technique is generally a very large-scale approach for sludges such as dredge spoils and usually does not produce a very stable, high load-bearing fill.

In certain areas of the United States, solar evaporation, or drying is practical. The climate must offer a substantial net evapotranspiration rate as compared to the precipitation rate in the area. By spreading or pouring the sludge or waste liquids into thin layers, usually several feet or less in thickness, much of the water can simply evaporate. The dried sludge may be left in place to build up a landfill area, or it may be plowed into the soil after drying, or scraped from the drying beds with earthmoving equipment, then deposited in a suitable landfill location. Other sludge dewatering technologies have been investigated, including solvent extraction, freeze-thaw cycling, and more exotic techniques, but to date these methods have found little commercial use in the United States.

Absorption or admixing of waste residues with dry solids is once again being investigated in the United States in the hope of finding less expensive methods for solidifying relatively nonleaching sludges, such as those produced from FGD. Absorbents range from material specifically produced to have high absorptive capacities to the admixing of liquid waste with other solid wastes such as municipal refuse. Perhaps the largest single absorptive medium used so far has been soil itself. For hazardous wastes, however, absorption has been eliminated as an acceptable solidification method, even when the waste is disposed in secure landfill.

Numerous industrial waste disposal operations, as well as sewage and mine waste disposal facilities, spray, spread, or plow their wastes into the soil to achieve a form of solidification. In certain cases, such as with sewage sludge, this technique is actually a treatment process. However, in most cases, the purpose is not to use the soil as a chemical or biological reactor, but simply as a water absorbent for the water phase of the liquid waste. On the other end of the scale, mineral absorbents, such as vermiculite, were used for many years for the drumming of radioactive wastes. Since no chemical reaction normally occurs between the waste and the absorbent, water so absorbed can in principle be squeezed from the waste–absorbent combination by application of physical pressure. In addition, such absorbent systems are highly leachable and are thus unsatisfactory for the final disposal of hazardous wastes.

A variation on absorption techniques is the combined process of using an absorbent to soak up water or chemisorb certain waste components such as metals or organics, and then to apply a chemical solidification technique. An example is that of chlorinated hydrocarbons that have been disposed of by first mixing them with coke or carbon absorbent and then solidifying that mixture with Portland cement. The solidified waste is subsequently disposed of by ocean dumping or in secured landfill. Certain clays are good sorbents for metals and some organics, and if dried before use, will absorb large quantities of water and result in a solid leach-resistant end product.

All absorption processes, however, must be very carefully controlled so that the absorbent is not oversaturated, thus producing an even larger quantity of sludge to be disposed of. Also, long-term stability under such environmental conditions as freeze–thaw cycling and wet–dry cycling must be established.

Vegetative stabilization, used for decades in the erosion control of consolidated soil structures, can also be used in cases of waste residues. The vegetative stabilization of lagoons can be effective if:

1. The waste contains no constituents incompatible with the vegetation to be grown.
2. The waste ideally contains some plant nutrients.
3. The original residue is high in solids content and sufficient land is available for this kind of use.

Many old sewage lagoons in this country have become naturally stabilized by the growth of vegetation on the ponds after they have been phased out of active use.

Much work has been done by the mining industry on the vegetative stabilization of mine tailings. In this case, because of the metal content in the tailings, such systems must be carefully managed with regard to how fertilizers will interact with the wastes and what kinds of vegetation are selected to be grown. Wheat, alfalfa, various grasses, and many shrubs have shown promise in this type of system. A very important consideration is how land so stabilized will ultimately be used. While the presence of contaminants such as heavy metals may not affect the vegetative cover to be grown, the vegetation nevertheless may take up the metals and concentrate them. Subsequent use of the vegetation for food crops or fodder may be environmentally unacceptable.

Organic Processes. With the exception of processes for the incorporation of waste residues into hot, liquid bitumen or gross encapsulation in polyethylene jackets, organic systems consist of monomers or prepolymers that are polymerized or cross-linked by the use of catalysts or accelerators after being mixed with the liquid waste. This kind of system is almost infinite in potential variety, but for practical purposes has been limited so far to those systems shown below.

Urea-formaldehyde
Polybutadiene
Polyester

Epoxy
Acrylamide gel
Polyolefin encapsulation
Bitumen (asphalt)

All kinds of combinations of the preceding and also of other organic polymer systems are conceivable. A main problem with organic systems is that many are solvent-based and are primarily hydrophobic in nature. For use on the water-based residuals that account for most of the disposal problem, the system usually must be formulated as an emulsion—not always easy to accomplish. In addition to these physical problems, the presence of water and other ingredients in the wastes often interferes with the polymerization reaction. These systems, however, could be useful in special situations for the solidification of certain organic waste streams, such as chlorinated hydrocarbons, and pesticides, which do not have a water base, assuming that these organics do not interfere with the solidification process.

Another difficulty with organic systems is that they may not be stable in the presence of such ambient environmental factors as microorganisms and ultraviolet light. In most cases, subsurface disposal eliminates exposure to ultraviolet light, but it increases the problem with microorganisms. Further, the presence of certain constituents in the waste can have long-term degradation effects on the polymer system, even though these may not show up immediately.

In most cases, the action of the organic polymer is primarily permeation limiting. Usually there is no direct reaction between the waste constituents and the polymer, nor does the system actually insolubilize, detoxify, or destroy the hazardous constituents. However, since these polymers generally have a very low order of permeability, they are more effective in this respect than most inorganic systems. Actually, the effect of most organic systems is microencapsulation of the waste material, which can be effective in separating the waste from its environment.

One of the technical advantages of many of the organic processes is that a given polymer can be applied to a wide variety of waste types, since direct chemical interaction between the polymer and the waste does not occur. This is helpful when dealing with a complex mixture of wastes or a wide variety of individual wastes in a given disposal situation.

Cost is a primary disadvantage of the organic processes. The urea-formaldehyde and bitumen processes are probably the least expensive, but are still costlier than the currently used inorganic processes. In fact, most organic processes at this time are of little interest commercially except in the solidification of extremely hazardous materials, such as radioactive wastes, and perhaps for hazardous organics that cannot be thermally destroyed. Organic systems are not likely to find widespread use in any high-volume waste disposal situation.

Inorganic Processes. Generic inorganic processes that have been used or proposed are listed below.

Portland cement
Portland cement–sodium silicate
Portland cement–clay
Portland cement–lime
Portland cement–flyash

Lime–pozzolan (lime flyash)
Cement kiln dust; lime kiln dust
Lime–clay
Lime
Calcium aluminate cement
Gypsum
Pozzolan–sodium silicate
Soil–lime

The most important systems at the moment involve either Portland cement alone (or with minor additives), cement or lime kiln dust, Portland cement–flyash, lime-flyash (or other pozzolan), lime-based, and Portland cement–sodium silicate. Portland cement has probably been the most widely used inorganic ingredient from the standpoint of diversity of use, especially when combined with sodium silicate or flyash. Lime–flyash and the kiln dust processes have seen the highest volume use for fluegas cleaning sludges and for solidification of liquid wastes, both in on-site remediations and in meeting the 1985 ban on hazardous liquids in landfills. Various combinations of lime and clay, sand, or Portland cement have been used for specific wastes throughout the world.

A problem with most inorganic systems is that they are often specific for a particular waste under a particular set of conditions. Unfortunately, most concentrated industrial wastes are more complex than specific. This is either because of the processes from which they are produced or because they are combined with other wastes in the waste storage area. The high content of dissolved inorganic ions or ion-formers often mixed with oils and other organic materials is probably the most serious problem a waste treatment process must overcome. The process chemistry of most of the inorganic processes is sensitive to the presence of those components that can act as setting retardants or, in some cases, can completely prevent solidification.

Another problem is that many processes require such large additions of solidification agents that volume increases significantly; this can affect both the technology and the cost of final disposal. However, in general, the inorganic systems are lower in cost than organic systems. The characteristics of inorganic systems are summarized as follows.

- Relatively low cost
- Good long-term stability, both physically and chemically
- Documented use on a variety of industrial wastes over a period of at least 10 years
- Widespread availability of the chemical ingredients
- Nontoxicity of the chemical ingredients
- Ease of use in processing (processes normally operate at ambient temperature and pressure and without unique or very special equipment)
- Relatively low volume increases with some of the processes
- Inertness to ultraviolet radiation
- High resistance to biodegradation
- Low water solubility
- Relatively low water permeability
- Good mechanical and structural characteristics

Design Parameters

Specific design parameters to determine before selecting a system include:

- Throughput rate of waste
- Type of mixing equipment required (which is a function of both the type of waste and also of the CFS system itself)
- Optimum size of the chemical storage required for the system
- Design of the chemical feed system (especially the accuracy required)
- Matching of the chemical feed system with the waste feed system
- Utility and power requirements for the unit

In a continuous system, the throughput rate is a function of the maximum output rate of the plant unless sufficient surge and holding capacity can be installed to average out the rate of waste being treated, and thus not require maximum-capacity design for the system. Surge capacity is also important in the case of dredging operations, where the output rate of the dredge is substantially higher than the input rate of the CFS system. If dredging is hired as an outside service, the cost per gallon of material dredged will be higher than normal if surge capacity is not provided.

One of the most critical elements of the mechanical design is the mixing system. There is a very large variety of mixer types and brands, and selection is heavily dependent on the waste to be treated and the CFS process used. Mixers and mixing will be examined in Chapter 9 and also as we discuss individual processes.

Requirements for chemical storage depend both upon the throughput rate and the chemical pricing and delivery situation in a specific locality. For example, with inorganic systems such as Portland cement, kiln dusts, and flyash, it is highly desirable to use tank truck or railroad car quantities to take advantage of sizable cost savings. This will determine the chemical storage required, although other factors, such as local delivery and possible shortage problems, must also be considered.

The accuracy of the chemical feed system, especially with dry chemicals, is important since the more accurate the system the higher the capital cost will be. This may be balanced out, however, in chemical cost savings by being able to better control the chemical feed rates. The chemical feed system must be matched accurately with the waste feed system and rated for the full range of waste throughput rates.

Power and water requirements become especially critical with mobile/portable operations in remote locations. Motor-generator units will be necessary in many instances, and fuel supplies must be provided for. Water for both process and cleanup requirements must be supplied at sufficient rate. In some locations, it may have to be trucked in.

Type of Delivery System

Chapter 16 explores the two categories of delivery system, fixed and mobile/portable. There are substantial differences in how these two types of installation operate, and where each is indicated. Fixed systems are normally used at waste-generating sites where the waste stream is both large and constant. However, there are many exceptions to this rule, based on economics, regulatory situations, and generator preference. Mobile/portable systems may be either of the throughput variety, or in situ. Throughput systems may be truly mobile—that is, on wheels—or portable, as in skid-mounted and site-erected units. Systems requiring extensive site erection and assembly are generally referred to as *transportable*.

Solids Handling

Once waste has been treated and solids produced, a solidification area must be provided. Its size will depend upon the solidification time required. In general, the longer the setting and/or curing time, the larger the temporary holding area must be. This may determine which system and throughput rate to choose. A major decision in system design is whether the treated waste will be allowed to solidify in its final resting place, or be transferred to a permanent disposal site after solidification in a temporary curing area or container. In principle, the former is usually more desirable because it results in lower handling costs, higher structural strength of the solidified material due to its low void structure, and better landfill space use for the same reason. However, other considerations usually take precedence over this factor in determining whether transfer of the solidified material is necessary. It is nevertheless one of the most important factors in analyzing the whole system design, and its cost.

Volume Increase

Finally, the question of volume (and weight) increase associated with a particular CFS process must be considered. This factor, one of the most important, is frequently neglected entirely in selecting a CFS system. This is probably because much CFS work has been done in on-site situations where the cost or value of the landfill space is either not considered or is believed, incorrectly, to be insignificant when choosing a system.

In the future, volume/weight increase may become the most important single factor in choosing the system, especially when the solidified waste must be placed in a secured or semisecure chemical landfill. The value of landfill disposal space, especially for hazardous waste, has been rising at a very rapid rate and this trend is likely to continue. Difficulties in obtaining new disposal sites is behind these escalating landfill costs. In fact, it is not unusual to find that the cost of transportation and disposal of the solidified material is higher, sometimes much higher, than the cost of running the solidification process. The hauler usually charges by weight, and the landfill site operator charges either by weight or volume. In any case, the final quantity of solidified material is what counts. (In solidified wastes, the weight/volume relationship normally falls within a narrow range.)

In some instances a large increase is acceptable or even desirable; for example, where the solidified material has sufficient structural strength to give it positive value as "clean" landfill. In situations where waste products are used as part of the chemical system, volume increase considerations become part of a complex relationship between the alternate disposal methods for the various waste streams involved.

Volume/weight increase is especially important for large-scale CFS projects. In such projects, the chemical, transportation and landfill costs account for most of a CFS system's operating expenses, comprising 75 percent or more of the total cost. These relationships will be examined in detail in Chapter 9 for three generic processes with two different general waste types: low-solids wastes, such as sludges and slurries (5–15 percent total solids), and high-solids sludges, filter ashes, etc. (25–50 percent solids). In most cases, costs tend to converge as landfill costs decrease, because the larger volume/weight increase of processes such as cement–flyash are offset by their lower unit chemical costs, and there is less penalty in terms of landfill and transportation costs.

An exception to this convergence is the case of Portland cement–soluble silicate systems, which were deliberately designed to treat low-solids waste streams. This exception illustrates the importance of having a CFS system custom-designed for a particular waste stream. Only then can optimum cost effectiveness be achieved.

THE DISPOSAL SITE

Most solidified wastes are disposed of in landfills, although there are exceptions, such as when they are stored for possible future recovery, used for road or other structural base, or ocean dumped. Landfills may range from so-called "clean" structural landfills, through municipal and private sanitary landfills designed for domestic refuse, to the highly secured landfills necessary for hazardous material. An example of the latter is shown in the photograph in Figure 7-14 while the landfill cell is being filled. Note the synthetic liner on the sides of the cell. The design of such landfills has become increasingly sophisticated in recent years. The state of the art is shown in the schematic diagram of Figure 7-15.

Some solidified waste may be used in direct water contact applications, such

Figure 7-14 Secure landfill cell being filled.

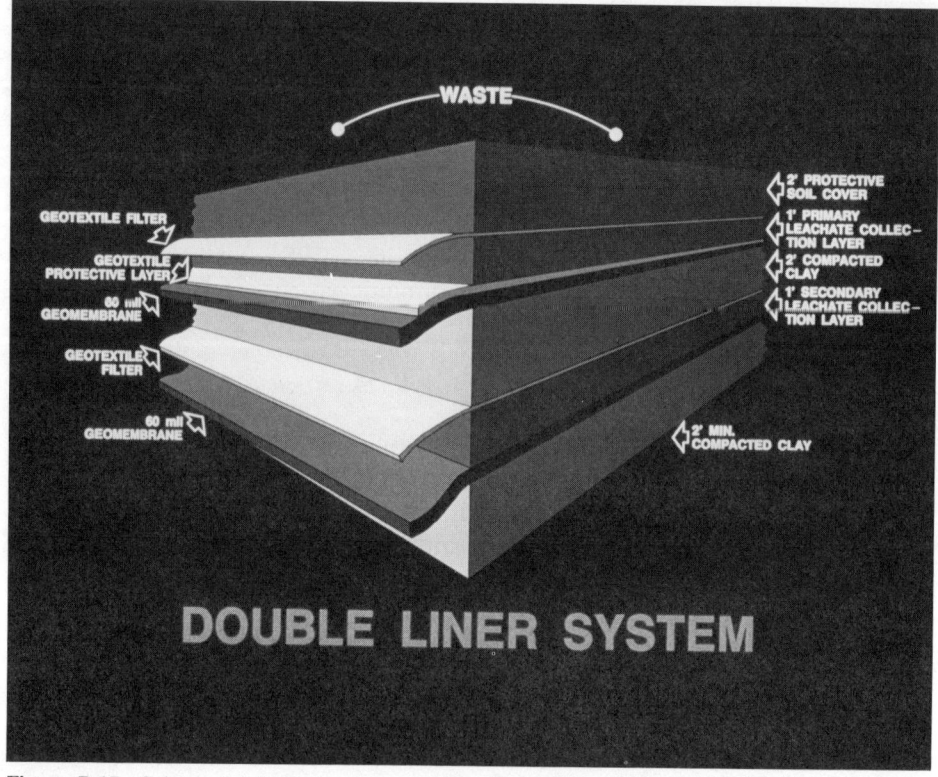

Figure 7-15 Schematic diagram of state-of-the-art design for secure landfill.

as for diking material, and for forming new land from lakes, streams, marine waterways, or low-lying swamp areas. In most cases, however, CFS-processed solidified waste is placed in landfills primarily used for waste disposal purposes. In such cases, the following factors should be considered:

- Distance from the waste source
- Location with reference to industrial and residential areas
- Geology and hydrology
- Available capacity
- Disposal cost per ton or cubic yard of waste disposed
- Overall environmental impact on the surrounding land, water, and air
- Attitude of neighbors
- Land use regulations (zoning, etc.)
- Political considerations
- Liability and insurance consideration

The whole subject of solid waste disposal sites, from choice of location to operation of the site, is a complex one dealing not only with technical, engineering, and cost factors, but also with the intangible political, public, and legal ones that are becoming increasingly sensitive. This book cannot address the subject in any detail, but it is important for those involved in CFS technology to recognize both the technical and nontechnical issues that affect the choices available to them. Several references on this subject are given in the Bibliography. At the moment, among the biggest problems in developing disposal sites are public reaction, financial responsibility, and liability.

Both the financial responsibility and liability issues are especially troublesome because insurance for waste disposal operations and facilities is difficult and expensive to obtain. Therefore, both the waste generator and the CFS processor must be as certain as possible that the waste treatment system has been properly chosen for the particular disposal site or, alternately, that the disposal site chosen is compatible with the characteristics of the treated waste.

The environmental impact of the solidified and/or chemically fixed waste is closely related to the disposal site or situation. There are two primary factors associated with environmental impact: leaching/chemical stability and land use. The reasons for using a CFS system at all generally come down to one or both of these factors.

COMMERCIAL CONSIDERATIONS AND COSTS

Cost is usually the final determining factor in choosing a treatment process and should therefore be kept in mind during every aspect of the decision-making process. The number of factors, variables, parameters, and considerations discussed in this chapter make it obvious that choosing a system is not a simple matter, and considerations of cost cannot be deferred until the technical and engineering choices have been made.

As with any other complex system, it is not possible to give flat costs or selling prices for a specific CFS process any more than it is possible for an engineering construction firm to bid on a water treatment plant without knowing the design parameters of that plant.

For purposes of comparison, price ranges for various commercial treatment and disposal alternatives, not including transportation, are given in Table 7-1 [2]. The large price spreads are due both to varying prices for different waste types, and to quantity pricing. In comparing costs and prices, it is important to determine exactly what operations are included; for example, dredging, chemical processing, transport, disposal. Typical cost ranges for the various operations involved in a complete disposal system based on CFS are given in Table 7-2 [2]. As indicated, transportation and landfill costs have a very wide range. Transportation cost is primarily dependent on distance, and can be readily determined on a "typical case" basis. Landfill cost, on the other hand, depends on many factors and can be accurately determined only by specific quote. Dredging costs are site-specific, and highly volume-dependent.

TABLE 7-1 Costs of Commercial Treatment and Disposal Alternatives

Method	Price ($/gallon)
CFS (treatment only)	0.10–0.75
Direct landfill or land spreading	0.05–0.30
Secure landfill (RCRA facility)	0.25–1.50
Incineration	0.25–3.00+
Ocean dumping	0.01–0.10

TABLE 7-2 Typical Costs of CFS Unit Operations

Operation	Cost ($/gallon)
Dredging	0.005–0.10
CFS (nonnuclear)	0.15 –0.75
Transportation	0.01 –1.00
Landfill	
Nonsecure	0.05 –0.30
Secure	0.25 –1.50

REFERENCES

1. Conner, J. R. *The Modern Engineered Approach to Chemical Fixation and Solidification Technology.* Chemfix Inc., Pittsburgh, 1981.
2. Conner, J. R. Fixation and solidification of wastes. *Chem. Eng.* 79–85 (Nov. 10, 1986).
3. Conner, J. R. Use of computer-based intelligent systems to select detoxification and solidification methods for residues and wastes. In *Proc. 5th International Symposium on Environmental Pollution.* Quebec City, P. Q., Canada, June 27, 1985.
4. Sax, I. R. *Dangerous Properties of Industrial Materials.* New York: Van Nostrand Reinhold, 1979.
5. Bretherick, L. *Handbook of Reactive Chemical Hazards.* Cleveland, OH: CRC Press, 1975.

Chapter 8

WASTES AND WASTE SOURCES

Confusion has always existed between treatment processes for dilute waste streams and those for concentrated residues. The primary application of CFS technology has been, and is, to concentrate residues. Dilute waste streams, usually waste waters, require other forms of treatment, or "purification," to allow discharge or reuse of the water. However, most purification is simply a way of concentrating the constituents. The further we purify these streams, the more residue we create. Therefore, even with the strong emphasis on waste reduction and concentration today, residues are continually being created from new sources.

Up until now, we have been discussing wastes, primarily hazardous wastes, without really defining them in any orderly and scientific manner. In Chapter 7, we discussed the properties and characteristics of wastes that categorize them as hazardous or otherwise of other interest in CFS technology. These attributes are hazardous constituents or properties that we would like to eliminate or neutralize, or other physical characteristics of importance in selecting a CFS process. Chapters 3–6 examined the properties of wastes that affect the fixation of hazardous constituents and the constituents themselves. Now, we will explore the wastes and where they come from. To do this in an orderly and rational manner, it is necessary to classify and categorize wastes by source, properties, regulatory framework, and generation rate.

WASTE CLASSIFICATION SYSTEMS

There are many systems that could be used to classify wastes and waste sources. Only two, however, have seen any significant use in the United States: the U.S. EPA (EPA) hazardous waste listing system of May 19, 1980 [1] and a system created by the University of California at Davis. The latter system has been used only in California and Ontario, Canada. The EPA listing system applies only to four groups of wastes determined by the EPA to require the most control because of their quantities and hazard potential to humans and the environment. The groups used by the EPA are (1) process specific (Section 261.32 [1]), (2) nonspecific (Section 261.31 [1]), (3) acutely hazardous discarded chemicals [Section 261.33(e)], and (4) hazardous discarded chemicals [Section 261.33(f)]. Any waste listed in this way, or specifically controlled under the TSCA [2] (for example, PCBs), or controlled under the various landbans, is hazardous by definition. Any other waste is determined to be hazardous only if it exhibits four basic criteria at levels above the limits set by the EPA. The criteria and limits are shown in Table 8-1 [1].

Interestingly, if a waste is not listed, the presence of hazardous constituents other than those specified in Table 8-1 does not make a waste hazardous. This is not an oversight by the EPA, but rather the result of insufficient information to set regulatory limits.* This deficiency is now being corrected in the landbans, in the CCW and CCWE tables (see description later in this chapter), and in the expanded list of constituents in the proposed rule utilizing the new TCLP test [3] to replace the present EPT for characterization of wastes as hazardous or nonhazardous. The new rule would add the 31 new organic constituents shown in Table 8-2.

U.S. EPA Listing Classification System

The EPA system divides wastes, other than discarded chemicals, into two groups: specific or "K-wastes," and nonspecific or "F-wastes." The listings in the original rule have been modified over the years, with some wastes being "delisted" and new ones added. In addition, Sections 261.33(e) and (f) of the latest revision [4] list discarded chemicals, otherwise known as the "P" and "U" wastes, that are also regulated. The latest lists for the four categories are given below. Each category has two or three groups, separated in the lists by extra space. This division is useful to

TABLE 8-1 Hazardous Characteristics of Wastes[a]

Ignitability (261.21) (D001)
Reactivity (261.22) (D002)
Corrosivity (261.23) (D003)
Toxicity (261.24)

Substance	Maximum Concentration (mg/l)
Arsenic (D004)	5.0
Barium (D005)	100.0
Cadmium (D006)	1.0
Chromium (D007)	5.0
Lead (D009)	5.0
Mercury (D009)	0.2
Selenium (D010)	1.0
Silver (D011)	5.0
Endrin (D012)	0.02
Lindane (D013)	0.4
Methoxychlor (D014)	10.0
Toxaphene (D015)	0.5
2,4-D (D016)	10.0
2,4,5-TP Silvex (D017)	1.0

[a] Taken from Reference [1]. Numbers in parentheses refer to section in the reference, or to the EPA hazardous waste number for that characteristic or constituent.

---

*Regulatory limits for the constituents stated in Table 8-1 were set based on the U.S. Interim Drinking Water Standards in effect at the time. No such standards existed for other hazardous constituents.

TABLE 8-2 Toxicity Characteristic Contaminants and Regulatory Levels

HWNO	Contaminant	CASNO	Regulatory Level (mg/l)
D004	Arsenic	7440-38-2	5.0
D005	Barium	7440-39-3	100
D019	Benzene	71-43-2	0.07
D006	Cadmium	7440-43-9	1.0
D021	Carbon disulfide	75-15-0	14.4
D022	Carbon tetrachloride	56-23-5	0.07
D023	Chlordane	57-74-9	0.03
D024	Chlorobenzene	108-90-7	1.4
D025	Chloroform	67-66-3	0.07
D007	Chromium	1333-82-0	5.0
D026	o-Cresol[a]	95-48-7	10.0
D027	m-Cresol[a]	108-39-4	10.0
D028	p-Cresol[a]	106-44-5	10.0
D016	2,4-D	94-75-7	1.4
D029	1,2-Dichlorobenzene	95-50-1	4.3
D030	1,4-Dichlorobenzene	106-46-7	10.8
D031	1,2-Dichloroethane	107-06-2	0.40
D032	1,1-Dichloroethylene	75-35-4	0.1
D033	2,4-Dinitrotoluene	121-14-2	0.13
D012	Endrin	72-20-8	0.003
D034	Heptachlor (and its hydroxide)	76-44-8	0.001
D035	Hexachlorobenzene	118-74-1	0.13
D036	Hexachlorobutadiene	87-68-3	0.72
D037	Hexachloroethane	67-72-1	4.3
D038	Isobutanol	78-83-1	36
D008	Lead	7439-92-1	5.0
D013	Lindane	58-89-9	0.06
D009	Mercury	7439-97-6	0.2
D014	Methoxychlor	72-43-5	1.4
D040	Methyl ethyl ketone	78-93-3	7.2
D041	Nitrobenzene	98-95-3	0.13
D042	Pentachlorophenol	87-86-5	3.6
D043	Phenol	108-95-2	14.4
D044	Pyridine	110-86-1	5.0
D010	Selenium	7782-49-2	1.0
D011	Silver	7440-22-4	5.0
D047	Tetrachloroethylene	127-18-4	0.1
D048	2,3,4,6-Tetrachlorophenol	58-90-2	1.5
D049	Toluene	108-88-3	14.4
D015	Toxaphene	8001-35-2	0.07
D052	Trichloroethylene	79-01-6	0.07
D053	2,4,5-Trichlorophenol	95-95-4	5.8
D054	2,4,6-Trichlorophenol	88-06-2	0.30
D017	2,4,5-TP (Silvex)	93-76-5	0.14
D055	Vinyl chloride	75-01-4	0.05

[a] o-, m-, and p-Cresol concentrations are added together and compared to a threshold of 10.0 mg/l.

know, since it is the timetable for the implementation for the last of the landbans. The first group was to have been implemented by August 8, 1988; land disposal rules for some of the wastes were implemented on that date, but others were either temporarily exempted or allowed to fall under the "soft hammer" provisions (see Chapter 1). The last two groups are scheduled for implementation on June 8, 1989, and May 8, 1990, respectively.

EPA F-Wastes
§261.31 Wastes
F006—Wastewater treatment sludges from electroplating operations except from the following processes: (1) Sulfuric acid anodizing of aluminum; (2) tin plating on carbon steel; (3) zinc plating (segregated basis) on carbon steel; (4) aluminum or zinc-aluminum plating on carbon steel; (5) cleaning/stripping associated with tin, zinc and aluminum plating on carbon steel; and (6) chemical etching and milling of aluminum.
F007—Spent cyanide plating bath solutions from electroplating operations.
F008—Plating bath sludges from the bottom of plating baths from electroplating operations where cyanides are used in the process.
F009—Spent stripping and cleaning bath solutions from electroplating operations where cyanides are used in the process.
F019—Wastewater treatment sludges from the chemical conversion coating of aluminum.

F010—Quenching bath sludge from oil baths from metal heat treating operations where cyanides are used in the process.
F011—Spent cyanide solutions from salt bath pot cleaning from metal heat treating operations.
F012—Quenching wastewater treatment sludges from metal heat operations where cyanides are used in the process.
F024—Wastes including but not limited to, distillation residues, heavy ends, tars and reactor clean-out wastes from the production of chlorinated aliphatic hydrocarbons, having carbon content from one to five, utilizing free radical catalyzed processes. [This listing does not include light ends, spent filters and filter aids, spent desiccants, wastewater, wastewater treatment sludges, spent catalysts, and wastes listed in §261.32.].

EPA K-Wastes
§261.32 Wastes
K001—Bottom sediment sludge from the treatment of wastewaters from wood preserving processes that use creosote and/or pentachlorophenol.
K004—Wastewater treatment sludge from the production of zinc yellow pigments.
K008—Over residue from the production of chrome oxide green pigments.
K011—Bottom stream from the wastewater stripper in the production of acrylonitrile.
K013—Bottom stream from the acetonitrile column in the production of acrylonitrile.
K014—Bottoms from the acetonitrile purification column in the production of acrylonitrile.
K015—Still bottoms from the distillation of benzyl chloride.
K016—Heavy ends or distillation residues from the production of carbon tetrachloride.
K017—Heavy ends (still bottoms) from the purification column in the production of epichlorohydrin.
K018—Heavy ends from the fractionation column in ethyl chloride production.
K020—Heavy ends from the distillation of vinyl chloride in vinyl chloride monomer production.
K021—Aqueous spent antimony catalyst waste from fluoromethanes production.
K022—Distillation bottom tars from the production of phenol/acetone from cumane.
K024—Distillation bottoms from the production of phthalic anhydride from naphthalene.

K030—Column bottom or heavy ends from the combined production of trichloroethylene and perchloroethylene.
K031—By-products salts generated in the production of MSMA and cacodylic acid.
K035—Wastewater treatment sludges generated in the production of creosote.
K036—Still bottoms from toluene reclamation distillation in the production of disulfoton.
K037—Wastewater treatment sludge from the production of disulfoton.
K044—Wastewater treatment sludges from the manufacturing and processing of explosives.
K045—Spent carbon from the treatment of wastewater containing explosives.
K046—Wastewater treatment sludges from the manufacturing, formulation and loading of lead-based initiating compounds.
K047—Pink/red water from TNT operations.
K048—Dissolved air flotation (DAF) float from the petroleum refining industry.
K049—Stop oil emulsion solids from the petroleum refining industry.
K050—Heat exchange bundle cleaning sludge from the petroleum refining industry.
K051—API separator sludge from the petroleum refining industry.
K052—Tank bottoms (leaded) from the petroleum refining industry.
K060—Ammonia still lime sludge from coking operations.
K061—Emission control dust/sludge from the primary production of steel in electric furnaces.
K062—Spent pickle liquor from steel finishing operations in chlorine production.
K069—Emission control dust/sludge from secondary lead smelting.
K071—Brine purification muds from the mercury cells process in chlorine production, where separately prepurified brine is not used.
K073—Chlorinated hydrocarbon waste from the purification step of the diaphragm cell process using graphite anodes
K083—Distillation bottoms from aniline production.
K084—Wastewater treatment sludges generated during the production of veterinary pharmaceuticals from arsenic or organo-arsenic compounds.
K085—Distillation of fractionation column bottoms from the production of chlorobenzenes.
K086—Solvent washes and sludges; caustic washes and sludges, or water washes and sludges from cleaning tubs and equipment used in the formulation of ink from pigments, driers, soaps, and stabilizers containing chromium and lead.
K087—Decanter tank tar sludge from coking operations.
K099—Untreated wastewater from the production of 2,4-D.
K101—Distillation tar residues from the distillation of aniline-based compounds in the production of veterinary pharmaceuticals from arsenic or organo-arsenic compounds.
K102—Residue from the use of activated carbon for decolorization in the production of veterinary pharmaceuticals from arsenic or organo-arsenic compounds.
K103—Process residues from aniline extraction from the production of aniline.
K104—Combined wastewater streams generated from nitrobenzene/aniline production.
K106—Wastewater treatment sludge from the mercury cell process in chlorine production.

K009—Distillation bottoms from the production of acetaldehyde from ethylene.
K010—Distillation side cuts from the productions of acetaldehyde from ethylene.
K019—Heavy ends from the distillation of ethylene dichloride in ethylene dichloride production.

K025—Distillation bottoms from the production of nitrobenzene by the nitration of benzene.
K027—Centrifuge and distillation residues from toluene diisocyanate production.
K028—Spent catalyst from the hydrochlorinator reactor in the production of 1,1,1-trichloroethane.
K029—Waste from the product steam stripper in the production of 1,1,1-trichloroethane.
K038—Wastewater from the washing and stripping of phorate production.
K039—Filter cake from the filtration of diethylphosphoro-dithioic acid in the production of phorate.
K040—Wastewater treatment sludge from the production of phorate.
K041—Wastewater treatment sludge from the production of toxaphene.
K042—Heavy ends or distillation residues from the distillation of tetrachlorobenzene in the production of 2,4,5-T.
K043—2,6-Dichlorophenol waste from the production of 2,4-D.
K095—Distillation bottoms from the production of 1,1,1-trichloroethane.
K096—Heavy ends from the heavy ends column from the production of 1,1,1-trichloroethane.
K097—Vacuum stripper discharge from the chlordane chlorinator in the production of chlordane.
K098—Untreated process wastewater from the production of toxaphene.
K105—Separated aqueous stream from the reactor product washing step in the production of chlorobenzenes.

K002—Wastewater treatment sludge from the production of chrome yellow and orange pigments.
K003—Wastewater treatment sludge from the production of molybdate orange pigments.
K005—Wastewater treatment sludge from the production of chrome green pigments.
K006—Wastewater treatment sludge from the production of chrome oxide green pigments (anhydrous and hydrated).
K007—Wastewater treatment sludge from the production of iron blue pigments.
K023—Distillation light ends from the production of phthalic anhydride from naphthalene.
K026—Stripping still tails from the production of methyl ethyl pyridines.
K032—Wastewater treatment sludge from the production of chlordane.
K033—Wastewater and scrub water from the chlorination of cyclopentadiene in the production of chlordane.
K034—Filter solids from the hexachlorocyclopentadiene in the production of chlordane.
K093—Distillation light ends from the production of phthalic anhydride from ortho-xylene
K094—Distillation bottoms from the production of phthalic anhydride from ortho-xylene.
K100—Waste leaching solution from acid leaching of emission control dust/sludge from secondary lead smelting.

EPA P-Wastes
§261.33(e) Wastes
P001—Warfarin, when present at concentration greater than 0.3%
P004—Aldrin
P005—Allyl alcohol
P010—Arsenic acid
P011—Arsenic (V) oxide

P012—Arsenic (III) oxide
P015—Beryllium dust
P016—Bis-(chloromethyl) ether
P018—Brucine
P020—Dinoseb
P030—Soluble cyanide salts not elsewhere specified
P036—Dichlorophenylarsine
P037—Dieldrin
P039—Disulfoton
P041—Diethyl-p-nitrophenyl phosphate
P048—2,4-Dinitrophenol
P050—Endosulfan
P058—Fluoracetic acid, sodium salt
P059—Heptachlor
P063—Hydrogen cyanide
P068—Methyl Hydrazine
P069—Methyllactonitrile
P070—Aldicarb
P071—Methyl parathion
P081—Nitroglycerine
P082—N-Nitrosodimethylamine
P084—N-Nitrosomethylvinylamine
P087—Osmium tetraoxide
P089—Parathion
P092—Phenylmercuric acetate
P094—Phorate
P097—Famphur
P102—Propargyl alcohol
P105—Sodium azide
P108—Strychnine and salts
P110—Tetraethyl lead
P115—Thallium (I) sulfate
P120—Vanadium pentoxide
P122—Zinc phosphide, when present at concentrations greater than 10%
P123—Toxaphene
P002—1-Acetyl-2-thiourea
P003—Acrolein
P007—5-(Aminoethyl)-3-isoxazolol
P008—4-Aminopyridine
P014—Thiophenol
P026—1-(o-Chlorophenyl)thiourea
P027—Propanenitrile, 3-chloro
P029—Copper cyanides
P040—O,O-Diethyl o-pyrazinyl phosphorothioate
P043—Diisopropyl fluorophosphate
P044—Dimethoate
P049—2,4-Dithiobiuret
P054—Aziridine
P057—Fluoracetamide
P060—Isodrin
P062—Hexaethyltetraphosphate
P066—Methomyl
P067—2-Methylaziridine

P072—Alpha-naphthylthiourea (ANTU)
P074—Nickel cyanide
P085—Octamethylpyrophosphoramide
P098—Potassium cyanide
P104—Silver cyanide
P106—Sodium cyanide
P107—Strontium sulfide
P111—Tetraethylpyrophosphate
P112—Tetranitromethane
P113—Thallic oxide
P114—Thallium (I) selenite

P006—Aluminum phosphide
P009—Ammonium picrate
P013—Barium cyanide
P017—Bromoacetone
P021—Calcium cyanide
P022—Carbon disulfide
P023—Chloroacetaldehyde
P024—p-Chloroaniline
P028—Benzyl chloride
P031—Cyanogen
P033—Cyanogen chloride
P034—4,6-Dintro-o-cyclohexylphenol
P038—Diethylarsine
P042—Epinephrine
P045—Thiofanox
P046—Alpha, alpha-Dimethylphenethylamine
P047—4,6-Dinitro-o-cresol and salts
P051—Endrin
P056—Fluorine
P064—Methyl isocyanate
P065—Mercury fulminate
P073—Nickel carbonyl
P075—Nicotine and salts
P076—Nitric oxide
P077—p-Nitroaniline
P078—Nitrogen dioxide
P088—Endothall
P093—N-Phenylthiourea
P095—Phosgene
P096—Phosphine
P099—Potassium silver cyanide
P101—Propanenitrile
P103—Selenourea
P109—Tetraethyldithiopyrophosphate
P116—Thiosemicarbazide
P118—Trichloromethanethiol
P119—Ammonium vanadate
P121—Zinc cyanide

EPA U-Wastes
§261.33(f) Wastes
U007—Acrylamide
U009—Acrylonitrile

U010—Mitomycin C
U012—Aniline
U016—Benz(c)acridine
U018—Benz(a)anthracene
U019—Benzene
U022—Benzo(a)pyrene
U029—Methyl bromide
U031—n-Butanol
U036—Chlordane, technical
U037—Chlorobenzene
U041—n-Chloro-2,3-epoxypropane
U043—Vinyl chloride
U044—Chloroform
U046—Chloromethyl methyl ether
U050—Chrysene
U051—Creosote
U053—Crotonaldehyde
U061—DDT
U063—Dibenz o (a, h) anthracene
U064—1,2:7,8 Dibenzopyrene
U066—Dibromo-3-chloropropane 1,2-
U067—Ethylene dibromide
U074—1,4-Dichloro-2-butene
U077—Ethane, 1,2-dichloro-
U078—Dichloroethylene, 1,1-
U086—N,N Diethylhydrazine
U089—Diethylstilbestrol
U103—Dimethyl sulfate
U105—2,4-Dinitrotoluene
U108—Dioxane, 1,4-
U115—Ethylene oxide
U122—Formaldehyde
U124—Furan
U129—Lindane
U130—Hexachlorocyclopentadiene
U133—Hydrazine
U134—Hydrofluoric acid
U137—Indeno(1,2,3-cd)pyrene
U151—Mercury
U154—Methanol
U155—Methapyrilene
U157—3-Methylcholanthrene
U158—4,4-Methylene-bis-(2-chloroaniline)
U159—Methyl ethyl ketone
U171—Nitropropane, 2-
U177—N-Nitroso-N-methylurea
U180—N-Nitrosopyrrolidine
U185—Pentachloronitrobenzene
U188—Phenol
U192—Pronamide

U101—Dimethylphenol, 2,4-
U106—Dinitrotoluene, 2,6-
U107—Di-n-octyl phthalate

WASTES AND WASTE SOURCES

U109—1,2,-Diphenylhydrazine
U110—Dipropylamine
U111—Di-N-Propylnitrosamine
U114—Ethylenebis-(dithiocarbamic acid)
U116—Ethylene thiourea
U119—Ethyl methanesulfonate
U127—Hexachlorobenzene
U128—Hexachlorobutadiene
U131—Hexachloroethane
U135—Hydrogen sulfide
U138—Methyl iodide
U140—Isobutyl alcohol
U142—Kepone
U143—Lasiocarpine
U144—Lead acetate
U146—Lead subacetate
U147—Maleic anhydride
U149—Malononitrile
U150—Melphalan
U161—Methyl isobutyl ketone
U162—Methyl methacrylate
U163—N-Methyl-N-nitro-N-nitrosoguanidine
U164—Methylthiouracil
U165—Naphthalene
U168—Napthylamine, 2-
U169—Nitrobenzene
U170—p-Nitrophenol
U172—N-Nitroso-di-n-butylamine
U173—N-Nitroso-diethanolamine
U176—N-Nitroso-N-ethylurea
U178—N-Nitroso-N-methylurethane
U179—N-Nitrosopiperidine
U189—Phosphorus sulfide
U193—1,3-Propane sultone
U196—Pyridine
U203—Safrole
U205—Selenium disulfide
U206—Streptozotocin
U208—Terachloroethane, 1,1,1,2-
U213—Tetrahydrofuran
U214—Thallium (I) acetate
U215—Thallium (I) carbonate
U216—Thallium (I) chloride
U217—Thallium (I) nitrate
U218—Thioacetamide
U235—Tris (2,3-Dibromopropyl) phosphate
U239—Xylene
U244—Thiram
U200—Reserpine
U209—Tetrachlorothane, 1,1,2,2-
U210—Tetrachloroethylene
U211—Carbon tetrachloride
U219—Thiourea
U220—Toluene

U221—Toluenediamine
U223—Toluene diisocyanate
U226—Methylchloroform
U227—Trichloroethane, 1,1,2-
U228—Trichloroethylene
U237—Uracil mustard
U238—Ethyl carbamate
U248—Warfarin, when present at concentrations of 0.3% or less
U249—Zinc phosphide, when present at concentrations of 10% or less

U002—Acetone
U003—Acetonitrile
U005—o-Acetylaminofluorene
U008—Acrylic acid
U011—Amitrole
U014—Auramine
U015—Azaserine
U020—Benzenesulfonyl chloride
U021—Benzidine
U023—Benzotrichloride
U025—Dichloroethyl ether
U026—Chlornaphazine
U028—Bis-(2-ethylhexyl)phthalate
U032—Calcium chromate
U035—Chlorambucil
U047—Beta-chloronaphthalene
U049—4-Chloro-o-toluidine, hydrochloride
U057—Cyclohexanone
U058—Cyclophosphamide
U059—Daunomycin
U060—DDD
U062—Diallate
U070—o-Dichlorobenzene
U073—Dichlorobenzidene, 3,3-
U080—Methylene chloride
U083—Dichloropropane, 1,2-
U092—Dimethylamine
U093—Dimethylaminoazobenzene
U094—Dimethylbenz(a)anthracene,7,12-
U095—Dimethylbenzidine,3,3'-
U097—Dimethylcarbamoyl chloride
U098—Dimethylhydrazine, 1,1-
U099—Dimethylhydrazine, 1,2-
U001—Acetaldehyde
U004—Acetophenone
U006—Acetyl chloride
U017—Benzal chloride
U024—Bis(2-chloroethoxy)methane
U027—Bis(2-chloroisopropyl)ether
U030—Benzene, 1-bromo-4-phenoxy
U033—Carbonyl fluoride
U034—Chloral
U038—Ethyl-4-4'-dichlorobenzilate
U039—4-Chloro-m-cresol

U042—Vinyl ether, 2-chloroethyl
U045—Methyl chloride
U048—o-Chlorophenol
U052—Cresols
U055—Cumene
U056—Cyclohexane
U068—Methane, dibromo
U069—Dibutyl phthalate
U071—m-Dichlorobenzene
U072—p-Dichlorobenzene
U075—Dichlorodifluoromethane
U076—Ethane, 1,1-dichloro-
U079—1,2-Dichlorethylene
U081—2,4-Dichlorophenol
U082—2,6-Dichlorophenol
U084—1,3-Dichloropropene
U085—2,2′-Bioxirane
U087—O,O,-Diethyl-S-methyl-dithiophosphate
U088—Diethyl phthalate
U090—Dihydrosafrole
U091—3,3′-Dimethoxybenzidine
U096—alpha,alpha-Dimethylbenzylhydroxyperoxide
U102—Dimethyl phthalate
U112—Ethyl acetate
U113—Ethyl acrylate
U117—Ethyl ether
U118—Ethylmethacrylate
U120—Fluoranthene
U121—Trichloromonofluoromethane
U123—Formic acid
U125—Furfural
U126—Glycidylaldehyde
U132—Hexachlorophene
U136—Cacodylic acid
U139—Iron dextran
U141—Isosafrole
U145—Lead phosphate
U148—Maleic hydrazide
U152—Methacrylonitrile
U153—Methanethiol
U156—Methyl chlorocarbonate
U160—Methyl ethyl ketone peroxide
U166—1,4-Naphthaquinone
U167—1-Naphthylamine
U181—5-Nitro-o-toluidine
U182—Paraldehyde
U183—Pentachlorobenzene
U184—Pentachloroethane
U186—1,3-Pentadiene
U187—Phenacetin
U190—Phthalic anhydride
U191—2-Picoline
U194—1-Propanamine
U197—p-Benzoquinone

U201—Resorcinol
U202—Saccharin and salts
U204—Selenious acid
U207—1,2,4,5-tetrachlorobenzene
U222—o-Toluidine hydrochloride
U225—Bromoform
U234—Sym-Trinitrobenzene
U236—Trypan blue
U240—2,4-D, salts and esters
U243—Hexachloropropene
U246—Cyanogen bromide
U247—Methoxychlor

University of California (UCD) Classification System

The University of California (UCD) Classification System classifies wastes by type, characteristics, and constituents, rather than by source. It is used in California, and has been adopted by the Ontario Waste Management Corporation (OWMC) [5] for use in Canada. The concept is useful because it provides another way of listing that can be cross-referenced to the EPA system and others. The primary categories and base codes used by the OWMC are shown in Table 8-3. The two-digit categories are further subdivided into smaller, more specific groups with the addition of another digit.

Standard Industrial Classification (SIC) System

The Standard Industrial Classification (SIC) codes are widely used for business, marketing, and financial purposes in the United States. While the categories are not really waste specific, using the complete four-digit code narrows the waste source and type to a useful degree in many cases. Combined with other systems, it provides worthwhile information about the possible characteristics and constituents of a waste.

TABLE 8-3 UCD Classifications Used by the OWMC

Description	UCD Code
Acidic solutions	11
Alkaline solutions	12
Metal finishing solutions	13
Other aqueous solutions	14
Inorganic solids	15
Other inorganic wastes	16
Spent solvents	21
Other organic liquids	22
Other organic wastes	24–25
Polymeric material wastes	26
Oily wastes	28
Relatively inert sludges	41
Metal sludges	43
Other sludges	44
Miscellaneous	50–53

Exempted Wastes

When the RCRA was passed, a number of wastes and waste types were exempted from the provisions of that law. While these wastes have been, and are being studied by the EPA to determine their degree of hazard, none have been removed by Congress from the exempted list to date. The exemptions are:

Mining wastes
Sewerage wastes
Domestic wastes (garbage, trash, etc.)
Nuclear waste (regulated by the NRC)
Waste water (covered by other laws and regulations)

These exemptions are for the RCRA system only. Other federal statutes and regulations may apply. Also, states may regulate these and any other wastes, so long as their regulations are at least as stringent as the RCRA's.

CONTEC System

The author developed a system of waste codes in the early 1980s to provide a reference system for wastes encountered in treatability studies and commercial jobs, which were not classified by the EPA [6]. This system uses the EPA codes where applicable, and then a random code using the letter W followed by three digits. It is used, along with the EPA codes, in the CFS Wastes Data Base (CWDB) in Appendix 1. Such a system is necessary to keep track of completed treatability studies in any CFS organization, because there is no standard waste name or description system. The CWDB reflects real-world wastes that have actually been examined, tested, and found to be treatable by CFS technology. It consists of a series of records, containing the following fields of information:

EPA waste code
CFS (CONTEC) code
UCD code
SIC code
Waste description
Generating industry
Generating process
Waste constituents
Applicable treatment technology

The CWDB cross-references all four classification systems and includes information about exempted wastes as well. In the case of EPA listed wastes, it provides more information than the RCRA listings, using both EPA background documents and the actual constituents found when testing the samples. Waste codes, both EPA and other, that are not amenable to CFS treatment have not been included. With non-RCRA wastes, the CWDB provides a reasonably comprehensive listing of industrial wastes that might be encountered by the CFS investigator, although new listings are being added periodically.

A complete printout of all CWDB listings is given below. This is ordered by

EPA/CONTEC CFS code, since it is useless to sort files by waste name for the reason previously given.

Waste Codes	Descriptions
F006	Waste water treatment sludges from electroplating, except for tin, zinc and aluminum on carbon steel, sulfuric acid anodizing, certain cleaning, stripping, etching, and milling
F007	Spent cyanide plating bath solutions from electroplating
F008	Plating bath sludges from the bottom of plating baths from electroplating where cyanides are used in the process
F009	Spent stripping and cleaning bath solutions from electroplating where cyanides are used in the process
F010	Quenching bath sludge from oil baths from metal heat treating where cyanides are used in the process
F011	Spent cyanide solutions from salt bath pot cleaning from metal heat treating
F012	Quenching waste water treatment sludges from metal heat treating where cyanides are used in the process
F013	Flotation tailings from selective floatation from minerals metal recovery
F014	Cyanidation waste water treatment tailing pond sediment from mineral metals recovery
F015	Spent cyanide bath solutions from mineral metals recovery
F016	Dewatered air pollution control scrubber sludges from coke ovens and blast furnaces
F017	Paint residues generated from industrial painting
F018	Waste water treatment sludges from industrial painting
F019	Waste water treatment sludges from the chemical conversion coating of aluminum
K001	Bottom sediment sludge from treatment of waste waters from wood preserving processes that use creosote and/or pentachlorophenol
K002	Waste water treatment sludge from the production of chrome yellow and orange pigments
K003	Waste water treatment sludge from the production of molybdate orange pigments
K004	Waste water treatment sludge from the production of zinc yellow pigments
K005	Waste water treatment sludge from the production of chrome green pigments
K006	Waste water treatment sludge from the production of chrome oxide green pigments (anhydrous and hydrated)
K007	Waste water treatment sludge from the production of iron blue pigments
K008	Oven residue from the production of chrome oxide green pigments
K021	Aqueous spent antimony catalyst waste from fluoromethane production
K031	By-product salts generated in the production on MSMA and cacodylic acid
K032	Waste water treatment sludge from the production of chlordane
K033	Waste water and scrub water from the chlorination of cyclopentadiene in the production of chlordane

K035	Waste water treatment sludges generated in the production of creosote
K037	Waste water treatment sludges from the production of disulfaton
K038	Waste water from the washing and stripping of phorate production
K040	Waste water treatment sludge from the production of phorate
K041	Waste water treatment sludge from the production of toxaphene
K044	Waste water treatment sludges from the manufacturing and processing of explosives
K045	Spent carbon from the treatment of waste water containing explosives
K046	Waste water treatment sludges from the manufacturing, formulation and loading of lead-based initiating compounds
K047	Pink/red water from TNT operations
K048	Dissolved air flotation (DAF) float from the petroleum refining industry
K049	Slop oil emulsion solids from the petroleum refining industry
K050	Heat exchanger bundle cleaning sludge from the petroleum refining industry
K051	API separator sludge from the petroleum refining industry
K052	Tank bottoms (leaded) from the petroleum refining industry
K053	Chrome (blue) trimmings generated by the leather tanning and finishing industry: hair pulp and save/chrome tan/retan/wet finish; no beamhouse; through-the-blue; shearing
K054	Chrome (blue) shavings generated by the leather tanning and finishing industry: hair pulp and save/chrome tan/retan/wet finish; no beamhouse; through-the-blue; shearing
K055	Buffing dust generated by the leather tanning and finishing industry: hair pulp and hair save/chrome tan/retan/wet finish; no beamhouse; through-the-blue
K056	Sewer screenings generated by the leather tanning and finishing industry: hair pulp and hair save/chrome tan/retan/wet finish; no beamhouse; through-the-blue; shearing
K057	Waste water treatment sludges generated by the leather and finishing industry: hair pulp and hair save/chrome tan/retan/wet finish; no beamhouse; through-the-blue; shearing
K058	Waste water treatment sludges generated by the leather tanning and finishing industry: hair pulp and hair save/chrome tan/retan/wet finish; through-the-blue
K059	Waste water treatment sludges generated by the leather tanning and finishing industry: hair save/nonchrome tan/retan/wet finish
K060	Ammonia still lime sludge from coking operations
K061	Emission control dust/sludge from the primary production of steel in electric furnace
K062	Spent pickle liquor from steel finishing operations
K063	Sludge from lime treatment of spent pickle liquor from steel finishing operations
K064	Acid plant blowdown/sludge resulting from the thickening of blowdown slurry from primary copper production
K065	Surface impoundment solids contained in and dredged from surface impoundments at primary lead smelting facilities
K066	Sludge from treatment of process waste water and/or acid plant blowdown from primary zinc production

K067	Electrolytic anode slimes/sludges from primary zinc production
K068	Cadmium plant leachate residue (iron oxide) from primary zinc production
K069	Emission control dust/sludge from secondary lead smelting
K070	Woven fabric dying and finishing waste water treatment sludges
K071	Brine purification muds from the mercury cell process in chlorine production, where separately prepurified brine is not used
K072	Waste water treatment sludge from the diaphragm cell process using graphite anodes in the production of chlorine
K074	Waste water treatment sludges from the production of titanium dioxide pigment using chromium-bearing ores by the chloride process
K075	Waste water treatment sludges from the production of titanium dioxide pigment using chromium-bearing ores by the sulfate process
K076	Arsenic-bearing sludges from the purification process in the production of antimony oxide
K077	Antimony-bearing waste water treatment sludge from the production of antimony oxide
K079	Water cleaning wastes from paint manufacturing
K080	Caustic cleaning wastes from paint manufacturing
K081	Waste water treatment sludges from paint manufacturing
K082	Air pollution control sludges from paint manufacturing
K084	Waste water treatment sludges generated during the production of veterinary pharmaceuticals from arsenic or organoarsenic compounds
K086	Solvent, caustic, and water washes and sludges from cleaning tubs and equipment used in the formulation of ink from pigments, driers, soaps, and stabilizers containing Cr and Pb
K088	Spent potliners (cathodes) from primary aluminum production
K089	Lead-bearing waste water treatment sludges from gray iron foundries
K090	Emission control dust/sludge from ferrochromium–silicon production
K091	Emission control dust/sludge from ferrochrome production
K092	Emission control dust/sludge from ferromanganese production
K098	Untreated process waste water from the production of toxaphene
K099	Untreated waste water from the production of 2,4-D
K100	Waste leaching solution from acid leaching of emission control dust/sludge from secondary lead smelting
K102	Residue from the use of activated carbon for decolorization in the production of veterinary pharmaceuticals from arsenic or organoarsenic compounds
K104	Combined waste water streams generated from nitrobenzene/aniline production
K105	Separated aqueous stream from the reactor product washing step in the production of chlorobenzene
K106	Waste water treatment sludge from the mercury cell process in chlorine production
W001	Cleanup wastes from rotogravure printing
W002	Thorium hydroxide filter cake
W004	Waste water treatment sludge, calcium phosphate
W005	See F006
W006	Incinerator dust

W006	Emission control residues from incineration
W007	Sludge from cathode coating process
W008	Sludge from industrial painting
W009	Digested sewage sludge
W011	Off-spec lotions
W012	Arsenic wood preservatives
W014	Waste water treatment residues from hazardous waste management facilities
W015	Sphincter cone sludge from emission control
W016	Sewage sludge, general
W017	Contaminated glass
W018	Waste water treatment residues—metal hydroxide
W020	Acid oil sludge
W021	Aluminum or tin dross
W022	Foundry sand
W023	Incinerator ash
W024	Barrel finishing baghouse dust
W025	Rolling mill sludges and mill scale
W026	Pickle liquor rinse water
W027	Waste automotive, hydraulic, and cutting oils that may contain solvents and metals
W031	Waste water-based paint
W032	Emission control baghouse dust
W033	Chromic rinse from metal finishing
W034	Waste electroplating baths, acid
W034	Waste electroplating baths, alkaline
W038	Arsenical herbicide wastes
W042	Oil/wax emulsion
W047	Waste water treatment residue—rag oil lagoon
W048	Waste water treatment residues—anaerobic lagoon
W050	Lime sludge from hydrofluoric acid neutralization
W051	Dry electrostatic precipitator dust
W052	Sludge from grinding of bearings
W053	Waste phosphoric acid rinse
W054	Cleanup wastes from latex paint manufacture
W055	Sludge from cold rolling
W057	Alodine waste from cleaning and refinishing
W058	Waste from BF production
W062	Electrochemical machining waste
W064	Waste sulfuric acid pickle liquor
W065	Drilling mud
W066	Baghouse dust from foundry cupola emission control
W067	Acid phosphate filter cake
W073	Dredging muds and sediments
W074	Spent pack hardening compound
W075	Waste water from organic coatings
W076	Waste water from pesticide manufacture
W077	Waste water treatment sludge from organic chemical manufacturing
W078	Sludge from waste water treatment

W078	Filter cake from waste water treatment
W080	Paint stripping waste
W080	Sludge from acid neutralization
W080	Neutralized pickle liquor
W080	Sludge from acid neutralization
W081	Cleaning waste
W082	Emission control dust
W084	Filter cake from waste water treatment
W085	Sludge from biological waste water treatment
W085	Sewage sludge
W086	Sodium metasilicate waste
W087	Contaminated soil
W088	Caustic waste
W089	Leachate from RCRA TSD landfills
W090	Spent catalyst
W092	CaF sludge from acid neutralization
W093	Coke dust
W095	Latex paint waste
W096	Dye sludge
W097	Spent nitric-hydrofluoric acid pickle liquor
W100	Waste water treatment residue from acid phosphating
W101	Sludge from alkaline stripping of nickel plate
W104	Carbon tetrachloride sludge
W105	Solder stripping waste
W109	Caustic paint kettle cleaning waste
W110	Aluminum powder
W113	Waste from caulk manufacture
W114	Pesticide cleanup
W115	Emission control dust containing pesticide
W117	Neutralized waste pickle liquor
W118	Waste oxalate solution
W119	Waste permanganate pickling solution
W120	Waste weld cleaning solution
W121	Zinc phosphate solution
W122	Adhesive waste
W123	Waste latex caulking compound
W125	Hazardous organic wastes, NOS
W133	Waste drawing lubricant
W135	Metal grinding dust
W136	Waste paint sludge
W137	Waste paint stripper
W138	Waste printed circuit etching bath
W139	Waste photoresist stripper
W141	Solder flux residues
W142	Sludge from spray painting
W145	Waste water treatment sludge
W146	Waste liquid
W147	Chromate waste water
W149	Oily water

W150   Organic containing waste water
W151   Waste water treatment residue from can manufacture
W151   Waste water treatment filter cake
W154   Dissolved air floatation sludge
W157   Paint sludge from spray painting
W158   Primary scum (skimmings) from POTW
W159   Primary grit screenings from POTWs and headworks
W161   Textile sizing waste
W162   Waste water treatment sludge from paint manufacturing
W165   Incinerator ash from sewage sludge incineration

Once an organization builds such a data base on a computer, searches and sorts can be made in many ways. One way to find a waste type, for example, is to sort by industry and/or process. Another is to search by constituents. Most database systems have a "wild card" search that allows searching of one or more fields by keyword. This is quite useful in picking out specific wastes from the waste description fields.

One other useful property of the CWDB is that it gives the applicable treatment technology actually used for the waste type, at least in broad terms. Each waste type is classified for treatment by one or more of three descriptors:

Pretreatment: Requires some form of pretreatment prior to CFS treatment.
CFS: Can be treated to meet current regulatory requirements for leachability
Solidification: Can be solidified for disposal in a secure landfill, but may not be treatable to a nonhazardous classification

ASSESSMENT OF WASTE QUANTITIES AND GROUPINGS

Total Waste Quantities

Although the total waste residuals market is not accurately known, it is now estimated that RCRA-controlled industrial manufacturing and pollution control hazardous wastes produced by all industry in the United States amounts to about 264 million metric tons per year [7] on a wet-weight basis. Another 322 million metric tons are exempt from the RCRA, because they are regulated under the Clean Water Act or some other regulatory system. And these totals do not include the exempted wastes, such as sewage sludge, fluegas cleaning sludge, and mining waste. For perspective, municipal sewage sludge is produced at the rate of 40 million metric tons per year.

Of the hazardous waste, about 10 percent is generated by small-quantity generators—those who produce less than 1000 kg per month—who are exempt from certain requirements of the RCRA. Of the remainder, more than 90 percent is waste water that is disposed of in injection wells and surface impoundments. Only about 6 percent goes to landfills, but much of this will fall under the landbans in one way or another by 1990. Thus, perhaps 16 million metric tons per year will be candidates for CFS treatment. This is more per year than the cumulative total of CFS-treated waste to date in the United States, not including fluegas cleaning wastes.

In addition, it is estimated that about $13 to $20 billion annually will be spent on Superfund and privately funded remedial action [8]. The General Accounting

Office estimates that the total costs for Superfund cleanup over the next 50 years could be more than $300 billion [9]. Of some 26,000 currently inventoried sites [10] only 14 have been fully cleaned up to date [11]. And there may be up to 425,000 Superfund sites across the United States [9]. Remedial action sites are good candidates for CFS on-site treatment, because of the cost and lack of capacity to remove these wastes to RCRA TSD facilities. In addition, much CFS work has been done, and will continue to be done, on wastes that are liquid, but not hazardous. In summary, the potential market for CFS services, technology, and equipment in the United States is estimated to be at least 5 to 10 million metric tons (wet) or about 1 to 2 billion gallons per year.

A reasonably up-to-date survey of the sources of all this waste [11] shows the distribution given in Table 8-4. Of this total, 256 million metric tons of waste are described as being handled, at least sometimes, by some treatment technology. Since much of this is waste water, volume reduction would be practiced before CFS in most cases. Nevertheless, the potential volume of waste amenable to CFS treatment is very large, and probably growing in spite of widespread waste-reduction efforts.

EPA Assessment Groups

To give a more detailed picture of the sources of hazardous wastes, their properties and constituents, it is instructive to look at a very extensive study on various industrial groupings done for the EPA in 1974–1976 by a number of different contractors [12]–[24]. This study led to many of the decisions made by the EPA about regulation of various waste streams. While the figures may be somewhat out of date, and some waste streams have disappeared, the information still gives valuable insights to industrial hazardous waste. The studies are summarized in the following sections.

TABLE 8-4 Sources of Industrial Waste

Industry	Number of Facilities	Quantity (million metric tons)	Description
Chemical	700	218	Contaminated waste waters, spent solvent residues, still bottoms, spent catalysts, treatment sludges, filter cakes
Fabricated metals	200	4	Electroplating wastes, sludges contaminated with metals and cyanides, degreasing solvents
Electrical equipment	240	1	Degreasing solvents
Petroleum refinery	100	20	Leaded tank bottoms, slop oil emulsion solids, others
Primary metals	150	4	Pickle liquor, sludge with metal contaminants
Transport equipment	150	3	Degreasing solvents, metals, sludges
National security	100	1	All types of wastes
Other	1360	24	All types of wastes
Total	3000	275	

Storage and Primary Batteries. [12] Waste residues from the storage battery industry are mainly high-volume sludges from water treatment along with some scrap cells. Waste from the primary battery industry consists almost entirely of scrap and rejected cells. By far the greatest amount of waste destined for land disposal, and therefore probably applicable to CFS processing, stems from lead–acid battery production. This residue consists of waste water treatment sludge containing suspended lead, lead sulfate, and lead hydroxide, or calcium sulfate sludge containing lead. The estimated quantity of such sludge in 1977 was about 160,000 metric tons (wet) or about 32 million gallons. This constitutes about 99 percent of the total waste from the storage battery industry, the remainder being small amounts of sludges containing cadmium and nickel compounds and scrap or rejected batteries, or residue containing lead, nickel, cadmium, zinc, manganese, mercury, magnesium, and silver.

Inorganic Chemicals. [24] Land-destined hazardous wastes from the inorganic chemical industry originate either directly from the process streams, or from air and water effluent treatment. The amount of water in these residual streams will usually vary between 30 and 80 percent, depending on whether the waste is a filter cake or a sludge going to a settling pond. Because of the variation in water content, these waste streams are reported on a dry basis. The inorganic chemical industry is divided into four major industry categories, which along with their hazardous waste generation rates, are as follows:

Alkalies and chlorine: 56,000 metric tons
Industrial gases: 0
Inorganic pigments: 230,000 metric tons
Inorganic chemicals: 2 million metric tons
   Total: 2.3 million metric tons

The primary hazardous constituents are asbestos, chlorinated hydrocarbons, cadmium, chromium, cyanide, fluoride, nickel, phosphorus, and zinc for the inorganic chemicals segment. In 1975, only about 12,000 metric tons per year, or less than 1 percent of the total, went to secured landfill. The remainder was handled by on-site storage and disposal, including deep-well disposal. Especially in this industry, it should be remembered that hazardous wastes account for only about 20 percent of the total liquid and semiliquid residuals generation, the other 80 percent being relatively inert sludges or liquids that were mostly lagooned or landfilled on-site.

Organic Chemicals, Pesticides, and Explosives. [13] Five major industries under the above heading are: gum and wood chemicals, cyclic crudes and intermediates, industrial organic, pesticides, explosives. The first three categories, which make up the organic chemicals group, generated an estimated 10.9 million metric tons (wet) of residuals in 1974. The pesticides group generated an estimated 584,000 metric tons, and the explosives category about 25,000 metric tons (wet).

Electroplating and Metal Finishing. Because this industry is composed primarily of small shops with few employees per shop, little good statistical information is available on the industry as a whole. The types of plants producing wastes in this industry are divided into small job shops and captive shops that are part of large manufacturing organizations. Estimated tonnages in 1977 from job shops are in the range of 130,000 to 230,000 metric tons (dry). Captive shops generated as estimated 1.3 million to 2.3 million metric tons (wet) in 1977.

The types of wastes generated are as follows:

Water Treatment Sludge: The predominant disposal practices are off-site landfill (usually uncontrolled using contractor hauling), on-site chemical fixation, and on-site landfill. Sludge or solutions containing large concentrations of cyanide may be incinerated.
Concentrated Chemical Solutions (Spent Baths): These are disposed of in the same manner as the water treatment sludges.
Organic Solvents: Except for reclaiming, the practices are the same as for water treatment sludges.

The sludges usually contain solids in the range of 2 to 20 percent, but are sometimes dewatered to the level of 20 to 25 percent solids. Hazardous constituents in these residuals are asbestos, chlorinated hydrocarbons, cyanide, arsenic, and the various heavy metals used in the plating operations: chromium, nickel, copper, cadmium, and zinc. More information about this waste category is given in the last section of this chapter.

Paint and Allied Products. [15] This industrial category includes three basic industries: paint and coatings manufacture, solvent reclaiming, and factory applied coatings operations. The paint and coatings industry generated about 100,000 metric tons per year (1974) of sludges and wastes from cleaning and from waste products and spills. This potentially hazardous part of the waste stream is also the total waste volume that could be approached by the CFS industry, since most of the remainder of the waste generated in this industry is nonhazardous solid waste. The vast majority of these wastes is discarded in off-site landfills by private contractors, but about 20 percent of the sludge undergoes some dewatering prior to disposal on landfill. The factory applied coatings part of this industrial grouping generates from 115,000 to 215,000 metric tons per year of paint waste. These wastes are disposed of along with the main stream of the manufacturing industry wastes and are therefore not separated out here for the type of disposal technology used.

Petroleum Refining. [16] There are 17 major waste streams in the petroleum refining industry. With their waste generation rates in wet metric tons per year, they are as follows:

Spent lime sludges from boiler feed water tanks	780,000
Lube filter clays	93,000
Lead product tank sludge	83,000
Waste biosludge	81,000
API sludge	69,000
Dissolved air floatation float	60,000
Cooling water sludge	38,000
Slop oil emulsion solids	34,000

FCC catalyst fines	31,000
Silt from storm water	30,000
Neutralized HF alkylation sludge	17,000
Kerosene filter clays	4,000
Coke fines	1,700
Lead tank sludge	1,700
Exchanger bundle cleaning sludge	1,300 crude
Tank sludge	800
Cooling tower sludge	500

The total of these waste streams is about 1.3 million metric tons (wet) per year.

The primary hazardous constituents of these wastes are oil, fluorides, and various heavy metals, including chromium, zinc, nickel, copper, vanadium, lead, arsenic, molybdenum, cobalt, selenium, mercury, silver, and cadmium. Minor hazardous components are phenol, cyanide, and benz-a-pyrene. Metallic constituents in refinery residuals are of a low order of hazard compared to other industry sources.

Widely used disposal methods for refinery wastes are various types of landfill, ranging from uncontrolled to secure landfill. Land spreading (soil farming) of certain sludges has also been practiced, especially during the last four years, and found to be relatively inexpensive. However, it can only be used effectively under the right climatological and environmental conditions. Little deep-well injection is used, and little or no ocean disposal. Refineries have been using CFS.

Pharmaceuticals. [17] Approximately 250,000 metric tons per year of wastes were produced by the pharmaceutical industry in 1974, of which about 25 percent or 60,000 metric tons per year were considered hazardous. The largest portion of these wastes comes from the production of antibiotics by fermentation. The filtered microorganisms plus the inorganic filter aid and nutrients makes up this residue. Other wastes are sludges from activated sludge systems used on the water effluent stream, solvents, and still bottoms. However, the major source of hazardous waste, and a significant source of nonhazardous waste, is the production of organic medicinal ingredients. These wastes consist of filter cakes, filter papers, carbon, sewage process sludge, and unrecoverable halogenated and nonhalogenated solvents and still bottoms.

Metal Smelting and Refining. [18] This group of industries disposes or stores large quantities of process and pollution control residues on land. These residuals are primarily inorganic slags and sludges containing silicates, oxides, sulfates, and chlorides. Sludges are often residues from FGD scrubbers or process waste water treatment with lime. Consequently, they contain calcium sulfate, calcium sulfite, calcium hydroxide, and calcium carbonate as well as metal oxides and hydroxides. Recycling of dust, especially electric arc furnace dust (K061), from emission controls is a fairly common practice, but much of it is disposed of on land. The latter practice will be curtailed when the first-third landban requiring recycling of high (>15 percent) zinc content dust becomes effective in 1990.

The principal potentially hazardous constituents found in the iron and steel, ferroalloy, and foundry residuals are heavy metals including lead, zinc, copper, manganese, nickel, and chromium as well as iron. Phenol and cyanide compounds are found in steel plant residuals from coking operations. Phenol is also present in some waste foundry sands. Oils and greases are present in steel mill scales and waste water treatment plant sludges, and fluoride salts used as fluxing agents are found in sludges. In the nonferrous smelting industry, the principal potentially hazardous constituents are heavy metals including arsenic, cadmium, lead zinc, copper, chromium, antimony, nickel, and mercury.

In 1974, the U.S. steel industry produced more than 16 million metric tons of pollution abatement residues. Of this, 7.5 million tons were dumped, or stored for "later reuse." However, the annual volume of waste pickle liquor generated in the United States exceeds 5 million tons annually, most of which originates in the steel industry, and this is in addition to the 7.5 million tons of other residues that are primarily nonhazardous dusts, solids, and sludges.

The nonferrous segment of this industry grouping generates the following potentially hazardous waste residues yearly (metric tons):

Primary copper smelting and tire refining	6,333,940
Primary copper electrolytic refining	6,655
Primary lead smelting and refining	542,400
Primary electrolytic zinc smelting and refining	22,946
Primary pyrometallurgical zinc smelting and refining	185,370
Primary aluminum	435,182
Primary antimony	8,313
Primary mercury	21,425
Primary titanium	5,100
Primary tungsten	236
Secondary copper	153,300
Secondary lead	164,500
Secondary aluminum	415,320

The predominant practices used in all of these industries for residuals disposal are lagooning and land disposal. Slags and other solid residues are generally open dumped on land. Residue from wet emissions and process waste water control with or without lime treatment is routed to settling pits or lagoons. Settled sludge is often dredged from ponds or lagoons and stored or disposed on land, but those plants producing small amounts of sludge often leave it permanently in lagoons. Unlike the other smelting and refining industries, the iron and steel industry generates considerable oily waste and acid pickle liquor. These are often removed from the plant site by contract disposers. Some pickle liquor is recycled, some deep-well injected. Neutralization, dewatering, and CFS are also practiced. Major increases in future sludge generation will be primarily from control of SO_2 emissions from primary copper and secondary lead smelters. It should be noted that more than 95 percent of residuals from primary copper smelting and fire refining, primary lead smelting, primary antimony smelting, primary mercury smelting, secondary copper smelting, and secondary lead smelting are solid slags and retort residues. These usually do not require treatment by CFS or other processes. On the other hand, processes employing electrowinning of metal produce predominantly sludge residues. For example, electrolytic copper, electrolytic zinc, and electrolytic antimony residues are 100 percent sludge, as are chlorination process residues from recovery of titanium.

Rubber and Plastics. [19] Rubber products manufacture does not generate much liquid waste. Most waste materials are solid, and while some of them are hazardous, they are not amenable to treatment by the CFS processes. For the plastic materials and synthetics industry, the following waste streams are important (given in metric tons per year):

Phenolic resin liquid fraction from reactor condensate	325,000
Styrene-butadiene still bottoms from monomer and solvent recovery	12,700
Polystyrene still bottoms	18,440
ABS/SAN still bottoms	2,700
Polypropylene still bottoms	22,000
Rayon zinc-containing sludge	7,200
Acrylic zinc-containing sludge	278,000

The disposal methods encountered in this industry are incineration, landfill, lagooning, and storage. Incineration is used for nearly all of the still bottoms and the liquid fraction from reactor condensate. Landfill or lagooning is used for the zinc-containing sludges.

Leather Tanning and Finishing. [20] Wastes from this industry are divided into two types: solid waste that consists of solids not applicable to CFS systems; and liquid residues from finishing operations and from waste water treatment, and potentially hazardous wastes containing lead, zinc, copper, and trivalent chromium. Process solid wastes in 1977 were generated in the amount of 214,000 metric tons on a wet basis, while potentially hazardous wastes were generated in the amount of 143,000 metric tons (wet).

Most disposal in the tannery industry is in off-site landfills and dumps. About 9 percent of the waste is disposed in lagoons, but this is nearly all off-site. A great deal of the water effluent from tanneries goes directly to municipal sewage treatment plants rather than being treated on-site; as the pretreatment requirements come into effect in the United States, more sludge will be generated on-site. Sludges produced from on-site waste water treatment plants have been considered to be suitable for disposal in sanitary landfills or secure landfills without chemical fixation.

Special Machinery. [21] Wastes in this classification come primarily from machine shops, heat-treating, electroplating, coating, and foundry operations in either industrial machinery or office computing and accounting machinery manufacture. Potentially hazardous constituents in these wastes are heavy metals, oil, acids and alkalies, cyanide, and flammable solvents. Total potentially hazardous waste in 1977 from the industrial machinery category was 86,000 metric tons (wet), while that from the office computing and accounting machinery category was about 67,000 metric (wet). Lubricating oils are often recycled or are burned with solvents or with fuel oil in plant boilers. Heavy metal sludges resulting from treatment of electroplating, heat-treating, and other metal finishing wastes are often drummed for landfill disposal. Acid and alkali cleaning solutions are neutralized or diluted before being discharged to a sanitary sewer. Off-site treatment and disposal by contractors is by far the most popular disposal method for these wastes. Most of the contract disposal goes to landfills, secure or otherwise.

Textiles. [23] The waste-generating operations in this industry are divided into the following categories: wool scouring, wool fabric finishing and dyeing, griege goods, woven fabric dyeing and finishing, and yarn and stock dyeing and finishing. It is important to note that the textile industry by definition does not include the manufacture of synthetic fibers, which is considered a part of the organic chemical industry in the United States.

As with most other industries, wastes in the textile industry originate directly from both the manufacturing processes and from the treatment of process waste water. Most of the direct manufacturing process wastes in the textile industry are largely innocuous, and land disposal in refuse landfills poses little problem. Only about 6 percent of the process-originating wastes are potentially hazardous, and these are primarily containers holding residual dyestuffs and chemicals. By far the greatest waste disposal problem in this industry is of the sludge, permeate, regenerate fluids, or other concentrates from water treatment process. The estimated yearly production (1974) of sludges from these processes in the United States is about 1.9 million metric tons (wet), or about 390 million gallons. Nearly all of this waste originates in two industry segments: wool scouring and woven fabric dyeing and finishing. Hazardous constituents in these wastes include the heavy metals—arsenic, barium, cadmium, chromium, cobalt, copper, iron, lead, manganese, mercury, nickel, and zinc—chlorinated organics, dyestuffs, and other chemicals.

Most of this waste currently does not show up in the U.S. hazardous waste disposal market due to the fact that only several categories of textile operations are currently disposing of excess sludge. In the remaining cases, the sludge is presumably still retained in the system in clarifiers, settling ponds, and lagoons, since water treatment systems in this industry are fairly recent. Eventually, however, this retained waste will appear in large quantities, either as a disposal problem for the generator or on the waste disposal market.

Most of the textile industry in all categories disposes of its solid wastes in sanitary landfills. In 1974, 35 percent of the waste stream in the industry was discharged to municipal sewage treatment plants, which practice will probably be significantly reduced in the future, thereby increasing the sludge generation rate at the plant level. Where sludge is disposed of, rather than being stored in accumulation ponds, the ratio of on-site to off-site sludge disposal is about 1:1.

Mining. [22] Discussion of this industry will be limited to metal mining, since the coal mining industry produces hazardous sludges or concentrated wastes only in the physically hazardous sense. The iron-containing sludges from the treatment of acid mine drainage water in the eastern United States is of such large volume and such a specialized situation that it is not an applicable source of business for the CFS industry, now or in the foreseeable future. There is a business in the stabilization of coal refuse piles and coal-fines holding ponds, but the technology and methodology used in this area are more in situ soil stabilization than CFS processing in the normal sense.

Information available in the United States on metal mining wastes is limited to five categories: copper, lead and zinc, mercury, uranium and vanadium, and miscellaneous ores. Except for the uranium and vanadium mining segment, only the concentrator tailings from copper mining operations were found to contain toxic materials above background levels and therefore considered to be potentially hazardous. For uranium and vanadium, both the waste rock and the concentrator tailings were considered to be potentially hazardous. Because of the immense volumes involved, it is unlikely that anything except potentially hazardous wastes in this industry will be treated by CFS methods, and even with these wastes there is some doubt about the application of the traditional CFS methods because of the total costs involved in the huge volumes. To give some idea of the volumes involved, an average of about 2 tons of wastes is produced for each ton of raw ore mined and about 52 tons of waste for each ton of product (concentrated ore). Of this total of nearly 800 million metric tons in 1974, about 24 percent is overburden, 47 percent waste rock, and 30 percent is concentrator tailings. Therefore, the concentrator tailings alone add up to about 234 million metric tons of waste. It is for this reason that the mining industry is separated from the other twelve industries in the EPA categorization.

In the copper mining area, approximately 217 million metric tons of tailings were generated in 1974, but this waste is considered to be only slightly hazardous. For lead–zinc mining, the amount of tailings generated was an estimated 9 million metric tons. These tailings are considered to be only slightly hazardous because the wastes are alkaline, and therefore the potential for water leaching of toxic metals is presumably minimized. Hazardous constituents include lead, zinc, pyrite, and other toxic metals in the ore. In mercury mining, the total potentially hazardous wastes amounted to 16,000 metric tons. In uranium and vanadium mining, the wastes generally contained significant amounts of radioactive substances such as uranium minerals and, in some cases, trace amounts of selenium, radium, and other toxic metals. The uranium wastes present a much greater potential hazard than do the vanadium wastes. The total potentially hazardous waste generated in this industry segment in 1974 was about 12 million metric tons, composed of about 3 million metric tons of waste rock and 9 million metric tons of tailings. Both the waste rock and the uranium concentrator tailings are considered moderately hazardous. The miscellaneous mining area includes antimony and beryllium, and the potentially hazardous wastes in 1974 amounted to only 310 metric tons.

The present methods used to handle waste residuals in the mining industry are either general fill or backfill for the overburden and waste rock, and tailings ponds for the concentrator wastes, except for mercury concentrator wastes, which are disposed of by covering with overburden and waste rock.

One other potential area for very low-cost CFS processes is in phosphate mining, where about 1.2 to 1.5 cubic feet of waste is generated for each cubic foot of phosphate matrix mined. This waste contains a great deal of water; for example, the slimes are 4 to 5 percent solids as they leave the plant, and rapidly concentrate by gravity to 10 to 15 percent in settling ponds. Even after years of settling, they reach only 25 to 30 percent solids and are still semiliquid in nature. This poses primarily a land reclamation problem rather than an environmental pollution problem, but the huge quantities involved require that any process used be very inexpensive.

Food Processing and Agriculture. These industries are not usually considered to generate hazardous wastes and, with two exceptions, are also not probable candidates in themselves for CFS processes. In the case of food processing wastes, waste water that is not discharged to municipal sewage treatment plants may be treated on-site in biological treatment processes that produce sludges similar to municipal sewage sludge. Therefore, opportunities for CFS processes exist here and are covered in the following subsection under "Municipal Sewage." However, it is believed that most food processing plants discharge their waste water to nearby municipal plants and that no on-site residues are generated. Other residues from food processing are not considered liquid or hazardous wastes and are usually handled along with plant solid waste. The other exception is in the possible use of some CFS processes for the solidification of animal manures to produce fertilizers or soil conditioners. This has been done experimentally, but has never been marketed by any CFS processor due to unfavorable economic conditions. In general, these manures are used for fertilizing or soil conditioning as generated or stockpiled, without any stabilization or solidification process being applied, with the possible exception of the use of lime for odor reduction. The wastes in agriculture are considered and handled as solid refuse, and are disposed of accordingly.

Municipal Sewage and Water Supply. In the past, sewage sludge disposal was the largest single waste disposal problem; it was really the only one that was considered important 20 years ago, for much of the present literature deals with this type of residue and a great deal of information is available on its composition, characteristics, disposal methods, and disposal costs. Because of this availability of information, and because the generation of sewage sludge is quite variable from country to country due to reasons already discussed in this book, we will try only to sum up the present situation in the United States.

The primary debate at this time in the United States concerns the metal content of such sludges and its effects on plant and animal life, including humans, when the sludges are used as fertilizers or soil conditioners, or disposed of by land farming on land used to grow fodder or food crops. Also, there are potential problems from various pathogens and the general environmental problem caused by the release of eutrophication-causing constituents in the sludges. Pollutants that have been found in primary sewage sludge include:

Cadmium: 100 ppm
Chromium: 700 ppm
Copper: 100 to 1000 ppm
Lead: 400 ppm
Mercury: 3 to 15 ppm
Nickel: 25 to 400 ppm
Zinc: 300 to 2000 ppm

In addition, pesticides (Aldrin, Chlordane, DDT, DDD, Dieldrin) and PCBs are found in sewage sludge from time to time.

The generation rate for sewage sludge in the United States is in the range of 10 to 30 billion gallons per year, containing about 7.5 million tons of dry solids. Sewage sludges range from 3 to 10 percent solids for nonfiltered sludge to 20 to 25 percent solids for vacuum-filtered sludge. By process, approximately 7400 gallons of lime sludge, 3000 gallons of alum sludge, or 3000 gallons of iron chloride sludge are generated per 1 million gallons of raw sewage treated, depending on which treatment method is used. Alternately, looking at the various treatment processes for sewage, the following generation rates (dry solids) are found per 1 million gallons of raw sewage treated:

Primary sludge	1000 pounds
Activated sludge	700 pounds
Trickling filter sludge	650 pounds
Aerated lagoon sludge	830 pounds
Extended aeration sludge	830 pounds
Aerated lagoon sludge	830 pounds
Filtration sludge	125 pounds
Algae removal sludge	125 pounds
Chemical additions to clarifiers	450 pounds

About 25 percent of municipal waste water treatment sludge produced in the United States is disposed of by land spreading; that is, by applying sludge to the land for agricultural benefits as well as for disposal purposes. The remainder of the sludge is disposed of by burial in sanitary landfills, lagooning on-site, incineration (in large urban areas), and by ocean dumping in coastal areas. Many of the large cities that are located in coastal areas have been using ocean dumping almost exclusively, but this practice is gradually being curtailed and is eventually planned to be entirely eliminated as a routine disposal method. Because of the lack of land in or around coastal cities, this sludge will presumably go to sanitary landfill or incineration in the future.

The cyclical shortages in fertilizer and the resulting price increases have periodically renewed interest in the use of sewage sludge as a fertilizer or soil conditioner. In a few instances, dried sewage sludge has been commercially marketed, but this is not a widespread practice in the United States. The main environmental objection to this approach in large cities is to the metal content. As advanced waste treatment processes and better biological techniques produce sludges of ever increasing inorganic content, the metal problem has been increasing, and this tendency is being compounded by discharges from industrial plants into municipal sewage systems. However, it is expected that the pretreatment requirements being imposed now on industrial plants will substantially reduce metal and other inorganic content in urban sewage sludges.

The cost for sewage sludge treatment and disposal ranges through a very wide scale. This is due to the diversity of practices as well as to the methods used for reporting costs. Traditionally, costs for disposal for sewage sludge have been reported on a dry ton basis, but very often the solids content of the sludge is not given. Since the CFS industry usually bases and calculates its costs on a volume basis, and on an actual volume rather than on a theoretical dry ton basis, it is difficult to correlate the two methods.

Wastes from water supply treatment processes are also municipal wastes, but are usually generated at a different location in the community and have very different characteristics. These residues are highly variable in composition and contain the concentrated substances removed from waste water, primarily sediment and the chemicals added in the treatment process itself. These wastes are usually considered to be nonhazardous and are a physical handling and disposal problem only in some situations. Gravity-thickened sludges range from about 2 to 6 percent solids by weight and mechanically dewatered sludges from 40 to 50 percent solids. The most widespread method for disposing of these wastes is lagooning. One method that is also favored by plant operators is disposal into the sanitary sewage system; this adds to the load on the sewage treatment plant and also changes the character of that sludge, sometimes for the better. However, this disposal method is not often permitted due to problems with deposition in the sewers and overloading of the sewage treatment plant's design capacity. Those sludges that are not lagooned are disposed of either in a solid waste landfill or in a separate landfill, but do not normally require a chemical landfill. Therefore, currently experienced costs for disposal of these residues are quite low.

Fluegas Cleaning. One of the major problems inherent in fluegas cleaning (FGC) systems based on lime or limestone scrubbing is the necessity to dispose of or somehow utilize, large quantities of sludge. At present, these systems generate a sludge with no commercial value. The sludge contains calcium sulfate, calcium sulfite, and calcium carbonate as well as trace concentrations of metal and other contaminants from the coal (or oil) being burned. For coal-fired installations where efficient particulate removal is not installed upstream of the absorber, the sludge may contain large quantities of flyash. The amount of sludge generated by a given plant is a function of the sulfur and ash content of the coal, the coal usage, the onstream hours for scrubber operation, the mole ratio of lime or limestone added to amount of SO_2 removed, the SO_2 removal efficiency, the ratio of sulfite to sulfate in the sludge, and the percent of moisture in the sludge. The yearly sludge generation of a typical 1000-megawatt (MW), coal-fired power station is 780,000 tons per year of wet limestone sludge at 50 percent solids, or about 0.25 gallon per minute per megawatt of design capacity. These figures are for FGC sludge only, without inclusion of flyash. Using the forecast demand for FGC in the United States, wet sludge disposed of annually in 1980 was estimated to be about 130 million tons, or about 20 billion gallons per year. This is based on the assumption that 90,000 MW of generating capacity would be controlled by lime–limestone scrubbing systems by that time. Another way of putting this is that about one ton of calcium sludge will be generated for each 3 tons of coal used.

The hazardous constituents in these wastes are the soluble dissolved solids of about 10,000 ppm and the elements arsenic, beryllium, cadmium, chromium, calcium, lead, mercury, selenium, and zinc present in the range of up to about one part per million. The nonmetal dissolved solids are composed of chloride, fluoride, sulfate, and sulfite, with sulfate normally being the largest single constituent. These residues are obviously not very hazardous, but because of the enormity of their generation rate, will be controlled as to disposal. In addition, these sludges in most areas pose a physical handling and disposal problem that is lessened by solidification, regardless of the chemical aspects of the need for CFS processes.

The present CFS methods used are primarily lime–flyash, which makes use to some extent of the lime or limestone and the flyash already present in the sludge, and the Calcilox™ process, which is more a settling and compaction additive system than a typical CFS process. Other systems that have been tested both experimentally and in the field include the Chemfix process and others using Portland cement. The most important issue in selection of a CFS process for FGD waste is cost. Extensive technical and cost evaluation work has been done for EPA by various contractors [25], and it has been determined that the processes tested all have total disposal costs ranging in the area of $7.30 to $11.40 (1977 dollars) per ton of sludge on a dry basis. The choice of process would be determined by generation rate, location, distance to disposal site, sludge water content, and other factors.

Dredging. In recent years, an ever increasing volume of dredging work has been done for the purpose of removing sludges, muds, and other materials from rivers, lakes, ocean bottoms, and sludge retention lagoons, either to prevent contamination of the water overlying the sludge beds or for ultimate disposal of the residues contained in lagoons. The problem of final deposal of the material so removed is becoming of critical proportion in certain areas, both from the environmental standpoint and from the standpoint of the dredging industry. In addition to the chemical and biological hazards posed by the various toxic constituents of such sludges, there is the physical problem of handling and stabilization of these semiliquid materials.

It is not possible to discuss the quantities of dredged materials produced, since most dredging work is done on noncontaminated areas for navigational purposes. To date, no CFS process has been applied in the United States, other than experimentally, for any dredging project. The less expensive processes might be usable for this purpose.

Hazardous Spill Residues. This is not a specific waste source, but it is a potential application of CFS technology. While the total yearly volume of hazardous spills in any county is small compared to the overall waste disposal problem, the potential danger resulting from such spills is a proportionately large menace due to the fact that spills are by definition unplanned and uncontrolled releases of hazardous material, often in concentrated form. CFS methods have good applicability because the spills, if reasonably contained, are usually concentrated and the CFS method can often be performed on-site much more efficiently and at lower cost than other treatment methods. On-site treatment of the spill eliminates transportation of the waste and all the associated problems. Ideally, the hazardous spill can be transformed by a mobile, fast-response CFS treatment plant into nontoxic, physically stable material at a reasonable cost. Even if this material should require further detoxification, it has at least been placed in a physical state that is cheaper, easier, quicker, and safer to transport to the final treatment location or to a secure disposal site. Often the spill results in the intermingling of the spilled material with water and soil, producing a contaminated or oily, muddy sludge that is not amenable to treatment by most other methods.

Other Countries

It is probable that the same general patterns of waste generation exist in other developed countries, or will exist there in the near future, although certain definite differences do exist. One of these differences is the emphasis on government involvement, even to the point of complete control of hazardous waste management facilities in certain countries. This is true to some extent in Germany, Scandinavia, and to a lesser extent in France and the United Kingdom. Except for the iron-curtain countries, Israel, South Africa, Australia, and Japan, there is little hazardous waste treatment activity in most other parts of the world; the waste management industry that does exist is composed of solid waste collection and disposal operations and sewage systems.

Canada seems to have a situation similar to that in the United States, which would be expected from its geography, population, and industry. In 1972, at least 50 million gallons of liquid wastes were known to have been disposed of in southern Ontario alone.

In the United Kingdom, one CFS contractor treated approximately 10 million gallons of waste between 1973 and 1977. The United Kingdom has a liquid waste generation rate proportionate (on a population basis) to that of the United States. However, there are some differences in traditional disposal techniques in that a larger proportion of wastes in the United Kingdom were ocean dumped than in the United States. There are a number of lagoons in the United Kingdom, although not as many as would be typical in the United States.

In France, lagoons at waste generators' plants are less common than they are in the United States, although large lagoons do exist. However, the overall CFS potential is considered to be very good—proportionate to that of the United States. One contractor, for example, treated about 6 million gallons of waste in 1976, 3 million gallons of which were on-site work at existing lagoons, the remainder at their central site.

In Japan, the situation is very much different. Industrial plants are much more likely to incinerate those wastes that can be burned, and to dewater all other sludges to as high as possible solids content and landfill the resulting sludges, often without treatment. It is expected that this will change to some extent in the future, requiring treatment of the dewatered sludges. Interestingly, a number of the CFS processes have originated in Japan. Apparently these have been used mostly for solidification of the top meter or so of the surface of dredge spoil impoundments and sewage lagoons, and for treatment of sewage sludge as it is produced. In the case of surface solidification, the lagoon is subsequently planted with vegetation. However, difficulties have been encountered in sinking and other unstable conditions in these areas, and that approach to CFS treatment may be altered in the future to require treatment of the complete lagoon. Ocean dumping in Japan is a very sensitive issue because of several disasters that have occurred due to release of toxic metal compounds (see Chapter 4). Thus, CFS treatment followed by landfill can be expected to increase in usage.

In recent years a considerable interest in CFS processes has been generated in Germany and in Scandinavia; however, it appears that much of this work will be done at government operated centers. Little has been published internationally about hazardous waste generation and treatment in Spain, in Italy, and in other major countries in Europe.

WASTE PROPERTIES AS RELATED TO CFS

The chemical properties of wastes, especially as they relate to fixation of hazardous constituents, were discussed exhaustively in Chapters 3–6. However, we have not yet said much about the chemical properties that affect the solidification part of the CFS process. Physical properties were briefly examined in Chapter 7, and will be explored in more detail in succeeding chapters. The testing of waste and CFS-treated waste forms, both chemical and physical, are described in Chapter 17. Here, we will examine some of these properties in a general way, as they relate to all wastes.

Physical Properties

Those properties that are important in CFS technology are listed below. Some have already been discussed (Chapter 7) and need no more elaboration here. Others deserve more attention in the context of general waste properties not specific to any individual process.

- Medium: aqueous, oil, solvent, etc.
- Physical state: solution, sludge, powder, etc.
- Phases/layering: none, bilayered, multilayered
- Total solids (%)
    Suspended solids (%)
- Type of solids: organic, inorganic, mixed
    Ease of wetting/surface tension
- Specific gravity
    Bulk density (for solids)
    Particle size distribution
    Particle morphology
    Viscosity
    Flash point
- Odor
- Color

Specific Gravity. If there is a significant difference in the densities of the various phases in a waste, especially between the suspended solid particulates and the medium, phase separation and layering will occur. This may be so rapid as to make homogenization difficult and cause excretion of fluid from the waste form as it cures. Specific gravity of the solid phase also affects viscosity, as discussed later in this section.

Particle Size Distribution. Particle size and distribution affect a number of waste properties, especially viscosity. Reactivity with the CFS system, wetability, inhibitory effects in some systems, and leachability of the final waste form are all influenced in one way or another by this factor. Small particles, because of the large surface-to-volume ratio, will generally leach more rapidly (on an equal weight basis) than large particles. If there is chemical interaction between the waste particles and the CFS reagents, small particles will tend to react faster. This is easily shown, for example, in Portland cement/soluble silicate systems. On the other hand, it has been demonstrated by the author that very fine particulates can coat the surfaces of reagents and inhibit the setting reactions of cements and pozzolans.

Particle size of the waste can affect the reaction rate of constituents inside the particles with fixation agents in the CFS formulation. This is especially important when the fixation agents have short lives in the CFS environment, or are immediately destroyed when the next process step takes place. The author found a case of the latter phenomenon in a field pilot study on "soil" contaminated with chromate (Cr^{+6}) solution. The "soil" was a combination of dry-well aggregate, shalelike natural rock, waste particulate material, soil, and the contaminant. The porous rock fragments, ranging from fine particles to large pieces, several inches or more in diameter, had, over many years, absorbed the chromate solution deep into the particle interior. Although the large-size fractions were screened out, the final size distribution before treatment ranged up to about one inch. The treatment process consisted of a first step that reduced the Cr^{+6} to Cr^{+3} using $FeSO_4 \cdot 7H_2O$ at pH <3.0, followed by neutralization and addition of the solidification reagents. After the chromium reduction step, the waste mixture was sampled and found to contain no detectable (colorimetric method) Cr^{+6}. The process was completed, but when the material was again tested after several weeks, a high Cr^{+6} level was detected. After investigation, this was attributed to slow diffusion of the chromate from the particle interior to the surface; no reducing agent remained in the system. The problem was solved by grinding the waste to fine particle size and reacting it with the reducing agent for a time sufficient for all of the chromium to diffuse out.

Physical properties of the CFS product are affected not only by the particle size but by the size distribution. A wide range of sizes generally contributes to a stronger product, as evidenced by the use of a well-graded aggregated distribution in concrete. The particle size distributions of different soil classifications are illustrated in Figure 8-1 [26]. Microscopically, the difference between poorly graded and well-graded particulate materials is shown in Figure 8-2. Particle size distribution also affects pumpability and mixability of wastes. Slurries of very fine materials, such as lime, are fluid only up to about 10 percent solids by weight, while coarser limestone slurries can remain fluid up to 50 percent solids [27]. Particle size distribution can be measured by laboratory or even field methods, but this is seldom done in practice.

Particle Morphology. Particle morphology, or shape, is also important, especially in the flow of pumpable wastes and in the mixing of wastes with a reagent in CFS systems. The effect of spherical particles, such as those from power plant flyash, in reducing viscosity and increasing pumpability of concrete mixes is well known. Typically, spherical particles produce slurries that are fluid up to about 50 percent solids, while platelet-shaped particle slurries are nonfluid at about 10 percent solids.

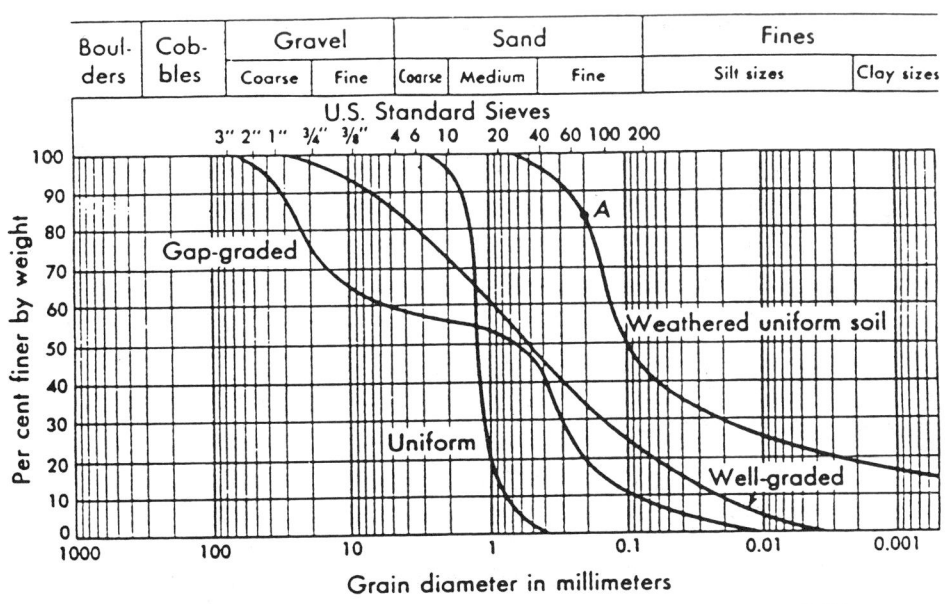

Figure 8-1 Particle size analysis grain size chart. (From Sowers and Sowers [26].)

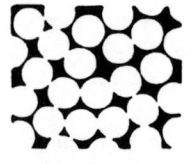

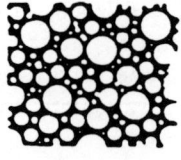

Poorly-graded   **Well-graded**   **Figure 8-2** Gradation of particles.

Mixing coarse sand into a viscous slurry, for example $Mg(OH)_2$, actually reduces the viscosity, even though the solids content rises. Acicular (needlelike) particles and fibrous wastes may produce nonpumpable mixtures at even lower solids content; for example, some paper industry wastes are virtual solids at about 1 percent by weight total solids. Particle morphology is difficult to measure except by laboratory techniques. Its effects are better determined by measuring viscosity.

Viscosity. It is evident from the previous discussion that viscosity is a very important characteristic of waste to be treated in CFS systems. It affects the handling of wastes and mixing them with reagents, to the degree that the choice of the mechanical system is often determined by waste viscosity alone. Solids content alone is not an adequate way of determining how the waste will act in the CFS system.

Viscosity is strongly related to particle size distribution and morphology, and to specific gravity of the particulates. Dallavalle [28] described both of these effects. Figure 8-3 shows the relationship between viscosity and suspension density for three suspensions, using different specific gravity materials (quartz, 2.65; ferrosilicon, 6.80; lead, 11.3) at the same particle size distribution. Each curve shows the same effect, but at different densities. As particulate is added to the water medium, viscosity increases gradually and nearly proportionately up to a certain point, after which the viscosity increases very rapidly with small increase in density. The critical

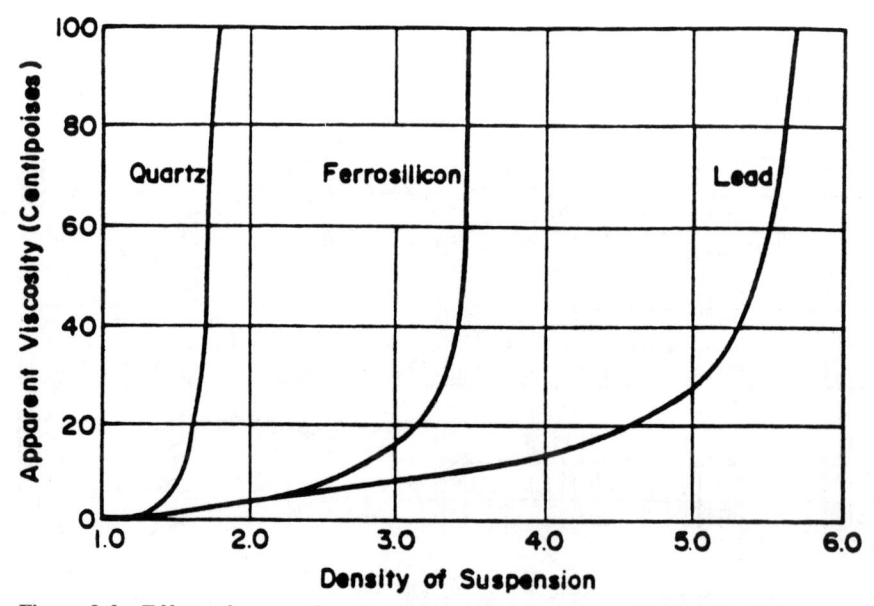

Figure 8-3 Effect of suspension density on viscosity. (From Dallavalle [28].)

points, where the slopes of the curves change suddenly, correspond to suspension densities of 1.38, 2.3, and 3.9 for quartz, ferrosilicon, and lead, respectively. Each critical point occurs at about the same volumetric solids content, 25 percent by volume.

Dallavalle [28] also determined the effect of particle size on viscosity. Figure 8-4 shows this relationship for four particulates with different particle sizes: Mg (OH)$_2$ at 10 μ, CaCO$_3$ at 12.6 μ, BaSO$_4$ at 18 μ, and SiO$_2$ at 40 μ. As the average particle size decreases, viscosity increases. The different relationships between weight percent solid and viscosity are also evident in Figure 8-4, and it becomes obvious why solids content is not a useful way to predict the viscosity of a waste, or of the mixed but unsolidified waste after treatment.

Viscosity is a rheological property. It can be defined as the force required to move a unit area of plane surface with unit speed relative to another parallel plane surface, both surfaces separated by a unit thickness of fluid. It is also defined and measured as the ratio of shear stress to shear rate.

Rheological types. Different wastes exhibit very different rheological properties, but they can be classified into two groups: Newtonian, meaning viscosity is constant (at constant temperature) and independent of shear rate; and non-Newtonian, meaning that viscosity changes in some way with shear rate and/or time. Newtonian materials include water, mineral oils, solvents, glycerol, and many clear solutions of substances in water. Non-Newtonian materials may be subdivided into the following types:

- *Pseudoplastic* materials are those in which viscosity decreases with increasing shear rate. This behavior is typical of many sludges and is exemplified in general by solutions of polymers, suspensions, dispersions, emulsions, paints and adhesives. Pseudoplasticity is probably the most common classification that we encounter with wastes.
- *Plastic* materials are very similar to pseudoplastic materials, except that a mini-

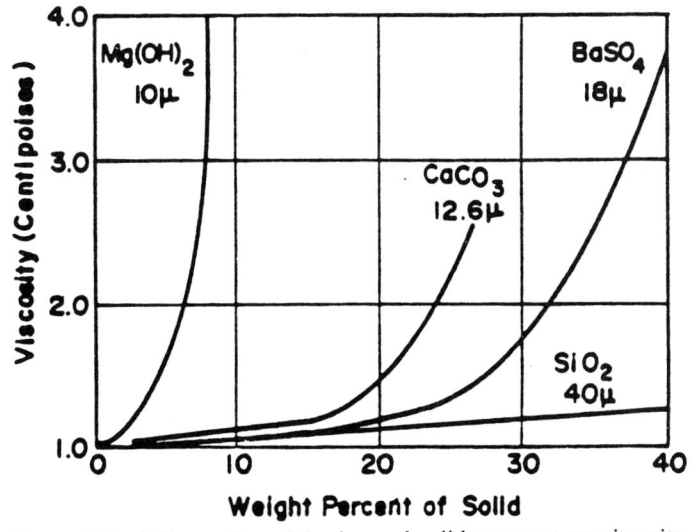

Figure 8-4 Effects of particle size and solids content on viscosity. (From Dallavalle [28].)

mum shear stress must be applied to the material before any movement takes place. This minimum stress (force) is called the *yield value*. Examples of plastic materials are catsup (the yield value must be exceeded by striking the bottle before the catsup will pour), mayonnaise, pastes, printing inks, and many suspensions with high solids content.
- *Thixotropic* materials are also similar to pseudoplastic materials, in that viscosity decreases with increasing shear rate, but in addition viscosity decreases with time at constant shear rate. This gives rise to the "memory" effect that eventually disappears, but may remain for days before the material returns to its original condition. The effect is due to the existence of a definite structure (gel) within the material. (There is also a very rare type of material known as *antithixotropic* or *rheopectic*, in which the viscosity increases with time at constant shear rate.) Examples of thixotropic materials include some coatings, bentonite suspensions, certain pharmaceuticals (ointments, cosmetics, gels, lotions, lipsticks, nail polish), and heavy printing inks.
- *Dilatent* materials are the reverse of pseudoplastic materials, that is, viscosity increases with increasing shear rate. This behavior is relatively rare, but may be encountered with highly concentrated suspensions such as plastisols, some coatings, ceramic and metal particle slurries, and some silicones. It is a serious problem because its occurrence can cause a pump to seize up when pumping an apparently fluid material. This can sometimes be detected in the laboratory by syringing a sample—the force applied will increase the viscosity to the point that the sample cannot be ejected.

Wastes may also be classified into rough viscosity categories. One such classification is:

- Liquid: Pumpable materials with low, generally less than several percent suspended solids.
- Pumpable: Pumpable fluid or sludge, usually less than 15 percent solids, but up to 50 percent with some wastes and as low as 5 percent with others.
- Flowable: Not pumpable, but flows and usually releases free liquid. More than 15 percent solids in most cases, but also includes dry dusts such as flyashes. Will not support a load.
- Non-flowable: Soils, many filter cakes.

Ease of Wetting/Surface Tension. In water-based sludges, solid particles are usually completely wetted, although not always. In emulsions, surface tension effects influence the droplet size and stability. Where the waste is multilayered, it may be difficult to homogenize the mixture for subsequent solidification. This is especially true with wastes containing oil and other hydrophobic organics. If the quantity of oil is small, homogenization may be accomplished with high shear mechanical mixing. If this is not sufficient, the use of surfactants may be indicated. The latter technique has been used effectively in the nuclear industry for incorporation of pump oils into a water-based CFS system, in this case, Portland cement/soluble silicate. The relative proportions and order of addition are often critical. There is a limit on how much oil can be incorporated into an aqueous system, although it is sometimes possible to use an inverse emulsion.

When dry powders are to be incorporated into CFS systems, either for physical stability or to fix a constituent, they must be wetted. The higher the surface area, the more difficult it is to wet the particles. Usually, wetting can be accomplished by intensive mixing, but occasionally a surfactant is helpful. Care should be taken to assure that the surfactant does not interfere with the solidification reaction, as some have been known to do.

Bulk Density (for solids). With unsaturated solids, such as soils, dusts, flyashes, and dried out filter cakes, it is important to know the bulk density so that weight-to-volume calculations may be carried out. Bulk density should be measured in two conditions, uncompacted and compacted, to provide complete information. Contaminated soils, for example, are often in the compacted state when excavated, then are in the uncompacted state as delivered to the mixer and transported through the system to the disposal site, then are recompacted in the disposal site. To properly design a CFS system, it is necessary to know all of these bulk density values. Measurement methods are described in Chapter 17.

Flash Point. Although usually not a factor with CFS systems, occasionally an aqueous waste will contain sufficient flammable solvent to be a potential hazard while being processed, or even in the final solid. Flash point determination is a standard test (Chapter 17) for liquids. No standard method is available for solids, but a useful quick test is to operate a spark generator (torch lighters work well) close to the surface. If the waste is a flammable hazard, it will ignite.

Chemical Properties

Chemical properties are important both for fixation, as we have seen in earlier chapters, and for solidification. Here we will be concerned only with chemical effects on solidification reactions and on the properties of the CFS product. The parameters of prime interest are listed below.

- pH: alkalinity/acidity (%)
- Waste analysis
    Metals, total
    Metals, leached
    Inorganics, total
    Inorganics, leached
    Organics, total
    Organics, leached
- Speciation of constituents—inhibition and acceleration
- Redox potential
- Zeta potential

Alkalinity/Acidity and pH. Most inorganic CFS processes can handle a wide pH range, but the extremes can cause problems. Very high pH, >12, is not counteracted, at least not immediately, by most processes. Very low pH is quickly neutralized by most reagents if the total acidity is not too high. In the latter case, it is advisable, and more economical, to partially neutralize the waste to pH 5 or greater with lime or other alkali before solidification. In the case of high pH, it may be important whether the pH comes from small amounts of strong alkali such as

NaOH, or from large quantities of lime. Actually, the most important parameter is not pH per se, but total alkalinity or acidity. The former can have an inhibitory effect; the latter is primarily of concern in the economics of the overall process. The anion associated with the acid may affect setting, for example, in the case of chloride from HCl, but again, the quantity is important. Small amounts will usually have no effect on inorganic systems.

Composition and Speciation. Soluble species in the waste may have considerable effect on setting, causing inhibition or even preventing setting entirely. These species can also act as accelerators. These effects are not always negative; sometimes it is desirable to speed up or slow down the normal setting reactions for practical reasons in the process design. For example, delayed setting may be necessary to prevent an excessive viscosity increase in the mixing or conveying equipment. This is especially true with fast-setting systems such as Portland cement/soluble silicate. On the other hand, a fast set may be required to get sensitive gelation reactions through that stage before the material freezes in cold weather.

The details of these effects vary considerably for different CFS systems, some of which include anti-inhibitors to counteract undesirable constituents in the waste. These will be discussed in Chapters 10–15 for the individual process types. Some general comments, however, can be made about component groupings.

Metals. Generally compatible with all processes, although they may temporarily inhibit setting in cement-based processes.

Soluble salts. Can decrease durability and inhibit setting in inorganic processes. Compatibility with organic processes varies with the process, from no effect to seriously decreasing durability.

Organics. Oils, greases, and other nonpolar organics may inhibit setting and decrease long-term durability in inorganic systems. Wastes with more than 10 to 20 percent nonpolar organics are usually not considered good candidates for the commercial inorganic CFS systems. Small amounts normally do not interfere with the solidification reactions, nor do they decrease durability, although they may volatilize from the waste during the process and afterward. Organic CFS systems may effectively immobilize nonpolar and polar organics, although some species may inhibit setting.

Polar organics generally have little effect on setting, although some alcohols may retard setting. There is some evidence that these compounds may decrease long-term durability [27]. Solid organics, such as tars and plastics, have no effect on setting and may increase durability.

Oxidizers. These components are generally compatible with inorganic systems, but not with organic processes. They may cause breakdown of the matrix, or even violent decomposition if the right type and quantity are present.

Redox Potential. The oxidation state of the system may affect solidification reactions in some cases, although there is no mention in the literature of direct effects, and the author has not observed any. Indirect effects can occur by mechanisms such as biological growth as determined by aerobic/anaerobic conditions. Also, valence states of metals are often determined by redox conditions, and inhibitory reactions can be altered by the valence state of the metal. Under extreme conditions of oxidation or reduction, organic CFS processes can be substantially affected and some inorganic and organic constituents in inorganic systems may be altered.

Zeta Potential. The electrochemical surface properties of very small particles, especially in the colloidal state, may have a pronounced effect on viscosity of wastes. This typically occurs in the size range of five microns and smaller. Zeta potential is a measurable quantity that correlates with these effects, and is widely used in waste water treatment for flocculation tests to determine the type and amount of polyelectrolyte to use. The properties of colloids do appear to influence setting behavior with some wastes and with some CFS processes, but this area has not been studied sufficiently to establish general cause–effect relationships.

WASTE STREAMS AND TYPES SUITABLE FOR CFS

We have already discussed an important source of information about the wastes that are, or are not, suitable for the application of CFS technology—the CONTEC Waste Data Base (CWDB). A summary of the waste codes and descriptions was given earlier in this chapter. The complete CWDB is presented in Appendix 1. Such a data base is vital to any individual or organization working in the CFS field. The CWDB will serve as a starting point, which the user can expand and enlarge to suit his or her own needs. As regulations change, and new ones come on the scene, the data base must change as this one has over the years. Wastes that were previously treatable with CFS methods may no longer be so under the upcoming landbans of 1989 and 1990.

In addition to the CWDB, there are other ways in which we can look at the results of years of testing in this field. Several of these are presented in the following lists and tables.

General Industrial Waste Categories

Wastes are listed below by general type under broad industry groupings.

Industry	*Waste Category or Source*
Automobile	Automobile assembly wastes, foundry plant wastes, neutralized pickle liquors, treated plating wastes, treatment plant wastes
Chemical	Acids, alkalies, metal-containing sludges, treatment plant sludge
Chemical cleaning	Spent cleaning solutions
Dredging	Contaminated dredge spoils
Food processing	Biological treatment sludges
Leather tanning and finishing	Biological treatment sludges, metal-containing sludges
Metal finishing and major appliance	Dissolved metal solutions, pickle li-

268   WASTES AND WASTE SOURCES

Industry	Waste Category or Source
	quors, rinse water neutralization sludge, treatment plant sludge
Municipal	Sewage sludges, water treatment sludges
Nonferrous metals	Air pollution control (APC) dust and sludges, lime/limestone wet scrubber sludge, waste pickle liquors, water treatment sludge
Paint and painting	Metal pickling and cleaning wastes, paint sludges
Pharmaceutical	Biological treatment sludge, filter cake, spent carbon
Plastic and rubber	Biological treatment sludge, metal-containing sludge
Pollution control	APC sludges, general spent activated carbon, spent resins, water treatment plant sludges
Power	Flyash, lime/limestone scrubber sludges, boiler cleaning solutions
Pulp and paper	Biological treatment sludges, spent clay and fibers
Refinery and petrochemical	API oil/water/sludge mixtures, biological treatment sludge, spent lime sludges
Sanitary landfill	Landfill leachates
Steel	APC dust and sludges, metal fines, scale pit sludge, waste pickle liquors, water treatment sludge
Textile	Biological treatment sludges, metal-containing sludges

Specific Waste Streams Suited to CFS

Table 8-5 lists a large number of waste streams that have been tested and were found to be amenable to CFS treatment, including fixation of constituents to render the waste nonhazardous as defined under the RCRA. This list does not include wastes currently listed under the RCRA [4], unless those wastes have been delisted at one time or another by use of CFS technology. It should be remembered in viewing this list that the forthcoming landbans may, by rule, remove some of these wastes from the possibility of CFS treatment. Also, some of these streams may not have been determinable to be hazardous by past, present, or future standards, but were solidified for other reasons.

Wastes Suited to Solidification Only

Many waste streams are not amenable to detoxification by fixation, either because of constituent types and levels that, by rule, do not allow that possibility or because nonhazardous leaching levels cannot be met in a practical way. An example of the

TABLE 8-5 Specific Waste Streams that Have Been Chemically Solidified and Detoxified

Waste Type	Industry or Process	Potentially Hazardous Constituents
Grease, soap, and vegetable oil in water		Oil and grease
Biosludge, activated sludge process	Chemical	Metals
Al and Na salt sludge	Metal finishing	—
Mixed plating waste sludge	Electroplating	Metals
Aluminum treatment waste sludge	Aluminum plating	Metals
Tannery waste	Tannery	Cr, organics
Sodium metasilicate waste	Minerals manufacturing	—
Contaminated soil	—	Cr
Caustic waste	Aluminum drawing	—
Filter cake	Aluminum drawing	—
DAF sludge	Railroad	Oil and grease
Landfill leachate	Hazardous waste management facility	Metals, organics
Copper catalyst or substrate	Organic chemicals—VCM process	Cu
Copper hydroxide sludge	Organic chemicals—VCM process	Cu
Clarifier sludge	Container plant	Metals
Centrifuge sludge	Container plant	Metals
Clarifier sludge	Brass plating	Cu, Zn
Filter press cake	Brass plating	Cu, Zn
Calcium fluoride sludge	Glass manufacturing	Metals, F
Metal hydroxide lagoon sludge	Machinery	Metals
Coke dust	Metal fabrication	Barium, organics
Battery waste sludge	Battery manufacturing	Lead compounds
Flue dust	Battery manufacturing	Lead, tin, antimony
Spent sulfite liquor	Pulping operation	Sulfite, oxalates, organics
Latex waste	Pulping operation	Organics
Mercury cell	Chlor-alkalie manufacturing	Hg
Dredging sludge	Chemical	—
Mixed refining wastes	Oil refining	F, organics
Landfill leachate	Marine terminal	Metals, organics
Acrylic/epoxy paint wash	Instrument manufacturing	Organics
Alodine conversion coating waste	Metal finishing	Metals
Electroplating WWT sludge	Metal finishing	Metals
Battery plant sludge	Battery manufacturing	Pb, Cd, Zn, and Ni
Mixed lagoon sludge	Manufacturing	Metals, organics
Boron fluoride waste	Pilot plant	B and F compounds
WWT filter cake	Chemical	Metals
Oily sludge containing < 500 ppm PCB	Manufacturing	PCB
Electroplating centrifuge sludge	Manufacturing	Metals
Neutralized acids	Manufacturing	Metals
Chromium plating waste	Manufacturing	Chromium, other metals
Cyanide plating waste sludge	Manufacturing	Metals, cyanide
Litho printing waste sludge	Printer	Metals, organics
Mixed sludge lagoon	Organic chemical manufacturing	Organics
Electrochemical machining	Metal fabrication	Metals
Neutralized waste aqua regia	Metal fabrication	Metals, inorganics
Waste sulfuric acid pickle liquor	Steel manufacturing	Metals
Drilling mud	Oil production	Cr, organics

(continued)

TABLE 8-5 Specific Waste Streams that Have Been Chemically Solidified and Detoxified (*Continued*)

Waste Type	Industry or Process	Potentially Hazardous Constituents
Foundry cupola dust	Foundry	Metals
Lagoon biosludge	Tannery	Metals, organics
Phosphoric acid filter cake	Phosphoric acid manufacturing	Arsenic
Kiln dust	Portland cement manufacturing	Lead
Buff wash	Metal finishing	Metals, oil
Mixed lagoon sludge	Oil refining	Metals, organics
Caustic waste	Oil refining	Organics
Latex paint WWT sludge	Paint manufacturing	Metals, organics
Acid oil sludge	Oil re-refining	Lead, oil, organics
Brass foundry sludge	Brass foundry	Copper, tin
Foundry sand	Brass foundry	Copper, tin
Latex paint sludge	Paint manufacturing	Metals
Caustic tank cleaning sludge	Paint manufacturing	Metals
Incinerator ash	Paint manufacturing	Metals
Lagoon sludge	Flatware manufacturing	Metals, organics
Barrel finishing baghouse dust	Metal finishing	Metals
Oil filter cake	Metal finishing	Metals, oil, and grease
Weak acid neutralization sludge	Hazardous waste management facility	Metals
Oil sludge	Hazardous waste management facility	Oil and grease
Anodizing sludge	Aluminum anodizing	Metals
Chromium sludge	Aluminum anodizing	Metals
Cupola dust	Foundry	Metals
Sludge	Foundry	Metals
Lead leach process residue	Mineral refining	Metals
Weathered oil waste	Petrochemical	Metals, organics
Sewage sludge	Small, private, WWT plant	Metals
Neutralized pickle rinse sludge	Stainless-steel finishing	Metals
Ink WWT sludge	Container printing	Metals
Ink WWT filter cake	Container printing	Metals
WWT sludge	Stainless-steel manufacturing	Metals
Chromium plating sludge	Stainless-steel manufacturing	Chromium
Metal salts slurry	Chemical manufacturing	Cadmium and zinc
Latex waste	Chemical manufacturing	—
Naphthalene waste sludge	Naphthalene-based chemical manufacturing	Naphthalene compounds
Metal hydroxide sludge	Hazardous waste management facility	Metals
Metal salts slurry	Hazardous waste management facility	Metals
Spent bleach clay	Oil re-refining	Metals, oil
Acrylic paint waste, solvent base	Auto manufacturing	Metals, organics
Primer paint waste, water base	Auto manufacturing	Metals, organics
Baghouse dust	Aluminum scrap remelting	Metals
Oil sludge	Oil refinery	Metals, oil
Chromic rinse	Auto manufacturing	Chromium
WWT sludge	Chemical manufacturing	Metals, organics
Flyash	Power company	Arsenic

TABLE 8-5 Specific Waste Streams that Have Been Chemically Solidified and Detoxified (*Continued*)

Waste Type	Industry or Process	Potentially Hazardous Constituents
WWT sludge	Hazardous waste management facility	Metals, organics
Paint waste filter cake	Instrument manufacturing	Chromium, organics
Spent plating bath	Instrument manufacturing	Metals, complexing agents
Organic acid process waste	Chemical manufacturing	Organics
Dye waste	Textile manufacturing	Copper, mercury, organics
Mixed lagoon sludge	Chemical manufacturing	Organics
Paint waste clarifier underflow	Auto manufacturing	Metals
River dredging mud	Hazardous spill	Pesticides
Waste nitric/hydrofluoric acid pickle liquor	Steel manufacturing	Metals
Mercuric sulfide sludge	Chemical manufacturing	Mercury
Barrel finishing sludge	Tool manufacturing	Metals, oil
Phenolic WWT lagoon sludge	Railroad car manufacturing	Phenol
Oil lagoon sludge	Railroad	Oil
Acid phosphating sludge	Tool manufacturing	Metals
Nickel filter cake	Tool manufacturing	Nickel, organics
Alkaline nickel stripping sludge	Tool manufacturing	Nickel, complexing agents
Filter cake	Steel manufacturing	Metals, oil
Lime neutralized waste pickle liquor	Metal finishing	Metals
NaOH neutralized waste pickle liquor	Metal finishing	Metals
Lagoon bottom sediment	Chemical manufacturing	Mercury
Rag oil lagoon sludge	Aluminium manufacturing	Oil
Anaerobic lagoon sludge	Tanning	Metals
Organic filter cake	Household products manufacturing	Organics
Electrochemical machining sludge	Fabrication	Metals
Oleum/hydrofluoric acid/iodine waste	Chemical manufacturing	Fluoride, iodine reactivity
Neutralized hydrofluoric acid sludge	Chemical manufacturing	Metals, fluoride
Electrostatic precipitator dust	Glass manufacturing	Borosilicates
Fine foundry sand sludge	Auto manufacturing	Phenol
Retention basin sludge	Oil refining	Metals, oil
DAF sludge	Oil refining	Metals, oil
Activated biosludge	Oil refining	Metals, oil
Bearing grinding sludge	Bearing manufacturing	Metals, oil
Phosphoric acid cleaner rinse	Metal fabrication	Metals
Latex-soap-resin waste	Chemical manufacturing	Organics
Water-based-paint waste	Paint manufacturing	Metals, organics
Rolling oil sludge	Aluminum manufacturing	Metals, oil

Note: These wastes are either toxic or of questionable toxicity before treatment, but are converted by solidification/chemical fixation into nontoxic products as determined according to Subpart C, 40 CFR Part 261, Federal Register, Vol. 45, No. 98, May 19, 1980.

TABLE 8-6 Specific Waste Streams that Have Been Chemically Solidified Only[a]

Waste Type	Industry or Process	Constituents
Rubber waste	Rubber manufacturing	Solvents, organics
Paint waste, solvent base	Paint manufacturing	Solvents, organics
Polymers and chlorinated hydrocarbons	Petrochemical manufacturing	Chlorinated hydrocarbons
Arsenical herbicide waste	Herbicide manufacturing	Arsenic compounds
Silicone waste	Silicone manufacturing	Silicones
Oil sludge	Tire manufacturing	Oil, organics
Rubber waste	Tire manufacturing	Synthetic rubber, organics
Waste lubricant	Electrical equipment manufacturing	Parafins, aluminum, water
Tall oil resin waste	Chemical manufacturing	Organics
Oil dispersion and wax emulsion waste	Chemical manufacturing	Organics
Organic resin waste	Chemical manufacturing	Organics
Resin soaps and resin waste	Chemical manufacturing	Organics
Landfill leachate, organic phase	Chemical landfill	Organics
PCB oil	Equipment manufacturing	PCB
Lacquer solvent still bottoms	Coatings manufacturing	Organics
Mixed waste resins, solvents, and pigments	Coatings manufacturing	Metals, organics
Synthetic resin waste	Chemical manufacturing	Organics
Alkyd resin gel waste	Chemical manufacturing	Organics
Acid-organic WWT sludge	Hazardous waste management facility	Metals, organics
Vegetable tanning waste	Tannery	Organics
Carbon tetrachloride sludge	Chemical manufacturing	Carbon tetrachloride, carbon disulfide
Solder stripping solution	Fabrication	Metals, organics
Vinyl plastisol waste	Wire manufacturing	Metals, organics
Phenolic resin wash	Chemical coatings manufacturing	Organics
Phenolic resin concentrate	Chemical coatings manufacturing	Organics
Paint kettle cleaning waste	Coatings manufacturing	Organics
Aluminum powder waste	Glass company	Reactive metal

[a] Some of these wastes are indeterminate as to toxicity under present RCRA standards. Others remain toxic after solidification and require secure landfill.

former is waste-containing PCBs at levels of more than 50 ppm, which makes the waste hazardous by rule under the TSCA. Table 8-6 lists a number of such waste streams that have been solidified, but not "detoxified."

EPA/RCRA Listed Wastes

The listed wastes [2,4] that are hazardous by definition under RCRA appear below. In this case, *detoxification* means that the constituents for which the waste was listed have been destroyed or fixed to levels that would meet the "characteristic" requirements of the RCRA if the waste were not listed. In some cases, these wastes have been "delisted" by means of a CFS process.† In others, CFS treatment may

---

†Delisting is a site-specific process; therefore, a waste delisted at one site is not automatically delisted at another, even though the same criteria have been met.

Hazardous Waste from Nonspecific Sources

Industry and EPA Hazardous Waste Number	Hazardous Waste
F006	Waste water treatment sludges from electroplating operations except from the following processes: (1) sulfuric acid anodizing of aluminum, (2) tin plating on carbon steel, (3) zinc plating (segregated bases) on carbon steel, (4) aluminum or zinc aluminum plating on carbon steel, (5) cleaning/stripping associated with tin, zinc, and aluminum plating on carbon steel, and (6) chemical etching and milling of aluminum
F007	Spent cyanide plating bath solutions from electroplating operations (except for precious metals electroplating spent cyanide plating bath solutions)
F008	Plating bath sludges from the bottom of plating baths from electroplating operations where cyanides are used in the process (except for precious metals electroplating plating bath sludges)
F009	Spent stripping and cleaning bath solutions from electroplating operations where cyanides are used in the process (except for precious metals electroplating spent stripping and clearing bath solutions)
F010	Quenching bath sludge from oil baths from metal heat-treating operations where cyanides are used in the process (except for precious metals heat-treating quenching bath sludges)
F011	Spent cyanide solutions from salt bath pot cleaning from metal heat-treating operations (except for precious metals heat-treating spent cyanide solutions from salt bath pot clearing)
F012	Quenching waste water treatment sludges from metal heat-treating operations where cyanides are used in the process (except for precious metals heat-treating quenching waste water treatment sludges)
F014	Cyanidation waste water treatment tailing pond sediment from mineral metals recovery operations
F015	Spent cyanide bath solutions from mineral metal recovery operations
F019	Waste water treatment sludges from the chemical conversion coating of aluminum

(continued)

Hazardous Waste from Nonspecific Sources (continued)

Industry and EPA Hazardous Waste Number	Hazardous Waste
Wood Preservation:	
K001	Bottom sediment sludge from the treatment of waste waters from wood preserving processes that use creosote and/or pentachlorophenol
Inorganic Pigments:	
K002	Waste water treatment sludge from the production of chrome yellow and orange pigments
K003	Waste water treatment sludge from the production of molybdate orange pigments
K004	Waste water treatment sludge from the production of zinc yellow pigments
K005	Waste water treatment sludge from the production of chrome green pigments
K006	Waste water treatment sludge from the production of chrome oxide green pigments (anhydrous and hydrated)
K007	Waste water treatment sludge from the production of iron blue pigments
K008	Oven residue from the production of chrome oxide green pigments
Organic Chemicals:	
K011	Bottom stream from the waste water stripper in the production of acrylonitrile
K021	Aqueous spent antimony catalyst waste from fluoromethanes production
K104	Combined waste water streams generated from nitrobenzene/anline production
K105	Separated aqueous stream from the reactor product washing step in the production of chlorobenzenes

WASTE STREAMS AND TYPES SUITABLE FOR CFS 275

Inorganic
Chemicals:
K071   Brine purification muds from the mercury cell process in chlorine production, where separately prepurified brine is not used
K106   Waste water treatment sludge from the mercury cell process in chlorine production

Pesticides:
K031   By-product salts generated in the production of MSMA and cacodylic acid
K032   Waste water treatment sludge from the production of chlordane
K033   Waste water and scrub water from the chlorination of cyclopentadiene in the production of chlordane
K034   Filter solids from the filtration of hexachlorocyclopentadiene in the production of chlordane
K035   Waste water treatment sludges generated in the production of creosote
K037   Waste water treatment sludges from the production of disulfoton
K038   Waste water from the washing and stripping of phorate production
K039   Filter cake from the filtration of diethylphosphorodithoric acid in the production of phorate
K040   Waste water treatment sludge from the production of phorate
K041   Waste water treatment sludge from the production of toxaphene
K096   Untreated process waste water from the production of toxaphene
K099   Untreated waste water from the production of 2,4-D

Explosives:
K044   Waste water treatment sludges from the manufacturing and processing of explosives
K045   Spent carbon from the treatment of waste water containing explosives
K046   Waste water treatment sludges from the manufacturing, formulation, and loading of lead-based initiating compounds
K047   Pink/red water from TNT operations

(continued)

276 WASTES AND WASTE SOURCES

Hazardous Waste from Nonspecific Sources (continued)

Industry and EPA Hazardous Waste Number	Hazardous Waste
Petroleum Refining:	
K048	Dissolved air flotation (DAF) float from the petroleum refining industry
K049	Slop oil emulsion solids from the petroleum refining industry
K050	Heat exchanger bundle cleaning sludge from the petroleum refining industry
K051	API separator sludge from the petroleum refining industry
K052	Tank bottoms (leaded) from the petroleum refining industry
Iron and Steel:	
K061	Emission control dust/sludge from the electric furnace production of steel
K062	Spent pickle liquor from steel finishing operations
Primary Copper:	
K064	Acid plant blowdown slurry/sludge resulting from the thickening of blowdown slurry from primary copper production
Primary Lead:	
K065	Surface impoundment solids contained in and dredged from surface impoundments at primary lead smelting facilities

WASTE STREAMS AND TYPES SUITABLE FOR CFS

Primary Zinc:
- K066 — Sludge from treatment of process waste water and/or acid plant blowdown from primary zinc production
- K067 — Electroytic anode slimes/sludges from primary zinc production
- K068 — Cadmium plant leach residue (iron oxide) from primary zinc production

Secondary Lead:
- K069 — Emission control dust/sludge from secondary lead smelting
- K100 — Waste leaching solution from acid leaching of emission control dust/sludge from secondary lead smelting

Veterinary Pharmaceuticals:
- K084 — Waste water treatment sludges generated during the production of veterinary pharmaceuticals from arsenic or organoarsenic compounds
- K102 — Residue from the use of activated carbon for decolorization in the production of veterinary pharmaceuticals from arsenic or organoarsenic compounds

Ink Formulation:
- K086 — Solvent washes and sludges, caustic washes and sludges, or water washes and sludges from cleaning tubs and equipment used in the formulation of ink from pigments, driers, soaps, and stabilizers containing chromium and lead

Coking:
- K060 — Ammonia still lime sludge from coking operations
- K087 — Decanter tank tar sludge from coking operations

(continued)

278 WASTES AND WASTE SOURCES

Hazardous Waste from Nonspecific Sources (continued)

Industry and EPA Hazardous Waste Number	Hazardous Waste
Formerly listed or tentatively listed wastes no longer listed	Chrome (blue) trimmings generated by the following subcategories of the leather tanning and finishing industry: hair pulp/chrome tan/retan/wet finish; hair save/chrome tan/retan/wet finish; retan/wet finish; no beamhouse; through-the-blue; and shearling
	Chrome (blue) shavings generated by the following subcategories of the leather tanning and finishing industry: hair pulp/chrome tan/retan/wet finish; hair save/chrome tan/retan/wet finish; retan/wet finish; no beamhouse; through-the-blue; and shearling
	Buffing dust generated by the following subcategories of the leather tanning and finishing industry: hair pulp/ chrome tan/retan/wet finish; hair save/chrome tan/retan/wet finish; retan/wet finish; no beamhouse; and through-the-blue
	Sewer screenings generated by the following subcategories of the leather tanning and finishing industry: hair pulp/chrome tan/retan/wet finish; hair save/chrome tan/retan/wet finish; retan/wet finish; no beamhouse; through-the-blue; and shearling

- Waste water treatment sludges generated by the following subcategories of the leather tanning and finishing industry: hair pulp/chrome tan/retan/wet finish; hair save/chrome tan/retan/wet finish; retan/wet finish; no beamhouse; through-the-blue; and shearling
- Waste water treatment sludges generated by the following subcategories of the leather tanning and finishing industry: hair pulp/chrome tan/retan/wet finish; hair save/chrome tan/retan/wet finish; and through-the-blue
- Waste water treatment sludges generated by the following subcategory of the leather tanning and finishing industry: hair save/non-chrome tan/retan/wet finish
- Sludge from lime treatment of spent pickle liquor from steel finishing operations
- Water or caustic cleaning wastes from equipment and tank cleaning from paint manufacturing
- Waste water treatment sludges from paint manufacturing
- Emission control dust or sludge from paint manufacturing
- Emission control dust or sludge from ferrochromium–silicon production
- Emission control dust or sludge from ferrochromium production
- Emission control dust or sludge from ferromanganese production

no longer be applicable by the publication date of this book, due to forthcoming landban regulations. Nevertheless, the listing demonstrates how useful CFS treatment can be.

Landban Wastes and BDATs

In Chapter 1, the effects of the various landbans were briefly discussed. This subject is so complex that it has been said that no one understands it fully, in all its ramifications. However, as it applies to CFS technology, it actually officially establishes CFS as a best demonstrated available technology (BDAT) for a number of wastes and residuals from other BDATs. Furthermore, the next two (and last) landbans that go into effect in 1989 and 1990, respectively, will certainly specify CFS as BDAT for a number of additional waste streams.

Table 8-7 lists the EPA-listed wastes or residuals for which CFS technology is required as a BDAT or may be used on residuals from other BDATs to meet the leaching standards. This regulation is for the so-called first-third of the landban waste list set forth in 1986 [4] by the EPA. This portion of the list involves only "listed" wastes at this time. The subsequent "thirds" will finally include the whole waste universe subject to the RCRA. Since many of the wastes yet to be subjected to landbans are inorganic, it is likely that many more CFS BDATs will be established.

CHARACTERIZATION OF TODAY'S WASTE STREAMS

We have seen how wastes have been characterized to date: the waste classification systems; paper studies of estimated waste types, sources, and industries; waste properties; waste listings and data bases. The perceptive reader may have noticed that there is a dearth of in-depth, hands-on, detailed studies of specific waste streams, even for those wastes "listed" by the EPA as deserving of special control. In fact, there are no published studies of individual wastes with extensive laboratory tests and analytical data (all done with the same techniques, in the same time period, by the same laboratory) on a full range of samples from representative sources. The reason for this is simple, if startling to the novice in this field. Regulatory agencies, having no practical access to a good sampling of any waste type, can only collect bits and pieces of data from various sources and try to put them together in a coherent and useful way.

Such a situation hampers attempts by the EPA and other regulatory agencies in making rational rules, especially with the time pressures of the landbans placed on them by Congress. Recognizing that this situation benefitted no one, and might result in the proposal of unreasonable rules on the basis of inadequate or incorrect information, Chemical Waste Management, Inc. (CWM) in 1987 carried out a novel program [29]. This program was designed to thoroughly characterize one waste stream of particular interest and importance in the upcoming first-third landban. The waste was F006—waste water treatment sludges from electroplating operations.

Within a period of several months, 100 samples were collected from two of the company's central treatment sites covering a large proportion of the U.S. industry that generates this waste. These samples were taken from commercial waste loads received at the sites from a large number and variety of generators. Each sample was completely characterized, using standard parameters and test methods. In addi-

TABLE 8-7 Landban BDAT Standards for CFS Treatment

Waste Code	BDAT Treatment Train	As	Cd	Cr	Pb	Hg	Ni	Se	Ag
F006	CFS		0.066	5.200	0.510		0.320		0.072
K001	Incineration + CFS				0.510				
K022	Incineration + CFS			5.200			0.320		
K046	CFS				0.180				
K048	Incineration + CFS	0.004		1.700			0.048	0.025	
K049	Incineration + CFS	0.004		1.700			0.048	0.025	
K050	Incineration + CFS	0.004		1.700			0.048	0.025	
K051	Incineration + CFS	0.004		1.700			0.048	0.025	
K052	Incineration + CFS	0.004		1.700			0.048	0.025	
K061	CFS (<15 percent Zn)		0.140	5.200	0.240		0.320		
K062	Metal fixation			0.094	0.370				
K071	Oxidation and treatment					0.025			
K086	Residue from incineration			0.094	0.370				
K087	Incineration + CFS				0.510				
K101	Incineration + CFS		0.066	5.200	0.510		0.320		
K102	Incineration + CFS		0.066	5.200	0.510		0.320		

TCLP Treatment Standard (mg/l)

tion to complete metals analyses, solids content and pH were measured and observations made about other subjective properties. The source of each sample was documented, so that information about generating process, treatment procedures, etc., could be gathered later if required. In addition to these analyses, each sample was subjected to the TCLP leaching test for all the metal parameters.

Eleven of these samples were then treated by one of CWM's standard CFS procedures, using cement kiln dust at various mix ratios. After curing, these treated samples were subjected to the TCLP procedure. The data from these stabilized samples were presented to the EPA, along with some of the characterization information and full quality control/quality assurance (QC/QA) documentation. Those data, with appropriate statistical factors derived from the QC/QA information, were used by the EPA to set the first-third landban standard for F006. The method used is described in the background document for F006 [30]. The complete leaching results are presented in Chapter 14. Statistical analyses of the distribution of the TCLP leachate concentrations for various metals of interest are given in Figures 8-5 and 8-6.

Unfortunately, this study cannot be released in its entirety because it contains a great deal of proprietary marketing information. However, some data of general interest are given in Figures 8-7 and 8-8. Figure 8-7 shows the distribution of total solids in the 100 waste streams sampled. The median solids range is 20 to 30 percent, with the distribution sharply skewed toward higher solids. This is not surprising in view of the strong trend, especially in the electroplating industry, toward dewatering. Ten years ago, this graph would have looked very different. Figure 8-8 shows the distribution of pH values for the as-received samples. These results are interesting because the pH range is lower than expected. For metal fixation, pH is usually

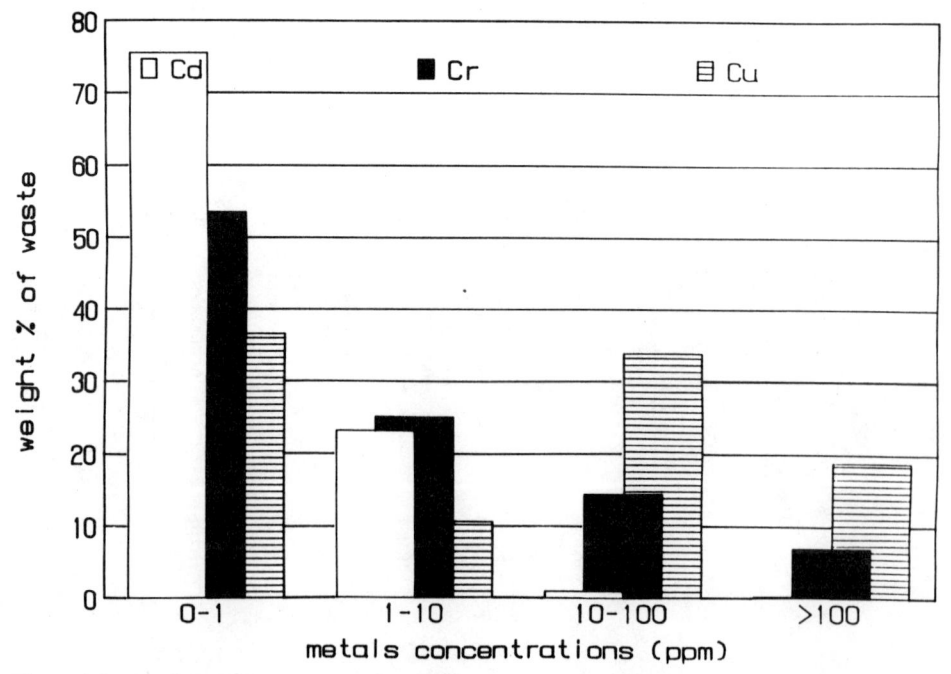

Figure 8-5 Statistical distribution of TCLP concentration in F006 waste streams: Cd, Cr, and Cu. 91 samples / 45000 tons. (From Chemical Waste Management [29].)

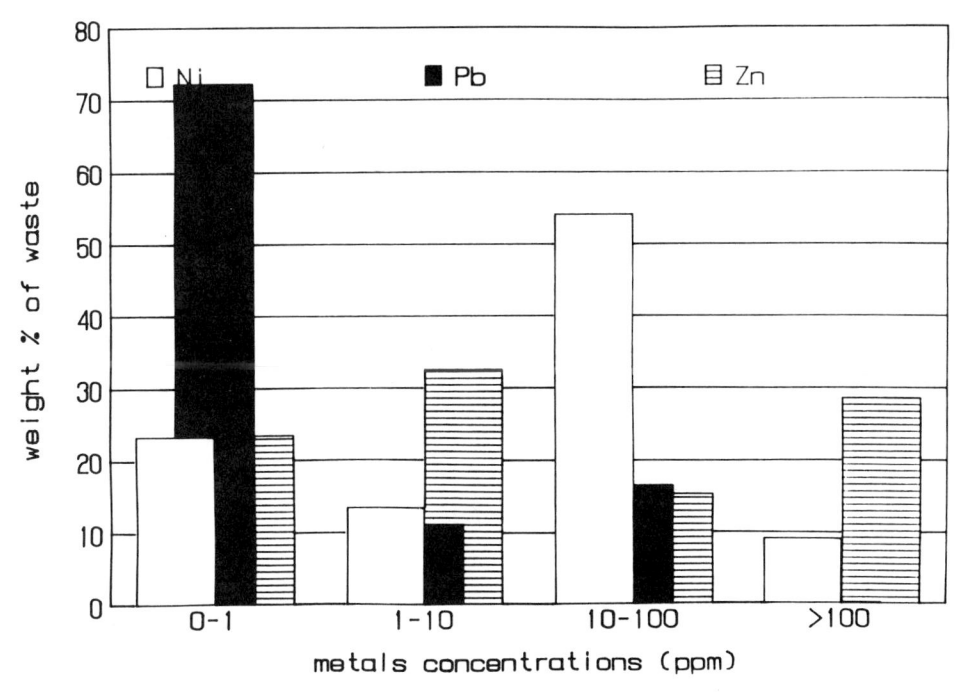

Figure 8-6 Statistical distribution of TCLP concentration in F006 waste streams; Ni, Pb, and Zn. 91 samples / 45000 tons. (From Chemical Waste Management [29].)

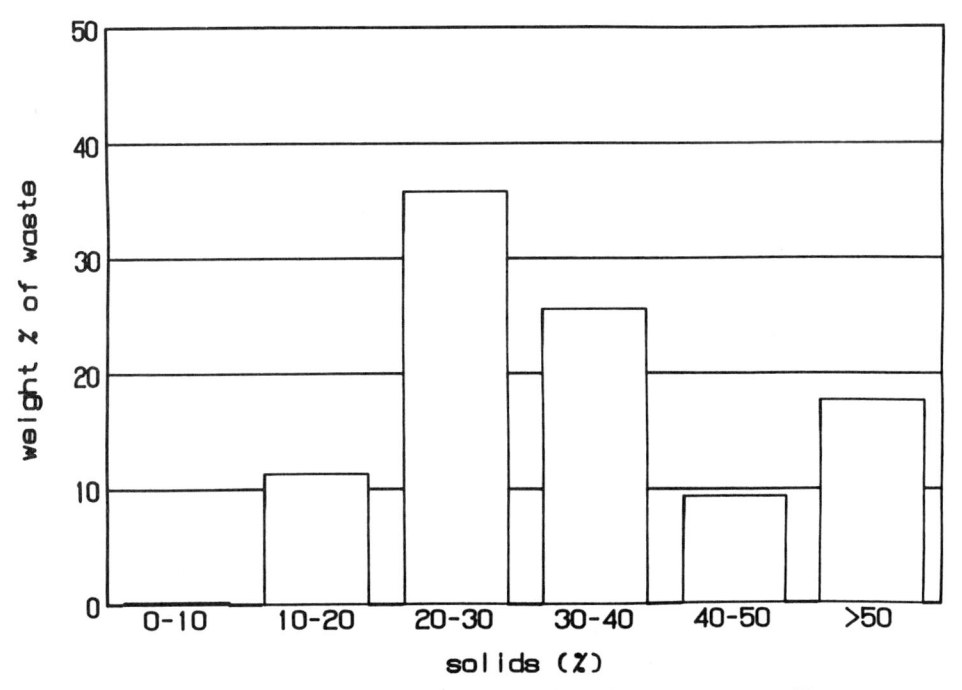

Figure 8-7 Statistical distribution of solids content in F006 waste streams. (From Chemical Waste Management [29].)

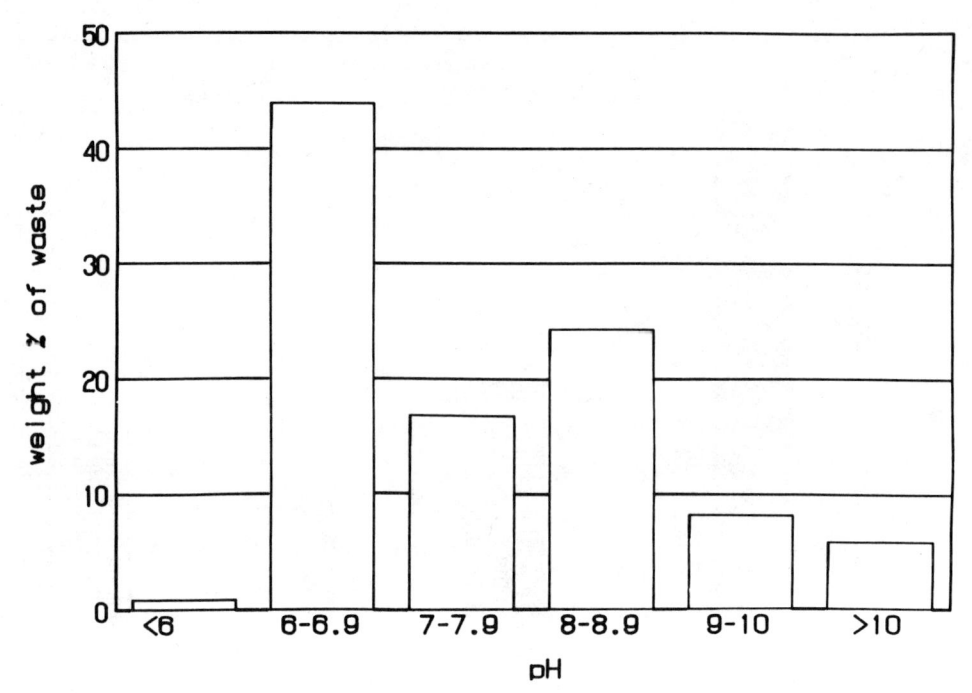

Figure 8-8 Statistical distribution of pH in F006 waste streams. (From Chemical Waste Management [29].)

set at about 9 or higher. Only 15 percent of the sample population had pH values in this range, and about 45 percent were in the below-7 range. On reflection, the lack of high values is not surprising, since the water phase associated with these sludges and filter cakes is usually discharged to POTWs or streams, where pH is limited to a narrow range around the neutral point. The low values may be evidence of the mixing of some of these residues with raw process or laboratory wastes in the disposal container. Other, metal speciation evidence also points to this conclusion.

Other studies such as this are badly needed to fully and accurately characterize important waste streams as they exist today. Hopefully, the CWM study will point the way for others.

REFERENCES

1. Federal Register. **45**(98): 33122–33133 (May 19, 1980).
2. U.S. Toxic Substance Control Act, Jan. 1, 1977.
3. U.S. EPA. *Solid Waste Leaching Procedure Manual,* SW 924. Cincinnati, OH, 1985.
4. Federal Register. **51**(102): 19305–19308 (May 28, 1986).
5. Ontario Waste Management Corporation. *Facilities Development Process, Phase I Report.* Ontario, Canada, 1982.
6. Conner, J. R. *CONTEC Data Base.* Conner Technologies, Atlanta, GA, 1985.
7. *Hazardous Materials Intelligence Report,* p. 3, Feb. 28, 1987.
8. *Wall Street Journal,* p. 33, Oct. 10, 1986.
9. *Inside EPA,* Sept. 2, 1988.
10. *Chem. Eng.* 95 June 22, 1987.
11. *Chem. Week:* 36 (Aug. 19, 1987).

12. Versar, Inc. *Assessment of Industrial Hazardous Waste Practices, Storage and Primary Battery Industries.* Springfield, VA, 1974.
13. TRW Inc. *Assessment of Industrial Hazardous Waste Practices, Organic Chemicals, Pesticides and Explosives Industries.* Redondo Beach, CA, 1975.
14. Battelle, Columbus Laboratories. *Assessment of Industrial Hazardous Waste Practices, Electroplating and Metal Finishing Industries.* Columbus, OH, 1975.
15. Wapora, Inc. *Assessment of Industrial Hazardous Waste Practices, Paint Industry.* Cincinnati, OH, 1975.
16. Jacobs Engineering Co. *Assessment of Industrial Hazardous Waste Practices, Petroleum Refining.* Cincinnati, OH, 1975.
17. Arthur D. Little Inc. *Hazardous Waste Generation, Treatment and Disposal in the Pharmaceutical Industry.* Cincinnati, OH, 1974.
18. Calspan Corp. *Assessment of Industrial Hazardous Waste Practices in the Metal Smelting and Refining Industry.* Buffalo, NY, 1975.
19. Foster D. Snell Inc. *Assessment of Industrial Hazardous Waste Practices, Rubber and Plastics Industry.* Florham Park, NJ, 1976.
20. SCS Engineers Inc. *Assessment of Industrial Hazardous Waste Practices—Leather Tanning and Finishing Industry.* Reston, VA, 1976.
21. Wapora Inc. *Assessment of Industrial Hazardous Waste Practices, Special Machinery and Manufacturing.* Cincinnati, OH, 1976.
22. Midwest Research Institute. *Assessment of Industrial Hazardous Waste Practices, Metals Mining Industry.* Cincinnati, OH, 1976.
23. Versar, Inc. *Assessment of Industrial Hazardous Waste Practices, Textiles Industry.* Springfield, VA, 1976.
24. Versar, Inc. *Assessment of Industrial Hazardous Waste Practices, Inorganic Chemical Industry.* Springfield, VA, 1974.
25. Leo, P. P., R. B. Fling, and J. Rossoff. Flue gas desulfurization waste disposal study at the Shawnee power station. Presented at the Symposium on Flue Gas Desulfurization, Research Triangle Park, NC, Nov. 11, 1977.
26. Sowers, G. B., and G. F. Sowers. *Introductory Soil Mechanics and Foundations,* New York: Macmillan, 1970.
27. JACA Corp. *Critical Characteristics and Properties of Hazardous Waste Solidification/Stabilization* (Draft Report). Cincinnati, OH: U.S. EPA, 1987.
28. Dallavalle, J. M. *Micrometrics.* New York: Pitman, 1948.
29. Chemical Waste Management, Inc. *Characterization of F006 Electroplating Wastes,* Technical Note 87-116. Riverdale, IL, 1987.
30. U.S. EPA. *Best Demonstrated Available Technology (BDAT) Background Document for F006,* EPA/530-SW-88-0009-1. Washington, DC, May 1988.

Chapter 9

CFS PROCESSES: GENERAL PROPERTIES AND COMPARATIVE EVALUATION

In Chapter 7, the basis was laid for discussion of the CFS process, removed from all of the other considerations involved in a total system or project. The CFS process was described in terms of its engineering features. Now, we will discuss CFS as a chemical process, focusing on the physical and chemical factors that affect it, and the generic and proprietary process categories and how they compare with each other. Examples of the results of laboratory, pilot, and commercial studies and projects will then be used to provide a comparative evaluation of how well they function with individual waste types and in CFS projects in the field.

Before going into further detail about processes, it is helpful for the reader to get a visual picture of what a CFS process really looks like. This is shown schematically in Figure 9-1. The process consists basically of waste and reagent feed systems, a mixer (reaction zone), a means of conveying the treated waste from the mixer, and an area for the mixture to harden. The same block diagram could be used for nearly all commercial inorganic systems, and for most organic ones as well. As a matter of fact, most CFS systems are quite simple conceptually [1].

When we move to the details, however, the system gets more complex. Figure 9-2 is the pictorial flow diagram of a major central CFS treatment facility that must handle all types of waste, delivered in a variety of containers and vehicles. Such a facility must allow for every contingency, mechanical as well as chemical.

The waste to be treated is conveyed by pump or mechanical conveyor into a surge tank or feed hopper, which in turn feeds the waste into the mixer where it is mixed with the CFS reagents. Depending on the system used, one or more dry or liquid reagent may be added to the waste in the mixer. The mixing process normally takes from 1 to 15 minutes in batch mode, or has a residence time of 1 to 3 minutes in a continuous system. Mixing time depends on the size of the batch, the type of waste, and the amounts and types of reagents used. After mixing, the waste, still in liquid or semisolid form, is removed from the mixer either by pumping, in the case of a liquid, or by mechanical conveyor or direct dump, in the case of semisolids. This same conveyance mechanism may move the waste to the curing area, or in some cases, directly to the disposal area. Alternately, the waste may be transferred to another means of conveyance—pump, screw or belt conveyor, or truck—for

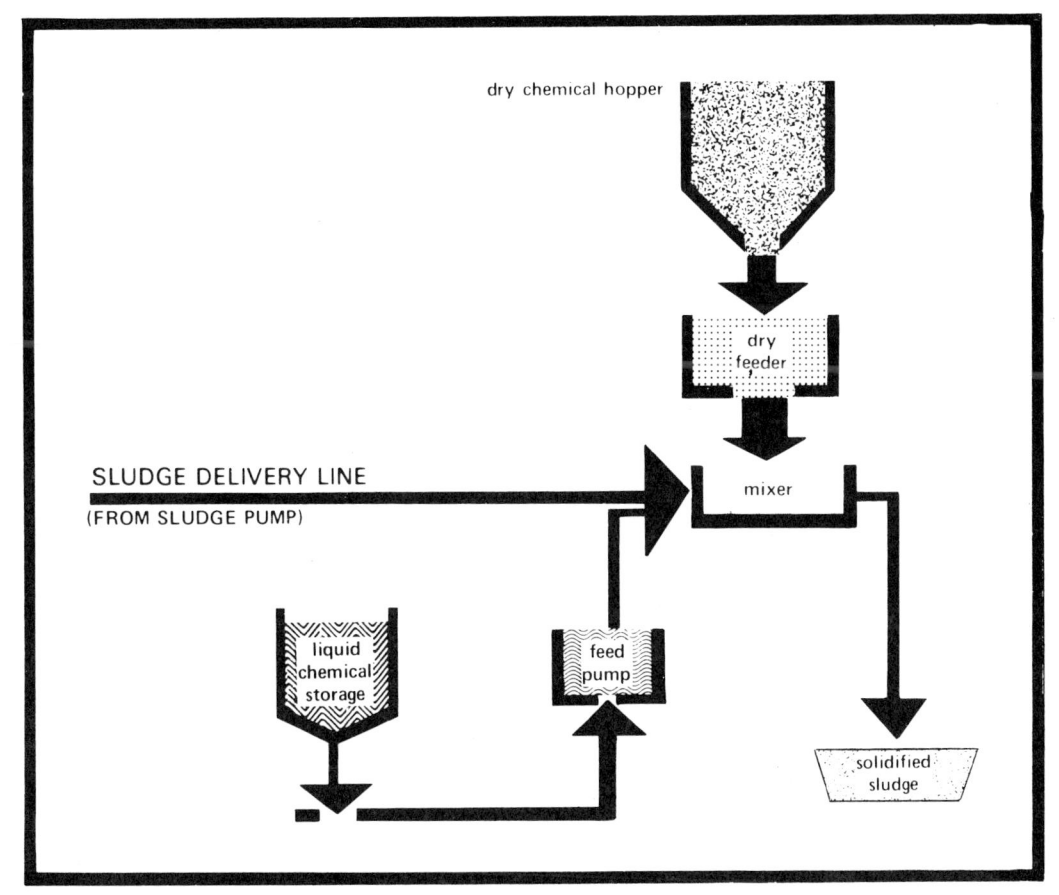

Figure 9-1 Schematic of CFS process.

movement to the curing area where it develops its final physical and chemical properties.

Photographs are even more able to give one a feeling for the "real world." Figure 9-3 shows one type of mobile CFS unit at its home base; Figure 9-4 shows the same unit set up at a remediation site and solidifying waste from a storage pond (behind the unit) into a curing area in the foreground. A close-up of treated, but still fluid waste being pumped into a test cell to cure is shown in Figure 9-5. The same activity is displayed for a full-scale commercial project in Figure 9-6. Note the earthen impoundment being used to contain the fluid waste temporarily until it can harden. When fully cured, the material has the appearance of hard clay after being excavated for removal to the disposal site (Fig. 9-7). Solidified waste, cast or compacted in place, can have a wide range of properties, depending on the waste, the CFS process, and the mix ratio (ratio of reagents added to waste, usually expressed as a decimal fraction on a weight-to-weight basis). A hard, strong product is shown in Figure 9-8.

When planning a solidification program, a pilot study is often essential. Laboratory samples and conditions do not completely duplicate the situation in the field. If nothing else, a pilot study is likely to develop data from a more representative

288 CFS PROCESSES

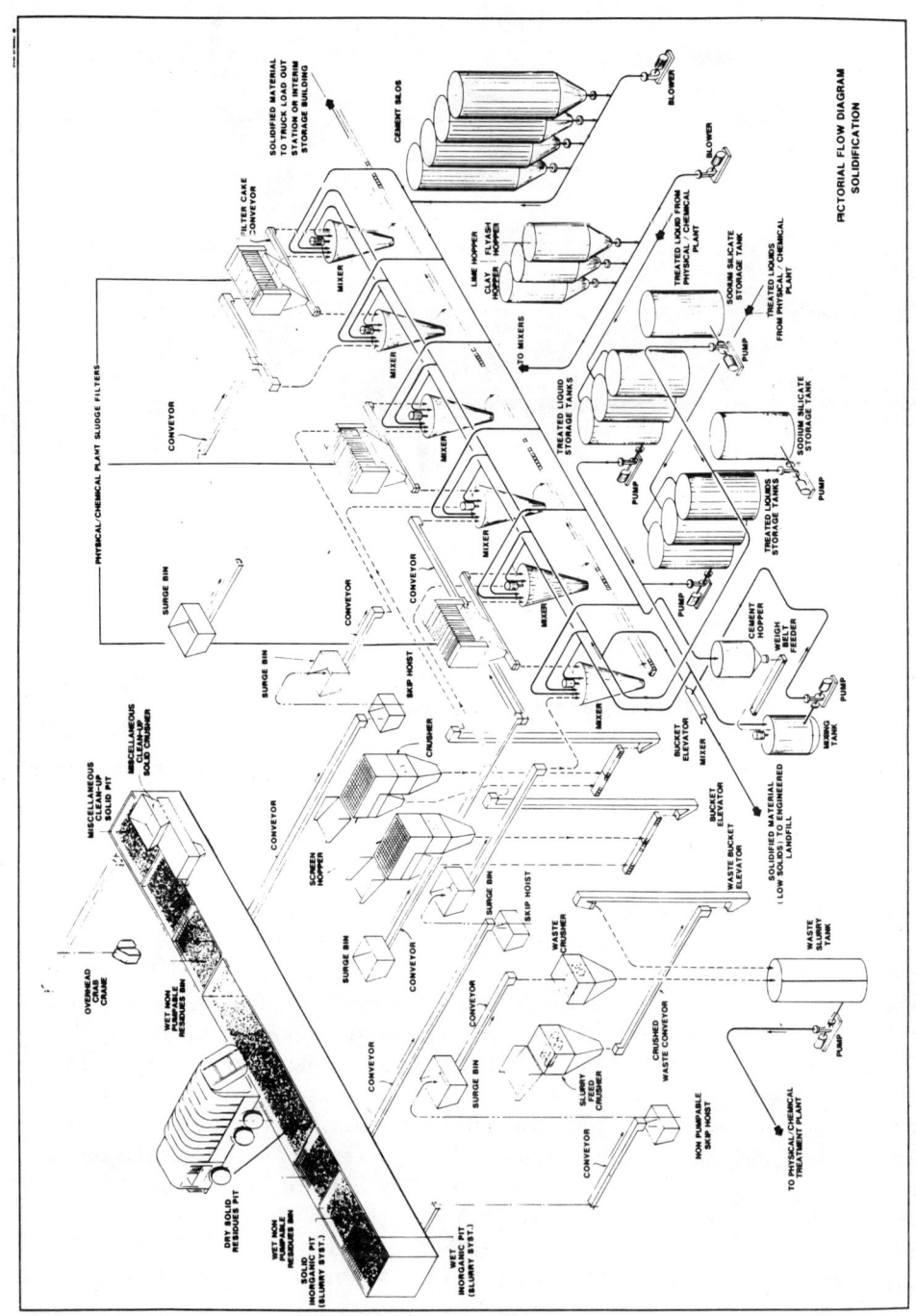

Figure 9-2 Pictorial flow diagram of a central CFS treatment facility. (From Ontario Waste Management Corp.)

Figure 9-3 A mobile CFS unit.

Figure 9-4 Mobile CFS unit in operation.

Figure 9-5 Treated liquid waste before setting.

Figure 9-6 Treated liquid waste being poured out into a prepared curing area. (Print by Associated Photographers, Inc.)

Figure 9-7 The fully cured CFS product. (Photo by George W. Sommer.)

sample of the waste, simply because so many more are required. It is also useful to train operators and technicians. More about this later, when we also see a wider variety of engineering designs, process layouts in the field, and operational photographs of specific projects and installation. For now, however, let us look at the process chemistry.

GENERAL CONCEPTS OF SOLIDIFICATION

The Setting and Curing of CFS-Treated Wastes

Different processes exhibit different setting and curing reactions. These will be explored in some detail in later chapters. Most of the commercial inorganic CFS systems, however, solidify by very similar reactions, which have been thoroughly studied in connection with the Portland cement technology used in concrete making. While the pozzolanic reactions of the processes using flyash and kiln dusts are not identical to those of Portland cement, the general reactions are alike.

One reason for this is presented in an interesting way by Cote [2]. The compositions of most of the primary reagents used in inorganic CFS systems were plotted on a ternary diagram using the three oxide combinations, SiO_2, $CaO + MgO$, and $Al_2O_3 + Fe_2O_3$. All of these reagents have the same active ingredients as far as solidification reactions are concerned; this is shown in Figure 9-9. The combinations

Figure 9-8 Cured CFS product, showing load support strength of the solid.

of these five oxides express the essential composition of any of these materials, even though the actual compounds are not all simple oxides, but more complex silicates and aluminates in many cases. And it is no accident that this is so. All of the reagents, with the exception of power plant flyash, have their origin in natural limestone and clay formations, and all are very inexpensive as industrial "chemicals" go.

The cementitious reactions that occur in these processes require the pH to be above 10. This is initiated by the dissolution of free lime from the solid, and contin-

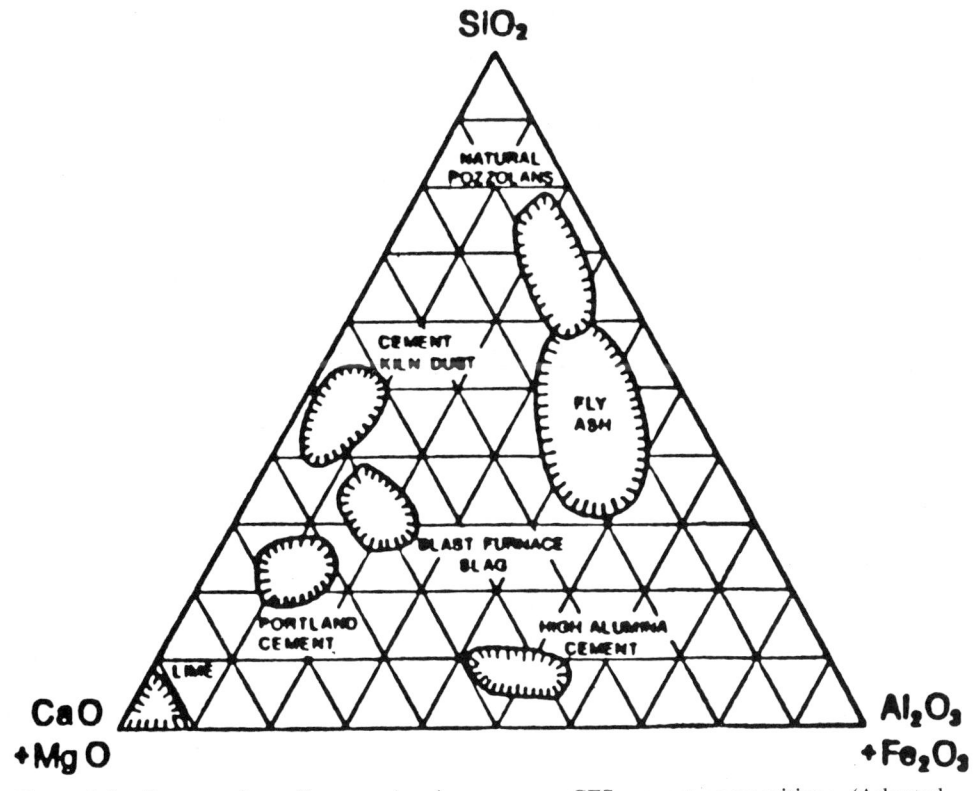

Figure 9-9 Ternary phase diagram showing common CFS reagent compositions. (Adapted from Cote [2].)

ues throughout the setting and curing stages of the mixture. A "false set" can happen when enough free lime is present initially to start the reactions, but availability decreases as the process proceeds, and the reactions stop or slow down. Therefore, any reaction that competes successfully for the calcium ion may inhibit setting. These reactions, and the physical and chemical properties that result from them, are described in more detail in Chapter 10. The cementitious or pozzolanic reactions also require sufficient free water if they are to go to completion. Water is used up in the hydration reaction of these chemical systems, so sufficient water must be provided. The water content as measured by total solids determination is not necessarily all available for the reaction.

Physical Factors Affecting Solidification

In Chapter 8 the important physical properties of wastes, as they related to handling and processing, were listed and discussed in some detail. Earlier, we had examined these and other waste properties as they relate to fixation of hazardous constituents. Here we will explore waste properties and attributes as they relate to the solidification process.

Particle Size and Shape. The characteristics of particles in waste, since most that are to be solidified are slurries, sludges, filter cakes, soils, and dusts, are some of the most important properties to CFS treatment. As we saw earlier (Chapter 8), they strongly affect viscosity and, therefore, handling properties of the waste. Particle attributes also affect both the physical and chemical aspects of solidification reactions, and the properties of the solid product. One of the physical factors is phase separation during the setting reaction. Particle properties, along with solids content, both suspended and dissolved, often determine whether a fast or slow setting process is to be used. (See the subsection, "Solids Content," below.)

Particle characteristics are also important in the final properties of the solid after curing. In many CFS systems, the product is a friable, soil- or claylike material that acts much like a soil when wetted and subjected to mechanical action, as in a leaching test. The product breaks down into a mud or sludge that then exhibits the particle size range associated with the combination of waste and reagents from which it was formed. In other systems, a rocklike structure is formed that does not mechanically degrade easily under the same conditions. In the former case, leaching becomes a function of the chemistry of the system only. In the latter example, leaching results may be determined by the deliberate size reduction performed on the material in preparation for the leaching test.

It has been widely assumed that the finer the particle size, the higher the leaching rate because of the larger surface area exposed to the leachant. Recent investigations by the author and from the literature [3,4] have shown that the opposite may be true in hard, monolithic CFS products such as those from some Portland cement processes. These findings are discussed in Chapter 10. This effect is attributed by all investigators to at least one factor: The influence of particle size on pH of the leachant at any given time. They found that smaller sizes release their alkalinity more rapidly, neutralizing the acid leachant more quickly and reducing metal leaching due to acid attack. Another possibility is that smaller particles provide greater surface area for sorption of metals and other species.

Another effect of particle size of the waste on the system was discussed in Chapter 8. This is the slow diffusion of soluble, hazardous constituents from the interior of large particles into the interparticle space after CFS processing has been completed. This may be no problem if the fixation reaction for that species can still occur in the water-saturated matrix of the solid, as is the case where pH control is dominant. However, in the example described, this was not the case. Often, CFS is a multistep process where the reactants in the early step(s) are destroyed or immobilized in the succeeding steps. This can occur in oxidation or reduction stages and in the use of soluble silicates, sulfides, and other fixants that have competing alternative reactions and are soon used up.

Free Water Content. The difference between total water and free water content is that "free" water is unbound chemically and can react anywhere in the system. It is determined analytically as the water that remains after drying to constant weight at 103°C (217°F). Generally, it is only the free water that is available for CFS setting reactions, although some bound water may become available later during the longer term curing process. Also, water may be generated during certain solidification reactions. In addition to chemically bound water, such as stoichiometric water of hydration, water may be immobilized in hydrating particle surfaces, in the formation of gels, and perhaps in other ways.

Solids Content. The total amount of solids in a CFS system affects both the setting and curing process, and the physical properties of the end product. This is due to two factors: (1) the solids, including the CFS reagents, may settle out, leaving a two-phase system that is generally undesirable, and (2) the final product may be weak mechanically, and thus unsatisfactory for the disposal scenario. Low solids, coupled with low viscosity and high specific gravity in a waste, requires either a fast-setting CFS process, or one that adds large amounts of solids as "bulking agents." In both cases, the overall cost of the treatment and disposal system is higher than it would be with a high solids waste, all other factors remaining equal.

High solids content does not assure that some phase separation will not take place during setting. Waste with high solids content consisting of high specific gravity solids or large, equiaxial particles may also settle very quickly. Conversely, materials with fine elongate or laminar particles of low specific gravity may not separate even at low solids loadings. Solids content, particle size and shape, and specific gravity (particle) are all inextricably intertwined in this respect.

While these latter two attributes are very important in the setting and curing stages of solidification, total solids content alone is more dominant in determining the physical properties of the product. It is nearly impossible to make a strong, durable solid with inorganic solidification systems at low total solids loading.* While good physical properties are not always necessary for fixation of hazardous constituents, they are often required for purposes of handling and disposal. In these cases, solidification of a low solids waste will require a high additive-ratio CFS process, usually one that uses the maximum possible of low-cost materials such as flyash. As we will see later, this comes at the cost of large volume and weight increases.

Specific Gravity/Density. Large differences between the waste and the reagent result in a tendency toward phase separation. Most reagents used in CFS have particle specific gravities much larger than the waste density as a whole: >2.0 versus 1.0 to 1.5 in most systems. Although reagent particle size is generally small, it is not usually sufficient to make up for this difference, and some settling will occur unless viscosity is sufficient to prevent phase separation until the initial setting reaction can physically immobilize all components in place. Alternatively, sufficient reagent can be added to quickly take up all the free water, or a viscosity-increasing agent can be added. Rheological agents that accomplish the latter technique are available, but add significantly to the cost of the process and sometimes can interfere with the reactions that must subsequently take place for proper curing. The Portland cement/soluble silicate system does this by causing an almost immediate thickening or gelling reaction between the free calcium ion from the cement and the silicate, creating an elastic silica gel that does not interfere with cement or pozzolanic reactions.

The problems with specific gravity differences do not occur in the solidification of filter cakes, soils, dusts, and some sludges. These wastes often require the addition of water for proper mixing and reaction of the reagents, so the system can be controlled to prevent phase separation.

Viscosity. In addition to all of the viscosity-related effects discussed earlier, viscosity per se may affect solidification reactions, as least their kinetics. An extreme example is the case of gels, which can be quite difficult to solidify [5]. This may be caused by the removal of free water from the system by the gel, but could also be

---

Solids loading, in this context, means the total resultant solids in the system from both waste and additives.

due to separation of reactive materials by a medium through which they diffuse only slowly.

A useful rough classification of viscosity categories is given below.

- Liquid: Pumpable materials with low, generally less than several percent suspended solids.
- Pumpable: Pumpable fluid or sludge, usually less than 15 percent solids, but up to 50 percent with some wastes, and as low as 5 percent with others.
- Flowable: Not pumpable, but flows and usually releases free liquid. More than 15 percent solids in most cases, but also includes dry dusts such as flyashes. Will not support a load.
- Nonflowable: Soils, many filter cakes.

Wetting. For the reagents to react, they must become wetted with the aqueous medium. In general, the higher the surface area of the particles, the more difficult they are to wet. Some additives, especially those that are themselves waste products, may even have hydrophobic surfaces initially. This is also true of many wastes, especially dry soils, emission control dusts, and some filter cakes. The presence of hydrophobic ingredients in the waste may also cause other problems, inhibiting the setting and curing reactions of the reagents. In some cases, it may be necessary to add surfactants to aid in the wetting of reagents and to allow thorough mixing of all components. Such additives should be pretested, since they may themselves interfere physically or chemically with process chemistry.

Mixing. This is a critical element of any CFS process. The mechanical aspects are discussed in later chapters, but the effects of mixing on the final chemical and physical properties of the CFS solid are also important. While it seems obvious that thorough dispersion of the CFS reagents in the waste is important, it is also possible to overmix certain systems. Fluid wastes such as low-solids sludges are usually easiest to mix, while sticky sludges and filter cakes with the consistency of peanut butter are the most difficult. Soils, solid particulates, and granular materials fall in between.

The choice of mixer is determined by the rheological properties of the waste, the reagents to be added (including how many and in what order), the mode of operation (batch, continuous, or in situ), the throughput rate, and the waste conveyance methods. For large-volume operations with a continuous, reasonably consistent feed, continuous mixing usually gives the best results at the lowest cost. For smaller projects, and where the waste feed is very variable, batch mixing may be the only feasible method. Sometimes the two are combined by accumulating a large batch of waste and homogenizing it to provide a uniform feed to a low throughput, continuous mixer until the batch is processed. In situ mixing is a special case unto itself. The equipment used is not generally thought of as a mixer, but has some other primary function as, for example, a backhoe. More recently, specialized equipment has been applied more and more to this area, most of it being constructed by or for the CFS contractor itself.

There is a myriad of mixer types, designs, and sizes, literally one or more for virtually any use. For fixed systems, the problem is simplified when only one waste is to be treated. This is also true for large remedial action projects where the mixer can be acquired specifically for that project. For central waste treatment sites receiving hundreds of different streams on a very frequent basis, the selection of a mixer is very difficult. Often, the best solution is to use two or more different types at the same location, combined with waste storage areas that can premix compatible wastes to provide larger batches of fewer types.

Selection of a mixer is not only determined by the physical characteristics of the waste, however. Some wastes and some CFS processes are quite sensitive to the energy input and shear rate of the mixer. Use of a high-shear mixer with some wastes may irreversibly alter their properties, causing phase separation. On the other hand, high-shear mixing may be required to completely homogenize other mixtures, for example, to incorporate nonpolar organics in aqueous systems. High-energy input may help with some plastic, pseudoplastic, or thixotropic wastes by decreasing their viscosity temporarily. In rare cases, dilatent materials may literally solidify in the mixer before reagents are even added.

Overmixing, either by using the wrong mixer or mixing too long, interferes with the initial gel formation of cementitious CFS systems, causing delayed set, slow curing, and even the loss of final physical properties. An extreme example of this is seen in the Portland cement/soluble silicate process. Overmixing irreparably destroys the silica gel structure, preventing the process from working properly at all. Overmixing can also go the other way. Many a processor has had to dig out a mixer because the blend set before it could be emptied. This usually happens in batch systems, but has also been observed in continuous mixers when the mixer is stopped while full.

One final comment about mixing. It is widely assumed that very thorough and intimate mixing is required to assure that a reaction will take place and fixation of hazardous constituents will be complete. However, it is known that this does not happen with most in situ techniques and still the end result may be satisfactory, at least from a physical viewpoint. Recently, Chemical Waste Management, Inc., has been investigating mixing from this perspective and has tentatively reached the conclusion that complete mixing at the microscopic scale is not always necessary [6] in commercial CFS systems. Other scientists have recently realized this as they began to look at the microstructures of solidified wastes [7]. At this level, apparently unreacted waste conglomerates are evident, as are areas of unmixed reagent. The question now remains: At what size level can inhomogeneities exist and still not prevent the process from achieving its function? There is probably no single answer; it may vary from waste-to-waste and process-to-process.

Temperature and Humidity. CFS processes are chemical reactions and would be expected to follow the general rules of such: that reaction rate increases with temperature, perhaps according to the old rule of thumb that reaction rate doubles with each 10°C (50°F) rise in temperature. CFS reaction rates do increase with temperature, but there are limitations. Below freezing, temperatures can cause irreversible changes in the final product by breaking gel structures at a critical stage, much as can happen with concrete. At very high temperatures, release of steam can actually break up the solid mass, and the loss of water can interfere with subsequent curing reactions. This latter phenomenon is of concern in CFS systems, most of which are exothermic. Some reagents, especially the kiln dusts and very active flyashes, generate a large exotherm and very rapidly, due to the presence of substantial quantities of CaO that can hydrate violently. This must be considered when designing the mechanical aspects of the CFS system. Also, the low heat transfer rate from a large mass of treated waste while curing can drive the internal temperature to unacceptable levels, even though the laboratory test did not indicate a potential problem. This must also be taken into consideration in the system design. Self-heating of the mass, however, is a valuable aid to curing if it is kept under control and within limits. To be assured that cementitious reactions will start, the temperature should be maintained above 27°F (-3°C) [5].

Ambient humidity in the curing area must be kept high if the product is to cure properly. Evaporation of water from the surface will inhibit or stop solidification, and perhaps some fixation reactions in the surface layer. In the laboratory, samples are normally cured at about 95 percent relative humidity to prevent this effect from masking the real solidification reactions. In very hot and dry field conditions, it may be necessary to keep the waste moistened as it cures, much as is done with concrete under the same conditions.

Chemical Factors Affecting Solidification

There are many factors that retard, inhibit, and accelerate the setting and curing of CFS systems. Some also affect the final strength, permeability, and other physical properties of the fully cured[†] CFS product. Others affect chemical properties, which we discussed in Chapters 3 and 4. Many of the compounds, materials, and factors that are known to have such effects are listed in Table 9-1. These have been accumulated from a variety of specific sources [1], [2], [5], [8], rather than [17], as well as the general knowledge in the field of cement technology. It is obvious that the effects are many and varied. In spite of the useful information given in Table 9-1, however, these effects are not simple to predict and sort out from knowledge of the composition of the waste. Most often, a number of species are present, sometimes with opposing effects. The same species may have opposite effects depending on concentration. This phenomenon has been found with calcium chloride, calcium sulfates, sodium hydroxide, sodium silicate, lead, copper and tin salts, amines, and hexachlorobenzene. And the effects vary in magnitude or even type with different CFS processes.

Some of the general effects are interesting. Ion exchange can inhibit or retard CFS reactions by removing calcium from solution, preventing it from entering into the necessary cementitious reactions. It can also accelerate the process by removing interfering metal ions from solution. Which of these occurs may depend on the selectivity of the ion exchange material. Other metals may retard and inhibit the reactions by substituting for calcium in the cementitious matrix, which may explain the effect of magnesium in dolomitic lime and lime products. Certain substances are natural or synthetic complexing agents that remove calcium from availability in the setting and curing reactions.

On the other hand, alcohols, amides, and specific surfactants can aid in wetting solids and dispersing fine particulates and oil that interfere with reactions by coating the reacting surfaces. Flocculants can also serve this purpose. Some of these materials and techniques are listed below. These are methods that can be used with

---

†The term *fully cured* is relative. Most pozzolanic and cementitious reactions continue for very long periods of time, perhaps forever. The changes in physical properties that take place in the long term are usually small, but may be significant in certain situations. Fully cured is usually taken to be periods ranging from 10 to 100 days, with 28 days being the most common.

TABLE 9-1 Factors Affecting Solidification

Compound or Factor	Effect	Mechanism Affected	CFS Processes
Fine particulates	I,P	P	PC, PZ
Ion exchange materials	I,A	I	AI
Metal lattice substitution	I,A	I	AI
Gelling agents		R,I,P−	P,I,M AI
Organics, general	I,P,R I,D	AI	
Acids, acid chlorides	P−	I	AI, Some O
Alcohols, glycols	R,P−	I,W	AI, Some O
Aldehydes, ketones	P−	I	C, Some O
Amides	R	I,W	Some O
Amines	R,A	I,F	Some O
Carbonyls	R	I,D	AI
Chlorinated hydrocarbons	P−,R	I,M	PC, Some O
Ethers, epoxides	P−	I	Some O
Grease	I,P	P	PC,PC/PZ,L
Heterocyclics	P−	I	C
Hydrocarbons, general	P−	I	C, Some O
Lignins	I	C	AI
Oil	I,P	P	PC,PZ
Starches	I	C	AI
Sulfonates	R	D	AI
Sugars	I,R	C	AI
Tannins	I	C	AI
Organics, specific			
Adipic acid			
Benzene			
EDTA			
Ethylene glycol	P	I PC	
Formaldehyde			
p-Bromophenol			
Hexachlorobenzene		P−,P+ I PC	
		P− I	PC/PZ
		P+,P− I	L
Methanol			
NTA			
Phenol	P−	I	PC,PC/PZ,L
Trichloroethylene	P−	I	PC,PC/PZ,L
Xylene			
Inorganics, general			
Acids	P−	I	PC, Some O
Bases	P−	I	PC/SS, C, Some O
Borates	R	M	PC,PZ
Calcium compounds			
Chlorides	R,P	I	AI
Chromium compounds	A	I	AI
Heavy metal salts		P−,A,R I	AI, Some O
Iron compounds	A	F,M	PC,PZ
Lead compounds	R	M	PC,PZ
Magnesium compounds	R	M	PC,PZ
Salts, general	P−,A,R	I	AI, Some O
Silicas	R	F	PC,PZ
Sodium compounds	I	I	AI
Sulfates	R,P	I	AI
Tin compounds	R	M	PC,PZ

(continued)

TABLE 9-1 Factors Affecting Solidification *(Continued)*

Compound or Factor	Effect	Mechanism Affected	CFS Processes
Inorganics, specific			
Calcium chloride		A,R M	PC,PZ
Copper nitrate	P+	I	PC
		P+,P− I	PC/PZ
		P− I	L
Gypsum, hydrate	R	I	PC,PZ
Gypsum, semihydrate	A	I	PC,PZ
Lead nitrate		P−,P+ I	PC
	P−	I	PC/PZ,L
Sodium hydroxide		P+,P− I	PC,PC/PZ,L
Sodium sulfate	P−	I	PC
		P+,P− I	PC/PZ,L
Tin			
Zinc nitrate		P+,P− I	PC
	P−	I	PC/PZ,L

Key: Effect: I = setting/curing inhibition (long term); A = setting/curing acceleration; R = setting/curing retardation (short term); P+ = alteration of properties of cured product, positive effect; P− = alteration of properties of cured product, negative effect. Mechanism: P = coats particles; I = interferes with reaction; C = complexing agent; M = disrupts matrix; F = flocculant; D = dispersant; W = wetting agent. Process: PC = Portland cement-based; PC/SS = Portland cement/soluble silicate; PC/PZ = Portland cement/pozzolan; PZ = pozzolanic (kiln dust, flyash); C = clay-based; L = lime-based; AI = all inorganic; O = organic.

Note: When the effect may be positive or negative, depending on concentration, the first symbol listed represents lower concentration, the last higher concentration.

a variety of inorganic systems. Many proprietary CFS processes have these additives built into the formulation for this purpose, as well as to fix metals and other hazardous constituents.

Method or Material	Mechanism
Flocculant	Aggregation of fine particles and film-formers
Wetting agent	Dispersion of oils and greases and fine particulates away from reacting surfaces
pH adjustment	Removal of interfering substances from solution; destruction of gels and filmformers
Fe^{+2}/Fe^{+3} addition	Precipitation of interfering substances
Ion exchange	Removal of interfering substances from solution
Sorbent addition	Removal of interfering substances from reacting surfaces
Redox potential	Destruction/conversion of interfering substances
Aeration	Alteration of biological status; removal of interfering volatiles
Temperature adjustment	Acceleration of reaction rate to counter retarding effect

Method or Material	Mechanism
Lime addition	Supplies additional calcium for reaction; reacts with certain interfering organics; pH adjustment
Sodium silicate	Reacts with interfering metals; causes acceleration of initial set
Calcium chloride	Accelerates set in Portland cement systems
Sodium hydroxide	pH adjustment; may solubilize silica for quicker reaction with calcium ion (low concentration)
Amines, other organics	Mechanism unclear
Metal ions	Mechanism unclear

A number of the studies just referenced will provide the reader with additional information about a specific CFS-waste system. One that is worthy of special note is that by Cullinane *et al.* [8]. Three CFS systems—Portland cement, cement/flyash, and lime/flyash—were used to treat sludge that was spiked with various known or suspected interfering compounds at three different levels. Samples were cured and evaluated for unconfined compressive strength by ASTM Method C-109. The results are shown in Figures 9-10 through 9-12, and are summarized in Table 9-2. Some important conclusion were drawn from this study:

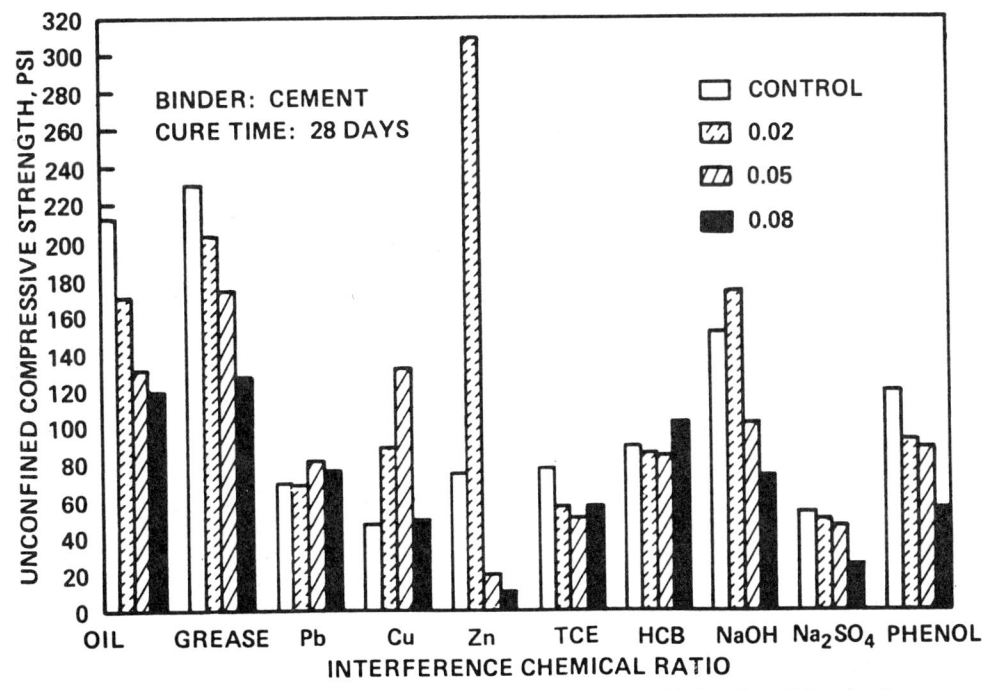

Figure 9-10 Effect of interference concentration on 28-day UCS for Type I Portland cement binder. (From Cullinane *et al.* [8].)

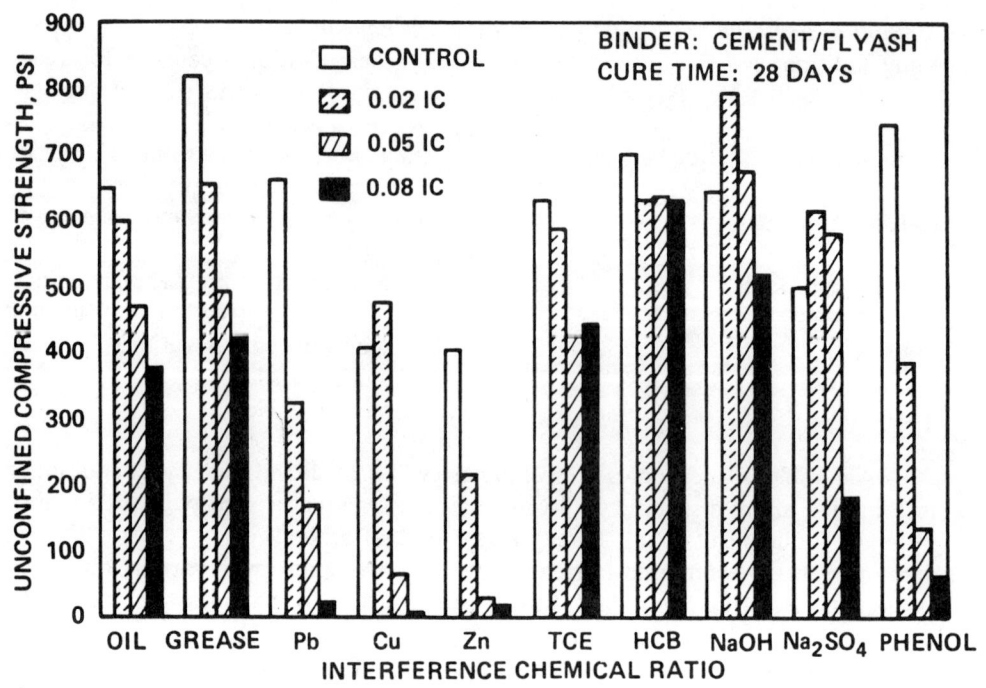

Figure 9-11 Effect of interference concentration on 28-day UCS for cement/flyash binder. (From Cullinane et al. [8].)

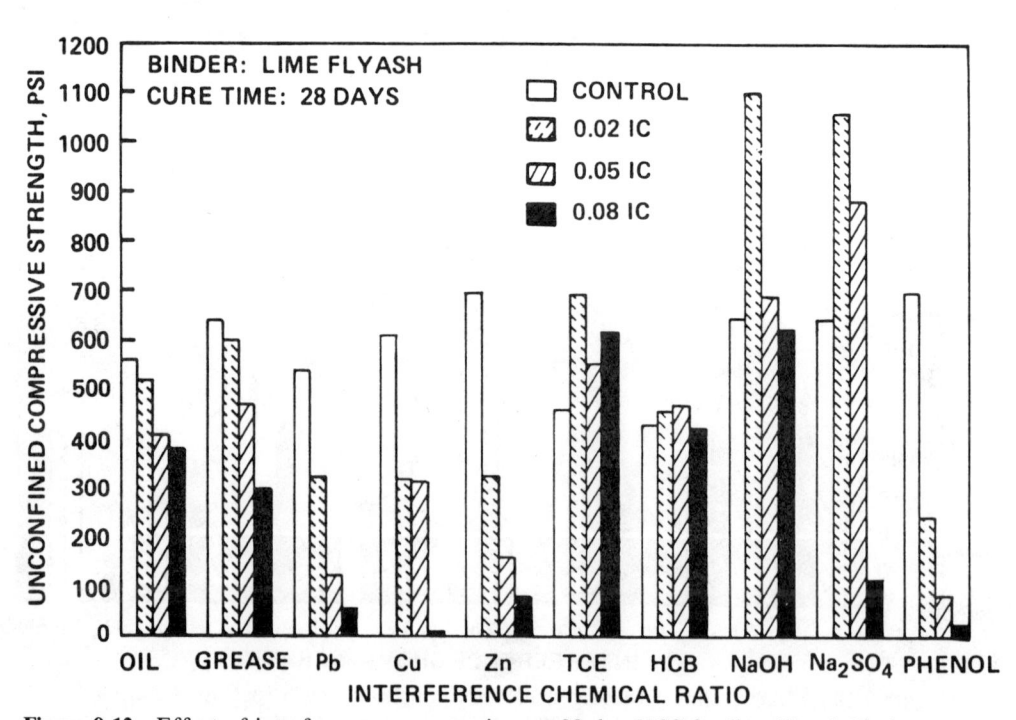

Figure 9-12 Effect of interference concentration on 28-day UCS for lime/flyash binder. (From Cullinane et al. [8].)

TABLE 9-2 Twenty-Eight-Day Unconfined Compressive Strength as a Percent of Control Specimen[a]

Interference Chemical	Portland Cement Binder I/B/S[b] Ratio 0.02	0.05	0.08	Cement/Flyash I/B/S Ratio 0.02	0.05	0.08	Lime/Flyash I/B/S Ratio 0.02	0.05	0.08
Oil	−20	−38	−44	−8	−28	−42	−7	−27	−32
Grease	−12	−25	−45	−48	−40	−20	−7	−27	−54
Lead nitrate	−2	+18	+10	−51	−75	−97	−40	−77	−90
Copper nitrate	+91	+181	+4	+17	−84	−98	−48	−48	−98
Zinc nitrate	+3	−73	−86	−47	−93	−95	−53	−77	−88
Trichloroethylene	−28	−36	−27	−7	−33	−29	+51	−20	−34
Hexachlorobenzene	−5	−6	+15	−10	−9	−10	+6	+9	−1
Sodium hydroxide	+14	−33	−52	+23	+5	−20	+1	−7	−4
Sodium sulfate	−7	−13	−53	+23	+16	−64	+65	+38	−82
Phenol	−22	−26	−54	−49	−82	−92	−65	−88	−96

[a] All results reported as percent increase (+) or decrease from the control specimen rounded to nearest whole percent.
[b] Interference to binder/sludge ratio.

- The effect of a given chemical was not necessarily the same with all three CFS processes. The effect was less for the Portland cement process than for the others, and depended on curing time.
- Oil and grease decrease strength development in all cases.
- Some chemicals, such as sodium hydroxide, increased strength at low concentration and decreased it at high concentrations. Hexachlorobenzene had the opposite effect with Portland cement.
- The chlorinated hydrocarbons, in general, had little effect.
- High (>5 percent) concentrations of zinc and phenol resulted in marked strength decreases with all processes.

The effects of chemical factors on solidification will also be discussed in later chapters when we look more closely at individual processes and systems.

GENERIC CATEGORIES OF CFS PROCESSES

Solidification systems are of two basic types, inorganic or organic, according to the nature of the solidification chemicals used, not the waste composition [18]. Organic systems have been little used for industrial wastes except in the area of radioactive waste solidification, where urea-formaldehyde, bitumen, and polymerization systems have been applied to some degree. These systems are sometimes hydrophobic in nature and difficulties are encountered when incorporating the waterbased wastes that make up the primary disposal problem. In addition, organic systems suffer from the problem of instability with regard to ambient environmental factors such as microorganisms and ultraviolet light.

Inorganic systems are used for the chemical fixation and solidification of complex wastes and/or mixtures thereof, with the aim of producing a nontoxic, environmentally safe material that can be used as landfill. The processes use inorganic reagents that react with certain waste components; they also react among themselves to form chemically and mechanically stable solids. These systems are based on reactions between binders, catalysts, and setting agents that occur in a controlled manner to produce a solid matrix. The matrix itself, as produced, is often a pseudomineral. This type of structure displays properties of stability, high melting point, and a rigid, friable structure similar to many soils and rocks.

One of the primary considerations and, in fact, a criterion in formulating CFS systems, is that the reagents used in these systems be basically nontoxic and safe to handle. Unless otherwise specified, this is the case of all systems discussed here. They are formulated from either natural or synthetic materials that have a long history of industrial use and are considered to be generally nonhazardous. Since many of the formulations are quite alkaline, and nearly all of them contain very fine powders that can be irritating to the mucous membranes, it is necessary to utilize normal caution in handling these materials, that is, eye protection in all cases, and in closed environments, dust masks.

Inorganic Processes

The inorganic processes can most meaningfully and conveniently be grouped into two categories: those that use bulking agents, such as Class F flyash, and those that do not. In this context, we mean by a bulking agent an addition that primarily adds to the total solids and viscosity of the waste, thus preventing settling out of the suspended waste components before solidification can occur; it may also help produce a solid with better physical properties. Examples of these two groups of processes are (1) cement-based or cement/soluble silicate systems (no bulking agents), and (2) cement/flyash, lime/flyash, cement/clay, or lime/clay (systems with bulking agents). There are two types of bulking agents: those that are essentially inert in the system and act as previously described, and those that also have reactive capacity or pozzolanic activity. A *pozzolan* is defined as a material that does not exhibit cementing ability when used by itself, but in combination with other materials, such as Portland cement or lime, will interact with these agents resulting in a cementitious reaction.

The most noticeable effect of the difference between systems with and without bulking agents is that systems with bulking agents often have lower chemical costs by virtue of replacing some of the more expensive cementing materials with the less expensive waste products such as flyash. Systems without bulking agents often have lower overall costs because of the lower weight and volume increase associated with them. This will be discussed in more detail later.

Generic inorganic chemical fixation and solidification processes that have been either used or proposed for use are shown below. Also listed are major references to each type of process.

- Portland-cement-based (major ingredient is cement): (Heacock [21], Landreth [24], Mahoney [25], Pojasek [27], Portland Cement Association [28], Skolny and Daugherty [29], U.S. Army Engineers Waterways Experiment Station [30])
- Portland cement/lime: (Pojasek [27], U. S. Army Engineers Waterways Experiment Station [30])
- Portland cement/clay: (Cote and Hamilton [19], Double and Hellawell [20], Heacock [21], Kupiec [22], Laguna [23], Landreth [24], Mahoney [25], National Academy of Science [26], Pojasek [27], Portland Cement Association [28], Skolny and Daugherty [29], U.S. Army Engineers Waterways Experiment Station [30], Rossoff *et al.* [31])
- Portland cement/flyash: (Clendenning [32], Cote and Hamilton [19], Laguna [23], Landreth [24], McClelland [33], Pojasek [27], Taub and Roberts [34])
- Portland cement/soluble silicate: (Chemfix [35], Conner and Gowman [36], Conner and Polosky [37], Conner [38], Conner [39], Conner [40], Conner [41], Conner [1], Conner [18], Conner [42], Cote and Hamilton [19], Falcone [43], Gowman [44], Heacock [21], Landreth [24], Leo [45], Mahlock [46], Pojasek [27], Spencer [47], U.S. Army Engineers Waterways Experiment Station [30], Rossoff [31])
- Lime/flyash: (Landreth [24], Mahoney [25], Pojasek [27], U.S. Army Engineers Waterways Experiment Station [30], I.U. Conversion Systems [48])
- Lime/clay: (Clendenning [32], Cote and Hamilton [19], Customized Waste Stabilization [49], Diamond [50], I.U.Conversion Systems [48], Landreth [24], Leo [45], Mahlock [46], Pojasek [27], Report on Leachate Test Comparison Programs [51], Rossoff [31], Thornton [52], U.S. Army Engineers Waterways Experiment Station [30])
- Cement or Lime Kiln Dust: (Cocozza [53], Landreth [24], Mahlock [46], Mineral

By-Products [54], Pojasek [27], U.S. Engineers Waterways Experiment Station [30])
- Gypsum-based: (Pojasek [27])

Many of the various permutations and combinations of these systems, including various types of additives, are patented or are covered with patents applied for, but most of the generic system types are believed to be in the public domain, at least in the United States and Canada.

The most important systems at the moment are:

1. Portland cement
2. Lime/flyash
3. Kiln dust (lime and cement)
4. Portland cement/flyash
5. Portland cement/lime
6. Portland cement/sodium silicate

All of these processes have been used commercially for solidification of water-based waste liquids, sludges, and filter cakes. A large body of technical information on leachability, physical properties, and general stability is readily available for all of these processes.

In volume of waste treated, the lime–flyash process has probably been the most used in the United States, although it has been more narrowly applied, primarily to flue gas desulfurization sludges. For other types of wastes and industrial sludges, the kiln dust and Portland-cement-based processes are the most widely used at the present time. In the United States, one of the most flexible techniques has been the Portland cement–sodium silicate process. This process has been applied to a variety of wastes and a good technical base, including leaching test information, is available. Another technique, the Portland cement/flyash process, has been used in Europe and Canada but has not been applied very broadly in the United States to date. This process works well with certain types of waste; like the lime/flyash process, it involves large additions of the solidifying agents and therefore large volume increases. In summary, the inorganic systems are characterized by:

- Relatively low cost
- Good long-term stability, both physically and chemically
- Documented use on a variety of industrial wastes over a period of at least ten years
- Widespread availability of the chemical ingredients
- Nontoxicity of the chemical ingredients
- Ease of use in processing (processes normally operated at ambient temperature and pressure and without unique or very special equipment)
- Wide range of volume increase
- Inertness to ultraviolet radiation
- High resistance to biodegradation
- Low water solubility
- Relatively low water permeability
- Good mechanical and structural characteristics

Organic Processes

Organic CFS processes consist of two basic types: thermoplastic, hot-melt systems, such as bitumen and polyethylene, and polymerization systems, such as polyester and urea-formaldehyde, where the waste is mixed with a monomer and one or more agents that cause the system to polymerize, thereby microencapsulating the waste throughout the matrix.

Polymerization Systems. Polymerization systems can be formulated to work with both organic solvent, oil, or water-base waste systems; most of the processes developed to date are designed to work with water-base wastes. The systems consist of monomers that are polymerized or cross-linked by the use of catalysts or accelerators after being mixed with the liquid waste. This kind of system is almost infinite in potential variety, but for practical purposes has been limited so far to urea-formaldehyde, polybutadiene, polyester–epoxy, acrylamid gel, urea-formaldehyde with plaster of Paris, polyolefin encapsulation, and polyurethane.

Many combinations of the preceding and of other organic polymer systems are conceivable. Some organic systems are solvent based and are primarily hydrophobic in nature. Difficulties are therefore encountered when incorporating the usual water-based residuals. In addition to these physical problems, the presence of water and other ingredients in the wastes often interferes with the polymerization reaction. These systems can be used for the solidification of certain waste organics—hydrocarbons, chlorinated hydrocarbons, pesticides—that do not have a water base. For use with water-base wastes, the systems are formulated as emulsions. The use of emulsion systems on water-base waste often produces excess water or "weep" water. This comes both from the reactions, and from being squeezed out of the matrix. The urea-formaldehyde process is especially prone to this problem, which has reduced its once extensive use in the nuclear waste area.

In most cases, the action of the organic polymer is primarily permeation limiting. Usually there is no direct reaction between the waste constituents and the polymer, nor does the system actually insolubilize, detoxify, or destroy the hazardous constituents. However, since these polymers generally have a very low order of permeability, they are more effective in this respect than many inorganic systems. The mechanism of most organic systems is microencapsulation of the waste material, which separates the waste from its environment. One of the technical advantages of many of the organic processes is that a given polymer can be applied to a wide variety of waste types, since there usually is not direct chemical interaction between the polymer and the waste. This is advantageous when dealing with a complex mixture of wastes or a wide variety of individual wastes in a given disposal situation. A primary disadvantage of the organic processes to date has been cost. In fact, most organic processes at this time are of little interest commercially except in the solidification of special hazardous materials such as radioactive waste, and for organic-based wastes that for some reason cannot be thermally destroyed.

Thermoplastic Systems. The hot-melt systems, primarily bitumen, work with either water-base or solvent/oil-base waste. In the case of water-base wastes, however, the water is usually evaporated before treatment, so smaller amounts will be evaporated during the hot-melt process. With organic-based wastes, the volatile organics are driven off during the hot-melt process and collected for either recycling or destruction in an incinerator.

The question of the biological stability of organic processes in the ground is still open. The very presence of natural tars, such as the Alberta tar sands, suggests that microbial attack is quite slow at best, at least under these anaerobic conditions. In the laboratory, microbes have been observed to oxidize bitumen [55]. Other tests showed no change in leach characteristics after inoculation with aggressive bacteria and subjection to severe disposal conditions. The only known commercial applications of the bitumen process in the United States are in the nuclear waste area.

Another possible thermoplastic process for the future is sulphur-based. To date, this technology is only in the experimental stage and has not at this time reached commercial use. Little information is available about the various mandatory and other parameters, including technical characteristics and the characteristics of the solidified product. It is likely that its operating characteristics would be similar to those of the bitumen process.

In summary, the characteristics of organic systems are:

- High cost
- Very low permeability
- Potential of using one system with a wide variety of wastes
- Wide-range volume increase
- Relative instability in the presence of ultraviolet light
- Questionable stability in the presence of microorganisms
- Potential long-term degradation effects on the polymer matrix by certain waste constituents
- Unproven long-term stability
- Very limited commercial use experience except with radioactive wastes, and no long-term commercial-scale field-test data
- Systems are more difficult to use and require more technical expertise than do the inorganic systems
- Some system components, especially catalysts, initiators, and accelerators, are hazardous

Nonchemical Processes

Glassification or Vitrification. Other than for use with high-level radioactive wastes, only one process has been brought to the point of commercial scale operation—the Battelle process. This will be discussed further in Chapter 15.

Gross Encapsulation Processes. No such processes are commercially or technically proven at the present time.

Sorbents. These processes or systems will be discussed briefly in Chapter 15. From a purist point of view, they are not really CFS systems at all. Nevertheless, some have present and/or potential values in the CFS field.

Self-Cementation. There are some wastes that have the property of reacting without chemical additions, usually over a long time scale, to solidify and/or fix hazardous constituents. Many flyashes, when considered as waste instead of as reagents, have this property.

PROPRIETARY CFS PROCESSES, PRODUCTS, AND SERVICES

Inorganic Systems

Many proprietary CFS processes that have been used, proposed, marketed, or discussed in the literature as of 1989 are listed in Table 9-3. A brief summary of their properties is also given in tabular form. This list is not exhaustive, nor are all of the processes still in existence. With the passage of the RCRA reauthorization in 1984, and especially because of the various landban rules, a plethora of processes and products have appeared on the market. Many of these are simply marketing phenomena, and represent nothing new or better in the technical sense. In nearly all cases, the more recent chemical products are basically generic or proprietary systems in new packages, although they may have utility in specific use circumstances. For example, the product may include additives to fix specific constituents, or anti-inhibiting agents to solidify wastes that are difficult.

Systems that have a sound technical basis and commercial applicability will be discussed in Chapters 10–15. Some will also be compared later on in this chapter.

Organic Systems

The same comments apply here as for the inorganic processes, except that the organic processes are primarily of technical, not commercial interest at this time. The one exception is in the nuclear waste treatment area. Proprietary organic processes are listed and briefly described in Table 9-4.

General Systems

A few systems that are not easily classified as either inorganic or organic *CFS* processes are described here. They are not, strictly speaking, CFS processes or systems at all, but broader waste treatment systems.

SoliRoc. SoliRoc is really not a solidification process but rather a complete hazardous waste treatment facility and system. While the system does use solidification as the final step, the solidification part of the process is performed using Portland cement. Therefore, in this respect it can be viewed as a cement-based process. The process consists of a fairly complex pretreatment process, which generates soluble silica species in situ at low pH, reaction of the soluble silica with metal ions, raising pH to precipitate the system as a sludge, dewatering of the sludge, and finally solidification with Portland cement.

Fersona. Fersona is really a very specialized fixation process that reportedly immobilizes sodium in a fairly complex structure.

310 CFS PROCESSES

TABLE 9-3 Proprietary Inorganic Processes

Process Name or Vendor	Location	Demonstrated?	Available?	Flexibility	Cost	Application	Delivery System[a]
Cement-based (with minor additives)							
Alsite	Japan	Yes	?	?	?	?	License in U.S.
ATCOR	Houston, TX	Yes	Yes	Low	High	Nuclear	F
Browning-Ferris (CECOS)	Various	Yes	Yes	Moderate	Medium	General	F, M, I
Chemical Waste Management	Columbia, SC	Yes	Yes	High	Low	General	F, M
Chem-Nuclear	Lima, OH	Yes	Yes	Low	High	Nuclear	License
Chem-Technics (EATA)	Japan	?	?	High	Varies	General	?
Ebara-Infilco	Japan	Yes	Japan	?	?	?	F
Fujibeton				Moderate	Medium	Sewage sludge	
Fujimasu	Japan	?	?	?	?	?	?
Hitman	Rockville, MD	Yes	Yes	Low	High	Nuclear	F, M
NUS Corp.	Japan	?	?	?	?	?	?
Onoda Cement	Oak Ridge, TN	?	?	?	?	?	?
ORNL (U.S. DOE)		Yes	Yes	Low	High	Nuclear	F
Silicate Technology Corp. (Fujibeton, U.S.)	Scottsdale, AZ	Yes	Yes	High	Varies	General	Chemicals
SolidTek Inc.	Morrow, GA	Yes	Yes	High	Varies	General	F, M, I
SoliRoc		Yes	?	Low	Low	Narrow	F
Stock Equipment		Yes	Yes	Low	High	Nuclear	?
Teledyne Energy Systems, Inc.	Timonium, MD	Yes	Yes	High	High	Nuclear	F
TJK, Inc.	N. Hollywood, CA	Yes	Yes	High	Low	General	License in U.S.
Toxco, Inc.	Claremont, CA	Yes	Yes	High	Varies	General	F, M
U.S. Waste, Inc.	Portland, OR	Yes	Yes	High	Varies	General	Franchise
Cement-lime							
Chemical Waste Management	Various	Yes	Yes	High	Low	General	F, M, I
Chem-Technics (EATA)	Lima, OH	Yes	Yes	High	Varies	General	License
Industrial Waste Management							
NuKem	Germany	?	?	?	?	?	?
Onoda Cement	Japan	?	?	?	?	?	?
Cement-clay							
Chem-Technics (EATA)	Lima, OH	Yes	Yes	Moderate	Varies	General	License
Environmental Technology Corp.		Yes	Yes	Moderate	?	?	?
ORNL (U.S. DOE)	Oak Ridge, TN	Yes	?	Low	High	Nuclear	F

Process/Company	Location					License
Cement/soluble silicate						
Chemfix Technologies Inc.	Kenner, LA	Yes	High	Low	General	F, M
Chem-Technics (EATA)	Lima, OH	Yes	Moderate	Varies	General	F
Delaware Custom Materiel	State College, PA	Yes	Low	High	Nuclear	?
French AEC	France	Yes	Low	High	Nuclear	?
Fujibeton	Japan	Yes	High	Medium	General	?
Kurita Water Industries	Japan	?	?	?	?	?
Nippon Synthetic Chemical Industry	Japan	?	?	?	?	?
United Nuclear Industries		Yes	Low	High	Nuclear	F, M
Cement/flyash						
FPL/QualTec	FL	Yes	High	Medium	General	F, M
ORNL (U.S. DOE)	Oak Ridge, TN	?	Low	High	Nuclear	F
Stablex Inc.	Blainville, Ont., Canada	Yes	High	Medium	General	F
Lime/flyash						
Conversion Systems	Horsham, PA	Yes	Moderate	Low	Power-FGD	F
Chem-Met Co.	MI	Yes	Moderate	Low	General	F
Velsicol Chemical Co.	Chicago, IL	Yes	Moderate	Low	General	Technology
Willis & Paul		Yes	Moderate	?	?	?
Lime-based						
Epes Lime Co.		?	Low	?	?	?
Ontario Liquid Waste Disposal	Ont., Canada	Yes	Moderate	Low	Limited	F
Sludgemaster		Yes	Low	Low	Oily waste	M, I
Wastetab		Yes	?	?	?	?
Lime or cement kiln dust						
Chemical Waste Management	Various	Yes	High	Low	General	F, M, I
Envirite		Yes	Moderate	Low	In situ only	I
Enreco		Yes	Moderate	Low	In situ only	I
Mineral By-Products	Marietta, GA	Yes	Moderate	Low	General	F, M, I
N-Viro		Yes	Low	Low	Sewage sludge	F
Research Cottrell		Yes	Low	Low	Power-FGD	F
Wehran Engineering		Yes	Low	Low	Power-FGD	F
Gypsum-based						
EnviroStone (U.S. Gypsum Co.)	Chicago, IL	?	Low	?	Nuclear	F
Other						
Petrifix	France	Pilot	?	?	?	?
Urritech, Inc.	Houston, TX	?	?	?	?	?

^{a} F = Fixed plant; M = Mobile plant; I = In situ.

TABLE 9-4 Proprietary Organic Processes

Process Name or Vendor	Location	Demonstrated?	Available?	Flexibility	Cost	Application	Delivery System[a]
Thermoplastic							
Sulphur-Based (Southwest Res. Inst.)		No	No	?	?	?	?
Waste Chem		Yes	Yes	Low	High	Narrow	F
Werner & Pfleiderer	Germany	Yes	Yes	Low	High	Nuclear	F
Polymerization							
Anefco (Urea/Formaldehyde)		Yes	?	Low	High	Nuclear	F, M
Dow Chemical Co. (Vinyl Ester)	Midland, MI	Yes	Yes	Low	High	Nuclear	F
Hittman (Urea/Formaldehyde)		Yes	?	Low	High	Nuclear	F, M
LSU	LA	No	?	?	?	?	?
NECO (Urea/Formaldehyde)		Yes	?	Low	High	Nuclear	F, M
Newport News Industries (Urea/Formaldehyde)	VA	Yes	?	Low	High	Nuclear	F, M
TRW (Polybutadiene)		No	?	?	High	?	F
Washington State Univ. (Polyester)	WA	No	?	Low	High	?	F

[a] F = Fixed plant; M = Mobile plant.

Calcilox. Calcilox is really not a solidification process, per se. It was designed to be used with fluegas desulfurization sludges and on coal waste fines. It is an improved thickening and densification process in which the Calcilox material (ground steel mill slag) is mixed with the waste and the waste then allowed to settle. The supernatant is drawn off, recycled, or discharged, while the settled material compacts and hardens. The process has been used commercially with some success at several installations in the United States.

COMPARISON OF PROCESSES

Perhaps the best way to understand the similarities and differences between the many CFS processes in existence today is to look at their chemical and physical properties, process parameters, costs, and general performances in pilot and field studies. Process parameters are discussed in later chapters. The remainder of this chapter is devoted to these comparisons, using a variety of laboratory, pilot, and field studies, as well as engineering and cost analyses from known data.

Physical Properties

Gel time and Physical State. In the selection of a CFS process, it is important to know the approximate gel time and the physical state of the product as it exits the solidification plant. These data are presented in Table 9-5 for both low and high solids wastes. The actual numbers will vary considerably with specific formulations and wastes, and with reagent mix ratios.

Strength and Durability. CFS processes develop a very wide variety of strength and durability values, depending on many factors: waste type, water content, reagent type, reagent addition ratio (mix ratio), curing time, and temperature. Many processes can adjust the final strength and durability values by changing reagent mix ratios. Others are designed to produce only a narrow range of physical properties. CFS products range from soft, soil-like materials to concretelike monoliths. Contrary to the opinion held by many nonprofessionals, rock-hard solids are not

TABLE 9-5 Gel Times and Physical States

Process	Set or Gel Time (hours) Solids Low	Set or Gel Time (hours) Solids High	Physical State Solids Low	Physical State Solids High
Cement-based	24–48	1–48	Heavy Slurry[a]	Soft Solid
Cement/flyash	48+	1–48	Heavy Slurry[a]	Soft Solid
Cement/soluble silicate	0.01–0.5	0.01–0.5	Pumpable Slurry	Soft Solid
Lime/flyash	48+	1–48	Soft	Soft Solid
Kiln dust	48+	24–48	Soft	Soft Solid

[a] May be pumpable with positive displacement pumps if distance and head are not too great.

always desirable. In a landfill operation, a friable, compactable material is usually preferable. And low permeability, while desirable from a leaching point of view, may make operation of a landfill difficult in wet weather.

A number of physical as well as chemical properties of solidified wastes were evaluated in an international cooperative test study during the period 1985–1987 [56]. Only a summary of that program had been published at the time of this writing, but the serious reader is encouraged to obtain the full report for useful details. Table 9-6 summarizes the physical properties found, as well as some determined in other work, for inorganic systems.

Another recent study by Weitzman *et al.* [57] evaluated a number of CFS processes, using a series of four synthetic analytic reference matrices (SARMs) prepared by the EPA. these SARMs of four types:

SARM 1: High organic, low metal; 31.4 percent water
SARM 2: Low organic, low metal; 8.6 percent water
SARM 3: Low organic, high metal; 19.3 percent water
SARM 4: High organic, high metal; 22.1 percent water

The analyses of these materials are given in the reference. The strength test results are shown in Table 9-7. Tests were run after 7, 14, 21, and 28 days curing at 90–100 percent relative humidity at 21°C ±6°C (70°F ± 43°F). Results are expressed in pounds/square inch. The comparison was made on a reasonably constant water-to-total-solids ratio (W/TS), at various binder-(reagent)-to-soil ratios (B/S). It is evident that Portland cement gives the strongest products through the range of mix ratios tested. However, nearly all samples tested gave adequate strength for most landfill disposal purposes.

Chemical Properties

The primary chemical property for comparison of different CFS processes is leachability, which has been very extensively studied by government, vendor, testing, and engineering organizations over the last 10 years. Prior to that, most information had been supplied by vendors, often using a variety of leaching tests and protocols, and was thus of limited use. Many recent studies have been funded by the U.S. EPA

TABLE 9-6 Ranges of Physical Properties

Test	Range of Properties
Bulk density	0.7–2.2 g/cm^3
Unconfined compressive strength	75–20,000 kPa
Water content (wet weight basis)	0.9–64%
Specific gravity	~2.5
Hydraulic conductivity (permeability)	10^{-4}–10^{-8} cm/sec
Freeze–thaw cycling	1–12 cycles
Wet–dry cycling	1–12 cycles
Thermal conductivity	0.8–1.7 W/m °C

Note: See Chapter 17 for test methods.

Source: Adapted from Stegemann *et al.*, Preliminary results of an international government/industry cooperative study of waste stabilization/solidification, *Hazardous Waste: Detection, Control, Treatment,* Elsevier, Amsterdam/New York, 1988; and other sources.

TABLE 9-7 Comparison of Product Strengths from Various CFS Processes

Sample Number	SARM Type	Binder Type	B/S Ratio	W/TS Ratio	\multicolumn{4}{c}{Days After Mixing}			
					7	14	21	28
1	I	Portland Cement Type 1	0.7	0.40	NA	977	NA	1093
2	I		1.2	0.40	NA	>1000	NA	>1000
3	I		2.3	0.40	NA	>1000	NA	>1000
4	II	Portland Cement Type 1	0.7	0.40	NA	>1000	NA	>1000
5	II		1.2	0.40	NA	>1000	NA	>1000
6	II		2.3	0.40	NA	>1000	NA	>1000
7	III	Portland Cement Type 1	0.7	0.40	NA	28	NA	>1000
8	III		1.2	0.40	NA	99	NA	>1000
9	III		2.3	0.40	NA	71	NA	>1000
10	IV	Portland Cement Type 1	0.7	0.40	NA	15.8	NA	16.2
11	IV		1.2	0.40	NA	167	NA	160
12	IV		2.3	0.40	NA	177	NA	300
13	I	Kiln Dust	1	0.40	5	72.9	93	113
14	I		2	0.45	5	51.8	54	241
15	I		3	0.40	176	211	215	81.1
16	II	Kiln Dust	1	0.40	37.5	59.7	78.3	85.1
17	II		2	0.40	128	190	164	216
18	II		3	0.40	183	225	275	252
19	III	Kiln Dust	1	0.40	32.9	36.6	37.1	38.5
20	III		2	0.40	33	38.4	40.8	39
21	III		3	0.40	45.7	44.7	43.7	79.8
22	IV	Kiln Dust	1	0.42	27.9	28.1	26.8	32.2
23	IV		2	0.43	38.9	55.7	52.4	52.2
24	IV		3	0.40	35.7	38.2	33	40.1
25	I	Lime/Flyash	1	0.45	24.1	27.3	26	32.3
26	I		2	0.45	22.2	29	33.9	40.4
27	I		3	0.45	19.5	30.4	32.9	46.6
28	II	Lime/Flyash	1	0.45	9.9	17.1	17.3	28.8
29	II		2	0.48	17.2	24.2	26.9	62.4
30	II		3	0.49	19.4	31.4	41.2	73.4
31	III	Lime/Flyash	1	0.48	21.9	28.7	29.1	30.7
32	III		2	0.49	30.3	33	36.4	36.5
33	III		3	0.50	34.8	48.8	48.2	50.9
34	IV	Lime/Flyash	1	0.45	34.9	36	34.8	37.9
35	IV		2	0.40	29.8	38.7	36.3	40.5
36	IV		3	0.45	36.9	35.7	37.9	42.2

or Environment Canada to provide data for regulatory purposes. The previously referenced study by Weitzman et al. [57] also reported leaching test results (TCLP method) for the same combinations of reagents and SARMs used in the physical test studies, at two different curing times 14 and 28 days. These results are presented in Table 9-8.

For cadmium and nickel, all processes gave essentially the same results for all wastes—very low leaching rates below or near analytical detection limits. For chromium, all processes appeared to slightly *increase* the leaching rate over that of the raw waste, but leaching levels were still very low, near or below the drinking water standard. With arsenic, copper, lead, and zinc, the Portland cement process was clearly superior for all wastes except SARM 2, where lime-flyash performed best on copper, lead, and zinc. Lime-flyash and kiln dust were also superior for lead fixation on SARM 1. SARM 1 and SARM 2 are both low metal content wastes.

TABLE 9-8 Summary of TCLP Results for Metals

(SARM) Sample Number	Binder (Day)	Arsenic a	Arsenic b	Cadmium a	Cadmium b	Chromium a	Chromium b
I	RAW	ND	—	0.53		ND	
1	PC(14)	ND	—	ND	100	0.06	+
14	KD(14)	ND	—	ND	100	0.06	+
27	LF(14)	ND	—	ND	100	0.02	+
1	PC(28)	ND	—	ND	100	0.06	+
15	KD(28)	ND	—	ND	100	0.09	+
27	LF(28)	ND	—	ND	100	0.02	+
II	RAW	ND		0.73		ND	
4	PC(14)	ND	—	ND	100	0.03	+
16	KD(14)	ND	—	ND	100	0.08	+
30	LF(14)	ND	—	ND	100	ND	—
4	PC(28)	ND	—	ND	100	0.03	+
16	KD(28)	ND	—	ND	100	0.05	+
29	LF(28)	ND		ND	100	ND	—
III	RAW	6.39		33.1		ND	
7	PC(14)	ND	—	ND		0.07	+
21	KD(14)	ND	—	ND	100	0.22	+
33	LF(14)	0.81	52	0.02	100	0.03	+
7	PC(28)	ND	—	ND	100	0.07	+
21	KD(28)	0.21	98	ND	100	0.12	+
33	LF(28)	0.79	51	0.02	100	0.07	+
IV	RAW	9.58		35.3		0.06	
10	PC(14)	ND	100	ND	100	0.06	+
23	KD(14)	0.16	95	ND	100	0.11	+
	LF(14)	1.61	50	ND	100	0.07	+
10	PC(28)	ND	100	ND	100	0.06	+
23	KD(28)	0.27	92	ND	100	0.12	+
	LF(28)	0.98	59	0.02	100	0.07	+
	Detection Limit	0.15		0.01		0.01	

[a] TCLP results in ppm.
[b] Percent reduction, corrected for dilution.
Note: ND = below detection limit; + = increase over raw SARM.

In the high metal content wastes, SARM 3 and SARM 4, only Portland cement produced acceptable fixation. Curing time had no significant effect on leachability, indicating that unconfined compressive strength is not a good indicator of metal fixation.

These results make a very important point about metal fixation in CFS systems: no one process is always superior for every waste tested by every leaching methodology. Subsequent data in this and later chapters will, again and again, reinforce this basic fact of CFS technology.

Another comparative study of importance is that done by the Waterways Experiment Station, U.S. Army Corps of Engineers, for the EPA [58]. This work was done to establish a data-base for setting the landban BDAT standards for K061, electric arc furnace dust. Three generic CFS process types were used: Portland cement, kiln dust, and lime–flyash. The results are shown in Tables 9-9–9-11, with the

TABLE 9-8 (*Continued*)

Copper		Lead		Nickel		Zinc	
a	b	a	b	a	b	a	b
0.61		0.49		0.27		9.2	
0.07	81	0.15	75	0.04	70	0.23	96
0.04	81	ND	100	ND	100	0.27	94
0.03	98	ND	100	ND	100	0.14	94
0.06	83	0.15	75	0.04	70	0.49	91
0.03	80	ND	100	ND	100	0.62	73
0.03	98	ND	100	ND	100	ND	100
0.89		0.7		0.4		14.6	
0.04	92	0.15	82	0.04	83	0.09	99
0.07	79	0.44	+	ND	100	0.25	97
ND	100	ND	100	ND	100	0.22	99
0.06	89	0.15	83	0.04	83	0.54	94
0.09	89	0.37	+	ND	100	0.78	89
0.03	90	ND	100	ND	100	0.02	100
80.7		19.9		17.5		359	
0.15	100	0.63	95	ND	100	0.58	100
1.02	96	13.3	+	ND	100	4.38	95
2.96	87	51	+	ND	100	3.81	96
0.09	100	ND	100	ND	100	0.69	100
0.85	96	18.3	+	ND	100	4.07	95
2.59	87	51	+	0.03	99	3.97	96
10		70.4		26.8		396	
0.14	100	0.39	99	ND	100	0.39	100
1.88	97	12.4	43	ND	100	4.57	97
1.92	96	91.8	+	ND	100	3.22	96
0.17	100	0.37	99	ND	100	0.74	100
1.67	97	21.4	9	ND	100	3.72	97
2.18	95	65	+	ND	100	3.64	96
0.02		0.15		0.04		0.01	

TABLE 9-9 Stabilization Testing[a]—EPA Collected Data

BDAT Constituents	Untreated Waste Total (ppm)	Untreated Waste TCLP (mg/l)	Treated Waste—TCLP (mg/l) Run #1	Treated Waste—TCLP (mg/l) Run #2	Treated Waste—TCLP (mg/l) Run #3
Antimony	294	0.040	<0.050	<0.050	<0.050
Arsenic	36	<0.010	<0.010	<0.010	<0.010
Barium	238	0.733	0.670	0.550	0.516
Beryllium	0.15	<0.001	<0.001	<0.001	<0.001
Cadmium	481	12.8	2.86	3.64	3.38
Chromium	1,370	<0.007	0.049	0.039	0.040
Copper	2,240	0.066	0.058	0.009	0.005
Lead	20,300	45.1	1.03	1.20	1.24
Mercury	3.8	0.0026	0.0013	0.0014	0.0012
Nickel	243	0.027	0.024	0.014	0.018
Selenium	<5.0	<0.050	<0.010	<0.010	<0.010
Silver	59	0.021	<0.003	<0.003	<0.003
Thallium	<1.0	0.038	<0.007	0.013	0.015
Vanadium	25	<0.006	0.084	0.091	0.290
Zinc	244,000	445	21.0	23.5	23.4

[a] Test #1: The binder was cement.

design and operating data for the tests given in Table 9-12. In this study, lime–flyash was found to be superior for fixation of cadmium and lead; however, the mix ratio for the lime–flyash process was twice that for the other two processes, so direct comparison is difficult. In a proprietary study, the author has found a modified Portland cement process to provide the best fixation (TCLP) for a wide range of K061 waste samples. It is likely that the poor results found in the referenced study were the result of too little alkali in the system to achieve a sufficiently high pH in the final leachate from the test.

The waste characteristics affecting performance for the parameters in Table 9-12 are listed on the next page.

TABLE 9-10 Stabilization Testing[a]—EPA Collected Data

BDAT Constituents	Untreated Waste Total (ppm)	Untreated Waste TCLP (mg/l)	Treated Waste—TCLP (mg/l) Run #4	Treated Waste—TCLP (mg/l) Run #5	Treated Waste—TCLP (mg/l) Run #6
Antimony	294	0.040	<0.050	<0.050	<0.050
Arsenic	36	<0.010	<0.010	<0.010	<0.010
Barium	238	0.733	0.516	0.454	0.552
Beryllium	0.15	<0.001	<0.001	<0.001	<0.001
Cadmium	481	12.8	2.92	1.80	0.508
Chromium	1,370	<0.007	0.027	0.035	0.034
Copper	2,240	0.066	0.019	<0.004	0.016
Lead	20,300	45.1	1.30	1.711	0.350
Mercury	3.8	0.0026	0.0005	0.0008	0.0009
Nickel	243	0.027	<0.012	<0.012	<0.012
Selenium	<5.0	<0.05	<0.010	<0.010	<0.010
Silver	59	0.021	<0.003	<0.003	<0.003
Thallium	<1.0	0.038	0.012	0.011	0.010
Vanadium	25	<0.006	0.083	0.088	0.091
Zinc	244,000	445	14.2	6.12	2.00

[a] Test #2: The binder was kiln dust.

TABLE 9-11 Stabilization Testing[a]—EPA Collected Data

BDAT Constituents	Untreated Waste Total (ppm)	Untreated Waste TCLP (mg/l)	Treated Waste—TCLP (mg/l) Run #7	Treated Waste—TCLP (mg/l) Run #8	Treated Waste—TCLP (mg/l) Run #9
Antimony	294	0.040	<0.050	<0.050	<0.050
Arsenic	36	<0.010	<0.010	<0.010	<0.010
Barium	238	0.733	0.462	0.431	0.500
Beryllium	0.15	<0.001	<0.001	<0.001	<0.001
Cadmium	481	12.8	0.033	0.049	0.073
Chromium	1,370	<0.007	0.093	0.072	0.053
Copper	2,240	0.066	0.015	<0.004	0.008
Lead	20,300	45.1	0.150	0.069	0.066
Mercury	3.8	0.0026	0.0016	0.0017	0.0018
Nickel	243	0.027	<0.012	<0.012	<0.012
Selenium	<5.0	<0.05	<0.025	<0.025	<0.010
Silver	59	0.021	<0.003	<0.003	<0.003
Thallium	<1.0	0.038	0.012	0.014	0.011
Vanadium	25	<0.006	0.080	0.089	0.085
Zinc	244,000	445	0.592	0.179	0.398

[a] Test #3: The binder was lime/flyash.

Fine particles— 90 percent of the waste composed of particles <63 μm or less than 230 mesh sieve size
Oil and grease— 282 ppm
Sulfates— 8440 ppm
Chlorides— 19,300 ppm
Total organic carbon— 4430 ppm

Cote [2] used a different leaching methodology, a dynamic leaching test (see Chapter 17 for test description) to study the comparative leaching of four processes: cement/soluble silicate, cement/clay, lime/flyash, and cement/flyash. He found that the cumulative amount leached from a synthetic waste (water solution of metal compounds) was a function of time, as would be expected, and varied greatly from one process to another. The degree of fixation also depended on the metal. Cement/flyash performed best with all the metals reported (arsenic, cadmium, chromium, and lead) and cement/soluble silicate performed worst except for lead. The other processes were intermediate except for lead leaching, where both lime/flyash and cement/clay were the poorest performers. Test results for all four metals are shown in Figures 9-13–9-16.

Cote attributed this, at least in part, to moderation of the initially high pH of these processes due to pozzolanic reactions that consume $Ca(OH)_2$. Figure 9-17 shows that Cote's time versus pH data fall into two bands: one of rapidly decreasing pH leveling off at pH about 9.5 for the flyash-based systems, and the other of high initial pH, leveling off at a much higher level (about pH 11) with the cement/clay and cement/soluble silicate systems. High pH can promote faster leaching with amphoteric metals, but there are many other factors involved. In this study, the leachant is distilled water, so the leaching system pH is directly determined by the pH and alkalinity of the waste form. In the TCLP and similar tests, the acid leachant alters the system so that leaching becomes a function of the total alkalinity of the waste, the amount of acid used in the test, the rate of acid addition, and the particle

TABLE 9-12 Stabilization Testing—EPA Collected Data

Design and Operating Data

Stabilization Process/Binder

Parameter	Cement Run #1	Cement Run #2	Cement Run #3	Kiln Dust Run #4	Kiln Dust Run #5	Kiln Dust Run #6	Lime and Flyash Run #7	Lime and Flyash Run #8	Lime and Flyash Run #9
Binder-to-waste ratio	0.05	0.05	0.05	0.05	0.05	0.05	0.10	0.10	0.10
Water-to-waste ratio	0.5	0.5	0.5	0.5	0.5	0.5	0.5	0.5	0.5
Mixture pH	10.9	11.5	10.5	11.5	11.6	11.1	12.1	12.0	12.0
Cure time (days)	28	28	28	28	28	28	28	28	28
Unconfined compressive strength (lb/in.^{2})	59.7	88.8	95.7	133.0	167.2	141.2	54.6	58.0	50.7

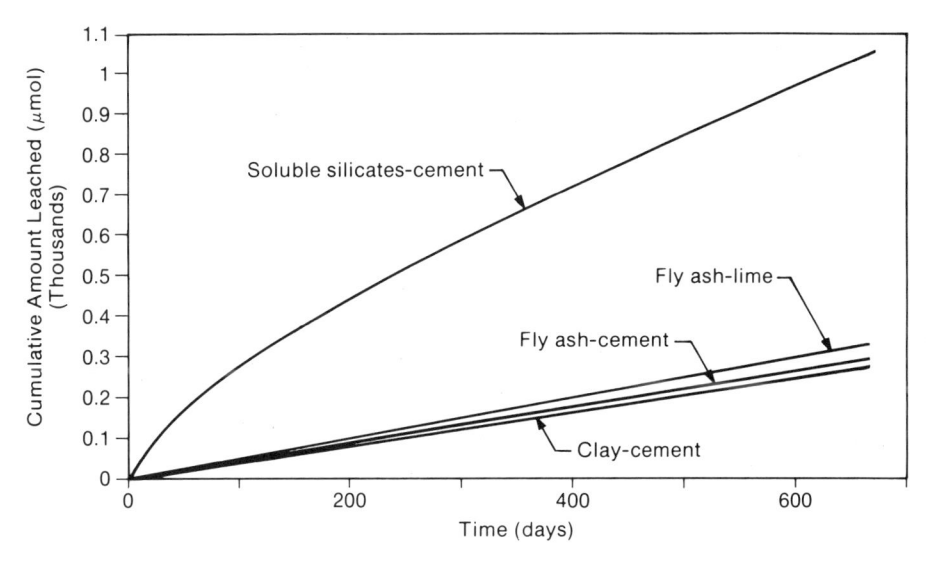

Figure 9-13 Arsenic leaching vs. time for various CFS processes. (From Cote [2].)

size distribution of the waste being leached. As we have seen, this creates a much more complex situation.

We will discuss the chemical properties of CFS-treated wastes again in the later chapters when the individual systems are discussed. There, the effects of leaching test and waste characteristics on process performance will be examined in detail. The discussion here, however, should have made one important point to the reader—that there is no one process, product, or system that is best for all waste treatment and disposal scenarios.

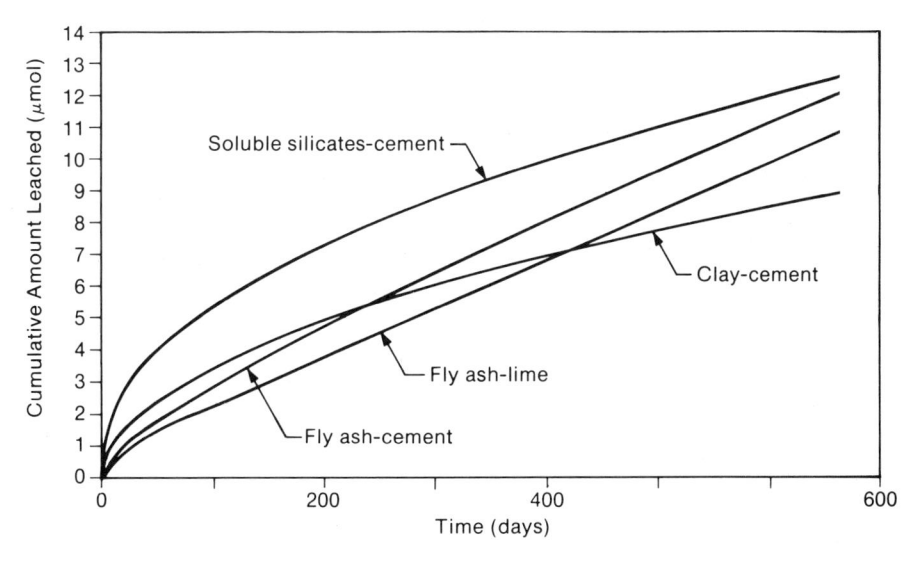

Figure 9-14 Chromium leaching vs. time for various CFS processes. (From Cote [2].)

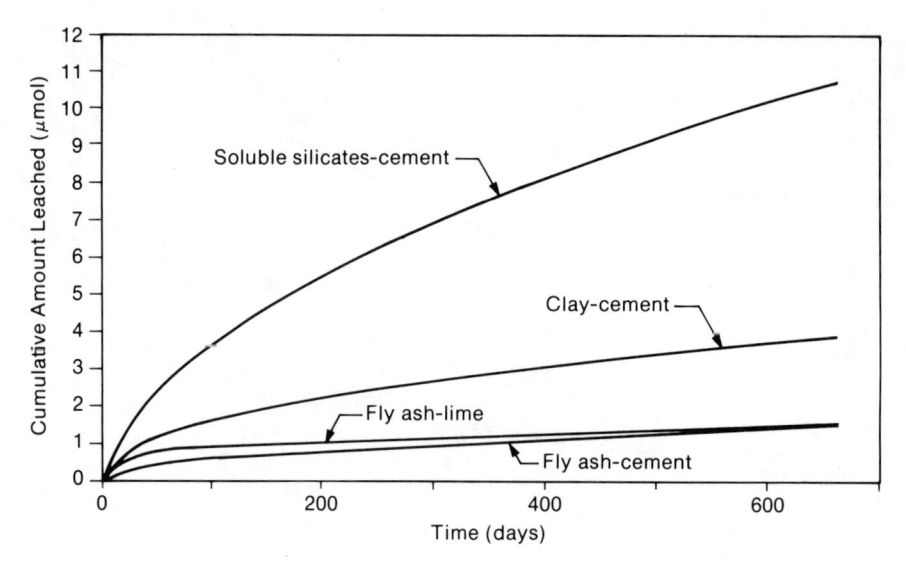

Figure 9-15 Cadmium leaching vs. time for various CFS processes. (From Cote [2].)

Costs

Some very general cost data assembled by the Waterways Experiment Station (WES) in 1982 [59] are given in Table 9-13. Typical costs for reagents are given in Table 9-14.

The prices given in Table 9-14 include local delivery, that is, within about 50 miles. These prices generally hold for industrial areas in the eastern United States. On the West Coast, prices are usually at the high end of the range. Transportation costs rise with distance, of course, but not in a linear fashion. Typical unit transpor-

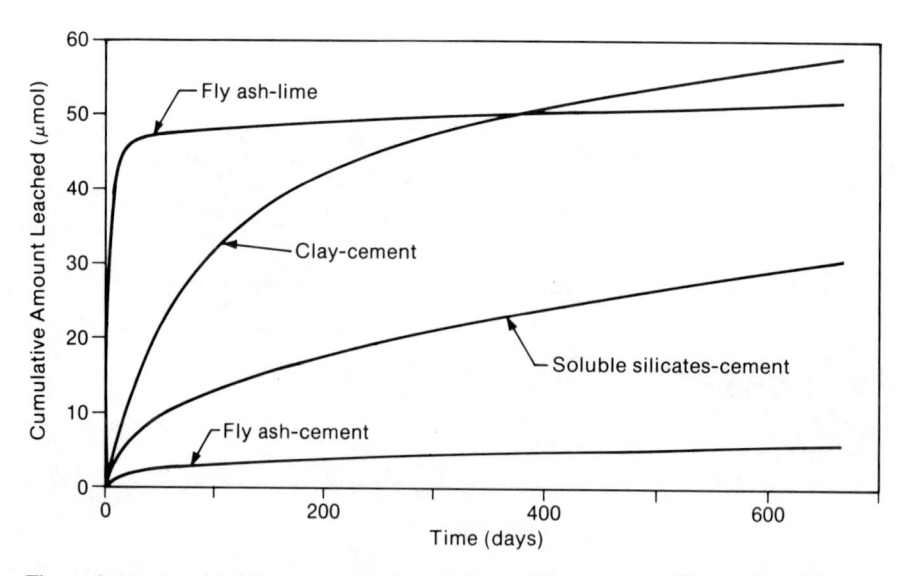

Figure 9-16 Lead leaching vs. time for various CFS processes. (From Cote [2].)

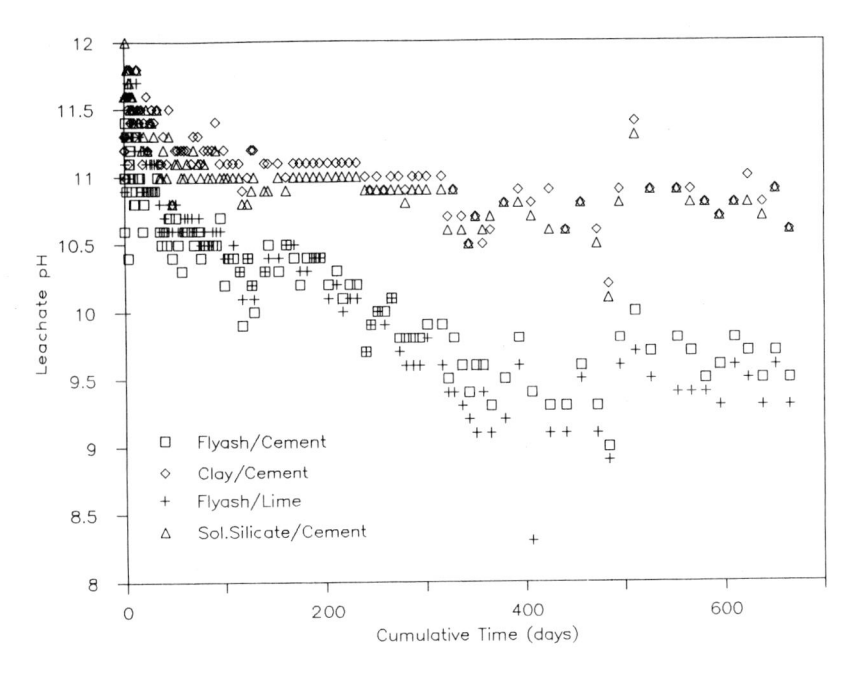

Figure 9-17 Comparison of pH data history: pH vs. time for various CFS processes. (From Cote [2].)

tation costs in the U.S. (1983 basis) are shown in Figure 9-18. Transportation costs often determine which CFS process will be used.

These data are useful for comparative purposes, but do not take into account many factors. To provide a more complete look at comparative costs of CFS processes, the author compared eight inorganic processes on the basis of total cost [60]. Since all have much the same low capital and operating (exclusive of chemical reagents) costs, the cost comparison was made on the basis of chemical cost and the cost associated with the increased volume and weight due to the chemical additions. Because transportation and landfill costs are generally calculated on a weight rather than volume basis, the increased weight cost is used here. Cost comparisons frequently ignore the effect of transportation and disposal of the treated waste. To be sure, these costs may be negligible in some on-site and in situ remediation projects. However, if the solidified waste is to be transported and disposed in a RCRA TSD facility, the added cost due to reagent weight and volume must be included, because the processes vary greatly in this respect. The following cost bases were used for chemical reagents. They were calculated from current, known delivered prices in typical U.S. northern industrial areas.

Reagent	*Price/Ton*
Portland cement	70
Flyash	15
Lime	50
Kiln dust (cement or lime)	20
Clay (regional)	70
Sodium silicate	140

TABLE 9-13 Present and Projected Economic Considerations for Waste Stabilization/Solidification Systems

Type of Treatment System	Major Materials Required	Unit Cost of Material	Amount of Material Required to Treat 100 lb of Raw Waste	Cost of Material Required to Treat 100 lb of Raw Waste	Trends in Price	Equipment Costs	Energy Use
Cement-based	Portland cement	$0.03/lb	100 lb	$ 3.00	Stable	Low	Low
Pozzolanic	Lime flyash	$0.03/lb	100 lb	$ 3.00	Stable	Low	Low
Thermoplastic (bitumen-based)	Bitumen drums	$0.05/lb $27/drum	100 lb 0.8 drum	$18.60	Bitumen prices are rising rapidly because of oil prices	Very high	High
Organic polymer (polyester system)	Polyester catalyst drums	$0.45/lb $1.11/lb $17/drum	43 lb of polyester–catalyst mix	$27.70	Price could rise rapidly due to oil shortage	Very high	High
Surface encapsulation (polyethylene)	Polyethylene	Varies	Varies	$ 4.50[a]	Price could rise rapidly due to oil shortage	Very high	High
Self-cementing	Gypsum (from waste)	[b]	10 lb	[b]	Stable	Moderate	Moderate
Glassification/ mineral synthesis	Feldspar	$0.03/lb	Varies	—	Stable	High	Very high

[a] Based on the full cost of $91/ton.
[b] Negligible, but energy costs for calcining are appreciable.

TABLE 9-14 Typical Costs for CFS Reagents

Delivered Prices, Industrial Areas in the United States

Reagent	Units	Cost Range ($)
Portland cement	Ton (bulk)	40–75
Portland cement	Ton (bag)	70–85
Quick lime	Ton (bulk)	45–90
Hydrated lime	Ton (bulk)	45–90
Hydrated lime	Ton (bag)	60–90
Cement kiln dust (local)[a]	Ton (bulk)	10–45
Lime kiln dust (local)	Ton (bulk)	20–50
Waste lime (local)	Ton (bulk)	15–22
Flyash (local)	Ton (bulk)	10–50
Sodium silicate solution	Pound (bulk)	0.05–0.20

[a] Reagents marked "local" may not be widely available at reasonable cost. Prices are for local delivery.

Costs attributable to weight increases were calculated as add-on costs for a number of transportation and disposal scenarios. The results of these calculations are given in Table 9-15. Typical weight increases for each process are shown for two solidification scenarios: low solids (10 percent) and high solids (35 percent) wastes. The calculation for add-on cost due to weight increases was done by first calculating dollar cost add-on factors per percent of weight increase for landfilling, and per mile of transportation distance, as follows.

Landfill Cost. High-quality engineered landfill is currently priced in the northern United States at $80–$120 per ton—a typical figure is $100 per ton. Capital and

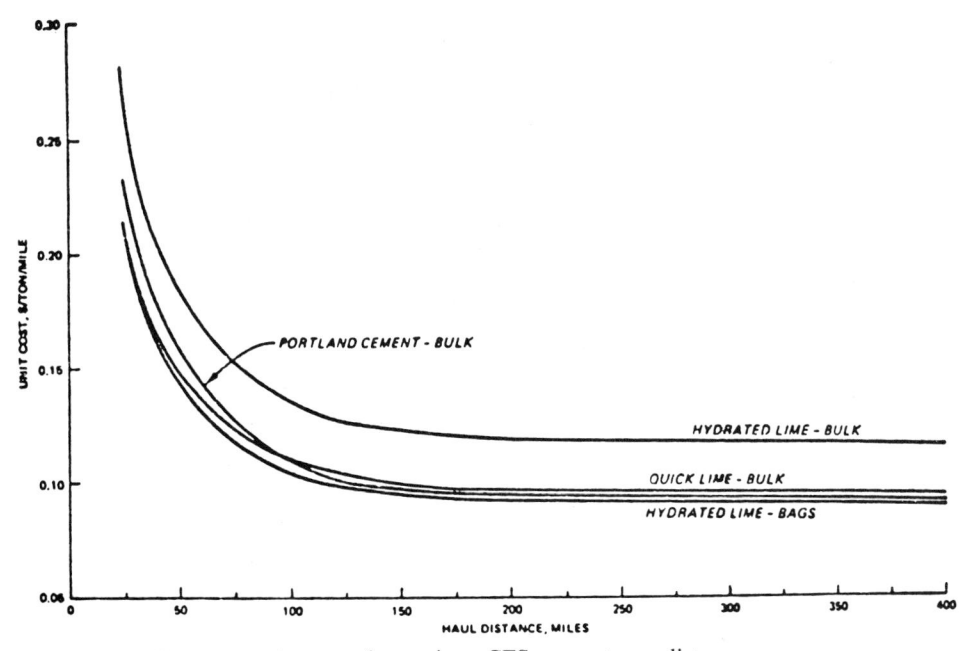

Figure 9-18 Transportation cost for various CFS reagents vs. distance.

TABLE 9-15 Cost Data for Various CFS Systems

		Percent of Weight Increase		Cost per Ton of Waste Treated and Disposed ($/ton)										
				Low Solids (5–15%) Wastes					High Solids (25–50%) Wastes					
					Landfill and Transportation Costs					Landfill and Transportation Costs				
Process	Reagent (delivered)	Low Solids Waste	High Solids Waste	Chemical Cost	0 mi.	50 mi.	200 mi.	500 mi.	Chemical Cost	0 mi.	50 mi.	200 mi.	500 mi.
Portland cement-based	70	100	20	70	100	115	160	190	14	60	69	105	150
Cement/lime (1:1)	60	100	25	60	150	172	240	270	15	80	92	140	200
Cement/clay (1:1)	70	100	35	70	66	76	106	156	24	60	69	105	150
Cement/soluble silicate (2.2:1)	96	33	20	32	100	115	160	194	19	62	71	108	156
Lime/flyash (1:9)	17	300	100	50	100	115	160	201	17	68	78	118	169
Lime/clay (1:4)	66	300	100	197	200	230	320	350	66	100	115	175	250
Cement/flyash (1:3)	27	200	60	54	200	230	320	350	16	100	115	175	250
Cement or lime kiln dust	20	200	75	40	150	172	240	281	15	88	101	154	219

operating cost for such a landfill is approximately half of the market price, or about $50 per ton. Therefore, each percent of weight increase costs $0.50, in addition to the $50 per ton associated with the raw waste.

Transportation Cost. Waste transportation in the northern United States costs about $3/loaded mile. Assuming a 20-ton load, the cost per loaded ton-mile is $0.15.

The calculated costs given in Table 9-16 include the cost of transporting and landfilling the total solidified product, which is the sum of the original waste and the solidification chemicals added. The distances chosen are arbitrary; they are useful only to illustrate the effects of transportation costs. Here, the total cost for each process is given for low and high solids wastes, and for transportation distances of 0 and 200 miles. The real cost of various solidification processes now becomes evident. For low solids wastes, a low volume-increase process such as cement/soluble silicate stands out because the other processes require very high addition ratios of reagents to form a solid that has no free water. In the high solids wastes scenario, three processes are about equal in full cost: cement-based, cement/soluble silicate, and cement/lime. When considering only chemical cost, all the processes except cement/clay and lime/clay are about equal.

In any such comparison, the assumptions made must be kept in mind if conditions change. For example, availability of dry flyash *as a waste stream,* such as is the case in the solidification of FGD sludges at power plants, could make the cement/flyash or lime/flyash processes more attractive. If the assumptions about landfill cost are not valid, then the cost rating must be recalculated.

Cost versus Solids Content. The total solids present in a waste strongly affects the amount of solidification chemicals used. To this point we have evaluated only two approximate cases: low-solids, for which we used a solids content range of 5 to 15 percent; and high-solids, for which we used a solids content range of 25 to 50 percent. To further refine cost comparisons of the preferred processes, the CONTEC data base was searched for actual detailed relationships between total solids content, chemical cost, and weight increase. This was done for the cement-based, cement/soluble silicate, and cement/flyash processes. The cement/lime process was searched, but insufficient detailed data were found. The results of the search are presented in Tables 9-17–9-19 for the three processes. Transportation cost for this comparison was taken as zero.

TABLE 9-16 Summarized Total Cost Data

	\$/Ton of Waste			
	Low Solids		High Solids	
Process	0 mi	200 mi	0 mi	200 mi
Cement-based	170	230	74	119
Cement/flyash	204	294	96	156
Cement/soluble silicate	98	138	79	124
Cement/lime	160	220	77	123
Cement/clay	170	230	92	142
Lime/flyash	250	370	117	192
Lime/clay	397	517	166	241
Cement or lime kiln dust	190	280	103	169

TABLE 9-17 Total Cost vs. Solids Content

Percent Solids	Portland Cement/Based		
	Chemical Cost ($/ton)	Weight Increase (%)	Total Cost ($/ton)
5	66	94	163
10	39	56	117
15	32	45	104
20	23	33	90
25	19	27	82
30	15	21	76
35	13	18	72
40	10	15	68
50 & higher	8	11	64

TABLE 9-18 Total Cost vs. Solids Content

Percent Solids	Portland Cement/Soluble Silicate		
	Chemical Cost ($/ton)	Weight Increase (%)	Total Cost ($/ton)
5	33	34	100
10	26	27	90
15	23	24	85
20	19	20	79
25	16	17	74
30	15	15	72
35	11	12	67
40	10	11	66
50 & higher	8	9	62

TABLE 9-19 Total Cost vs. Solids Content

Percent Solids	Portland Cement/Flyash		
	Chemical Cost ($/ton)	Weight Increase (%)	Total Cost ($/ton)
5	58	212	214
10	46	170	181
15	37	135	154
20	27	98	126
25	23	86	116
30	20	73	106
35	16	60	96
40	13	49	88
50 & higher	10	37	78

These results are consistent with the results shown in Table 9-16, given the assumptions stated previously. Chemical cost and weight increase decrease with increasing solids content. Total cost, from the 5 percent solids level to the 50 percent plus level, falls by factors of 2.8 for cement-based, 1.7 for cement/soluble silicate, and 2.8 for cement/flyash. At high total solids, the cement-based and cement/soluble silicate processes have nearly identical costs. As the solids content falls, the cement/soluble silicate process becomes more cost effective. The cement/flyash process costs about 25 percent more than the cement-based process at all solids levels, and like the latter process, is much less cost effective than the cement/soluble silicate process at low solids content.

Total cost versus total solids is presented in graphic form in Figure 9-19 for the three preceding processes, for the data previously given. From these curves, several conclusions can be drawn.

1. The relationship between total cost (which in turn is directly related to the quantity of solidification chemicals added) and total solids in the waste is regular for all three wastes; that is, the curves are smooth and fit the points well.
2. The cement/soluble silicate process is least sensitive to fluctuations in total solids; the cement/flyash is most sensitive.

Pilot and Field Studies

Until very recently, there were virtually no data available on the chemical and physical properties of CFS-treated wastes from large pilot or commercial-scale projects, except the limited information put out by vendors and others doing the treatment

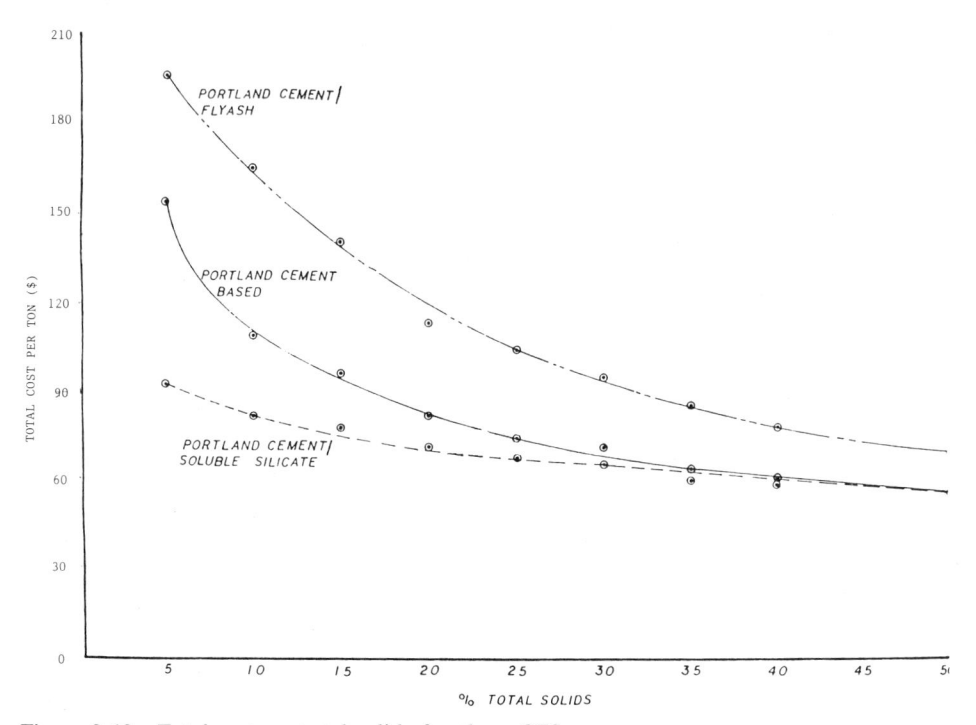

Figure 9-19 Total cost vs. total solids for three CFS processes.

work. The hundreds, perhaps thousands, of CFS projects completed over the last 20 years could have provided a wealth of information. If they had been recorded in such a way as to allow sampling and testing of the solids and the contiguous soil and groundwater environment after years of weathering and leaching, CFS technology would be far more widely accepted than it is currently. In most cases, however, the treated waste was sent to municipal or other landfills and comingled so that analysis is now impossible. In other projects, the waste was used for structural fill on-site, and is now covered by roads or structures, or cannot be identified.

One exception to this situation is in a study done by the Wasterways Experiment Station (WES) [61] in 1981. Four locations where CFS-treated wastes had been deposited in monofills on-site were examined and sampled, and the present groundwater quality was also determined. All four sites had wastes that had been treated with the same cement/soluble silicate process in the period 1973 to 1974; one site's waste was removed to a municipal landfill after about one year. The characteristics of the sites are given in Table 9-20. Sites "W" and "Y" were large automobile assembly plants; site "X" was an electroplating plant; site "Z" was a refinery.

TABLE 9-20 Summary of Characteristics of the Four Sites Selected

Characteristic	Site W	Site X	Site Y	Site Z
Geographic area within the United States	Central	North Central	North Central	South Central
General geologic setting	Glacial drift	Glacial outwash	Pleistocene—lake terrace	Deltaic—fluvial deposits
Mean annual rainfall	102 cm	93 cm	88 cm	117 cm
Mean annual air temperature	12°C	11°C	10°C	21°C
Nature of waste	Paint, putty	Electroplating	Paint, putty	Refinery sludge
Major pollutants detected in sludge analyses	B, Cr, Fe, Pb, Mn, Ni, Zn	Cd, Cr, Cu, Mn, Na, Zn	Cr, Fe, Pb, Mn, Zn	Pb, Mn, and phenol
Liner used below fill	None	None	None	None
Thickness of waste	1.22–3.05 m (avg. 2.14 m)	0.91–1.22 m (avg. 1.07 m)	Thin[a]	1.83–3.20 m (avg. 3.79 m)
Nature of material in unsaturated zone	Sandy clay	Sandy clay	Clayey sand	Clay
Thickness of unsaturated zone	3.05–8.60 m (avg. 5.60 m)	2.16–2.93 m (avg. 2.41 m)	1.46–11.80 m (avg. 8.79 m)	1.04–6.63 m (avg. 3.79 m)
Average hydraulic conductivity below waste	1.1×10^{-7} cm/sec	3.45×10^{-7} cm/sec	2.63×10^{-4} cm/sec	4.2×10^{-7} cm/sec
Character of covering material	None	None	None	Clay
Average thickness of cover	0.0	0.0	0.0	0.5 m
Dates of emplacement of fixed sludge	1974	1973	1974[a]	1974
Type of operation	Diked fill	Diked fill	Fill	Diked fill and cover

[a] Fixed waste was placed on the ground in April and May 1974, but the major portion of the fixed material was removed to a landfill in January 1975 when the area was regraded.

General waste characterization is given below, and chemical analyses are given in Table 9-21.

Site	Waste Characterization
W	Treated paint waste, oil, other residues. Contains iron hydroxides, calcium, sulfates, silica, and lime putty.
X	Electroplating waste containing oxides, hydroxides, and salts of Cd, Cr, Ni, Sn, and Zn.
Y	Waste contained paint pigments, oil, iron, zinc, phosphate, and chromium.
Z	Oil refinery waste containing 8–15 percent organics, 16–28 percent sediment. Metals are Cd, Cu, Fe, Pb, Ni, and Zn. Also contained phenol.

Waste "soil" samples were analyzed and extracted with distilled water, which is more realistic for the actual site "leaching" than a test such as the EPT. Groundwater analyses were also made. The general conclusions drawn from the study were as follows.

- No consistent physical changes in the underlying soil were found.
- Toxic metal contamination in groundwater was not detected at levels "that would present a serious pollution problem."
- The underlying soils did not show consistent increases in major toxic metals that were traceable to the wastes, and the overall composition of these soils was within the range of compositions for natural soils in those areas.

TABLE 9-21 Concentrations of Major Chemical Contaminants in the Fixed Sludge at Each Site[a]

Constituent	Site W (mg/kg)[b]	Site X (mg/kg)	Site Y (mg/kg)	Site Z (mg/kg)
B	43	35	10	
Cd	4.5	105	0.1	1.8
Cr	230	1600	30	960
Cu	28	1650	18	130
Fe	47,000	4000	23,500	16,800
Pb	40	20	40	3,500
Mn	850	110	572	275
Ni	75	50	30	25
Se	1	1.3	0.62	0.83
Na	350	1281	2,000	1,340
Zn	1,300	450	200	520
Hg	0.05	0.20	0.02	6.2
Phenol				Present

[a] Data obtained from analyses of sludge samples.
[b] On a dry weight basis.

- Distilled water extracts of underlying soils showed some increases in metal content.
- "No major contamination of groundwater or soil could be detected under the stabilized wastes; but the sub-waste soils do contain larger quantities of leachable toxic metals than do background soil samples"‡

It appears from this study that these CFS-treated wastes are stable and have not caused groundwater pollution in the disposal area, at least through an approximately six year period. Since leaching rate (i.e., concentration of constituents in the leachate) is generally found to decrease with time, there is no reason to believe that the situation would become worse with time. In fact, the opposite should be true.

A number of projects on CFS treatment are now being done as part of the EPA's Superfund Innovative Technology Evaluation (SITE) program, but detailed data are not available as of this writing. Information from these projects, and others like them being done under other programs, will finally provide the much needed technical and engineering data to establish CFS technology as a major element in environmental control of hazardous wastes.

REFERENCES

1. Conner, J. R. *The Modern Engineered Approach to Chemical Fixation and Solidification Technology*. Pittsburgh: Chemfix, Inc., 1981.
2. Cote, P. *Contaminant Leaching from Cement-based Waste Forms Under Acidic Conditions*. Ph.D. Thesis, McMaster Univ., Hamilton, Ont., Canada, 1986.
3. Bishop, P. L. Leaching of inorganic hazardous constituents from stabilized/solidified hazardous wastes. *Hazardous Wastes Hazardous Mater.* **5**(2): 129–143 (1988).
4. Prange, N. E., and W. F. Garvey. *The Impact of Particle Size on TCLP Extraction of Cement-Stabilized Metallic Wastes*. St. Louis, MO: Monsanto Co., 1988.
5. JACA Corp. *Critical Characteristics and Properties of Hazardous Waste Solidification/Stabilization* (Unpublished Draft). Cincinnati, OH: U.S. EPA, 1987.
6. Chemical Waste Management, Inc. *Chemical Waste Management Internal Study,* Riverdale, IL, 1987.
7. Eaton, H. C., M. B. Walsh, M. E. Tittlebaum, F. K. Cartledge, and D. Chalasani. Microscopic characterization of the solidification/stabilization of organic hazardous wastes. Presented at the Energy Sources and Technology Conference and Exhibition (ASME Paper No. 85-Pet-4). Cincinnati, OH, 1985.
8. Cullinane, M. J., R. M. Bricka, and N. R. Francingues, Jr. An assessment of materials that interfere with stabilization/solidification processes. In *Proc. 13th Annual Research Symposium, U.S. EPA:* Cincinnati, OH, pp. 64–71, 1987.
9. U.S. EPA. *Best Demonstrated Available Technology (BDAT) Background Document For F006, Electroplating*. Washington, DC, 1988.
10. Ashworth, R. Some investigations into the use of sugar as an admixture to concrete. *Proc. Institute of Civil Engineering*. London, 1965.
11. Chalasani, D., F. K. Cartledge, H. C. Eaton, M. E. Tittlebaum, and M. B. Walsh. The effects of ethylene glycol on a cement-based solidification process. *Hazardous Wastes Hazardous Mater.* **3**(2): (1986).

‡It should be pointed out that the treated wastes were deposited on the site of old lagoons or burning areas, so that the slightly increased levels may be due to past waste management practices at the sites. It was not possible to obtain baseline data on the condition of the soils before the treated wastes were deposited.

12. Roberts, B. K. *The Effect of Volatile Organics on Strength Development in Lime Stabilized Fly Ash Compositions.* M.S. Thesis, Univ. of Pennsylvania, Philadelphia, 1973.
13. Rosskopg, P. A., F. J. Linton, and R. B. Peppler. Effect of various accelerating chemical admixtures on setting and strength development of concrete. *J. Testing Eval.* **3**(4): (1975).
14. Smith, R. L. *The Effect of Organic Compounds on Pozzolanic Reactions.* Horsham, Pa.: I.U. Conversion Systems Report No. 57, 1979.
15. Walsh, M. B., H. C. Eaton, M. E. Tittlebaum, F. K. Cartledge, and D. Chalasani. The effect of two organic compounds on a Portland cement-based stabilization matrix. *Hazardous Wastes Hazardous Mater.* **3**(1): (1986).
16. Young, J. F. A review of the mechanisms of set-retardation of cement pastes containing organic admixtures. *Cement and Concrete Research 2,* No. 4, 1972.
17. Young, J. F., R. L. Berger, and F. V. Lawrence. Studies on the hydration of tricalcium silicate pastes. III Influences of admixtures on hydration and strength development. *Cement and Concrete Research 3,* No. 6, 1973.
18. Conner, J. R. Ultimate disposal of liquid wastes by chemical fixation. In *Proc. 29th Annual Purdue Industrial Waste Conference.* Purdue Univ., West Lafayette, IN, 1974.
19. Cote, P. L., and D. P. Hamilton. Leachability comparison of four hazardous wastes fixation processes. In *Proc. 38th Annual Purdue Industrial Waste Conference.* Purdue Univ., West Lafayette, IN, 1984.
20. Double, D. D., and A. Hellawell. The solidification of cement. *Sci. Am.* 82–90 (1977).
21. Heacock, H. W. Alternative nuclear waste solidification processes. In *Proc. Symposium on Waste Management.* Tucson, AZ, pp. 177–206, 1975.
22. Kupiec, A. R. U.S. Patent 4,149,968 (1979).
23. Laguna, W. Radioactive waste disposal by hydraulic fracturing. *Ind. Water Eng.* 32–37 (Oct. 1970).
24. Landreth, R. E. *Guide to the Disposal of Chemically Stabilized and Solidified Waste.* Cincinnati, OH: U.S. EPA, 1980.
25. Mahoney, J. D. *Radiochemical Studies of Leaching of Metal Ions from Sludge Bearing Concrete.* American Electroplating Society, 1980.
26. National Academy of Sciences. *Solidification of High-Level Radioactive Wastes.* Washington, DC: U.S. Nuclear Regulatory Commission, 1979.
27. Pojasek, R. B. *Toxic and Hazardous Waste Disposal,* Vols. I and II. Ann Arbor, MI: Ann Arbor Science Publishers, 1979.
28. Portland Cement Association. *Design and Control of Concrete Mixtures.* Skokie, IL, 1979.
29. Skolny, J., and K. E. Daugherty. Everything you always wanted to know about Portland cement. *ChemTech:* 38–45 (Jan. 1972).
30. NTIS. *Survey of Solidification/Stabilization Technology for Hazardous Wastes.* Springfield, VA, 1979.
31. Rossoff, J., *et al. Disposal of By-Products from Non-regenerable Flue Gas Desulfurization Systems,* Final Report No. EPA-600/7-79-046. U.S. EPA, 1979.
32. Clendenning, T. G., *et al. Current Technology in the Utilization and Disposal of Coal Ash.* Toronto, Ont., Canada: Ontario Hydro Corp., 1975.
33. McClelland, N. I. *Leachate Testing of Hazardous Chemicals from Stabilized Automobile Wastes.* Ann Arbor, MI: National Sanitation Foundation, 1979.
34. Taub, I., and K. Roberts. Leach testing of chemically stabilized waste. In *Proc. 3rd Annual Conference on Treatment and Disposal of Industrial Wastewaters and Residues.* 1978.
35. Chemfix Inc. *Chemfix Data Package I.* Pittsburgh, PA, 1972.
36. Conner, J. R., and L. P. Gowman. Chemical fixation of activated sludge and end use applications. In *Proc. National Conference on Complete Water Reuse.* Cincinnati, OH, 1976.

37. Conner, J. R., and R. J. Polosky. *The Use of Chemically Fixed Solids to Improve Sanitary Landfills.* Pittsburgh, PA: Chemfix, Inc., 1974.
38. Conner, J. R. Testing of chemically solidified waste for disposal in sanitary landfills or on-site monofills. In *Proc. Workshop Environmental Assessment of Waste Solidification.* Vegreville, Alta., Canada, Nov. 1983.
39. Conner, J. R. Chemical fixation and solidification of dredged materials. In *Proc. 7th World Dredging Conference.* San Francisco, July 1976.
40. Conner, J. R. *Disposal of Concentrated Wastes from the Textile Industry.* Pittsburgh, PA: Chemfix, Inc. (Dec. 1976).
41. Conner, J. R. *Ind. Water Eng.* (July/Aug. 1977).
42. Conner, J. R. Ultimate liquid waste disposal methods. *Plant Eng.* (Oct. 1972).
43. Falcone, J. S., R. W. Spenser, and E. P. Katsanas. *Chemical Interactions of Soluble Silicates in the Management of Hazardous Wastes.* Lafayette Hill, PA: The PQ Corp., 1982.
44. Gowman, L. P. Chemical stability of metal silicates vs metal hydroxides in ground water conditions. In *Proc. National Conference on Complete Water Reuse.* Cincinnati, OH, 1976.
45. Leo, P. P., et al. Flue gas desulfurization waste disposal field study at the Shawnee power station. In *Proc. Symposium on Flue Gas Desulfurization.* Research Triangle Park, NC, Nov. 1977.
46. Mahlock, J. L., et al. *Pollutant Potential of Raw and Chemically Fixed Hazardous Industrial Wastes and Flue Gas Desulfurization Sludges.* Vicksburg, MS: U.S. Army Engineer Waterways Experiment Station, July 1976.
47. Spencer, R. W., et al. Applications of soluble silicates and derivative materials for the management of hazardous wastes. *Res. and Devel.* (1972).
48. I.U. Conversion Systems. *Simulated Field Leaching Test for Evaluation of Surface Runoff from Land Disposal of Waste Materials.* Horsham, PA, 1977.
49. Customized waste stabilization. *Waste Age* (April 1981).
50. Diamond, S., et al. Transformation of clay minerals by calcium hydroxide attack. In *Proc. 12th National Conference on Clay and Clay Minerals.* pp. 359–379.
51. ASTM. *Report on Leachate Test Comparison Program.* Philadelphia, PA, Nov. 1977.
52. Thornton, S. I., et al. *Fly Ash As Fill and Base Material in Arkansas Highways.* Washington, DC: Federal Highway Administration, 1975.
53. Cocozza, E. P. U.S. Patent 4,049,462 (1977).
54. Mineral By-Products. *Pozzolime.* Marietta, GA, 1984.
55. *Handbook for Stabilization/Solidification of Hazardous Wastes.* Cincinnati, OH: U.S. EPA, 1986.
56. Stegemann, J., P. L. Cote, and P. Hannak. Preliminary results of an international government/industry cooperative study of waste stabilization/solidification. *Hazardous Waste: Detection, Control, Treatment.* Amsterdam/New York: Elsevier, 1988.
57. Weitzman, L., L. E. Hamel, and E. Barth. Evaluation of solidification/stabilization as a best demonstrated available technology. In *Proc. 14th Annual Research Symposium.* U.S. EPA, Cincinnati, OH, 1988.
58. U.S. EPA. *Onsite Engineering Report for Waterways Experiment Station for K061* (Draft Report). Washington, DC, 1988.
59. U.S. Army Engineer Waterways Experiment Station. *Guide to the Disposal of Chemically Stabilized and Solidified Waste.* Cincinnati, OH: U.S. EPA, 1982.
60. Conner, J. R. Private study, 1988.
61. Larson, R. J., P. G. Malone, J. H. Shamburger, J. D. Broughton, D. W. Thompson, and L. W. Jones. *Field Investigation of Contaminant Loss from Chemically Stabilized Industrial Sludges.* Springfield, VA: NTIS, 1981.

Chapter 10

PORTLAND CEMENT-BASED SYSTEMS

As we begin to discuss the individual CFS processes in detail in Chapters 10–19, it is appropriate that the first process to be addressed is that based on Portland cement. This process, and variations on it, was the first to be used for actual CFS work—in the nuclear waste field in the 1950s. Since then, it has been the most widely applied ingredient in CFS systems, although bulk agents and pozzolans probably constitute the largest total mass of reagents used. Some systems that incorporate Portland cement—cement/soluble silicate and cement/flyash—are discussed in later chapters because they represent special cases either technically or commercially. The other variations on Portland cement, including additives, are addressed in this chapter. We will use the terms "Portland cement" and "cement" interchangeably here; *cement* will be taken to mean a Portland cement unless otherwise specified. Other cement types, such as alumina cement and sorel cement, have been relatively little used in CFS technology (except in the nuclear waste area) in the United States, primarily due to cost. They may, however, find more future use in special situations as the requirements for fixation become more stringent and the allowable costs therefore become higher.

Some of the specific reasons why Portland cement processes are so important in CFS technology, and so useful for studying the reactions of inorganic CFS systems, are as follows.

1. Its composition is much more consistent from source to source, eliminating some of the many variables in studying CFS processes.
2. Much more is known about the reactions of Portland cement in setting and hardening, and more recently, in fixation of metals.
3. Good data on modeling of environmental effects from the leaching of cement-based waste forms are available from the nuclear waste field.
4. Most of the recent studies at Louisiana State University [1] and elsewhere [2,3] have been done on this process type.
5. Portland cement processes may also emulate many of the reactions of pozzolanic processes, which make up much of the current commercial CFS technology; however, the extent of emulation has not been determined.

Another reason for starting out our coverage of individual processes with Portland cement is that its reactions are similar in many ways to the pozzolanic reactions of kiln dusts, flyashes, and other materials used in CFS. Ancient Egyptian, Greek, and Roman buildings and other structures used the latter cementitious materials as mortars. These "mortars" were based on unsintered calcerous compounds combined with sand and/or other silicious materials, often of volcanic origin [4]. These mortars and concretes were made from various materials of the time: burnt gypsum, calcined limestone, sand and gravel, volcanic ash, and crushed tiles [5]. Roman concrete work reached a surprising level of quality. A poured concrete dome supported by walls 20 feet thick is still in excellent condition in the Roman Pantheon.

The Portland cement that we know today is made by heating together limestone and clay (or some other source of silica) at about 2700°F, forming a mass called clinker. A small amount of gypsum is added and the clinker is ground to a fine powder. Its invention is attributed to an English mason, Joseph Aspdin, who called his invention (patented in 1824) Portland cement because it produced an artificial stone resembling the natural stone from the famous quarries of Portland, England. Portland cement is basically a calcium silicate mixture containing predominantly tricalcium and dicalcium silicates, which in cement shorthand are called C_3S and C_2S, respectively,* with smaller amounts of tricalcium aluminate and a calcium aluminoferrite with the approximate formulas C_3A and C_4AF, respectively [6]. Typical weight proportions in an ordinary cement are 50 percent C_3S, 25 percent C_2S, 10 percent C_3A, 10 percent C_4AF, and 5 percent other oxides. High alumina cement consists mainly of $CaO \cdot Al_2O_3$, or CA. This cement hardens very rapidly compared to ordinary cement, but its strength may deteriorate over long time periods, especially in a high-temperature environment and high water content in the cement mix.

There are many types of cement used for many purposes, but only those classified as Portland cement have seen substantial use in CFS technology. The primary reason for this is cost. Portland cement processes can be divided into two general types: Portland cement alone, and Portland cement plus other additives, such as lime and clay. Portland cement has been used alone or with minor additives for many years for solidification of radioactive wastes [7], especially for subsequent ocean disposal.

With the Portland cement process, water in the waste reacts chemically with Portland cement to form hydrated silicate and aluminate compounds. Solids in the waste act as an aggregate to form a "concrete," although the types of solids encountered in wet wastes may produce a concrete of low strength. The optimum combination of waste and Portland cement, the type of Portland cement chosen, and any additives, will vary with the waste type and its composition. Portland cement requires a minimum amount of water to obtain workability. This minimum water to cement ratio is approximately 0.40 by weight for Portland cement, but will also depend on the waste itself, since some waste solids may absorb large amounts of water. The addition of too much water may result in a layer of freestanding water on the surface of the solidified product, as well as reduction in strength and increase in permeability of the final product.

Optimum formulations should take each waste into consideration individually, because of possible interactions between the waste constituents and the cement. One

---

*This notation system represents calcium, silicon, aluminum, and iron oxides by C, S, A, and F, respectively. The subscripts denote the relative mole ratios of each component, for example, $2CaO \cdot SiO_2$ is C_2S.

such interaction is the effect on setting of the cement matrix. Acceleration of the set can result in the cement hardening in the processing equipment. On the other hand, boric acid and borate salts, for example, retard the setting of Portland cement; if sufficient quantities are added, the set may be retarded to the extent that the cement never hardens. Other substances in wastes that may be deleterious to the setting and curing reactions of Portland cement include organic compounds, silt, coal and lignite, some inorganic salts, and metal compounds, which may prevent setting for several days. Very fine particles such as are present in silt and clay may weaken the bond between the cement paste and the particulates by particle coating action. Salts of manganese, tin, zinc, copper, and lead are active in reducing strength, and other salts such as sodium phosphate, sodium iodate, sodium sulfide, and sodium borate are potential retarders. In some cases, the addition of the lime to the Portland cement has been found to counteract the retarding action and allow a proper set to be achieved. Sodium silicate has been used for the same purpose [8].

CHEMICAL AND PHYSICAL CHARACTERISTICS OF CEMENT

Cement Types and Compositions

The ASTM provides for eight types of Portland cement, while the Canadian Standards Association (CSA) provides for five. In addition, there are a number of other cements of interest in CFS. These types are listed and briefly described below.

Type	*Description*
ASTM Type I	General-purpose Portland cement, and usually the least expensive. Used most commonly in CFS work (CSA "Normal").
ASTM Type IA	Same as Type I, but contains air-entraining agents for improved resistance to freeze–thaw and scaling.
ASTM Type II	General use where moderate sulfate attack is expected, or where moderate heat of hydration is required (CSA "Moderate").
ASTM Type IIA	Same as Type II, but with air-entraining agents.
ASTM Type III	Used where high early strength is required, and in cold weather (CSA "High early strength").
ASTM Type IIIA	Same as Type III, but with air-entraining agents.
ASTM Type IV	Low heat of hydration, used in massive structures where temperature rise must be controlled. Develops strength more slowly than Type I (CSA "Low heat of hydration").

Type	Description
ASTM Type V	Used where soils and groundwaters have high sulfate content. Develops strength slowly (CSA "Sulfate resisting").
ASTM Type IS, IS-A	These are Portland/blast furnace slag cements, the latter designation containing air-entraining agents. The slag reacts in the presence of $Ca(OH)_2$ and gypsum in the cement paste. Low early strength, but hardens slowly to the same values as Type I. It is more common in Europe than in the United States.
ASTM Type IP, P	Pozzolan-containing cements made by intergrinding of suitable pozzolan with the cement clinker. Similar uses and properties as Type IS; also, air-entraining grades.
Masonry cements	Used as mortar in bonding brick and masonry. Usually contains one or more of hydrated lime, limestone, chalk, shell, talc, slag, or clay. Good workability, plasticity, and water retention.
Natural cements	Made from same materials as Portland cement, but at lower temperatures below the sintering point. More like a hydraulic lime. Not restricted in magnesia content.
Expanding cement	Contains slag and "calcium sulfoaluminate cement" (calcined mixture of 50 percent gypsum, 25 percent bauxite and 25 percent chalk). Expands slightly on hydration, due to the formation of ettringite.
High-alumina	Not a Portland cement. Made by fusing limestone and bauxite with small amounts of silica and titania. Develops very high strength quickly, but may have long-term stability problems. Sets slowly.
Waterproofed	Made by intergrinding small amounts of calcium, aluminum, or other stearates with cement clinker.

Type
Sorel cement

Description
Also known as magnesium oxychloride cement. Not a Portland cement. Made by adding MgO to a solution of magnesium chloride. The reaction product is hard, but not water resistant.

Sources of the raw materials used in the manufacture of cement are shown in Table 10-1. While the properties of Portland cements are specified, and much more highly controlled than most of the other ingredients used in commercial CFS systems, there can be noticeable differences due mostly to the varied raw materials used in different locations. These may show up in CFS work mostly in terms of leaching and other chemical properties. Most of the cement types just shown are not used in commercial CFS work in the United States, but they appear rather consistently in patent literature, especially in Japanese patents. Because of differences in raw materials availability and energy costs in various parts of the world, other cements may find practical utilization in CFS processes in specific locations, or with particular wastes.

The chemical compositions of some of the different cement types are given in Tables 10-2 and 10-3. Composition may be expressed in the more familiar form, as elements or stoichiometric compounds, or in terms of cement nomenclature, which is more useful in discussing the hydration reactions of cement. Both are given in Table 10-2. Cement nomenclature is expressed as follows:

$$C_2S = 2\ CaO \cdot SiO_2 = \text{dicalcium silicate}$$

$$C_3S = 3\ CaO \cdot SiO_2 = \text{tricalcium silicate}$$

TABLE 10-1 Sources of Raw Materials Used in Manufacture of Portland Cement

Lime CaO	Iron Fe_2O_3	Silica SiO_2	Alumina Al_2O_3	Gypsum $CaSO_4 \cdot 2H_2O$	Magnesia MgO
Alkali waste	Blast furnace	Calcium	Aluminum	Anhydrite	Cement rock
Calcite	flue dust	silicate	ore refuse	Calcium	Limestone
Cement rock	Clay	Cement rock	Bauxite	sulfate	
Chalk	Iron ore	Clay	Cement rock	Gypsum	
Clay	Mill scale	Flyash	Clay		
Fuller's earth	Ore washings	Fuller's earth	Copper slag		
Limestone	Pyrite cinders	Limestone	Flyash		
Marble	Shale	Loess	Fuller's earth		
Marl		Marl	Granodiorite		
Seashells		Ore washings	Limestone		
Shale		Quartzite	Loess		
Slag		Rice hull ash	Ore washings		
		Sand	Shale		
		Sandstone	Slag		
		Shale	Staurolite		
		Slag			
		Traprock			

TABLE 10-2 Typical Compositions of Portland Cements: Chemical Composition (%)

Cement Type	Designation or Characteristics	Ignition Loss	Insoluble Material	SiO$_2$	Al$_2$O$_3$	Fe$_2$O$_3$	CaO	MgO	SO$_3$
I	Normal Portland—general use	0.6	0.1	22	5.1	3.2	65	1.4	1.6
II	Moderate heat evolution							2.5	
III	High early strength							2.0	
IV	Low heat evolution							1.8	
V	Sulfate resistance							1.9	
	Rapid hardening	0.9	0.2	21	4.9	2.8	66	1.1	2.5
	Super rapid hardening	0.9	0.1	19.7	5.1	2.7	65	2.0	3.0
	Jet cement	0.6	0.1	13.8	11.4	1.5	59	0.9	10

TABLE 10-3 Typical Compositions of Portland Cements: Mineral Composition (%)

Cement Type	Designation or Characteristics	C_3S[a]	C_2S[b]	C_3A[c]	C_4AF[d]	Free CaO	$CaSO_4$	Specific Gravity	Specific Surface Area[e] cm^2/g
I	Normal Portland—general use	45	27	11	8	0.5	3.1	3.17	3220
II	Moderate heat evolution	44	31	5	13	0.4	2.8		
III	High early strength	53	19	11	9	0.7	4.0		
IV	Low heat evolution	28	49	4	12	0.2	3.2		
V	Sulfate resistance	38	43	4	9	0.5	2.7		
	Rapid hardening	66	11	8	9			3.13	4340
	Super rapid hardening	68	5	9	8			3.14	5950
	Jet cement	52	0	22[f]	5			3.04	5300

[a] $3CaO \cdot SiO_2$.
[b] $2CaO \cdot SiO_2$.
[c] $3CaO \cdot Al_2O_3$.
[d] $4CaO \cdot Al_2O_3 \cdot Fe_2O_3$.
[e] Blaine value.
[f] $C_{11}A_7 \cdot CaF_2$.

$C_3A = 3\ CaO \cdot Al_2O_3$ = tricalcium aluminate

$C_4AF = 4\ CaO \cdot Al_2O_3 \cdot Fe_2O_3$ = tetracalcium aluminoferrite

$C\bar{S} = CaSO_4$

$H = H_2O$

One of the best ways to visualize the various cement types in terms of their compositions is to plot the compositions on a ternary diagram, as shown in Figure 10-1 [4]. The individual minerals that make up the various cements all lie along the axes since they are binary mixtures of lime/silica/alumina (ferrite is not included in this diagram). The complex minerals lie in the various interior regions of the diagram as shown. It is instructive to compare Figure 10-1 with Figure 9-9, which includes all of the main ingredients (with the exception of sodium silicate) of the commercial CFS processes. Another property of cements, and all other reagents of importance in CFS, is to what degree they leach in standard tests. TCLP test results

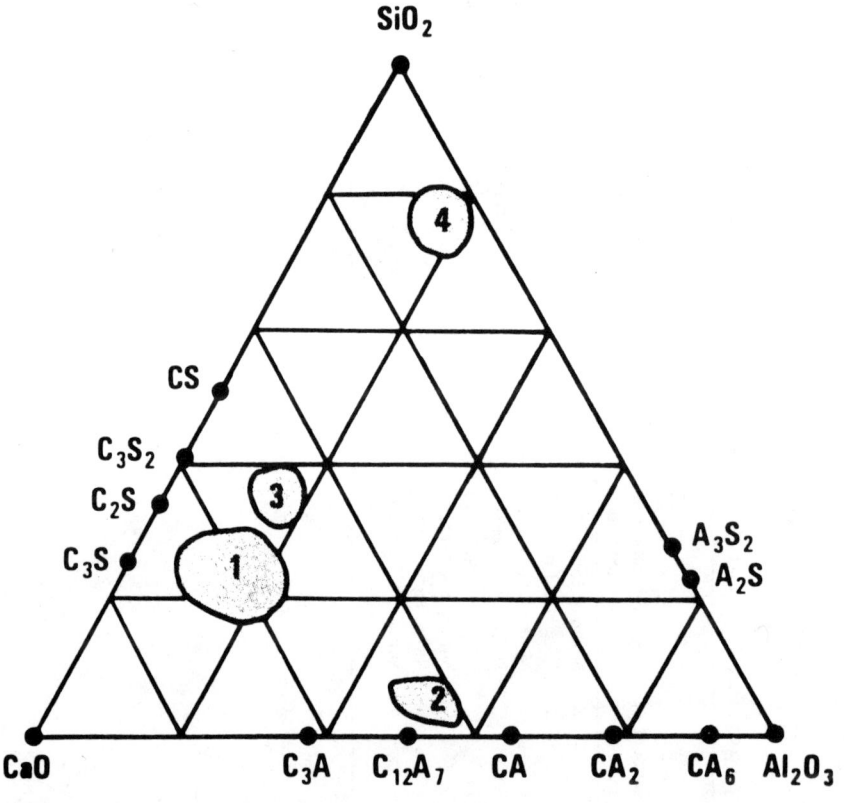

Figure 10-1 Ternary diagram of the system $CaO - Al_2O_3 - SiO_2$ with areas of basic cement and related materials compositions. 1. Portland cement. 2. Aluminous cement. 3. Basic blast furnace slag. 4. Volcanic and artificial pozzolanas. Reprinted with permission from J. Skalny and K. E. Daugherty, Everything you always wanted to know about Portland cement, but did not ask. *ChemTech,* p. 38, Jan. 1972. Copyright The American Chemical Society.

for a number of samples from different locations are shown in Table 10-4. It must be remembered, however, that these data are for the unhydrated cement directly leached with the acid TCLP solution; results might be greatly different for a fully hydrated and cured cement. The significance of the data in Table 10-4 is that there are limiting factors in the reagents themselves that will control the ultimate leaching resistance that it is possible to attain with any commercial system. As we will see later, cement is one of the best reagents in this regard.

Chemistry of Cement Setting

Discussion of the chemistry of cement setting is important to an understanding of cement-based CFS processes. The references give more detailed examination of cement chemistry. Especially useful is Cote's discussion of the subject as it applies to CFS [3]; it is used extensively in this section.

Normal Portland Cements. In spite of the seemingly simple composition of cement (limestone and clay) and its long history of use, there is still a surprising amount of disagreement among scientists in this field about how cement combines with water. Two models, the crystalline and the osmotic or gel, emphasize different mechanisms. The former is more widely accepted in the United States; the latter was developed by Double *et al.* [9] at Oxford, England. In both models, the same basic reactions occur. In the presence of water, each of the major crystalline compounds hydrates, but the products are different and their contributions to the final waste form are different. Tricalcium aluminate and sulfates react almost immediately to form hydrates. If sufficient sulfate is present, the reaction product is hydrated calcium aluminate sulfate, which coats the surfaces of the particles, preventing rapid further hydration. This is why gypsum retards setting. If no gypsum is present, calcium aluminum hydrates form immediately and the system sets. The overall contribution of aluminate and ferrite hydrates to the strength of the cement paste is thought to be minor [4].

Strength development after setting is primarily due to C_3S and b-C_2S, both of which give the same reaction products, that is, $Ca(OH)_2$ and calcium silicate hydrate, or tobermorite gel (CSH). All of the reaction products, with the exception of lime, have low solubilities in the lime-saturated medium of the cement paste compared to the anhydrous minerals in the original cement. The basic hydration reactions [4] are given in Table 10-5. Reaction starts when the cement powder and water are mixed together. First C_3A hydrates, causing the rapid setting that produces a rigid structure. Setting rate is controlled by the amount of gypsum added in the cement manufacture. The ettringite that forms does not contribute to setting, but coats the cement particles and retards the setting reactions. If a large amount of gypsum is present, secondary gypsum precipitation results, causing quick setting. Hydration of C_3S and C_2S, which account for approximately 75 percent of the cement by weight, is responsible for strength development after the initial set. The reaction products in both cases are the same—$C_3S_2H_3$, or tobermorite, and crystalline calcium hydroxide.

Four stages in the hydration of Portland cement are illustrated in Figure 10-2 [4]. Initially, the cement grains dispersed in water appear as in Figure 10-2(a); after two minutes, calcium sulfoaluminate hydrate begins to form on the surfaces of the grains [Fig. 10-2(b)]. Two hours later the sulfoaluminate hydrates, and possibly

TABLE 10-4 Leaching of Portland Cement

Concentration of Constituent (mg/l) in Waste or Leachate

Reagent Name	Test Type	Arsenic	Barium	Cadmium	Chromium	Copper	Lead	Mercury	Nickel	Selenium	Silver
Portland cement, Type I	Total	26.300	109.000	1.380	36.700		70.100	<0.025		<0.230	4.140
Portland cement, Type I	TCLP-2	<0.420	0.840	0.010			0.250	<0.001		<0.230	<0.010
Portland cement, Type I	Total	31.600	104.000	1.110	33.700	47.800	57.500	<0.033	24.900	<0.25	3.490
Portland cement, Type I	TCLP-2	<0.010	1.010	<0.010	0.260	0.180	0.260	<0.001	0.020	<0.010	0.010
Portland cement, Type I	Total	26.300	109.000	1.380	36.700		70.100	<0.025		<0.230	4.140
Portland cement, Type I	TCLP-2	<0.420	0.840	<0.010			0.250	<0.001		<0.230	<0.010
Portland cement, Type II	Total										
Portland cement, Type II	TCLP-2										
Portland cement, Type II	Total	21.400	215.000	2.050	35.800	64.200	85.900	<0.032	9.210	<0.280	7.780
Portland cement, Type II	TCLP-2	<0.010	1.770	<0.010	0.240	0.210	0.340	<0.001	0.020	<0.010	0.020
Portland cement, Type III	Total										
Portland cement, Type III	TCLP-2										

TABLE 10-5 Basic Hydration Reactions

Reactants	Products	Heat Evolved (cal/g)
$C_3A + 6H$	C_3AH_6	207
$C_3A + 3C\bar{S} + 32H$	$C_6A\bar{S}_3H_{32}$ (ettringite)	347
$2C_3S + 6H$	$C_3S_2H_3 + 3CH$ (tobermorite gel)	120
$2C_2S + 4H$	$C_3S_2H_3 + CH$ (tobermorite gel)	62
$C + H$	CH	279

Source: Adapted from J. Skalny and K. E. Daugherty, Everything you always wanted to know about Portland cement, *ChemTech,* pp. 38–45, Jan. 1972.

other hydrates, begin forming an intermeshing network that causes setting [Fig. 10-2(c)]. After two days, the network has developed further due to the hydration of calcium silicates, forming tobermorite and causing hardening [Fig. 10-2(d)].

In the crystal model, initial hydration proceeds by nucleation and growth of hexagonal crystals of calcium hydroxides that eventually fill up all the spaces between the cement grains, leaving a silicate-rich layer on the surfaces of the grains. This layer slows the movement of water to the cement surface and the release of calcium and silicate ions from the cement. Meanwhile, particles of calcium silicate hydrate grow outward from the cement grains in the form of needles or spines [5], followed by a different crystalline morphology of the same chemical nature that pushes the spines outward. As they come into contact with other spines on neighbor-

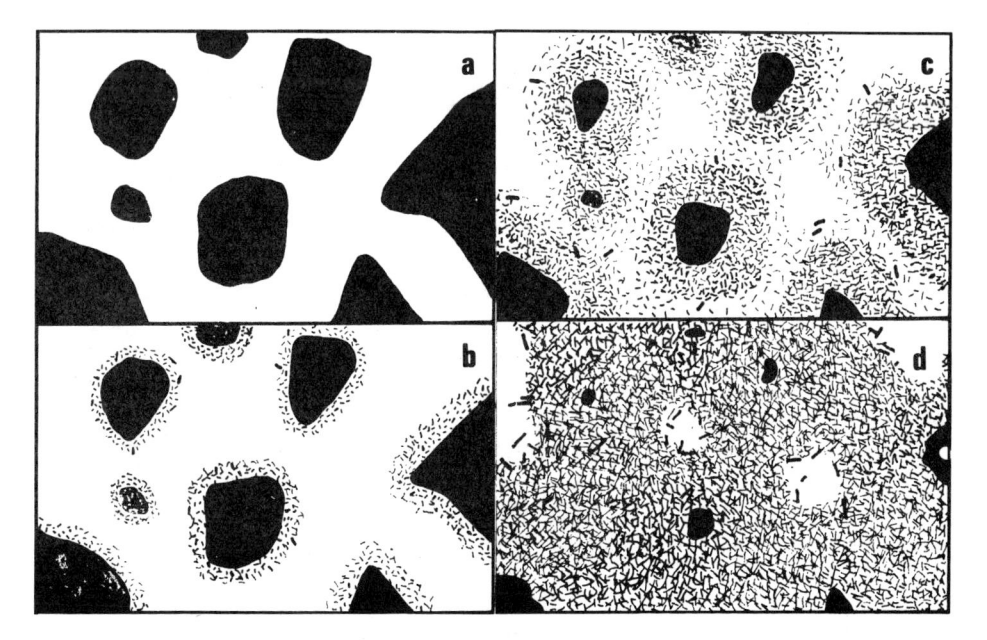

Figure 10-2 The phases of cement setting. Reprinted with permission from J. Skalny and K. E. Daugherty, Everything you always wanted to know about Portland cement, but did not ask. *ChemTech,* p. 38, Jan. 1972. Copyright The American Chemical Society.

ing cement grains, the spines crumple into sheets of tobermorite, which accounts for the final strength and durability of cement.

In the osmotic model [6], calcium silicate in the cement hydrates in the presence of water to coat each grain with a hydrate gel. The initial coating develops into protuberances which become tubular (hollow) fibrils—the spines of the crystalline theory. At the same time, calcium hydroxide (called "Portlandite") precipitates and forms large crystals between the grains. The major difference between the theories is that, in the osmotic model, the fibril growth occurs by a mechanism similar to the familiar silica "garden".† The membrane initially formed around the grain "acts as an osmotic pump, the salt solution flowing up the tube and precipitating at the tip by an almost steady-state process" [6]. This happens because the semipermeable membrane allows the inward flow of water and the outward flow of small ions—Ca^{+2} and OH^-—by diffusion, but prevents the outward diffusion of the large hydrated silicate ions. The result is an excess of $Ca(OH)_2$ on the fluid side of the membrane causing precipitation of Portlandite crystals, and an excess of silicate ions on the cement grain side of the membrane, producing an osmotic pressure differential across the membrane. The latter results in fibril growth as the membrane alternately ruptures and reforms by extruding concentrated silicate solution into the saturated $Ca(OH)_2$ pore solution where it reacts to form a new CSH gel membrane.

Both models end in the same result—the cement grains become interlocked, first setting the cement and finally hardening it. There are, however, still some significant unsolved problems in Portland cement chemistry [4]: the role and effect of minor components; the exact mechanisms of hydration of the cement minerals in the presence of other components and admixtures; the exact structure of the cement paste and its effect an the engineering properties of waste forms; how the properties of cement-based waste forms can be improved. For example, in the latter case, modification of the microstructure, which has been so successful in metallurgy, has not been tried with cement.

Several major factors affect the morphology and performance of cement-based waste forms. Fineness affects the rate of hydration, with finer grinds showing accelerated strength development during the first seven days [10, 11]. Considerable heat is liberated by the exothermic hydration reactions. This is generally of no consequence, in fact, it helps speed up the curing process. However, in massive pours of concrete, or waste forms, heat transfer limitations may cause the development of temperatures that affect the final physical properties of the solid. The rate of heat development is determined by the cement composition, and also by certain waste constituents. C_3A evolves heat more rapidly than the other cement compounds—50 percent more than at two days than C_3S [11].

The water-to-cement (W/C) ratio is very important, even though it cannot always be optimized in solidification work. The volume of the cement itself approximately doubles upon hydration [12], creating a network of very small gel pores. The volume originally occupied by the added water forms a system of much larger capillary pores. As the water–cement ratio increases, the percentage of larger pores increases, substantially increasing the permeability of the waste form [3]. In a pure

---

†So-called because of the common experiment in which crystals of water-soluble metal salts are placed in a solution of sodium silicate, producing a treelike growth of metal silicates as the salt first forms a silicate membrane, which then ruptures due to osmotic pressure, forming a new membrane, which again ruptures, and so on.

water–cement system, the permeability is essentially zero at W/C ratio of 0.32, but increases exponentially as the W/C ratio reaches 0.6 to 0.7. These effects are shown in another way in Figure 10-3, for two different W/C ratios, 0.31 and 0.48. At W/C ratio of 0.32, the initial volume is split equally between water and cement, but after full curing, all the water is used up, leaving some unreacted cement and air voids. At a W/C ratio of 0.48, the cement is fully hydrated, leaving free water (pore water), gel water, and the air voids. Very high W/C ratios will leave "bleed water," or water that appears as standing water on the surface of the solid mass. This occurs at a W/C ratio of about 0.5 or greater. As we will see later (Chapters 11 and 13), the addition of gelling agents like sodium silicate and/or bulking agents such as flyash allow much higher W/C ratios to be achieved without generation of bleed water. They may, however, alter the final physical and chemical properties of the waste form and they do affect set time.

The long-term durability of concrete and cement mortars is well proved.

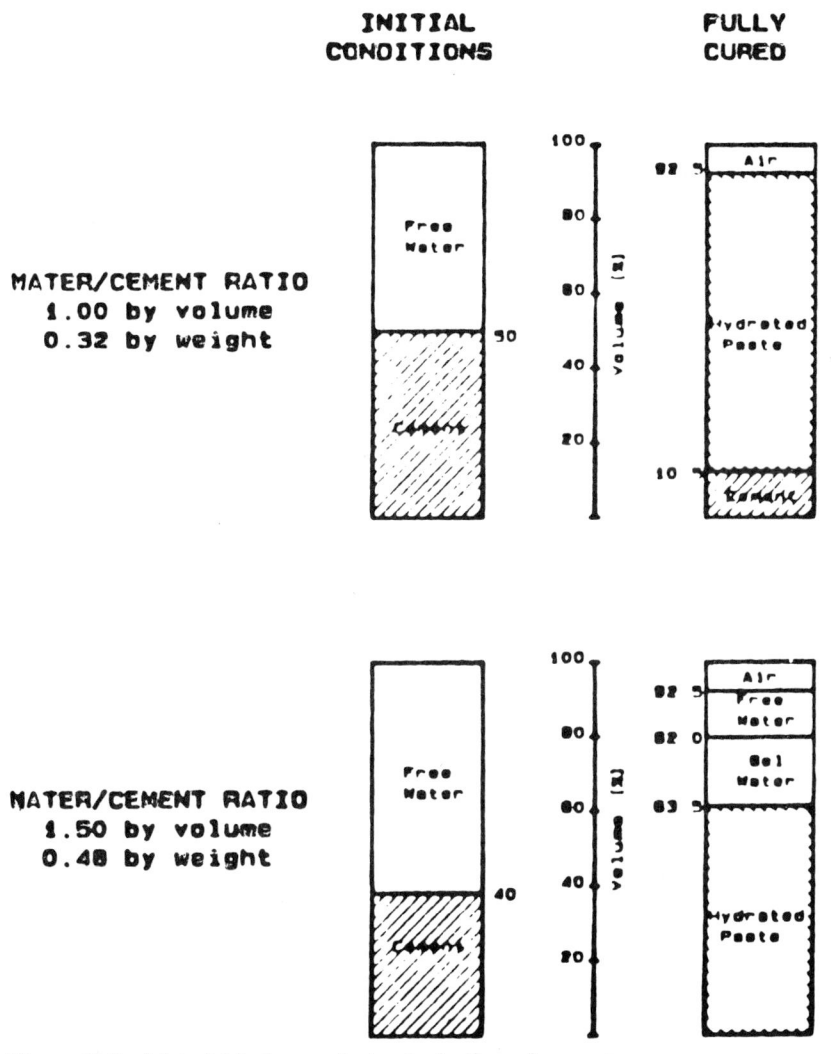

Figure 10-3 Material balances during hydration of cement.

Nevertheless, chemical and physical attacks due to environmental conditions do occur. The chemical attacks "are interfacial phenomena which take place through complex mechanisms which include ion exchange, dissolution of hydrated solids or formation of new insoluble compounds" [3]. Sulfates are the most destructive compounds normally present in the environment because they react with aluminates in the cement structure to form the expansive sulfoaluminate, ettringite, as well as acids. The former break down the physical structure of the waste form. Acids leach lime from the waste form until 10 to 15 percent of the original weight of the wet cement has dissolved, but this has little effect on the strength of the waste form [13], because the Portland cement matrix will resist attack from solutions with pH as low as 5 and the high-alumina cement as low as pH 4. Strong acids, however, can completely dissolve the matrix.

In CFS work, the amount of lime produced in cement hydration is of importance. A typical cement composition will produce about 30 percent $Ca(OH)_2$ (0.3 gram per 1.0 gram of cement), which corresponds to an acid neutralization capacity of 8 milliequivalent/gram of dry cement [3]. This produces a pore solution with pH in the range of 12 to 13. While cement can be used as an alkali for the neutralization of acid wastes, it is not cost effective for this purpose when compared to lime. Also, the $Ca(OH)_2$ used in neutralization is not available for the normal cement reactions, so the cement will not serve a dual purpose.

High-Alumina Cements. The principal anhydrous constituent of these cements is $CaO \cdot Al_2O_3$, or CA, and the development of strength is derived primarily from the hydration products of this compound, which at normal room temperature consist mainly of the decahydrate, CAH_{10}, with smaller amounts of C_2AH_8 and alumina gel. These initial compounds are unstable, and eventually age to the more stable C_3AH_6 and gibbsite, AH_3. This aging is accompanied by a substantial reduction in volume due to the different crystal structures—the initial compounds are hexagonal and C_3AH_6 is cubic. These changes are the basis for the loss of long-term strength and sensitivity to corrosive attack in high-alumina cements. High temperatures and high W/C ratio facilitate these aging effects [6].

Inhibitors and Accelerators

Inhibitors. As we saw in Chapter 9, many chemicals and substances act either as inhibitors or accelerators for the setting and curing of inorganic CFS systems. Many of these can act as accelerators in cement setting, an often desirable effect, and some actually improve the properties of the product. These will be discussed in the next section, since they may be considered as admixtures. With respect to Portland cement processes specifically, constituents in wastes that have a retarding or inhibiting effect, or that affect the final properties of the waste form, are shown in Table 10-6. Many aqueous wastes contain large concentrations of electrolytes, as well as soluble organics, particulates, and other materials that are not inert with respect to cement reactions. Some of the inhibitors can also act as accelerators, the nature of the effect depending on concentration.

Dispersants such as sulfonates and carboxylate compounds retard the formation of cementitious gels in cement-based systems. Flocculants can have either positive or negative effects, depending on concentration. Acids and alkalies affect cement setting, since an alkaline pH is required. Oil and grease in small concentrations, especially in emulsified form, can be incorporated into the cement matrix with relatively little effect, but some soluble organics cause more difficulty. Many of the substances listed in Table 10-6 can reduce the ultimate mechanical strength of the waste form by producing cracking and spalling. One especially deleterious group is the sulfates, which cause the expansive compound ettringite to form. If substantial quantities of sulfates are present, it may be desirable to use one of the sulfate-resistant grades of cement.

A great deal of work has been done on inhibition and durability effects in concrete technology. One study by Kantro [18] measured the hydration of C_3S in the presence of various salts by conduction calorimetry. He found that the additives can be divided into four groups.

Soluble calcium salts
Salts with soluble hydroxides
Salts with insoluble hydroxides
   No subsequent reaction with $Ca(OH)_2$
   Subsequent reaction with $Ca(OH)_2$
Salts with anions that form insoluble calcium salts

The effect on hydration varies with the amount of salt added, and the extent of hydration affects the activation energy of the system. There is a stage in the hydration at which the apparent activation energy becomes nearly constant, and the presence of various salts causes a broadening of this stage in terms of the amount of hydration that occurs during the stage, thus changing the rate of hydration. While these kinds of data cannot usually be applied to specific CFS systems to predict their behavior, they may be a useful tool in studying the reactions that occur.

TABLE 10-6 Substances Affecting Cement Reactions: Inhibition and Property Alteration

Substance or Factor	Inhibition	Property Alteration
Fine particulates	X	X
Clay	X	
Silt	X	
Ion exchange materials	X	
Metal lattice substitution	X	
Gelling agents	X	X
Organics, general	X	X
Acids, acid chlorides		X
Alcohols, glycols	X	X
Aldehydes, ketones		X
Carbonyls	X	
Carboxylates	X	
Chlorinated hydrocarbons	X	X
Grease	X	X
Heterocyclics		X
Hydrocarbons, general		X
Lignins	X	
Oil	X	X
Starches	X	
Sulfonates	X	
Sugars	X	
Tannins	X	

(continued)

TABLE 10-6 Substances Affecting Cement Reactions: Inhibition and Property Alteration (*Continued*)

Substance or Factor	Inhibition	Property Alteration
Organics, specific		
Adipic acid		
Benzene		
EDTA		
Ethylene glycol		X
Formaldehyde		
p-Bromophenol		
Hexachlorobenzene		X
Methanol		
NTA		
Phenols	X	X
Trichloroethylene		X
Xylene		
Inorganics, general		
Acids		X
Bases		X
Borates	X	
Calcium compounds		
Anions that form insoluble Ca salts		
Chlorides	X	X
Copper compounds	X	
Heavy metal salts	X	X
Hydroxides, insoluble		
Hydroxides, soluble		
Lead compounds	X	
Magnesium compounds	X	
Phosphates	X	
Salts, general	X	X
Silicas	X	
Sodium compounds	X	
Sulfates	X	X
Sulfides	X	
Tin compounds	X	
Zinc compounds	X	
Inorganics, specific:		
Calcium chloride	X	
Copper hydroxide	X	
Copper nitrate		X
Gypsum, hydrate	X	
Lead hydroxide	X	
Lead nitrate	X	X
Sodium arsenate	X	
Sodium borate	X	
Sodium hydroxide		X
Sodium iodate	X	
Sodium sulfate		X
Sulfur	X	
Tin		
Zinc nitrate		X
Zinc oxide/hydroxide	X	

Sources: This book, Chapter 9; ASTM, *Temperature Effects on Concrete,* Philadelphia, 1985; U.S. Patent 3, 947, 283; U.S. Patent 3, 947, 284; M. J. Cullinane, R. M. Bricka, and N. R. Francingues, Jr., An assessment of materials that interfere with stabilization/solidification processes, *Proc. 13th Annual Research Symposium,* Cincinnati, OH, pp. 64–71, 1987.

Another study done by Cullinane et al. [17] described the effect of ten possible interfering agents on three CFS processes: Portland cement, lime/flyash, and cement/flyash. The results were shown in Table 9-2. This work showed that the magnitude of the effect depended on the process type, curing time, and the additive used. Oil and grease cause decreases in final strength development as a function of concentration. Zinc and phenol additions resulted in marked decreases in strength, while NaOH increased in strength at low concentration and decreased at high concentration. Chlorinated hydrocarbons had little effect. In general, the interfering material had less effect on Portland cement systems than on the others.

Butler et al. [1] found little permanent effect on cement hydration for cadmium salts, although there was some shortening of the dormant period due to the formation of solid $Cd(OH)_2$ that acts as a nucleating agent. However, lead salts caused "extreme retardation of the hydration reactions" that "applied to the aluminate phases as well as the silicate phases." This delay lasted as long as three days. Even after 28 days, the salt-doped mixtures still had a higher proportion of exchangeable water and a lower proportion of both CSH and $Ca(OH)_2$ compared to cement alone.

Accelerators/Additives. In Chapter 9, we discussed methods of countering the inhibitory effects of waste constituents (see the listing therein). These methods can also be used in many instances to accelerate the setting and/or hardening of uninhibited cement systems. The additives and anti-inhibitors that have been reported in the literature or found to be useful by the author are listed below [this book, Chapter 9], [14]–[17], [19]–[30].

Compound or Factor	Mechanism
General	
Absorbents	Remove interfering ions and organics
Clay‡	
Diatomaceous earth	
Flyash‡	
Activated carbon‡	
Biocides‡	Destroy organisms that coat reactive reagent particles
Ion exchange materials	
Metal lattice substitution	
Oxidants‡	Destroy interfering organics
Seawater‡	Provides NaCl
Organic Compounds	
Acrylamide/formaldehyde/catalyst	
Calcium tetraformate	
Ethylene glycol diacetate	
Hydroxylated organic acids	
Tartaric, citric, gluconic, malic, lactic, salicylic	
Mono-, di-, or triacetin	
Mono-, di-, or triethanolamine	

‡Known use in pilot or commercial CFS processes or products.

Compound or Factor	Mechanism
Mono-, di-, or triethylene glycol	
Polyethylene glycol	
Starch hydroxylate	
Surfactants‡	
Inorganic compounds	
Alkali metal aluminates‡	
Alkali metal bicarbonate‡	
Aluminum sulfate	
Ammonium chloride	
Calcium aluminate	
Calcium chloride‡	
Calcium haloaluminate cement‡	
Chromium compounds‡	
Cobalt chloride	
Copper nitrate	
Copper sulfate	
Heavy metals‡	
Gypsum, hemihydrate‡	
Iron compounds, Fe^{+2}, Fe^3‡	Coprecipitation of interfering metals, for example, Sn and Pb
Jet cement‡	
Lime‡	
Magnesium chloride‡	
Magnesium oxide‡	
Metal salts‡	
Silica, amorphous‡	Forms gels, reacts with alkali from cement
Sodium chloride‡ (low concentration)	
Sodium hydroxide‡	
Sodium silicate‡	
Trisodium phosphate‡	Precipitates metals
Zinc nitrate	
Zinc oxide	Inhibitor, but reportedly increases final compressive strength

Little is known about the detailed mechanisms of hydration acceleration [31]. Kondo et al. [32] postulated that mobile anions diffuse inward through the silicate gel coating on the cement grains, forcing calcium ion to diffuse out to maintain electrical neutrality. This driving force accelerates the formation of more gel.

Other additives and ingredients also accelerate setting and hardening and contribute to final physical and chemical properties of the waste form. These will be discussed in the subsequent sections of this chapter.

GENERIC VS. PROPRIETARY PROCESSES

A listing of various proprietary, cement-based CFS process, products and systems was given in Table 9-3. Many of these neither are, nor ever will be commercial processes, largely because they are either impractical or are marketing variations on what are really generic systems. Because of this, and the fact that new offerings are appearing almost weekly at this time, it is of little value to attempt coverage in this way. It would be outdated almost by the time this book is published. Furthermore, most of the promoters of these proprietary systems offer little reliable technical information. Where adequate information is available, it will be discussed elsewhere in this book. Some of these specific processes will be discussed later in this chapter because they represent the commercial or technical state of the art, or are of special technical interest. Also, information about operating mobile/transportable and fixed operations will be presented in Chapter 16.

Portland cement processes can be divided into two general types: Portland cement alone,§ and Portland cement plus other additives such as lime, clay, and sorbents. Several systems that merit separate treatment—cement/soluble silicate and cement/flyash—are discussed in Chapters 11 and 13, respectively.

Portland Cement

Portland cement was used alone for many years for solidification of radioactive wastes and for solidification of other wastes containing hazardous constituents, especially for subsequent ocean disposal. In Japan it is reported that in 1974 about 47,000 tons of sludges containing mercury were subjected to concrete solidification before ocean disposal [33]. As a result, the Japanese Environmental Agency has worked out a unified standard for concrete solidification methods to be used in that country to assure that when this type of process is used, it is used properly. With the Portland cement process, water in the waste reacts chemically with the cement to form hydrated silicate and aluminate compounds. Solids in the waste act as an aggregate to form a "concrete," although the types of solids encountered in wet wastes may produce a concrete of low strength.

The optimum properties of waste and cement, and the type of cement chosen, will vary with the waste type and its composition. Cement requires a minimum amount of water to obtain workability. This minimum water-to-cement ratio is approximately 0.3 by weight for Portland cement alone, but will also depend on the waste itself, since some waste solids may absorb large amounts of water. The addition of too much water may result in a layer of freestanding water on the surface of the solidified product. Optimum formulations must take each waste into consideration individually because of the possible interactions between the waste constituents and the cement that we have already discussed.

Portland Cement Plus Additives

In addition to cement/soluble silicate (Chapter 11) and cement/flyash (Chapter 13), there are a number of cement-based processes that use various additives in substantial amounts: cement/lime, cement/clay, cement/sorbent, cement/polymer, and mixtures of various kinds of cements.

Cement/Lime. In some cases, the addition of lime to the Portland cement has been found to counteract the retarding action of various waste constituents and to

---

§ This category includes minor additions, such as accelerators, set retarders, additives to fix specific waste constituents, redox agents, biocides, and surfactants.

allow proper set and hardening to be achieved. Iffland et al. [34] found this to be useful in solidifying radioactive, borate-containing liquids. Others [35]–[37] have used cement/lime and mixtures of cement, lime, and ferrous sulfate to produce solid, low-leaching products with lead-, chromium-, and other heavy-metal-containing wastes. Another additive commonly used along with cement and lime is aluminum sulfate or alumina [38]–[40], which reportedly counters the retarding effect of organic constituents in the waste. Lime is probably the most used additive with cement, and this probably comes from its long-standing use in other cement products.

Cement/Clay or Sorbent. A cement/bentonite composition [41], containing a large proportion of bentonite, has been used commercially for CFS treatment (see Chapter 15). Others have used smaller amounts of clays to control viscosity and reduce available water [42], as well as to sorb organics and metals [43]–[45]. Other sorbents such as diatomaceous earth [46], vermiculite, rice hull ash [47], and various expanded mineral absorbents have been tested, and some used in commercial CFS formulated chemicals. Such mixtures have considerable utility, especially for low solids wastes where large cement additions would otherwise be required.

Mixtures of Cements. Many formulations for CFS use contain more than one cement to balance the effects of waste constituents and excess water, and to fix specific constituents, especially metals. Usually, the base is Type I Portland cement to reduce cost, using the minimum amount of the more expensive cement to achieve the project goals. Some vendors sell the special cement as a proprietary product, recommending the proper blend with Portland cement for a specific use.

Miscellaneous Cement Additives. A very large variety of additives have been described in the literature, especially in the patent literature. Most of these have either not been used commercially, or have found only narrow use for specific projects, often for the developer's own CFS undertaking. A number of these are listed below, along with a brief description of their function or purpose. Others were listed in the tables of proprietary CFS processes (Table 9-3) in Chapter 9. Commercial systems are described later in this chapter.

Additive	Function or Purpose
Sodium lignosulfonate [48]	Surfactant
Calcium lignosulfonate [49]	Surfactant
Polyoxyethylene [50]	Surfactant
Calcium aluminate [51]	Accelerator
Silica fume	Hardener
Polyvinyl acetal [52]	Decreases permeability
Aluminum stearate [53]	Makes product hydrophobic
Magnesium oxide [54]	Accelerator
Sodium bicarbonate [55]	Accelerator
Isocyanate-Thiourea polymer [56–61]	Decreases permeability, improves mechanical properties
Polyvinyl alcohol [62]	Hardener
Paraffin (C12–C16) [63]	

Additive
Milling or high-speed mixing [64,65]
Slag/gypsum [66]
Vermiculite [67]
Gypsum/sulfate [68]

Function or Purpose
Improved mechanical properties

Accelerator; improved strength
Sorbent
Anti-inhibitor

PHYSICAL PROPERTIES

Waste forms can have a variety of physical forms: monolithic, concretelike masses; soil-like granular materials; soft, wet, claylike solids; even powders. The form will vary to some extent with the waste, but usually can be designed to meet the requirements of the project. In general, the harder and more monolithic the product, the higher the cost with cement-based systems. Test methods and product specifications must be appropriate to the end use or disposal scenario for the product.

The bulk density of cement-based waste forms varies between 1.25 and 1.75 g/cm^3 [69], with water contents ranging from about 15 to 60 percent. Porosity values range from 0.25 to 0.75, with the voids only partially filled with water [3]. Porosity can be calculated from specific gravity of the solid matrix, bulk density, and water content [70].

Reported values for unconfined compressive strength vary greatly, depending on the process, additive type and ratio, and overall mix ratio. They also vary according to the test method used. Most products range from 15 to about 1000 lb/in^2, but much stronger products can be prepared if desired, up to the strength level of concrete, by addition of slag, aggregates, and additives. Volume increase factors (ratio of final to initial volume) range from 1.05 to about 1.5. The bearing strengths of a number of CFS-treated wastes at 1, 3, and 10 days are shown in Table 10-7 [71]. These strength values are reported in tons/square foot as measured by penetrometer. Also given are total solids contents and volume increase (%) for each of the waste-product combinations. In general, volume increase for cement-based CFS systems equals about 0.4 times the weight increase; that is, 10 percent volume increase equals 25 percent weight increase (mix ratio of 0.25). This applies to water saturated wastes only. Dry ashes and dusts may actually see a volume decrease due to lowered void space when the material is wetted.

Permeabilities are strongly influenced by the waste being solidified. They normally fall in the area of 10^{-5} to 10^{-8} cm/s, depending to some extent on how they are measured. Permeability of cement-based waste forms are comparable to those of clay. Other properties, such as resistance to weathering, wetting/drying, and freeze/thaw, can be measured, but are usually not very important in typical waste disposal scenarios. If the product is to be used in some other way, however, such properties may be critical.

Water-to-cement ratio is very important in determining the final waste form physical properties [69], but unlike concrete, the water content cannot usually be controlled independently, since it depends on the original waste. An exception is the solidification of ashes and dusts, with essentially zero water content. Economics usually dictate the amount of cement and other additives that can be used to balance the waste's water content. If a stronger product is required, while keeping the

356 PORTLAND CEMENT-BASED SYSTEMS

TABLE 10-7 Penetrometer Hardness vs. Cure Time in Cement-Based Processes

Waste	Physical State	pH	Total Solids	Volume Increase	Bearing Strength (Penetrometer) (tons/ft²) 1 day	3 days	10 days
F006	CAKE	8.2	54	5	2	2	2.5
F006	CAKE	8.0	43	5	1	2	2.6
F006	CAKE	6.5	44	5	1	2	3.5
F006	SLDG	7.0	33	10	0	1	2
F007	SOLN	12.0	40	25	0	2	3.3
K006	SLDG	6.7	50	5	>4.5	>4.5	>4.5
K006	SLDG	6.5	18	10	1	4	>4.5
K006	CAKE	7.2	2	10	>4.5	>4.5	>4.5
K062	SOLN	1.8	30	25	0	1	2.7
Calcium phosphate sludge	SLDG	12.8	80	9	2	3	>4.5
Paint sludge, NOS	SLDG	7.4	59	15	1	2	
Arsenic wood preservative	SLDG	0.5	35	5	0	1	2.6
Arsenic wood preservative	SLDG		55	25	>4.5	>4.5	>4.5
Metal hydroxide sludge, NOS	SLDG	9.0	3	10	0	3	
Sulfuric acid pickle liquor	SOLN	0.5	30	20	0	0	>4.5
Waste water treatment residue	SLDG	7.8	28	16	0	>4.5	>4.5
Lime neutral, residue, NOS	CAKE	9.5	24	7	1	2	2.4
Lime neutral, residue, NOS	SLDG	12.1	36	10	3	4	
Lime neutral, residue, NOS	CAKE	12.4	25	7	0	3	>4.5
Waste filter cake, NOS	SLDG	12.7	75	7	>4.5	>4.5	>4.5
Calcium fluoride sludge	CAKE	11.9	36	9	2	4	>4.5
Weld cleaner	SOLN	0.8	10	15	0	4	4.2
Printed circuit etch bath	SOLN		25	25	3	>4.5	>4.5
DAF sludge	SLDG	6.4	7	15	0	1	4.5
F009	SOLN	13.3	1		0	0	1.1
Thorium hydroxide filter cake	CAKE	2.5	20	15	2	4	>4.5
Cleaning waste, NOS	SUSP	10.8	50	50	0	0	0.7
Textile sizing waste	SUSP	8.7	3	25	0	1	1.2

cement addition ratio low, it may be necessary to go to another process, such as cement/flyash. With low-cost flyash, this is an effective method if the volume/weight increase does not impose too severe an economic penalty. This is sometimes the case in on-site remediation projects.

In general, Portland-cement-based systems produce stronger matrices than other inorganic binder systems, and they do it at lower mix ratios, which results in a smaller volume of waste requiring ultimate disposal [72]. This is important both for economic reasons and because of the limited availability of landfill space in the United States.

CHEMICAL PROPERTIES

Chemical properties of Portland-cement-based processes, like other CFS processes, are described primarily in terms of leachability. In Chapter 4, we explored the reactions and leachability of various metal species in CFS systems. In this section, we discuss some general findings on the leachability of cement-based waste forms, as well as conclusions and speculations on fixation mechanisms. For ease of comparison and reference, the results of leaching studies and tests are summarized in Table 10-8, rather than being presented for each study or case separately. Table 10-8 lists the information by waste type, stressing laboratory studies and general data from the technical literature; this same information is also used in the individual metal sections of Chapter 4, but presented in a different way. The data sources are referenced at the end of Chapter 4.

Studies have shown that heavy metal compounds—oxides and hydroxides, chlorides, sulfates, nitrates—interact in the hydration reactions of cement both during setting and later during the hardening process. In addition to affecting the setting and hardening rate, these interactions also may function to fix the metals chemically or physically in the microstructure. Bishop [2] attributed the immobilization of cadmium, chromium and lead in cement-based waste forms, even after the alkali had all been leached out, to chemical complexes formed during such interactions with the siliceous cement matrix. Other work has shown the same sort of immobilization for metals, such as mercury, that do not form low-solubility hydroxides.

Bhatty [73] approached the immobilization of metals by Portland cement in a different way. Using pure tricalcium silicate (C_3S) instead of Portland cement per se, he reacted it with metal salt solutions in dilute solution so that cemented solids were not formed, then filtered the slurry and analyzed the filtrates. In this system, the immobilization of metals would be caused by the C_3S hydration reactions alone, not by physical entrapment, although sorption could occur. The metals studied were cadmium, chromium, lead, mercury, and zinc. Metal retention in the solid (immobilization) was found to increase with C/S ratio and with time. In most of the experiments, the retention was almost complete even after one day. The metal partition ratio (filtrate concentration/solid concentration) was of the order of 10^{-5} or less in most cases. The anion (chloride, nitrate, sulfate) was partitioned primarily into the filtrate, except for sulfate, where the solubility was limited by the formation of calcium sulfate.

Studies of X-ray diffraction of the solids after hydration showed unidentified lines apparently belonging to complex compounds other than calcium silcate

TABLE 10-8 Summary of Metal Fixation Data, Portland-Cement-Based Processes

Reference[a]	Waste Description	Waste Code	CFS System	Mix Ratio	Treatment Scale	Leaching Test
4:101	Electroplating sludge	F006	Cement-based	0.10	Lab	TCLP
4:101	Electroplating sludge	F006	Cement-based	0.15	Lab	TCLP
4:101	Electroplating sludge	F006	Cement-based	0.20	Lab	TCLP
4:103	Arc furnace dust	K061	Cement-based	0.20	Lab	TCLP
4:90	Arsenic sludge		Portland cement	1.00	Lab	Static Water
4:117	Cadmium hydroxide sludge		Portland cement	0.45	Lab	TCLP
4:117	Cadmium nitrate sludge		Portland cement	0.41	Lab	TCLP
4:117	Lead hydroxide sludge		Portland cement	0.45	Lab	TCLP
4:117	Lead hydroxide sludge		Portland cement	0.41	Lab	TCLP
4:119	Arc furnace dust	K061	Portland cement	0.05	Lab	TCLP
4:122	Electroplating waste	F006	Jet cement/Fe(NO$_3$)$_2$	0.20	Lab	Water
4:123	Incinerator ash		Portland cement + Al$_2$(SO$_4$)$_3$ + CaSO$_3$	0.83	Lab	Water
4:125	Sodium chromate waste		Cement + additives	0.50	Lab	Water
4:132	Electroplating waste	F006	Cement + zeolite + FeSO$_4$ + CaCO$_3$/CaCl$_2$	1.05	Lab	Water
4:135	Electroplating waste	F006	Unstabilized		Lab	EPT
			Cement-based		Lab	EPT
	CCA wood preserving waste		Unstabilized		Lab	EPT
	CCA wood preserving waste		Portland cement	0.10	Lab	EPT
	CCA wood preserving waste		Portland cement	0.20	Lab	EPT
4:137	Pigment sludge		Cement + Al$_2$(SO$_4$)$_3$		Lab	Water
4:140	Cl-caustic brine sludge	K071	Cement + thiourea	0.65	Lab	Water
4:140	Cl-caustic brine sludge	K071	Cement + NaS	0.65	Lab	Water
4:141	Acetylene sludge		Cement + Fe^{+2} + thiourea	2.5	Lab	Water

[a] References are listed at the end of Chapter 4.
[b] Maximum concentration.
[c] Sequential batch test, CO$_2$-saturated water.

hydrate or the original metal compounds. Bhatty speculates that four mechanisms may be involved in the fixation of metals by C$_3$S:

$$\text{Addition:} \quad CSH + M \longrightarrow MCSH \qquad (10\text{-}1)$$

$$\text{Substitution:} \quad CSH + M \longrightarrow MSCH + Ca \qquad (10\text{-}2)$$

Formation of new compounds

Multiple mechanisms

Addition [Eq. (10-1)] is most likely to be favored by hydrates containing low C/S mole ratios. Cadmium, chromium, and lead are likely to be immobilized by this mechanism. Substitution [Eq. (10-2)] is found in high C/S ratio hydrates. As calcium is removed by dissolution from the C$_3$S, more metallic ions may be substituted in the hydrate. The effects of this mechanism are limited by the number of calcium ions that can be removed from the CSH structure.

TABLE 10-8 *(Continued)*

| Metal Content (mg/kg or mg/l) ||||||||||
| Antimony || Arsenic || Barium || Cadmium || Chromium ||
Total	Leachate	Total	Leachate	Total	Leachate	Total	Leachate	Total	Leachate
							0.010		0.010
							0.010		0.010
							0.080		0.090
							<0.010		0.610
		20000	0.760						
						15340	0.026		
						18980	0.043		
294	<0.050	36	<0.010	238	0.580	481	3.290	1370	0.043
								3000 (Cr^{+6})	0.010
								25000 (Cr^{+6})	0.200
								1050 (Cr^{+6})	<0.010
								(Cr^{+6})	0.120
						6700	12.000		
							0.030		
		11500	1.800					16000	90.000
		11500	13.8					16000	13.390
		11500	4.3					16000	4.110

(continued)

LABORATORY AND PILOT STUDIES

Because cement-based processes have been so widely used in CFS treatment of hazardous wastes, most commercial vendors and users have concentrated on adapting and modifying formulations to meet the everchanging regulatory requirements under the RCRA and CERCLA and their amendments and rulemaking procedures. Little attention has been paid to the underlying mechanisms until quite recently, when the government and academic sectors started to explore the basic principles of CFS processes in general, and cement-based waste forms in particular. Cement-based processes offer certain advantages: reagent composition is comparatively consistent; basic concepts have been explored to some degree in the nuclear waste area; a considerable body of scientific knowledge is already available from the cement industry; the effects of bulk dilution and sorption on particle surfaces are minimized.

As we have already seen, some good basic groundwork has been laid by those working in the development of test methods, especially leaching test methods. Cote [3] and others have explored the mechanisms to some degree. Unfortunately, data

TABLE 10-8 (*Continued*)

	Copper		Lead		Mercury	
Reference[a]	Total	Leachate	Total	Leachate	Total	Leachate
4:101		73.000		20.000		
4:101		0.160		0.220		
4:101		0.450		0.100		
4:103				0.190		
4:90						
4:117						
4:117						
4:117			17200	70		
4:117			51600	490		
4:119	2240	0.024	20300	1.157	3.8	0.001
4:122						
4:123						
4:125						
4:132						
4:135	11000	20.000				
		0.350				
	8000	13.000				
	8000	9.910				
	8000	7.410				
4:137				<0.010		
4:140					15	<0.010
4:140					15	0.017
4:141					381	<0.001

Metal Content (mg/kg or mg/l)

from the current regulatory test procedures yield little information about the basic chemistry of fixation or solidification. Therefore, investigators have turned to other methods, especially sequential batch leaching with various leachants. Data from protocols such as ANS 16.1 (see Chapter 3, "Molecular Diffusion," and Chapter 17 for details) yield single-number figures of merit like the leachability index (LX) for diffusion-controlled systems. Cote and Hamilton [74] developed LX values for As, Cd, Cr, Pb, nitrate, and phenol from three different synthetic wastes for several CFS generic processes, among them Portland cement.

One of the more interesting and useful studies was published recently by Bishop [2]. Using a modified form of the EPT, samples of a synthetic waste sludge containing Cd, Cr, and Pb were solidified with Type II Portland cement, ground to minus 9.51 mm (0.375 in.), and sequentially extracted with fresh batches of 0.04 M acetic acid. Extractions were continued until the alkalinity due to the waste and the cement was essentially gone and the extract pH approached that of the extraction solution (pH 4)—fifteen sequential extractions on the same sample. The EPT is essentially equal to the first extraction in this series. Each leachate batch was then analyzed for the metals. The relationships between pH or alkalinity and extraction number are shown in Figure 10-4.

TABLE 10-8 (*Continued*)

Nickel		Selenium		Silver		Zinc	
Total	Leachate	Total	Leachate	Total	Leachate	Total	Leachate
	3.000				0.010		1.000
	0.110				0.010		0.010
	0.090				0.050		0.110
							0.040
243	0.019	<5.0	<0.010	59	<0.003	244000	22.633
240	1.100					33000	220.000
	0.130						0.320

A number of interesting observations resulted from these tests.

(1) Metal leaching *rate* decreased with decreasing particle size, just the opposite of the usual expectation, but confirming some prior work [75–77]. This result has been ascribed to the more rapid leaching of alkalinity from smaller particles with larger surface area, which therefore more quickly neutralizes the leachant acid and decreases the metal hydroxide dissolution rate. Another possible mechanism is that the increased surface area of the smaller particles provided more sorption sites for metal ions from the liquid phase to the solid phase. These leaching results for the three metals are shown graphically in Figure 10-5. However, small particles eventually leached more of all the metals than the larger ones, after the leachant pH dropped below 6.0.

(2) It appeared that metals in the system were leaching and then reprecipitating or sorbing onto particles where alkaline conditions still prevailed. Again this supports previous observations by the author and others.

(3) Cadmium leached out when the pH and total alkalinity dropped, eventually releasing about 75 percent of the total cadmium content. This is the expected result for a fixation mechanism dependent on metal hydroxide speciation and surface sorption of metal ions.

362   PORTLAND CEMENT-BASED SYSTEMS

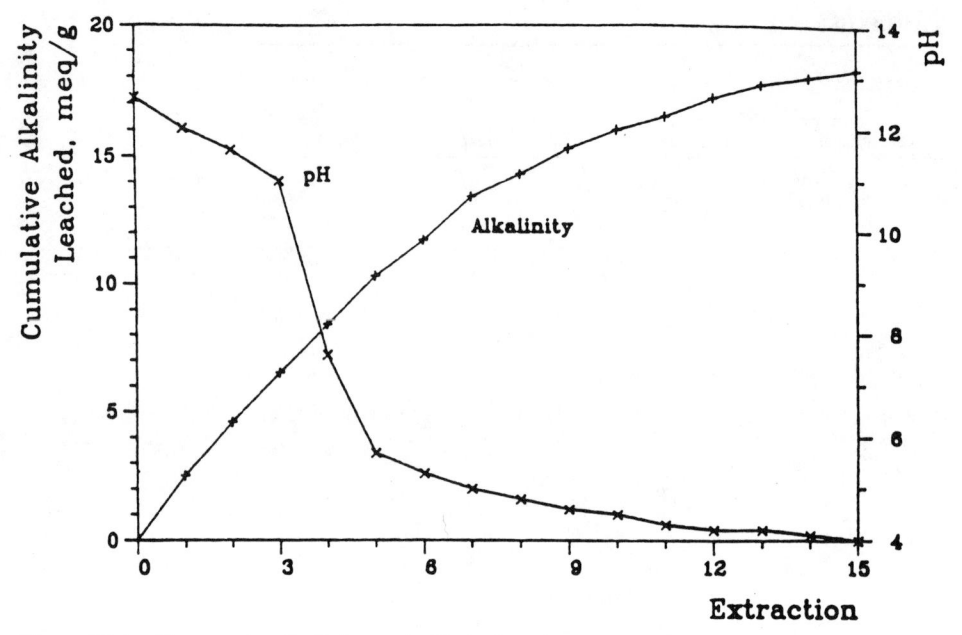

Figure 10-4 Changes in alkalinity and pH during the sequential batch leaching of a cement-based CFS product.

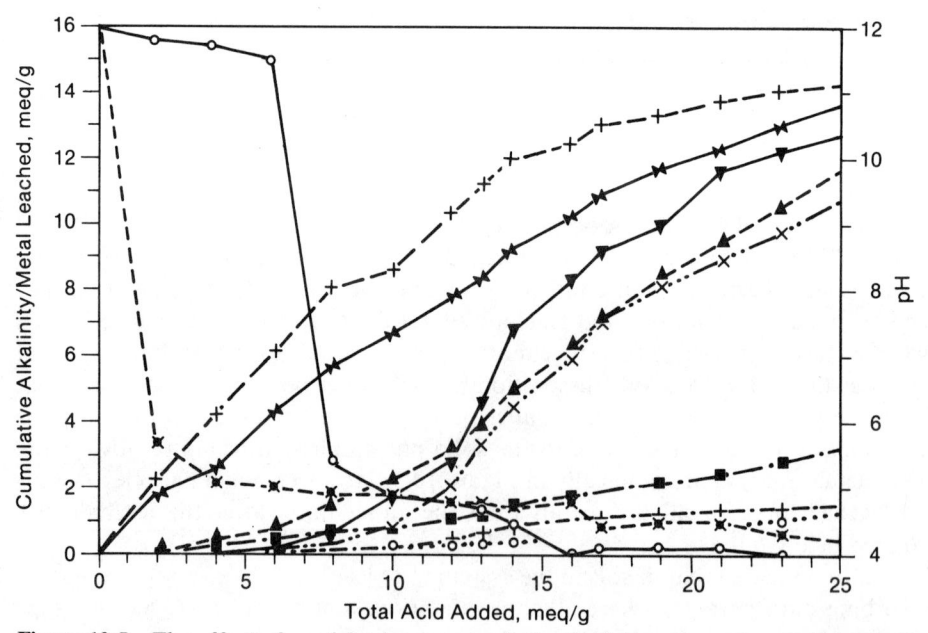

Figure 10-5 The effect of particle size on cumulative leaching of metals and changes in alkalinity and pH. —○— pH—small particles; --×-- pH—large particles; —⋈— Alkalinity—large particles; —+— Alkalinity—small particles; —▼— Cd—small particles; --▲-- Cd—large particles; ---+--- Cr—small particles; ⋯○⋯ Cr—large particles; ⋯×⋯ Pb—small particles; —■— Pb—large particles.

(4) Chromium and lead leached at cumulative amounts well below those expected for metal hydroxide/surface sorption fixation mechanisms (15 and 25 percent, respectively). This indicates that these metals are bound into the silica matrix itself, and are immobilized as long as that matrix is intact. This is borne out by the leaching curve for silicon, which follows closely those of lead and chromium.

These results indicate that certain metals, such as lead and chromium, are bound into the silica matrix of cement-based systems (and, likely other similar systems) as "silicates" rather than as free hydroxides. As we shall see later (Chapter 11), such complexes are probably not stoichiometric silicates, but rather are variable composition species of the general formula yMe(OH)$_x$·zSiO$_2$. In forming such complexes, the original metal compounds actually respeciate in the waste form. If this is so, it would likely occur gradually as the cementitious reactions occur. Bishop found that the metal leaching rate was essentially the same after about six week's curing as after two years. Cure time is important when doing leaching studies; however, in practice, little leaching would be expected during this early time period in normal treatment/disposal scenarios, so that waste forms probably achieve their maximum resistance to leaching without special precautions in the field.

The findings of Bishop and others have some serious implications in CFS work. Metals such as cadmium, which show increased leachability with time in an acid leaching environment, should be placed in disposal environments that minimize contact with acids, or that control permeation rate and movement into groundwater, so that the concentration of metal in the ground- or surface water environment never reaches unacceptable levels. On the other hand, chromium, lead, and other metals that do become complexed in the waste form require less control.

Another interesting study was done by Campbell *et al.* [78] using seawater as a leachant in a sequential batch extraction protocol. Sample cement-based waste forms were cured for three years before testing. They also correlated leaching with macroscopic and microscopic [scanning electron microscope (SEM)] observations of the waste forms. Increased release of cadmium correlated well with cracking of the solid, which was connected with the formation of crystalline microstructures, especially the expansive ettringite, due to sulfate attack. Lead hydroxide does not appear to cause these effects on the solid waste form. Total leaching over a 50-day period (about 16 cycles), however, leached only about 1 percent of the total cadmium in the solid, and no detectable lead.

Shin *et al.* [79] investigated a number of factors in cement-based CFS process design in a system that included addition of sand: sand/cement (S/C) ratio; water/cement (W/C) ratio; sludge content and dosage of a "precipitating agent" (an organic sulfide compound). They found that the S/C ratio had the greatest effect on Cr^{+6} leaching, while the W/C ratio affected Zn leaching and compressive strength the most. The use of the "precipitator" and sand was not economically practical or necessary.

The stabilization of contaminated soil from a superfund site was studied by Barich *et al.* [80]. The Solid Waste Leaching Procedure (see Chapter 17) was used to provide engineering estimates of leachate that might be expected from treated wastes landfilled on-site. Several proprietary processes were compared: vitrification (Chapter 15), cement-based, and two that were not identified, but were believed to be cement-based. The chemical methods produced a one-to-two order of magnitude reduction in metal (Cr, Pb, Cd, Zn) leachate concentrations over those of the untreated waste. Simulated leaching to a nearby watercourse indicated ambient concentrations well below acceptable levels for the treated waste, but unacceptable levels if the soil were to be left in place untreated. Unexpectedly, the vitrified soil did not perform as well as the chemically treated soil in these tests.

CASE STUDIES, COMMERCIAL PROJECTS, AND PROCESSES

Commercial Process Examples

Onoda Cement Company Ltd. This so-called *chemicolime* process consists of a mixture of Portland cement and lime. The system has been used by engineering and construction companies in Japan to solidify the surface layer of dredged sludge. While the process is not known to have been used in the United States or other countries, it has been promoted in English publications for solidification of wastes having a high content of hazardous metals.

Onoda has two pertinent patents, U.S. 3,947,283 and U.S. 3,947,284, which, respectively, show the solidification of sludges containing substances having a harmful influence on the setting of Portland cement by mixing the sludge with either a rapid hardening cement containing calcium haloaluminates, or a substance from the group consisting of alkali sulfate, gypsum dihydrate, and gypsum anhydrase. Presumably, in these patents either the calcium haloaluminate or the sulfate interacts in some way with retarding agents in the waste and prevents their retarding action on the setting of the cement. Neither of these patents mentions the use of lime, but our information from Japan is that, in practice, Onoda is using a Portland cement–lime mixture which, as we have previously seen, is known to help prevent the retardation of setting in cement.

Fujisash Industries Ltd. Described as an "alumina setting system" that uses Portland cement and a cement hardener called "Alset." Alset is described as a complex alumina-based hardener of cement, which is supposed to allow the solidification of industrial wastes and dredged spoils with much smaller quantities of cement, thus lowering cost. Fujisash states costs for treatment of industrial wastes in Japan as being about $20 to $30 per ton, which would be roughly 10 to 18 cents per gallon. It is believed that the alumina-based cement hardener being used is either sodium aluminate or sodium aluminosilicate, but this is not known for certain. This system has been in use at some ten fixed installations in sewage plants in Japan since 1978. One plant, in Tokyo, has been treating about 720,000 tons of sludge annually since 1978. It is claimed to be low-cost and to reduce odor and fix heavy metals [81].

Ebara-Infilco Company Ltd. In Japan, Ebara-Infilco applied for a CFS process patent for using a hydraulic cement such as Portland cement and inorganic chemicals containing a sulfate group. This process requires the addition of large amounts of slaked lime as a filter aid plus an organic polymer flocculent to produce a more filterable sludge, which is then dewatered and subsequently mixed with Portland cement and an inorganic chemical containing a sulfate group such as gypsum, plaster of Paris, sodium sulfate, aluminum sulfate, or ammonium sulfate. It is stated that this mixture hardens completely after one to two days and does not revert back to sludge. In this kind of process it is difficult to determine which of the agents is really doing the solidification and what type of process it really represents. The large amounts of lime, along with the Portland cement and the inorganic sulfate, could represent a lime-based process, a gypsum-based process, or a cement-based process. In any case, this process has not been promoted or commercialized outside of Japan.

Fujibeton. This technology was one of the earliest to be used for the CFS treatment of hazardous wastes. It was developed in Japan in the early 1970s (private communication to the author, September 1972) and [82], and has been used there since that time. Basically a cement-based process, Fujibeton uses a proprietary additive to aid in solidification and to help fix various constituents including, reportedly, organics. Various vendors in the United States have at one time or another represented the process for licensing. These vendors include Environmental Technology Corp., Schaumberg, Illinois, and Priority 1 Systems, Scottsdale, Arizona, and New Materials Technology Corporation. The process is claimed to have been used to detoxify more than 140 sites in Japan (private communication to the author from Environmental Technology Corp., September 2, 1988). With a mix ratio of 0.2 part Fujibeton plus 0.1 part Portland cement, the following results on an unspecified waste were claimed:

Metal	Total Analysis (ppm)	TCLP Leachate Concentration (ppm)
Zinc	12,000	1.0
Cadmium	4,850	<0.01
Chromium	4,840	0.30
Lead	30	<0.1

Silicate Technology Corporation. This company markets a series of reagents for CFS work with the trade names of Toxsorb, Soilsorb™, and Chemisorb™ [83]. These compounds are claimed to fix metals and organics (see Chapter 6). In the case of metals, Toxsorb HM, a "cementitious waste treatment reagent," is claimed to produce the following results with electroplating sludge (Table 10-9) and electric arc furnace dust (Table 10-10) [83].

Toxco, Inc. This remediation company uses cement-based processes custom-designed for the specific project at hand. It has done a number of small projects in California and the western United States, especially in the fixation of lead- and

TABLE 10-9 Electroplating Sludge

		Concentration in ppm	
Metal	Total Analysis (ppm)	Raw Waste TCLP Leachate	Treated Waste TCLP Leachate
Cadmium	6,700	12	0.03
Copper	11,000	20	0.35
Zinc	33,000	220	0.32
Nickel	240	1.1	0.13

TABLE 10-10   Electric Arc Furnace Dust

		Concentration in ppm	
Metal	Total Analysis (ppm)	Raw Waste TCLP Leachate	Treated Waste TCLP Leachate
Lead	410		0.5
Cadmium	78		0.2
Zinc	1023		1.3

chromium-contaminated soils and automobile shredder waste (W. J. McLaughlin, private communication, 1988). In one project on chromium (Cr^{+6})-contaminated soil [84] results of the fixation treatment, that included a chromium reduction step, were as follows:

		Concentration in mg/l	
Metal	Total Analysis (ppm)	Raw Waste WET Leachate	Treated Waste WET Leachate
Chromium +3	2200	67	15
Chromium +6	530	33	0.55

These results are interesting because the leaching tests were run using the California Waste Extraction Test (WET) rather than the TCLP or EPT. The WET is a much more severe test, usually yielding leachates with metal concentrations of from one to two orders of magnitude greater than those obtained from the latter tests. Thus, these results are quite good even though the leaching of Cr^{+3} is at 15 mg/l, since the California soluble threshold concentration limit is 560 for Cr^{+3}. The leaching of Cr^{+6} is quite low, indicating that chromium reduction was successful.

Waste Service Technologies, Inc.   This vendor supplies cement-based CFS technologies through a system of franchisees in the United States. Pilot-scale treatment projects have been conducted on contaminated soils at Superfund sites in the northwestern United States.

Takenaka Sludge Treatment System.   According to the developer, 26 projects have been completed in Japan since 1973, 17 of which involved deposits under water (46,000 m^3) and 7 of which treated factory discharges (14,500 m^3) [69]. The system is marketed in the United States by TJK, Inc., North Hollywood, California.

Other Large-Scale Projects

One CFS system of unusual type is the disposal of low- and intermediate-level radioactive waste at Oak Ridge National Laboratory (ORNL) in Oak Ridge, Tennessee. Here, the waste was mixed as a cement-based grout that was then pumped into a hydrofractured, underground geological formation where it solidified and became physically immobile. This process operated successfully over a period of 20 years. It required very special formulation to produce a grout with the right combination of properties [85]. It had to be, and remain fluid enough to pump over some distance and time interval. The formulation was as follows.

Type II Portland cement: 2.5 lb/gal
Flyash: 2.5 lb/gal
Attapulgite clay: 1.0 lb/gal
Illite clay: 0.5 lb/gal
d-Gluconolactone (a sugar): 0.002 lb/gal

The cement, of course, is the primary solidification agent. Flyash is added because it acts as an inexpensive pozzolanic binder along with the cement and improves the retention of strontium in the solidified waste. The illite clay is added to sorb cesium-137, which is not immobilized by the solidification agents, but is specifically sorbed by this type of clay. Attapulgite clay prevents phase separation of the mixture until it sets after introduction into the underground formation. The sugar was originally added to delay cement setting, but was later found to reduce viscosity of the mix without countering the effect of the attapulgite. Depending on the specific waste stream, if foaming is a problem, a small amount of tributyl phosphate is used.

This CFS composition is an excellent example of the art and science of formulation, balancing components that each perform some function without having them interfere with each other. While few commercial CFS projects or systems to date have used this many components, it is very likely that we will see increasing complexity in CFS formulations as requirements become more stringent in the future.

Remediation Projects: Case Studies

Filter Cake and Contaminated Earth. An old plant site in Massachusetts contained some 10,000 yd^3 of filter cake and earth mixture from former operations. The waste was acidic and contained naphthalene-based compounds. Not until its removal for a mass transit system station was the filter cake and earth mixture found to be hazardous. The options were CFS treatment on-site with disposal at local sanitary landfills, or transport to a remote, secure landfill. The CFS option was chosen because the total cost was about one-half that of transporting it to a secure landfill (approximately $1 million was saved). Because the waste was not pumpable, a solids type of treatment unit was used. The job parameters were as follows.

Waste analysis:
    DAXAD (polymer of naphthalene sulfonic acid)—2.5–10 percent
    Miscellaneous naphthalene-based hydrocarbons—10–12.5 percent
    β-naphthalene sulfonate—0.5–2 percent
    Sodium sulfate—0–1.5 percent
    Filter aid (perlite)—20–30 percent
Volume treated: 10,000 yd^3
Treatment rate: 150 yd^3/day, average
Time: Spring 1981
Treatment cost: $75.00 per yd^3
Disposal method: Sanitary landfill, daily cover material
Properties of treated solid:
    Unconfined compressive strength: 2.5–5.0 tons/ft^2
    Leachate results:
        Organic priority pollutants—0.6 ppm

Naphthalene—7.5 ppm
RCRA metals (except barium)—<0.06 ppm
Barium—1.0 ppm
Phenol—0.22 ppm
Total cyanides—0.050

Superfund-type Cleanup. One of the tasks in a "superfund-type" project (run much like a Superfund job, but not paid for out of Superfund) in Ohio in 1982 was the cleanup of a so-called "polymer pit." This concrete block impoundment contained three layers of waste: a floating, gummy, organic layer containing a wide range of organic priority pollutants, an intermediate contaminated-water layer, and a heavy sludge layer on the bottom. Because of the variety and high concentration of priority pollutants in the waste, secure landfill was chosen as the disposal method. Solidification was required before landfill, and the options were solidification either on-site, or at the landfill. On-site solidification was chosen for economic and public safety reasons.

Because there were essentially three distinct waste streams, three different formulations were chosen to minimize volume increase and the associated transportation and disposal costs. A cement-based process was used for the sludge and water layers, and a lime kiln dust formulation for the organic layer. The volume was relatively small and because three different formulations were used, a small, portable, batch-type treatment unit was used. The job parameters were as follows.

Waste analysis:
  Organic polymers
  Organic priority pollutants
  Metals
  Inert inorganic solids
  Water
Volume treated: 210,000 gal
Treatment rate: 20,000 gal/day
Time: May 1982
Treatment cost: $0.41/gal
Disposal method: Secure landfill (off-site)
Properties of treated solid:
  Unconfined compressive strength: >1.0 ton/ft^2

Uranium Sludge. This very unusual project involved the solidification of waste from a nuclear fuel processing plant that contained minute amounts of uranium. Because of the uranium content, the waste was classified as a nuclear waste, even though its radioactivity was essentially at background level, and was therefore subject to the Nuclear Regulatory Commission and its system rather than the RCRA. The project plan called for the solidification of the sludge in drums that would later be shipped to the nuclear waste repository at Barnwell, South Carolina. Solidification with no free water had to be confirmed before the drums could be closed, labeled, and shipped. Leachability was not a consideration.

Figure 10-6(a) through (h) shows the progression of the operation. In Figure 10-6(a), the waste pit and the equipment setup are shown. The sludge was removed from the pit by clamshell and delivered to the hopper directly above the mixer

(a)

(b)

Figure 10.6 Stages in the operation of a commercial CFS process in the field. (a) Waste pit and equipment set-up. (b) Transport of sludge to the mixer. (c) Mixing of waste and reagents. (d) Filling of drums with the treated waste. (e) Transport of filled drums. (f) Cleaning of filled drums. (g) Capping of drums. (h) Wrapping of palletized drums for shipment.

(c)

(e)

Figure 10-6 (*Continued*)

CASE STUDIES, COMMERCIAL PROJECTS, AND PROCESSES 371

(d)

(f)

Figure 10-6 (*Continued*)

372 PORTLAND CEMENT–BASED SYSTEMS

(g)

(h)

Figure 10-6 (*Continued*)

[Fig. 10-6(b)]. After mixing with the cement formulation contained in the silo [Fig. 10-6(c)], the waste was discharged into a batch hopper below the mixer and metered into drums [Fig. 10-6(d)]. Filled drums were rolled by conveyor away from the mixing area [Fig. 10-6(e)] and the exterior cleaned with a high-pressure steam jet [Fig. 10-6(f)]. After testing for free water and solidity, the drums were capped [Fig. 10-6(g)], labeled, palletized, and wrapped [Fig. 10-6(h)] for storage until they could be shipped to the disposal site.

This project is a good example of a commercial CFS project because it has its own particular characteristics. Every CFS project has some combination of product requirements, wastes to be treated, and physical/mechanical considerations that makes it unique. That is why it is so important to understand the principles and practices of CFS, rather than attempt to fit every project to some preconceived, inflexible process and set of concepts. Later, we will see a number of other physical layouts that have been used in the field.

REFERENCES

1. Butler, L. G., F. K. Cartledge, D. Chalasani, H. C. Eaton, F. Frey, M. E. Tittlebaum, and S. L. Yang. Immobilization mechanisms in solidification/stabilization using cement/silicate fixing agents. Louisiana State Univ., 1988.
2. Bishop, P. L. Leaching of inorganic hazardous constituents from stabilized/solidified hazardous wastes. *Hazardous Wastes Hazardous Mater.* **5**(2): 129–143 (1988).
3. Cote, P. *Contaminant Leaching from Cement-Based Waste Forms Under Acidic Conditions.* Ph.D. thesis, McMaster Univ. Hamilton, Ont., Canada, 1986.
4. Skalny, J., and K. E. Daugherty. Everything you always wanted to know about Portland cement. *ChemTech:* 38–45 (Jan. 1972).
5. Hansen, J. The delicate architecture of cement. *Science:* 49–55 (Dec. 1982).
6. Double, D. D., and A. Hellawell. The solidification of cement. *Sci. Am.* **237**: 82–90 (1977).
7. Heacock, H. W. Alternative nuclear waste solidification processes. In *Proc. Symposium on Waste Management.* Tucson, AZ, pp. 177–206, 1975.
8. Portland Cement Association. *Design and Control of Concrete Mixtures.* Skokie, IL, 1979.
9. Double, D. D., W. L. Thomas, and D. A. Jameson. The hydration of Portland cement. Evidence for an osmotic mechanism. In *Proc. 7th International Congress on the Chemistry of Cement.* 1980.
10. Portland Cement Association. *Concrete Information.* Skokie, IL, 1974.
11. Kirk-Othmer. *Encyclopedia of Chemical Technology,* 3rd ed. New York: Wiley, 1979.
12. Popovics, S. *Concrete Making Materials.* New York: McGraw-Hill, 1979.
13. Lea, F. M. *The Chemistry of Cement and Concrete.* London: Arnold, 1970.
14. ASTM. *Temperature Effects on Concrete.* Philadelphia, 1985.
15. U.S. Patent 3,947,283.
16. U.S. Patent 3,947,284.
17. Cullinane, M. J., R. M. Bricka, and N. R. Francingues, Jr. An assessment of materials that interfere with stabilization/solidification processes. In *Proc. 13th Annual Research Symposium.* Cincinnati, OH, pp. 64–71, 1987.
18. Kantro, D. L. Tricalcium silicate hydration in the presence of various salts. *ASTM J. Test. Eval.* 312–321 (1975).
19. Monnel, B. U.S. Patent 3,656,985 (Apr. 18, 1972).
20. U.S. Patent 4,060,425.
21. U.S. Patent 3,642,503.

22. Kire, K., T. Inoue, and T. Tsutsumi. Japan Kokai 80 48,282.
23. Shinoki, K., T. Katayama, and S. Yamaguchi. Japan Kokai 80 44,355.
24. Nakagawa, H. Japan Kokai 80 44,309.
25. Simosono, C., Y. Togawa, and Y. Taguchi. Japan Kokai 80,126,557.
26. Japan Kokai 80 136,166.
27. Kajioka, H., H. Tadani, T. Shimizu, and T. Kanamoto. *Kure Kogyo Shikenjo Honoku* **23**:46–48 (1980).
28. Bonnel, B. U.S. Patent 3,591,542 (July 6, 1971).
29. Galer, R. E., and P. C. Webb. U.S. Patent 4,286,991 (Sept. 1, 1981).
30. Berry, D. U.S. Patent 4,191,584 (Mar. 4, 1980).
31. Ramachandran, J. S. *Calcium Chloride in Concrete.* London: Applied Science Publishers, 1976.
32. Kondo, R., M. Daimon, E. Sakai, and H. Ushiyama. Influence of inorganic salts on the hydration of tricalcium silicate. *J. Appl. Chem. Biotechnol.* **27**: 191 (1977).
33. *Bulletin of the Environmental Technical Information Center* **1**(4): 1 (1976).
34. Iffland, N., H. Isensee, G. Wagner, and H. Witte. U.S. Patent 4,122,028 (Oct. 24, 1978).
35. Okada, T., I. Kanai, T. Sato, and M. Mizuno. Japan Kokai 75 87,964 (July 15, 1975).
36. Kitsugi, K., M. Shimoda, and T. Hita. Japan Kokai 78,117,677 (Oct. 14, 1978).
37. Ando, T., T. Shimotori, T. Kida, and H. Udagawa. *Rev. Gen. Meet. Tech. Sess. Cem. Assoc. Jpn.* **31**: 65–67 (1977).
38. Senda, K. Japan Kokai 75 43,048 (Apr. 18, 1975).
39. Uemura, T., and E. Hirotsu. Japan Kokai 80 47,251 (Apr. 3, 1980).
40. Suzuki, K., Y. Fujimori, and K. Amagai. *Rev. Gen. Meet. Tech. Sess. Cem. Assoc. Jpn.* **33**: 73–75 (1979).
41. Kupiec, A. R., and E. D. Escher. U.S. Patent 4,149,968 (Apr. 17, 1979).
42. Nakagana, H. Japan Kokai 80 44,309 (Mar. 28, 1980).
43. Marvi, A. Japan Kokai 77 100,744 (Aug. 24, 1977).
44. Rudolph, G., and R. Koesten. *Reaktor Tag.*:482–485 (1978).
45. Conner, J. R. U.S. Patent 4,518,508 (May 21, 1985).
46. Uchida, Y., and Y. Nagasawa. Japan Kokai 78 102, 868 (Sept. 7, 1978).
47. Durham, R. L., and C. R. Henderson. U.S. Patent 4,460,292 (July 17, 1984).
48. Kokusai, G. Japan Kokai 81 53,796 (May 13, 1981).
49. Yokouchi, H. Japan Kokai 75 44,977 (Apr. 22, 1975).
50. Veda, S., and M. Ito. Japan Kokai 77 38,468 (Mar. 25, 1977).
51. Yoshida, A. Japan Kokai 75 93,264 (July 25, 1975).
52. Nakanishi, M. Japan Kokai 79 107,923 (Aug. 24, 1979).
53. Takeshita, I. Japan Kokai 79 162,844 (Dec. 24, 1979).
54. Kawasaki Steel Corp. Japan Kokai 80 136,166 (Oct. 23, 1980).
55. Shinoki, K., T. Katayame, and S. Yamaguchi. Japan Kokai 80 44,355 (Mar. 28, 1980).
56. Nippon Soda Co. Japan Kokai 80 29,119 (Aug. 1, 1980).
57. Bonnel, B., P. Allemand, and P. Versmee. U.S. Patent 3,591,542 (July 6, 1971).
58. Wagner, H. B. Polymer modifications of Portland cement systems. *ChemTech:* 105–108 (Feb. 1973).
59. Donato, A., and G. Giacani. *Atti Congr. Naz.—Assoc. Ital. Fis. Sanit. Prot. Radiaz.* **20**: 543–549 (1978).
60. Nippon Soda Co. Japan Kokai 80 29,119 (Aug. 1, 1980).
61. Columbo, P., R. M. Neilson, and W. W. Becker. U.S. Patent 4,174,293 (Nov. 13, 1979).
62. Shiraishi, M., T. Moritami, and T. Tasaka. Japan Kokai 77 111,260 (Sept. 17, 1977).
63. Wakimura, Y. Japan Kokai 80 31,475 (Mar. 5, 1980).
64. Kitsugi, K., and M. Shimoda. Japan Kokai 78 12,770 (Feb. 4, 1978).
65. Munster, L. U.S. Patent 4,338,134 (July 6, 1982).
66. Chudo, A., T. Sugi, and K. Katsoka. U.S. Patent 4,266,980 (May 12, 1981).
67. Chen, K. S., and H. W. Majewski. U.S. Patent 4,113,504 (Sept. 12, 1978).
68. Onoda Cement Co. U.S. Patent 3,947,284 (Mar. 30, 1976).

69. Waterways Experiment Station. *Guide to the Disposal of Chemically Stabilized and Solidified Waste,* Pub. SW-872. Cincinnati, OH: U.S. EPA, 1980.
70. Peck, R. B., W. E. Hanson, and T. H. Thronburn. *Foundation Engineering.* New York: Wiley, 1974.
71. Conner, J. R. CONTEC Data Base, 1985.
72. Weitzman, L., L. E. Hamel, and E. Barth. Evaluation of solidification/stabilization as best demonstrated available technology. In *Proc. 14th Annual Research Symposium.* U.S. EPA, Cincinnati, OH, 1988.
73. Bhatty, M. S. Y. Fixation of metallic ions in Portland cement. Portland Cement Assoc., Skokie, IL, 1986.
74. Cote, P. L., and D. P. Hamilton. Leachability comparison of four hazardous waste fixation processes. In *Proc. 38th Annual Purdue Industrial Waste Conference.* Purdue Univ., West Lafayette, IN, May 10-12, 1983.
75. Brown, T., W. Shively, P. Bishop, and D. Gress. Use of an upflow column leaching test to study the release patterns of heavy metals from stabilized/solidified heavy metal sludges. *Hazardous Industrial Solid Waste Testing and Disposal,* Vol. 6. ASTM STP 933, p. 79, 1986.
76. Brown, T., and P. Bishop. The effect of particle size on the leaching of heavy metals from stabilized/solidified waste. In *Proc. International Conference on New Frontiers in Hazardous Waste Management.* 1985.
77. Conner, J. R., and S. Cotton. CWM Unpublished Study, 1988.
78. Campbell, K. M., T. El-Korchi, D. Gress, and P. Bishop. Stabilization of cadmium and lead in Portland cement paste using a synthetic seawater leachant. *Environ. Progr.* **6**(2): 99-103 (1987).
79. Shin, H., N. Her, and J. Koo. Design optimization for solidification of hazardous wastes. *Hazardous Wastes Hazardous Mater.* **5**(3): 239-250 (1988).
80. Barich, J. J., J. Greene, and R. Bond. Soil stabilization treatability study at the Western Processing superfund site. *Superfund '87:* 198-203 (1987).
81. Alset Process Brochure. Tokyo: Fujisash Co., 1987.
82. *Fujibeton . . . Public Nuisance Material.* Tokyo: Fujimasu Synthetic Chemical Laboratory Co. Ltd., 1972.
83. STC Technology. *Hazardous Waste Management Solutions from Silicate Technology Corporation.* Scottsdale, AZ, 1988.
84. Toxco, Inc. *Treatment of Chromium Contaminated Soils* (Final Report). Claremont, CA, Aug. 1988.
85. *Ind. Water Eng.* (Oct. 1970).

Chapter 11

PORTLAND CEMENT/SOLUBLE SILICATE PROCESSES

Sodium silicates have been used for more than a century for the production of commercial products such as special cements, coatings, molded articles, and catalysts. In these mixtures, the soluble silicate is mixed with cement, lime, slag, or other sources of multivalent metal ions that promote the gelation and precipitation of silicates. Katsanis *et al.* [1] have shown that the solubility of systems containing multivalent ions in combination with soluble silicates differs significantly from precipitation of solutions composed of metal salts or silica. The presence of calcium or magnesium ions, even at low concentrations, can reduce the solubility of silica by several orders of magnitude. These results suggest that soluble silicates reduce the leachability of toxic metal ions by formation of low-solubility metal oxide/silicates and by encapsulation of metal ions in a silicate- or metal silicate-gel matrix. These characteristics are one basis for the use of soluble silicates in CFS systems.

The history of silicate use in the CFS treatment of industrial wastes dates back to about 1970 [2], although its roots go further back to similar technology used for soil stabilization [3], grouting [4], mine back filling, and production of stabilized base courses for road construction, and there was also earlier work done in the field of nuclear wastes. Insoluble silicates in the form of Portland cement, flyash, kiln dusts, and clays were used in many of these processes, but the only recorded use of soluble silicates was in injection grouting of unstable soils and in the nuclear waste area. When the author first introduced the use of soluble silicates for CFS treatment of wastes in 1970 [2], solidification of industrial residues was done primarily, if at all, to make sludges and liquids physically suitable for handling in landfills. Nearly all of the process development work in this area of CFS technology has been done on a proprietary basis, primarily by one vendor [5,6], until recently.

In the case of soluble silicate processes, there appears to have been more misunderstanding and confusion than with the other major CFS process types. This was due not only to the greater complexity of the process, but also because it is used for several distinct and different purposes. The process uses Portland cement and soluble silicates, usually sodium silicate in the liquid (38 percent solution) form, to solidify a wide range of wastes. Because the system is quick gelling, it was designed to be used with low solids, rapid-settling waste streams to prevent the formation of freestanding water on the surface. It is also useful in fixing certain toxic metal species in a form believed to be less soluble than the hydroxides, especially under very high or low pH leaching conditions. To date, this process has been used to treat at least 500 million gallons of industrial and municipal liquid wastes [7]. Much of the work was performed using mobile treatment vans at on-site lagoons near the point of waste generation [8].

CHEMICAL AND PHYSICAL PROPERTIES

The Portland cement/soluble silicate (PCSS) system is based on the reactions between soluble silicates and Portland cement to produce a solid matrix in a controlled manner. The gel matrix itself, as produced, is based on tetrahedrally coordinated silicon atoms alternating with oxygen atoms along the backbone of a linear chain. The charged side groups—in this case, oxygen—when reacted with polyvalent metal ions result in strong ionic bonding between adjacent chains to form a crosslinked, three-dimensional, polymer matrix that is much like many of the natural pyroxene minerals. This type of structure displays properties of high stability, high melting point, and a rigid, friable structure similar to many soils. It is not possible here to discuss all of the complex chemistry that can take place in such a system. However, we can define three classes of interactions. First are the rapid reactions between soluble silicates and most polyvalent metal ions, producing low-solubility metal silicates. These compounds are nontoxic and cannot easily be resolubilized later on. In some cases they are similar to the minerals from which the metals were originally extracted, and are quite resistant to breakdown under most environmental conditions [9,10,11].

The second set of reactions occurs between the soluble silicate and reactive components of the Portland cement, which has limited solubility of the cross-linking calcium ion but a high reserve capacity, so that the reaction can take place slowly under controlled conditions. In addition to operational requirements that are served by such controlled reaction rates, the gel structure thus formed is more suitable to producing good solid properties, especially in waste that has a high water content. The gel has the unique property of being able to hold very large quantities of water while acting like a solid. The gel reaction can occur quickly enough, for example, in seconds or minutes, to prevent the settling out of solids that are to be contained in the structure. Because of its properties, the gel holds ions in place by various bonding mechanisms (see Chapters 3 and 4). The third class of reactions occurs between Portland cement, the waste, and water as it undergoes a series of hydrolysis, hydration, and neutralization reactions. This group of reactions was discussed extensively in Chapter 10.

As is evident from the preceding discussion, the reactions taking place are not easily set forth theoretically. Therefore, our present knowledge of such systems is largely empirical and, while we can often predict what general reactions will occur and what the results will be, it is still necessary to test a waste that has not been previously handled.

Gelation and Solidification

Soluble silicates are one of the oldest classes of industrial chemicals known. They have been used for myriad purposes and their reactions and uses are well described in the literature [12,13]. In the CFS field, there are many possible silicates that can be used and a number of reactants or setting agents. However, the overriding consideration in formulating any system for CFS use has been cost. In view of this and other factors, the most commercially important system to date has been that which uses Portland cement, Type 1, and a 38 percent solution of sodium silicate (silica-to-sodium oxide ratio of 3.22). The latter is a syrupy liquid with density of 11.6 lb/gal and pH of 11.3; it is commonly known as *water glass*. These two chemicals are the most widely used in their respective categories, and are the lowest priced grades. The chemical compositions of the sodium silicate and Portland cement used in PCSS processes are as follows.

Silicate: Sodium silicate, grade "N"
  Weight ratio SiO_2/Na_2O = 3.22
  % Na_2O = 8.90
  % SiO_2 = 28.7
  % Solids = 37.6
  Specific gravity = 1.39

Portland cement, Type I
  % $3CaO \cdot SiO_2$ = 45
  % $2CaO \cdot SiO_2$ = 27
  % $3CaO \cdot Al_2O_3$ = 11
  % $4CaO \cdot Al_2O_3 \cdot Fe_2O_3$ = 8
  % MgO = 2.9
  % Free CaO = 0.5
  % $CaSO_4$ = 3.1

These chemicals are available in all developed countries, although the pricing situation may vary.

Each of the materials used in this system is capable of forming a solid, cemented mass under the right conditions. This property of Portland cement is obvious, and a number of CFS processes use it alone or in various other combinations. Sodium silicate solution (SS) can also act to form a solid, in some cases with components of the waste itself. In the PCSS system, the cement is often referred to as the setting agent for the SS. At the same time, sodium silicate is used as an accelerator or anti-inhibitor for the setting of cement, and the hardening that takes place as the solidified waste cures is due primarily to the cement. There is no contradiction in this; both processes take place at different times during solidification, although there is no sharp delineation between them and they may overlap on the time scale. The chemistry of the system is more easily understood by examining the reasons (in addition to fixation of metals) for using soluble silicates in solidification work: (1) to quickly gel low solids wastes, and (2) to aid in the cement hardening process.

Gelation. Most industrial waste streams, especially those stored in lagoons, contain large amounts of water. When the PCSS process was first introduced, most wastes had solids contents of less than 20 percent. Today, with the emphasis on volume reduction, wastes are usually dewatered where possible to produce semisolid "cakes," but there are still many instances where dewatering is not feasible. For example, in remedial action work solidification is commonly practiced on the lagoon waste without any pretreatment. The difficulty with solidification of these low solids, low viscosity wastes is that they settle rapidly when not agitated. Since most CFS processes solidify slowly—over a period of hours or days in most cases—a contaminated watery layer remains on top of the hardening mass. This is usually undesirable. One of the reasons for developing the PCSS process was to eliminate this effect by quickly gelling the mass, increasing viscosity to the point where settling could not occur. As a result, all constituents are kept in a homogeneous suspension in the gel while the mixture hardens to a true solid. The alternative is to add large amounts of inexpensive ingredients that act primarily as fillers or bulking agents, artificially increasing the solids content and viscosity of the waste so that it cannot settle while it hardens.

The gelling ability of SS is unique, and should not be confused with various gelling or thickening agents, usually organic gums or polymers, that act by physical means only. In the PCSS system, gelation occurs chemically by reaction of the soluble silicate with calcium hydroxide (lime) produced by the hydration of PC when mixed with water. Sodium silicate is not a simple solution, because it contains both polymeric silicates and colloidal silica [10,11]. These species interact with the divalent calcium ion from cement hydration to form a silica/silicate gel that holds all of the available system water, as well as the nondissolved solids, suspended in a nonmobile (physically) state.

The initial gel is quite delicate, and forms properly under only the right conditions. Addition of soluble silicate to a concentrated solution of polyvalent metal ion will result in a hard, dense precipitate, not a gel. Too little metal ion will result in slow gel formation, or none at all. To be effective, the gel must immobilize the suspended solids, including the cement, within a period of several seconds to several minutes, depending on the waste and the mechanical aspects of the CFS system.

An uncommon property of the PCSS system is that cement provides just the right amount of calcium ion in solution at any given time. It does this by hydration of the cement's components, which ocurs slowly enough so that hard precipitation of the silicate does not take place. As the calcium ion from the cement is used up, more is generated by hydration to further react with the remaining SS, strengthening the gel. At the same time, the PC begins its own series of reactions that first produce setting and then hardening. In the early stages of reaction, there is competition for the calcium ion between the soluble silicate and hydrosilicate residues on the cement grains. Since the cement setting reaction is substantially slower than the soluble silicate gel formation, the latter occurs at the expense of the former. As the soluble silicate is used up, calcium ion becomes available to enter into the cement's own gel reactions that result in the cement setting phase. The cement then continues to harden through its complex set of reactions, which continue almost indefinitely. Considering this complex, but fairly well-defined, set of reactions, it is evident why the ratios of cement and soluble silicate must be properly proportioned. If too much soluble silicate is used, the cement setting and curing reactions, which rely on the generation of calcium hydroxide, may become calcium-starved and never go to completion.

Obviously, the presence of metal ions from the waste itself, as well as many other common waste components, will affect the gelling properties of soluble silicate in this system. One common example is pH; acid wastes will precipitate the soluble silicate before the gel can be formed. Given all of the various problems that are encountered with different waste streams, it is surprising that the system works so well with so many and varied waste types, especially in view of the complexity and sensitivity of the reactions that must occur.

Cement Hardening. Many potential waste constituents can act as retarders or accelerators in the setting of cement. The presence of accelerators is not usually a problem, but retarders can slow or even prevent the hydration of the calcium silicate components of cement. There are compounds or ions that form precipitates on the surface of the cement grains, for example, gypsum and salts of heavy metals such as zinc, copper, and lead. Another category of retarders adsorb on the cement grains, or coat them so that reaction with water is prevented. These retarders are exemplified by fine particulates such as silt and clay, and by organics like oil and algae. Soluble silicates have been used both as accelerators and as anti-inhibitors for concrete, and have the same function in PC-based CFS systems. In the case of the precipitant-type retarders, such as heavy metals, soluble silicate probably works by removing the metal from solution before it can precipitate on the cement grains. With retarders that operate by coating the grains, soluble silicate may function as a surfactant, emulsifying oils and flocculating fine particulates so that they remain suspended in the water phase. In any case, soluble silicate has been useful in many instances for this purpose rather than as a fixant or gellant.

Metal Fixation

The reactions of polyvalent metal salts in solution with soluble silicates have been studied extensively over many years [10,12]. Nevertheless, the "insoluble" precipitates that result from such interactions are not usually well characterized, especially in the complex systems representative of most wastes.

Reactions in Solution. This situation is best summarized by Vail [10], who states:

> The precipitates formed by the reaction of the salts of heavy metals with alkaline silicates in dilute solution are not the result of the neat stoichiometric reactions describing the formation of crystalline silicates, but are the product of an interplay of forces which yield hydrous mixtures of varying composition and water content.

These reaction products are usually noncrystalline and therefore very difficult to characterize structurally. They are most often described as hydrated metal ions associated with silica or silica gel. Iler [12] mentions "that many ions are held irreversibly on silica surfaces by forces still poorly understood in addition to ionic attraction."

The composition and form of the metal "silicates" formed from metal ions and soluble silicates are functions of the conditions under which they are formed: temperature, concentration, addition rate, metal ion speciation, presence of other species, etc. For example, in dilute solutions, the species may remain in colloidal suspension, while at higher concentration or in the presence of other destabilizing conditions, it will precipitate. Metals are often present as soluble complexes or as negatively charged anions that should not bind to the silica surface. PH is very important, for, as Iler [12] has pointed out:

> ... silica suspended in a solution of most polyvalent metal salts begins to adsorb metal ions when the pH is raised to within 1–2 pH units below the pH at which the

polyvalent metal hydroxide is precipitated ("silica" here can refer to either colloidal silica or polymeric silicates).

If the pH is not raised, the metal is not precipitated; if the pH of the system is already above the adsorption point, the metal may precipitate as the hydroxide instead. Different metals have different precipitability by soluble silicate solutions. According to Vail [10], the order, beginning with the most precipitable, is: copper, zinc, manganese, cadmium, lead, nickel, silver, magnesium, calcium.

Real Systems. The preceding discussion deals only with the complexities of reactions in solution, and only with those between the metal species and the soluble silicate. With real wastes, the components may be present also as suspended solids or in immiscible liquid phases. Metals may already be speciated as relatively insoluble hydroxides or other solid phase compounds. Nearly always, cement or other solidification reactant is added to the system; these materials themselves often interact with the metal species, as well as with the soluble silicate. If too little soluble silicate is added to the system, the reaction products that form will depend on which competing reactions are most successful, and also on the order of addition of the reactants. If an excess is added, unwanted leachable metal ion silica complexes may form (J. S. Falcone, private communication, 1986). The rate of addition, degree of agitation, and temperature will also affect the nature of the reaction products.

The aspect of soluble silicate fixation of metals that seems most confusing to workers in the field involves the speciation of the cation before the silicate is introduced into the system. Most wastes that are treated by CFS are sludges, filter cakes, and other residues from waste water treatment systems. The metals have been precipitated with lime or other alkali, or with other agents such as sulfide, to produce metal hydroxides, sulfides, etc., that have low solubility under the conditions of precipitation, usually in the pH range of 6 to 8 and a dilute water medium. In this state, there is little short-term reaction between the metal species and the soluble silicate. If the metal hydroxide is more soluble than the silicate, there may be gradual respeciation of the hydroxide at the particle surface, but total respeciation would not be expected to occur except over a very long period of time. Furthermore, the soluble silicate will not be available as such for long, because it will react rapidly with other components of the CFS system. The ultimate result is a mixture of metal hydroxides, sulfides, etc., dispersed in a cementitious matrix. In this case, the degree of fixation of the metals will be similar to that of a low-ratio Portland cement CFS system.

The leachability of the whole will be determined by the solubility of the metal compounds and by the permeability of the monolith, which in turn is a function of pore structure and the amount of water present in the waste [14]. In this system, the primary function of the soluble silicate (other than in the gelation/solidification reactions) is to reduce permeability. It does this by forming precipitates in the matrix that block pores, reducing the effective pore volume and slowing the movement of any mobile species throught the matrix into the environment [10,15]. This effect probably is also responsible for the reduced leachability of species that would not be expected to react with soluble silicates: monovalent cations, anions, and organics. It accounts for the fact that monoliths produced by CFS usually (but not always) show higher levels of leachability when subjected to mechanical degradation, either naturally or during leaching test procedures. Grinding or crushing the monolith increases the surface area in contact with the leaching medium, thereby reducing the contribution of impermeability produced by the CFS process.

The preceding problem of speciation may be avoided when the metal species to be fixed is in solution. Unless it is complexed in soluble, stable form, the metal can usually be precipitated from solution as a "silicate" that exhibits low solubility through a wide pH range. This is especially important for the heavy metals of environmental interest that exhibit amphoterism, such as chromium, zinc, nickel, and lead. The hydroxides of these metals exhibit minimum solubility through a narrow pH range, usually in the area of pH 7.5 to 9; it increases rapidly above pH 9. Because most CFS systems are quite alkaline, usually above pH 11 (at least initially), the solubility of the metal species in the CFS-treated waste may actually be higher than in the original, untreated sludge. This problem tends to be exacerbated with increased surface area. While the soluble silicate, if available, will respeciate the metal hydroxide as it dissolves, it usually will not be available because it will have been used up in other reactions. There are three solutions to this problem: (1) reduce permeability, (2) add a fixant other than silicate that remains available in the matrix as needed, or (3) pretreat the waste by reducing pH to dissolve the metal hydroxide, then respeciate as silicate.

Another set of real-world complexities is introduced by the way in which industrial metal hydroxide or sulfide residues have been handled in the past. Typically, these materials have been stored in lagoons where other residues or waste products may be added. Some of the added residues may also be the result of water treatment systems—for example, biological processes for organic wastes—while others may be concentrated, untreated process wastes. Frequently, water immiscible organics such as oily lubricants will be carried along with one or another of the waste streams into the lagoon. Thus, an extremely complex mixture forms, and further reactions among the various wastes and as a result of weathering and aging occur. Often, the additional components will interfere with otherwise possible reactions between metal species and soluble silicates, or will themselves precipitate the silicate from solution before it can achieve either of its purposes: fixation or gelation/solidification. Oily constituents, for example, may coat reactive particles, preventing reaction with soluble silicate. When the oil is removed as a first step in the EPA Oily Waste Extraction Procedure (see Chapter 17), the unreacted material will then leach at higher than expected rates. The presence of biological organisms or their by-products can interfere with the solidification process; it may also have an effect on metal fixation.

Properly applied, PCSS CFS systems can reduce the leachability of heavy metals to levels below 0.1 mg/l in most wastes. Often, much lower levels can be achieved. Unfortunately, we cannot predict the leachability in these systems from theoretical models, an outcome that is not surprising in view of the complexities described previously.

General Chemical Properties

Chemical properties of Portland cement/soluble silicate processes, like other CFS processes, are described primarily in terms of leachability. In Chapter 4, we explored the reactions and leachability of various metal species in CFS systems. We have discussed some general findings on the leachability of PCSS waste forms. For ease of comparison and reference, the results of a number of leaching studies and tests are summarized in Table 11-1, rather than being presented for each commercial process or case study separately. In addition to this, some older but useful data are presented later, using different leaching tests than are currently in use. Table 11-1 lists the information by waste type, stressing laboratory studies and general data from the technical literature. The data sources are referenced at the end of Chapter 4.

Chemical theory and principles relative to PCSS systems have also been discussed previously in this chapter. Many of the fixation and matrix formation principles are very similar to those of cement-based processes discussed in Chapter 10, especially in the long-term hardening processes. Other soluble silicate-based systems (those not using Portland cement) might have very different properties and chemistries, but at the present time none of these are important commercially.

Physical Properties

The addition of soluble silicates to a Portland cement (or other cementitious or pozzolanic) system allows the solidification of low-solids wastes without the addition of massive amounts of bulking agents. This can be a cost effective approach, but it carries with it certain penalties and product characteristics that are quite different from those resulting from most other inorganic CFS processes. This is primarily due to the much greater water content in the solid, which increases the porosity of the solid. Cote [14] shows this in terms of increased pore diameter, which he calls "macropores." Both increased pore diameter and cumulative porosity are shown in Figure 11-1. Higher water content also causes reduced physical strength and relatively higher permeability in the solid.

Reported values for unconfined compressive strength vary greatly, depending

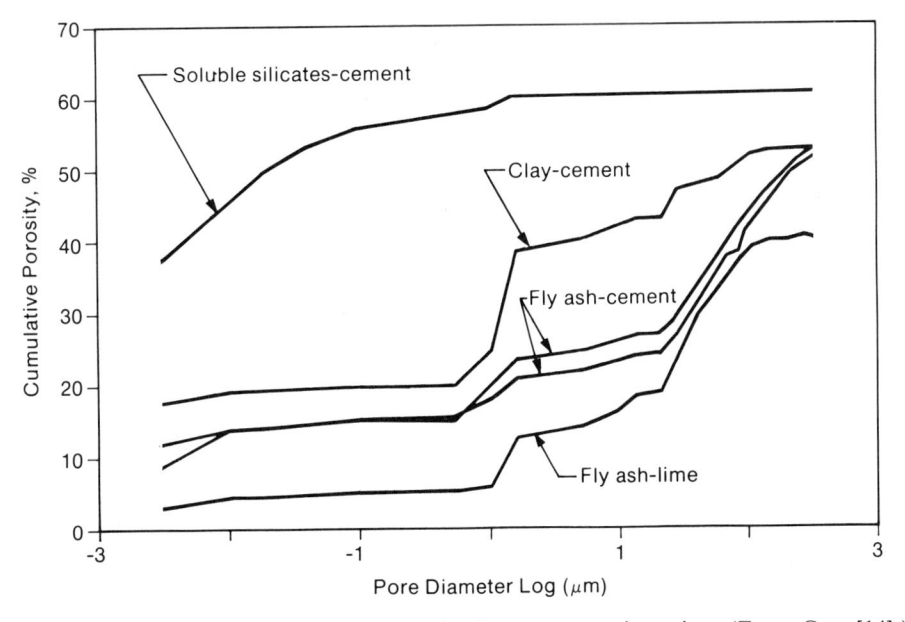

Figure 11-1 Corrected cumulative porosity from mercury intrusion. (From Cote [14].)

TABLE 11-1 Summary of Metal Fixation Data: Soluble Silicate Processes

Reference[a]	Waste Description	Waste Code	CFS System	Mix Ratio	Treatment Scale
4:103	Arc furnace dust	K061	Unstabilized raw waste		Lab
4:103	Arc furnace dust	K061	Unstabilized raw waste		Lab
4:103	Arc furnace dust	K061	Cement/silicate	0.20	Lab
4:33	Mixed		Cement, sodium silicate, EDTA	0.20	
4:117	Cadmium hydroxide sludge		Cement/silicate	0.45	Lab
4:136	Flyash A		Unstabilized		Lab
4:136	Flyash A		Gelled silicate	0.26	Lab
4:136	Flyash B		Unstabilized		Lab
4:136	Flyash B		Gelled silicate	0.26	Lab
4:136	Plating waste		Unstabilized		Lab
4:136	Plating waste		Gelled silicate	0.16	Lab
4:136	Plating waste		Cement/silicate	0.24	Lab
4:136	Mixed municipal/industrial waste		Unstabilized		Lab
4:136	Mixed municipal/industrial waste		Gelled silicate	0.16	Lab
4:136	Mixed municipal/industrial waste		Cement/silicate	0.24	Lab
	CCA wood preserving waste		Unstabilized		Lab
	CCA wood preserving waste		Cement/silicate	0.40	Lab
	CCA wood preserving waste		Potassium silicate		Lab
	Arc furnace dust	K061	Potassium silicate		Lab
4:138	Ceramic slip waste		Cement/silicate	0.30	Lab
4:138	Ceramic slip waste		Cement/silicate	0.12	Lab
4:138	Ceramic slip waste		Cement/silicate + Na$_2$S	0.22	Lab
4:138	Ceramic slip waste		Cement/silicate + Na$_2$S	0.14	Lab
4:138	Ceramic slip waste		Cement/silicate + (NH$_4$)2HPO$_4$	0.22	Lab
4:138	Ceramic slip waste		Cement/silicate	0.14	Lab
4:139	Ceramic glaze waste		Cement/silicate	0.30	Lab
4:139	Ceramic glaze waste		Cement/silicate	0.12	Lab
4:139	Ceramic glaze waste		Cement/silicate + (NH$_4$)2HPO$_4$	0.22	Lab
4:139	Ceramic glaze waste		Cement/silicate + (NH$_4$)2HPO$_4$	0.14	Lab
4:139	Ceramic glaze waste		Cement/silicate + Na$_2$S	0.22	Lab
4:139	Ceramic glaze waste		Cement/silicate + Na$_2$S	0.14	Lab
4:143	Synthetic sludge		Cement/silicate + hexadecyl mercaptan	0.2	Lab

[a] References are listed at the end of Chapter 4.

on the process, additive type and ratio, and overall mix ratio. They also vary according to the test method used. Most products range from 15 to about 100 lb/in.^{2}, but stronger products can be prepared if desired. Volume increase factors (ratio of final to initial volume) range from 1.05 to about 1.2. The bearing strengths of a number of PCSS-treated wastes at one, three, and ten days are shown in Tables 11-2–11-4 for three types of PCSS processes: cement/liquid silicate, cement/dry silicate, and cement/clay/dry silicate. These strength values are reported in tons/square foot as measured by penetrometer. Also given are total solids contents and volume increase (percent) for each of the waste–product combinations. In general, volume increase for PCSS systems equals about 0.5 times the weight increase; that is, 10 percent

TABLE 11-1 (*Continued*)

Leaching Test	Metal Content (mg/kg or mg/l)							
	Antimony		Arsenic		Barium		Cadmium	
	Total	Leachate	Total	Leachate	Total	Leachate	Total	Leachate
EPT							600	1.700
TCLP								1.090
TCLP								<0.010
			40	0.016				
TCLP							15340	0.014
EPT								0.860
EPT								0.190
EPT								64.000
EPT								24.000
EPT							3.2	0.33
EPT								0.050
EPT								0.050
EPT							0.9	0.200
EPT							0.9	0.060
EPT							0.9	0.100
EPT			11500	1.800				
EPT			11500	2.300				
EPT			11500	<0.010				
EPT				<0.400		2.800		0.050
EPT								
EPT								
EPT								
EPT								
EPT								
EPT								
EPT								
EPT								
EPT								
EPT								
EPT								
EPT								
TCLP		1.310		2.280		0.900		0.074

(*continued*)

volume increase equals 20 percent weight increase (mix ratio of 0.20). This applies to water-saturated wastes only. Dry ashes and dusts may actually experience a volume decrease due to lowered void space when the material is wetted.

As we saw previously, permeabilities are strongly influenced by the waste being solidified. They normally fall in the area of 10^{-5} to 10^{-7} cm/s, depending to some extent on how they are measured. Permeability of PCSS-based waste forms is comparable to that of clay. Other properties, such as resistance to weathering, wetting-drying, and freeze–thaw, are usually not very good for PCSS products, but are not very important in typical waste disposal scenarios. If the product is to be used in some other way, however, such properties may be critical.

TABLE 11-1 (Continued)

	Chromium		Copper		Lead		Mercury	
Reference[a]	Total	Leachate	Total	Leachate	Total	Leachate	Total	Leachate
4:103	1100	0.900			38000	139.000		
4:103		0.060				14.300		
4:103		0.730				0.170		
4:33								
4:117								
4:136		0.140		2.000		3.200		
4:136		0.090		0.130		1.200		
4:136		0.130		37.000		170.000		
4:136		0.060		14.000		74.000		
4:136	56	2.500	10.4	2.600	1048	8.600		
4:136		0.420		0.140		2.700		
4:136		1.400		0.480		12.000		
4:136	353	0.130			823	4.500		
4:136	353	0.150			823	2.500		
4:136	353	0.190			823	3.200		
	16000	90.000	8000	13.000				
	16000	16.000	8000	0.200				
	16000	0.500	8000	<0.050				
		0.120				0.26		<0.001
4:138					10000	0.18		
4:138					10000	2.7		
4:138					10000	0.18		
4:138					10000	5.6		
4:138					10000	0.73		
4:138					10000	0.21		
4:139					450000	43		
4:139					450000	87.8		
4:139					450000	38.1		
4:139					450000	81.7		
4:139					450000	41.8		
4:139					450000	1290		
4:143		<0.050		0.110		0.300		<0.001

Metal Content (mg/kg or mg/l)

LABORATORY AND PILOT STUDIES

A great deal of laboratory and pilot work has been done on the PCSS process—until recently, most of it by Chemfix, Inc. (CI) or Chemfix Technologies Inc.(CTI). The CI and CTI work will be discussed later in this chapter. More recently, the PQ Corporation has published the results of laboratory testing of a number of soluble silicate-treated waste streams [9,13,15,16]. Falcone et al. [9] have explained the apparent negative effects of soluble silicates on metal leachability experienced by some other investigators. When an excess of soluble silicate over available metal ion is present in such a system, the silica species with adsorbed metal ion may remain suspended, not be filtered in the leaching protocol, and give the appearance of increased leachability. Also, large excesses of soluble silicate can so decrease the set

TABLE 11-1 (*Continued*)

Nickel		Selenium		Silver		Zinc	
Total	Leachate	Total	Leachate	Total	Leachate	Total	Leachate
						1460.000	
							0.030
	1.600						36.000
	0.280						11.000
	0.400						8600.000
	0.230						4200.000
23.4	44.000						20.000
	1.400						6.200
	1.900						4.100
564	4.700						30.000
564	1.700						8.300
564	2.200						22.000
			<0.400		<0.030		
	0.190		0.066		0.020		0.054

time that it is impossible to obtain a homogeneous mass, and the resulting pore size is much larger. This counters the reduction in permeability that is the real benefit of soluble silicate in systems where the metal has already been precipitated as another species. Davis *et al.* [15] found that "treatment of a waste with sodium silicate reduced the degree to which leaching acid can penetrate the waste matrix by a factor of five to seven." The sum result of these effects is demonstrated graphically in Figure 11-2. In this system, a silicate content range of 6 to 10 percent gives optimum leachability. Interestingly, this is precisely the range that has long been used empirically by workers in this CFS specialty.

The author also has found that many investigators not familiar with soluble silicate processes have difficulty reproducing in the laboratory the results that are routinely obtained in the field. The reason for this is simple. Field operations

TABLE 11-2 Penetrometer Hardness vs. Cure Time Cement/Liquid-Silicate-Based Processes

Waste	Physical State	pH	Total Solids	Volume Increase (%)	Bearing Strength (Penetrometer) (tons/ft^2) 1 day	3 days	10 days
F006	SLDG	6.7	6	7	1	2	2.5
F006	CAKE	7.5	42	7	3	>4.5	>4.5
F006	SLDG	9.8	12	9	0	1	4.5
F006	SLDG	2.0	4	13	0	1	2
F006	SLDG	7.5	2	13	0	1	2
F006	SLDG	9.6	6	13	1	2	4
F006	CAKE	8.6	46	13	4	>4.5	>4.5
F006	CAKE	7.5	50	13	>4.5	>4.5	>4.5
F019	SLDG	7.5	3	14	2	3	3.2
F019	SOLN	7.5	3	14	2	3	3.2
K058	SLDG	6.9	19	14	1	2	
K058	SUSP	7.1	1	14	2	2	2.7
K058	SLDG	7.1	41	14	>4.5	>4.5	
K058	SLDG	7.4	3	14	2	>4.5	>4.5
K106	SLDG	12.4	50	6	1	2	3
Sewage digestor sludge	SLDG	5.8	2	13	1	2	
Off-specification lotions	SUSP	6.8	12	13	0	2	2.1
Municipal sewage sludge	SLDG	7.1	4	7	1	1	2
Municipal sewage sludge	SLDG	7.4	2	9	2	3	3.8
Acid oil sludge	SLDG	1.2	40	14	0	3	>4.5

Waste	Form					
Organic coating waste	SUSP	8.1	1	13	0	1	3
Organic chemical waste, NOS	SLDG	7.3	21	13	3	3	3.2
Waste water treatment sludge	SLDG	12.7	70	7	2	4	4.3
Lime neutral. waste, NOS	SUSP	12.9	5	10	0	0	1.2
Lime neutral. waste, NOS	CAKE	7.3	5	10	1	2	
Biosludge, NOS	SLDG	7.0	30	7	2	3	4.5
Biosludge, NOS	SLDG	12.2	25	10	1	2	>4.5
Caustic waste	SUSP	6.5	8	32	1	2	
Latex waste	SOLN	12.4	1	16	4	4	3.3
Paint kettle washdown	SUSP	6.1	4	13	1	2	2.4
Caulk manufacturing washdown	SUSP	8.4	18	14	1	2	2.2
Drawing lubricant residue	SOLN	8.0	6	13	1	2	3
Paint stripping waste	SOLN	10.5	1	13	0	1	3.3
Photo-resist stripping solution	SOLN	10.6	10	9	2	2	1.7
Solder flux residue	SOLN	10.3	3	13	0	1	>4.5
Amine waste	SUSP	13.0	1	18	4	>4.5	2.5
Chromate waste	SUSP	8.0	8	18	2	2	3.9
Organic waste, NOS	SOLN	3.4	35	7	1	2	1.1
Manufacturing sludge, NOS	SLDG	10.4	13	14	0	1	2
Paint manufacturing waste, NOS	SLDG	7.8	25	10	1	1	2.5
Paint manufacturing waste, NOS	SLDG	6.8	26	10	0	2	
Paint manufacturing waste, NOS	SLDG	12.0	25	10	>4.5	>4.5	>4.5
Paint manufacturing waste, NOS	SLDG	12.0	15	10	>4.5	>4.5	>4.5
Organic chemical manufacturing waste	SLDG	3.8	20	16	2	4	

TABLE 11-3   Penetrometer Hardness vs. Cure Time Cement/Dry-Silicate-Based Process

Waste	Physical State	pH	Total Solids	Volume Increase (%)	Bearing Strength (Penetrometer) (tons/ft^2) 1 day	3 days	10 days
F006	CAKE	8.5	26	10	0	1	1.1
F006	SLDG	6.6	38	10	0	1	2.1
F006	SLDG	10.5	15	10	1	2	2.3
F006	SLDG	6.6	8	10	0	3	4
F006	SLDG	11.2	24	10	3	4	4.5
F006	SLDG	7.4	12	15	3	3	3.8
F006	CAKE	9.5	40		0	4	4.5
F006	SLDG	8.9	3	15	2	1	2.5
K006	SLDG	6.7	7	10	0	2	2.5
K071	CAKE	8.8	71	5	4	>4.5	>4.5
K106	CAKE	7.5	27	5	3	>4.5	>4.5
K106	CAKE	11.0	50	10	0	1	3
K106	CAKE	11.0	50	10	4	>4.5	>4.5
Electroplating waste, NOS	CAKE	9.6	38	8	2	3	4
Cathode coating sludge	SLDG	8.0	60	8	3	4	4.5
HWM facility sludge	SLDG	8.9	37	10	1	1	4.5
Municipal sewage sludge	SLDG	6.0	3	15	0	0	1
Metal hydroxide sludge, NOS	SLDG	11.8	10	15	0	>4.5	>4.5
Waste water treatment residue	CAKE	9.5	40	5	1	3	
Lime neutralization sludge	SLDG	7.4	20	10	1	2	3.8
Waste pickle liquor sludge	SLDG	12.0		5	1	2	4.5
Zinc phosphate bath	SOLN	11.6	2	10	0	1	2.3
Latex caulking cpd.	CAKE	7.2	83	5	1	3	4.5
Latex caulking cpd. washwater	SUSP	10.5	36	10	1	2	
Paint sludge, NOS	CAKE	10.0	90	10	0	0	1.5
Manufacturing waste, NOS	SUSP	6.8	21	8	3	4	4.5
F006	SLDG	10.2	16	10	0	1	2.6
F006	SLDG	10.3	19	15	0	3	3.1
K106	SLDG	8.3	62	10	0	0	4.5
K106	CAKE	11.0	50	5	4	>4.5	>4.5
Sphincter cone sludge	CAKE	8.2	79		2	3	
Waste water treatment sludge	SLDG	12.4	55	15	0	1	1.4
Lime neutral. sludge, NOS	SLDG	11.7	27	5	2	4	4.5
Pesticide cleanup waste	SLDG	12.3	8	15	0	>4.5	>4.5
Production residue, NOS	SLDG	8.8	5	10	0	0	1.9
Oily water	SUSP	5.9	54	25	2	4	4.3
DAF Sludge	SLDG	4.5	12	15	0	1	2.5

390   PORTLAND CEMENT/SOLUBLE SILICATE PROCESSES

TABLE 11-4 Penetrometer Hardness vs. Cure Time Cement/Clay/Dry-Silicate-Based System

Waste	Physical State	pH	Total Solids	Volume Increase (%)	Bearing Strength (Penetrometer) (tons/ft^2) 1 day	3 days	10 days
F006	SLDG	10.2	16	10	0	1	2.6
F006	SLDG	10.3	19	15	0	3	3.1
K106	SLDG	8.3	62	10	0	0	4.5
K106	CAKE	11.0	50	5	4	>4.5	>4.5
Sphincter cone sludge	CAKE	8.2	79		2	3	
Waste water treatment sludge	SLDG	12.4	55	15	0	1	1.4
Lime neutral. sludge, NOS	SLDG	11.7	27	5	2	4	4.5
Pesticide cleanup waste	SLDG	12.3	8	15	0	>4.5	>4.5
Production residue, NOS	SLDG	8.8	5	10	0	0	1.9
Oily water	SUSP	5.9	54	25	2	4	4.3
DAF Sludge	SLDG	4.5	12	15	0	1	2.5

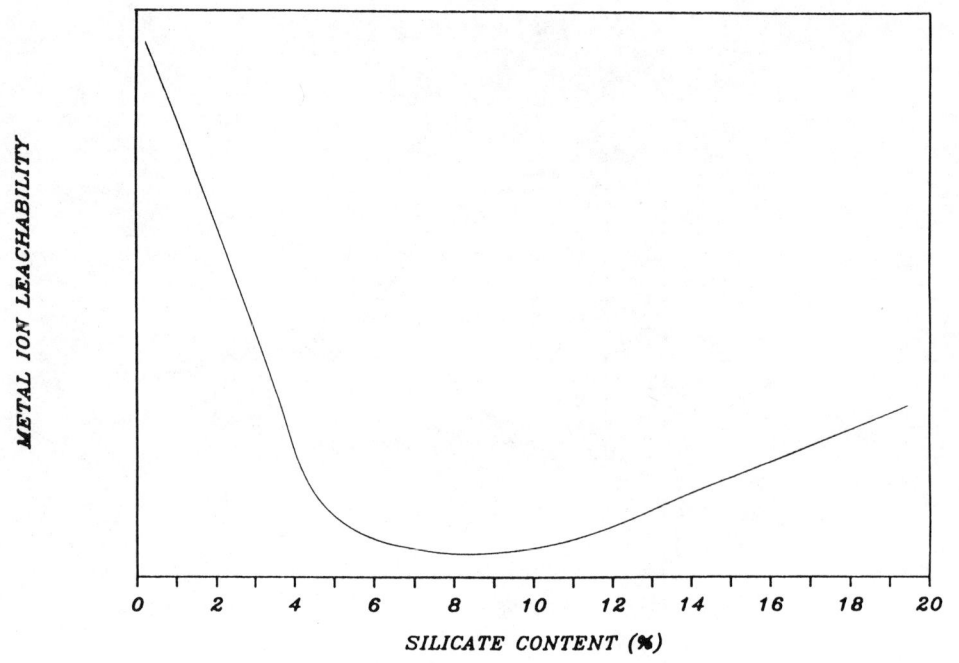

Figure 11-2 Solidified waste leaching as a function of silicate content. (From Falcone et al. [9].)

introduce the soluble silicate after the cement (or other reagent) has been thoroughly mixed into the waste, usually in the discharge pump from the mixer. This produces an intense, but very rapid mixing action that results in a homogeneous product before the setting reaction can begin. In the laboratory, this is difficult to reproduce without the use of a high-intensity mixer, which may overmix the formulation and, in any case, is difficult to clean out after the test. Laboratory workers therefore tend to mix by hand or by slow-speed mixer, which does not produce a homogeneous product. In attempting to get thorough mixing by these systems the formulation may also be overmixed, breaking the delicate initial gel. In the PCSS process, this gel often cannot be reformed, as it usually can in straight Portland cement processes, because the soluble silicate has been reacted. To counter this problem, the novice experimenter will sometimes add excess soluble silicate, causing the problems discussed previously.

Spenser et al. [16] studied the effects of several types of soluble silicate CFS on flyashes and mixed municipal-plating wastes. They compared the use of soluble silicate alone, gelled (with acid) sodium silicate, and PCSS with standard lime treatment. The results can be seen in Table 11-1. The silicate methods produced solids able to meet the EPT characteristics standards for the mixed waste and for one flyash, but they would not have met the present landban standards. In the case of the PCSS treatment, it appears that the cement ratio would have to be increased substantially to accomplish the latter purpose. This experiment points out again the importance of designing the system to meet the goal of treatment. In this case, the work was done when the landban standards had not yet been established.

Falcone et al. [9] did some interesting work using a column leaching test with various leachant compositions, as well as a long-term batch leaching test. They studied the leachability of untreated waste and the same waste treated with several CFS processes as a function of time. The leachants used were a buffered citric acid solution (pH 5.5) and a 5 percent acetic acid solution adjusted to pH 5.5 with NaOH. The waste was a lime-precipitated, plating waste sludge containing zinc and chromium. The CFS processes used were:

Ungelled sodium silicate
Freshly gelled sodium silicate
Aged, gelled sodium silicate
Portland cement/flyash
Portland cement/flyash/sodium silicate

The results of the batch tests are shown in Table 11-5. The cement/silicate product gave the lowest leaching levels for both zinc and chromium after 720 hours.

The results of the column tests are shown in Figures 11-3 and 11-4 for the gelled-silicate sludge and the cement/flyash with and without silicate, respectively. The gelled-silicate sludge leached zinc at very high levels early in the tests, with the leaching decreasing as zinc in the sample was depleted. The chromium leached at much lower levels, but was present at lower levels in the sludge. It was observed that the structure of tested waste forms broke down with time in the column. In the case of the cement/flyash matrices, the structure remained intact and the leaching curves showed a different behavior, which correlates well with the change in pH of the system, shown in Figure 11-5. However, when the cement/flyash waste form included sodium silicate, the leaching rate remained low even after the pH fell to below 7.

It was concluded that three mechanisms could be involved in the solubility reduction.

1. Formation of insoluble chromium and zinc hydrous oxide silicates
2. Encapsulation of chromium and zinc involving the formation of calcium and magnesium hydrous oxide silicates on the sludge particle surfaces
3. Encapsulation of chromium and zinc involving the formation of a gelled-silicate

TABLE 11-5 Total Amount of Zinc and Chromium Leached with Buffered Citric Acid from Treated Plating Sludges Under Batch Conditions for 720 Hours

Batch Number[a]	Sludge Treatment	Zinc Amount Leached (g)	Zinc Percent of Total	Chromium Amount Leached (g)	Chromium Percent of Total
—	Total metal ion present per 100 g sludge[b]	1.27		0.98	
A	None	1.2	95	0.16	16
B	Ungelled sludge	1.2	94	0.13	13
C	Fresh gel	1.1	84	0.10	10
D	Aged gel	0.84	66	0.08	8
E	Cement	0.44	35	0.05	5
F	Cement/silicate	0.07	5	0.02	2

Note: All values estimated to be within ± 5 percent.
[a] For treatment details, see Reference 9.
[b] Determined by chemical analysis.

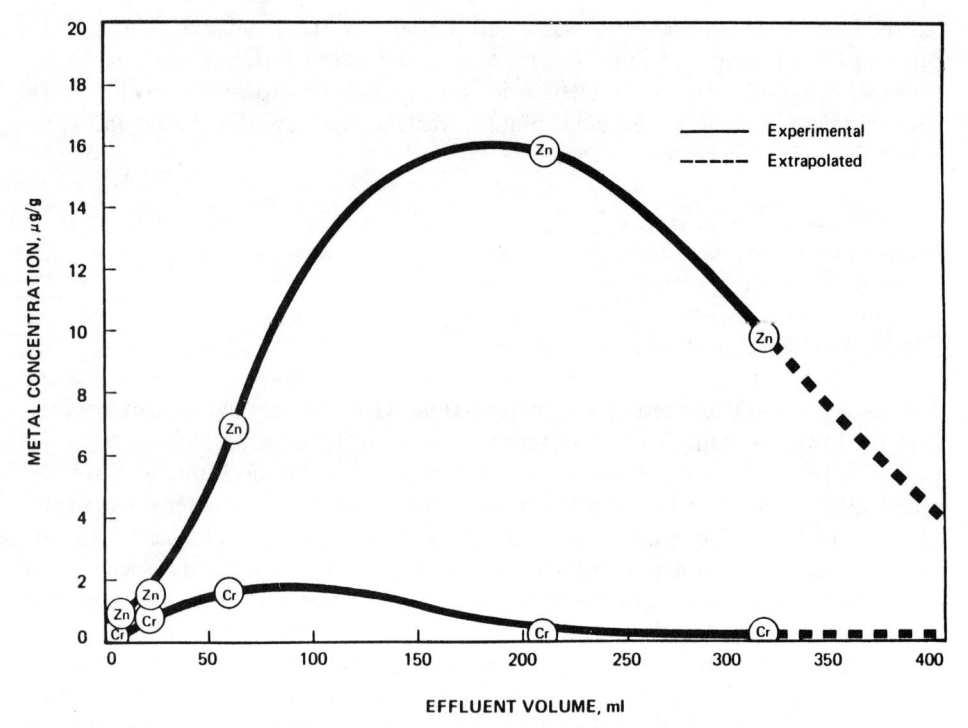

Figure 11-3 Leaching of silicate gelled sludge. (From Falcone et al. [9].)

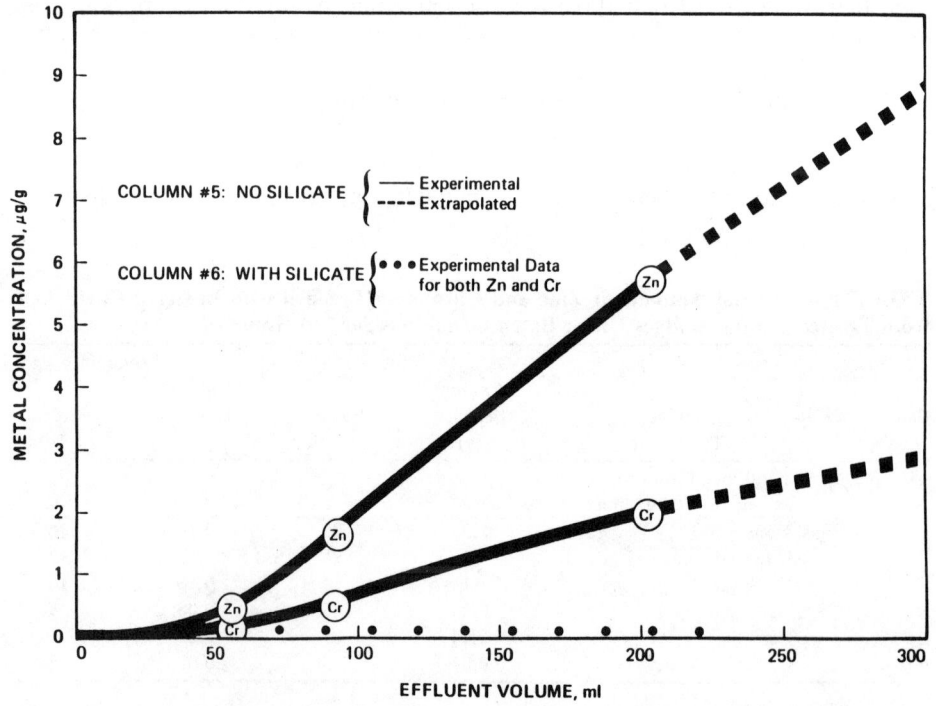

Figure 11-4 Leaching of Portland cement/flyash treated sludge with citric acid. (From Falcone et al. [9].)

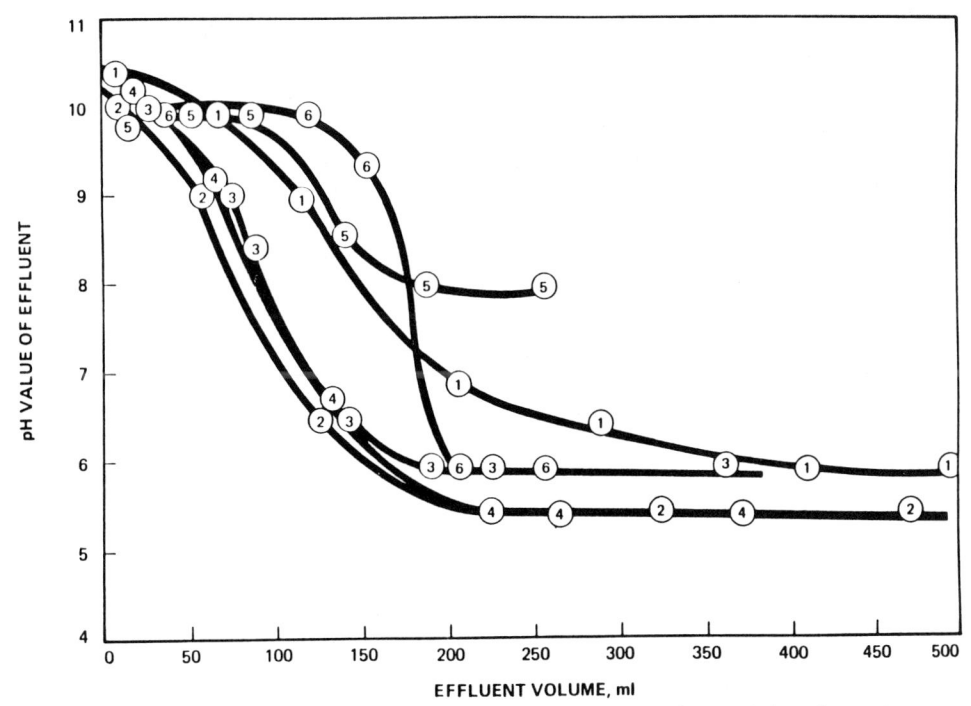

Figure 11-5 PH of waste solids treated in various ways. ○Experimental data for each column treatment. 1. Untreated sludge, acetic acid. 2. Untreated sludge, citric acid. 3. Acid/silicate gelled sludge, acidic acid. 4. Acid/silicate gelled sludge, citric acid. 5. Cement/flyash treated sludge, citric acid. 6. Cement/flyash/silicate treated sludge, citric acid. (From Falcone et al. [9].)

sheath over the sludge particles, which then is insolubilized by calcium ions from the cement

The investigators were unable to determine which mechanism or mechanisms were operative in this system.

GENERIC VS. PROPRIETARY PROCESSES

Any present or potential vendors or users interested in using soluble silicate systems should review very thoroughly both the patent and technical literature on this subject, especially in their own countries. So many compositions for the setting of soluble silicates have been patented or desribed in the literature that it is rather unlikely that a new one will be sufficiently novel to obtain patent protection. Also, a number of these patents are sufficiently new to be in force and infringement may result. However, most of the systems patented or otherwise described are really not applicable to waste solidification either for technical reasons or for reasons of cost. This is true, for example, for most of the organic gelling agents that are described in the patent literature. It is for this reason that the Chemfix® process and several of the others using all inorganic chemicals, and especially those using Portland cement, lime, and mixtures thereof, are the most suitable for commercial CFS use.

One interesting adaptation has been reported: the use of a mixture of about 50 percent lime and 50 percent Portland cement in place of either agent alone, in situations where setting inhibition of the sodium silicate–Portland cement system was being experienced.

The following companies are known to offer solidification processes using soluble silicates. A comparative evaluation of the processes was presented in Table 9-3. The commercial systems are discussed further in this section.

Chemfix® Technologies, Inc., Kenner, Louisiana (Chemfix)
EATA, Lima, Ohio
Fujimasu Synthetic Chemical Laboratory Company, Ltd., Japan
Kurita Water Industries, Ltd., Japan
Lopat, Wanamassa, New Jersey
Nippon Synthetic Chemical Industry, Japan
ProTek (enviroGuard), Houston, Texas
SolidTek, Inc., Atlanta, Georgia
Hitachi, Japan
Japan Organo Co., Japan
PD Pollution Control, England
CGE-SARP, France
Trezek Group, Caltrans, California
California Cast Metals Assoc./Jim Furness

This list does not include uses of soluble silicates in the nuclear waste field. These will be described briefly in Chapter 15. In addition, there are a number of vendors and others who offer or describe various products and processes for CFS work that use soluble silicates.

The Chemfix® Process

This soluble silicate process was developed by the author in the late 1960s, and first put into commercial operation in 1970. It is currently owned by Chemfix Technologies, Inc., who purchased the technology and other assets of the original Chemfix, Inc. The nature of the chemistry was maintained as a trade secret until 1974, when the first patents were issued [17,18], followed by additional patents in 1977 [19,20] and a reissue of the basic Chemfix® patent in 1986 [21]. Since then, a number of vendors have marketed CFS processes that profess to use soluble silicates in various ways and forms, and with setting agents other than those claimed in the Chemfix® patents. This has created a great deal of controversy, and various claims and counterclaims in the legal realm, which subject is not appropriate for discussion here. However, the potential user or vendor of CFS processes using soluble silicates is advised to explore the legal as well as the technical aspects of any such system before proceeding.

The Chemfix® process was first demonstrated in 1970 and first applied on a full-scale, commercial CFS job in 1971. To date this patented process has been used to treat approximately 500 million gallons of industrial and municipal liquid wastes. Nearly all of the industrial treatment work has been performed using mobile treatment units at on-site lagoons at or near the point of waste generation. The solid product of Chemfix® treatment is obtained with relatively small amounts of sodium silicate and cement. In general the volume added to the waste by the treatment chemical is less than 10 percent by volume, or about 20 percent by weight.

A typical operation uses a mobile or transportable treatment unit, which contains the chemical storage, metering, and mixing equipment to operate the process at flow rates of 300 to 500 gal/min. The process is continuous and occurs at ambient temperature and pressure. Process control is maintained by automatic or semiautomatic equipment that meters the required ratios of chemical reactants into the waste as it flows through the unit. More detail on the physical and mechanical aspects of this type of operation is given in Chapter 16.

The process features the following important characteristics:

1. Relative low cost
2. Controlled rate of solidification
3. High continuous throughput rate
4. Mobility
5. Small volume increase due to chemical additives
6. Ability to react with complex waste mixtures
7. Ability to process low solids wastes without water discharge from the process
8. Controlled range of physical and mechanical properties for use as landfill
9. Nontoxicity of the solid material

The process has also been used in stationary plants for applications where such a mode of operation is indicated or desired by the waste producer. The Chemfix® process has been used on waste residues from the automotive, chemical, petrochemical, and refining, electronics, electric utility, metal finishing, and primary metals industries, and has application in the mining, food processing, municipal sewage, pulp and paper, sanitary landfill, and dredged areas. While there is no such thing as a universal waste treatment process, this soluble silicate process works on an unusually large variety of materials and is especially useful in handling mixed wastes with low solids content.

The Chemfix® process has only recently been applied commercially to the treatment of municipal sewage sludge. While not classified as a hazardous waste, sewage sludge poses a number of potential environmental threats: toxic metals, odor, pathogens and nutrient runoff, as well as the physical problems in handling and disposal. With the possible exception of certain pathogenic effects, treatment can solve all of these problems at costs competitive with land spreading and lower than incineration. Table 11-6 gives leaching results for a typical aged sludge from a large municipal lagoon, using a column leaching test. While this leachate is not of drinking water quality (esthetics aside), it constitutes an improvement over the raw sludge. As usual, the metal problem is completely resolved; COD, chloride, nitrogen forms, phosphate, and sulfate levels are reduced to acceptable levels for general landfill.

In general, this type of process is not suited to batch processing. The set time is usually quite short—of the order of 10 to 60 seconds. It is not an independent variable, but a function of the optimum formulation, which in turn is a function of waste composition and desired properties of the product, as well as temperature.

Leaching properties of this process, and other soluble silicate processes, were given in Table 11-1, as well as on a metal-by-metal basis in Chapter 4. However, many early column leaching tests were run on solids produced by the Chemfix®

TABLE 11-6 Chemfix®-Treated Sewage Sludge

		Inches of Leachate Water[a]			
Constituent	Raw Sludge	0–25 in.	25–50 in.	50–75 in.	75–100 in.
Biological oxygen demand (BOD)	5500	430	<100	<50	<25
Cadmium (Cd)	9.2	<0.10	<0.10	<0.10	<0.10
Calcium (Ca)	>400	11	34	35	32
Chemical oxygen demand (COD)	>200,000	2,500	600	1,300	900
Chloride (Cl)	>75,000	200	100	50	25
Total chromium (Cr)	210	<0.10	<0.10	<0.10	<0.10
Copper (Cu)	190	7.9	1.5	<0.50	<0.25
Cyanide (CN) and related	10	<0.50	<0.10	<0.10	<0.10
Iron (Fe)	920	<0.10	<0.10	<0.10	<0.10
Lead (Pb)	14	<0.10	<0.10	<0.10	<0.10
Mercury (Hg)	<0.10	<0.10	<0.10	<0.10	<0.10
Nickel (Ni)	10	1.5	<0.50	<0.10	<0.10
Nitrogen (NH$_3$)	1,500	2.9	2.5	3.5	4.5
Nitrogen (NO$_2$)	1.0	<0.10	<0.10	<0.10	<0.10
Nitrogen (NO$_3$)	300	5.0	4.0	<0.50	<0.25
Phenol (ϕ) and related	3.5	<0.50	<0.10	<0.10	<0.10
Phosphate (PO$_4$)	7,000	16	6	4.5	2.0
Potassium (K)	>100	43	46	16	7.5
Sulfate (SO$_4$)	10	180	40	25	20
Total organic carbon (TOC)	17,000	330	<100	<25	<25
Zinc (Zn)	410	<0.10	<0.10	<0.10	<0.10

Note: All results in ppm.
[a] Each 25 inches of leachate represents approximately 800cm^3 of distilled water.

process before the current leaching tests were developed and promulgated as standards. Tables 11-7–11-9 give some of these data for general interest purposes.

Fujimasu Process (Fujibeton)

Fujimasu is known to have several solidification processes, one using various combinations of lime, Portland cement, and waste siliceous materials, while the other, which is of interest here, uses sodium silicate, Portland cement, and a surface active agent. Fujimasu's interest seems to be in selling the chemicals to processors who wish to solidify sludges or other wastes. Typically, Fujimasu recommends additions of about 5 percent of a sodium silicate–surfactant mixture and 10 percent Portland cement by weight to the waste being treated. Presumably, the process could be operated in any of the modes in which other PCSS processes are operated, but Fujimasu has apparently not attempted to market the process or chemicals anywhere but in Japan. However, it is known that at least one company, Toa-Kohatsu Company Ltd., uses the Fujibeton solidification chemicals for treatment of the surface layer of dredged sludge in Japan.

Lopat Process

Lopat has been actively marketing a process using potassium silicate and setting agents. It has been used in California to reduce lead leaching from automobile shredder waste, a material that is normally not classified as hazardous elsewhere, but frequently will not pass the STLC leaching limits with the California WET

TABLE 11-7 Chemfix®-Treated Dredge Mud

		Sludge with Supernatant		Sludge without Supernatant	
Constituent	Raw Sludge	CF Test[a]	Japan Test	CF Test[a]	Japan Test
Cadmium (Cd)	<0.10	<0.10	<0.10	<0.10	<0.10
Total chromium (Cr)	42	<0.10	<0.10	<0.10	<0.10
Iron (Fe)	170	<0.10	<0.10	<0.10	<0.10
Nickel (Ni)	6.0	<0.10	<0.10	<0.10	<0.10
Zinc (Zn)	55	<0.10	<0.10	<0.10	<0.10
Copper (Cu)	<0.10	<0.10	<0.25	<0.10	<0.10
Lead (Pb)	13	<0.10	<0.10	<0.10	<0.10
Mercury (Hg)	<0.10	<0.10	<0.10	<0.10	<0.10
Arsenic (As)	11	[b]	<0.10	<0.10	[b]
Phenol	<0.25	<0.10	<0.10	<0.10	<0.10
Cyanide (CN⁻)	<0.10	<0.10	<0.10	<0.10	<0.10
COD	37,500	1800	[b]	[b]	[b]
Total inorganic phosphate (PO$_4$)	580	2.2	[b]	3.0	[b]

Note: All results in ppm.
[a] The Chemfix® leachate test used 25 in. of distilled water or 800 cm^3 through the Chemfixed sludge.
[b] Not analyzed.

leaching procedure. It has been used by several vendors and processors, such as the Trezek Group. Little technical information has been published, although claims are made for the immobilization of lead, dioxins, and PCBs [22]. Results of several tests provided by the company were given in Table 11-1.

SoliRoc

This process, discussed earlier in this book, is not really a soluble silicate process per se [23], but it is mentioned here because it inevitably comes up in any discussion

TABLE 11-8 Chemfix®-Treated Chemical Waste

	Raw Sludge (2/15/73)	Inches of Leachate Water[a]			
Constituent		0–25 in.	25–50 in.	50–75 in.	75–100 in.
Aluminum (Al)	<1.0	<0.10	<0.10	<0.10	<0.10
Iron (Fe)	160	<0.25	<0.10	<0.10	<0.10
Cadmium (Cd)	0.3	<0.10	<0.10	<0.10	<0.10
Nickel (Ni)	2.2	<0.25	<0.10	<0.10	<0.10
Arsenic (As)	280	<1.0	<0.25	<0.25	<0.10
Total chromium (Cr)	0.8	<0.10	<0.10	<0.10	<0.10
Copper (Cu)	11	<0.10	<0.10	<0.10	<0.10
Zinc (Zn)	1.1	<0.10	<0.10	<0.10	<0.10
Lead (Pb)	[b]	<0.10	<0.10	<0.10	<0.10
Cyanide (CN⁻)	<0.10	<0.10	<0.10	<0.10	<0.10
Phenol	<0.10	<0.10	<0.10	<0.10	<0.10

Note: All results in ppm.
[a] Each 25 in. of leachate represents approximately 800 cm^3 of distilled water.
[b] Not analyzed.
Source: Lab Leachate of 3/2/73 Field Chemfix® Product.

TABLE 11-9 Chemfix®-Treated Electronics Waste

Constituent	Company Analysis (10/19/71)	Chemfix Analysis (1/17/72)	0–25 in.	25–50 in.	50–75 in.	75–100 in.
Total chromium (Cr)	100–1,000	134	<0.10	<0.10	<0.05	<0.05
Iron (Fe)	5,000–50,000	106	<0.10	<0.10	<0.025	<0.025
Zinc (Zn)	100–1,000	137	<0.10	<0.05	<0.025	<0.025
Nickel (Ni)	5–100 10–100	32.3	<0.10	<0.10	<0.10	<0.10
Cadmium (Cd)	500–5,000	19.6	<0.10	<0.10	<0.10	<0.10
Magnesium (Mg)	5–100	232	<0.20	<0.10	<0.10	<0.10
Copper (Cu)	1,000–10,000	5.1	<0.10	<0.10	<0.05	<0.05
Aluminium (Al)	5–50	[b]	<0.10	<0.10	<0.10	<0.10
Lead (Pb)	500–5,000	[b]	<0.10	<0.10	<0.10	<0.10
Manganese (Mn)	[b]	[b]	<0.10	<0.10	<0.10	<0.10
Cyanide (CN⁻)	[b]	[b]	<0.01	<0.01	<0.01	<0.10
Phenol		[b]	<0.10	<0.10	<0.10	<0.10

Inches of Leachate Water[a]

Note: All results in ppm.
[a] Each 25 in. of leachate represents approximately 800 cm³ of distilled water.
[b] Not analyzed.

Source: Lab Leachate of 3/22/72 Field Chemfix Product.

of silicate-based processes. The reader is referred to Chapter 9 for further discussion.

Environmental Technology Associates (ETA)

An investment organization, ETA owns or has license rights for several patents in the soluble silicate CFS area [24,25]. One of these is interesting because it uses a combination of liquid and dry sodium silicates to solve the problem of too-rapid setting of PCSS systems. In a batch-type process, which is usually the preferred approach for small quantities of wastes, the liquid silicate system sets too rapidly. The ETA process uses small amounts of silicate solution to thicken (but not gel to a nonflowable state) the waste so that low solids wastes will not separate before they can set, along with a powdered sodium silicate and Portland cement, which react more slowly to set and harden the mixture without the use of large amounts of reagents.

SolidTek Process

SolidTek operates a fixed CFS site near Atlanta, Georgia, and also sells formulated CFS chemicals for waste generators use under the trade name Fix-Sorb®. The company also has a patented process [25] that uses cement, a dry water-absorbent material such as clay, and a powdered soluble silicate. The clay acts to reduce the free water content of the waste, thickening the mixture and preventing phase separation until the cement and soluble silicate can react to set and harden it.

enviroGuard/ProTek/ProFix™

This company sells a number of sorbents and formulated chemicals under the name enviroGuard (sorbent), enviroGuard Plus™ (CFS agent), and ProFix™. All of these products are based on rice hull ash, an amorphous, biogenetic silica [26]. Because of its sorptive and alkali-reactive nature, rice hull ash has some unusual properties. Its sorptive nature will be discussed alongside other sorbents in Chapter 15, but its ability to react with alkalies to form soluble silicates is of primary interest here. Under alkaline conditions, the amorphous silica reacts slowly to produce soluble silicates that can then react with toxic metals ions to form low-solubility metal silicates. At the same time, the soluble silicate can react with available calcium or other polyvalent metal ions to set and harden the system in a controlled manner. The advantage of this method over that of most soluble silicate processes is that the slow, continuous generation of soluble silicate provides a reserve capacity analogous to the action of buffers in a pH-control system. As metal hydroxides and other species slowly dissolve in the alkaline environment of the waste form, they can then become respeciated as the "silicate." The process has patents applied for in the United States [27] and elsewhere.

Other Soluble Silicate Processes

In addition to the preceding companies that actually offer systems, there are many patents and other examples of technical literature that describe the use of soluble silicate processes for soil solidification, grouting, and various compositions of matter for commercial use. These technological areas will not be discussed in detail here, since it would be impossible to do so in view of the volume of literature. However, it should be realized that from this area new CFS companies could arise at any time. Most of the commercial soluble silicate processes use sodium silicate and Portland cement as the chemical system. Some consider it a Portland cement system with sodium silicate as an additive, while others view it as a soluble silicate system that uses Portland cement as one of a number of possible setting agents for the soluble silicate. If we look at these processes as being based on soluble silicates along with a setting agent, then we find that a variety of setting agents have been enumerated in the technical and patent literature. Some of the setting agents used are:

Glycolides
Glyoxal
Polyalcohol esters
Boric–phosphoric acid condensation products
Sequestered metal ion complexes
Succinic acid diesters; e.g., dimethyl succinate
Methyl acetate, methyl propionate, methyl formate mixtures
Formaldehyde or paraformaldehyde
Diacetin and triacetin
Formamide and ethyl acetate
Phosphates
Amides
Glycerin–glacial acetic acid reaction products
Chlorides, sulfates, and nitrates of aluminum, magnesium, and iron
Soluble polyvalent metal compounds in general
Silicon polyphosphate
Potassium silicofluoride and sodium silicofluoride acids
Mono-, di-, and triacetate acid esters of glycerol

There is an almost infinite variety of reagents that react with sodium silicate to produce gels or solids in addition to the specific setting agents just mentioned. The known chemistry of most such systems is discussed in detail in Vail [10] and also in other references. One company that has done a great deal of research in this area is Progil de Miromesnil. Progil has developed the use of various gelling and setting agents for soluble silicates in applications such as soil stabilization, grouting, and manufactured compositions. The two Progil patents relating directly are U.S. 3,493,406, which describes the use of succinic acid diesters as gelling agents for alkali metal silicates in soil stabilization, and U.S. 3,558,506, which describes the use of a mixture of methyl formate, methyl acetate, and methyl propionate as a gelling additive for the same use. Neither of these processes is known to have been offered commercially outside of France for CFS work, although there was one report of their commercial availability for this purpose in France. Progil's patents do not specifically suggest the use of these gelling mixtures for CFS processing but, like all other gelling or setting agents for soluble silicates, they have that potential use.

CASE STUDIES, COMMERCIAL PROJECTS

Cost Effectiveness

Soluble silicates have been widely used in CFS work because of several important properties: their ability to fix metals, to quickly gel low solids wastes, and to prevent inhibition of cement-based systems. The gelation function of the PCSS system usually makes it the process of choice for low-solids wastes. This was shown graphically in Figure 9-19, where the total cost (solidification plus landfill) is plotted against percent of solids for three processes: PCSS, PC alone (plus minor additives), and PC plus flyash. The cost assumptions used for this comparison were given in Chapter 9. At 5 percent solids, the PCSS process costs about half that of PC/flyash and about 60 percent that of PC alone. At 20 percent solids, the numbers are 63 and 87 percent, respectively. At high solids (40–50 percent), the cost of the PCSS and PC processes are nearly identical, and both are about 75 percent that of PC/flyash. The explanation for this, of course, is that the gelling ability of SS allows the use of much smaller quantities of the CFS chemicals, even though they have a higher unit cost. For example, the weight increase due to CFS treatment for a 20 percent solids sludge was only 20 percent for PCSS, compared to 33 percent for PC and 98 percent for PC/flyash. As landfill costs continue to escalate, minimizing weight, volume increase will become even more important.

Commercial Projects

Waterways Experiment Station (WES) Evaluation. In 1981, the WES evaluated the soil and groundwater around and beneath four disposal sites where PCSS-treated wastes had been deposited seven to eight years earlier. They found no deterioration of soil or groundwater attritable to the CFS treatment or product. This study was described in more detail in Chapter 9 and Tables 9-20 and 9-21.

Case Studies

Tanked Glycol Waste. This waste originated at a chemical plant in West Virginia. It had been transported routinely to New Orleans, tanked temporarily, and then ocean dumped. When ocean dumping was stopped, 4 million gallons remained in storage tanks. After treatment, it was considered suitable for local, sanitary landfill as a nontoxic solid. The toxicity problem had been caused by 234 ppm antimony in dissolved form. After CFS treatment, antimony leaching from the waste (column leaching test with water) was less than 0.1 ppm, which allowed the solid to be used as daily cover in the sanitary landfill.

One requirement of the project was that all 4 million gallons had to be contained on-site after solidification, until the state could test it. Because only limited space was available, and the groundwater situation precluded digging a deep pit for curing, the job was done in stages, using the solid produced in one stage to build up the impoundment for the next stage. In this manner, a 35-foot-high "hill" was created.

The detailed job parameters were:

Waste analysis: 6.6 percent ethylene glycol
              0.2 percent diethylene glycol
              2.5 percent sodium terephthalate
              8.3 percent sodium chloride
              0.9 percent sodium sulfate
              1.1 percent ammonium chloride
              234 ppm antimony
Volume treated: 4 million gallons
Treatment rate: 130,000 gal/day, average
Time: Spring 1976
Treatment cost: $0.167/gallon
Disposal method: Sanitary landfill, daily cover material
Properties of treated solid:
  Unconfined compressive strength: 4–5 tons/ft^2
  Leachate results: 0.1 mg/l antimony
                     5.0 mg/l sulfate
                   160.0 mg/l chloride
                   350.0 mg/l COD

Hazardous Waste Landfill Leachate. At this secure landfill in Ohio, leachate was stored in ponds on-site. Because the landfill had no water treatment or discharge capability, on-site CFS treatment was found to be the most economical method of treatment at the time. The solid produced was then used as daily cover material in the landfill. The treatment unit with treated waste in the foreground is shown in Figure 9-4. Approximately 3.3 million gallons of leachate were treated over a period of about 15 months. In at least one instance, the CFS mobile service responded within ten days; however, the basic CFS testing (see Chapter 23) had been done earlier and only a verification test was required before startup. Job parameters were:

Waste analysis: 99+ percent water
              10–200 mg/l phenol
              0–10 mg/l metalsalts
              100–30,000 mg/liter COD
  (The waste was considered hazardous by definition, not by test.)
Volume treated: 3.3 million gallons
Treatment rate: 50,000–200,000 gal/day
Time: March 1980–June 1981
Treatment cost: $0.11–0.165/gallon
Disposal method: Secure landfill, on-site
Properties of treated solid:
  Unconfined compressive strength: 0.5–5 tons/ft^2
  Leachate results: <1.2 mg/l phenol
                    <1.0 mg/l metals

Dredging Pilot Project

Dredging operations are performed for maintenance of navigational channels, for land reclamation, and sometimes for pollution control in lakes and seabed areas. An

TABLE 11-10 Leaching from Seabed Mud

	Concentration in Leachate (mg/l)[a]	
Constituent	Raw Mud	Treated Mud
Mercury, total	1.71	0.0002
Cadmium	8.64	<0.001
Lead	516.0	<0.001
Arsenic	3.80	0.003
Cyanide	0.08	<0.001
PCBs	18.8	<0.005

[a] Distilled water leach test; combination of two samples.

experimental project of particular interest to the dredging industry was conducted in February 1974 as part of an undertaking by the Association for Protection of Japanese Aquatic Resources, comprising tests for the commercial application of technologies for control of the red tide problem in the Inland Sea (Japan). The Chemfix® process was chosen as one of the test methods for the purpose of developing a practical, commercial system of seabed mud treatment, so that treated mud could be disposed of on land or used in reclamation projects.

The project was conducted using a pneumatic dredge to minimize contamination of the surrounding water while the dredging operation was underway. Mud accumulation on the sea bottom in the area dredged ranged from several centimeters to several meters in thickness. Solids content of the dredged mud ranged from 14.5 to 24.6 percent that was ideally suited for treatment by a mobile unit. Solidification tests were conducted under a variety of weather conditions including snow, rainfall, and temperatures as low as 35°F (1.7°C). Solidification was not affected by rainfall, and the rain did not cause any breakdown or washing away of this particular material. In general, the solidified material became solid enough to support foot traffic within 24 hours. Samples of solidified material were taken 24 hours after it was poured into the solidifying pond.

Leaching tests were done to determine the environmental properties of the solidified seabed mud, in this case by the standard methods specified in the Japan EPA Notification #13. The results are shown in Table 11-10. Extremely small concentrations of heavy metals were detected in the leachate, indicating that the solidified, treated material is safe for use as landfill. PCB and hexane extract values were also extremely small, indicating that harmful organic substances were entrapped in the solid.

REFERENCES

1. Katsanis, E. P., P. H. Krumrine, and J. S. Falcone, Jr. Chemical reactions in an alkaline flood. *Proc. Am. Chem. Soc.* (Sept. 12-17, 1982).
2. Conner, J. R. *Chemical Fixation and Solidification.* CHEMFIX, Inc. Pittsburgh, PA, 1970.
3. Joosten, H. U.S. Patent 2,081,541 (1937).
4. Karol, R. H. *Chemical Grouting.* New York: Dekker, 1983.
5. Conner, J. R. A critical comparison: Ultimate waste disposal methods. *Plant Eng.* (Oct. 1972).
6. Chemfix, Inc. *Chemfix Data Package I.* Pittsburgh, PA, 1972.

7. Conner, J. R. *The Modern Engineered Approach to Chemical Fixation and Solidification Technology.* Chemfix, Inc, Pittsburgh, PA, 1981.
8. Conner, J. R. Ultimate disposal of liquid wastes by chemical fixation. In *Proc. 29th Annual Purdue Industrial Waste Conference.* Purdue Univ., West Lafayette, IN, 1974.
9. Falcone, J. S., R. W. Spencer, and E. P. Katsanes. *Chemical Interactions of Soluble Silicates in the Management of Hazardous Wastes.* ASTM Special Technical Publication #851, 1983.
10. Vail, J. G. *Soluble Silicates.* New York: Reinhold, 1952.
11. Gowman, L. P. Chemical stability of metal silicates vs. metal hydroxides in ground water conditions. In *Proc. 2nd National Conference on Complete Water Reuse.* Chicago, IL, May, 1975.
12. Iler, R. K. *The Chemistry of Silica.* New York: Wiley, 1979.
13. Falcone, J. S. *Soluble Silicates.* New York: Reinhold, 1982.
14. Cote, P. *Contaminant Leaching from Cement-Based Waste Forms Under Acidic Conditions.* Ph.D. thesis, Hamilton, Ont., Canada, McMaster Univ., 1986.
15. Davis, E. L., J. S. Falcone, S. D. Boyce, and P. H. Krumrine. *Mechanisms for the Fixation of Heavy Metals in Solidified Wastes Using Soluble Silicates.* Lafayette Hill, PA: The PQ Corp., 1987.
16. Spenser, R. W., R. H. Reifsnyder, and J. C. Falcone, Jr. *Applications of Soluble Silicates and Derivative Materials in the Management of Hazardous Wastes.* Lafayette Hill, PA: The PQ Corp., 1986.
17. Conner, J. R. U.S. Patent 3,837,872 (Sept. 24, 1974).
18. Conner, J. R., and R. J. Polosky. U.S. Patent 3,841,102 (Oct. 15, 1974).
19. Conner, J. R., E. A. Zawadzki, and R. J. Polosky. U.S. Patent 4,012,320 (Mar. 15, 1977).
20. U.S. Patent 3,893,656 (July 8, 1975).
21. Conner, J. R. B1 3,837,872 (Feb. 25, 1986).
22. *Environ. Sci. Technol.* **20**(2): 110 (1986).
23. Rousseaux, J. M., and A. B. Craig, Jr. *Stabilization of Heavy Metal Wastes by the Soliroc Process.* Brussels, Belgium: Cemstobel, S.A., 1982.
24. Conner, J. R. U.S. Patent 4,600,514 (July 15, 1986).
25. Conner, J. R. U.S. Patent 4,518,508 (May 21, 1985).
26. Durham, R. L., and C. R. Henderson. U.S. Patent 4,460,292 (July 17, 1984).
27. Conner, J. R. U.S. Patent Applied For (1987).

Chapter 12

LIME/FLYASH AND OTHER LIME-BASED PROCESSES

Lime is a general term usually used to cover the various chemical and physical forms of quick lime, hydrated lime, and hydraulic lime. Typical chemical and physical properties were given in Table 4-7. Combining lime and flyash with water forms a cementitious material. The reaction product formed is initially a noncrystalline gel, but eventually becomes calcium silicate hydrate, a compound found in hydrated Portland cements. One of the problems in lime/flyash processes is that flyash is a by-product of coal-burning power plants and its composition is dependent upon not only the composition of the coal burned, but also how the plant is operated.

For those readers interested in more detail on lime/flyash/sulfate processes and their chemical and physical properties, considerable additional information on flyash and lime/flyash compositions can be found in publications from the flyash associations in different countries, and from the references given at the end of this chapter. Further discussion of flyash itself is also provided in Chapter 14. Not all of this information is directed specifically toward solidification, but it does provide some insight into the chemical properties and reactions, the physical properties attainable with such mixtures, and the variability of flyash sources.

Lime-based processes can be divided into several groups for more meaningful discussion and analysis:

Lime/flyash processes
Lime/clay processes
Lime processes
Lime/kiln dust processes

The latter group will be discussed in Chapter 14.

LIME/FLYASH PROCESSES

Without question, compositions including lime and flyash have comprised the largest volumes of wastes treated in the United States. This is largely due to the use of this technology in the electric power industry to solidify fluegas cleaning (FGC) sludges from fossil fuel-burning power plants. However, lime/flyash processes have also been widely used in other industrial applications at generators' plants, at central treatment facilities, and for remedial action projects [1].

Chemical and Physical Properties

Cementitious reaction products typically are mixtures of gel, semicrystalline, and crystalline structures. Many of the reactions are analogous to those of Portland cement, and so do not need further elaboration here. In general, however, these reactions are slower than those of cement and do not produce exactly the same products in terms of chemical and physical properties. Unconfined compressive strengths are in the range of 20 to 1000 psi, with permeabilities between 10^{-5} and 10^{-8} cm/s.

Because of the nature of flyash, other differences from cement chemistry are introduced. For example, unburned organics reduce the cementing action by covering reactive surfaces and preventing contact of the cementitious materials. Smaller particles are more reactive (due to larger surface area) than large particles. Some flyashes (Type C) have self-hardening characteristics (contain large quantities of calcium oxide), so that the addition of water alone will produce a cementing reaction without the addition of lime. This kind of process is claimed to be usable for solidification. However, it depends strongly on the particular flyash being used, and therefore processes combining lime and flyash, and also additions of sulfate, are often necessary to produce a reliable solidification mixture.

Many of the applications, and most of the volume of lime/flyash CFS technology have another difference: The wastes being treated contain sulfates and sulfites that enter into the pozzolanic reactions in complex ways that are not fully understood. It is known, however, that the expansive mineral, ettringite, is formed in the presence of sulfate [2]. Smith and Larew [3] discuss this in some detail, and further information can be obtained from a number of patents in this area [4–11].

Lime-based and lime/flyash processes are able to accommodate large quantities of organics as well as the more common inorganic sludges. Chestnut et al. [12] describe the solidification of oily wastes and other water-insoluble organic materials at organic levels above 20 percent. In fact, this is a major use for these processes. On the other hand, it has been the author's observation that the lime processes, in general, are not as effective in reducing leachability of metals as the cement-based systems. One reason for this is the very high pH that usually results from lime-based systems; another may be that these pozzolanic processes do not bring metals such as lead and chromium into the silica matrix as effectively as does cement. However, as with any CFS process, each waste and disposal scenario must be evaluated separately, and lime/flyash may provide acceptable leaching results in many cases. Some data to this effect are given in Table 12-1 for lime-based processes. Materials that interfere with cement setting and hardening may do the same in lime/flyash systems, but the effects are not as common. As we saw in Chapter 10, lime is often used as an anti-inhibitor for cement setting. Materials that have been found to inhibit lime/flyash processes include sodium borate, calcium sulfate, potassium bichromate, and carbohydrates.

In view of the extremely widespread use of this technology, it is not possible here to cover all of the applications or results to date. In fact, most have never been published. One reason for this is that the process is so often an integral part of a water or waste treatment system or facility that it is not considered on its own as a CFS process. Another is that much of the volume of lime/flyash treatment has been in "nonhazardous" (or, perhaps more accurately, non-RCRA) waste applications such as FGC sludge, municipal sewage sludge, mine wastes, and road stabilization. It is more instructive to examine the activities of one CFS vendor that pioneered and still is the primary factor in this technology—Envirosafe Services Inc., formerly Conversion Systems, Inc. (CSI), I.U. Conversions Systems, Inc. (IUCS), and G&W.H. Corson, Inc. Before doing that, however, it is important to note that a number of other companies offer lime/flyash CFS treatment. In fact, most of the larger CFS vendors, hazardous waste management companies, and engineering and construction firms use the technology at one time or another.

Commercial Processes

The following companies are known to offer, or have offered, lime/flyash processes:

Conversion Systems, Inc., Horsham, Pennsylvania
Envirosafe Services, Inc., Valley Forge, Pennsylvania
Research-Cottrell, Bound Brook, New Jersey
Willis and Paul Corporation, Denville, New Jersey
Chem-Met Services, Inc., Detroit, Michigan
Sludgemaster, Santa Barbara, California.

Envirosafe Services, Inc. (ESI) (Poz-O-Tec). ESI and/or its predecessors CSI and IUCS has been, and probably still is, the most prominent CFS company in the United States in terms of total volume treated. The patent situation with the Poz-O-Tec process is somewhat complex. A number of patents have been issued to ESI and its predecessors. The first patent issued, U.S. 3,720,609 [4], appears to be the basic Poz-O-Tec patent. This patent describes the combination of sludge with materials producing aluminum ion, lime-, and/or sulfate-bearing compounds, wherein the preferred source of aluminum ions is flyash. In other words, Poz-O-Tec is a basic lime/flyash process that also requires the presence of sulfates that are stated to aid in the hardening and strength development reactions. The other patents listed refer primarily to the use of lime/flyash for in situ treatment of roadways, treatment of coal mine refuse, and an improvement of the lime/flyash process caused by mechanical size reduction on the ingredients.

The Poz-O-Tec process is highly commercial and has been used widely for FGC sludges and for certain other wastes. As of April 1978, IUCS had contracts at nine power plants in the United States for the stabilization and disposal of approximately 5.5 million tons per year of scrubber sludge and flyash. These contracts are shown in Table 12-2, and give some idea of the magnitude of lime/flyash usage even before the implementation of the RCRA.

In addition to work on FGC wastes, IUCS did considerable work on stabilized road bases using lime and flyash to produce what was previously called Poz-O-Pac. Also, large-scale field experiments were done on sulfuric acid waste and on battery plant residues [13]. ESI reported that laboratory and pilot plant operations have indicated the successful stabilization of:

TABLE 12-1 Summary of Metal Fixation Data, Lime-Based Processes

Reference[a]	Waste Description	Waste Code	CFS System	Mix Ratio	Treatment Scale
4:98	Metal/cyanide	F006-8	Lime	—	Commercial
4:98	Metal finishing	F006-9, D002-3	Lime/sulfide	—	Commercial
4:98	Metal finishing	F006-9, D002-3	Lime/sulfide	—	Commercial
4:98	Metal finishing	F006-9, D002-3	Lime/sulfide	—	Commercial
4:98	Metal finishing	F006-9, D002-3	Lime/sulfide	—	Commercial
4:92	Organic arsenical sludge		Sulfate, ferrous/ferric ion, lime	1.00	Lab
4:114	Electroplating sludge	F006	Lime, sulfide		Commercial
4:119	Arc furnace dust	K061	Lime/flyash	0.10	Lab
4:136	Flyash A		Unstabilized		Lab
4:136	Flyash A		Lime	0.10	Lab
4:136	Flyash B		Unstabilized		Lab
4:136	Flyash B		Lime	0.10	Lab
4:158	FeCl$_3$ etching waste		Lime	1.5	Lab
4:158	FeCl$_3$ etching waste		Lime	0.7	Lab

[a] References are listed at the end of Chapter 4.

TABLE 12-1 (Continued)

Metal Content (mg/kg or mg/l)

Reference[a]	Chromium Total	Chromium Leachate	Copper Total	Copper Leachate	Lead Total	Lead Leachate	Mercury Total	Mercury Leachate
4:98	11200	0.380			970	<0.100		
4:98	16300	<0.010	330	0.050	375	0.080	<1	<0.002
4:98		<0.050				<0.10		<0.002
4:98	10000	<0.010	432	0.010	42	0.080	<1	<0.002
4:98		<0.050				<0.100		<0.002
4:92								
4:114	16300	<0.010	775	0.030	2800	0.180	<1	<0.002
		<0.050				<0.100		<0.001
4:119	1370	0.073	2240	0.008	20300	0.095	3.8	0.002
4:136		0.140		2.000		3.200		
4:136		0.080		0.480		0.440		
4:136		0.130		37.000		170.000		
4:136		0.260		20.000		350.000		
4:158		0.004						
4:158		0.026						

TABLE 12-1 (*Continued*)

Leaching Test	Metal Content (mg/kg or mg/l)								
	Antimony		Arsenic		Barium		Cadmium	
	Total	Leachate	Total	Leachate	Total	Leachate	Total	Leachate
EPT						0.130	7200	0.260
EPT			<1	<0.500	30	<0.020	20	0.008
TCLP				<0.100		0.180		<0.020
EPT			1	<0.050	<10	<0.020	<5	0.007
TCLP				<0.010		<0.100		<0.020
Stirred water			6850	16.000				
EPT	<10	0.010	5	<0.050	40	0.030	53	0.031
TCLP				0.016		0.280		<0.020
TCLP	294	<0.050	36	<0.010	238	0.464	481	0.052
EPT								0.860
EPT								0.270
EPT								64.000
EPT								26.000
EPT								
EPT								

TABLE 12-1 (*Continued*)

Metal Content (mg/kg or mg/l)							
Nickel		Selenium		Silver		Zinc	
Total	Leachate	Total	Leachate	Total	Leachate	Total	Leachate
1700	0.100	<10	<0.010	<2	0.010	375	0.010
			<0.010		<0.020		
1600	0.060	<10	<0.010	<2	0.090	68	0.010
			<0.010		<0.020		
4700	0.090	<10	<0.010	14	0.060	2100	0.100
			<0.040		<0.020		
243	<0.012	<5.0	<0.010	59	<0.003	244000	0.390
	1.600						36.000
	0.840						7.000
	0.400						8600.000
	0.280						4100.000
	0.024						
	0.030						

TABLE 12-2 Lime/Flyash Power Plant Installations (As of 1978)

Plant	MW	Tons/Year Waste
Columbus & Southern Ohio Electric, Conesville, OH	830	650,000
Duquesne Light Co., Elrama, PA	500	505,000
Duquesne Light Co., Phillips, PA	400	410,000
Indianapolis Power & Light Co., Petersburg, IL	515	550,000
Indianapolis Power & Light Co., Unit #4, IL	515	550,000
Commonwealth Edison, Powerton, MA	450	561,000
Central Illinois Public Service Co., Newton, IL	1,200	800,000
Texas Municipal Power Agency, Gibbons Creek, TX	400	1,050,000
Big Rivers Rural Electric, Reid Cooperative	500	800,000

Aluminum oxide sludges
Iron oxide sludges
Electroplating waste
Calcium sulfate sludge from hydrofluoric acid production
Lead acid battery production sludges
Pickle liquor sludges
Acid mine drainage
Steel mill process sludges

IUCS financed, engineered, designed, constructed, operated, and maintained CFS treatment and disposal facilities under long-term agreements. It also designed, engineered, and constructed facilities for customer operation and maintenance. IUCS spent considerable effort on attempts to get several large sewage treatment jobs in the eastern United States. The Poz-O-Tec process works well with sewage sludges when flyash is available nearby, but is less than economic when the flyash must be transported for considerable distances. IUCS installations at power plants were always accompanied by dewatering devices beyond the clarifier, usually by vacuum drum filtration. This enhanced the marketability of CFS processes, since the smaller volume of sludge to be treated means lower overall costs for the generator, especially in the industrial waste area.

An interesting project was paving the parking lot of the TRANSPO 1972 exhibition at Dulles airport near Washington, D.C., with a five-inch-thick mixture of flyash, waste sulfate sludge, and hydrated lime. The sludge came from neutralized sulfuric acid pickle liquor. The cost of doing this job was equivalent to about $0.03 per gallon of waste treated. After the road base was laid, a layer of liquid asphalt was spread on the surface, but no regular "wearing course" was applied to the base. The results, at least after several years, were reported to be good.

LIME/CLAY PROCESSES

The oldest processes that exemplify the reaction between lime and various silicas and silicates are the sand/lime mortars used for many centuries before Portland cement was invented. Sand/lime mortars are slow setting and develop their strength only very gradually. However, there are examples of such mortars many centuries old that are still in existence today, as are sand/lime stuccos, plasters, and bricks.

Soils engineers, and others in that field, have done some work on the consolidation of loose sand by the reaction of aqueous solutions of calcium hydroxides or other alkalies. However, these cementation reactions are very slow compared with most solidification processes at ambient temperature.

Chemical and Physical Properties

The reaction of lime with clays, which are silicates, causes the slow formation of calcium silicates with bonded properties. There is considerable literature on the transformation of clay minerals by calcium hydroxide reaction [14], since lime has been used as a practical soil stabilizing agent for a long time. The reaction of lime with quartz produces a calcium silicate hydrate gel, or just calcium silicate hydrate, depending upon whether the lime reacted with quartz, with kaolin, or with montmorillonite, and also depending on reaction conditions such as temperature.

Commercial Processes

Only one company appears to have really explored this area in the sense of using this kind of chemistry for solidification work: Ontario Liquid Waste Disposal (OLWD) in Canada. OLWD did treatment work at a central treatment site some years ago in the Province of Ontario, Canada [15]. It is believed that this operation, and the technology, was subsequently sold to one of the waste management companies in Canada, and it is not known if the technology is still being used. The OLWD process included certain pretreatment steps that resulted in their being issued a patent [16]. OLWD has taken advantage of these reaction possibilities to develop an approach to solidification and, in some cases, to fixation of hazardous wastes based on the fact that many waste sludges already contain large amounts of silica or silicates from other sources. In this case, the simple addition of lime can allow cementation of the mass at very low cost. If the waste product does not contain sufficient silica or silicate, it can be added in the form of locally obtained soil or clay which, in the case of most land disposal situations, is available at little cost on the site. Where applicable, this approach works well for central site operations that do have land disposal options, since the bulk of the reactants added to the waste will be dug from the disposal site and refilled afterward so that utilization of the disposal site area is optimized.

This process also includes another step that may or may not be necessary, depending upon the waste. This is the use of acid and ferrous sulfate to reduce or otherwise detoxify certain waste constituents before solidification takes place. For example, hexavalent chromium is reduced under these conditions to the trivalent form that can then be precipitated by neutralization with lime or other alkali as the hydroxide. Other reactions can also take place under these conditions, according to OLWD, although care must be taken when acidifying industrial wastes not to release toxic gases, such as hydrogen cyanide, when large quantities of cyanide are present.

When the waste is subsequently neutralized with lime, the process is known as the *ferro-lime* process and has been used for many years in waste treatment. It should be noted that this is not necessarily an essential step to the solidification reactions, although in some cases with dilute solutions, the precipitated ferrous and ferric hydroxides formed from the iron salt add to the solids content of the waste and aid solidification in this respect. Actually the first step of addition of sulfuric acid and ferrous sulfate can be considered as a pretreatment step in a solidification process.

The OLWD process has been used at the central treatment site run by OLWD in the Province of Ontario, Canada, and also on-site. It is a highly customized system that must be engineered for each individual treatment job. It is thus not a "packaged" process or equipment system in the same sense that many of the other CFS processes are. Operations at OLWD's central site were assessed technically by the Ontario Ministry of the Environment in 1976 [17]. Its conclusion was that the process appeared to hold and stabilize most heavy metals contained in the waste to a level of below one part per million in the leachate. As is usual with such processes, the leachate does contain fairly high concentrations of nontoxic dissolved solids.

A typical application of the OLWD process to on-site treatment of industrial wastes is one involving ferric hydroxide sludges in a brine solution. Analysis of the process in this case indicated that proper solidification could be conducted using a mixture composed of 1 cubic yard of sludge, 2 cubic yards of soil (clay), and 100 pounds of lime as CaO. This mixture required as much as two months curing time to achieve the desired properties, and cost in the neighborhood of $0.06 per gallon, if done by OLWD.

Another application of this approach was for the solidification of tailings from oil tar sand mining [18]. This makes use of the fact that the basic chemical reagents already exist in the waste, and that it is necessary only to add sufficient quantities of chemicals to trigger reactions that will then lead to solidification. In this case, it was necessary to achieve a process that had a cost of less than $0.01 per gallon because of the very large quantities of wastes being generated. This was reportedly accomplished, using small quantities of acid or of alkaline earth-metal-containing compounds. The mechanisms for this process are not clearly known, although it is stated in the patent that the small amount of sulfuric acid reacts with kaolinite to produce soluble aluminum compounds that then react to form a cementitious product, calcium aluminosilicate, that is similar to Portland cement. It is questionable whether this explanation would elucidate the action of the alkaline magnesium and calcium compounds that can also be used.

LIME-BASED PROCESSES: GENERAL

Lime is used by literally hundreds, perhaps thousands of waste generators in the United States and elsewhere, for CFS treatment in the broad sense. Originally used for neutralization of acidic waste water and to precipitate metals before discharge, many generators discovered that they could meet the leaching test standards simply by using excess lime. In most cases, this becomes a separate step in the process, because the addition of sufficient lime for fixation raises the pH of the effluent above that acceptable for discharge or recycle. However, the lime can be added before the filtration step in the sludge conditioning tank, or after filtration to the filter cake. In the latter case, a separate mixing step is required.

Most central waste treatment facilities use lime as a fixation agent, along with other additives as required. Lime is widely used in the "stabilization" of sewage sludge [19], primarily for odor control and pathogen reduction. Some specific CFS applications of lime are listed in Table 12-3.

TABLE 12-3   Lime Applications in CFS

Process	Waste	Reference
Lime	Steel pickle liquor	[20]
Lime	Ferric chloride etching waste	[21]
Lime + hydrophobizing agent	Oily waste	[22]
Lime + special reagent	Hydrocarbon waste	[23]
Lime	Incinerator ash	[24]
Lime + gypsum	Petroleum sludge	[25]
Lime	Phosphoric acid residue	[26]
Lime or lime/flyash	Various	[27]

Chemical and Physical Properties

Leaching properties of lime-based CFS systems were given in Table 12-1. Lime operates, as we have seen before, by controlling the pH of the waste form so that the metals are kept in a range where they are least soluble. When several metals are present in significant quantities, pH control must be a compromise. Also, the pH of interest is not that of the waste form, but of the leaching test solution after the test is complete. For these reasons, adjustment and control of pH with lime alone is difficult, and is a primary reason for the addition of other reagents, such as cement, flyash, and carbonate ion.

Commercial Processes

For the reasons just discussed, it is not feasible to name individual processes or their properties. If the process uses lime only, there is little difference between one and another. Some, however, include additives such as hydrophobing agents, surfactants, or silicates to improve properties, limit permeability, control pH, or provide better mixing with the waste. The main "additive," however, is flyash, which we have already discussed.

REFERENCES

1. Pojasek, R. B. *Toxic and Hazardous Waste Disposal,* Vol. 1. Ann Arbor, MI: Ann Arbor Science Publishers, 1979.
2. Cote, P. L. *Contaminant Leaching from Cement-Based Waste Forms Under Acidic Conditions.* Ph.D. thesis, McMaster Univ. Hamilton, Ont., Canada, 1986.
3. Smith, L. M., and H. G. Larew. *User's Manual for Sulfate Waste in Road Construction,* Report N. FHWA-DR-76-11. Springfield, VA: NTIS, 1975.
4. Smith, C. L., and W. C. Webster. U.S. Patent 3,720,609 and Reissue 29,783 (Mar. 13, 1973 and Sept. 26, 1978).
5. Minnick, L. J. U.S. Patent 3,753,620 (Aug. 21, 1973).
6. Minnick, L. J. U.S. Patent 3,854,968 (Dec. 17, 1974).
7. Minnick, L. J. U.S. Patent 3,870,535 (Mar. 11, 1975).
8. Webster, W. C., and C. L. Smith. U.S. Patent 4,018,619 (Apr. 19, 1977).
9. Webster, W. C., R. G. Hilton, and R. F. Cotts. U.S. Patent 4,028,130 (June 7, 1977).
10. Smith, R. H. U.S. Patent 4,342,732 (Aug. 3, 1982).
11. Minnick, L. J. U.S. Patent 4,397,742 (Aug. 9, 1983).
12. Chestnut, R., J. J. Colussi, D. J. Frost, W. E. Keen, Jr., and M. C. Raduta. U.S. Patent 4,514,307 (Apr. 30, 1985).

13. IU Conversion Systems, Inc. *The World of IU Conversion Systems.* Philadelphia, 1978.
14. Diamond, S., *et al.* Transformation of clay minerals by calcium hydroxide attack. In *Proc. 12th National Conference on Clay and Clay Minerals,* Lafayette, IN, pp. 359-379, 1963.
15. Ontario Ministry of the Environment. *An Assessment of a Process for the Solidification and Stabilization of Liquid Industrial Wastes.* Ont., Canada, 1977.
16. Krofchak, D. U.S. Patent 4,229,295 (Oct. 21, 1980).
17. Ontario Ministry of the Environment. *An Assessment of a Process for the Solidification and Stabilization of Liquid Industrial Wastes.* Ont., Canada, 1976.
18. Krofchak, D. Canadian Patent 1,003,777 (Jan. 18, 1977).
19. Roediger, H. Using quicklime. *Oper. Forum:* 18–21 (Apr. 1987).
20. Sandesara, M. D. *Toxic Hazardous Waste Disposal* **4:** 127-133 (1980).
21. Oberkrom, S. L., and T. R. Marrero. Detoxification process for a ferric chloride etching waste. *Hazardous Waste Hazardous Mtl.* **2**(1): 107-112 (1985).
22. Bolsing, F. U.S. Patent 4,018,679 (Apr. 19, 1977).
23. SRS fixation process handles up to 80% hydrocarbons. *HazTech News* 3(13): 95 (1988).
24. Japan Patent 63 35,469 (Feb. 16, 1988).
25. Nippon Mining Co. Japan Kokai 80 116, 497 (Sept. 8, 1980).
26. Schroeder, K., R. Gradl, K. P. Ehlers, and W. Scheibitz. Ger. Offen. 2,909,572 (Sept. 25, 1980).
27. DuPont, A. *Lime Treatment of Liquid Waste Containing Heavy Metals, Radionuclides and Organics,* Topic No. 43. Arlington, VA: NationalLime Assoc., 1986.

Chapter 13

PORTLAND CEMENT/FLYASH PROCESSES

Portland cement and flyash have been used in flyash concrete for many decades [1,2], and a whole area of technology has grown up around this usage. In fact, the most extensive nonlandfill utilization of flyash is in flyash concrete. Flyash is also used directly in the making of flyash cement by mixing with the other raw materials before forming the cement clinker [3] (see also Chapter 10 for a description of the different cement grades). Flyash imparts many desirable properties to concretes for certain usages, the most important of which is significant economy because it replaces 25-35 percent of the Portland cement normally used. Because of such wide usage and the fact that flyash is a waste in itself, it was fairly obvious to use cement/flyash mixtures as a solidification system in the early days of solidification, especially in the nuclear waste industry. One of the early usages of cement and flyash is described in 1970 in an article about solidification work on nuclear waste at Oak Ridge [1]. Flyash and Portland cement were used together in this application for a grouting formula that was produced from nuclear waste and subsequently pumped underground into fractured shale zones where it solidified and became immobile. This work was started in the 1960s and used cement/flyash mixtures, both with and without additives such as clay to prevent settling, sugar to delay setting time and reduce viscosity, and tributyl phosphate to reduce foaming. The additive ratio was about 0.6 part of a 1:1 mixture of Portland cement and flyash to one part of waste by weight; it was also necessary to add 1.5 pounds of clay per gallon of waste to prevent settling in this type of mixture.

A leaner mixture of cement to flyash would require a higher total ratio of the additives to waste. Our laboratory experience has shown that in relatively low solids wastes, that is, 5-15 percent solids, additive ratios of one to two parts Portland cement/flyash per part of waste by weight were required to prevent settling and achieve good solidification, using cement to flyash ratios of one part of cement to three or four parts of flyash. In high solids wastes, where settling is not a problem, cost savings can often be achieved by substituting flyash for part of the Portland cement in the mixture. However, this results in larger volume and weight increase than with Portland cement alone, and thus is usually only justified where low handling, transportation, and disposal costs are encountered.

CHEMICAL AND PHYSICAL PROPERTIES

Flyash (Coal Ash)

Flyash in a Portland cement/flyash process acts both as a bulking agent and as a pozzolan. From a chemist's viewpoint, flyash is composed primarily of silica, alumina, iron oxide, and calcium oxide. Physically, flyash consists of finely divided spheroids of siliceous glass, about 1 to 50 microns in diameter, plus minor fractions of larger, irregularly shaped particles. Compositions of the two main grades of flyash according to ASTM C 618 are given in Table 13-1. Actual flyash compositions vary greatly from these specifications. Many flyashes themselves leach substantial amounts of metals, and have been the subject of a number of investigations in this respect (Western Ash Co., private communication, April 29, 1987) [4]. Table 13-2 shows the results of leaching tests on several flyashes.

While most flyashes do not harden significantly by themselves (there are self-hardening flyashes that contain large amounts of free lime), in the presence of a sufficiently large amount of Portland cement they will exhibit cementing activity [2]. This is believed to be caused by the release of calcium hydroxide by the Portland cement as it crystallizes, which then reacts with the flyash to form calcium aluminum silicates that harden much as does the Portland cement itself.

Chemical Reactions and Properties

In Chapter 12, we touched briefly on the pozzolanic properties of flyash as a component in lime/flyash processes. When mixed with Portland cement, the gypsum and calcium hydroxide (as formed from cement hydration) components of the cement are the activators for pozzolanic reactions of the flyash [5]. The resulting cement paste may have higher strength than Portland cement alone [6]. The flyash also

TABLE 13-1 Chemical and Physical Properties of Flyash

Chemical Composition (%)	Class F	Class C
Silica (SiO_2)[a]	35.0	35.0
Alumina (Al_2O_3)[a]	20.0	20.0
Iron Oxide (Fe_2O_3)[a]	6.0	6.0
Total $SiO_2 + Al_2O_3 + Fe_2O_3$	70.0 min	50.0 min Sulfur
Sulfur Trioxide (SO_3)	5.0 max	5.0 max
Calcium oxide (CaO)[a]	5.0	15.0
Magnesium oxide (MgO)	5.0 max	5.0 max
Moisture content	3.0 max	3.0 max
Loss on ignition	12.0 max	6.0 max
Available alkali as Na_2O	1.5 max	1.5 max
Physical Properties		
Fineness (retained on #325 Sieve, %)	34 max	
Pozzolanic activity index		
with Portland cement @ 28 days (ratio)	75 min	
with lime @ 7 days (psi)	800 min	
Water requirement (% of control)	105 min	
Soundness (autoclave expansion, %)	0.8 max	
Drying shrinkage (increase @ 28 days)	0.03 max	
Specific gravity[a]	2.75	

[a] Typical values.

TABLE 13-2 Leaching of Flyash

Concentration of Constituent (mg/l) in Waste or Leachate

Reagent Name	Test Type	Arsenic	Barium	Cadmium	Chromium	Copper	Lead	Mercury	Nickel	Selenium	Silver
Class C flyash	Total	122.000	348.000	2.970	75.400	352.000	221.000	0.500	52.500	<0.280	9.650
	EPT	<0.010	1.070	<0.010	0.140	0.220	0.350	<0.001	0.020	<0.010	0.020
Class F flyash	Total	9.000	1000.000	<0.500	20.000	28.000	14.000	<0.300	10.000	12.000	<1.000
Class F flyash	WET									0.580	
Class F flyash	Total	55.000	87.000	0.500	8.000	114.000	<5.000	<0.400	4.500	2.700	0.200
Class F flyash	WET	4.8									
Class F flyash	Total	22.100	1410.000	2.440	59.600	100.000	122.000	0.251	25.500	3.540	14.300
Class F flyash	EPT	<0.010	1.360	<0.010	0.400	0.180	0.280	<0.001	0.020	0.120	0.020
F006 BDAT				0.066	5.200		0.510		0.320		0.072
K061 BDAT				0.140	5.200		0.240		0.320		
EPT characteristic level			100.000	1.000	5.000		5.000	0.200		1.000	5.000
Delisting level			6.300	0.063	0.315		0.315	0.013		0.063	0.315
Drinking water standard			1.000	0.010	0.050		0.050	0.002		0.010	0.050

helps to bind additional water, decrease the pore pH, and act as an adsorbent for metal ions. Waste components that interfere with cement setting and hardening will likely also interfere with those functions in the cement/flyash system. The flyash itself may retard setting of cement.

Cote [5] studied the leachate pH history of several CFS systems over a period of about two years, and found that the flyash-containing processes had lower initial pH and the pH decreased faster with time. This shown graphically in Figure 13-1, and is attributed to the pozzolanic reactions that consume lime produced from cement hydration. At the same time, the cement/flyash process produced the lowest leaching of metals with time in the dynamic leaching test, as shown in Table 13-3. Cote attributes this to the moderation of pore water pH by the flyash, as well as to other effects such as adsorption.

Cote has done extensive work on cement/flyash systems [5,7]. In a study using the modified ANS 16.1 test at different renewal frequencies, he found the cumulative leaching rate for arsenic to be as shown in Figure 13-2. In this study, he used

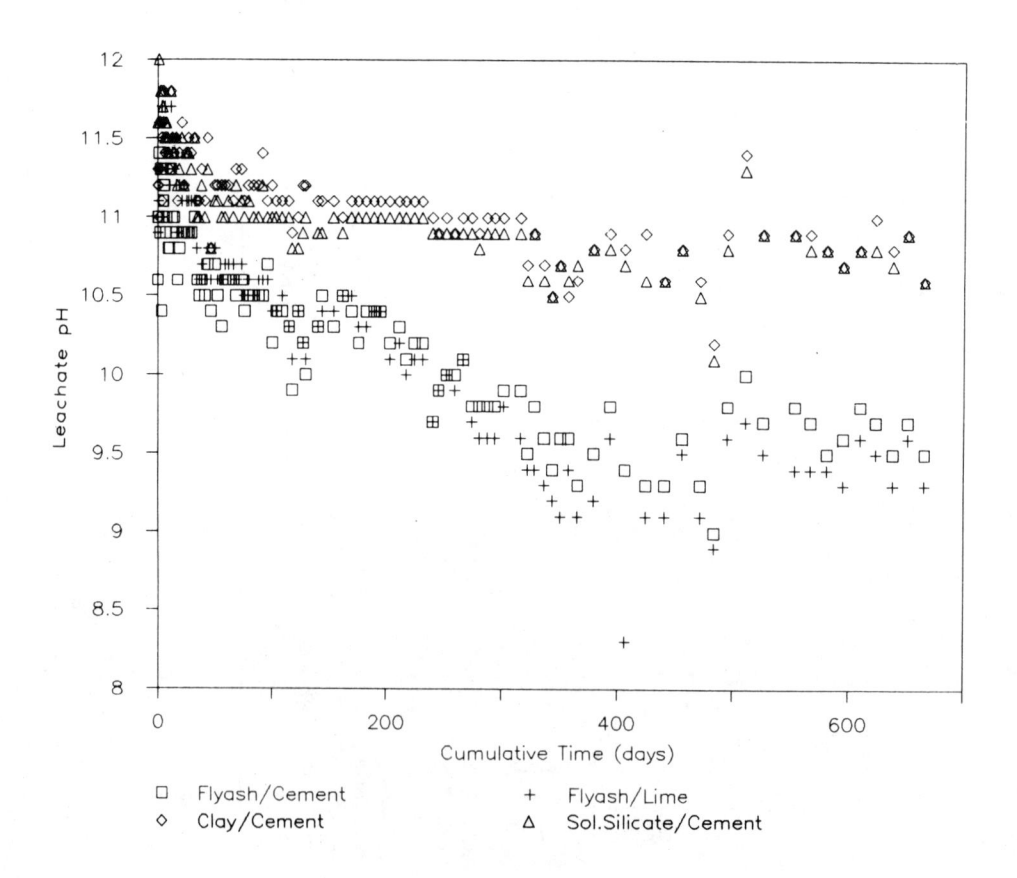

Figure 13-1 Long-term leaching—pH versus time. (From Cote [5].)

TABLE 13-3 Leaching of Cement/Flyash vs. Other Processes

	Cumulative Fraction Leached (%)			
Process	Arsenic	Cadmium	Chromium	Lead
Cement/flyash	3.97	0.02	0.19	0.11
Lime/flyash	4.82	0.03	0.18	0.85
Cement/clay	3.42	0.04	0.11	0.68
Cement/silicate	15.16	0.14	0.19	0.42

Source: Adapted from P. L. Cote, *Contaminant Leaching from Cement-Based Waste Forms Under Acidic Conditions,* Ph.D. Thesis, McMaster Univ., Toronto, Ont., Canada, 1986.

two parts flyash, one part of Portland cement (Type I) to one part synthetic waste, a fairly high reagent ratio but not atypical for low solids wastes. For additional reading on cement/flyash systems, the reader is referred to Cote's thesis [5].

The greatest disadvantage of the cement/flyash process is the volume increase associated with large additions of flyash. The flyash:cement ratio, by weight, is

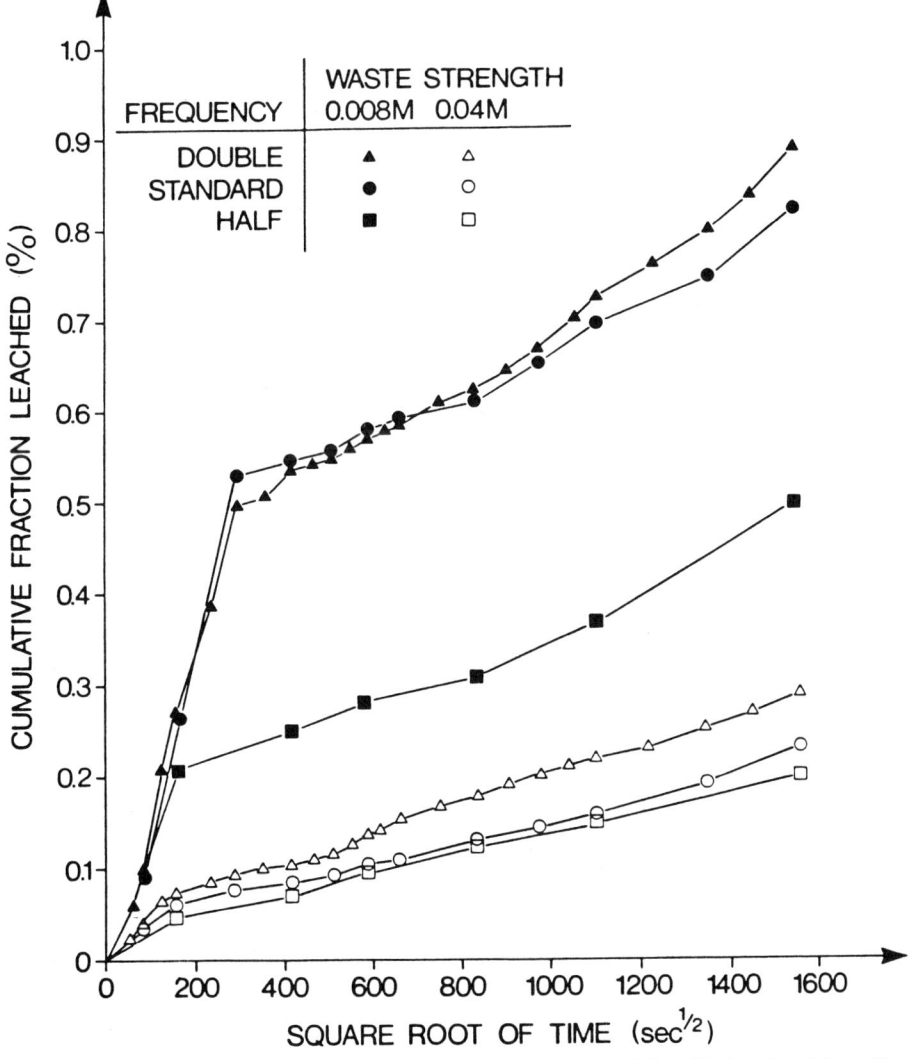

Figure 13-2 Arsenic data following a simple diffusion model for 28 days leaching. (From Cote [5].)

TABLE 13-4 Summary of Metal Fixation Data, Cement/Flyash Processes

						Metal Content (mg/kg or mg/l)					
						Arsenic		Barium		Cadmium	
Reference[a]	Waste Description	CFS System	Treatment Scale	Leaching Test		Total	Leachate	Total	Leachate	Total	Leachate
4:159	Mixed waste	Cement/flyash	Lab	EPT		635	0.022	1730	0.073	79.1	<0.050
4:159	Mixed waste	Cement/flyash	Lab	MEP[b]		635	0.172	1730	0.158	79.1	<0.050
4:160	Nickel waste	Cement/flyash	Lab	ELT, 1 day							
4:160	Nickel waste	Cement/flyash	Lab	ELT, 28 days							
4:160	Chromium waste	Cement/flyash	Lab	ELT, 1 day							
4:160	Chromium waste	Cement/flyash	Lab	ELT, 28 days							
4:160	Copper waste	Cement/flyash	Lab	ELT, 1 day							
4:160	Copper waste	Cement/flyash	Lab	ELT, 28 days							
4:160	Zinc phosphate waste	Cement/flyash	Lab	ELT, 1 day						1.9	<0.001
4:160	Zinc phosphate waste	Cement/flyash	Lab	ELT, 28 days							
4:160	Waste treatment plant sludge	Cement/flyash	Lab	ELT, 1 day		0.8	<0.005			1.9	<0.001
4:160	Waste treatment plant sludge	Cement/flyash	Lab	ELT, 28 days		0.8	0.020				
4:160	Zinc phosphate waste	Cement/flyash	Lab	ASTM,[c] 2 days						1.9	0.001
4:160	Zinc phosphate waste	Cement/flyash	Lab	ASTM,[c] 10 days						1.9	<0.001

[a] References are listed at the end of Chapter 4.
[b] Maximum concentration.
[c] Sequential batch test, CO_2-saturated water.

TABLE 13-4 Summary of Metal Fixation Data, Cement/Flyash Processes (Continued)

Metal Content (mg/kg or mg/l)

| Chromium || Copper || Lead || Mercury || Nickel || Selenium || Silver || Zinc ||
Total	Leachate	Total	Leachate	Total	Leachate	Total	Leachate	Total	Leachate	Total	Leachate	Total	Leachate	Total	Leachate
1100	1.09	1740	0.173	3320	0.003	0.377	0.003	645	0.168	1275	0.096	18.1	<0.050	4300	<0.050
1100	0.08	1740	0.112	3320	0.029	0.377	0.027	645	0.114	1275	0.36	18.1	<0.050	4300	<0.050
		220	<0.002					2900	<0.005						
1750	0.090	220	<0.002					2900	<0.005					100	<0.005
								40	<0.005						
1750	0.005	7500	0.020					40	<0.005	0.36	0.020			100	<0.005
26	0.300	7500	<0.002	23.4	<0.005	2.6	<0.002	1040	<0.005	0.36	<0.001			27625	<0.005
26	<0.002			23.4	<0.005	2.6	<0.002	1040	0.030					27625	0.200
136	0.070			79	0.010									10735	<0.005
136	0.002			79	0.006									10735	0.200
26	0.060			23.4	0.030	2.6	0.003	1040	<0.005					27625	<0.005
26	0.002			23.4	<0.005	2.6	0.002	1040	0.030					27625	0.700

typically two to four, with total weight increases of 50 to 150 percent corresponding to volume increases of 25 to 75 percent. This is especially true with low solids waste streams, where the flyash acts as a bulking agent to increase viscosity and prevent phase separation until the mass sets. Where the increase is not important, as in some remedial action projects, cement/flyash may be the optimum choice. Increasingly, however, at RCRA TSD facilities the value of the landfill space is such that minimal volume increase is a necessity. The economic effects of this consideration were discussed in detail in Chapter 9, and are shown graphically in Figure 11-6.

Leaching data for various cement/flyash processes and studies from the literature are given in Table 13-4. The best and most complete information has been developed by the Stablex Corporation, and is discussed in the next section.

COMMERCIAL PROCESSES

It is known that a number of CFS vendors and central treatment facilities use some version of the cement/flyash process, sometimes with other additives for specific uses [1]. These are listed in Table 9-3, along with other CFS processes and systems. Central treatment facilities often allow for a number of reagents to be used—cement, lime flyash, lime dust, clay, etc.—and so it is difficult to get a good idea of the full extent of use of any of the pozzolanic processes. In addition to Stablex's SealOSafe™ process (discussed below), several others have been described specifically in the literature. One of these, the ORNL nuclear waste process [1], has already been discussed. Ray [8] describes a process for using acid industrial waste to create a soil cement with the addition of cement and flyash. Pichat [9] has patented a process using flyash/lime-containing materials for neutralization, and lime or cement for solidification. The requirement is that the waste have a starting pH not greater than two, which leads one to believe that soluble silicates are formed first by combination of acid and flyash. This step in the process is termed *activation* of the flyash. When neutralized, the system should precipitate metals as silicates and/or hydroxides.

Stablex™ Process

The Stablex Company has done a great deal of technical work on the Portland cement/flyash process and has been involved in a considerable amount of regulatory testing by the U.S. EPA and regulatory agencies in Canada and the United Kingdom [10,11]. Also known as SealOSafe, the process has been used at two locations in England, one in Japan, and is now being used at a central treatment facility in the Canadian Province of Quebec. In 1973, Crossford Pollution/Services Ltd. (predecessor of Stablex) of the United Kingdom introduced this process in England. Crossford described the process as being "a mechanical-chemical polymerization process which uses a polymer believed to outlast the contained wastes." This process is claimed to be suitable for materials containing lead, cadmium, copper, nickel, tin, maganese, zinc, and sulfides, and other materials, such as arsenic, cyanide, chromium, ammonia, and acidic wastes, can be handled with preteatment. In 1974, the cost of the SealOSafe process was stated as being between $7.50 and $15.00 per ton of waste treated, or about $0.04 to $0.08 per gallon—about the same price range as vendors of other processes were charging in the United States.

Stablex has three U.S. patents on the process [12-14]. The basic patent [13]

TABLE 13-5 Summary of Waste Types Received at the Thurrock Plant[a]

Type of Waste	Percent of Waste Received	Type of Waste	Percent of Waste Received
Sulphuric acid	4.7	Neutral sludges	10.4
Hydrochloric acid	4.6	Lime sludge	14.0
Chromic acid	0.6	Other sludges	0.8
Mixed acids/other acids	6.6	Filter cakes	1.1
Aluminum chloride solution	16.2	Paint stripper washings	1.7
Ferric chloride solution	0.9	Ferrous sulphate	1.0
Solid/liquid cyanides	2.7	Others	3.1
Caustic solutions	30.5		

[a] Waste classifications during the period 7/80 to 12/80.

is for a process using a mixture of Portland cement and an aluminum silicate or aluminosilicate, that is, flyash. It also shows a claim for the use of a powdered alkali metal silicate with flyash. The other two patents describe the addition of activated carbon to sorb organic contaminants [11], and an oil sorbent such as calcium carbonate [12].

Stablex has been in business for some time in the waste treatment and disposal field. In 1974, it announced a license to Polymeric Treatments Ltd. in the United Kingdom who planned to build a 50,000 metric ton per year plant using the process to treat metal plating wastes. According to Stablex in 1979 [15], this plant had a current throughput of 200,000 metric tons per year. The original facility was at Brownhills (Birmingham) and another was built from a reconditioned cement plant at West Thurrock. This latter plant was stated to have a throughput of 400,000 metric tons per year, and to handle a variety of wastes, as shown in Table 13-5. Stablex attempted to site several central treatment plants in the United States, but was unsuccessful in this. Finally, a plant was sited in Blainville, Quebec, with a capacity of 70,000 metric tons per year [16] at a cost of about $20 million. In 1977, Stablex announced that it had licensed the SealOSafe process to the Japan Environmental Protection Company. It was also announced that a SealOSafe installation would be done for Sanei Gumi, a Toyota subsidiary. Typical leach test results from the Stablex process are given in Table 13-6 for several leaching tests; additional data were presented in Table 13-4.

TABLE 13-6 Worst-Case Results from West Thurrock Testing Program

Parameter	Maximum Bulk Sample Concentration (mg/kg dry wt)	Corresponding EP Toxicity Concentration (mg/l)	Maximum MEP Concentration	EP Toxicity Standard	Drinking Water Standard
Ag	18.1	<0.05	<0.05	5.0	0.05
As	635	0.022	0.172	5.0	0.05
Ba	1,730	0.073	0.158	100	1.0
Cd	79.1	<0.05	<0.05	1.0	0.01
Cr	1,100	1.09	0.08	5.0	0.05
Cu	1,740	0.173	0.112	—	1.0
Fe	29,800	0.179	0.132	—	0.30
Hg	0.3772	0.0032	0.0271	0.2	0.002
Ni	645	0.168	0.114	—	—
Pb	3,320	0.003	0.029	5.0	0.05
Se	1,275	0.096	0.360	1.0	0.01
Zn	4,300	<0.05	<0.05	—	5.0

REFERENCES

1. Laguna, W. Radioactive waste disposal by hydraulic fracturing. *Ind. Water Eng.:* 32–132 (Oct. 1970).
2. Clendenning, T. G., A. E. Dalrymple, and T. W. Klym. Current technology in the utilization and disposal of coal ash. *EngCon '75,* 1975.
3. Power plant ash disposal a growing problem. *Chem. Eng. News*: 26 (Nov. 6, 1978).
4. Young, H. W., J. C. Young, M. Boybay, and T. Demirel. The effect of stabilization of flyash on the leaching of metals. Presented at the International Symposium on Industrial and Hazardous Solid Wastes, Mar. 7–10, 1983.
5. Cote, P. L. *Contaminant Leaching from Cement-Based Waste Forms Under Acidic Conditions,* Ph.D. thesis, McMaster Univ., Hamilton, Ont., Canada, 1986.
6. Popovics, S. *Concrete Making Materials.* London: Arnold, 1970.
7. Cote, P. L., and D. Isabel. Application of a static leaching test to solidified hazardous waste. Presented at the ASTM International Symposium on Industrial and Hazardous Solid Wastes, Mar. 7–10, 1983.
8. Ray, L. F. U.S. Patent 4,353,749 (Oct. 12, 1982).
9. Pichat, P. U.S. Patent 4,375,986 (Mar. 8, 1983).
10. Taub, I., and K. Roberts. Leach testing of chemically stabilized waste. In *Proc. 3rd Annual Conference on the Treatment and Disposal Industrial Wastewaters and Residues.* (1978).
11. Stablex Corp. *Testing and Approvals,* Information Bull. No. 102. Radnor, PA.
12. Chappell, C. L. U.S. Patent 4,230,568 (Oct. 28, 1980).
13. Chappell, C. L. U.S. Patent 4,274,880 (June 23, 1981).
14. Chappell, C. L. U.S. Patent 4,116,704 (Sept. 26, 1978).
15. Schofield, J. T. *SealOSafe*™. Ann Arbor, MI: Ann Arbor Science Publishers (1979).
16. Quebec site of new facility to treat hazwaste. *World Wastes:* 26 (Aug. 1983).

Chapter 14

KILN DUST AND FLYASH-BASED PROCESSES

Lime kiln dust, cement kiln dust, and various coal flyashes have been used on dozens, perhaps hundreds of solidification projects in the United States, primarily in remedial actions ("Superfund"). They have also been used extensively at central hazardous waste management facilities (see the section titled Commercial Projects below). When so used, they function primarily as absorbents or bulking agents or, in the case of lime kiln dust, as neutralizing agents for acidic wastes. The kiln dusts are especially effective in both roles because of their calcium oxide content. This gives them high alkalinity and the ability to remove free water by the hydration of CaO to Ca(OH)$_2$, an advantage over inert sorbents. They also use up water in their pozzolanic reactions, described earlier in Chapters 10–13. Physical and chemical properties of a number of flyashes and kiln dusts were determined by Collins [1] and are shown in Tables 14-1–14-6. These reagents vary widely in chemistry from source to source, and their sources are uncertain and variable as to both quality and cost. Cement kiln dust has been well characterized, and a good description given of its variability by Haynes and Kramer [2].

In recent years, several suppliers in the United States have brought some order into the marketplace for kiln dusts [3,4]. They supply the reagents from various sources throughout the United States, and can guarantee price and delivery, at least on a short time scale. Flyash has been marketed in this way for many years, and has a much more established technological base due to its use in concrete and in other construction areas. The general properties of the two flyash types were given and discussed in Chapter 13. Much of the usefulness of these reagents is due to their availability and low prices compared to Portland cement, lime, and other manufactured products. In recent years these prices have been rising, and if the trend continues, they will likely be replaced in many applications by more expensive but more efficient reagents. The availability is also decreasing as energy costs and air pollution control requirements are forcing companies to recycle the dusts through the kilns. However, in 1985 between 20 and 24 million tons of lime and cement kiln dusts were generated in the United States, 80 to 90 percent of which was cement kiln dust [1]. It is estimated that there are well in excess of 100 million tons of dust in stockpiles.

TABLE 14-1   Physical Characterization Test Results for Cement Kiln Dusts

		Particle Size Distribution Percent Passing								Other	
Sample Number	Top Size	60 Mesh	200 Mesh	325 Mesh	0.02 mm	0.01 mm	0.006 mm		SG	Blaine Fineness	pH[a]
CD-1	#50	100	99	97.5	87.5	3			2.76	9,090	11.9
CD-2	#20	99.6	96.7	93	11	0.5			2.78	4,890	11.9
CD-3	#50	97.5	83.2	71	2.5	0			2.78	5,870	11.9
CD-4	#50	98.7	84	70	28				2.76	7,190	11.9
CD-5	#14	97.2	71	50	14				2.87	7,550	12.1
CD-6	#50	99.9	84	61	19				2.74	9,980	12.1
CD-7	#50	98.7	86	76.5	35	0.3			2.84	5,120	11.9
CD-8	#50	98.6	72	52	21.5				2.84	4,610	11.9
CD-9	#10	96.5	85.7	77	1.2	0.3			2.85	4,940	12.0
CD-10			100	97.5	76		0		2.79	10,760	12.0
CD-11	#50	99.5	95.2	91.5	55	4.4	2.2		2.76	8,130	11.8
CD-12	#50	99.9	94	89.5	77.5	44	19		2.74	10,370	11.9
CD-13	#10	97.8	79.5	59.5	21.5				2.84	7,110	11.9
CD-14	#50	98.6	80	68	38				2.78	6,810	12.0
CD-15	#50	99	89	82	67	27	19.3		2.69	7,090	11.5
CD-16	#50	99.5	94	88	63				2.77	8,700	12.0
CD-17	#50	98.5	79.3	63	34.5				2.79	10,010	12.1
CD-18	#50	99.5	97.4	96.3	80	3	1.3		2.74	9,120	12.0
CD-19	#50	99.5	88.5	81.2					2.85	5,680	12.1

CD-20	#50	99.5	90	80	3.5		2.75	6,090	12.1
CD-21	#50	97	78	66	4.5	0	2.79	4,460	12.1
CD-22[b]	1/2 in. (99.5%)	32	12						
CD-22[c]	#10	79	57.5	46.5	28	14	2.60	3,740	11.8
	1/2 in. (95%)	11							
CD-23[b]	#10		2				2.78	13,900	11.2
CD-23[c]	#50	76	57	41	29.5	22	2.93	5,630	12.0
CD-24		99.5	85.5	68	86.5	0	2.70	13,060	12.0
CD-25			100	97.5	9	2.5	2.48	6,950	12.0
CD-26	3/4 in.	29.5	17.5	12.5	25	0	2.79	5,520	12.0
CD-27	#50	99.5	86	72	4.7	3	2.86	7,370	12.1
CD-28	#50	99.6	98	95.6			2.83	5,600	12.0
CD-29	#50	99	86	79	80	2.5	2.76	8,780	11.9
CD-30			98.8	97.7					
Calcitic Hyd. Lime	#100		100	98	91	0	2.31		

[a] 15 g in 150-ml distilled, deionized water, stirred 1 min, pH of slurry (pH of water 7.8).
[b] As received.
[c] Ground.

Source: R. J. Collins, DOT/DOE evaluation of kiln dust/flyash technology, presented at the 6th Annual International Conference on the Economic and Environmental Utility of Kiln Dust & Kiln Dust/Fly Ash Technology, Feb. 19–21, 1985.

TABLE 14-2 Physical Characterization Test Results for Lime Kiln Dusts

| Sample Number | Top Size | Particle Size Distribution Percent Passing ||||||| SG | Other ||
		60 Mesh	200 Mesh	325 Mesh	0.02 mm	0.06 mm	0.001 mm		Elaine Fineness	pH[a]
LD-1	#20	99.5	89.3	73	0			2.96	2,920	12.0
LD-2	#50	99.7	86.5	40				2.85	5,180	12.0
LD-3	#10	90.5	78.3	65				2.87	3,650	11.9
LD-4			100	98	49			2.73	10,350	12.0
LD-5	#10	92.3	49.5	4				2.89	1,310	12.0
LD-6	#10	98.9	88.5	69				2.83	2,180	12.0
LD-7	#10	97	79	56				2.93	1,870	12.0
LD-8[b]										
LD-9[b]										
LD-10	#50	98.5	62.5	43	23.3	13	7.6	2.98	2,090	12.0
LD-11	#10	84.8	74	56				3.04	1,750	12.0
LD-12	#10	80.5	57	43	19.5	4.5		2.83	2,630	11.9
	(99.5%)									

[a] 15 gm in 150 ml distilled, deionized water, stirred 1 minute, pH of slurry, (pH of water 7.8).
[b] Testing to be completed.

Source: R. J. Collins, DOT/DOE evaluation of kiln dust/flyash technology, presented at the 6th Annual International Conference on the Economic and Environmental Utility of Kiln Dust & Kiln Dust/Fly Ash Technology, Feb. 19–21, 1985.

TABLE 14-3 Physical Characterization Test Results for Flyashes

Sample Number	Top Size	\multicolumn{7}{c}{Particle Size Distribution Percent Passing}							SG	Other Blaine Fineness	pH[a]
		60 Mesh	200 Mesh	325 Mesh	0.02 mm	0.06 mm	0.001 mm			
FA-1	#10	99	94	85	50	17.7	10.3	2.65	4,170	11.2
FA-2	#10	98.2	92.8	84	43.9	11.2	7	2.74	3,550	11.6
FA-3		100	95	85	66.8	26.3	11	2.41	3,360	10.4
FA-4	#10	99.4	92.8	83	54	27	18	2.31	3,790	11.4
FA-5	#10	99.5	91	81.5	61.2	34	18.1	2.18	2,620	10.0
FA-6	Not tested							2.62		
FA-7	#10	98.9	87.2	78	54	30	17	2.28	2,820	8.7
FA-8	#10	99	89	80	54.5	29.7	17	2.45	2,820	9.8
FA-9		100	96.3	89	62	37	22.3	2.44	3,150	11.0
FA-10	#10	98.8	92	84	7	2	1	2.70	3,900	11.4
FA-11	#10	99.2	90	78.5	53.7	31	17.5	2.52	2,880	11.3
FA-12		100	96.7	94	70			2.48	10,060	11.6
FA-13	#10	98	88.7	76	47.8	25	14	2.45	2,360	4.2
FA-14	#10	97.2	84	72	47	25.8	14	2.30	2,870	5.5
FA-15	#10	99	93.5	83	62.5	42.7	26.8	2.38	3,510	5.0
FA-16	#50	99.8	99	96	82.5	60	35	2.49	5,080	3.5
FA-17	#50	99.9	88	76	50.3	28.8	16.3	2.27	3,160	8.2
FA-18	#50	99.9	93.7	83	42	22	11.8	2.24	3,360	4.3
FA-19	#50	99.9	93	84	48	22	13.2	2.38	3,640	10.0

[a] 15 g in 150-ml distilled, deionized water, stirred 1 min, pH of slurry (pH of water 7.8).

Source: R. J. Collins, DOT/DOE evaluation of kiln dust/flyash technology, presented at the 6th Annual International Conference on the Economic and Environmental Utility of Kiln Dust & Kiln Dust/Fly Ash Technology, Feb. 19–21, 1985.

TABLE 14-4 Chemical Characterization Test Results for Cement Kiln Dusts (Percent)

Sample Number	CaO	Free Lime CaO	SiO$_2$	Al$_2$O$_3$	MgO	Na$_2$O	K$_2$O	Fe$_2$O$_3$	SO$_3$	LOI 1050°C
CD-1	38.3	3.3	13.2	4.61	2.49	0.15	3.96	2.32	6.74	25.3
CD-2	44.5	4.1	17.1	4.84	1.15	0.27	2.91	1.97	3.32	22.8
CD-3	39.1	3.72	15.4	2.93	2.63	0.55	3.52	2.13	8.56	22.8
CD-4	37.2	0.3	12.5	4.18	2.02	0.68	4.70	1.51	6.79	27.3
CD-5	38.0	2.3	15.3	4.25	0.91	0.32	7.30	1.83	7.34	19.6
CD-6	25.8	0.2	9.71	2.21	1.13	1.35	15.3	1.77	17.40	19.5
CD-7	41.9	1.7	16.2	4.11	1.64	0.34	3.22	2.39	4.79	22.9
CD-8	39.4	1.1	17.7	4.07	0.92	1.20	3.90	2.84	3.47	22.7
CD-9	41.6	4.2	20.0	5.76	2.22	0.41	3.76	2.46	6.59	12.7
CD-10	45.9	4.8	11.9	2.92	1.39	0.07	1.54	2.04	6.24	28.2
CD-11	40.8	2.2	13.3	4.85	1.02	0.27	2.90	2.26	6.24	25.6
CD-12	44.4	0.4	12.0	3.13	1.66	0.08	2.86	1.27	3.30	31.8
CD-13	45.2	6.7	16.8	3.88	1.37	0.18	1.78	2.11	3.72	23.2
CD-14	34.6	0.3	15.1	4.24	1.83	0.58	7.05	2.06	8.54	22.9
CD-15	19.4	0.20	22.4	10.0	0.64	1.34	14.1	4.06	10.14	13.2
CD-16	37.4	1.2	15.2	4.75	1.96	0.48	5.03	2.78	6.37	24.0
CD-17	26.8	1.2	13.0	4.50	0.54	1.47	12.4	2.04	16.93	13.5
CD-18	47.6	4.2	9.91	3.08	1.33	0.11	1.08	1.21	2.92	31.6
CD-19	41.1	13.15	15.2	3.92	1.30	0.20	3.39	2.19	13.76	11.7
CD-20	45.5	4.8	14.0	3.39	1.16	0.28	2.50	1.26	2.40	28.4
CD-21	42.9	4.3	14.9	4.62	0.89	0.14	3.16	2.31	5.54	22.2
CD-22	39.6	0.0	17.6	4.42	2.04	0.20	2.60	2.04	3.75	26.6
CD-23[a]	31.4	0.0	11.7	3.18	0.97	0.13	1.65	2.16	8.24	40.4
CD-24	57.1	16.0	9.70	4.18	1.81	0.00	0.22	0.24	2.67	21.1
CD-25	37.9	0.0	8.85	2.98	1.15	0.36	7.94	1.51	11.74	27.5
CD-26[a]	44.2	0.0	11.9	3.24	1.73	0.27	2.92	1.45	2.40	30.2
CD-27	42.5	7.8	14.3	3.34	2.09	0.44	5.21	1.82	3.10	23.8
CD-28	49.7	21.3	13.2	3.24	1.73	0.40	4.03	1.48	3.02	18.7
CD-29	47.5	9.1	14.3	3.03	1.20	0.30	2.02	1.93	3.20	24.1
CD-30	43.0		16.0	3.97	3.28	0.28	2.09	2.20	2.15	27.1

[a] Stockpiled kiln dust sample.

Source: R. J. Collins, DOT/DOE evaluation of kiln dust/flyash technology, presented at the 6th Annual International Conference on the Economic and Environmental Utility of Kiln Dust & Kiln Dust/Fly Ash Technology, Feb. 19-21, 1985.

TABLE 14-5 Chemical Characterization Test Results for Lime Kiln Dusts (Percent)

Sample Number	CaO	Free Lime CaO	SiO$_2$	Al$_2$O$_3$	MgO	Na$_2$O	K$_2$O	Fe$_2$O$_3$	SO$_3$	LOI 1050°C
LD-1	28.5	3.4	9.19	5.27	20.5	0.21	0.49	6.82	6.37	18.2
LD-2	31.2	2.6[a]	2.46	0.74	23.5	0.00	0.09	0.94	2.80	37.4
		5.1					0.22			
LD-3	54.5	26.68[a]	9.94	4.16	0.49	0.03		1.98	7.97	14.2
		26.4								
LD-4	44.3	2.1	10.1	4.92	3.56	0.14	0.38	1.36	4.84	27.5
LD-5	66.1	40.8	1.92	0.48	2.16	0.00	0.13	0.43	1.72	19.6
LD-6	56.7	13.1	3.45	1.83	1.11	0.00	0.21	0.80	0.27	34.4
LD-7	58.0	14.5	3.21	1.18	0.43	0.00	0.10	3.48	2.20	27.6
LD-8[b]	31.2	0.0	1.74	0.71	23.3	0.05	0.03	1.30	3.50	39.7
LD-9	53.9	8.8	5.90	2.57	2.23	0.09	0.33	1.79	3.50	27.9
LD-10	35.6	2.9	0.62	0.06	23.8	0.00	0.00	0.56	0.87	36.2
LD-11	62.4	29.40[a]	12.7	4.85	0.70	0.09	0.19	1.36	2.05	8.47
		29.7								
LD-12	35.1	2.0	6.41	1.41	21.5	0.02	0.12	0.75	0.05	35.3

[a] Comparison by wet chemistry–glycol methanol extraction.
[b] Stockpiled kiln dust sample.

Source: R. J. Collins, DOT/DOE evaluation of kiln dust/fly ash technology, presented at the 6th Annual International Conference on the Economic and Environmental Utility of Kiln Dust & Kiln Dust/Fly Ash Technology, Feb. 19–21, 1985.

TABLE 14-6 Chemical Characterization for Flyashes and Miscellaneous Test Materials

Sample Number	CaO	SiO$_2$	Al$_2$O$_3$	MgO	Na$_2$O	K$_2$O	Fe$_2$O$_3$	SO$_3$	LOI 1050°C
Flyashes									
FA-1[a]	28.21 to 36.4	35.94 to 50.03	13.9 to 20.3	5.0 to 6.1			5.1 to 9.4	1.5 to 3.2	0.36
FA-2	14.6	38.4	13.0	1.36	0.40	2.04	20.6	3.30	1.62
FA-3[a]		47.3	25.7	1.2			14.1	0.66	2.90
FA-4	6.69	55.1	12.1	1.61	1.73	1.24	5.16	0.52	0.62
FA-5[a]	0.57 to 0.86	55.44 to 59.37	28.00 to 32.31	0.45 to 1.0	0.26 to 0.30	2.46 to 2.90	3.21 to 4.96	0.10 to 0.47	0.66
FA-6[a]	4.51	43.16	20.75			1.12	19.82	1.26	
FA-7	2.41	50.9	25.3	0.95	0.28	2.83	8.42	0.27	2.08
FA-8[a]	2.2 to 2.46	48.3 to 50.24	19.4 to 26.28	0.73 to 1.16	2.12 to 2.33		11.8 to 14.74	0.56 to 0.68	2.33
FA-9[a]	13.40	45.03	20.68	3.86	3.67		6.68	1.80	0.35
FA-10	26.3	34.8	17.3	4.73	1.72	0.42	5.31	3.15	0.23
FA-11	15.8	50.5	17.2	3.06	0.49	0.82	5.91	0.95	0.39
FA-12	20.6	30.3	13.0	3.16	0.36	0.39	4.76	16.36[b]	16.0
FA-13	1.53	44.1	19.6	0.70	0.33	2.39	21.5	0.77	4.93
FA-14	1.81	46.2	31.3	0.67	0.25	1.99	8.53	0.45	4.47
FA-15	1.44	50.9	28.9	0.92	0.32	2.52	5.43	0.42	2.93
FA-16[c]	1.93 to 1.70	44.0 to 48.9	26.8 to 28.2	0.74 to 0.91	0.28 to 0.36	2.14 to 2.55	14.6 to 7.43	0.90 to 0.45	4.39 to 4.47
FA-17	7.48	48.9	22.8	1.54	2.86	0.67	6.09	.30	0.54
FA-18	1.67	48.9	28.3	0.92	0.38	2.51	7.48	0.40	4.23
FA-19	3.26	42.8	22.4	0.95	0.61	2.09	14.8	1.15	8.70
Miscellaneous Test Materials									
Calcitic Hyd. Lime	65.2	0.98	0.42	1.31	0.00	0.28	0.22		26.9
Dolomitic Hyd. Lime	38.0	1.73	0.27	28.0	0.00	0.01	0.54		30.9
Type 1 PC	57.9	19.8	5.81	2.64	0.22	0.82	2.47		1.23

[a] Data provided by flyash supplier, except for LOI values.
[b] May be in partly sulphide form.
[c] Two separate XRF samples prepared for FA-16.

Source: R. J. Collins, DOT/DOE evaluation of kiln dust/fly ash technology, presented at the 6th Annual International Conference on the Economic and Environmental Utility of Kiln Dust & Kiln Dust/Fly Ash Technology, Feb. 19–21, 1985.

CHEMICAL AND PHYSICAL PROPERTIES

Physical Properties

Both lime and cement kiln dusts can produce hard, reasonably strong solids due to pozzolanic reactions, and they continue to harden with time. Some typical bearing strengths for short curing times are shown in Table 14-7. Collins [1] gives bearing strengths for a number of kiln dust/flyash mixtures; these are presented in Table 14-8. He found a strong relationship between reactive oxide content (CaO + MgO)

TABLE 14-7 Penetrometer Hardness vs. Cure Time for Lime-Kiln-Dust-Based Processes

Waste	Physical State	pH	Total Solids	Volume Increase (%)	Bearing Strength (Penetrometer) (tons/ft^2) 1 day	3 days	10 days
Water-based paint	SUSP	8.1	23	50	0	0	0.8
Pesticide manufacture	SUSP	5.7	12	35	0	2	2
Organic chemical manufacture, NOS	SLDG	7.8	9	30	1	3	3
Waste water treatment residue	SLDG	11.1	15	35	1	4	>4.5
Permanganate solution	SOLN	12.5	23	50	3	>4.5	>4.5
Mixed wastes, NOS	SLDG	5.8	39	50	0	0	1.2
Spray painting sludge	SUSP	8.8	47	40	0	1	1

TABLE 14-8 Comparison of Kiln Dust/Flyash–Aggregate Reference Mix Strengths vs. Total Reactive Oxides of Kiln Dusts

Cement Dust Number	Optimum KD-FA Ratio	Maximum 7-Day Strength	Group of Dust	CaO	MgO	LOI	Net Reactive Alkaline	Alkalies	Total Reactive Oxide
1. CD-28	2:1	2,001	I	49.7	1.73	18.7	32.7	4.43	37.2
2. CD-19	1:1	1,948	I	41.1	1.30	11.7	30.7	3.59	34.3
3. CD-29	2:1	1,444	III	47.5	1.20	24.1	24.6	2.32	26.9
4. CD-13	1:1	1,441	III	45.2	1.37	23.2	23.4	1.96	25.4
5. CD-8	2:1	1,301	III	39.4	0.92	22.7	17.6	5.10	22.7
6. CD-20	2:1	1,282	I	45.5	1.16	28.4	18.3	2.78	21.1
7. CD-30	2:1	1,281	I	43.0	3.28	27.1	19.2	2.37	21.6
8. CD-3	2:1	1,213	IV	39.1	2.63	22.8	18.9	4.07	23.0
9. CD-10	2:1	1,159	I	45.9	1.39	28.2	19.1	1.61	20.7
10. CD-18	2:1	989	II	47.6	1.33	31.6	17.3	1.19	18.5
11. CD-16	2:1	845	IV	37.4	1.96	24.0	15.4	5.51	20.9
12. CD-17	2:1	801	III	26.8	0.54	13.5	13.8	13.87	19.8
13. CD-1	2:1	734	II	38.3	2.49	25.3	15.5	4.21	19.7
14. CD-14	2:1	719	IV	34.6	1.83	22.9	13.5	7.63	19.5
15. CD-4	2:1	581	IV	37.2	2.02	27.3	11.9	5.38	17.3
16. CD-6	2:1	562	IV	25.8	1.13	19.5	7.4	16.65	13.4
17. CD-12	2:1	477	II	44.4	1.66	31.8	14.3	2.94	17.2
18. CD-15	2:1	115	IV	19.4	0.64	13.2	6.8	15.44	12.8

Lime Dust Number	Optimum KD-FA Ratio	Maximum 7-Day Strength	Group of Dust	CaO	MgO	LOI	Net Reactive Alkaline	Alkalis	Total Reactive Oxide
1. LD-3	1:1	1,669	H	54.5	0.49	14.2	40.8	0.25	41.1
2. LD-1	1:1	973	L	28.5	20.5	18.2	30.8	0.70	31.5
3. LD-4	1:1	888	L	44.3	3.56	27.5	20.4	0.52	20.8
4. LD-2	1:1	606	L	31.2	23.5	37.4	17.3	0.09	17.4
5. LD-11	1:1	564	H	62.4	0.70	8.47	54.6	0.28	54.9
6. LD-7	1:2	191	H	58.0	0.43	27.6	30.8	0.10	30.9

Note: Strength values given in psi; 1 lb/in.^{2} = 0.0703 kg/cm^2.

Source: J. R. Collins, DOT/DOE evaluation of kiln dust/fly ash technology, presented at the 6th Annual International Conference on the Economic and Environmental Utility of Kiln Dust & Kiln Dust/Flyash Technology, Feb. 19–21, 1985.

and compressive strengths of the solids that had been compression molded and oven cured for seven days. It has been shown that even the traditional 28 day maximum curing time is not sufficient to develop full strength in these systems. Often, however, kiln dust and flyash solidification techniques result in friable, even granular products, which are usually desirable from the operational point of view at the landfill. Also, kiln dusts have been used most often in operational scenarios where immediate "solidification" was required, usually to pass the paint filter test before placement in a landfill. This results in "overdosing" the waste, which uses all free water immediately, leaving none for the subsequent pozzolanic reactions. As a result, the product cannot harden further and becomes quite friable because of the unreacted components of the reagent.

Chemical Properties

Lime kiln dust usually has good neutralizing properties, depending upon free lime content from the particular source. Since it is a fine dust, it also acts as a bulking agent or absorbent. Cement kiln dust has less neutralizing capacity, although it is alkaline, but it has more cementing ability in most cases, even acting as a somewhat poor substitute for cement. It is also a very fine dust and thus acts as a bulking agent or absorbent. The relatively large volume increase associated with these materials—the weight addition percentage ranges from 50 to 200 percent—makes them unattractive in some applications. However, there are instances, especially with highly organic waste streams, where these systems are the most efficient for solidification, even on a weight-to-weight basis [5]. This is apparently due to the anhydrous lime content, which may range up to 40 percent or so.

The actual setting reactions of kiln dusts and flyashes are pozzolanic, and resemble those of Portland cement in many ways. Kiln dusts contain silica and silicates from their natural rock and clay genesis, with cement kiln dust generally having a much higher silica content than the lime dust. Many factors—particle size, silica/silicate content, composition and crystal structure of the silica/silicate, degree of hydration, calcium/magnesium ratio, and others—determine the rate and extent of the long-term chemical reactions.

A "fresh," that is, unhydrated, dust will normally have a large exotherm when mixed with an aqueous waste. This is often enough to generate steam, sometimes explosively, if large volumes are mixed quickly. Therefore, caution must be exercised in the use of these materials in both the laboratory and the field. Also, the rapid hydration of CaO by reaction with sweat on the skin may cause severe irritation, even burns. The exothermic reaction, however, contributes to the rate of solidification in two ways: evaporation of water and increased chemical reaction rate.

While by far the largest use of kiln dusts and flyashes has been for solidification only, they have also been used for fixation of metals in much the same manner as the cement-based systems discussed previously [6-8]. A great deal of work in this area has been done by Cotton, Hammock, Fochtman, Swanstrom, and Conner [9] at Chemical Waste Management (CWM). In 1987, in anticipation of the upcoming first-third landban, CWM launched an extensive program to define the properties, especially composition and leachability, of electroplating sludge (F006) waste streams at two of its sites that received major quantities of F006. This included a treatability program on selected F006 streams that represented the typical and extreme materials received by its sites. The results are summarized in Table 14-9, along

TABLE 14-9 Comparison of CFS Processes, Kiln Dust Type

Reference Number	Proprietary Process or Additives	Mix Ratio	Mixture pH	UCS (tons/ft^2 @ 24 hr)	Cure Time (days)	Waste Type
6	None					Electric arc furnace dust—K061
	None					
	Kiln dust, unspecified	0.05	11.4	147.0	28	
	Water/waste ratio = 0.5					
7	None		6.0			Electroplating sludge—F006
	None					Autopart manufacturing
	Cement kiln dust	0.20				
	Cement kiln dust	0.50				
7	None		8.0			Electroplating sludge—F006
	None					
	Cement kiln dust	0.20				
	Cement kiln dust	0.50				
7	None		6.5			Electroplating sludge—F006
	None					
	Cement kiln dust	0.20				
	Cement kiln dust	1.00				
7	None		6.5			Electroplating sludge—F006
	None					
	Cement kiln dust	0.20				
	Cement kiln dust	0.50				
7	None		6.5			Electroplating sludge—F006
	None					
	Cement kiln dust	0.20				
	Cement kiln dust	0.50				
7	None		8.0			Electroplating sludge—F006
	None					
	Cement kiln dust	0.20				
	Cement kiln dust	0.50				
7	None		7.0			Electroplating sludge—F006
	None					
	Cement kiln dust	0.20				
	Cement kiln dust	0.50				
7	None		6.0			Electroplating sludge—F006
	None					
	Cement kiln dust	0.20				
	Cement kiln dust	0.20				
8	None					Electric arc furnace dust—K061
	None					
	Kiln dust/flyash				0	
	Kiln dust/flyash				7-98	
	Kiln dust/flyash				7	

KEY: (1) = reserved
    * = Average of three runs

BDAT proposed 1st-3rd limits
RCRA EPT limits
Delisting limit
Drinking water

TABLE 14-9 (*Continued*)

| % Moisture | Physical State | Test Type | \multicolumn{4}{c}{Concentration of Constituent (mg/l) in Waste or Leachate} |
			Antimony	Arsenic	Barium	Beryllium
0.0	Powder	Total	294.000	36.000	238.000	0.150
		TCLP-2	0.040	<0.010	0.733	<0.001
		TCLP-2	<0.050	<0.010	0.507	<0.001
70.9	Filter cake	Total				
		TCLP-2				
		TCLP-2				
		TCLP-2				
32.4	Filter cake	Total			85.500	
		TCLP-2			1.410	
		TCLP-2			0.330	
		TCLP-2			0.310	
31.8	Filter cake	Total				
		TCLP-2				
		TCLP-2				
		TCLP-2				
19.4	Filter cake	Total			14.300	
		TCLP-2			0.380	
		TCLP-2			0.310	
		TCLP-2			0.230	
19.0	Filter cake	Total				
		TCLP-2				
		TCLP-2				
		TCLP-2				
30.7	Filter cake	Total				
		TCLP-2				
		TCLP-2				
		TCLP-2				
20.5	Filter cake	Total			15.300	
		TCLP-2			0.530	
		TCLP-2			0.320	
		TCLP-2			0.270	
38.4	Filter cake	Total			19.200	
		TCLP-2			0.280	
		TCLP-2			0.190	
		TCLP-2			0.080	
0.0	Powder	Total		50.000	<1000.000	
		EPT				
		EPT				
		EPT		<0.020	<1.000	
		MEP (day 9)		<0.020	<0.500	
		Standards	(1)	(1)	(1) 100.000 6.300 1.000	

(*continued*)

440 KILN DUST AND FLYASH–BASED PROCESSES

TABLE 14-9   Comparison of CFS Processes, Kiln Dust Type (*Continued*)

Concentration of Constituent (mg/l) in Waste or Leachate

Reference Number	Cadmium	Chromium	Copper	Lead	Mercury	Nickel
6	481.000	1370.000	2240.000	20300.000	3.800	243.000
	12.800	<0.007	0.066	45.100	3.800	0.027
	1.740	0.032	0.013	0.790	<0.001	0.012
7	31.300	755.000	7030.000	409.000		989.000
	2.210	0.760	368.000	10.700		22.700
	0.500	0.400	5.400	0.400		1.500
	0.010	0.390	0.250	0.360		0.030
7	67.200	716.000		257.000		259.000
	1.130	0.430		2.260		1.100
	0.060	0.080		0.300		0.230
	0.020	0.200		0.410		0.150
7	1.310		1510.000	88.500		374.000
	0.020		4.620	0.450		0.520
	0.010		0.300	0.300		0.100
	<0.010		0.150	0.210		0.020
7	720.000	12200.000	160.000	52.000		701.000
	23.600	25.300	1.140	0.450		9.780
	3.230	0.250	0.200	0.240		0.530
	0.010	0.300	0.270	0.340		0.030
7	7.280	3100.000	1220.000	113.000		19400.000
	0.300	38.700	31.700	3.370		730.000
	0.020	0.210	0.210	0.300		16.500
	0.010	0.380	0.290	0.360		0.040
7	5.390	42900.000	10600.000	156.000		13000.000
	0.060	360.000	8.690	1.000		152.000
	0.010	3.000	0.400	0.300		0.400
	0.010	1.210	0.420	0.380		0.100
7	5.810		17600.000	1.690		23700.000
	0.180		483.000	4.220		644.000
	0.010		0.500	0.310		15.700
	0.010		0.320	0.370		0.040
7			27400.000	24500.000		5730.000
			16.900	50.200		16.100
			3.180	2.390		1.090
			0.460	0.270		0.020
8	200.000	1600.000		15000.000	<2.000	<1000.000
	1.400	0.050		55.000		
	0.020	0.050		0.010		
	0.020	0.080		0.020	<0.002	<0.100
	<0.020	<0.050		<0.010	<0.002	<0.100
	0.066	3.800	0.710	0.530		0.310
	1.000	5.000		5.000	0.200	
	0.063	0.315		0.315	0.013	
	0.010	0.050		0.050	0.002	

TABLE 14-9 *(Continued)*

Concentration of Constituent (mg/l) in Waste or Leachate

Selenium	Silver	Thallium	Vanadium	Zinc	Cyanide
<5.000	59.000	<1.000	25.000	244000.000	
<0.050	0.021	0.038	<0.006	445.000	
<0.010	0.003	0.011	0.087	7.440	
	6.620			4020.000	
	0.140			219.000	
	0.030			36.900	
	0.050			0.010	
	38.900			631.000	
	0.200			5.410	
	0.200			0.050	
	0.050			0.030	
	9.050			90200.000	
	0.160			2030.000	
	0.030			32.000	
	0.030			0.010	
	5.280			35900.000	
	0.080			867.000	
	0.040			3.400	
	0.040			0.040	
	4.080			27800.000	
	0.120			1200.000	
	0.030			36.300	
	0.060			0.030	
	12.500			120.000	
	0.050			0.620	
	0.030			0.020	
	0.050			0.020	
	8.110			15700.000	
	0.310			650.000	
	0.030			4.540	
	0.050			0.020	
				322.000	
				1.290	
				0.070	
				<0.010	
70.000	<100.000			107000.000	<50.000
<0.050	<0.050				<0.100
0.020	<0.050				
(1)	0.260			0.086	(1)
1.000	5.000				
0.063	0.315				
0.010	0.050				

with other data on the use of kiln dusts for fixation of metals to landban standards. These data were submitted to the EPA, which subsequently used them in development of the first-third landban BDAT for F006. Information on kiln dust leaching is presented in Table 14-10.

Flyash alone is less frequently used for anything but solidification. However, as we saw in Chapters 12 and 13, it becomes quite important in CFS technology when mixed with lime or Portland cement, and has also been used effectively in combination with kiln dusts. In the latter case, the kiln dust replaces all or part of the cement or lime that would otherwise be used.

One limitation on the use of any of these materials is that they contain significant amounts of metals, which they leach at levels at or above the latest landban BDAT standards. This is demonstrated by the leaching test results presented in Table 14-10. It should be noted, however, that these results were obtained by leaching the dust or ash in the raw, powder form, not the solidified mass that would result from wetting and curing the materials. The latter experiments have not been performed to date per se, but test results from wastes treated with these reagents often equal or better the metal fixation obtained with either the raw reagent or the waste alone.

COMMERCIAL PROCESSES

Kiln dusts and flyashes have been so widely used at central treatment facilities, remedial action projects, and at the point of waste generation that it is impossible to list all of the uses, projects, and "processes." In fact, most of these uses are not considered processes in themselves any more than, for example, would be the use of lime in a waste water treatment plant. They are simple ingredients in a system in most cases. There are, however, specific users and vendors whose systems are based on kiln dusts and/or flyashes, along with additives in some instances. Several have done considerable technical and engineering development on the systems; a few of these are even patented for narrow uses or reagent combinations. The more important are described here.

Chemical Waste Management (CWM)

In 1985, CWM undertook a major research effort to develop a company-wide system for the use of kiln dusts to stabilize liquid wastes to meet the May 8, 1985 "Ban on Disposal of Bulk Liquids in Landfills" [10]. This effort included a procedure for establishing mix ratios for various wastes and a quality control system for assuring the proper treatment of a large variety of wastes with highly variable reagent supplies [11–13]. This effort was successful in determining and setting forth the factors involved in solidification, the reactions and properties of kiln dusts, and the tests and observations that would assure consistent results in their use. CWM's specifications are summarized as follows.

Appearance
  Must be a free-flowing powder
  No lumps larger than $3/8$ in.
  No foreign material (rocks, plant fragments, etc.)

Chemical
- pH of 10 percent slurry — > 10
- Alkalinity as CaO — > 15 percent
- Moisture [105°C (221°F)] — < 1 percent
- Loss on ignition [1100°C (2012°F)] — < 30 percent
- Temperature rise — 20°C (66°F)

Physical
- Apparent bulk density (loose) — 35–60 lb/ft^3
- Apparent bulk density (packed) — 60–100 lb/ft^3
- Mesh size
  - + 8 mesh — 1 percent maximum
  - − 325 mesh — 90 percent maximum

This specification also points out the following fundamental basis for selection of stabilization reagents: "... price alone cannot be used to determine the best stabilization reagent. Mix ratio, volume expansion, delivered price of the reagent, and the cost of cell (i.e., landfill) space must all be considered."

These specifications were established for the disposal of stabilized wastes in RCRA TSD facilities, or secure landfills. At that time, chemical fixation was not a requirement for this disposal scenario. The conversion of liquids into stable solids, with certain limitations on organics and reactive components, was the goal of stabilization. The basic characteristics or properties that define the presence of a chemical reaction* in kiln dust processes are:

(a) Temperature change
(b) Hardening over time
(c) Presence of materials *known* to react
(d) Resistance to desorption
(e) Chemical transformation of toxic compounds

As we saw previously, CWM has more recently completed a major project on the fixation properties of kiln dust processes, as well as other systems for specific purposes. One finding of this research was that the basic CFS reagents themselves, whatever the type, are not always sufficient to immobilize metals to the levels required under either first-third landban or delisting regulations. A number of proprietary additives have been developed for use with problem wastes or where special properties are required for the specific disposal scenario. With the rapidly changing, and ever more severe regulatory requirements, the present limits of chemical technology are being challenged to the extent that more fundamental knowledge is essential.

Wehran Engineering Corporation

Wehran Engineering Corporation, a consulting-engineering firm, has patented [14] a solidification process based on the use of cement kiln dust in the presence of

---

*The EPA requires that the solidification of bulk liquids to be placed in landfills be done by chemical reaction, not absorption.

444 KILN DUST AND FLYASH-BASED PROCESSES

TABLE 14-10 Leaching of Kiln Dusts

Concentration of Constituent (mg/l) in Waste or Leachate

Reagent Name	Test Type	Arsenic	Barium	Cadmium	Chromium	Copper	Lead	Mercury	Nickel	Silver	Zinc
Cement kiln dust	Total	38.000	92.700	3.140	31.900	44.800	156.000	<0.033	12.600	4.130	65.600
Cement kiln dust	TCLP-2	<0.010	2.740	<0.010	0.050	0.160	0.290	<0.010	0.020	0.020	0.040
Cement kiln dust	Total	10.600	104.000	10.100	59.200	58.700	432.000	<0.034	24.100	14.200	145.000
Cement kiln dust	TCLP-2	<0.010	2.740	<0.010	0.100	0.110	0.390	<0.001	0.020	0.030	<0.010
Cement kiln dust	Total	1.460	9.020	2.200	13.100	22.800	94.400	<0.034	13.000	6.700	9.560
Cement kiln dust	TCLP-2	<0.010	0.210	<0.010	0.070	0.120	0.430	<0.001	0.020	0.030	<0.010
Cement kiln dust	Total	22.900	157.000	10.700	54.900	55.600	227.000	<0.034	28.000	10.500	171.000
Cement kiln dust	TCLP-2	<0.010	0.610	<0.010	0.090	0.130	0.490	<0.001	0.030	0.040	0.040
Cement kiln dust	Total	50.000	133.000	7.590	40.900	49.500	151.000	<0.003	16.300	3.980	249.000
Cement kiln dust	TCLP-2	<0.010	0.450	<0.010	0.030	0.060	0.130	<0.001	<0.010	<0.010	0.010
Cement kiln dust	Total	10.600	63.200	4.030	24.700	28.900	155.000	0.197	15.800	7.090	34.100
Cement kiln dust	TCLP-2	<0.010	0.580	<0.010	0.080	0.100	0.290	<0.001	0.020	0.010	0.030
Cement kiln dust	Total	4.450	93.800	1.880	26.500	33.100	107.000	<0.034	13.100	5.170	41.400
Cement kiln dust	TCLP-2	<0.010	0.770	<0.010	0.200	0.120	0.390	<0.001	0.020	0.030	0.030
Cement kiln dust	Total	10.100	122.000	1.180	20.500	23.500	58.700	<0.033	8.990	1.420	18.000
Cement kiln dust	TCLP-2	<0.010	0.710	<0.010	0.330	0.120	0.240	<0.001	0.010	0.010	0.010
Cement kiln dust	Total	11.800	55.600	2.210	15.100	21.700	86.000	0.092	10.800	1.450	19.900
Cement kiln dust	TCLP-2	<0.010	0.400	<0.010	0.060	0.140	0.280	<0.001	0.010	0.010	0.010
Cement kiln dust	Total	28.000	388.000	29.900	23.700		114.000	<0.025		3.040	
Cement kiln dust	TCLP-2	<0.010	0.760	<0.010	0.070		0.300	<0.001		<0.010	
Cement kiln dust	Total		94.500	6.160	26.600		329.000		9.350	1.030	470.000
Cement kiln dust	Total	20.100	109.000	7.400	30.500	37.000	394.000	<0.033	10.900	2.230	839.000
Cement kiln dust	TCLP-2	<0.010	0.850	<0.010	0.060	0.160	0.160	<0.001	<0.010	<0.010	0.060
Cement kiln dust	Total	2.860	10.800	3.620	193.000	32.400	1390.000	<0.033	90.900	1.460	266.000
Cement kiln dust	TCLP-2	<0.010	0.300	0.010	0.300	0.090	3.260		0.020	0.020	0.040

Cement kiln dust	Total	38.100	153.000	3.950	29.000	38.300	191.000	<0.033	11.000	3.770	36.700
	TCLP-2	<0.010	1.410	<0.010	0.070	0.170	0.490	<0.001	<0.010	<0.010	0.060
Cement kiln dust	Total	24.400	61.900	1.140	18.700	28.000	60.300	<0.033	13.600	1.610	19.500
	TCLP-2	<0.010	0.680	<0.010	0.060	0.200	0.390	<0.001	0.010	0.010	0.060
Cement kiln dust	Total	8.590	66.400	7.700	21.700	37.800	274.000	<0.037	19.400	4.550	158.000
	TCLP-2	<0.010	0.600	<0.010	0.050	0.160	0.310	<0.001	<0.020	0.020	0.020
Cement kiln dust	Total	1.530	14.400	2.560	9.230	11.700	32.500	<0.033	26.500	0.120	29.200
	TCLP-2	<0.010	0.440	0.010	0.060	0.180	0.380	<0.001	0.020	<0.020	0.050
Cement kiln dust	Total	28.900	137.000	3.220	19.900	30.100	85.000	<0.033	12.600	1.100	49.400
	TCLP-2	<0.010	0.720	<0.010	0.070	0.150	<0.140	<0.001		<0.010	0.080
Cement kiln dust	Total	14.200	42.100	12.200	23.700	29.500	312.000	0.069	29.600	2.390	99.400
	TCLP-2	<0.010	0.380	<0.010	0.090	0.120	0.210	<0.001	0.010	0.020	0.020
Cement kiln dust	Total	38.400	198.000	7.230	22.300		500.000	0.440		2.760	
	TCLP-2	<0.010	0.860	<0.010	0.040		0.750	<0.001		<0.010	
Cement kiln dust	Total	<0.450	82.700	19.400	17.700	35.900	367.000	0.749	8.680	2.670	50.700
	TCLP-2		0.420	<0.010	0.130	0.070	0.240	<0.001	0.020	0.020	<0.010
Cement kiln dust	Total		<10.000		<10.000	<10.000	200.000		<10.000		5000.000
Cement kiln dust	Total	24.000	<55.000	21.000	41.000	30.000	253.000	<0.130	22.000	5.400	462.000
	TCLP-2										

KEY:

F006 BDAT	(1)	(1)		0.066	5.200			0.320	0.072	0.086
K061 BDAT				0.140	5.200			0.320		
EPT characteristic level		100.000	1.000	5.000		5.000	0.200		5.000	
Delisting level		6.300	0.063	0.315		0.315	0.013		0.315	
Drinking water standard		1.000	0.010	0.050		0.050	0.002		0.050	

sufficient water along with the waste to provide a workable consistency, then adjusting the pH of the alkaline mass with acid, such as sulfuric or phosphoric acid, to a value from about pH 5 to about pH 7, and finally drying the pH-adjusted mass to produce an environmentally stable and hardened solid product. The Wehran process has been included in the EPA study program done by the U.S. Army Corps of Engineers at Vicksburg, Virginia [15].

N-Viro Energy Systems Ltd.

N-Viro Energy Systems Ltd. has developed a flyash/cement kiln dust process that may also incorporate Portland cement [16]. It has been commercially developed primarily for the solidification of sewage sludge, but appears to be applicable for other wastes as well. The combination of reagents is interesting, but no information is available on its properties as applied to hazardous wastes.

Research Cottrell

This kiln dust process was developed only for the stabilization of fluegas desulfurization sludges, and apparently has not been commercially used for other wastes.

CASE HISTORY: SUPERFUND-TYPE CLEANUP

One of the tasks in a superfund-type project (run much like a Superfund job, but not paid for out of Superfund) in Ohio in 1982 was the cleanup of a so-called "polymer pit." This concrete block impoundment contained three layers of waste: a floating, gummy organic layer containing a wide range of organic priority pollutants; an intermediate contaminated-water layer; and a heavy sludge layer on the bottom. Because of the variety and high concentration of priority pollutants in the waste, secure landfill was chosen as the disposal method. Solidification was required before landfill, and the options were solidification either on-site or at the landfill. On-site solidification was chosen for economic and public safety reasons.

Because there were essentially three distinct waste streams, three different formulations were chosen to minimize volume increase and the associated transportation and disposal costs. A cement-based process was used for the sludge and water layers, and a lime kiln dust formulation for the organic layer. The volume was relatively small, and because three different formulations were used, a small, portable, batch-type treatment unit was used. The job parameters were as follows.

Waste analysis
  Organic polymers
  Organic priority pollutants
  Metals
  Inert inorganic solids
  Water
Volume treated: 210,000 gal
Treatment rate: 20,000 gal/day
Time: May 1982
Treatment cost: $0.41/gal

Disposal method: Secure landfill (on-site)
Properties of treated solid
  Unconfined Compressive Strength: >1.0 ton/ft^2

REFERENCES

1. Collins, R. J. DOT/DOE evaluation of kiln dust/fly ash technology. Presented at the 6th Annual International Conference on Economic and Environmental Utility of Kiln Dust & Kiln Dust/Fly Ash Technology, Feb. 19–21, 1985.
2. Haynes, B. W. and G. W. Kramer. *Characterization of U.S. Cement Kiln Dust,* Information Circular 8885. Washington, DC: U.S. Bureau of Mines, 1982.
3. JTM Industries, Inc. *Making Industrial By-Products Work For You.* Marietta, GA, 1985.
4. Marblehead Lime Co., Chicago, IL, 1988.
5. Jach, Z. Czech. Patent 180,864 (Sept. 15, 1979).
6. U.S. EPA. *Best Demonstrated Available Technology (BDAT) Background Document for K061,* Vol. 1. EPA/530-SW-88-031D. Washington, DC, Aug. 1988.
7. U.S. EPA. *Best Demonstrated Available Technology (BDAT) Background Document for F006,* Vol. 13. EPA/530-SW-88-0009-1. Washington, DC, May 1988.
8. Bethlehem Steel Corp. *Petition to Delist Chemically Stabilized Electric Arc Furnace Dust,* Petition Number 0681. Bethlehem, PA, Nov. 1986.
9. Chemical Waste Management. *Stabilization Treatment of Selected Metal-Containing Wastes,* CWM TN 87-117, 1987.
10. Chemical Waste Management. *Documentation in Support of CWM Procedures to Meet the May 8, 1985 Ban on Disposal of Bulk Liquids in Landfills,* CWM Tech. Rep. 85-102. Riverdale, IL, 1985.
11. Chemical Waste Management. *Identifying Reactions When Stabilizing Bulk Liquids for Landfilling,* CWM TN 85-101. Riverdale, IL, 1985.
12. Chemical Waste Management. *Reactivity Evaluation of Lime Based Stabilization Reagents,* CWM TN 85-104, Riverdale, IL, 1985.
13. Chemical Waste Management. *Specifications for Stabilization Reagents,* CWM TN 86-129, Riverdale, IL, 1986.
14. Cocozza, E. P. U.S. Patent 4,049,462 (Sept. 20, 1977).
15. Larson, R. J., P. G. Malone, J. H. Shamburger, J. D. Broughton, D. W. Thompson, and L. W. Jones. *Field Investigation of Contaminant Loss from Chemically Stabilized Industrial Sludges.* Springfield, VA: NTIS, 1981.
16. Nicholson, J. P. U.S Patent 4,101,332 (Jul. 18, 1978), Reissue 30,943 (May 25, 1982).

Chapter 15

OTHER CFS AND NONCHEMICAL PROCESSES AND SYSTEMS

There are a great many processes, reagents, and systems that claim to accomplish solidification and/or fixation of hazardous wastes. Most of the commercial forms of CFS processes fit into the categories discussed in Chapters 10–14. Others, such as those used in the nuclear waste area and in solidification of nonhazardous wastes such as sewage sludge, are important commercially and technically, but are not within the purview of this book.* With these limitations in mind, let us look at a number of nonclassified processes and systems, a few of which have had commercial application. To provide some kind of context, these have been grouped as follows:

Miscellaneous inorganic processes
Nuclear waste CFS: introduction
Organic-based processes
Nonchemical solidification
Systems combining dewatering and CFS
In situ processes and systems

MISCELLANEOUS INORGANIC PROCESSES

Gypsum-Based Processes

In general, gypsum-based processes have been suggested or experimented with in the laboratory, but do not appear to have been used in hazardous waste treatment. The only "name" process of this type, U.S. Gypsum's Envirostone® process, uses calcium sulfate hemihydrate, $CaSO_4 \cdot \frac{1}{2}H_2O$, (plaster of Paris) plus a proprietary polymer [1]. The process is described in detail in U.S. Gypsum's Topical Report [2]†

---

*Because of the increasing interest in "mixed wastes," that is, those that both have the RCRA hazardous properties and contain radionuclides, a short discussion of this area is included in this chapter.
†The Topical Report system, per the U.S. Nuclear Regulatory Commission's Branch Technical Position Report (May 1, 1983-Rev.O) on 10 CFR 61, is a required performance testing protocol for qualification to use a CFS process for certain purposes that fall under NRC jurisdiction.

for use on nuclear waste, for which purpose it was developed. Leaching tests were conducted according to ANS 16.1 on very specific solidified nuclear wastes, and so the data have little direct applicability to RCRA hazardous waste CFS work. However, the Topical Report is recommended as further reading, since procedures of this sort may someday be required by the EPA, states, or other governments to qualify to use CFS technology on hazardous wastes.

The author has used gypsum in certain CFS formulations where a nonalkaline system is required, for example, to prevent the evolution of ammonia. The primary problem is the expected lack of water resistance of gypsum-based formulations. This problem may be addressed by the addition of hydrophobing agents or other additives that make the waste form less permeable and/or more water-repellent. The proprietary polymer used in the Envirostone process may be added for this purpose; information given does not clarify the purpose of the additive.

Other processes that use gypsum are listed below. Most of these use it along with other reagents, or it is a necessary ingredient found in the waste itself.

Process or Ingredients	*Reference*
Gypsum	[3], [4]
Slag + gypsum	[5]
Ground slag	[6] – [8]
Self-cementation (calcined $CaSO_4$ or $CaSO_3$)	[9]

Calcilox Process

Calcilox is fundamentally a settling or compaction process that relies on the addition of a proprietary ingredient to slowly form a cementitious mass under the overlying water layer. In normal operation, this process requires the water layer for proper compaction and reaction, and in this respect is much different from other CFS processes. In fact, in some ways it is more related to conventional waste treatment than to CFS work. However, it has been marketed as a CFS process and competes in certain areas with other CFS processes.

The proprietary ingredient, Calcilox, is actually finely ground blast furnace slag obtained as a waste product from basic steel-producing plants. The process developer, Dravo Lime Co., produces Calcilox in quantity at a plant in the Pittsburgh, Pennsylvania, area. It is reportedly sold at a price of about $25 per ton. Chemical details of the process are given in the patents referenced in the preceding list. Calcilox is typically added in the amount of about 5 to 10 percent by weight to the waste [10]. The composition of the sludge being treated is quite critical, both in solids content and in other chemical factors. The process may use other ingredients, such as flyash, kiln dust, lime, or Portland cement as well as the ground slag.

This process has been used to stabilize fluegas desulfurization sludges and coal waste fines in very large projects. The ground slag is mixed with the dilute waste slurry, which is then allowed to settle and compact in an impoundment where the supernatent water is drawn off and recycled or discharged. One of these was at the Bruce Mansfield power station near Pittsburgh, in 1975. The capital cost for this system was more than $70 million. The planned capacity was 18,000 tons of sludge per day, at a reported price of $2.50 per ton, or a daily cost of approximately $45,000; the project is scheduled to run for 30 years before the disposal site is filled.

More recently, Dravo reported that the process would be used to stabilize 90,000 tons per year of coal fines refuse. Dravo also reported that they are testing the use of Calcilox on sludges from smelters, chemical, and pharmaceutical plants. The process has applications where very large volumes of dilute waste are produced and where large areas for settling and compaction ponds can be constructed and operated over long periods of time.

Other Inorganic Processes

A variation on CFS techniques is the combined process of using an absorbent to soak up water or chemisorb certain waste components, such as metals or organics, and then to apply a chemical solidification technique. An example is that of chlorinated hydrocarbons that have been disposed of by first mixing them with coke or carbon absorbent and then solidifying that mixture with Portland cement. The solidified waste is subsequently disposed of by ocean dumping or in secured landfill. Certain clays are good sorbents for metals and some organics, and if dried before use, will absorb large quantities of water and result in a solid, leach-resistant end product [11].

Self-cementation can be applied to wastes containing large amounts of calcium sulfate or sulfite, such as fluegas cleaning sludges. A portion of the waste itself, about 10 percent, is calcined and then remixed with the remainder of the waste along with other additives [9]. Along similar lines, a process was developed for sealing coal refuse piles by mixing with hydrated lime and then treating in place with carbon dioxide to form cementitious calcium carbonate crystals.

The literature contains a number of references to surface coating, sealing, or hardening methods, where the cementitious waste form is surface treated to improve its hardness, abrasion resistance, permeability, or leaching resistance. One of these uses fluorosilicate solutions, such as a mixture of zinc and magnesium fluorosilicates, to react with free lime in the solid and form low-solubility fluorides and silicates. There is no record of these techniques ever being used commercially.

There is one other type of solidification process that is worthy of note, even though it is not "chemical" per se. This is glassification or vitrification, conducted by melting the waste alone or with glass-forming substances to produce a glasslike product. There are two approaches that have been tested and are in pilot or small-scale commercial use at this time: glassification of radioactive wastes by mixing the waste with glass frits or glass-forming substances, and in situ vitrification. Both are discussed later on in this chapter.

NUCLEAR WASTE CFS: INTRODUCTION

Radioactive waste residuals can be classified into three categories: low level, intermediate level, and high level. High-level wastes from nuclear fuel manufacture and reprocessing and from weapons operations are not a pertinent part of CFS technology, since their control, handling, and disposal are outside of the hazardous waste management industry. Laboratories and fuel processing operations also produce quantities of essentially nonradioactive sludges that can be handled by normal CFS processing applicable to the particular type of sludge. The purpose of this section is to discuss a unique area—that of the solidification of liquid or semiliquid waste from nuclear power plants—that was one of the first applications of CFS processes in the United States and elsewhere.

There were more than 60 nuclear reactors operating in the United States in 1977 and more were under construction or planned. All wet wastes from nuclear power plants are solidified before shipment to one of the several active commercial burial grounds in the United States. Solidification is required because licensing conditions of the burial grounds prohibit the burial of liquids as such, and because local, state, and federal regulations control the shipment of liquid radioactive material and encourage the solidification of wet wastes.

Wet wastes from nuclear power plants can be classified into four basic types: spent resins, filter sludges, concentrated salt solutions, and evaporator concentrates, plus miscellaneous liquids. Spent resins result from demineralization of dilute liquid wastes, reactor coolant cleanup, etc. Filter sludges consist of spent filter aid material from precoat filter and powdered resin from precoat filter-demineralizers. Concentrated salt solutions result from the treatment of more dilute liquid radwaste by evaporators. Evaporator concentrates consist primarily of concentrated boric acid solutions and sodium sulfate solutions. These concentrates may also contain other salts and chemicals from simultaneous evaporation of floor drain, laboratory, and decontamination wastes. Miscellaneous liquid wastes include more concentrated wastes from laboratories and other sources that are to be solidified directly without concentration.

The basic hazardous constituents in these residues are radioactive isotopes of various types, usually present in relatively small quantities, and toxic, nonradioactive, heavy metals.

The quantity of these wastes generated per year (1977) was between 10 million and 20 million gallons. At that time, nearly all of the systems in use in the United States utilized either Portland cement-based or urea-formaldehyde polymerization systems as the basic solidification reagent. Absorbents such as vermiculite were also used but were eliminated due to increasingly stringent regulatory requirements at disposal sites. More recently, the use of urea-formaldehyde systems has been discontinued due to their tendency to throw off free water from the condensation reaction. The Nuclear Regulatory Commission requires each reactor to have available a solidification system for all wastes leaving the plant, and no freestanding water is allowed in any container of wet waste leaving the plant. Therefore there is, and will continue to be, a substantial market for both on-site and mobile solidification systems and services.

In the nuclear waste area, the allowable cost for solidification and disposal is between one and two orders of magnitude higher than in the nonnuclear area. Since major costs in the total treatment and disposal process are transportation, burial, and the cost of the container, packaging efficiency in the container is very important, and more expensive processes and reagents are cost effective as long as they give better packaging efficiencies [12]. For example, the actual volume of waste contained in a finished 55-gallon drum will vary from about 25 gallons with Portland cement systems to about 42 gallons for the cement–silicate processes. Essentially nonradioactive sludges in large lagoons at fuel processing and reprocessing plants are, on the other hand, treated in bulk with systems that are more in the cost and technology range of conventional CFS processes. Organic processes are also used in nuclear waste solidification, which is really their only use in the United States. They are discussed in the next section of this chapter.

Processes and vendors in the nuclear waste area were listed in Table 9-3 for inorganics, and Table 9-4 for organics. Brief comments on several of these are made here.

Chem-Nuclear Systems, Inc. (CNSI)

Chem-Nuclear Systems, Inc. (CNSI) is the largest of the nuclear waste CFS vendors. CNSI operates mobile solidification plants, installs fixed units if required, and performs transportation and disposal services at its Barnwell, South Carolina, facility. CNSI uses cement-based CFS technology with the addition of proprietary additives for specific waste streams [13,14].

ATCOR, Inc.

The ATCOR, Inc., system uses Portland cement as the solidification agent and is offered as an engineered equipment system that is designed to be purchased and operated by the waste generator.

Hittman Nuclear and Development Corporation

Hittman uses both a Portland cement system and a polymer system. In the cement process, Hittman offers either an engineered, fixed system installed for customer operation, or a mobile cement solidification and packaging service for use in filling 55-gallon drums or larger containers up to 200 cubic feet in volume.

United Nuclear Industries Inc. (UNI)

UNI uses cement-based processes, including cement/soluble silicate [15]. They sell on-site drum solidification systems to nuclear power plants and other sources of radioactive waste.

ORGANIC-BASED PROCESSES

This section is divided into three parts; the first discusses urea-formaldehyde systems; the second, other organic polymerization processes; and the last, thermoplastic systems. The reason for this is that the urea-formaldehyde (UF) systems are a subclass in themselves, having at one time attained common usage for the solidification of radioactive wastes in drumming operations. Other organic polymer processes, on the other hand, are relatively new to the CFS business, and except for the Dow system, are not really commercial at this time. It is important to note that the only large-scale commercial use of organic processes, at least in the United States, has been in the nuclear waste area. Organic system vendors were listed and briefly described in Table 9-4.

Urea-Formaldehyde (UF) Systems

It is believed that UF processes are obsolete, at least in the United States. This is due to their cost, their tendency to throw off free water during the condensation reaction, and recent environmental concerns over formaldehyde. However, various nuclear plants that have installed and used such systems may still have them in use. Typically, these systems use a partially polymerized UF dispersion that is capable of mixing in all proportions with water and that can be fully polymerized by the addition of a curing agent [16]. The polymer produced during curing is thermosetting and cannot be redissolved by normal means. Urea-formaldehyde systems require relatively large quantities of the resin dispersion for solidification and, compared to Portland cement or soluble silicate processes, are relatively expensive in use for drumming radioactive wastes.

Urea-formaldehyde resin is an aqueous emulsion of urea and formaldehyde chemically combined to form linear polymeric chains. The resulting liquid is a viscous emulsion with water containing approximately 65 weight percent solids that is completely miscible with water, but immiscible with nonpolar solvents. Upon the addition of an acid catalyst, cross-linking polymerization by a condensation mechanism occurs and a solid is obtained. During polymerization, water added to the resin before the addition of the acid catalyst is physically entrapped in the polymer matrix. The UF emulsion may generally be mixed with aqueous waste in volume ratios of from 1:1 to 1:3, with 1:2 typically recommended as a compromise between product properties and cost. The optimum formulation must be determined for each waste to be solidified. A saturated solution of sodium bisulfate is usually employed as the catalyst and is typically added in the ratio of 2–3 percent by volume of the waste–UF mixture to initiate polymerization. Polymerization is pH dependent and the amount of catalyst necessary to produce a waste–UF mixture pH of 1.5 ± 0.5 should be determined for each waste. The formulation will begin to gel within several minutes after addition of the required quantity of catalyst and will generally form a freestanding solid within 30 minutes. The polymerization reaction may continue for several hours, however, during which small quantities of acidic, freestanding water may be released. Certain wastes, such as concentrated sodium sulfate and soap solutions, are difficult to entrap in UF, although such difficulties can be minimized by proper selection of the UF/waste ratio. Various clays can be added to UF to decrease product leachability.

As in the case with cement, the UF mixing process can be carried out either in an in-line mixer or directly in the burial container. In both cases, the catalyst is added last to avoid the problems associated with premature resin setting. With an in-line mixer, the catalyst is added to the premixed UF/waste mixture as it enters the solidification container. Perhaps the main problem encountered in using UF systems is the standing water that remains on the surface of the drum being solidified. This has been cited as being a problem by various users and suppliers of such systems. In addition, the resin dispersion has poor shelf life at elevated ambient temperatures. Urea-formaldehyde systems have good tolerance for variations in the liquid waste, have the ability to solidify a fairly wide variety of wastes, and have good physical properties after curing. They were widely used for the drumming of radioactive wastes in the United States and have also reportedly been applied to some other types of hazardous wastes, although commercial activity in this area has been very small.

Other Organic Polymerization Systems

A number of other organic polymerization systems were proposed in the 1970s, but the interest seems to have diminished in recent years. There are several reasons for this, but the primary one is cost. Even in the nuclear industry, the use of polymers is expensive compared to the inorganic systems or even the UF processes. Potentially, organic polymers have wide applicability to liquid waste solidification since they can be formulated in an almost endless variety of compositions to meet different requirements. They also have one major advantage over most of the inorganic systems: once cured, they have a much higher degree of impermeability and tend to remain in monolithic form due to the strength and elasticity properties of the polymer. Also, most of these systems are hydrophobic after curing and tend to resist leaching even if crushed to a small particle size. Thus these systems are good at retaining highly toxic metals and organic compounds whose leaching must be limited to the parts per billion (ppb) range. Another characteristic of organic polymer systems is that very quick gelation and development of physical strength can be obtained if desired.

Most processes use polyester or vinyl ester resins. Aqueous waste is dispersed into the basically hydrophobic, liquid, monomer, or prepolymer resin using high-shear mixing. This forms a water-in-resin emulsion to which is added a catalyst to polymerize the system, forming a hard monolith. The resultant thermoset polymer is resistant to solvents, moderate heat, and water. No reaction takes place with waste components insofar as a fixation is concerned, but certain waste constituents can interfere with polymerization by reacting with the catalyst or scavenging free radicals.

The properties, characteristics, and chemistry of these systems are discussed below under the individual processes. However, virtually no information about the leachability of toxic metal species from these processes is available.

Dow Process. A full description of the Dow solidification system is given by Filter [17], including equipment and a schematic diagram of a treatment system operating on a continuous basis for the solidification of highly radioactive material. The process, which used vinyl ester resin, has been approved by the NRC and various states, and is reportedly in commercial use for nuclear waste solidification.

Washington State University (WSU) Process. Washington State University developed their process under a research grant from the National Science Foundation. It is thus available to users who wish to commercialize it, although this should be done in cooperation with the originators at WSU. The process is a polyester system using water extensible polyester resins and a polymerization initiator to produce a solid [18]. The author observed the process in operation at the WSU pilot plant. It seems to work very smoothly, producing hard, strong, elastic monoliths after curing. Since the process requires about 25 percent by weight of the polyester resin in the mixture with the waste, and assuming chemical costs of approximately $0.45 per pound for this resin in bulk, chemical costs alone would be in the neighborhood of $0.75 to $1.50 per gallon of waste processed, which is quite expensive by today's CFS standards. While further work on these systems may allow reduction in the required resin amount, or lead to less expensive resin systems, it is unlikely that these processes will be widely used for CFS work in the near future.

Another Organic Polymer System. Another system for the solidification of radioactive wastes using polyvinyl resins produced from monomers and catalysts was developed by Bahr *et al.* [19].

Thermoplastic Processes

Bitumen Processes. These systems have been marketed only for radioactive waste solidification in the United States, except for isolated circumstances where bitumen has been generated as a waste product and mixed with other waste products for disposal. Therefore, we will discuss bitumen processes in terms of their present use in the nuclear waste area, with the understanding that they could also be applied to other waste types.

Bitumen or asphalt is a mixture of high molecular weight hydrocarbons obtained as a residue in petroleum or coal-tar refining. Several types of bitumen are available, but the direct distillation product is the one most widely suggested for radioactive waste solidification. The properties of the material vary from batch to batch, but this variation can be minimized by specifying named products or grades of bitumen. Bitumen systems appear to be able to handle most reactor waste streams and permit a wide latitude of waste proportions. Substances that decompose at the working temperature of the bitumen process should not be added. Bitumen does burn and there is some evidence that the incorporation of oxidizing agents increases the fire risk.

There are four main types of processes that use bitumen for radioactive waste solidification: stirred evaporation, film evaporation, the emulsified bitumen process, and screw extrusion. With the exception of the emulsified bitumen process, these techniques rely upon the tendency of heated bitumen to flow and simultaneously allow the evaporation of water from the wastes. Screw extrusion appears to have gained the most widespread use. In this process, bitumen and either liquid or sludge wastes are continuously pumped into one end of a screw extruder that may contain one or multiple screws. The design and operation of the extruder is such that the bitumen and waste are intimately mixed and spread into a thin film on the heated surface of the extruder barrel [20]. This mechanical processing together with the maintenance of a temperature of approximately 200°C (392°F) affects the almost complete (99.5 percent) evaporation of the water contained in the waste and provides a homogeneous product. The evaporated water is vented through large disengaging sections called steam domes and passed through an oil separator before being condensed. Carryover of activity to the condensate is reported to be less than 0.1 percent. The bitumen waste mixture is discharged directly into solidification containers at the end of the extruder and allowed to cool. Average residence time in the extruder is a few minutes.

It is believed that one or two power reactors in the United States, and several in Canada, use bitumen solidification processes, although they are used more extensively in Europe. The Werner and Pfleiderer Company [21,22] has offered a system for use in the United States and has promoted it widely in the nuclear industry. Werner and Pfleiderer (WP) is a manufacturer of mixing and extruding equipment. In this process the liquid waste goes to an evaporator-compounder containing intermeshing, self-cleaning, multiscrew extruders that effectively remove the water and incorporate the radioactive residue into the matrix of bitumen. The hot mixture is poured into drums where it cools and solidifies. This system has been used at the West German Nuclear Center at Karlsruhe and elsewhere in Europe and in Argentina. The WP process is reported to be able to operate with a wide variety of waste, to eliminate the supernatent water, and to have excellent leaching properties. It is also considered to be expensive.

One problem that has been encountered with this process is with the solidification of soluble salts, such as those encountered from nuclear power plants. Water does slowly diffuse through asphalt, and when it encounters a hygroscopic salt, it forms hydrates that build osmotic pressure within the waste form, causing it to swell and eventually fall apart. This has been observed in several instances, and places some limitation on the use of hot-melt bitumen systems. Concern has also been raised about the biodegradability of asphalt.

Another approach to the use of bitumen or asphalt in CFS processing is to use an asphalt–water emulsion (which is available commercially for use in road surfacing and roofing) to mix with the waste. Subsequently, the water is volatilized by heating the mixture to about 160°C (320°F) and the residue containing the radioactive solids is drained into the disposal container where it solidifies. This has been shown to provide good leaching results and is sufficiently resistant to radiation for use with intermediate-level radioactive wastes [23]. A variation on this is the process used by Nicherecki Chemical Company in Japan—an asphalt emulsion to settle solids from a slurry, with subsequent solidification by Portland cement and sodium silicate. This process has not been very widely used, but does have the advantage of reducing sludge volume before solidification.

A different concept for the use of asphalt was set forth by the author in 1986 [24]. In this approach, no heat is used. Emulsified asphalt is caused to coalesce into a hydrophobic mass by the interaction with counterions in the waste or by additions to the mixture of asphalt and waste. When the asphalt emulsion breaks, the water is released and the organic phase forms a continuous matrix of hydrophobic asphalt around the waste solids. This process is intended especially for use with high solids wastes, contaminated soils, emission control dusts, and the like.

Other Thermoplastic Polymers. Although some laboratory work has been done with thermoplastic polymer systems for CFS work, nothing in this area has been commercialized or even reported on a pilot scale other than one project conducted by TRW Systems Group [25]. This system consists of cementing the waste with 3 to 4 percent by weight of a polybutadiene binder and then encapsulating the cemented waste in a quarter-inch-thick polyethylene jacket. Obviously, this encapsulated material exhibited excellent retention of contaminants under leaching conditions by a broad spectrum of aqueous solutions. It also withstood considerable mechanical stress. The engineering study on this process indicated a cost of approximately $100 per ton of waste treated at a throughput rate of 20,000 tons per year. However, it was required that the waste be dried before being cemented with the polybutadiene binder, and this would be an additional cost for liquid wastes. The resins used in this process accounted for approximately 50 percent of the total cost.

Other macroencapsulation processes include those by Bondico, Inc. [26] and the author [27]. The former describes a system in which the waste is placed inside a thermoplastic container and the cover permanently fused on. The latter is a continuous process of encapsulation combined with solidification, and is intended for use primarily with wastes that have volatile components or emanate odors. The CFS-treated, but not yet hardened, waste is continuously extruded into a casing of polyethylene or other flexible, low-permeability membrane where it is isolated from the environment while it solidifies. The process is described graphically in Figures 15-1 to 15-3. As shown in Figures 15-2 and 15-3, the extruded product can also be laid in place in a landfill, or it can be placed in a shipping device or other temporary area for later disposal.

Before using a thermoplastic process, an assessment should be made of the

Figure 15-1 CFS continuous encapsulation system—concept.

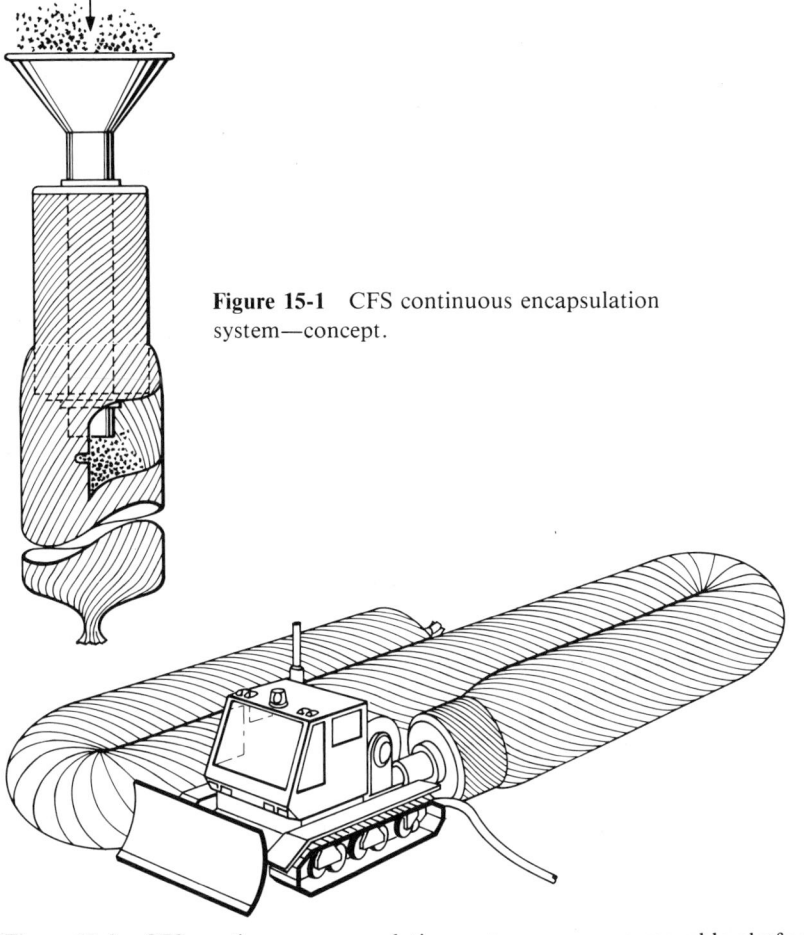

Figure 15-2 CFS continuous encapsulation system—use on a movable platform.

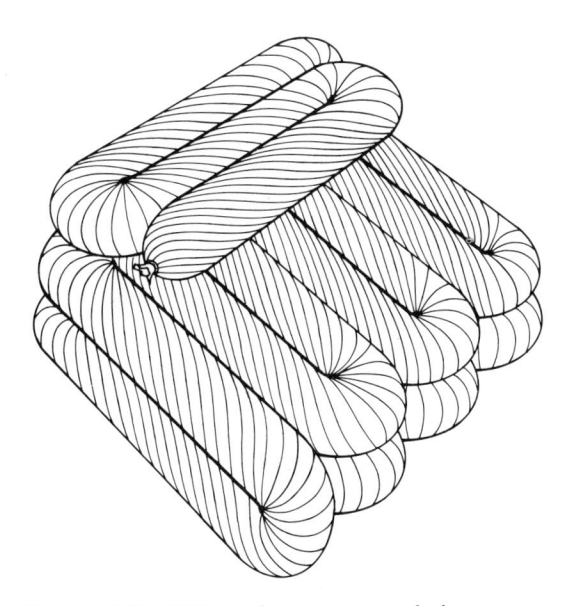

Figure 15-3 CFS continuous encapsulation system—emplacement of encapsulated waste in a landfill.

TABLE 15-1 Long-term Chemical Resistance of Organic Polymers (Resins) Used in Solidification

	Resistance of Resins		
Chemical	Conventional Polyethylene	Linear (High-density) Polyethylene	Polyvinyl Chloride
Acetic acid 50%	Excellent	Excellent	Moderate
Benzene	Poor	Moderate	Not resistant
Butadiene	Not resistant	Not resistant	Not resistant
Carbon tetrachloride	Poor	Moderate	Moderate
Chloroform	Poor	Moderate	Not resistant
Chromic acid	Excellent	Excellent	Excellent
Cresol	Poor	Poor	Poor
Dichlorobenzene	Poor	Poor	Not resistant
Diethyl ether	Not resistant	Not resistant	Not resistant
Gasoline	Poor	Moderate	Poor
Metallic salt solution	Excellent	Excellent	Excellent
Sulfuric acid (concentrate)	Moderate	Moderate	Not resistant
Trichloroethane	Not resistant	Not resistant	Not resistant

possible interaction of the polymer with waste components, especially organics and strong oxidizing agents. Table 15-1 lists the long-term resistance of several polymers with a number of chemicals.

NONCHEMICAL SOLIDIFICATION

The use of sorbents that do not chemically react with the waste or with other additives was fairly prevalent until the 1985 landban on bulk liquids in landfill [28]. Inexpensive materials such as clay and vermiculite were used to solidify both bulk and containerized waste by generators, treaters, and landfill operations. Sorbents alone are still allowed for certain containerized wastes, but it is expected that this use will also disappear, at least for hazardous wastes. Since no chemical reaction normally occurs between the waste and the absorbent, liquids so absorbed can in principle be squeezed from the waste–absorbent combination by application of physical pressure. In addition, such absorbent systems may be highly leachable and thus unsatisfactory for use with hazardous wastes where immobilization of constituents is required. In practice, however, some sorbents are capable of containing liquids even under considerable pressure. The author, for example, has demonstrated that pressures up to 10,000 psi will not cause the release of a variety of aqueous and organic liquids from several proprietary sorbent systems [29]. Sorption processes, however, must be very carefully controlled so that the absorbent is not oversaturated.

Such systems are used extensively in spill control work where quick, foolproof action by nontechnical personnel is required for safety reasons. The residues from such spills, both large and small, are generally considered a special category of waste, are usually not treated further, and are often landfilled locally in municipal refuse landfills unless extremely toxic materials are involved. Examples of the latter case are spills of PCBs or dioxin-containing materials. Industrial operations usually have stocks of granulated clay, sawdust, vermiculite, and various proprietary sorbents on hand for spill-control use. Fire services, HAZMAT teams, and laboratories also make extensive use of sorbents, and both the products and the delivery systems are becoming more sophisticated. One example of this is the use of pressurized containers of sorbent for more convenient and safer application to spills. Chem-Technics Inc. pioneered this use with a product known as the Spill Arrester™ [30] shown in Figure 15-4. The pressurized applicator is used to apply the sorbent from a distance, providing easier, more uniform, and safer operation, as shown in Figure 15-5. For small spills, such as those in laboratories, a smaller hand-pressurized applicator is used (Fig. 15-6).

There are literally hundreds of spill-control products in the market. Some time ago, the author conducted an extensive evaluation of the sorptive properties of a number of generic and proprietary agents used on a variety of chemicals and solvents that might be spilled. The results of some of this work are given in Table 15-2 with permission of Environmental Technology Associates Inc. The sorptive ability is expressed in pounds of sorbent required to just immobilize a gallon of the liquid with no free liquid remaining. Since cost is an important factor, and the costs of these agents vary considerably, the resulting cost ($) per gallon of liquid sorbed is

Figure 15-4 Pressurized spill sorbent application device. (From Environmental Technology Associates [30].)

TABLE 15-2 Characteristics of Selected Spill Absorbents

Code	Product Name	Chemical Cost ($/lb)	Chemical Type	Neutralizes Acids?	Selectively Absorbs Organics?	Ease of Use[a,b]	Typical Final Product Characteristics
910KAX	CT hydrophobic absorbent	0.200	Organic	No	Yes	1A	Particulate
710AAA	CT SPILL ARRESTER absorbent	0.076	Inorganic	Yes	No	3C	Solid Cake
710AAN	CT all-purpose absorbent	0.183	Mixed	No	No	2B	Particulate
900MAS	Fumed silica, AEROSIL 200	2.310	Inorganic	No	No	5E	Gel
900KAM	Hydrophobic polymer, O-SORB	2.600	Organic	No	Yes	2C	Gel
900KAW	Expanded sodium silicate, HAZORB	1.750	Inorganic	No	No	1E	Particulate
900KAF	Polymer, EMCO IMBIBER BEADS	6.700	Organic	No	Yes	1B	Gel
900KAU	Spill pillow, FISHER	5.600	Organic	No	No	1E	Pillow
900EAN	Rice hull ash, ENVIROGUARD	0.230	Inorganic	No	No	2C	Particulate
900KAL	Starch copolymer, WATER-LOCK	6.000	Organic	No	No	4D	Gel
900MBB	Polymer, CARBOPOL 940	4.000	Organic	No	No	4D	Gel
900MAA	Guar gum, JAGUAR HP-11	1.450	Organic	No	No	3C	Gel
900KAN	Granular clay, OIL-DRI	0.050	Inorganic	No	No	1B	Particulate
900CAA	Kaolin clay, powder	0.055	Inorganic	No	No	3C	Cake
900AAA	Portland cement	0.046	Inorganic	No	No	3C	Solid Cake
900DAA	Hydrated lime	0.046	Inorganic	Yes	No	3C	Cake
900EAG	Lime kiln dust	0.068	Inorganic	Yes	No	3C	Cake
900EAM	Fly ash, Type C	0.026	Inorganic	No	No	3C	Cake

[a] *Dustiness*: (1) Nondusting granules, medium to high bulk density, (2) low-dusting powder and/or granules, medium to high bulk; (3) medium-dusting powder, high bulk density, settles rapidly; (4) high-dusting powder, low bulk density, settles slowly; (5) very high-dusting powder, very low bulk density, settles very slowly.

[b] *Ease of Application and Absorption Speed*: (A) Very good in all situations; (B) good in most situations; (C) satisfactory outdoors in low wind; (D) difficult in most situations, tends to float, may require mixing; (E) very difficult in any situation, floats away, requires mixing.

TABLE 15-2   Characteristics of Selected Spill Absorbents (*Continued*)

Acids		Alkalies		Gasoline		Motor Oil		Chlorinated Oganics	
(lb/gal)	($/gal)	(lb/gal)	($/gal)	(lb/gal)	($/gal)	(lb/gal)	($/gal)	(lb/gal)	($/gal)
5.26	1.05	5.88	1.18	2.63	0.53	5.26	1.05	1.89	0.38
6.67	0.51	17.50	1.33	16.40	1.25	24.40	1.85	17.50	1.33
				3.45	0.63	6.67	1.22	3.33	0.61
0.58	1.34	0.83	1.92	0.42	0.97	0.50	1.16	0.54	1.25
				2.86	7.44	4.17	10.84	1.77	4.60
0.65	1.14	0.50	0.88	0.58	1.02	0.58	1.02	0.92	1.61
13.60	91.12	12.80	85.76	2.63	17.62	8.33	55.81	4.76	31.89
1.06	5.94	1.08	6.05	0.75	4.20	1.85	10.36	1.16	6.50
3.51	0.81	6.25	1.44	3.23	0.74	3.12	0.72	3.57	0.82
1.14	6.84	0.46	2.76						
1.91	7.64	1.00	4.00	1.85	7.40	3.12	12.48	2.44	9.76
5.74	8.32	5.56	8.06	12.00	17.40	14.10	20.45	11.60	16.82
14.90	0.75	12.50	0.63	14.70	0.74	13.30	0.67	16.70	0.84
7.69	0.42	7.69	0.42	4.17	0.23	2.33	0.13	5.13	0.28
5.06	0.23	16.40	0.75	12.30	0.57	22.20	1.02	13.70	0.63
5.40	0.25	7.14	0.33	6.67	0.31	7.14	0.33	5.00	0.23
6.56	0.45	9.52	0.65	7.14	0.49	9.09	0.62	8.33	0.57
11.00	0.29	19.20	0.50	13.70	0.36	25.00	0.65	15.90	0.41

Figure 15-5 Use of the pressurized spill sorbent device on a spill. (From Environmental Technology Associates [30].)

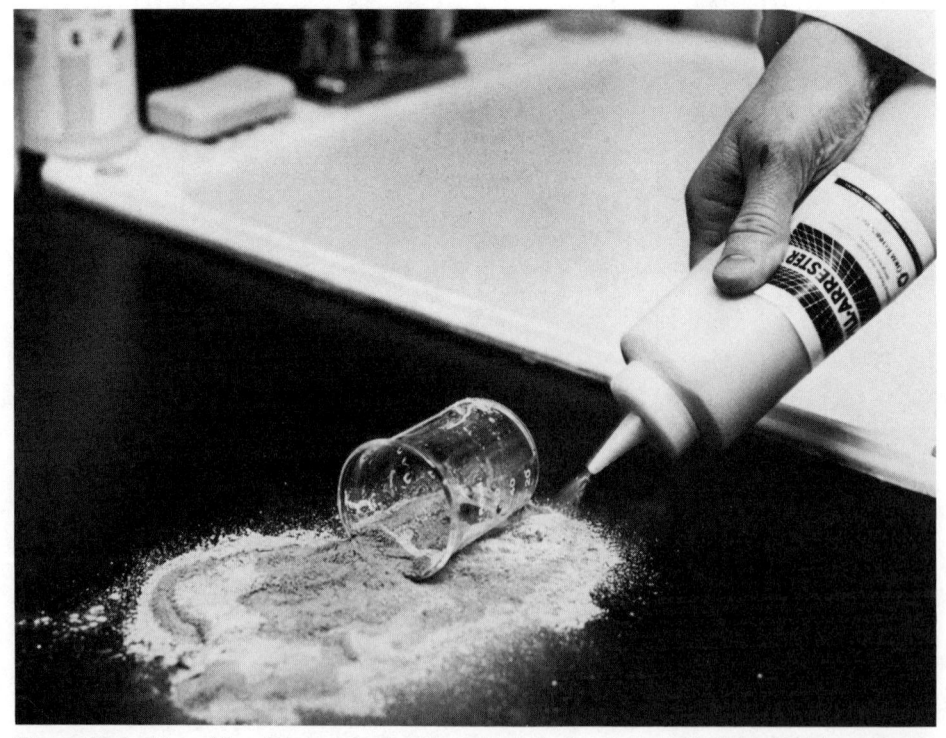

Figure 15-6 Use of hand-pressurized spill sorbent device on a spill. (From Environmental Technology Associates [30].)

also given. In controlling spills, factors other than cost and efficiency are important: acid/alkali neutralizing ability, ease of use, ability to selectively sorb organics (for spills of water-immiscible organics on water), and properties of the sorbent–liquid product. These are also given in Table 15-2. While discussion of spill control and sorbents is not a direct subject of this book, it is a closely associated field in which many of the products have potential applicability as additives to CFS systems.

It is important to understand, however, that the landbans do not rule out the use of sorbents as part of a reactive system, such as any of those already described in this and earlier chapters. In fact, at least part of the functionality of systems using kiln dusts and flyash is due to absorption of water or organic liquids from the waste. One vendor of sorbents has marketed a product, EnviroGuard Plus, combining an effective sorbent, rice hull ash, with various cementitious agents [31]. This combination meets the requirements of the liquid landban while providing an easy-to-use product that has high sorption efficiency. Another similar combination, Pro-Fix, has already been discussed in Chapter 11. ProFix reacts in a different way than EnviroGuard Plus, forming soluble silicates that react with metal-containing sludges to form a solid, low-leaching product. It is very likely that we will see more pre-blended products of this sort in the future, especially for use on small quantities of wastes and by small generators who have neither the equipment nor the technical resources to design and operate a full CFS system.

SYSTEMS COMBINING DEWATERING AND CFS

Dewatering/volume reduction is frequently a companion process to CFS, and can take many forms. Water may be removed using thermal drying processes, but more commonly, dewatering is done by filtration or centrifugation. Sludges especially amenable to filtration are those that are not gelatinous in nature and that can be dewatered to less than about 75 percent water, although the water content of "solid" residues varies anywhere from zero to 80 percent or more. Clarifiers, thickeners, and settling basins can also be used for dewatering. The dewatering process, whatever the mechanical or thermal means used to accomplish it, is important to CFS technology. Dewatering means volume reduction, which equates to cost reduction when treatment and disposal costs are high. It is being practiced to a high degree today, and its use can be expected to increase in the future. What this means to CFS technology is higher solids wastes with little free (i.e., mechanically separable) water. This, in turn, partially obviates the need for processes that were designed to handle low solids wastes, such as the cement/soluble silicate, unless they serve some other purpose.

There is another connection between dewatering and CFS technology. The high solids sludges and filter cakes from dewatering operations are often hard to mix with CFS reagents, requiring expensive, high-energy mixing equipment. In some cases, CFS operators even add water to aid mixing, which is an undesirable practice for obvious reasons. Another approach is to add the CFS reagents before dewatering, where they are easy to mix with the more dilute slurry or sludge. This approach is workable within the following limitations.

1. The CFS reagent is slow reacting, so that it does not interfere with the dewatering process.

2. The CFS reagent has low water solubility, so that it is not lost into the filtrate or decant stream.
3. The CFS reagent does not contaminate the filtrate or decant stream, so that its reuse or discharge is adversely affected.
4. The physical system has a conditioning tank or other element where the reagent can easily be added and mixed in, or this element can easily and inexpensively be added.
5. The mix ratio of CFS reagent is not so high that it produces so much more solids that the design capacity of the dewatering system is exceeded.

Another desirable property of the CFS reagent is that it aids in filtration, producing a residue with higher solids and/or increasing the dewatering rate. Some reagents do have this property.

The author has done several laboratory- and pilot-scale tests of this concept, using filter-press dewatering and a cement-based CFS system, and it is feasible in practice. Others have reported success using lime-based systems (private communication). Cement-, lime-, or pozzolan-based systems seem to work best, with the primary limitation being the pH of the filtrate. It may be necessary to adjust filtrate pH if it is to be discharged directly into surface water or POTWs. For recycle use, the pH may be satisfactory as is. The reagent should be introduced into the system as close to the dewatering device as possible to minimize reaction in this area. Mixing should be thorough, and may be attained with moderate-speed impeller mixers in the conditioning tank (in the case of filter-press dewatering). Care must be taken not to interfere with the conditioning process, so the CFS reagent is usually introduced after flocculation is complete. It seems likely that certain CFS reagents may aid in conditioning, replacing all or part of the conditioning agents, but this has not been verified.

IN SITU PROCESSES AND SYSTEMS

In situ literally means *in place,* and in the context of CFS means that the waste is *not* removed from the storage or disposal area to be processed through the system. Usually, treatment is accomplished by mixing the reagent into the waste storage zone by some mechanical means such as a backhoe, auger, or rotary tilling device. However, in many cases there are other options. For example, permeable soils may be treated to immobilize contaminants by injection or irrigation of a reagent solution into the contaminated area. In situ solidification was one of the earliest techniques used to physically stabilize sludge lagoons and is still being used for this purpose. Most of the commercial CFS processes that we have discussed can be used with in situ delivery systems, and the characteristics of the product are much the same if the reagent is mixed in properly. The latter, however, is easier said than done.

Most of the hazardous waste management firms in the United States have used this method, and a number of construction and engineering organizations have developed both equipment and expertise for in situ treatment. The EPA has encouraged this technology for use in the Superfund and other remedial action programs, and several evaluations have been done under the Superfund Innovative Technology

Evaluation (SITE) program. A number of papers have been published describing the results of these and other studies [32–34].

Perhaps the largest such project ever undertaken in the United States was done at one of Chemical Waste Management's facilities in 1985. Seven hundred and fifty thousand cubic yards of lagoon sludge were solidified with a kiln-dust-based system using in situ mixing with backhoes. The solid product was then retained in a pile on-site awaiting regulatory approvals for on-site landfill. Many similar projects have been done during the last decade at numerous sites in the United States. The most common device used for mixing is the backhoe, because it is widely available and does a satisfactory job within limits. A large-scale backhoe operation is shown in Figure 15-7. The reagent is introduced by pneumatic or mechanical conveyance to the surface of the waste, where it is then mixed in with the backhoe. This may be repeated until sufficient reagent has been added to produce an acceptable solid.

Obviously, this arrangement is rather crude. It produces a great deal of dust (from the reagent) and the mixing is generally not very thorough. For the latter reason, the method is not recommended for CFS work where fixation is required. Also, because of the relatively poor mixing and loss of reagent due to dusting, the method is limited to low-cost systems such as kiln dusts and flyash. This has caused CFS vendors and their contractors to move toward more sophisticated reagent feed and mixing systems. One of these is shown schematically in Figure 15-8, and in Figure 15-9 in actual operation at a sludge lagoon. The hollow tubes convey the reagent underneath the waste surface, minimizing dust and maximizing reagent

Figure 15-7 Large-scale in situ CFS project using backhoes. (From National Lime Association.)

Figure 15-8 Injector-type in situ solidification. (Schematic). (From Enreco, Inc.)

Figure 15-9 Injector-type in situ solidification in operation. (From Enreco, Inc.)

usage. The remainder of the assembly serves to mix the reagent into the waste as it is moved back and forth. Another injector device is shown in Figure 15-10. In this case, each injection tube has a motor-driven mixer assembly attached to the end for better mixing.

All of the injectors shown in Figures 15-8–15-10 are mounted on backhoes, and their reaches and working depths are limited only by the size and power of the backhoe. They can work to depths of 15 feet or more in sludge ponds. For shallow ponds and contaminated soil areas, a different type of injector/mixer can be used. Two variations on this are shown in Figures 15-11 and 15-12. The first is a hollow

Figure 15-10 In situ injection system. (Harmon Environmental Services, Inc.)

Figure 15-11 Used for high solids stabilization not requiring high-energy mixing. (Harmon Environmental Services, Inc.)

Figure 15-12 Mixer for high solids stabilization where high-energy mixing is required. (Harmon Environmental Services, Inc.)

tine injector mounted on a tractor and pulled through the waste, much like tilling a field. The second uses a rotary tiller, again tractor mounted, to achieve more thorough mixing. These devices can work in waste depths from several inches to about five feet. They are depth-limited by the tractor's size and traction ability.

These in situ injectors and mixers are limited both in the depth of the waste impoundment (or soil contamination) and in the reagents that can be used in the system. Another approach, described by Stinson and Sawyer [34], uses the GeoCon Deep Soil Mixing System. This device, shown schematically in Figure 15-13, consists of a drilling auger containing one set of cutting blades and two sets of hollow mixing blades attached to a vertical drive shaft. It works much like an auger post-hole driller, but is much larger and instead of removing the soil, it churns and mixes it in place. In operation, the auger is advanced to the desired maximum depth, the reagents are then injected in slurry form through the mixing blades, and mixing is continued during withdrawal of the auger.

The working end of the rig is shown in operation in Figure 15-14; in close-up, the auger is shown advancing downward into the soil (Fig. 15-15), and introducing reagents at depth (Fig. 15-16). Since each boring creates a column of treated material circular in cross section, the columns must be overlapped to assure full treatment. This can be done to provide area treatment, or to treat small blocks, create grids, or form walls, as shown in the diagrams in Figure 15-17. The sequence of operation may be visualized more accurately by referring to the explanatory drawing and schematic for a specific project (the one described by Stinson and Sawyer [34]) shown in Figures 15-18 and 15-19, respectively. In the operation, the cement-based reagent was slurried and injected as the auger advanced, then a sodium silicate solution was injected at the maximum depth to speed solidification in that portion of the column,

Figure 15-13 Schematic of a vertical auger mixer/injector operation. (From GeoCon, Inc.)

Figure 15-14 General view of a vertical auger mixer/injector operation. (From GeoCon, Inc.)

after which the auger was withdrawn without injection. This project appeared to be successful; final evaluation by the EPA will be completed in 1989. Reportedly, the system can also be modified for use on sludge ponds and for the treatment of contaminated bottom layers underwater (Fig. 15-20). In soil applications, depths of up to 120 feet can be treated.

In general, a number of grouting techniques and chemicals are potentially applicable to CFS in situ treatment. One of these, a grout system used at Oak Ridge, Tennessee, for radioactive waste disposal, was described in Chapter 13. Others use cement, bentonite/cement, and other materials to solidify unstable formations underground and also can be used to physically and chemically stabilize contaminated soils. These systems are generally applicable only to fissured rock and coarse-grained sandy or gravel areas, since the grout will not penetrate fine-grained, low-permeability materials. One possible exception to this is the use of soluble silicates to immobilize soluble metals; because the silicate is a solution that can be thinned to low viscosity, it can be pressure-permeated into many contaminated soil structures. If necessary, physical stability can also be enhanced by use of a two-part system that causes the silicate to gel in place. This was one of the earliest uses of soluble silicates in construction.

Chemicals other than soluble silicates have been used in several instances to decontaminate soils. The chemical is injected into the soil and allowed to react with the contaminant to immobilize or destroy it. This approach was used to destroy cyanide with dilute hypochlorite solution in a demonstration project in Georgia. One caution in such treatment schemes is that the reagent used must either be nontoxic, such as sodium silicate, or unstable in the soil environment, so that any excess

Figure 15-15 Close-up view of a vertical auger mixer/injector operation. (From GeoCon, Inc.)

does not cause secondary pollution. Also, potential reaction products must be determined so that they will not cause pollution. For shallow soil applications, in situ fixation/destruction may be accomplished simply by spray or trickle irrigation of the reagent solution at the surface, allowing it to permeate the contaminated area by gravity flow. Soil permeability, groundwater conditions, and rainfall are all factors that must be considered when designing such a system.

Recent interest in in situ soil washing systems will probably enhance the use

Figure 15-16 Close-up view of a vertical auger mixer/injector operation, with injection of CFS reagents. (From GeoCon, Inc.)

of in situ treatment as well. If the soil can be washed and the permeate recovered for treatment, then in situ treatment should be even easier. Also, if the permeate can be recovered, then a wider range of chemical systems can be used, since excess chemical can be recycled for both economic and environmental reasons, and any toxic reaction products recovered for separate treatment. As a general rule, in situ CFS systems of any of the types discussed to this point are less costly than removing

WALL TYPE   GRID TYPE

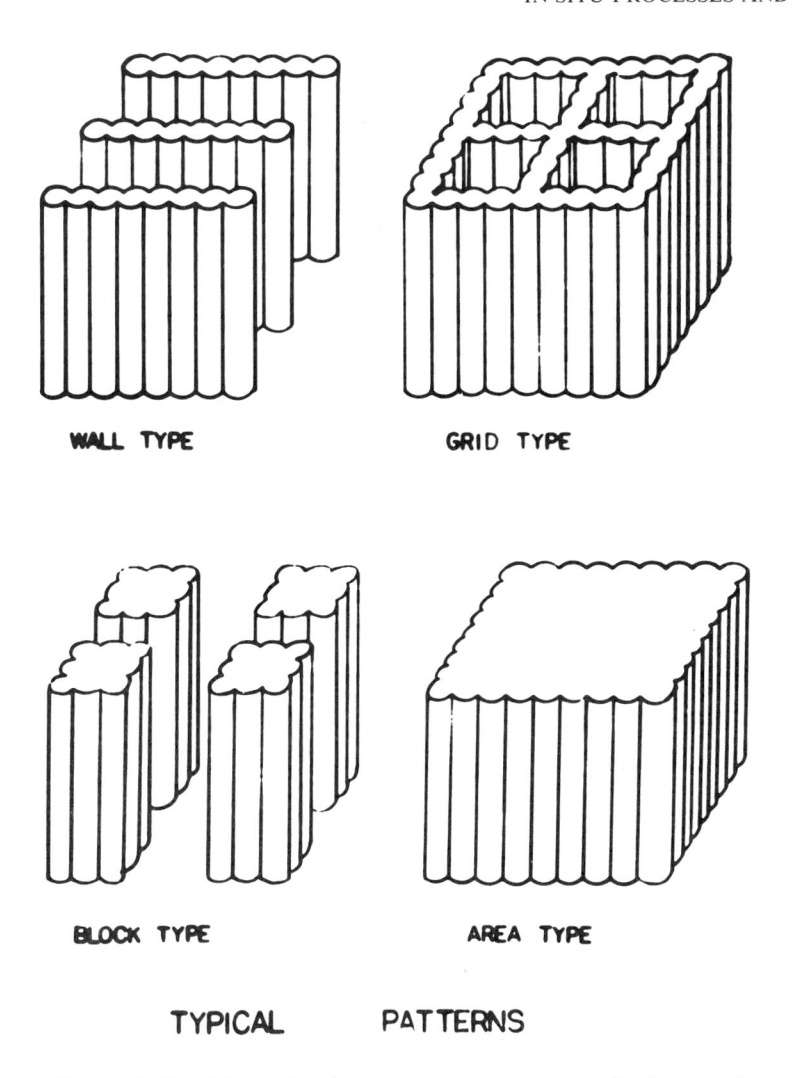

BLOCK TYPE   AREA TYPE

TYPICAL   PATTERNS

Figure 15-17 Schematic of typical treatment patterns in the use of the vertical auger mixer/injector operation. (From GeoCon, Inc.)

the waste for treatment and replacement. The primary question is whether the in situ method accomplishes the requirements of the project.

One other in situ method has recently received considerable attention. This is the vitrification technique developed by Battelle Pacific Northwest Laboratories for the U.S. Department of Energy [35]. In this method, electrodes are inserted into the soil or other waste, and an electric current is passed between them. This heats the soil to about 3600°F (1984°C) and causes it to melt, producing a solid, glasslike material that immobilizes waste constituents within the matrix. The process description and layout are shown in Figure 15-21. Since the "glass" product is formed from the waste itself, its properties are a function of that waste, unlike glassification

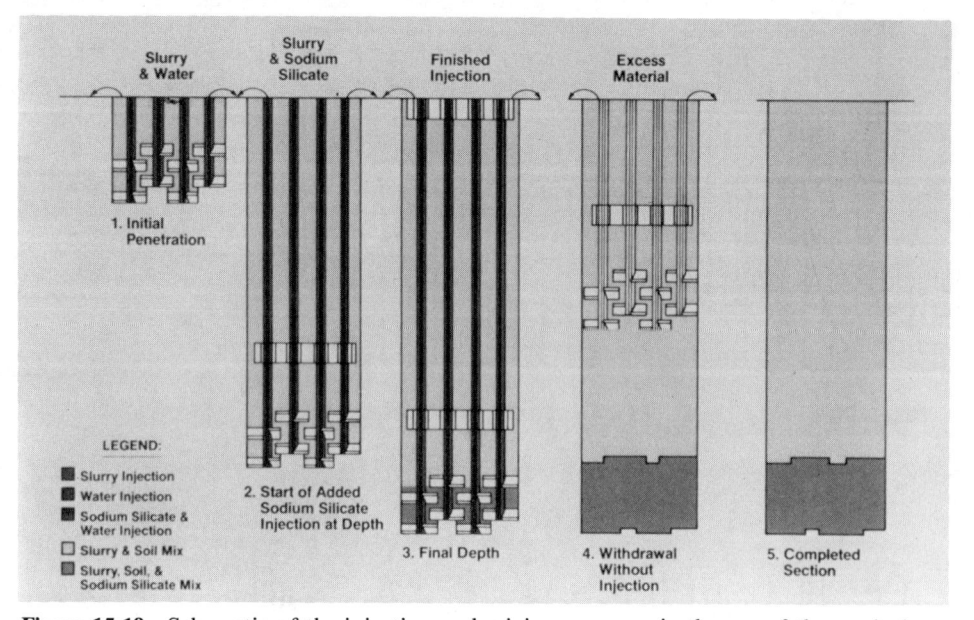

Figure 15-18 Schematic of the injection and mixing sequence in the use of the vertical auger mixer/injector operation. (From GeoCon, Inc.)

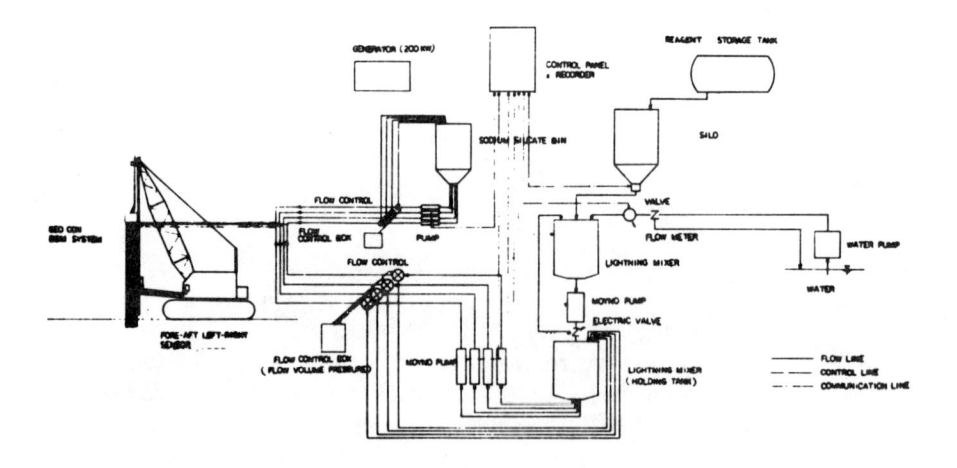

Figure 15-19 Flow diagram of the vertical auger mixer/injector operation. (From GeoCon, Inc.)

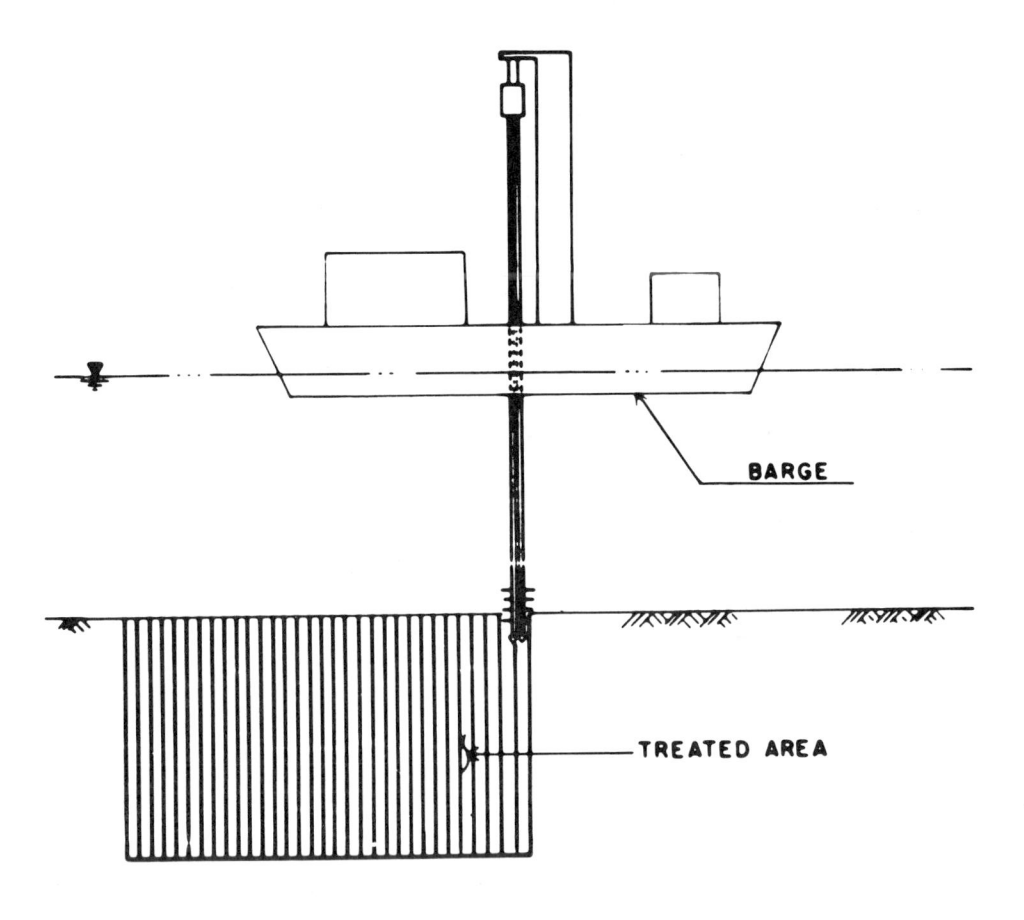

Figure 15-20 Schematic of the use of the vertical auger mixer/injector operation in the treatment of underwater contaminated areas. (From GeoCon, Inc.)

processes used in the nuclear waste industry where glass frits are added. However, it is expected that it will generally have low leachability and high structural integrity. The primary drawback to the process is cost, which is a function of the electrical rate and the water content of the waste or soil. Total costs have been reported to range from about $100 to $250 per ton for soil treatment. The process has been tested at both pilot and larger scale, but has not been truly commercialized as of this writing. The normal processing rate for a large-scale system is 4 to 5 tons per hour [36].

Figure 15-21 View of the set up and operation of an in situ vitrification plant. (From Battelle Pacific Northwest Laboratories [35].)

REFERENCES

1. U.S. Patent 4,424,148 (Jan. 3, 1984).
2. Rosenstiel, T. L., S. P. Bodett, and R. G. Lange. *ENVIROSTONE® Gypsum Cement, 10 CFR 61 Compliance Testing,* Topical Rep. Libertyville, IL: U.S. Gypsum Co., May 1984.
3. U.S. Patent 3,855,391
4. U.S. Patent 4,151,940
5. U.S. Patent 4,124,405
6. U.S. Patent 3,920,795 (Nov. 18, 1975).
7. U.S. Patent 4,015,997 (Apr. 5, 1977).
8. U.S. Patent 3,859,799 (Jan. 14, 1975).
9. U.S. Patent 4,108,677
10. Pojasek, R. B. In *Toxic and Hazardous Waste Disposal,* Vol. 4. Ann Arbor, MI: Ann Arbor Science Publishers, 1980.
11. Conner, J. R. U.S. Patent 4,518,508 (May 21, 1985).
12. Heacock, H. W. Alternative nuclear waste solidification processes. In *Proc. Symposium on Waste Management.* Tucson, AZ, pp. 177–206, July 1975.
13. Carlson, J. E. U.S. Patent 4,620,947 (Nov. 4, 1986).
14. Carlson, J. E. *Containment of Radioactive Wastes Using Improved Cementitious Binders.* Barnwell, SC: Chem-Nuclear Systems Inc., 1987.
15. Heacock, H. W. In *Proc. Symposium on Waste Management.* Tucson, AZ, pp. 177–206, 1975.
16. U.S. Patent 4,010,108

17. Filter, H. E. Polymeric solidification of low-level radioactive wastes from nuclear power plants. In *Toxic and Hazardous Waste Disposal,* Vol. 1. Ann Arbor, MI: Ann Arbor Science Publishers, 1980.
18. Mahalingham, R., P. K. Jain, and R. K. Biyani. Mixing alternatives for the polyester microencapsulation process for immobilization of hazardous residuals. *J. Hazardous Mater.* **5:** 77–91 (1981).
19. Bahr, W., S. Drobnik, W. Hild, R. Kroebel, A. Meyer, and G. Naumann. U.S. Patent 4,009,116 (Feb. 22, 1977).
20. Meir, G. U.S. Patent 3,971,732 (July 27, 1976).
21. U.S. Patent 3,305,894.
22. Doyle, R. D. Use of an extruder/evaporator to stabilize and solidify hazardous wastes. In *Toxic and Hazardous Waste Disposal,* Volume 1. Ann Arbor, MI: Ann Arbor Science Publishers, 1980.
23. Godbee, H. W., J. H. Goode, and R. E. Blanco. Development of a process for incorporation of radioactive waste solutions and slurries in emulsified asphalt. *Environ. Sci. Technol.* **2**(11): 1034–1040 (1968).
24. Conner, J. R. U.S. Patent 4,623,469 (Nov. 18, 1986).
25. Lubowitz, H. R., and C. C. Wiles. Encapsulation technique for control of hazardous wastes. In *Toxic and Hazardous Waste Disposal,* Vol. 1. Ann Arbor, MI: Ann Arbor Science Publishers, 1980.
26. Shaw, M. D. *Macroencapsulation: New Technology Provides Innovative Alternatives for Hazardous Materials Storage, Transportation, Treatment and Disposal.* Jacksonville, FL: Bondico, Inc., 1986.
27. Conner, J. R. U.S. Patent 4,518,507 (May 21, 1985).
28. U.S. EPA. *Prohibition on the Disposal of Bulk Liquid Hazardous Waste in Landfills—Statutory Interpretive Guidance.* Washington, DC, Mar. 19, 1985.
29. Conner, J. R. Unpublished treatability studies, 1985–1987.
30. Patents applied for by Environmental Technology Associates, Lima, OH. Licensed by Priority 1 Systems, Inc.
31. EnviroGuard, Inc. *EnviroGuard Data Pack.* Houston, TX, 1988.
32. Barich, J. J., J. Greene, and R. Bond. Soil stabilization treatability study at the Western Processing superfund site. *Superfund '87:* 198–203 (1987).
33. Kuhn, R. C., and K. R. Piontek. A site-specific in situ treatment process development program for a wood preserving site. *Superfund '87:* 182–186 (1987).
34. Stinson, M. K., and S. Sawyer. In situ treatment of PCB-contaminated soil. *U.S. EPA Site Papers:* 504–507 (1988).
35. Battelle Pacific Northwest Laboratories. *Battelle Technology Transfer Bulletin: In Situ Vitrification.* Richland, WA, 1987.
36. *The Hazardous Waste Consultant:* 4-7–4-10 (May/June 1988).

Chapter 16

DELIVERY SYSTEMS

Delivery systems are the means by which CFS processes are carried out. They include not only equipment but the complete process of developing, designing, planning, permitting, operating, controlling, and financing a CFS project. The project may be remedial or continuing; the equipment may be mobile/portable or fixed; the regulatory structure under the RCRA or the CERCLA. The equipment may be designed, built, owned, and operated by a CFS vendor, a generator, a central RCRA TSD facility, or even a public entity, or any combination of these elements. The ultimate disposal of the CFS product may be part of the delivery system, or it may be totally separate. Many of these elements are outside the scope of this book. They are, however, considerations that affect every phase of CFS technology, as we saw in Chapter 7, and should be kept in mind when designing equipment or conducting a laboratory treatability study. If the process cannot operate or the equipment be used because of limitations in other areas, then the work has gone for naught.

Here, we will consider the two primary types of delivery systems from the mechanical point of view. These are mobile/transportable operations and fixed installations. The latter category is more familiar to most readers, since it is the most common method of running any chemical process; therefore, we will consider it first.

FIXED INSTALLATIONS

CFS system operation at a fixed installation is usually only one of a number of technologies used in a central treatment facility. For example, a typical large central treatment facility will receive nearly every kind of hazardous and nonhazardous liquid waste with the exception of radioactive materials, explosives, or certain military wastes. Very often, facilities will have their own fleet of trucks for the transportation of waste, and will also accept wastes from other transportation firms. A well-run central treatment facility will require a complete analysis of the waste to be treated before it is accepted at the plant site.

A typical treatment plant will have equipment and facilities for storage, neutralization, CFS, and landfill, and perhaps also biological treatment, dewatering, and incineration. An example of a typical system is shown in the process flow diagram of Figure 16-1. The actual CFS work at the central facility can be done either with fixed equipment installed specifically for that purpose, with temporarily set up mixing, pumping, and holding equipment for infrequent batches, or with a mobile CFS unit. One early licensee for a cement–soluble silicate process in France chose to use a fixed unit at their plant. The process flow diagram for this operation is shown in Figure 16-2, and photographs of the facility are shown in Figure 16-3. A 50,000-gal waste holding tank feeds a mixer that also receives cement from a storage hopper and feeder. The liquid soluble silicate is fed into the suction side of the discharge pump, usually a 4- to 6-in. centrifugal pump, where it is thoroughly, but quickly mixed into the waste/cement combination. This arrangement is standard with fast-setting systems such as cement–soluble silicate, where introduction of the silicate causes a thickening/gelling reaction, often within seconds. By the means shown, even if the mixture set prematurely, only one pump and a section of discharge line is affected.

In this particular facility, a sludge thickener [Fig. 16-3(a)] is included, as well as a sludge storage and homogenization tank and a solidification area [Fig. 16-3(b)]. The homogenization operation is shown in Fig. 16-3(d) and (e). This allows a variety of compatible small loads to be dumped into the holding tank over a period of time (design capacity is two days in this case) and homogenized to provide a continuous waste feed of reasonably consistent composition. After homogenization, the waste is pumped to the mixing chamber [Fig. 16-3(f)] and mixed with waste from the cement storage silo [Fig. 16-3(c)]. This operation was designed for, and is only suitable for pumpable wastes.

For wastes that are not pumpable, a typical process flow and schematic may look like the one shown for a municipal sewage sludge solidification facility using flyash as the CFS reagent (Fig. 16-4). In this case, the sludge at about 20 percent

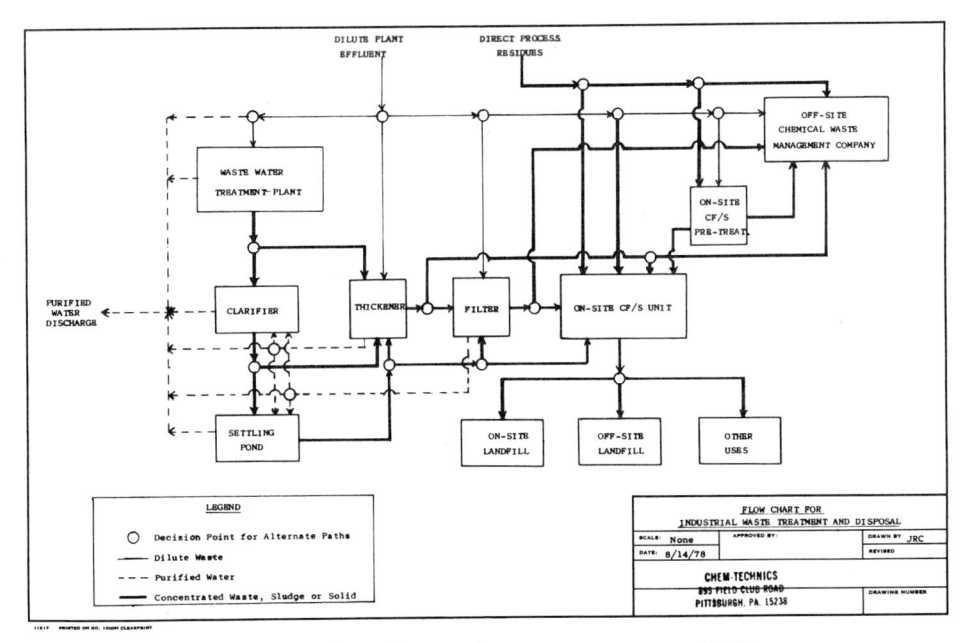

Figure 16-1 General process flow diagram for a waste treatment facility.

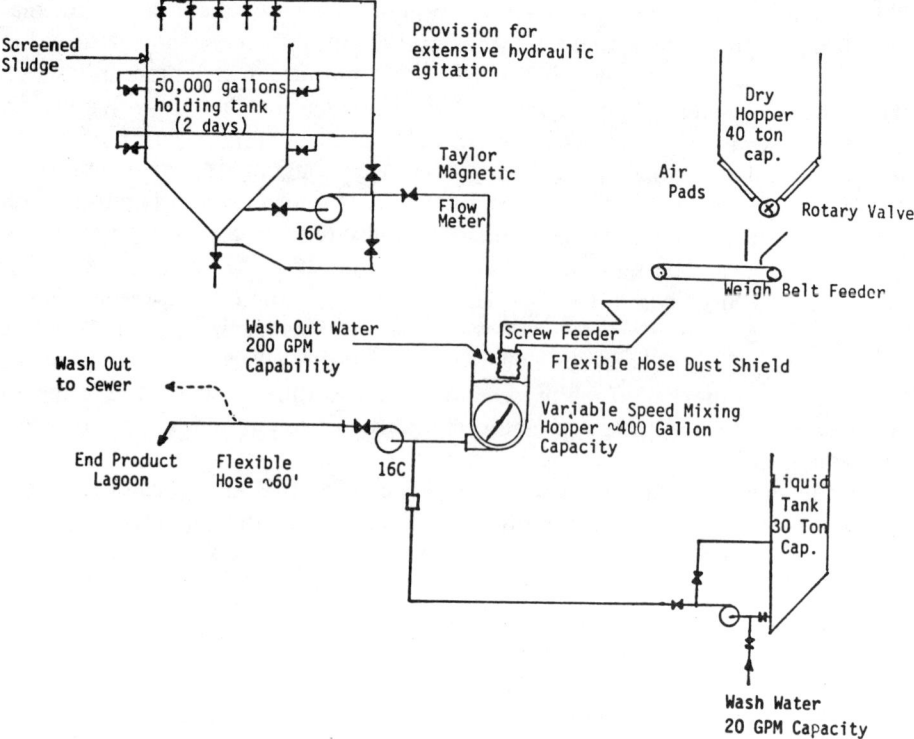

Figure 16-2 Process flow diagram for a central CFS facility in France.

solids is fed into the system by a solids loader, which feeds the sludge hopper, which in turn "pumps" the sludge to the pug mill as a semisolid. Another facility, designed by Chemical Waste Management for the stabilization of electric arc furnace dust (K061) is shown in the isometric drawing in Figure 16-5. This facility conveys the dry dust pneumatically to the weigh-batcher above the mixer, where it flows by gravity into the mixer and, after treatment, discharges by gravity into trucks for transport to the landfill. Solid reagents are also conveyed pneumatically, though liquid reagents are pumped, into the mixer. The reader should have noticed by this time that all of the systems discussed use different mixers. This emphasizes a basic rule in mixing: There is no such thing as an all-purpose mixer. When a specific waste stream is to be treated, as in the last two examples, the most efficient mixer can be chosen or designed simply by using good mixer engineering. In the first example of a central treatment facility, however, the mixer must be designed to handle a wide variety of wastes, or the waste must be homogenized, as was done in this case, to provide a uniform, pumpable stream. The latter approach works well where the average water content of the wastes received is high. Semisolids can be dispersed into liquid streams to produce a pumpable slurry.

More and more, however, the average solids content of wastes received at large, central TSD facilities such as those run by Chemical Waste Management is so high that the waste cannot be made pumpable unless non-waste water is added. This is generally undesirable, so a facility that can handle a wide range of viscosities and solids contents is required. One approach, used almost exclusively at these

FIXED INSTALLATIONS 481

(a)

(b)

Figure 16-3 Operations in a central CFS facility in France. (a) Sludge thickener and holding tank. (b) General view of CFS area of the plant. Sludge lagoon is in the background (in front of large tanks), cement and silicate tanks to the right, and solidification area in the foreground. (c) Close-up view of cement silo, silicate tank, and CFS treatment area. (d) Sludge holding lagoon with floating dragline and discharge pump. (e) Close-up view of floating pontoon dragline, with CFS area in background. (f) CFS mixing chamber. Cement is added here and silicate is added at suction side of discharge pump, as with a mobile treatment unit.

(c)

(d)

Figure 16-3 (*Continued*)

facilities since 1985 (the advent of the liquid-in-landfill ban) has been the backhoe pit. In this system, shown schematically in Figure 16-6, the waste is unloaded directly into a large, steel-lined pit where the solidification reagent is fed in and the mixture mixed with a conventional backhoe with modified bucket. While seemingly somewhat crude, the method has the great advantage of being extremely tolerant of de-

(e)

(f)

Figure 16-3 (*Continued*)

bris, including the plastic liners used in many trucks and roll-off boxes to contain any free liquid. Otherwise, the system must be able either to remove trash down to a level compatible with the mixer, or grind it to small size before mixing. Either option has disadvantages. The backhoe pit's main disadvantage is that thorough mixing to meet the new landban standards takes considerable time, which makes

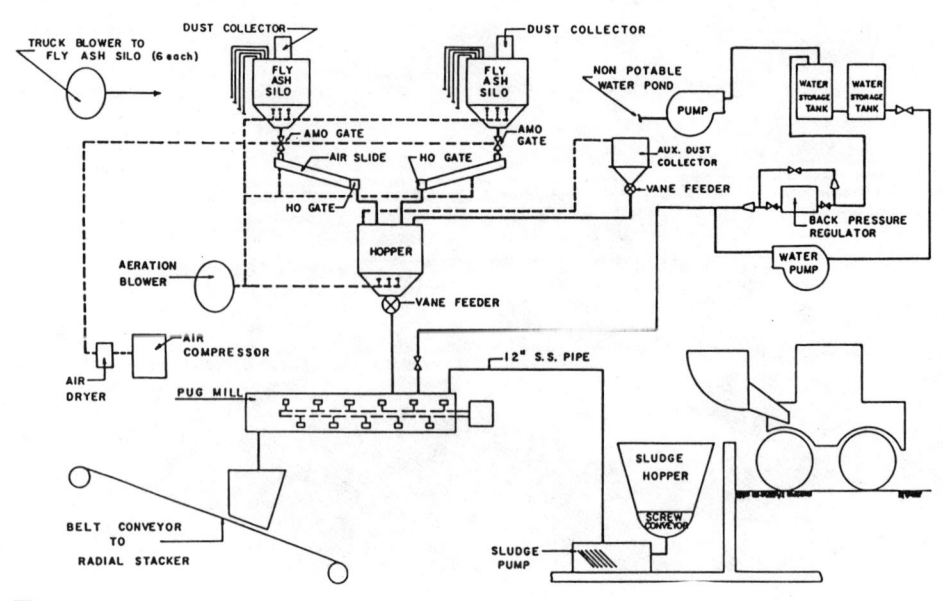

Figure 16-4 Process flow diagram for a sewage sludge solidification plant.

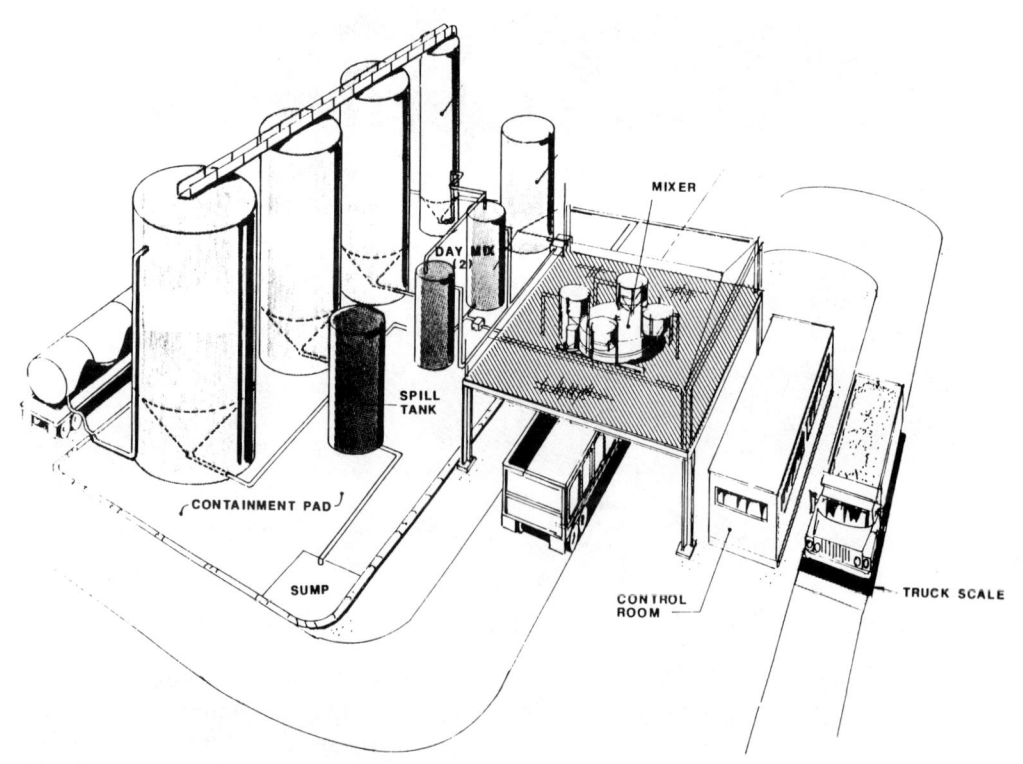

Figure 16-5 Artist's drawing of a CFS plant for electric arc furnace dust treatment. (From Chemical Waste Management, Inc.)

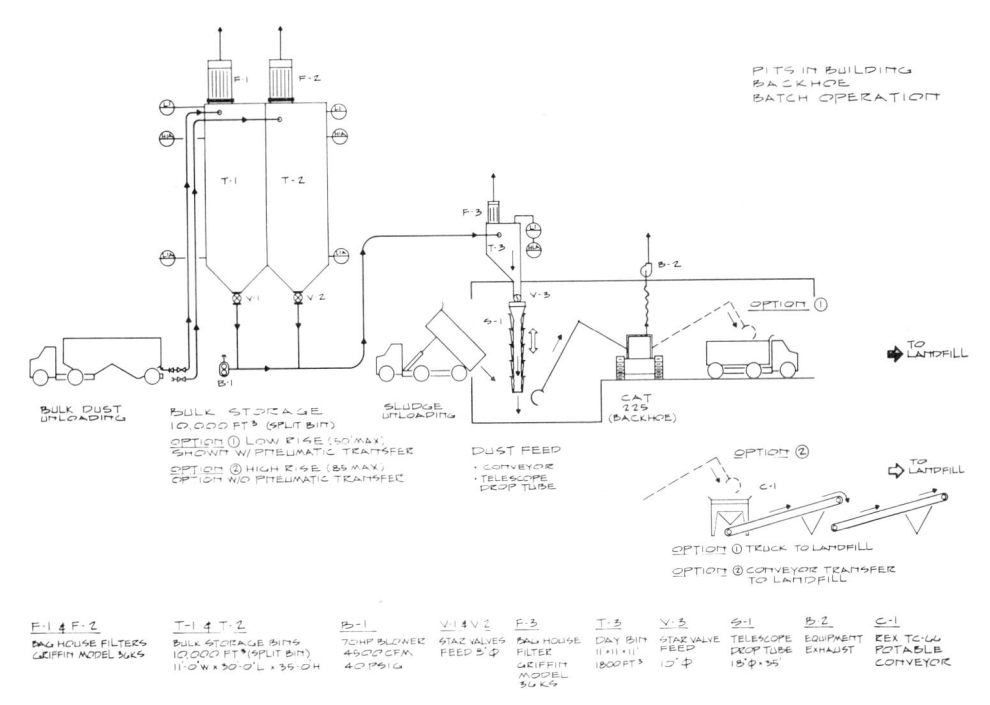

Figure 16-6 Process flow diagram for CFS at a central hazardous waste management facility—Backhoe operation. (From Chemical Waste Management, Inc.)

the system inefficient from a cost point of view. Dust and volatile vapor control is also more difficult, requiring complete enclosure of the building under negative pressure in some circumstances.

If the trash problem is not too severe, the system shown in Figure 16-7 can be used. Here, the receiving facility consists of separate areas for pumpable sludges and for solids. Both are fed into a batch mixer, the former by pumping and the latter by frontloader, as shown in the inset. Large trash, such as liners, can be picked out by a mechanical grapple or other device, or separated by the loader operator. There are a great many variations possible on such a system, depending on throughput rate, topography of the site, and the waste mix to be treated.

Type of System

Fixed CFS systems can be categorized into three groups: onstream, (normally) continuous systems; semicontinuous systems using intermediate storage; and batch systems.

On-Stream Systems. The most traditional way to utilize any waste treatment process is to apply it continuously to the waste as it is generated. This is common in waste water treatment, but its application in CFS has been primarily with streams such as fluegas desulfurization sludge and sewage sludge, both of which are not defined as hazardous wastes. This is the easiest type of system to design and operate, especially if the scale is large enough to permit automation and the waste feed is

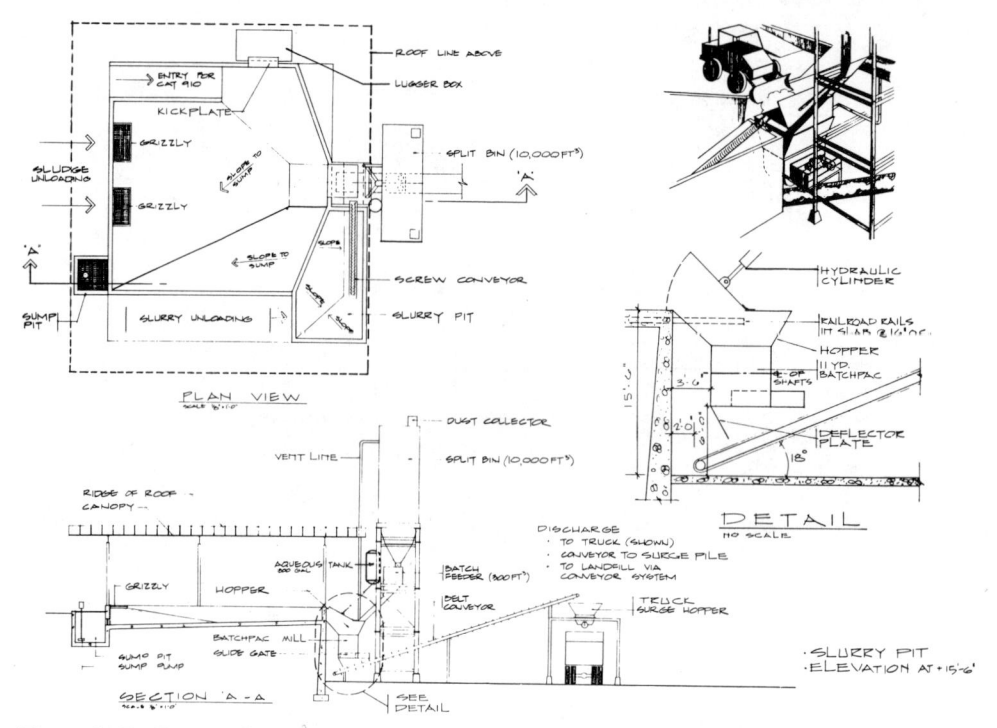

Figure 16-7 Process flow diagram for CFS at a central hazardous waste management facility—Throughput mixer operation. (From Chemical Waste Management, Inc.)

consistent. One such system will be discussed later. Given the necessary conditions, this system generally provides the lowest possible unit cost for waste treatment.

Semicontinuous Systems Using Intermediate Storage. Some discussion of this kind of system was presented previously, when we discussed the cement/silicate operation in France. It allows the wastes to be stored and mixed together, if desired, until sufficient material is available to allow economical processing. The processing itself can be continuous for the "batch" being treated; generally, the processing may take hours to days to complete. This is a very flexible approach—it can be operated on a variety of scales and with a wide diversity of wastes. However, complete automation is rarely practical, so the system has higher labor costs and the resultant unit costs are higher than for continuous systems. One significant advantage is that the waste mixture can be sampled and analyzed, and even subjected to a fast treatability study, before the "batch" is processed.

Most mobile treatment systems really operate this way, as do some central TSD facilities. An option on this approach is to use one mobile treatment unit to service several central treatment facilities, thus maximizing utilization of equipment. One operation in the United Kingdom uses the mobile treatment unit both for on site jobs at the point of waste generation and for treatment of stored wastes at their own central sites. There will be more discussion of this later.

Advantages of the semicontinuous approach using intermediate storage are:

- Ability to use either fixed or mobile equipment
- Flexibility to handle a variety of waste streams

- Allows homogenization of a variety of wastes, allowing the optimum combinations to be used, permitting the neutralization of one waste with another, and generally minimizing reagent costs
- Minimizes the redundancy necessary with an on-stream system

Disadvantages are:

- Space requirements for storage are large
- Regulatory complexities may be increased; where different treatment standards are in effect for different waste streams, the most severe standard will be applied to the whole "batch"

Batch Systems. One type of batch system was shown in Figure 16-7. Another is displayed in Figure 16-8. In the latter case, the CFS system is tied into a plate-and-frame filter press, which is itself a batch-type operation. This sort of matchup is ideal, with the CFS batch sized to that of the filter press or other device generating the waste. The unit shown in Figure 16-8 was designed to use bagged reagents rather than bulk, and deposits the waste in 55-gal drums for final disposal. Advantages of batch systems are:

- Low capital investment
- Minimal space requirement
- High degree of control over the operation
- Can be operated part-time by personnel with other duties, saving on the cost of dedicated personnel
- Allows pretreatment or other operations in the same mix tank

Disadvantages are:

- High labor cost per unit waste processed
- High reagent costs if bagged reagents are used

Size of the System

System size and type are independent of each other in principle, but not in practice. Small systems are generally batch, very large systems continuous. In between, they may be either.

Small. A small system for nonpumpable waste was presented in Figure 16-8. Another small system, this one for pumpable waste, is shown in Figure 16-9. Both use bagged reagents, but the latter unit uses manual bag opening, while the larger unit (Fig. 16-8) has a bag-opening machine. Such individual options depend primarily on the trade-off between capital cost and operating cost, and the required throughput rate. For example, if the unit shown in Figure 16-9 is to have relatively high throughput, it would be equipped with a bag-opening machine or even bulk storage and feeding of reagents. It could then be converted into a medium-size unit, because the mixing cycle is fast and simple with pumpable wastes. In the case of the Figure 16-8 unit, however, throughput rate is determined by the filter press, not the CFS system.

488 DELIVERY SYSTEMS

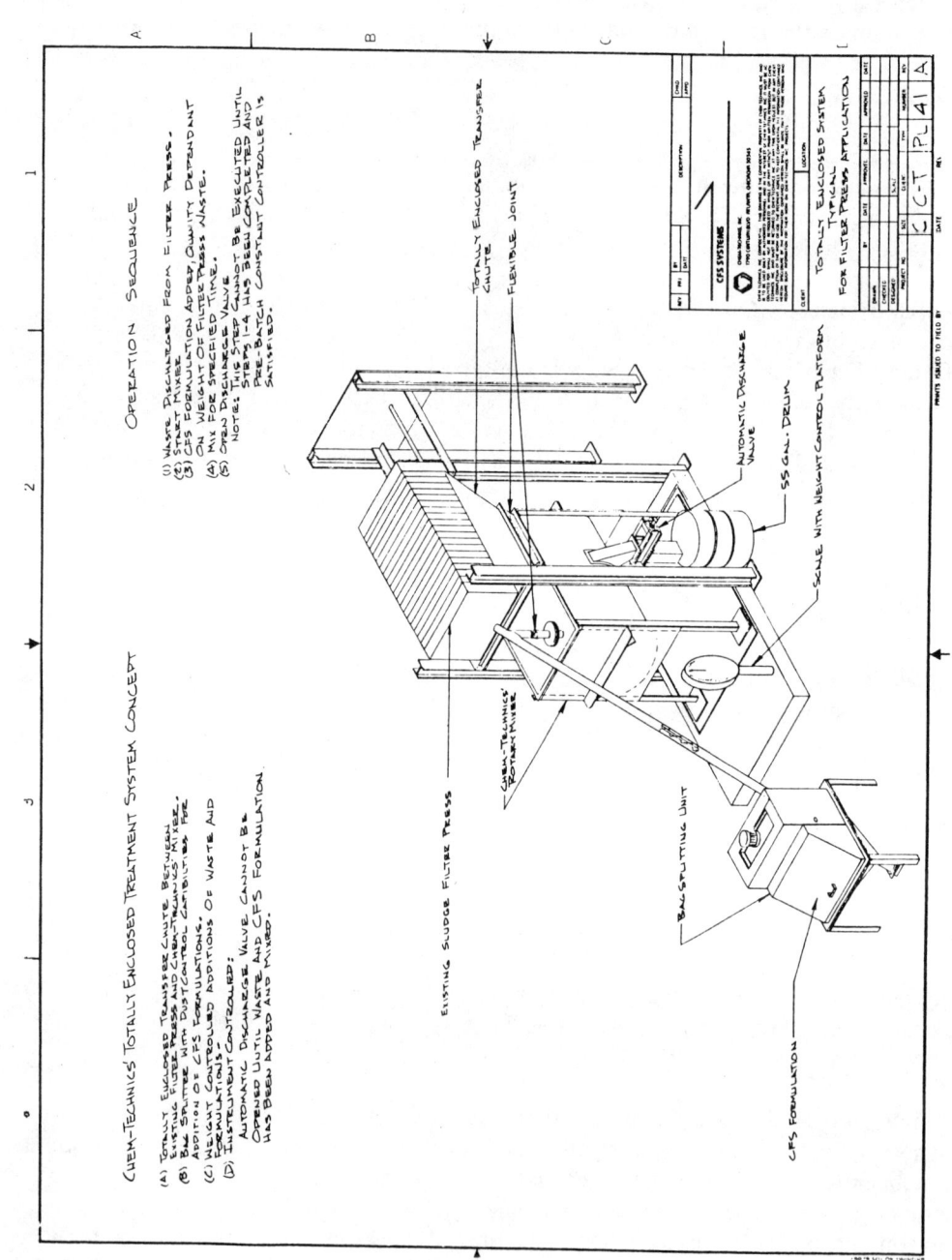

Figure 16-8 Artist's drawing of a batch-type CFS system combined with a filter press.

FIXED INSTALLATIONS 489

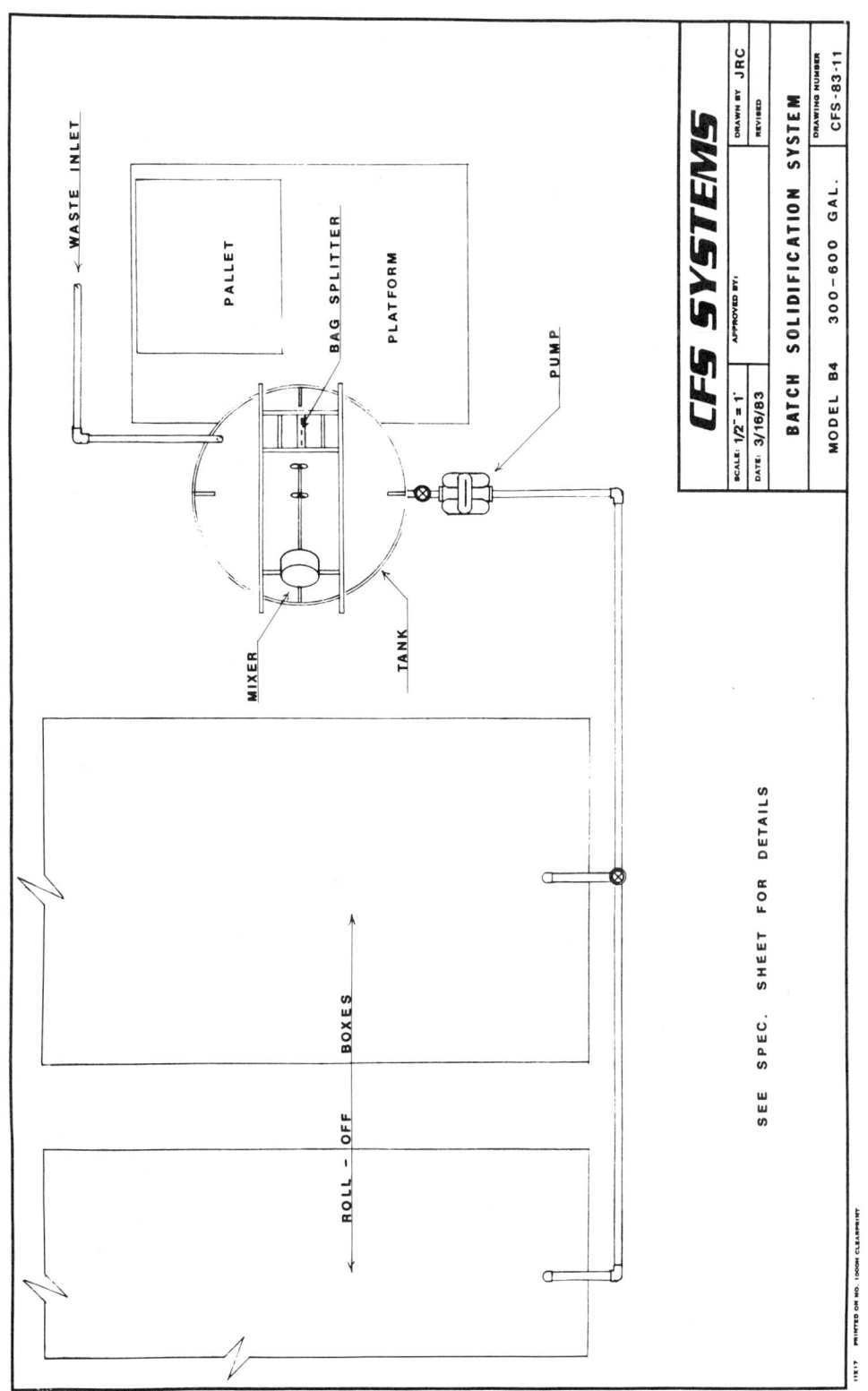

Figure 16-9 Layout of a small batch-type system for pumpable wastes, using bagged reagents.

Medium to Large. Figures 16-10 and 16-11 diagram medium- to large-scale units operating in batch and continuous modes, respectively. The former system, designed for a specific waste stream, uses only one reagent addition, although additional storage and feeding subsystems could be added. The latter system is a more general design that can use any combination of a variety of liquid and powdered reagents. Liquid wastes are stored in Tanks A, B, and C, and pumped into the continuous mixer. Solid and semisolid wastes are fed directly into the surge hopper above the mixer; waste-handling subsystems are not shown.

Both of these systems require either size reduction or prescreening of the waste if trash is present.

Very Large. Figure 16-12 is the process flow diagram for a large, continuous, on-stream CFS operation of about 1000-gal/min design capacity. This system was designed for the solidification of fluegas desulfurization sludge from a coal-burning power plant, and was based on actual pilot-scale field tests with a mobile CFS unit. Cost estimates for 40 and 50 percent solids sludges, respectively, were $9.63 and $4.16 per ton (1975 dollars). The substantial effect of solids content can be seen in these costs, which reflect primarily the difference in the amount of reagents used. At these high throughput rates, labor and equipment amortization costs are a very small part of the total. Therefore, the optimization of the CFS chemistry is of prime importance, as is very close quality control over reagent addition rate.

Commercial Installations

There have been a number of fixed, commercial-scale CFS systems in operation in the United States and Canada since the early 1970s, not including those facilities dedicated to nuclear wastes and materials outside the hazardous waste regulatory schemes. In addition, facilities have been in operation in Great Britain, Europe, and Japan for some time. Some of these installations are listed in Tables 16.1 through 16.6, categorized by CFS process type. The data in these tables are believed to be correct, but may not be up to date as of this publication. It does not include central RCRA TSD facilities in the United States; these are difficult to analyze because they usually include operations other than CFS and information on throughputs and capacities is usually not available.

MOBILE OPERATIONS

While the fixed delivery system is more familiar, mobile or transportable systems are more complex. They consist not only of equipment, installation, labor, and support services, but also involve a number of elements that do not come into play in fixed operations. The difference between "mobile" and "transportable" is more than just semantics, but these words are very similar in many respects. *Mobile* is generally taken to mean that the equipment is on wheels, and that the whole operation can be rapidly moved, set up, and put into operation at a new site, often within a few days. *Transportable*, or *portable,* operations, on the other hand, may be broken down into a number of segments that must be separately transported and are assembled at the operational site. This process may require weeks, even months, to complete. An example is a transportable incinerator that may require five or more

MOBILE OPERATIONS 491

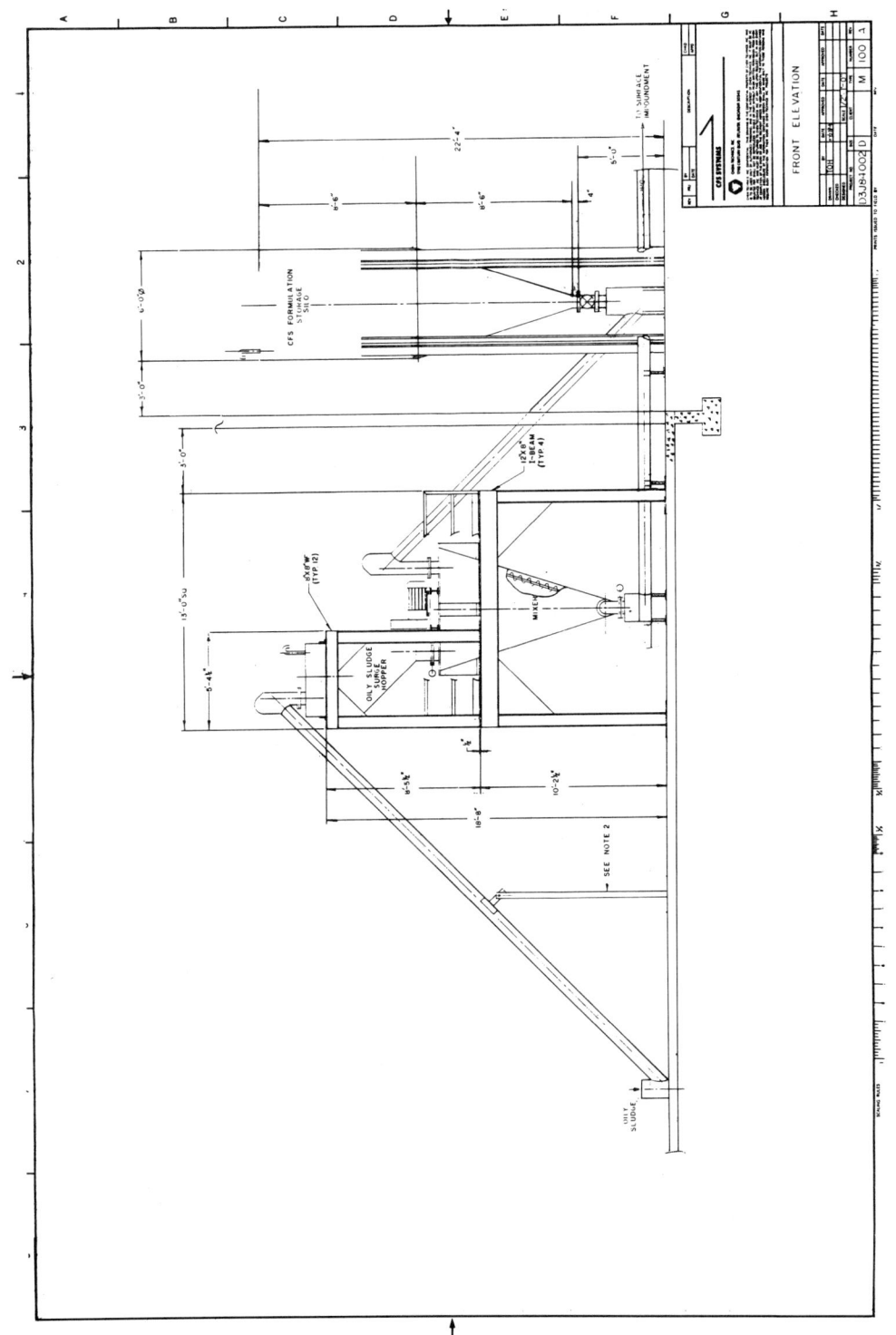

Figure 16-10 Elevation drawing of a medium-scale CFS plant operating in batch mode.

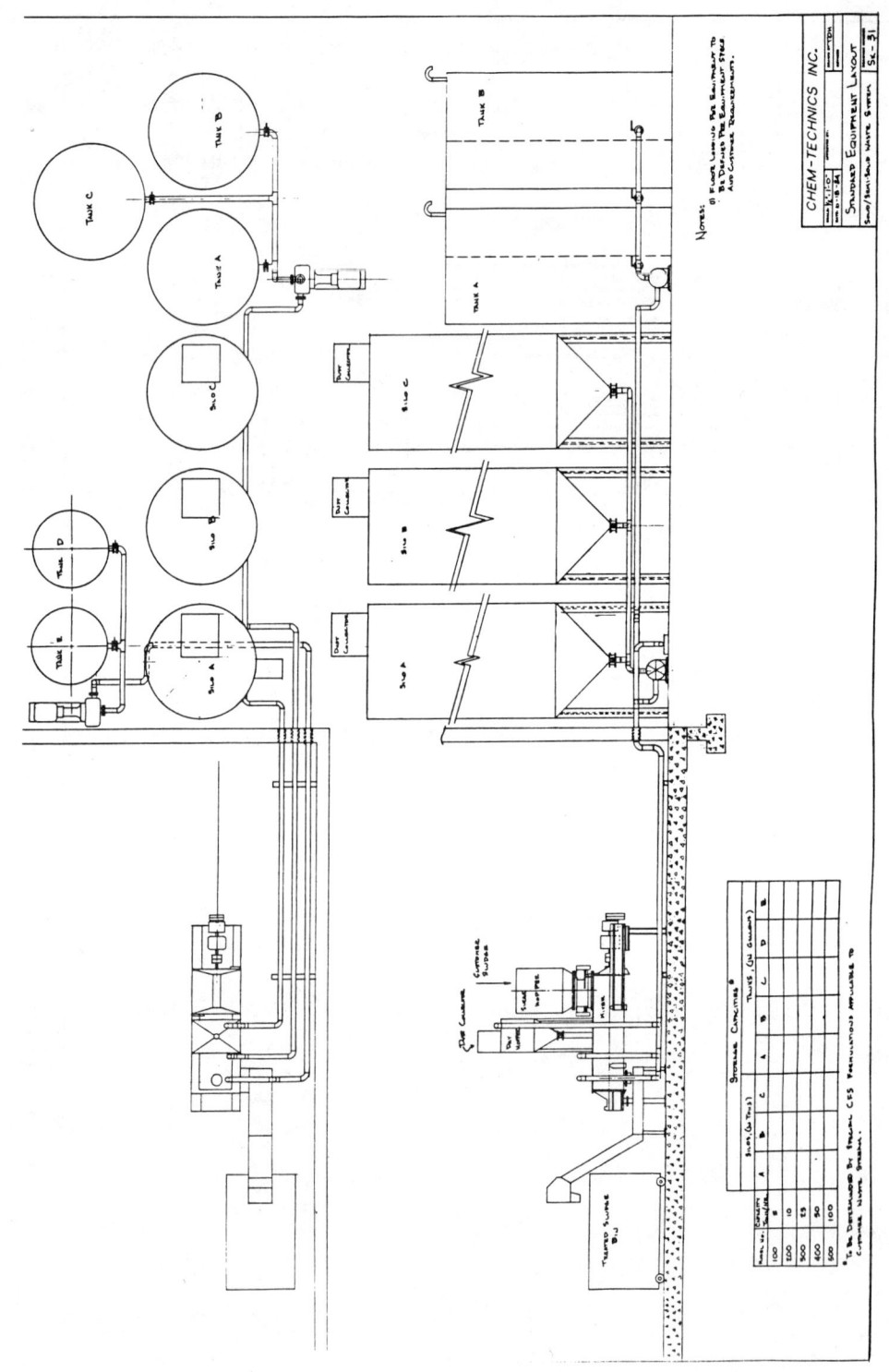

Figure 16-11 Elevation drawing of a large-scale CFS plant operating in continuous mode.

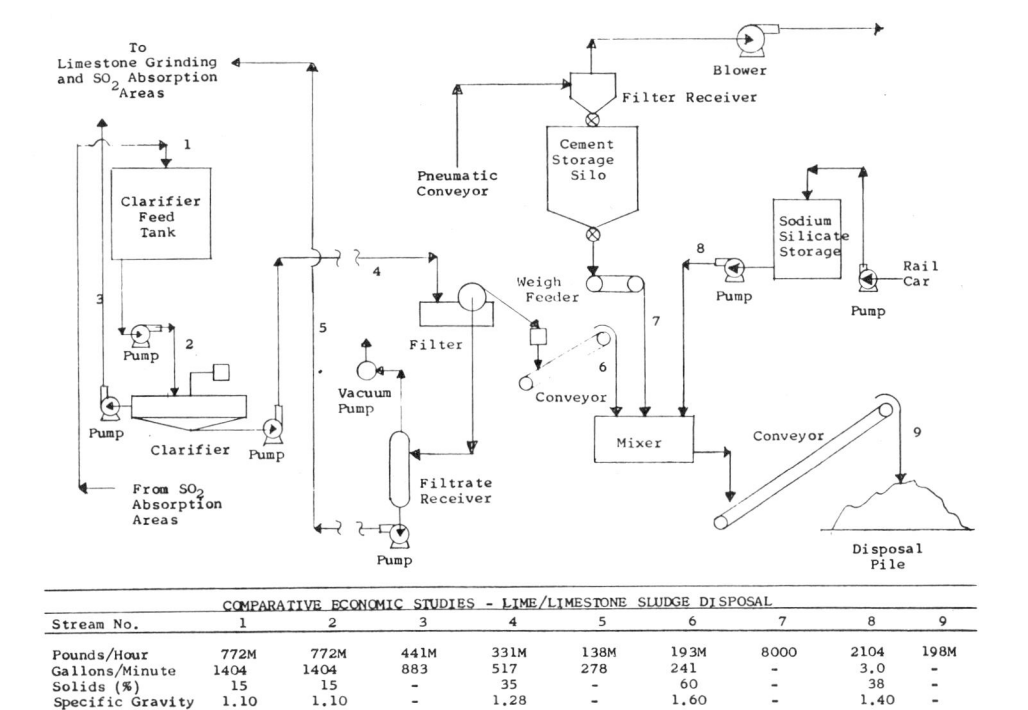

Figure 16-12 Process flow chart for a large-scale, continuous CFS plant treating FGC sludge.

	COMPARATIVE ECONOMIC STUDIES - LIME/LIMESTONE SLUDGE DISPOSAL								
Stream No.	1	2	3	4	5	6	7	8	9
Pounds/Hour	772M	772M	441M	331M	138M	193M	8000	2104	198M
Gallons/Minute	1404	1404	883	517	278	241	-	3.0	-
Solids (%)	15	15	-	35	-	60	-	38	-
Specific Gravity	1.10	1.10	-	1.28	-	1.60	-	1.40	-

trailers and long assembly and shakedown times. The two appellations are often used interchangeably; in the case of CFS systems, there often is little difference because the mechanical aspects are simple relative to processes such as incineration. Here, we will use the word *mobile* unless there is reason to do otherwise, and expand the discussion to include some elements other than just the equipment and its set up and operation; hence, the word *service* rather than system.

The mobile service concept has been mentioned a number of times, but it has not been discussed in detail. Most of this discussion will be based on the author's experience with two different mobile service operations operated in the United States from 1970 to 1982. The reason for dwelling so heavily on mobile service is that more information is available on this CFS delivery system than on any other, due to its early use on a wide variety of wastes and waste sources.

At a minimum, a CFS mobile service utilizes a treatment unit containing the chemical storage, metering, and mixing equipment necessary to mix the waste with the CFS reagents and discharge it to a holding or disposal area. It may also provide the means of removing the waste from its storage containment, homogenize it, and convey it to the treatment unit. Alternately, these latter elements may be leased at the site, or subcontracted to another party. The service is usually used when customers reach storage capacity, wish to eliminate storage for land-use reasons, or are required by a regulatory body to treat and dispose of the waste. Unless the CFS contractor is engaged to empty the storage facility on a periodic basis, the mobile service is usually bought with one-time budget appropriations that tend to be canceled or delayed at times of economic turndown. Therefore, it is not unusual for a CFS contractor to do jobs bid one or two years previously.

TABLE 16-1 Solidification Plant Installations Technology Employed: Portland Cement-Based[a]

Organization	Location	Type of Wastes	Capacity (t/a)	Total Quantity Treated (t/a)	Years of Operation
Browning Ferris Industries	Landfill sites around the U.S.	Various industrial	Very large	>1,000,000	Since 1970
Chemical Waste Management	Landfill sites around the U.S.	Various industrial	Very large		Since 1985
GSX (Formerly SCA Chemical)	Landfill sites around the U.S.	Various industrial	Very large		
NUS Corp. (Formerly Environmental Technology Co.)	Pittsburgh, PA	Pickle liquors and sludges		Reported to be installed as part of a central waste treatment facility in western Pennsylvania. This process now owned and supplied by NUS Corp.	
Fujisash Co. (Formerly described as Fujimasu Industries) "Alset Process," supplied by Kenko Co., Ltd., in U.S. (Portland cement plus fine-divided alumina setting agent)	Various Japanese cities, including Tokyo, Hirakata, Ohtsu, Kobe, Kita-Kyushu, Tsu, Hiroshima, Sakado, and Nagaoka	Sewage sludge and industrial wastes (primarily sewage sludge)	720,000 t/a at Tokyo site alone	>5,000,000 t to date	1973 to present
TJK Inc. (Takenake Sludge Treatment Process)	Japan. Also, office in N. Hollywood, CA, U.S.A.	Sewage sludge and industrial waste	Variable on a project basis	>60,000 m^3 in 26 projects	1973 to present

[a] Including Portland cement/lime and Portland cement/clay.

TABLE 16-2 Solidification Plant Installations Technology Employed: Portland Cement/Soluble Silicate

Organization	Location	Type of Wastes	Capacity (t/a)	Total Quantity Treated (t/a)	Years of Operation
Browning Ferris Industries	Houston, TX	Industrial	150,000		1974 to present
S.A.R.P. Industries	Baton Rouge, LA France	Industrial	50,000		1974 to present
Burlington Northern Railroad	W. Burlington, IA	Railroad maintenance shop sludge	5,000		1976 to present
AMAX Corp.	Climax, CO				
South Essex Water Pollution Control Plant	Salem, MA	Sewage sludge with industrial waste	40,000	80,000	1984 to present
W. R. Grace Co.	Cambridge, MA	Organic waste filter cake	20,000		1981 to 1983

TABLE 16-3 Solidification Plant Installations Technology Employed: Lime/Flyash

Organization	Location	Type of Wastes	Capacity (t/a)	Total Quantity Treated (t/a)	Years of Operation
Conversion Systems Inc. (Industrial division is called Envirosafe Services)	11 power plants in the U.S.	SO2 scrubber sludge	8,000,000 at present	75,000,000 estimated 5,500,000 as of 1978	1975 to present
	Honey Brook, PA	Industrial waste			1980 to present
	Mountain Home, ID	Industrial waste			More than 3
Chem-Met Services	Wyandotte, MI	Acids, caustics, sludges, oils and paints			1966 to present
Stabatrol Corporation Purchased by Chemical Waste Management about 1980 or 1981; closed by PA DER shortly thereafter. Terra-Tite Process.	Norristown, PA	Industrial	Closed now	>65,000	Mid-1970s to 1980 or 1981
Industrial Waste Mgt., Inc. Enviroclean Process.	Japan?	Industrial			
Chemline Corp.	New Brighton, PA	Acids and caustics			Early 1970s to ?

MOBILE OPERATIONS   497

TABLE 16-4 Solidification Plant Installations Technology Employed: Lime/Clay

Organization	Location	Type of Wastes	Capacity (t/a)	Total Quantity Treated (t/a)	Years of Operation
Canadian Waste Technology Licensed by Ontario Liquid Waste Disposal to Laidlaw Transport, Ltd. in 1976	Hamilton, Ontario	Industrial	125,000	>1,000,000	1970 to ?

TABLE 16-5 Solidification Plant Installations Technology Employed: Portland Cement/Flyash

Organization	Location	Type of Wastes	Capacity (t/a)	Total Quantity Treated (t/a)	Years of Operation
Stablex Corporation All plants now reported to have been acquired by other entities	Blainville, P.Q., Canada	Industrial	70,000		1983 to present
	Birmingham, U.K.	Industrial	200,000	80,000 t/a in 1977	1974 to present
	London (West Thurrock), U.K.	Industrial and mixed	400,000	30,000 t/a recently	1978 to present
	Japan—plants for Kanagawa Prefecture (Tokyo area) and Toyota (Nagoya area)		Originally scheduled to begin operation in 1979. No further information since that time		
Industrial Waste Management. Enviroclean Process	Japan?	Industrial			

498 DELIVERY SYSTEMS

TABLE 16-6 Solidification Plant Installations Technology Employed: Cement Kiln Dust/Lime Kiln Dust

Organization	Location	Type of Wastes	Capacity (t/a)	Total Quantity Treated (t/a)	Years of Operation
Browning Ferris Industries Chemical Waste Management GSX (Formerly SCA Chemical)	Processing and landfill sites around the U.S.	Industrial	Very large	>1,000,000	1970 to present Since 1985
Envirite Field Services Formerly American Resources, now part of Envirite Corp. Permix Process	Thomaston, CT York, PA Canton, OH Harvey, IL	Industrial	100,000 at Thomaston, CT		1976 to present at Thomaston, CT
Various waste generators and waste management companies	Various in U.S.	Industrial		>1,000,000	1970 to present

In addition to the mobile treatment units described in this chapter, an array of other equipment listed below may be required, depending on the nature of the project.

Pumps
Hoses
Valves
Fittings
Agitation and homogenization equipment
Conveyance devices
Portable offices
Portable laboratories
Testing equipment

In addition, provision must be made to transport the equipment from one site to another. This may be done either with hired traction units or operator-owned equipment, whichever is the most economically desirable. For example, in a situation where a small number of treatment units is covering a very large territory, as is common in the United States, it is more feasible to use hired trucks and drivers for pulling the vans.

However, in a more compact geographical area, such as much of Europe, it may be economically preferable for a company to have its own traction units. In addition to transporting the equipment, it is necessary to transport the personnel who run the job. This is usually done with company-owned trucks that can serve a dual purpose—to transport the crews and their belongings as well as tools and other equipment, and to pull small trailers carrying pumps, hoses, valves, and other field supplies and equipment.

The technical and operational activities involved in operating a CFS mobile service are, in normal chronological sequence, the following:

1. Obtaining samples of the waste
2. Preliminary laboratory testing
3. Preliminary quote
4. Meeting with customer, field sampling, and preliminary meeting with the regulatory agency
5. Final laboratory solidification, leaching, and physical tests
6. Firm quotation to customer
7. Regulatory approval
8. Mobilization
9. Set up at job site
10. Treatment of the waste
11. Close down and clean up at job site and return to home base
12. Final laboratory leaching and physical tests on solid produced in job to satisfy contract requirements and protect warranty
13. Possible follow-up sampling and laboratory testing of solidified waste at various times, if required by contract or desired by CFS contractor for information or warranty protection.

Each of these steps requires, of course, the successful completion of the previous step and the customer's go-ahead. Several of the steps are discussed in more detail below. Others have already been examined in earlier chapters.

REGULATORY APPROVAL PROCESS

The regulatory approval process is an integral part of doing a CFS project. Usually, after the CFS contractor has submitted a preliminary quote and the customer has expressed interest in having the work done, the next step is a joint visit by the customer and the contractor's technical people to the regulatory agency to obtain permission to do the project. This normally requires a presentation that describes the process to be used and its past history, and provides specific technical information required by the permitting agency. At this point the preliminary leaching test results on the waste to be treated are submitted to the regulatory agency along with a request for a permit or letter of approval for treatment of a specific quantity of the waste at a specific site, usually within a specified time period. If the regulatory agency approves, it will eventually grant either a letter of permit or a letter of no objection, stating that the customer and the CFS processor may proceed with the work. If there are questions that the regulatory agency needs answered, it may be necessary to provide additional leaching or other data, or for the waste owner to provide more information about the waste disposal site. Of course, this regulatory approval process will vary considerably from country to country, as it does from state to state in the United States, and it is rapidly changing with the evolution of laws and regulations governing the treatment and disposal of residuals, especially those designated as hazardous.

Operations

Any discussion of mobile service field operations must include logistics, permits and licenses, chemical purchasing and supply, equipment necessary to do the work, transportation, maintenance and field service, personnel, quality control, safety and health considerations, interaction with the customer during the job, startup and closedown, cost control, and final evaluation of the job results. This will be done by going through the actual steps involved, from the initial site evaluation to the final cleanup. The major steps are as follows:

Site evaluation and analysis
Job costing—generation of data necessary to quote the job
Preparation of a detailed operational plan
Mobilization
Set up of equipment and facilities at the job site
Treatment
    Lagoon agitation
    Treatment
    Technical control
    Chemical acquisition
    General field purchasing
Closedown
Cleanup and maintenance

Site Evaluation. If the job is of significant size, site evaluation will be done by operations and/or technical service personnel; smaller jobs may be evaluated by field sales personnel if they are sufficiently experienced. A complete site analysis is done according to a formal, written checklist, which describes in detail all of the information usually needed by the operations department to do the job costing and to prepare a detailed operational plan if the job is acquired. Site analysis includes evaluation not only of the lagoon or storage space itself, but also of the area in which the treatment unit is placed, the area available for discharge of the treated waste and subsequent solidification, a general terrain analysis, pump and hose requirements, communications analysis, and analysis of protective devices and clothing needed for crew members. When this site analysis is complete, the necessary information is forwarded to the marketing department for quotation and the remainder of the information is filed for preparation of the operational plan.

Operational Plan. Additional information may be necessary for the preparation of an operational plan. A complete operational plan for a large treatment job requires detailed planning of every phase of the operation.

Mobilization. When the starting date for the job is set, mobilization begins. From the operational plan the project engineer and his or her assistants begin accumulating and testing the equipment necessary for the job and planning the details of transportation and setup at the job site. At the same time, the operations manager and the salesperson involved communicate with the customer concerning preparation of the job site, access to the site, clearances for personnel, acquisition of equipment to be supplied by the customer, and the running of the necessary utilities and services to the site.

Concurrently, the purchasing department makes arrangements for chemical deliveries according to the schedule in the operational plan, and also makes any necessary arrangements for temporary chemical storage facilities at the job site. The purchasing department at this time ascertains whether there is likely to be chemical shortages and makes contingency plans in the event that the selected chemical supplier is unable to deliver. The chemical scheduling and delivery is probably the most critical logistical aspect that is encountered after the CFS equipment arrives at the site. Without on-site chemical storage capacity, job scheduling will be controlled by the chemical deliveries, and if these are not on time, job costs increase and the job completion deadline may not be met. Frequently, therefore, it is desirable to establish on-site storage, especially where the job is large. Storage of liquid reagents is usually not a problem, requiring only portable tanks. In the case of dry reagents such as Portland cement, it may be necessary to set up a portable storage silo at the site. These facilities are available for use with various types of dry chemicals in construction sites. Another method is to use a mobile storage unit, which is like a large cement tank truck. It is delivered empty to the site and when filled will hold 4 to 6 truckloads of Portland cement; it has the necessary equipment to transfer the reagent to the CFS treatment unit. These storage units are usually available from the larger suppliers and are not expensive to rent.

Another aspect of the mobilization step is the making of necessary reservations, travel arrangements, housing, and eating accommodations for the personnel on the job. Arrangements should also be made for paying field personnel on long jobs, as well as providing cash to them for living expenses. The purchasing department should set up the necessary field purchasing with suppliers, such as hardware stores, clothing stores, and equipment parts suppliers, since it is difficult for field personnel to make these financial arrangements at the site. Cash purchasing is undesirable from the standpoint of cost control.

As the date approaches for transportation of equipment and personnel to the site, the project manager will obtain the necessary transport vehicles for both equipment and personnel and the crew will begin to pack the supplies for the job. It is desirable to have an enclosed staging area where this can be done efficiently in time to avoid problems as the departure date approaches. For a major field job, this staging operation is a fairly large undertaking and must be well planned and scheduled. Normally, the field crew that will do the job will also pack their equipment, and unpack, clean, and provide the necessary maintenance to it at the end of the job.

Setup. The operational plan provides for the necessary time to set up the treatment unit, the pump and hose runs, connect utilities, and do the other work required before treatment ban begins. It is important that the plan provide sufficient time for this to be done properly, especially in the case where concrete footings may have to be poured to anchor draglines, and where earthmoving work is required to lay out the hose lines and construct the discharge area where solidification will occur. At this point all the local arrangements concerning supplies and equipment, housing, eating, etc., will be confirmed with the appropriate suppliers so that any problems can be corrected before treatment begins.

Consideration must be given to the health and safety aspects of the job, both those required by regulatory authority and those used by the CFS processor. A complete safety manual should be written for operation of the treatment equipment, and this should be understood by the various crew personnel. The crew members must be required to wear the specified safety clothing and equipment, whether or not they wish to. Mechanical safety with equipment is usually well established, and it is up to the crew chief to make certain that his or her personnel follow procedures. In the case of waste treatment, there is also the problem of occupational health considerations due both to the chemicals being added and to the waste being treated. The occupational health requirements due to the CFS chemicals are, of course, known in advance and field personnel should be well educated in this aspect before starting on the job. However, each job will require different health and safety considerations according to the waste being treated. In some cases, wastes containing volatile organics will require special breathing equipment for those working at the treatment unit; in all cases, dust masks must be supplied and used. In other situations the waste may be corrosive or irritating to mucous membranes or skin, and the appropriate safety clothing must be worn. This equipment should be made available during the setup process and the crews reeducated in the hazards of the waste being treated.

Treatment Operations. Treatment operations are of two general types: demonstration or pilot jobs, and full-scale on-site treatment jobs. Since some full-scale jobs are no larger than pilot jobs, the difference between them can be rather vague from the operational standpoint. However, the main difference is in the interactions with the customer and other visitors to the site, since a demonstration program will usually draw large numbers of personnel from the customer's plant as well as regulatory people, and sometimes people from other companies. Therefore, in a pilot program special care may be necessary to keep the job site neat and clean and to have everything working smoothly even at the expense of high production rates (the expectation of lower production rates is usually factored into the price quoted for the pilot work). A typical layout for a pilot project is shown in Figure 16-13.

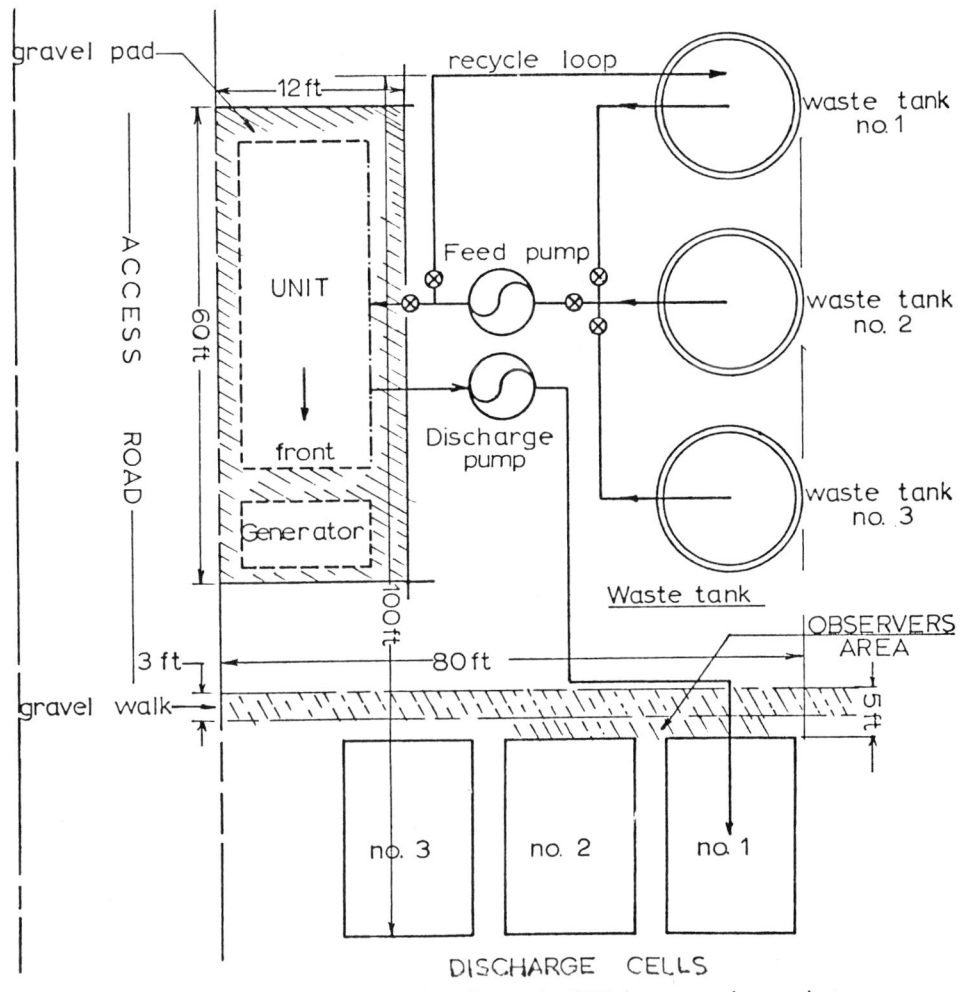

Figure 16-13 Typical layout for a mobile pilot-scale CFS demonstration project.

The actual job usually begins with agitation and homogenization of the lagoons or waste holding tanks, both to obtain a uniform material for treatment throughout the job and to prepare the waste physically for pumping. One layout for a large project of this sort is presented in Figure 16-14. Hydraulic agitation is normally used for the smaller lagoons and for the more fluid wastes. On large lagoons, or with difficult-to-pump wastes, draglines, clamshells, and floating dredges may be necessary for the transport of the waste to the treatment unit. Some of these methods were discussed in Chapter 7.

After all of this preparation, the actual treatment work comes almost as an anticlimax and is often the most uneventful part of the job. Once the waste has been prepared in the lagoon or pumped from the lagoon into a temporary surge pit or holding tank, operation of the treatment unit usually goes smoothly if all of the preparations have been properly done. In most of the author's experience with mobile operation, the treatment unit is operated as a continuous process during a one-shift operation, usually 8 to 10 hours per day.

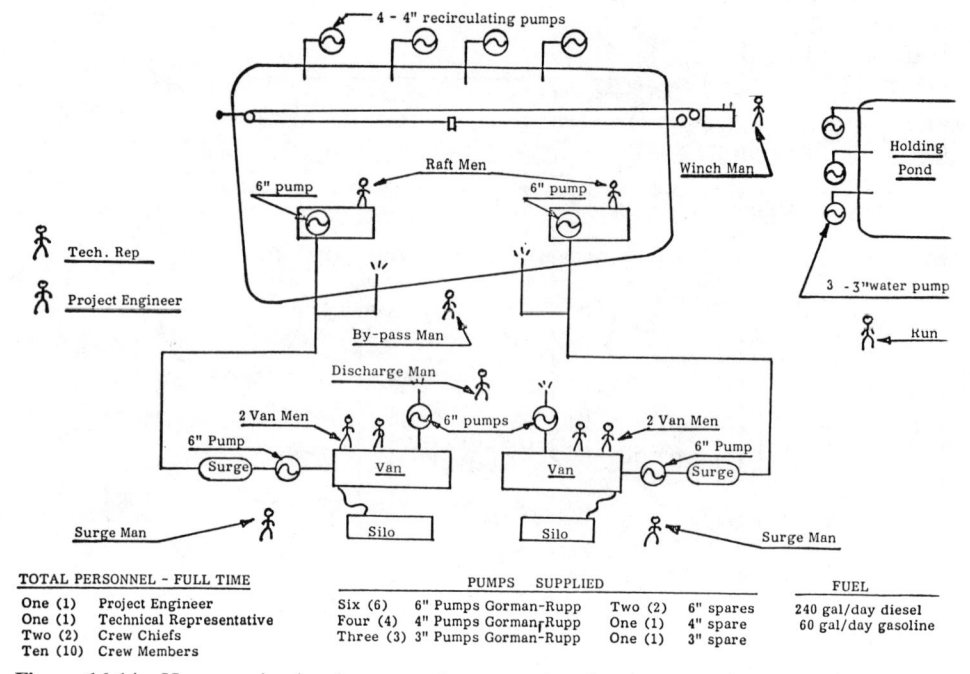

Figure 16-14 Homogenization layout and personnel assignments on a large lagoon treatment project.

Technical Control. To operate a mobile service operation, it is necessary to have a well-equipped laboratory capable of doing not only the relatively simple solidification tests necessary to determine proper chemical additive ratios, but also the more complex leaching tests and analytical work necessary to characterize both the samples and the leachate from the leaching tests. A large CFS operation would consist of the following.

- The technical director, who supervises the laboratory and the field technical service personnel.
- Field technical service engineers, who travel to the jobs, take samples, perform quality control functions, and supervise the jobs technically.
- Laboratory people, chemists, and technicians, to run solidification and leaching tests and to do analytical work.

The extent of the laboratory depends to a great degree on the size of the CFS operation and, especially, on whether it is part of a larger waste treatment firm that already has laboratory facilities. In the latter case, most of the laboratory equipment and personnel would already be available; the additional outlay in money and personnel would be small in the beginning. However, it is necessary to have at least one technical person thoroughly trained and competent in the CFS process to be used, and who has full-time responsibility for the technical aspects of the area.

Technical service engineers are normally in the field at the beginning of and at other times during every major job. These engineers are responsible for assuring that the waste to be treated is the same waste that was tested initially, that the CFS processor's operating personnel are adding the proper quantities of chemicals and otherwise properly operating the job technically, and that the final product meets the specifications set forth in the quote and in the leaching test warranty. Frequently, the technical service engineer must interact not only with the customer, but with representatives of the regulatory agency inspecting the job. At the beginning of the job, a description of what is to be expected is given to the customer's project engineer and to others. This is often helpful in avoiding lengthy explanations of what is taking place on the job. In addition to checking for proper solidification, making quick tests for leachability, and collecting samples for shipment to the laboratory for more complete testing, the technical service engineer's main job in the field is to keep the chemical usages within the control limits specified in the job plan.

This is extremely important, since the profitability of the job is predicated on the quantities of chemicals used. Because it is difficult to pump large quantities of sludge through units such as these and still keep the chemical addition rates accurate, an allowable tolerance of $\pm$ 10 percent is initially established, although one should expect to do better than this. During one year's operation in the author's experience, the average chemical variances from laboratory ratios were 4.5 percent for cement and 3.8 percent for liquid reagents. In the following year, this had been improved to 1.8 percent for cement and 0.7 percent for liquid, a substantial improvement. Often, the reagent quantities used are less than the laboratory ratios. This is desirable from an economic standpoint, but it is important that the operating personnel not be permitted to deliberately cut chemical ratios in order to improve their cost performance, since this can result in disaster if the waste being treated does not solidify or meet the leaching requirements.

During treatment, the technical service engineer will take samples, measure addition rates of chemicals, and ascertain at all times that the proper chemical ratios are being maintained and that the treated waste is solidifying as designed. As the job runs, it is also important that the project engineer and/or the technical service engineer maintain good relationships with the customer's representative, usually a plant engineer. This includes reporting progress on the job and any problems that are encountered.

Cost Control. Cost control on the job is an important function of the project engineer. It should be done on a running basis as well as at the end of the job, so that any cost problems can be corrected, insofar as possible, before the situation gets out of hand. This is especially true on long jobs. The required procedures for any company will, of course, depend on that company's policies and experience with field operations. Another aspect of the cost control program is the establishment of definite field purchasing procedures. In mobile operations, this is especially important, since considerable money in the form of cash may be required in the field to purchase the necessary running supplies and equipment. As we have just seen, the major cost item in a CFS project, mobile or fixed, is normally that of reagents. Therefore, control of reagent usage, as previously described, is the highest priority for the project engineer.

Closedown, Cleanup, and Maintenance. When the job has been completed, as determined by the contract provisions and/or agreement between the customer's representative and the project engineer, it is important to leave the job site in good condition and to properly prepare the equipment for return to home base. The job site must be cleaned up according to the provisions of the contract and in keeping with good workmanship practice. The site should be checked carefully to make sure that all tools, clothing, supplies, and other equipment are packed. All trash, broken equipment, and other material to be discarded should be placed in the area designated by the customer for that purpose. The utility services to the treatment unit should be terminated as requested by the customer. At this point, the technical service engineer will take the final samples of the solidified waste and the project engineer will check the job site after the crews and equipment have departed. The project engineer will leave the job last and only after agreement with the customer representative that the job has been completed according to the contract.

The job is not complete until the equipment and personnel have returned to the home base, the equipment has been unloaded and cleaned as necessary, and the required repair and routine maintenance work reported and scheduled to be done. At this point the shop supervisor should have checked in all the equipment, determined what is missing or broken, and made the necessary reports to the operating manager. Only after this is completed can the final job profitability analysis be done.

Job Evaluation and Final Report. The final job cost analysis sums up the various cost factors: labor, materials, equipment, and miscellaneous, and compares these with the budgeted amounts. These data can then be analyzed by the operations, marketing, technical, and general management staffs. More importantly to this discussion, a final technical report sums up the job briefly and provides necessary technical data so that the customer can judge compliance with the warranty and provide required information to the appropriate regulatory authority. This final report, along with the final invoice and support documents, will be the basis for

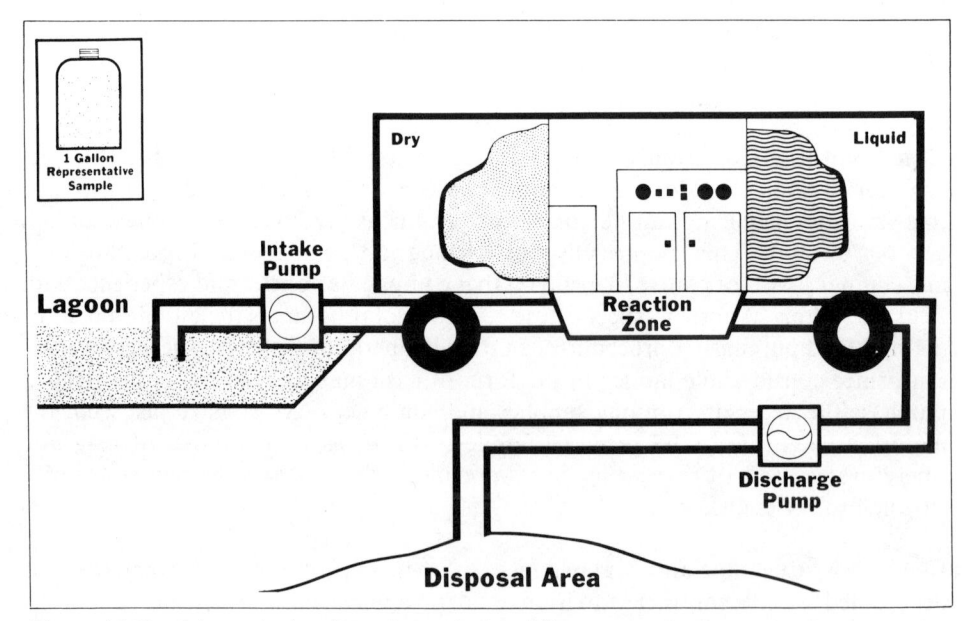

Figure 16-15 Schematic drawing of a typical mobile treatment unit.

Figure 16-16 A self-contained mobile CFS unit.

Figure 16-17 A self-contained mobile CFS unit operating at a treatment site.

508 DELIVERY SYSTEMS

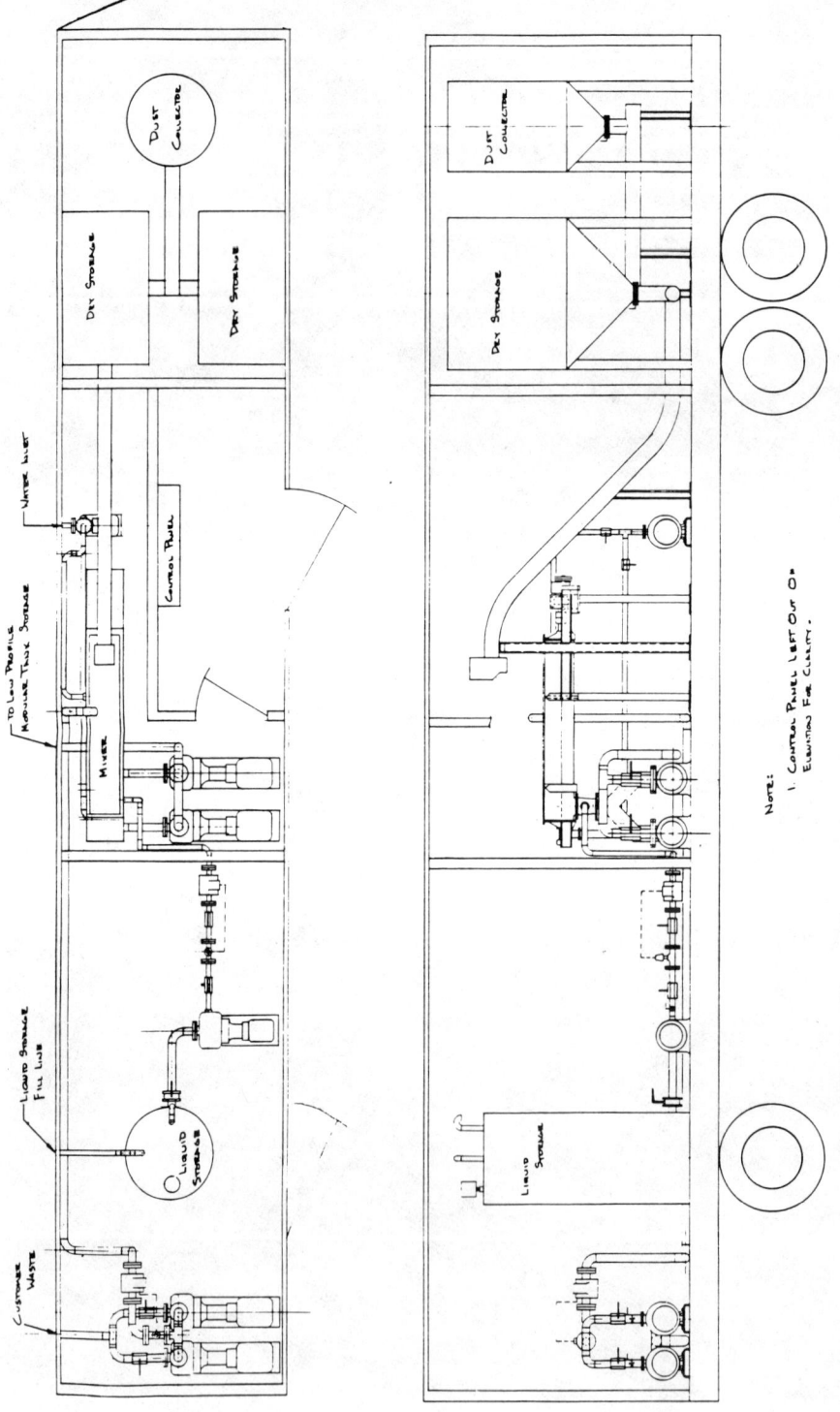

Figure 16-18 Plan and elevation drawings of a mobile CFS treatment unit for pumpable wastes.

Figure 16-19 A mobile CFS treatment unit for nonpumpable wastes.

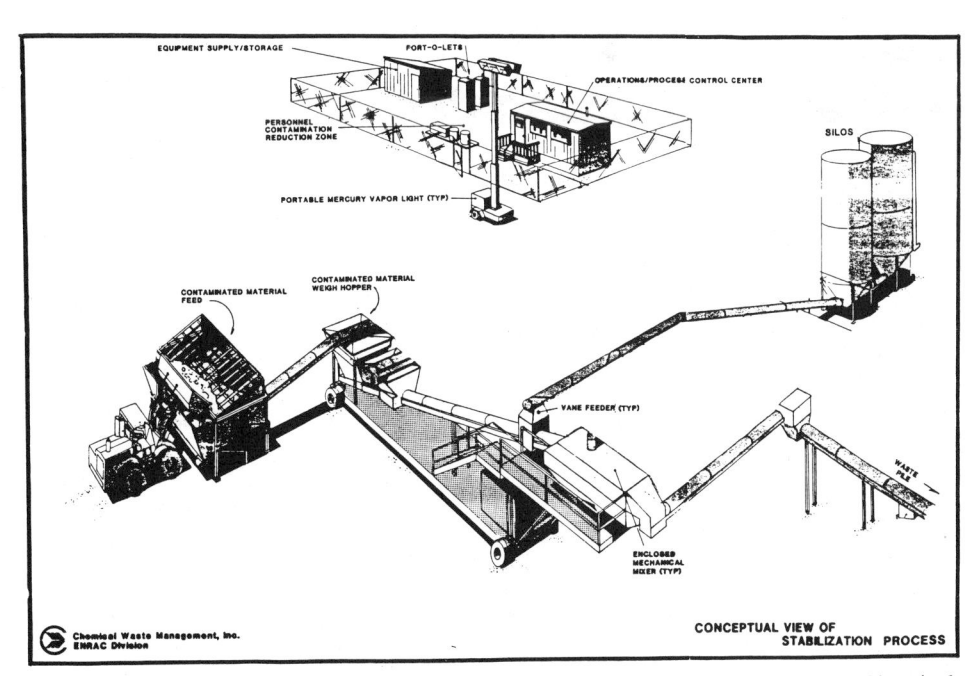

Figure 16-20 Artist's conceptual drawing of a transportable CFS system. (From Chemical Waste Management, Inc.)

Figure 16-21 A "standard" mobile CFS treatment plant.

Figure 16-22 Outpour of treated, pumpable waste before it sets up. (Print by Associated Photographers, Inc.)

payment by the customer for the work done. It is also important for the CFS processor's records, for both legal and financial purposes.

As can be seen by the discussion in this and other sections of this book, the operation of a mobile CFS job is a complex operation requiring the coordination of a number of people and considerable experience to achieve a smoothly functioning, profitable project. No matter how experienced the CFS contractor may be in other areas of waste handling or treatment, there will be a learning period during which mistakes will be made and problems encountered. The important thing is to have properly planned, organized, staffed, and prepared the operation so that these mistakes and problems are minimized, can be detected quickly, and rectified before too much harm has been done.

Equipment

To this point, the *operation* of a mobile CFS process has been briefly outlined. This was necessary to provide a framework for discussing the equipment that is used, because it can be quite different in some ways from the fixed plants described earlier in this chapter. Schematically, a mobile treatment unit in its simplest form is as shown in Figure 16-15. One manifestation of Figure 16-15 looks like the actual mobile unit shown in Figures 16-16 and 16-17. This was the first mobile unit for the CFS treatment on industrial wastes in the United States and, it is believed, in

Figure 16-23 Placement of a treated, pumpable waste into an impoundment while it is still liquid. (Photo by George W. Sommer.)

512 DELIVERY SYSTEMS

Europe, Great Britain, and Japan as well. It was used to treat more than 100 million gal of industrial wastes at more than 100 different sites and projects from 1970 through 1977. A later model of this general type is shown schematically in Figure 16-18. Both of these designs were intended for the treatment of pumpable liquid wastes only.

When nonpumpable wastes are to be treated, the requirements are quite different. Figure 16-19 shows a completely different type of unit for the processing of contaminated soils. In this kind of operation, the waste is delivered to the unit by frontloader, conveyor, backhoe, clamshell bucket (shown here), or other means, after which it is handled as a solid. Solids require different feed and materials han-

Figure 16-24 Waste of Figure 16-24 after it has hardened. (Print by Associated Photographers, Inc.)

dling devices and often different mixer types than liquids. A conceptual view of another transportable system is shown in Figure 16-20. Until recently, such mobile/transportable plants were custom-built by CFS contractors, but a number of equipment manufacturers now build them in more or less standard form with custom variations specified by the buyer. One of these units is shown in Figure 16-21. It will handle either liquid or solid wastes with some modifications for the waste type in each specific project.

The peripheral equipment used in mobile operations is especially important, and is frequently overlooked by engineers when assembling a mobile treatment system. It will vary greatly with the project size, location, waste characteristics, and other job parameters. Some of the general equipment types were listed previously. Not only must this equipment be provided, but it must be transported, maintained, and stored between jobs, if owned, or rental arrangements must be set up in advance, if leased.

Waste Handling, Curing and Disposal

Before leaving the subject of delivery systems, a quick visual look at the handling and curing of wastes may be interesting. Contaminated soils and many semisolids such as filter cakes undergo little obvious physical change during the CFS process, but the change with liquids is often dramatic. Figure 16-22 shows the output of a CFS process for liquid waste after the chemicals have been added and before the material has set and hardened. This liquid is being poured into a bermed area (Fig. 16-23), where it quickly sets and then begins to harden. After several days, the liquid

Figure 16-25 Landfilling of solidified waste after hardening.

has become a firm, friable solid (Fig. 16-24) that can easily be handled and landfilled. Such a landfilling operation is shown in Figure 16-25, where the waste has been cured, graded to a flat surface, and compacted. Interestingly in Figure 16-25, the solid product of CFS treatment is being used to displace rainwater from an old quarry, filling in the quarry, disposing of the waste, and returning the land to some beneficial use, all at the same time.

Chapter 17

CFS TESTING AND FORMULATION

We have seen that, while much has been learned about the chemistry of CFS in the last ten years, it is still largely an empirical science. The "real-world" operating rule has been "if it works, use it." But how do we know that "it works"? In most cases we cannot reliably prescribe a treatment formulation based on knowledge of the characteristics of the waste and the properties desired in the final product. Therefore, formulation requires a treatability study to select and optimize the CFS system for a particular waste problem. The treatability study, in turn, requires such adjuncts as proper sampling, problem definition, waste handling, waste characterization, and testing of the product. In this chapter, we will discuss all of these areas; however, the subject of leaching tests will not be covered in minute detail, since a number of excellent reference works are available.

WASTE SAMPLING AND HANDLING

Very often, the most difficult part of CFS testing is to obtain a representative sample of the waste. This is especially true in remediation work where large sludge ponds are involved. Recently, a number of articles and manuals have been written on this subject specifically for use in hazardous waste work [1-6], so it is unnecessary to go into detail here as to overall techniques, equipment, and statistical methods. However, we will discuss some of the special considerations in sampling of pits, ponds, and lagoons for remedial action. The official methods for sampling are given in the EPA's manual, better known as SW-846 [1]. Exner describes a sampling strategy that is useful [2]. Other methods and techniques have been described by Brantner [3] and in various ASTM draft documents [4-6].

The most difficult sampling problems arise when attempting to characterize large and, especially, old pits, ponds, and lagoons filled with waste. The sampling plan is generally based on a grid system such as that taken from EPA SW-846 [1] and shown in Figure 17-1. Such plans are straightforward; the actual sample taking is not. Waste lagoons are usually not neatly rectangular, nor are the sides vertical. In many cases it is difficult to determine where the lagoon ends and the surrounding native soil or fill begins. In fact, a grid sampling program must often be completed

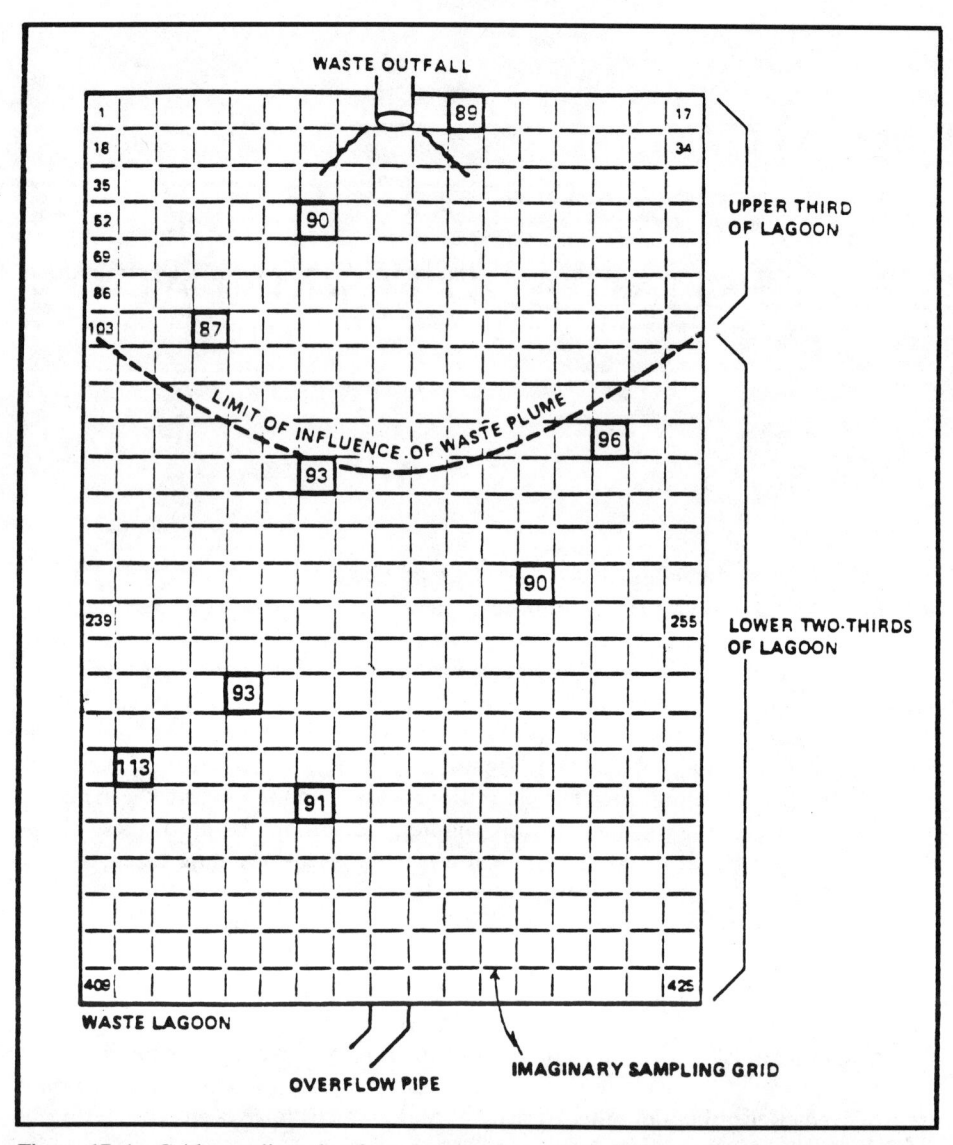

Figure 17-1 Grid sampling plan for a lagoon. Legend: 1–425 = units of sampling grid. $\boxed{90}$ = barium concentrations (ppm) asociated with nine samples of sludge. (From SW-846 [1].)

to determine by analyses of the samples the waste or contamination boundaries within which remediation is necessary. Lagoons are also normally stratified vertically, and are nonuniform horizontally due to flow patterns from the waste outfall.

To make matters worse, many lagoons are neither covered with water on which a boat can float, nor solid enough to support the weight of sampling equipment and personnel. Smaller impoundments can be reached from the perimeter by backhoes or equivalent devices that can take samples at different grid locations and at various depths. With large lagoons, special platforms must be devised, often custom-made for a particular situation. Health and safety considerations also limit the techniques that can be used; for example, placing sampling personnel on flat-bottom boats in water-covered lagoons, a practice common some years ago, is now often forbidden. Fortunately, there are now a number of consulting and engineering organizations that specialize in the sampling of lagoons and contaminated soil areas. They have the necessary equipment and expertise to do a professional job, and are listed in various hazardous waste services and equipment guides.

When undertaking a sampling program, even with a professional and experienced consulting firm, it is important to make clear at the outset the purpose of the sample taking. Much of this work has been done for Superfund Remedial Investigation and Feasibility Studies (RIFS), where the primary purpose is waste characterization. Samples taken for this purpose are often not suitable for CFS treatability testing, due to lack of sufficient quantity or the use of preservatives for analytical purposes. For treatability work, at least five gallons of waste should be taken from every sampling location, and more if a wide variety of treatment technologies is to be investigated or if it would be very difficult and expensive to get additional samples later on.

It is a sad but common occurrence to find a multimillion dollar remediation project predicated on one or two grab samples not representative of the problem as a whole. And this often happens where a very thorough and expensive characterization had been done, but the samples were not saved or were insufficient for treatability testing. The type of tests conducted and data generated in characterization cannot substitute for actual treatability testing. If the sample subjected to CFS treatability testing is not representative of the waste to be treated, extrapolation of results to a different waste composition is very risky.

Sampling Techniques

There are three basic techniques for obtaining the sample: coring, pumping, and digging.

Coring. The core sampler is basically a length of pipe or tube, generally steel or plastic, such as that shown in Figure 17-2. The inside diameter should be a function of the amount of sample required as well as the viscosity of the waste; it is generally between one and two inches. The sampler is forced vertically into the lagoon until the bottom is reached. The upper end is then sealed with a cork or threaded cap and the filled sampler pulled out. The end cap is then removed and the sample forced out with a plunger into the sample container. If it is desired to determine properties at different vertical locations, a series of sample containers can be used to collect aliquots of the extruded material as it is forced out, thus creating a vertical profile. This method is useful for viscous sludges and soft soils; the more viscous the material, the larger the corer diameter that can be used, but two inches is about the maximum for hand-operated sampling. Low viscosity wastes and hard-packed solids cannot be sampled by this method, at least not to any depth. Core-sampling difficulty increases rapidly with depth, and is generally limited to about ten feet or less. For greater depths, or with hard-packed materials, a drilling rig is required.

A word of warning for those contemplating core sampling. It is not unusual for the sampling person to stop after reaching a false bottom, usually a hard interlayer of crystalline material or other dense substance overlying another, often softer or even liquid, waste layer. This premature stopping can be prevented by doing a quick "fingerprinting" evaluation of the bottom plug in the sampling tube, compar-

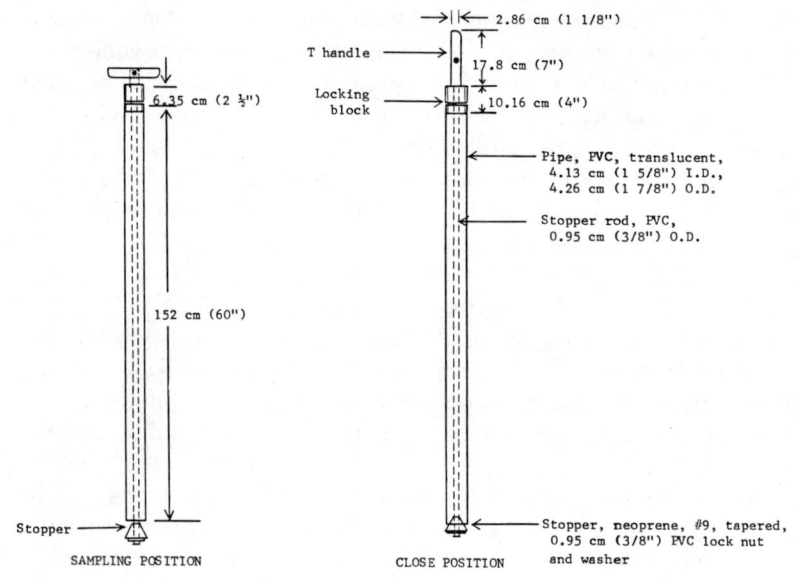

Figure 17-2 Core sampler.

ing it with the known native soil or pond bottom liner. Also, core samples should not be taken too close together if a vertical profile is required, because removal of the first core causes an upwelling of bottom material in the vicinity of the sampling point, distorting the profile in the immediate vicinity.

Pumping. If the waste is fluid enough to pump, a variety of pump types can be used to get good samples, even samples representing different vertical horizons, if care is taken. Diaphragm pumps are good for this purpose because their positive displacement characteristic allows the sampling of quite viscous materials, the pump can be small and light, and it can be air driven for even lower weight and where there are possible ignitable gases present. In some cases, a hand pump can be used; in others, a small powered pump may be required. Another way in which a pump can be used is to recirculate the waste, pumping from the bottom and discharging at the surface until the vertical column of waste is homogenized and a final sample taken. With very viscous wastes, this "column" will not intermix significantly with adjacent material; with low viscosity wastes, the column can be isolated by first sinking a caisson into the bottom, and working within it. This can be done at various grid locations to yield a very good representation of the impoundment, particularly if it is to be homogenized before treatment. It is also a good way to take large samples when required.

Digging. For very solid wastes, sampling with a backhoe, clamshell, or dragline may be used. Samples can be taken at different depths as the overlying layers are removed and sampled, or each sampling area homogenized from top to bottom. This is usually the best method for contaminated soil sites and old, dried-up lagoons.

The number of sampling points is determined by the size of the impoundment, its age, expected degree of stratification, etc. For CFS treatability study purposes, a minimum of five horizontal locations is recommended. Unless there are data to indicate otherwise, one sample should be taken from the center and one near each corner, about one-third of the way from the corner to the center. The sample should be either vertically homogenized, or a complete core taken. For larger lagoons, a sample should be taken for each 10,000 ft^2 of surface area; for very large lagoons, the area per sample can be increased for practical reasons.

Sampling for CFS treatability purposes involves considerable time, effort, and expense to properly characterize the waste. However, to skip this step can, and usually will, cost a great deal more in the long run. Responsible waste treatment firms will refuse to guarantee results when the impoundment has not been properly sampled, or will base their warranties on the premise that the impoundment when homogenized is identical to the sample given them. The only way to get an unconditional guarantee is to involve the CFS vendor in the characterization process at the earliest possible time.

SAMPLE CONTAINERS, STORAGE, SHIPPING, AND HANDLING

Most CFS treatment is done for the purpose of solidifying nonsolids and/or immobilizing metals and other nonvolatile species. For this purpose, the best sample container is a five-gallon plastic (polyolefin) pail with gasketed lid. This type of container can usually be shipped without a secondary container if the waste meets certain DOT requirements. If volatile organics are present, glass containers with Teflon-lined closures may be required. Information about proper shipping containers, labeling, shipping manifests, chain-of-custody records, and other state and federal regulatory requirements may be obtained by consulting reference guides such as the Keller manuals [7]. These regulations and safety practices change from time to time, and it is important to use the proper procedures to protect both shipper and testing laboratory. Most CFS companies and testing laboratories will not accept samples without the proper packaging and paperwork.

Several special considerations are necessary for samples that are to be CFS tested, as opposed to those for chemical analysis. No preservatives are to be used for CFS samples, and the samples should be protected from freezing. If these conditions are not met, the CFS test results may be meaningless, or worse, misleading. Shipping is best done by the United Parcel Service or Federal Express, and should be as rapid as possible to avoid changes in the sample before testing. Biologically active samples such as sewage sludges must be shipped refrigerated, but not frozen, (there is special packaging available for this purpose) and the container should be provided with a pressure-release cap to vent any gas pressure build-up. Containers, labels, and other supplies that meet regulatory requirements and the other needs previously specified are available from companies such as LabelMaster [8].

INFORMATION SYSTEM: CHAIN OF CUSTODY

The proper and timely transfer of information concerning the waste sample is just as important as the sample itself. Furthermore, certain manifest, chain-of-custody, and representative sample documents are a legal requirement for hazardous materials shipping and sample-custody transactions, and are normally required by the testing laboratory or CFS firm as a condition for processing the sample. Examples of typical forms are given in Figures 17-3, 17-4 and 17-5. In addition, most CFS companies require some sort of waste profile form, such as that shown in Figure 17-6. The information contained in the latter form is vital for designing a proper treatability study, as we discussed in Chapters 7 and 8. It also becomes part of a valuable resource—a waste treatability data base—as we shall see in Chapter 18.

When the sample is received, it is promptly logged in and given a sample number that will be its identification from that point on. It is then stored in a sample-receiving area until processing begins. Both of these procedures should be under the supervision of a sample-control custodian who understands the procedures and also the regulations concerning the receipt, possession, control, storage, and disposal of hazardous waste samples. Figure 17-7 is the flow chart for information throughout the testing cycle in a CFS testing laboratory that uses a computer-based laboratory information management system (LIMS). The routing of the physical sample through the laboratory is shown in Figure 17-8. These flow charts are also decision trees that show the decisions necessary in the processing of a typical sample.

When the necessary sample-control paperwork has been completed, the next step is to fill out a laboratory work sheet that contains the information necessary for beginning sample characteristics and treatability testing. A format that has been extensively used by the author for this purpose is shown in Figure 17-9. It is laid out in three sections.

Waste Description. This is primarily information furnished by the generator, but which may require verification or elaboration as discussed later in this chapter.

Waste Analysis. Often, analysis of the raw waste is supplied by the generator, but it may require verification. Provision is made to include leaching test results on both the raw waste and the treated end product.

Test Results. Results from the treatability testing are written in the space provided on the form, which also includes a summation area for results on the optimum sample.

This form is set up so that the final data can be entered from it in a logical manner onto the computer entry form shown in Figure 17-10. If the computer entry is to be made by a technician familiar with the testing protocol and the interpretation of results, entry may be made directly from the laboratory work sheet. In many laboratories, however, computer entry is done by clerical staff, in which case the entry form is necessary.

DETERMINATION OF WASTE CHARACTERISTICS

Sample Preparation

The first step in the determination of waste characteristics is to prepare the sample for testing. A step-by-step procedure for this is as follows.

1. If the sample is in a small container (one gallon or less), take it directly to the CFS laboratory or workplace. Larger containers are best broken down in the

Figure 17-3 Manifest form. (From Chemical Waste Management.)

CHAIN OF CUSTODY RECORD
Hazardous Materials

Location of Sampling: ___ Producer ___ Hauler ___ Disposal Site
　　　　　　　　　　✓ Other: _ANALYTICAL LAB_
Company's Name _CHEMICAL WASTE MGMT._ Telephone (_312_) _841-8360_
Address _150 WEST_ _137TH STREET_ _RIVERDALE_ _IL_ _60627_
　　　　　number street　　　　　city　　　　　state　　　zip
Collector's Name _____ Telephone (___) _____
Date Sampled _____ Time Sampled _____ hours

Collector Sample No.	No. of Containers	Sample Description/Source

Sample(s) Submitted to:

1. _____
　　　　　　　　　　name of organization
2. _____
　　　　　　　　　　name of organization

Chain of Possession:

1. _____　　_____　　_____
　　　signature　　　　　　title　　　　　inclusive dates
2. _____　　_____　　_____
　　　signature　　　　　　title　　　　　inclusive dates
3. _____　　_____　　_____
　　　signature　　　　　　title　　　　　inclusive dates
4. _____　　_____　　_____
　　　signature　　　　　　title　　　　　inclusive dates
5. _____　　_____　　_____
　　　signature　　　　　　title　　　　　inclusive dates

"Certificate of Representative Sample" attached; ___ Yes ___ No/Explain

Figure 17-4 Chain-of-custody form. Note: the apparent gaps/breaks in the "inclusive dates" section of the "chain of possession" section are covered by site sample shipping/receiving logs. (From Chemical Waste Management.)

Chemical Waste Management, Inc.
RECERTIFICATION OF GENERATOR'S WASTE MATERIAL PROFILE SHEET
PLEASE PRINT IN INK OR TYPE (Elite, 12-pitch).

Waste Profile Sheet Code

CWM Location of Original: _____ (SHADED AREAS FOR CWM USE ONLY) CWM Sales Rep. #: _____

INSTRUCTIONS FOR COMPLETING THIS FORM ARE FOUND ON THE OPPOSITE SIDE. ANSWERS MUST BE MADE TO ALL QUESTIONS. A copy of the previously completed Generator's Waste Material Profile Sheet which is being recertified must be attached to this recertification form.
Return this form and attachments to: _____
_____
_____ Zip Code: _____

A. GENERAL INFORMATION
1. Generator Name: _____ 2. Generator USEPA ID: _____
3. Facility Address: _____ 4. Generator State ID: _____
_____ 5. Zip Code: _____
6. Technical Contact: _____ 7. Title: _____ 8. Phone: ( ) ___ - ___

B. 1. **NAME OF WASTE** _____
2. **PROCESS GENERATING WASTE** _____

C. CHANGES OR ADDITIONS SINCE LAST PROFILE SHEET PREPARATION OR RECERTIFICATION
1. Have you obtained any laboratory analysis of this waste? ☐ Yes ☐ No. If yes, please attach copies.
2. Have you changed the raw materials used in the waste-generating process? ☐ Yes ☐ No
3. Have you changed the waste-generating process itself? ☐ Yes ☐ No
4. Are you aware of any facts or circumstances which have, or reasonably could have, altered the physical characteristics or chemical composition of the waste? ☐ Yes ☐ No
5. Are you aware of any human health effects of exposure to the waste not previously conveyed to Chemical Waste Management, Inc.? ☐ Yes ☐ No
6. If you answered "Yes" to questions 2, 3, 4 or 5, please provide details by attaching additional pages.

D. SPECIFIC CONSTITUENTS
1. Is this waste a listed Dioxin waste as defined by 40 CFR 261.31 (e.g., F020, F021, F022, F023, F026, F027, or F028)? ☐ Yes ☐ No
2. Is this waste a listed solvent waste as defined by 40 CFR 261.31 (F001, F002, F003, F004, or F005)? ☐ Yes ☐ No
3. Does this waste contain greater than 1000 ppm total halogenated organic compounds? ☐ Yes ☐ No

E. REPRESENTATIVE SAMPLE
1. ☐ I have obtained and will send a representative sample of this waste to Chemical Waste Management, Inc. as outlined on the attached Certification of Representative Sample.
2. ☐ I hereby authorize Chemical Waste Management, Inc. to obtain a representative sample of this waste from any shipment arriving after the receipt of this recertification form. **DO NOT COMPLETE** the attached Certification of Representative Sample form.

F. TRANSPORTATION INFORMATION Has the transportation information for this waste changed from that on the attached Generator's Waste Material Profile Sheet? ☐ No ☐ Yes If Yes, complete the remainder of this Part F.
1. Is this a DOT Hazardous Material? ☐ Yes ☐ No 2. Anticipated Annual Volume/Units: _____ / _____
3. Proper Shipping Name: _____
4. Hazard Class: _____ 5. I.D. #: _____
6. Additional Description: (_____)
7. Method of Shipment: ☐ Bulk Liquid ☐ Bulk Solid ☐ Drum (Type/Size): ___ / ___ Other: _____
8. CERCLA Reportable Quantity (RQ): _____ 9. RQ Units (lb/kg): _____
10. USEPA Hazardous Waste? ☐ Yes ☐ No 11. USEPA Hazardous Waste Number(s): _____
12. State Hazardous Waste? ☐ Yes ☐ No 13. State Hazardous Waste Number(s): _____

G. RECERTIFICATION The information provided in this document, the attached Chemical Waste Management, Inc. Generator's Waste Material Profile Sheet, and all other attached documents contain true and accurate descriptions of this waste material. All new information regarding known or suspected hazards in the possession of the generator has been disclosed.

1. Signature _____ 2. Title _____
3. Name (Type or Print) _____ 4. Date _____

Form CWM 50-B © 1987 Chemical Waste Management, Inc.

Figure 17-5 Representative sample form. (From Chemical Waste Management.)

Waste Management, Inc.
GENERATOR'S WASTE MATERIAL PROFILE SHEET
PLEASE PRINT IN INK OR TYPE (Elite, 12-pitch).

H 98837
Waste Profile Sheet Code

WMI Location of Original: _____ (SHADED AREAS FOR WMI USE ONLY) WMI Sales Rep. #: _____

A. GENERAL INFORMATION
1. Generator Name: _____ 2. Generator USEPA ID: _____
3. Facility Address: _____ 4. Generator State ID: _____
_____ 5. Zip Code: _____
6. Technical Contact: _____ 7. Title: _____ 8. Phone: ( ) ___ - ___

B. MAIL WASTE MANAGEMENT, INC. INVOICES TO
1. ☐ Generating Facility (A, above), or
2. Company Name: _____ 3. Phone: ( ) ___ - ___
4. Address: _____
_____ 5. Zip Code: _____

C.
1. **NAME OF WASTE** _____
2. **PROCESS GENERATING WASTE** _____
3. Is this waste a Dioxin listed waste as defined in 40 CFR 261.31 (e.g., F020, F021, F022, F023, F026, F027, or F028)?
   ☐ Yes ☐ No   If yes, **DO NOT COMPLETE** this form. Contact your Waste Management, Inc. sales representative for assistance.

D. PHYSICAL CHARACTERISTICS OF WASTE

| 1. Color: | 2. Does the waste have a strong incidental odor? ☐ No ☐ Yes If known, describe: | 3. Physical State @ 70°F: ☐ Solid ☐ Semi-Solid ☐ Liquid ☐ Powder Other: | 4. Layers: ☐ Multilayered ☐ Bi-layered ☐ Single Phased | 5. Specific Gravity: Range: ___ - ___ | 6. Free Liquids: ☐ Yes ☐ No Volume: ___ % |

7. pH: ☐ ≤ 2   ☐ > 2-4   ☐ 4-7   ☐ 7   ☐ 7-10   ☐ 10- < 12.5   ☐ ≥ 12.5   ☐ Range ___ - ___   ☐ NA

8. Liquid Flash Point: ☐ < 73°F   ☐ 73-99°F   ☐ 100-139°F   ☐ 140-199°F   ☐ ≥ 200°F   ☐ None   ☐ Closed Cup   ☐ Open Cup

E. CHEMICAL COMPOSITION
RANGE MIN. - MAX.
1. _____ ___ - ___ %
   _____ ___ - ___ %
   _____ ___ - ___ %
   _____ ___ - ___ %
   _____ ___ - ___ %
   _____ ___ - ___ %
   _____ ___ - ___ %
   _____ ___ - ___ %
   _____ ___ - ___ %

Please note: The chemical composition total in the maximum column must be greater than or equal to 100%. **TOTAL:** ___ %

2. Indicate if this waste contains any of the following:

	NONE	or	LESS THAN	or	ACTUAL
PCB's	☐		☐ < 50 ppm		___ ppm
Cyanides	☐		☐ < 50 ppm		___ ppm
Phenolics	☐		☐ < 50 ppm		___ ppm
Sulfides	☐		☐ < 50 ppm		___ ppm

F. METALS
Indicate if this waste contains any of the following:
1. ☐ EP TOX/TCLP   or   2. ☐ Total

METAL	LESS THAN	or	ACTUAL (Parts Per Million)
Arsenic	☐ < 5	☐ < 500	___
Barium	☐ < 100		___
Cadmium	☐ < 1	☐ < 100	___
Chromium	☐ < 5		___
Lead	☐ < 5	☐ < 500	___
Mercury	☐ < 0.2	☐ < 20	___
Selenium	☐ < 1	☐ < 100	___
Silver	☐ < 5		___
Chromium-Hex	☐ < 5	☐ < 500	___
Copper	☐ < 5		___
Nickel	☐ < 5	☐ < 134	___
Thallium	☐ < 5	☐ < 130	___
Zinc	☐ < 5		___
_____	☐ <		___
_____	☐ <		___
_____	☐ <		___

Side 1 of 2   TURN PAGE AND COMPLETE SIDE 2
Form WMI-6000 (Rev. 05/87) © 1980 Waste Management, Inc.

Figure 17-6 Waste profile form. (From Chemical Waste Management.)

DETERMINATION OF WASTE CHARACTERISTICS 525

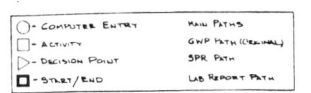

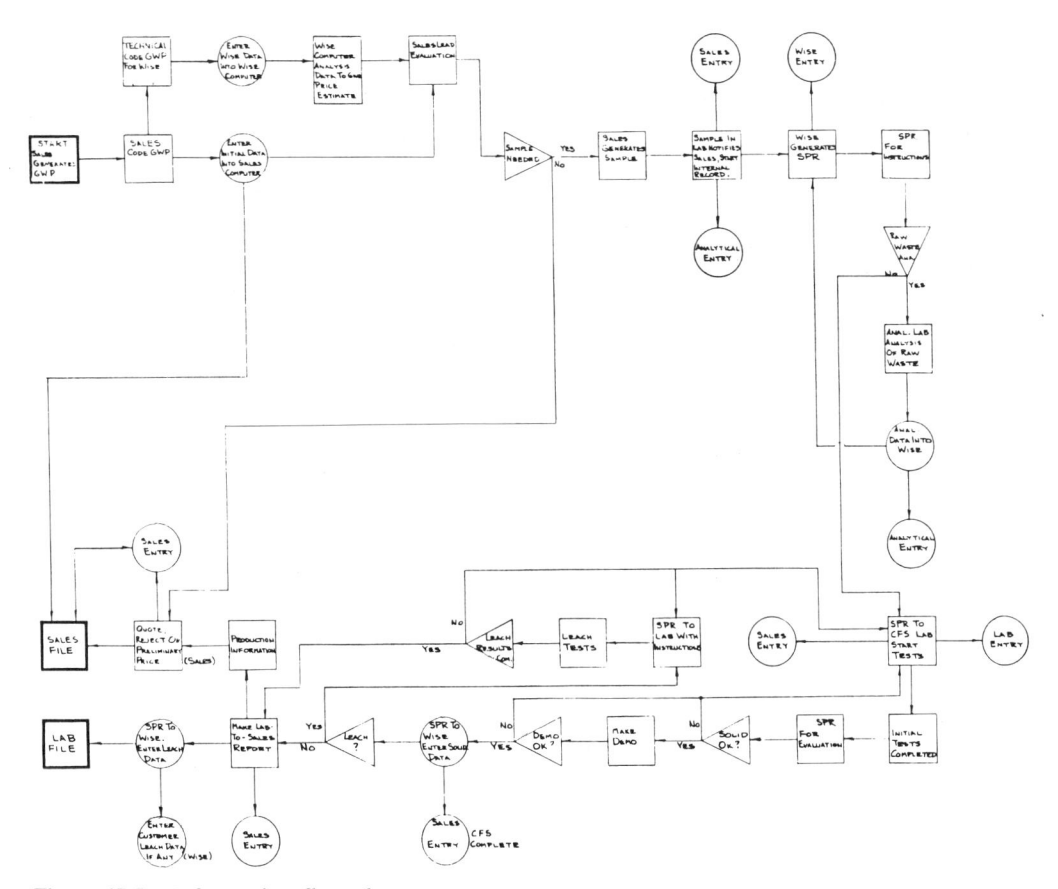

Figure 17-7 Information flow chart.

sample storage area, if it is equipped for this purpose, since the process is often messy and prone to spillage. The sample storage area or laboratory area used for this purpose should have spill-containment provisions.

2. The container should be opened under a chemical fume hood. Alternately, personnel should wear respirators suited for the purpose. In any case, lab personnel should be protected with rubber gloves, lab coat and/or apron, safety glasses, and (when opening a sample container for the first time) face shield.

3. The sample must be homogenized so that any aliquots taken for analysis, physical testing, or CFS treatability testing are representative of the whole sample. The following methods can be used.

    (a) For single phase, clear liquids shaking or simple stirring will normally suffice. Make sure that there is no sediment, or if there is, that it is evenly distributed by stirring.

526 CFS TESTING AND FORMULATION

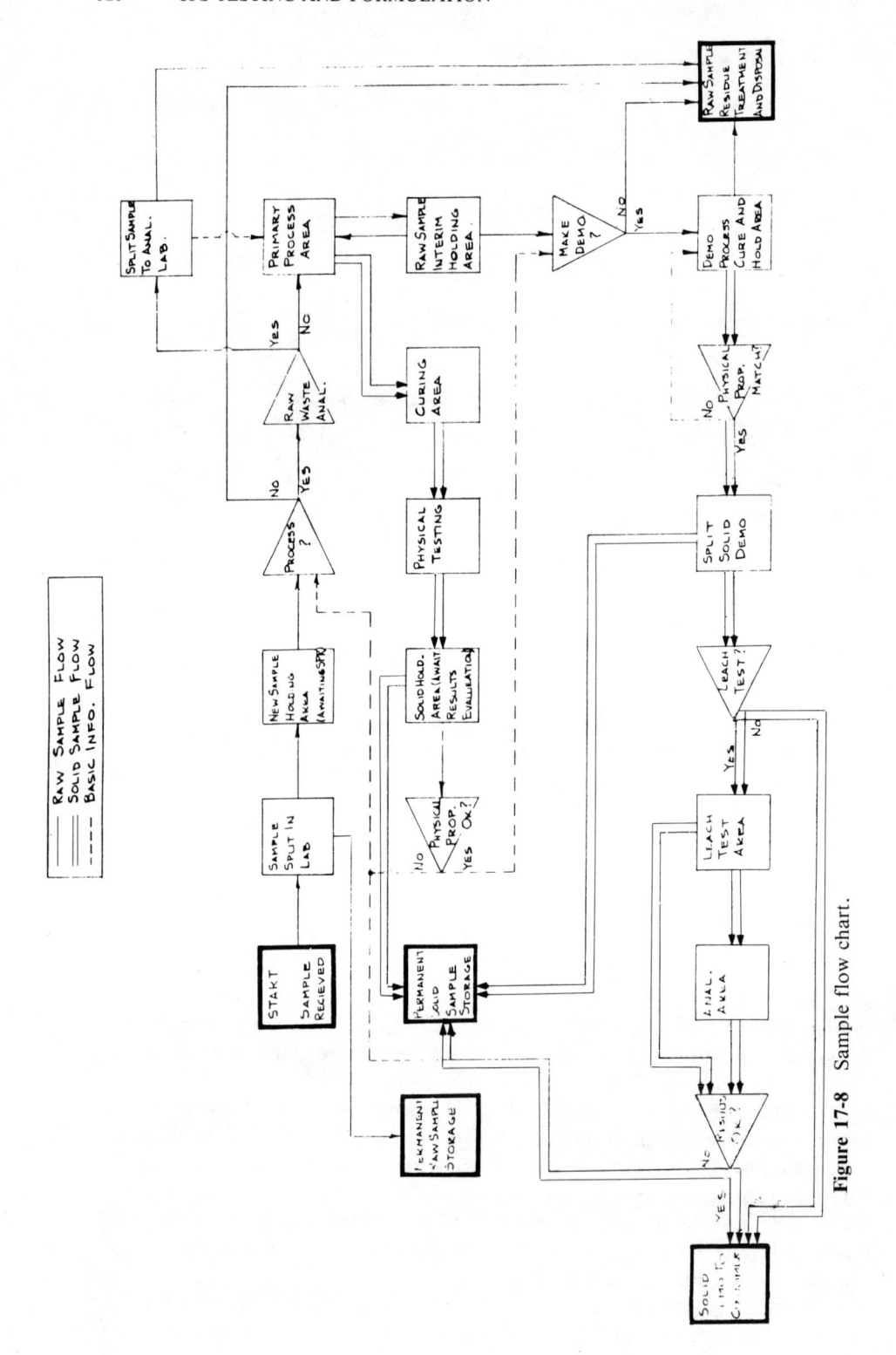

Figure 17-8 Sample flow chart.

DETERMINATION OF WASTE CHARACTERISTICS 527

Figure 17-9 CFS laboratory work record sheet.

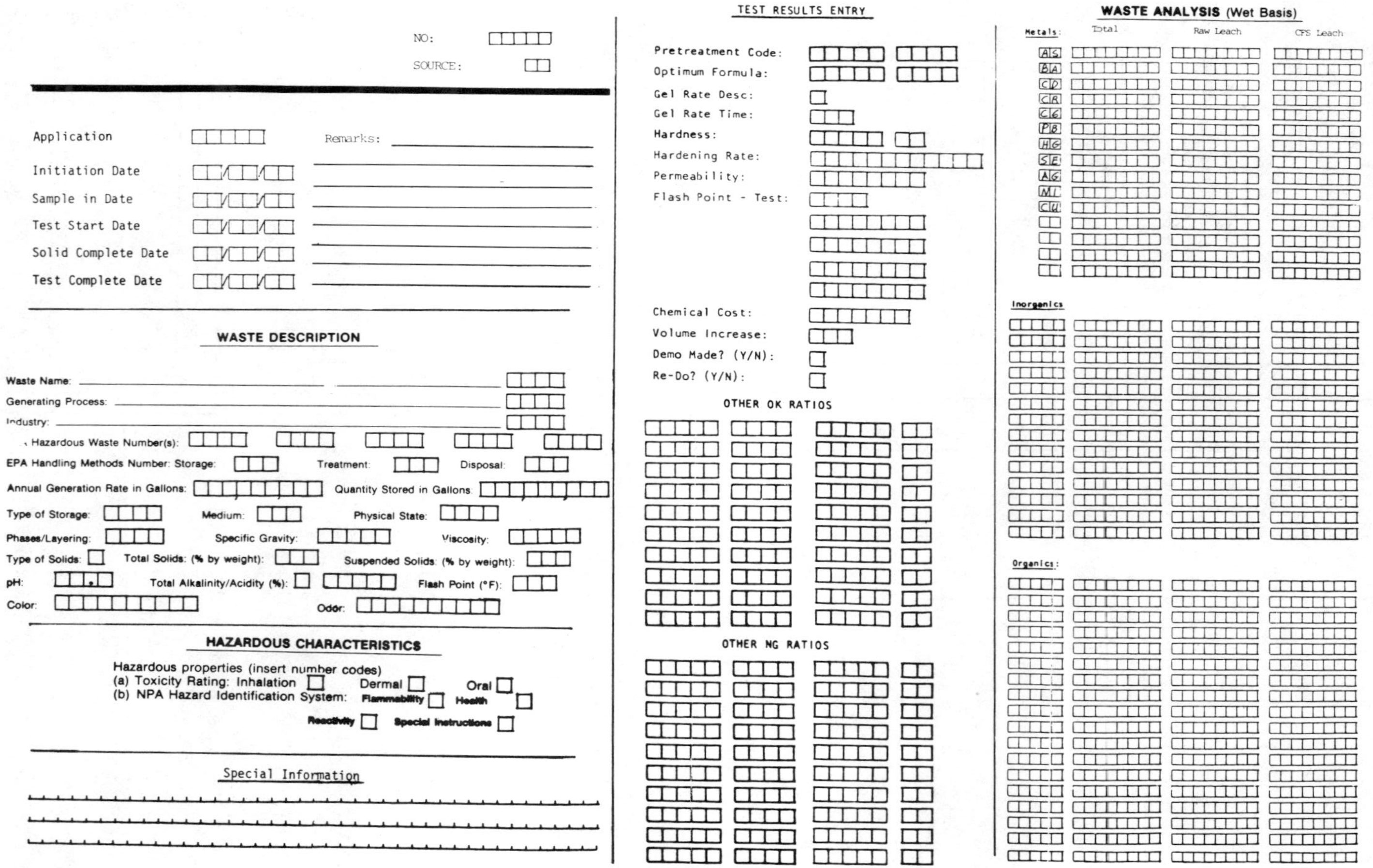

528 CFS TESTING AND FORMULATION

Figure 17-10 CFS computer data entry form.

(b) Sludges or slurries can be homogenized by shaking or tumbling of small containers, or the use of a long shaft, propeller-type stirrer driven by an electric drill. The stirrer should be stainless steel or Teflon-coated. This method is much preferred over hand stirring, because the latter takes so much time and effort that thorough homogenization is usually not accomplished. Alternately, the whole sample can be emptied into a large laboratory mixer such as a Hobart mixer. While this is the best method, it is time-consuming because of the necessity to thoroughly clean the mixer afterward. Typically, sludge samples from waste impoundments contain debris, such as rocks, wood, and metal. For testing purposes, any debris over about a quarter inch in size should be removed. It may be necessary to screen the whole sample to accomplish this. Any debris removed should be weighed and the proportion of debris to treated waste recorded. The debris should be retained for possible analysis later.

(c) Aliquots from layered samples that are difficult to homogenize can be taken by one of two methods.
  (1) *Coring.* For viscous materials, a simple tube is used, pushing it vertically to the bottom of the container, inserting a cork into the top, and withdrawing the tube with a twisting motion. The sample is placed in the lab container by removing the cork and using a plunger (if necessary) to force out the sample. For low viscosity materials, bottom valve samplers are available that will keep the aliquot in the tube until the valve is opened. In either case, cores should be taken at three to five horizontal locations to assure a representative aliquot.
  (2) *Separation.* Sometimes it is easier and more useful to separate phases, measuring the relative volume of each, and recombining in the proper proportions to make up an aliquot. This is especially useful with immiscible oil/water systems, or where a solid phase settles out very rapidly.

(d) Solid waste samples, such as emission control dusts, contaminated soil, and solid process residues can be homogenized either by the standard methods used for this purpose [9] or by tumbling or shaking the sample. The preferred method will be determined by the characteristics of the waste itself. Materials with fairly uniform particle size can usually be most easily handled by tumbling or shaking for small samples. Other methods are best for large samples, or where the particle or aggregate size distribution is wide.

Aliquots should be directly placed in the appropriate sample container. Samples for chemical analysis can be preserved at this point; samples for physical and CFS testing must remain unchanged. During the preceding preparatory steps, loss of volatiles, including water, that might change the properties of the waste must be minimized. Great care must be taken to avoid cross contamination arising from equipment (stirrers, core samplers, etc.).

The purpose of the preceding is to divide the original sample into aliquots that are usable for various testing procedures. One- or two-liter containers can then be shaken as required to keep the contents homogenous, and poured or spooned into the actual test container or apparatus. The preparation steps used should be recorded on the lab bench sheets, along with observations about the waste's physical characteristics. These observations—color, odor, debris content and type, non-homogeneity, ease of handling, viscosity/pumpability—are very important in characterization, often more important than the chemical and physical characterization test results themselves.

In treatability testing, a sample of the raw waste should always be retained for archival purposes. The size of the sample will vary depending on the storage space available, but should be at least 500 ml. Archival samples should be recorded in the waste-tracking system and kept in locked storage cabinets under supervision of the sample-control custodian. These samples are to be used for legal and reference purposes in case a dispute ever arises as to the nature of the test material upon which the CFS formulation was based. Archival samples should be kept at least until the CFS project is completed and accepted by the customer and the appropriate regulatory agencies; if space allows, the samples should be retained indefinitely for possible future comparative test work.

Waste Characterization Tests

The general information that is required for proper characterization of a waste stream depends not only on the waste itself but also on the purpose of the CFS (or any other) treatment program. This was discussed in depth in Chapter 7, and needs no further elaboration here. Some of the information is obtainable from observation alone, or is specified in the transmittal documents. A complete list of waste characteristics appropriate to CFS treatability testing is given below.

Characteristic	*Reference*
Waste name	(1)
Generating process	
Industry	
EPA hazardous waste number(s)	(1)
EPA handling methods number: Storage	(1)
Treatment	
Disposal	(1)
NFPA hazard identification: Flammability	(2)
Health	(2)
Reactivity	(2)
Special	(2)
Toxicity rating: Inhalation	(3)
Dermal	(3)
Oral	(3)
Annual generation rate in gallons	
Quantity stored in gallons	
Type of storage	
Medium: (aqueous, oil, solvent, etc.)	
Physical state: (solution, sludge, powder, etc.)	
Phases/layering: (none, bilayered, multilayered)	
Total solids (%)	(4) Part A
Suspended solids (%)	(4)
Type of solids: (organic, inorganic, mixed)	
Bulk density	
Grain size	
Specific gravity	ASTM D854

Characteristic	Reference
Bulk density (for solids)	
Grain size distribution	
Viscosity	(5)
Flash point	(6) Method 1010
pH	(6) Method 9040
Total alkalinity/acidity (%)	
Odor	
Color	
Waste analysis: Metals, total	(6) Method 6010
Metals, leached	(7)
Inorganics, total	(6) Methods 9030
Inorganics, leached	and 9010
Organics, total	(7)
Organics, leached	(6) Methods 8240, 8250, and 8080
	(7)

Key: (1) EPA, various publications in the Federal Register; (2) National Fire Prevention Association; (3) OSHA; (4) *Standard Methods of Examination of Water and Waste Water* [10]; (5) see discussion later in this chapter; (6) EPA SW 846 [1]; also, appropriate methods in *Standard Methods of Examination of Water and Waste Water* [10] for non-RCRA constituents; (7) Appendix I to 40 CFR Part 268, 51 Fed. Reg. 40643-40652 (Nov. 7, 1986).

It is often useful to organize all or part of the waste characteristic information just listed into a check list for use by persons who obtain the information. One such partial check list used by customer programs personnel at a major hazardous waste treatment firm is shown in Figure 17-11.

Each parameter in the profile is defined by a number specifying the level (e.g., ppm for constituents) or by an alphanumeric code defining type or level. The chemical analytical methods are defined by the EPA [1] in the case of RCRA priority pollutants, or by *Standard Methods* [10] for non-RCA constituents, and need no elaboration here except for one point. In doing waste-residual analyses, the interferences and matrix effects are greatly exaggerated over those problems in water or waste water analysis. This is taken into account in SW 846, but not always in other methodologies such as those in *Standard Methods,* which were often designed for much more dilute streams.

If the waste is hazardous due to toxic characteristics, its leachability should be determined using the appropriate test or tests. If ignitable, the flash point should be determined; otherwise, flash point is usually not necessary. Total solids, pH, and specific gravity should always be determined before doing the treatability testing. These tests are fast and easy to do, and often reveal inconsistencies between the waste profile information given by the generator and the actual waste received. When this situation occurs, the discrepancy must be cleared up before testing begins. The subjective observations, especially those made during waste preparation, are also valuable in determining whether the material is really that described on the waste profile form. In addition, they provide valuable information for the person doing the treatability testing, and can help determine what formulations are to be used and in what mix ratios.

WASTE INFORMATION NEEDED

GENERAL AND PHYSICAL PROPERTIES

Waste Name   (EPA No., Etc.)
Generating Process
Industry   (SIC Code or Description)
Hazard Identification   (Flammability, Reactivity, Etc.)
Toxicity Rating   (Inhalation, Dermal, Oral)
Size of Project   (Gallons, Tons, Cubic Yards)
Purpose of Project   (Solidification, Stabilization, Etc.)
Regulatory Framework/Location   (CERCLA, California, Etc.)
Time Frame
Medium   (Aqueous, Oil, Solvent, etc.)
Physical State   (Solution, Sludge, Powder, etc.)
Phases/Layering   (None, Bilayered, Multilayered)
Total/Suspended Solids (%)
Type of Solids   (Organic, Inorganic, Mixed)
Bulk Density/Specific Gravity
Viscosity
pH
Alkalinity/Acidity (%)

CHEMICAL PROPERTIES

Waste Analysis: Metals, Total
                  Metals, Leachable
                  Inorganics, Total
                  Inorganics, Leachable
                  Organics, Total
                  Organics, Leachable

Figure 17-11 Information check list.

In addition to the standard parameters shown in the preceding list, it is very useful to know the speciation of various constituents, especially the metals. The reasons for this were discussed in Chapter 3 and elsewhere, but the subject is worth reviewing here when discussing test methods. Elemental analysis only is standard procedure for the metals. Determination of the actual metal species is difficult and expensive, especially in these very complex systems; nevertheless, it is becoming more necessary as we try to understand the mechanisms of both fixation and solidification. Recent work in this area has shown the value of species identification, even on a qualitative basis. A combination of scanning electron microscopy/energy dispersive X-ray analysis (SEM/EDX) and Fourier transform infrared analysis (FTIR) has, in one instance, determined that lead in a particular flyash is primarily associated with chlorine, that is, lead chloride (W. Hubble, EnerGroup, Inc., private communication, 1987). Other work on speciation is also underway [11]. There are also indirect methods of species determination, such as the sequential chemical extraction approach used by Cote [12].

Measurement of Viscosity.   One characteristic of great importance in CFS treatment is viscosity, which is directly related to pumpability (or other handling methods) of the waste. Unfortunately, no simple, objective measure of viscosity as it relates to pumpability and mixability of wastes has been found. To understand the reasons for this, and at the same time why it is so important in this field, it is necessary to discuss viscosity and its science, rheology.

Viscosity is a flow-related property of materials, in our case, waste "fluids." More specifically, viscosity is the material's resistance to flow; thus, the higher the viscosity number, the more resistance the waste has toward flow or pumping, or the "thicker" the waste. The force required to shear the waste, or "get it moving," is called the *shear stress*. The measure of the degree of shearing the waste experiences, or the velocity of the liquid layers relative to each other, is called the *shear rate*. *Absolute viscosity* is defined as the ratio of shear stress to shear rate and is expressed in a unit called the *poise* (or more frequently, as the *centipoise,* which is 0.01 poise). Another way of expressing viscosity is as *kinematic viscosity,* which is the ratio of absolute viscosity to the density of the waste, and is expressed in stokes or centistokes.

To develop a frame of reference, the following list gives the absolute viscosities in centipoises of various substances at 20°C (68°F).

Material	Viscosity in Centipoise
Acetone	0.33
Water	1
Concentrated sulfuric acid	24
Olive oil	84
Sodium silicate solution	200
Castor oil	990
Glycerol (100%)	1500

Viscosity is highly dependent on temperature, usually exponentially. For example, #50 motor oil changes viscosity 10 percent for each 1°C (33.8°F) change in temperature.

Viscosity may be measured in many different ways, with a variety of test equipment and methods, depending on the purpose of the measurement. This is a complex subject that cannot be covered in depth here, but it is necessary to understand the basics of rheology (the study of flow) to be able to communicate between lab and field personnel, and with waste generators. A basic explanation of rheological types was given in Chapter 8.

A rating system. While viscosity can be measured in various ways, the different materials encountered in CFS cover every rheological type. To objectively assign a single viscosity number for use in waste characterization is not feasible, because instruments that measure viscosity are different for different types of material. Therefore, it is necessary to use a largely subjective evaluation system that relies on experience, but that can be meaningfully communicated from one experienced person to another.

The system presented in Table 17-1 uses two descriptors, one for viscosity level and one for rheological type. Occasionally, additional comment may also be necessary; for example, when the presence of foreign objects (rocks, wood, plastic film, or the like) or fibers would significantly affect field performance.

Examples of the use of this viscosity rating system are given in the following list.

TABLE 17-1 Viscosity Rating System for Wastes

		Viscosity	
Code	Meaning	Examples	Pumpability/Pump Type
L	Low	Water	Yes/centrifugal
LM	Low/medium	Light motor oil, pickle liquor	Yes/centrifugal
M	Medium	Heavy motor oil, glycerol, oil-based paint, low solids lagoon sludge	Yes/centrifugal
MH	Medium/high	Moderate solids lagoon sludge, soft filter cake	Yes/centrifugal or positive Displacement
H	High	Lime paste, high solids lagoon sludge, filter cake	Yes/positive Displacement
VH	Very high	Wet clay, peanut butter, tars	Sometimes/positive Displacement
S	Solid	Flyash, hard clay, dry filter cake	No
PP	Pseudoplastic	Latex Paint	Yes
P	Plastic	Catsup	Yes/positive Displacement
N	Newtonian	Water	Yes/any pump
T	Thixotropic	Lotions, filter cake	Sometimes/positive Displacement
D	Dilatent	Ceramic slips	No
G	Gel	Jello	Sometimes

Waste	Rating
Landfill leachate	L/N
Filter cake	H/T or H/P
Lagoon sludge, typical	MH/PP
Pickle liquor	LM/N
Solvent-based paint waste	M/N
Water-based paint waste	M/PP
Tarry tank bottoms	MH/N or MH/PP

The system requires some experience on the part of both the rater and the person using the information, and is subject to the variability of any subjective rating system. Nevertheless, it works well when used by experienced people.

INTERROGATION OF DATA BASE

When the problem has been defined, the data collected and the waste characterized physically and chemically, the next step is to analyze this information in light of previous experience. After many years of testing, an organization accumulates an enormous amount of data. Experimentation with dozens of different formulations on thousands of distinct waste streams results in a large body of unique knowledge. This knowledge may include data from:

- Actual experimental and field test results and knowledge gathered over many years
- Commercial-scale operations
- Current and retrospective worldwide information on CFS technology: patents, technical literature, unpublished papers and data, private communications

- Comparative laboratory evaluation of many CFS system types on many distinct waste streams and individual waste samples

It became evident long ago that this mass of data was nearly useless without some sort of automated data processing system. Conversely, it was recognized that if such a system could be developed for use in a practical way, it would not only eliminate much repetitive laboratory testing, but could help lead to a better understanding of the complex chemistry of CFS systems.

A first approach to this problem was made by the author some years ago with a microcomputer (PC) data-base system called WISE [13]. One record from this data-base, representing a single treatability study on one waste, is shown in Figure 17-12. Due to the limited memory and data-base capacity of PCs at that time, it was necessary to combine parameters and use codes for long parameter descriptions

```
Record Number : 189
                          ***** WISE DATA BASE *****

CT Number: Z4STN0010                            Processing Date: 01/04/84
                          WASTE CHARACTERISTICS
Waste Name: W032         Generating Process: G041          Industry: 3325

EPA Codes: D006, D007, D008

Annual Generation Rate: 46000

Medium: SLD    Physical State: PWDR   Phases: NONE  Specific Gravity: .62

Viscosity: 8       Type of Solids: I  Total Solids: 100  Suspended Solids: 100

pH: 6.2   Color: DK BROWN       Odor: NONE      Toxicity Rating: 1/1/1

Comments: SEE Z4STN0006 - REPORTED TO BE SAME MATERIAL.

                          ******************
                              TEST RESULTS

Pretreatment: 901WAA033   Optimum: 900AAA006  Gel Rate: 0-60    Hardness: >>4.5

Hardness Description/Rate: 9G-A0SB9GC9G                Permeability:

Chemical Cost: .118    Volume Increase: -40   Passes EP Tox (Raw/Fixed)?: Y/?

Other OK Ratios: 901WAA025/110AAL006,110ADL006;901WAA050/120ABD006,310AAN004
Other NG Ratios: 901WAA050/110AAL006,110ADL006

Metals: AS,BA,CD,CR,PB,FE

Inorganics 1: I0082
Inorganics 2:
Organics 1:
Organics 2:

Analyses 1: AS>0; BA>0; CD>0,>0; CR>0,>0; PB>0,>0, FE>1000
Analyses 2: I0082>1000
Analyses 3:
Analyses 4:
```

Figure 17-12 WISE™ data-base record.

such as the names of organic compounds. Present systems have much more flexibility and power and, in relational data bases, can contain lookup tables so that manual coding and decoding are unnecessary. However, even the system exemplified in Figure 17-12 is enormously useful. The data base can be searched for specific parameter values or ranges to quickly determine whether a waste with a certain set of characteristics has been tested before, and if so, what the results were. It will be noted that the WISE record contains essentially the same information about waste characteristics listed in the previous section, and also a summary of the test results described in more detail on the laboratory test sheet (Fig. 17-9). CFS formulations are coded for two reasons: to minimize writing in the laboratory and space in the computer memory, and to protect proprietary information.

Another use for a treatability data base such as WISE is to develop quick selection charts for use by marketing and technical service personnel in estimating the applicability of different formulation systems to actual treatment cases or waste types. One such chart is shown in Table 17-2, where general waste categories are matched up with general formulation types, or "series." These series were developed for use in supplying preformulated CFS chemicals for use by generators and others doing their own treatment work. More detailed data sheets are then developed, again using the computer data base to furnish the information; examples of these data sheets are given below.

Series 110AAL
Description of Wastes Handled
    Water-based sludges, filter cakes, and contaminated soils with medium to high solids content and viscosity, and pH range of 5 to 11
Typical Wastes
    Mercury cell brine mud from chlorine production
    General waste water treatment residues
    Mercury cell waste water treatment sludge from chlorine production
    Waste water treatment sludge or filter cake from electroplating and metal finishing
    Latex caulking compound waste
    Lime–neutralized waste pickle liquor
    Plant sludge
    Sodium metasilicate waste

TABLE 17-2 Quick Selector Chart CFS Systems

General Waste Categories	CFS System Series
Dilute, watery wastes, sludges, and slurries	300
Low viscosity solutions with medium to high dissolved solids	270AAB; 300
Sludges with low to medium solids and viscosity	120ABD
Sludges, filter cakes, and soils with medium to high solids and viscosity	110AAL; 110ADL; 910AAA
Powders, soils, and other solids	910AAA
Aqueous sludges with medium oil content	300
Sludges and filter cakes with high organic content	270AAB; 210AAB
Oil- and solvent-based wastes	910EAG; 230AAE
Caustic wastes	300; 230AAE; 910EAG
Acid wastes	210AAB; 230AAE

Zinc phosphate waste from conversion coating cathode, coating sludge from electronics manufacturing
Municipal sewage sludge
Metal hydroxide sludge from storage battery manufacturing

Typical Industries
Inorganic chemicals
Organic chemicals
Electroplating and metal finishing
Paint manufacturing
Industrial painting and coating
Steel finishing
Electronics
Municipal sewage
Storage batteries
Automotive transportation
General manufacturing
Waste management

Series 120ABD
Description
Water-based sludges with low to medium solids content and viscosity, and pH range of 5 to 12

Typical Wastes
Lime-neutralization sludge from chemical manufacturing
Emission control sludge from brass foundry
Water treatment sludge
Waste water treatment sludge from electroplating and metal finishing
Dissolved-air-flotation sludge from textile manufacturing
Cleaning water sludge from pesticide manufacture

Typical Industries
Inorganic chemicals
Organic chemicals
Electroplating and metal finishing
Electronics
General manufacturing
Leather tanning and finishing
Textile
Pulp and paper
Pharmaceutical
Plastic and rubber

Series 210AAB
Description
Water-based sludges, solutions, and suspensions with low to medium solids content and viscosity, and pH range of 2 to 11

Typical Wastes
Spent stripping and cleaning bath solutions from electroplating where cyanide is used (pretreated to destroy cyanide)
Filter sludge from can manufacturing

Textile sizing waste
Cleaning waste from chemical manufacturing
Thorum hydroxide sludge from rare earth manufacturing

Typical Industries
Inorganic chemicals
Organic chemicals
Electroplating and Metal Finishing
Electronics
General manufacturing
Textile

Series 230AAE

Description
Oil- and solvent-based wastes (solidification only); concentrated caustic waste solutions and sludges

Typical wastes
Organic chemical treatment residues
Distillation bottoms
Solvent-based paint residues
Caustic stripper solutions and sludges
Pesticide manufacturing waste water
Permanganate solution
Solvent washup from printing
Rag-oil lagoon sludge
Silicone oil
Oil sludges, general
Rubber solvents and wastes
Synthetic organic resin solutions
Resin soaps
Natural oils and resins
Oil dispersions
Wax emulsions
Plastisols

Typical Industries
Organic chemicals
Paint manufacturing
Industrial painting and coating
Primary metals
Nonferrous metals
Electronics
General manufacturing
Textile
Refining and petrochemical
Plastic and rubber
Waste management

Series 270AAB

Description
Water-based sludges and solutions containing large amounts of dissolved or layered organics, having low to medium solids and viscosity, and pH range of 4 to 13; certain oil- and solvent-based wastes

Typical Wastes
  Cleaning waste from specialty chemical manufacture
  Paint kettle cleaning waste
  Paint sludge, solvent-based
  Paint sludge, spray painting
  Waste water sludge from organic chemical manufacturing
Typical Industries
  Inorganic chemicals
  Organic chemicals
  Paint manufacturing
  Industrial painting and coating
  Electronics
  General manufacturing
  Refining and petrochemical
  Power
  Plastic and rubber
  Waste management

Series 300
Description
  Water-based solutions and sludges, with low solids content and/or low viscosity; emulsions and latices of oils or other organics in water; low to medium dissolved solids content; pH range of 2 (low total acidity) to 13
Typical Wastes
  Chromate waste water from chemical manufacturing
  Waste water treatment residue from rare earth manufacturing
  Amine waste water from chemical manufacturing
  Latex waste from cleaning paint manufacturing equipment
  Waste water treatment sludge from chemical conversion coating of aluminum
  Drawing lubricant from wire manufacturing
  Landfill leachate
  Waste water treatment sludge from leather/tanning
  Waste water treatment sludge from electroplating and metal finishing
  Wash water from paint and allied product manufacturing
  Municipal sewage sludge
  Wash water from organic coating
  Oxalate pickling solution from nonferrous metal finishing
  Solder flux residue wash
  Acid oil sludge from petroleum refinery
  Paint solutions and sludges
  Caustic washes and stripping solutions
  Biosludges, scum, and grit
Typical Industries
  Inorganic chemicals
  Organic chemicals
  Electroplating and metal finishing
  Paint manufacturing
  Industrial painting and coating
  Steel finishing
  Primary metals

Nonferrous metals
Electronics
Municipal sewage
Storage batteries
Automotive/transportation
General manufacturing
Leather tanning and finishing
Textile
Pulp and paper
Refining and petrochemical
Power
Pharmaceutical
Waste management
Food processing
Dredging

Series 910AAA
Description
   Water-based filter cakes and heavy sludges with high solids content and viscosity, having pH range of 4 to 13; powders, contaminated soils, and other solids
Typical Wastes
   Emission control dusts and sludges from electric furnace steel production
   Emission control dusts, general
   Contaminated scrap glass
   Waste water treatment filter cake from electroplating and metal finishing
   Arsenic-containing sludge from wood preservatives
   Chrome oxide sludge
   Incinerator dust
   Sludge/filter cake from storage battery manufacturing
   Calcium fluoride and calcium phosphate sludges
   Grinding and polishing dusts
   Spent pickle liquor
Typical Industries
   Inorganic chemicals
   Organic chemicals
   Electroplating and metal finishing
   Paint manufacturing
   Industrial painting and coating
   Steel finishing
   Nonferrous metals
   Electronics
   Municipal sewage
   Storage batteries
   Automotive/transportation
   General manufacturing
   Leather tanning and finishing
   Power
   Pharmaceutical
   Waste management
   Dredging

A more detailed type of selection chart that can be developed using the data base is shown in Table 17-3. Here, various individual characteristic ranges are given for each formulation type, along with the characteristics of the formulation itself: number of components, typical additive ratios and associated volume increases, etc.

The examples just given are only a few of the many uses of computer data bases in treatability testing and CFS system selection. Another approach to using the data generated from treatability studies is the "expert system" described in some detail in Chapter 18.

CFS TREATABILITY STUDY: FORMULATION

Treatability studies are carried out on solid or liquid hazardous wastes for the purpose of developing chemical, physical, biological, or thermal methods to eliminate or reduce the hazard potential. Some treatability work, such as neutralizing the corrosivity characteristic of acids and bases, is simple and straightforward. In thermal destruction the testing is more complex and expensive, with the results being used primarily in engineering design and emission testing. Physical methods such as dewatering are also involved primarily with engineering and mechanical considerations, although the step of conditioning involves complex chemical and electrochemical phenomena more akin to CFS mechanisms.

CFS formulation procedures are deceptively simple in appearance, involving little in the way of sophisticated measuring devices or complex equipment. The method used in this phase of the treatability study is to make up a number of formulations on small aliquots of the waste, choosing the mixtures on the basis of waste characteristics, experience from previous CFS treatability testing and the judgment of the formulator. These experiential and judgmental inputs can be greatly aided by the use of a computer data-base system as described earlier in this chapter. The goal at this point of testing is to minimize the number of formulations to be made and tested, especially if the testing is extensive and expensive, as it is when evaluating leachability. Often, the experimental design deliberately contemplates an iterative process, establishing the best general formulation type in the first round of formulations and tests, and then optimizing it in the second or third rounds. If the formulation system involves two or more components, it is desirable to use a statistical experimental design method such as a factorial or Simplex design [14]. In addition to meeting the project specifications listed below, some of the other factors that may be important in establishing boundary conditions for the formulation work are:

- Allowable curing time
- Maximum volume increase
- Allowable cost
- Type of mixing: batch or continuous
- Heat generation due to chemical reactions
- Gas generation by volatilization or chemical reaction
- Necessary contact time in the mixer: minimum or maximum
- Restrictions due to rapid gel reactions
- Handling characteristics and hazards of reagents
- Handling characteristics of the waste

TABLE 17-3 CFS System Selection Chart

	110AAL	110ADL	120ABD	210AAB	230AAE	270AAB	300	900AAAA	900EAG
Number of components	1	1	1	1	1	1	2	1	1
Powder or liquid	Powder	Powder	Powder	Powder	Powder	Powder	Powder and liquid	Powder	Powder
Volume increase (%/lb of additive/gal of waste)	4	4	5	5	6	5	5	4	6
Typical additive ratio (lb/gal of waste)	1–2	1–2	1–3	2–5	5–8	2–5	1–3	1–3	6–10
Typical volume increase (%/gal of waste)	4–8	4–8	5–15	10–25	30–48	10–25	5–15	11–12	36–60
Bulk density (not compacted)	0.94	1.00	0.83	0.53	0.75				

Compatible Waste Characteristics

	110AAL	110ADL	120ABD	210AAB	230AAE	270AAB	300	900AAAA	900EAG
Total solids content (%)	15–40	15–40	5–25	10–50	10–50	5–40	1–15	50–100	10–50
Dissolved solids content, max (%)	5	5	5	25	50	25	15	5	100
Viscosity	Med-high	Med-high	Low-med	Low-med	Low-med	Low-med	Low-med	High	Low-med
pH range	5–11	6–12	5–12	2–12	2–13+	4–13	2–13	4–13	2–13+
Allowable total acidity (for wastes with pH below 4)	—	—	—	High	High	—	Low	—	High
Oil content, max (%)	2	2	5	25	50	25	25	5	100
Solvent content, max (%)	1	1	2	10	50	25	10	5	100
Description									
solution				X	X	X	X		X
suspension	X	X	X	X	X	X	X		X
sludge	X	X	X	X	X	X	X	X	X
filter cake		X	X					X	X
emulsion or latex		X			X	X	X		
powder		X						X	
soil	X	X						X	

After curing, these formulations are then tested by a variety of methods as prescribed by the specifications for the final product. These specifications may include some or all of the following:

Physical strength and stability
Reactivity
Ignitability
Corrosivity
Resistance to biodegradation
Leachability
Permeability
Special requirements

Formulation

There are no "standard" (i.e., EPA, ASTM, etc.) test methods for CFS work. However, most workers in the field use similar techniques for CFS testing and development, modified or varied to suit the specific process and application. The following method is used in a number of laboratories and is representative of the technique for inorganic reagent systems.

Preparation. A series of mixes, using a different quantity of each reagent in each, is made with small aliquots of the waste to be tested, normally at least 100 g. The amount used is based on experience with that waste and reagent, and is best expressed as the "mix ratio." Mix ratio (MR) is defined as follows:

$$MR = \frac{\text{weight of reagent}}{\text{weight of waste}}$$

Before beginning the mixes, a small quantity of the waste should be tested under a fume hood with the reagents to be evaluated to ascertain whether there are any hazards associated with the mixture; for example, violent reaction, gas evolution, or rapid heat generation.

Aliquots are weighed out into 250-ml polyolefin jars with tight-fitting screw caps of the same material. Alternately, jars or disposable beakers with no lids may be used, in which case the containers may be individually sealed with laboratory film or commercial plastic wrap, or cured without covering in a controlled humidity environment. The latter procedure may be preferable where exact humidity control is specified for curing. However, it has the disadvantage of making possible the transfer of volatiles among the samples, especially where different wastes are cured in the same atmosphere. The jar should have a heavy wall, a flat bottom, straight sides, and no lip. It should also be translucent, so that the waste level can be observed from the outside. The expected number of aliquots, as determined by the experimental design, are all weighed at the same time and then covered until the mixes are complete. Each jar should be identified with a waterproof label or with permanent marker. Aliquots should be of sufficient size so that the jar will be about half full after reagents are added. This allows enough freeboard for mixing, and provides sufficient depth of treated waste so that subsequent penetrometer tests are not affected by proximity of the jar bottom.

The waste should be leveled in the jar bottom without void space, and the level marked on the outside of the jar. After treatment and compaction of the mixture, the level should be marked again. The two markings can then be compared to a calibrated, marked jar to measure the approximate volume increase due to the treatment. This method is usually easier than the method described later, which requires measurement of density of the treated waste, and is sufficiently accurate for most purposes.

Mixing. With most CFS systems and wastes, optimum mix ratios fall in the range of 0.1 to 2.0. A typical first-cut experimental scheme might use four reagents or reagent mixtures, each at three mix ratios. The reagents are usually weighed directly into the jars onto the waste, and immediately mixed in. Mixing is done with a wide-blade, fairly stiff, stainless-steel spatula. Mixing action depends on the waste and reagents used, and will range from a stirring action to cutting and scraping. Completion of mixing is based on the judgment and experience of the technician, and usually requires several minutes.

When mixing is complete, the sample is compacted into the bottom of the plastic test jar after scraping the mixed wastes from the sides of the jar. It is here that the reason for the requirement of straight-sided jars with no lip becomes evident. It is difficult to cleanly scrape jars with a lip, and this problem can result in residual unmixed waste under the lip and on the sides that will give false leaching test results when the cured waste is removed. During and after mixing, observations are made about the sample (difficulty of mixing, appearance, viscosity, presence of excess fluid, color or odor changes, heat or gas evolution, physical properties). The sample is then sealed to maintain a relative humidity (RH) of 95 to 100 percent during curing.

If the treated waste is still fairly fluid, it can be compacted by tapping the jar sharply on the counter top; this will also remove most large voids and air bubbles. Use of a vibrator or vibrating table is often helpful, especially when the waste is to be molded into special forms for specific test procedures. However, this procedure sometimes causes separation of fluid at the top surface, which is to be avoided in most cases. Treated samples that are solid or semisolid normally should be compacted into the jar bottom with a steel or plastic rod of about one inch diameter. Sufficient force must be used to assure that the treated waste has minimum void space so that it can cure as a homogeneous mass. Voids will result in false penetrometer and volume increase readings and a mass that is not monolithic, preventing proper application of some test procedures. The level of treated waste is marked on the outside of the jar for subsequent calculation of approximate volume increase.

Occasionally, it is desirable to produce a solid with granular properties directly from the mixing stage without grinding the cured waste. In this case, the sample should not be compacted; obviously, strength tests such as the penetrometer will be meaningless, and so are not run on such samples. Care should be taken to prevent compaction during handling of the jar.

Another special case happens when the waste to be treated is a dry dust or soil. For such wastes, water addition is usually required in all commercial inorganic systems that consume water in the chemical solidification reactions (although, there are fixation reactions in which water is generated). However, the addition of water will not cause a volume increase, since it fills in the void spaces in the unsaturated waste. In fact, water addition usually allows compaction of the waste, with a resulting volume *decrease*. Therefore, the jar should be marked at three different levels: the compacted dry waste level, the level following water addition and compaction, and the final compacted level of the treated waste. The information so gained will be valuable in later commercial application of the formulation.

Evaluation. Samples are usually evaluated after one, three, and ten days, but shorter and longer times may also be specified. Some slow curing systems, such as lime–flyash, are evaluated after 28 days, or even longer. Between evaluations, the samples are resealed. The following evaluation parameters are standard; others may be specified for individual scenarios.

Hardness/structural strength. A rough, comparative measure of bearing strength (incorrectly called unconfined compressive strength by the penetrometer manufacturer) is made using the SoilTest Model CL-700 penetrometer, or equivalent. This method gives good, consistent results when used on larger samples, 100 grams or more. Small samples with low agent addition ratios do not provide enough test material to minimize wall and bottom effects. If actual UCS numbers of use for engineering purposes are required, tests of freestanding formed cubes or cylinders should be made on a mechanical tester according to procedures such as ASTM D-2166 or D-1633. For the purpose of determining the suitability of a material for landfill, however, penetrometer tests are satisfactory. In addition to the penetrometer test, a subjective test is done by pressing the sample with the tip of the index finger to give a "feel" for the solidity of the material. The results are expressed as shown in Table 17-4.

The required level of strength or hardness is determined by the individual application. In general, however, a minimum bearing strength of 1 ton/ft^2 is acceptable for waste landfill purposes. This value represents the footprint pressure of a typical tracked vehicle used on landfills. For wheeled vehicles and other more stringent requirements, 2.5-ton/ft^2 bearing pressure is sufficient.

Another approach to evaluating the suitability of formulation is by strength *increase*, rather than any absolute value of the final product strength. This approach demonstrates that a chemical reaction has taken place during the treatment, compared with only physical absorption of fluid. Such a demonstration is required by the EPA [15] for any hazardous, noncontainerized liquid that is to be treated and landfilled. A decision matrix proposed for verification of chemical reaction [16] is shown in Figure 17-13.

TABLE 17-4 Subjective Evaluation vs. Penetrometer

Subjective "Finger Test" Result	Penetrometer Result (ton/ft^2)
Soft	
Sample presents little resistance and deforms completely; consistency of mud	0 to 0.5
Firm	
Sample deforms but presents resistance; consistency of compacted soil	1.0 to 3.5
Hard	
Sample deforms little or not at all; consistency of hard clay	3.5 to >4.5
Very hard	
Sample does not deform even under maximum pressure of a penetrometer; consistency of natural rock	»4.5

546 CFS TESTING AND FORMULATION

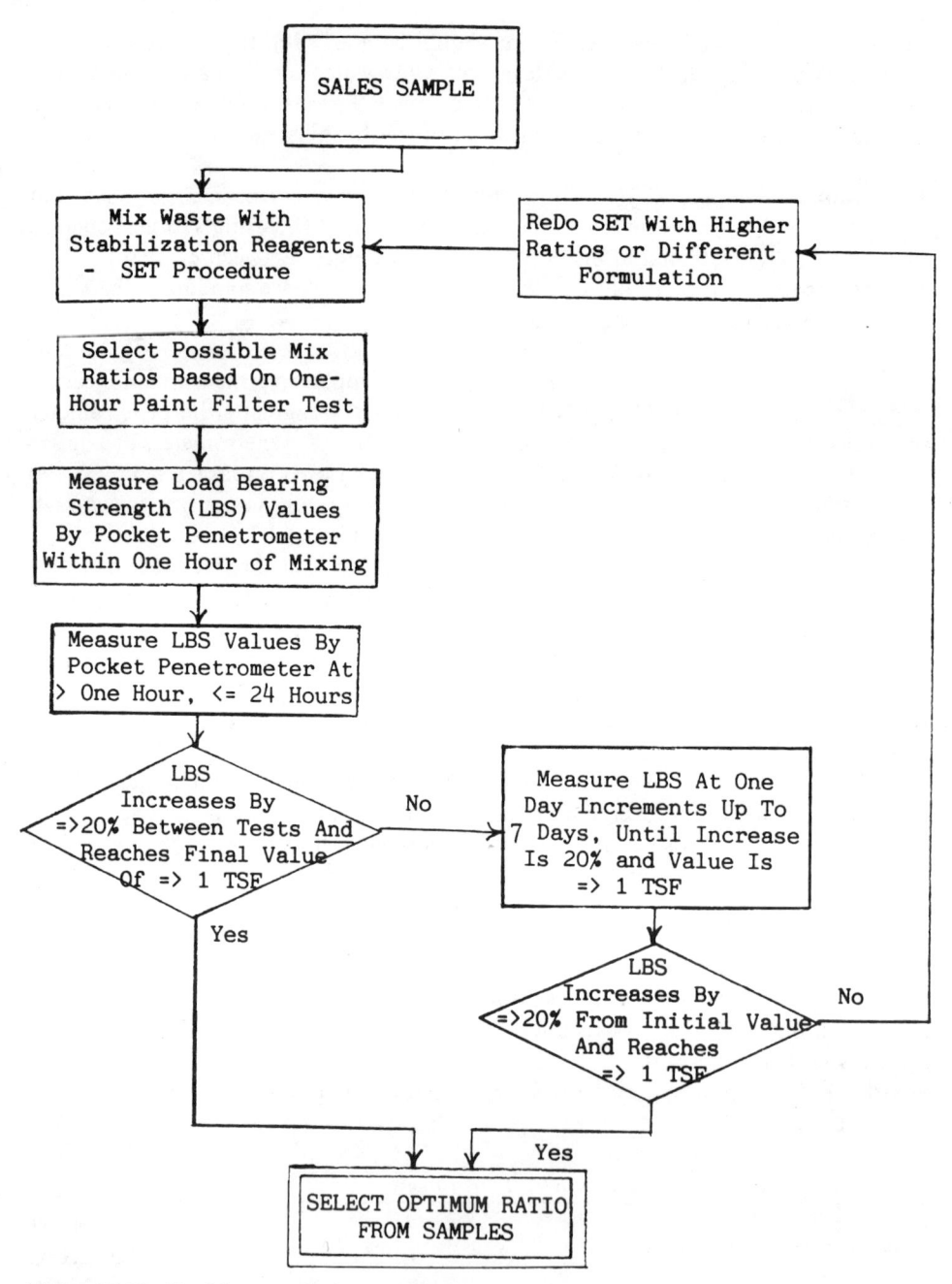

Figure 17-13 Decision matrix for verification of chemical reaction during waste stabilization by the increasing-strength method. (From Chemical Waste Management.)

Other physical properties. In addition to these results, observations about the following attributes of the sample are recorded.

Is the sample spongy, powdery, granular, etc.?
Is the surface softer than the underlying material, or vice versa?
Is there excess water on the surface, even if the underlying sample is firm, or is the surface wet?

Does the sample exude fluid when subjected to pressure from the finger or penetrometer, and if so, is the fluid reabsorbed when the pressure is released?
Has the color or odor changed?
Has the sample expanded, shrunk, evolved gas, etc.?

Chemical properties. There are many chemical properties of a product that could be measured. Some of these have already been discussed and others are listed later in this chapter. However, the primary property of interest is leachability. The initial leaching evaluation is performed on the treated waste from the jar tests, after the physical evaluations have been completed. The usual leaching tests require only 100 grams of test material, but additional QC procedures required under SW 846 [1] usually dictate that samples of 200 grams be provided.

Using the information from all of these test results, either a tentative optimum formulation is selected or the jar testing is reiterated using different reagents or mix ratios. When the jar test procedure appears to have produced acceptable results, the optimum result is usually that which meets the specification at the lowest projected cost, taking into account chemical costs, transportation, disposal, ease of treatment in the field, and any other applicable factors. These factors and their relative weighting will vary from project to project. This is one reason for obtaining all the information listed in the fourth section of this chapter, so that the factors important in test optimum selection can be ascertained.

Confirming Tests. If the first set of jar tests produces an acceptable formulation, the next step is confirmation of the results obtained. This is usually done on a larger sample, using a mechanical mixer, to scale-up the process and simulate more closely an actual field operation. The sample size is increased to 1000 grams, typically, and a motorized laboratory mixer with intermeshing, counterrotating ribbon or blade beaters is used. The product is poured or compacted into a large, plastic container whose height and diameter are about equal, and cured for the specified period. Evaluations are made, as previously described, at the appropriate time intervals to determine if scale-up adjustments are required.

One evaluation parameter that is especially important in the scale-up is the presence of free liquid associated with the cured sample. Small samples often do not display this phenomenon, which nevertheless occurs at larger scale. Another is temperature rise. Large samples with smaller surface-to-volume ratios than the jar test samples often exhibit much larger temperature increases in the interior, due to slower heat transfer out of the material. This may be beneficial in the sense that curing occurs more rapidly at higher temperature, but excessive heat build-up can cause hazards from heat and emission of volatiles. It can also break down the physical structure of the product due to pressure cracking from steam generation.

After curing, leaching tests (if appropriate) are run on the scale-up sample and the results compared with those from the original optimum jar sample. If the results match within acceptable limits, the formulation process is considered complete. If not, the process is reiterated until the results are both acceptable and repeatable. Widely varying results from repeated testing indicate the need to examine factors such as the homogeneity of the sample or the high sensitivity of the CFS chemistry to process variability. In the latter case, the formulation may not be suitable for commercial application, and another approach should be considered.

TESTING OF CFS PRODUCTS

Volumes have been published on leaching and physical test methods for hazardous waste, including the testing of CFS treated waste. Theory, equipment, and protocols have been proposed or established by the EPA, ASTM, states, and various associations and vendors. The connection between leaching tests and chemical fixation was explored in Chapters 3 and 4. Both physical and chemical tests have been mentioned in Chapters 9–15 when discussing the characteristics of specific CFS processes. The bibliography provides many reference sources for those wishing to delve deeper into testing methodologies. Therefore, the remainder of this chapter will be devoted to listing and briefly describing the methods and their applications. For clarity and ease of reference, test methods have been broken down into the following eight categories.

Physical Testing

The EPA has concentrated on the establishment of leachability testing procedures rather than physical testing, since the former relate directly to protection of human health and the environment. Generators, treaters, disposers, and vendors have paid more attention to physical properties, because they relate to handling and disposal characteristics of the solidified waste. In addition, there is a wide selection of ASTM, Department of Transportation (DOT), U.S. Army Corps of Engineers (USACOE), and other agencies' methods that could be applied to physical testing in the hazardous waste area.

Existing methods such as those developed by the ASTM were designed to test the properties of a material for a *specific* purpose, usually not related directly to hazardous waste. Therefore, such methods must be applied with great caution, and modifications are usually required both in the methodology and in the specification values that may be set. Unfortunately, EPA Regional Offices, state regulatory agencies, and some users have sometimes specified inappropriate methods or achievement levels. This may have a number of serious ramifications: excessive cost, unnecessary volume increase resulting in excessive use of limited landfill space, and production of landfill with undesirable operating characteristics.

The following methods, categorized into types, have been or might be used in CFS testing.

Compressive or Bearing Strength. These methods must be applied with great caution to CFS testing. They were developed for construction purposes—buildings, roads, dams, etc. In a hazardous waste landfill, however, the important factors are workability, minimum settling, and ability to support landfill equipment during the operational period. Unconfined compressive strength as measured by procedures such as ASTM D-2166 is meaningful only on cohesive materials. The test requires the preparation of a monolithic, unsupported test specimen, which is then loaded axially to failure in a compression machine. Obviously, materials such as sand and aggregate, which are commonly used as load-bearing road and building foundation base, would yield zero strength in such a test. Therefore, the test's use for specification of suitable bearing strength in a waste landfill is improper. If, however, the treated waste were to be used for embankments, retaining walls, or other applications requiring a certain level of shear strength, then this kind of test for unconfined compressive strength would be applicable, along with other test methods listed below.

ASTM D-2166	Unconfined compressive strength of cohesive soil
ASTM D-1633	Unconfined compressive strength of noncohesive soil
ASTM C-109-86	Method for molding samples for UCS testing [17]
EM 1110-2-1906	USACOE—unconfined compressive strength
—	California bearing ratio
—	Pocket penetrometer
ASTM D-3080	Shear test
ASTM D-2573	Shear test
ASTM D-4318-84	Atterberg limits
ASTM D-698	Compaction
ASTM D-422	Grain size analysis
EM-1110-2-1906	USACOE—grain size analysis
ASTM D-698	Water content, moisture-density
ASTM D-2206	Water content, moisture-density
ASTM D-2216	Water content, moisture-density
ASTM D558-82	Water content, moisture-density
EM-1110-2-1906	USACOE—porosity/void ratio

Physical Stability. A very controversial area of physical testing of CFS products involves the evaluation of how well a material holds up under repeated wet-dry and freeze-thaw cycling. While this property is obviously important in road and building construction, its relevance to waste treatment and disposal is questionable. Hazardous wastes today, even after CFS treatment, are normally disposed of in landfills specifically designed for this purpose. Properly designed and located landfills are subjected to such cycling only for a limited period during the filling of the cell, if at all. The effect of wet-dry or freeze-thaw cycling would be the breakdown of physical integrity, a possibility that is already taken into account in the design of regulatory standard leaching tests such as the TCLP. Therefore, such tests are only meaningful if: (1) the product is to be reused, not landfilled, or (2) the landfill is designed or located in such a way that such cycling is a valid operating mode. In these cases, specific physical and mechanical tests would be applied according to the end use. However, the use or disposal of hazardous wastes outside of the RCRA and state regulatory systems is becoming increasingly rare. Tests that have been used to measure these properties are ASTM D-559-82 (wet-dry durability test) and ASTM D-560-82 (Freeze-thaw durability test).

Permeability. This property is frequently specified in CFS technology. Contrary to general belief, low permeability is not necessarily a desirable property in a landfill, RCRA or otherwise. Very low permeability can result in standing water in the landfill cell while it is being filled, making workability very difficult. And, as we have seen, permeability is not necessarily related to leachability. Several test methods are ASTM D-2434 (permeability), EM-1110-2-1906 (USACOE—permeability), and SW-846 [1] (permeability, Method 9100).

Miscellaneous Physical Test Methods. A variety of other physical tests are applicable to the characterization of hazardous waste, solidified or not. Many of these are commonly used in engineering and economic calculations, especially in doing materials balances for a system design or comparing the relative total costs of various technologies. The most common methods are listed below.

EM-1110-2-1906	USACOE, Appendix II—dry density and bulk density
ASTM D854-83	Bulk density
APHA [18], ASTM D854	Specific gravity
APHA [18]	Total solids
APHA [18]	Moisture (Method 209A)
SW-846 [1]	Free liquid (paint filter test, Method 9095)
—	Water soak tests
ASTM D2216-80	Water content

Chemical (Nonleachability) Testing

The tests listed below are, with several exceptions, standard methods familiar to those in the environmental fields. Their uses in CFS have been discussed elsewhere in this book. The exceptions are tests for biological activity, which have not been applied in CFS but may be in the future, and heat of hydration, which is applicable to the testing of reagents such as kiln dusts, flyashes, and quick lime. The latter application is discussed in Chapters 12–14.

APHA [18]	pH
SW-846 [1]	pH (Method 9045)
40CFR 261.21	Ignitability
40CFR 261.22	Corrosivity
40CFR 261.23	Reactivity
ASTM G-21, 22	Biological activity
ASTM C-186	Heat of hydration
APHA [18]	Total solids and ash (Method 209 G)
—	Acidity/alkalinity
[19]	Sequential chemical extraction
ASTM D1498-76	Oxidation-reduction potential
ASTM C-114	Major oxides

Leachability Testing

The primary objective of chemical fixation is to immobilize (or destroy, in the case of certain inorganic anions and organics) constituents in the waste. Leachability testing is used to predict the degree to which this objective has been accomplished. In view of the variety of possible landfill scenarios, it is not surprising that no single leachability testing procedure or protocol can duplicate all possible field conditions. Ideally, the treated waste would be leach-tested with the actual surface, ground- or rainwater that is present at the site. In practice this is rarely possible, both because of lack of definitive knowledge about what the conditions really are, and because of regulatory philosophy. Therefore, standard leachability tests have been developed and promulgated by the EPA and several states.

Also, Canada and other countries have developed their own methodologies. Leaching/extraction test methods are listed below. A decision tree for selecting leach tests is given in Figure 17-14.

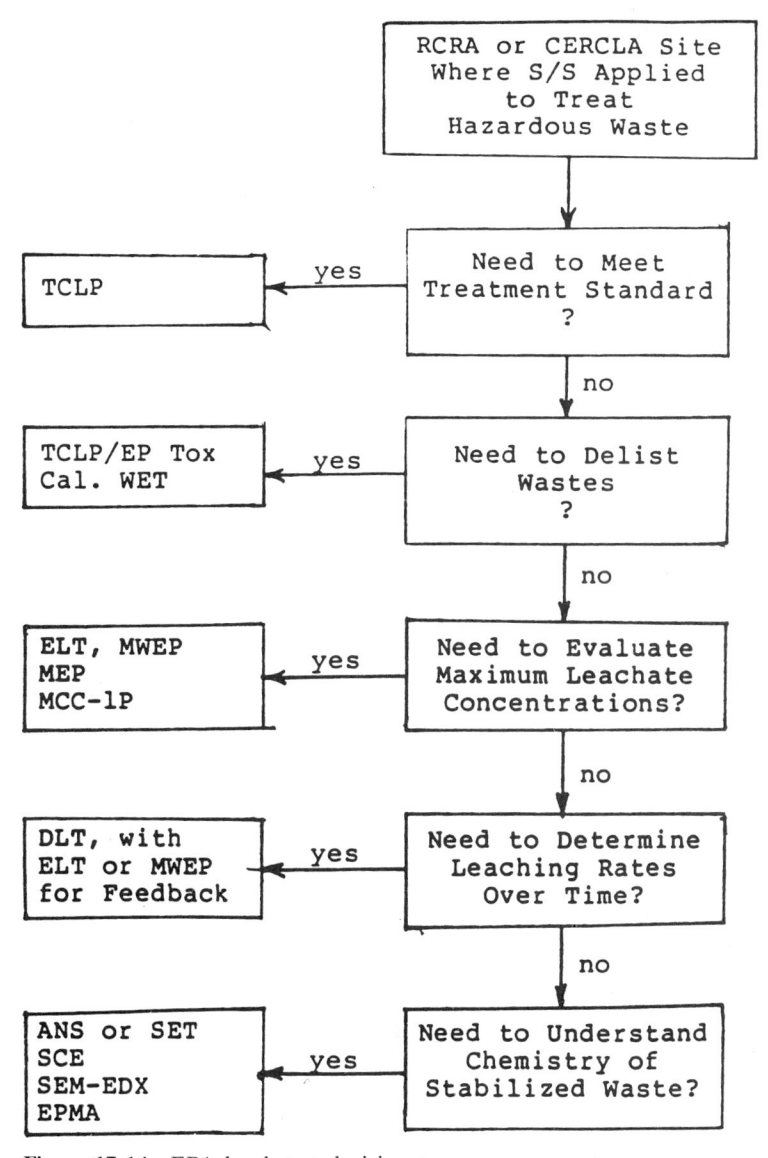

Figure 17-14 EPA leach test decision tree.

U.S. EPA test procedures
  EP Toxicity Test (EPT) [20]
  Toxicity Characteristic Leaching Procedure (TCLP) [21]
  Multiple Extraction Test (MEP) [22]
  Oily Waste Extraction Test (OWEP) [23]
State and Province Test Procedures
  California Waste Extraction Test (WET) [24]
  Ontario Province Test
ASTM Test Procedures
  D3987
  Other
Diffusion Modeling Tests
  ANS 16.1
  MCC Tests
  Static Leaching Test (DLT) [25]
Sequential Batch Extraction Tests
  Solid Waste Leaching Procedure (SWLP)
Flow-through Tests
European and Japanese Test Procedures
Special and Experimental Tests
  Equilibrium Leach Test (ELT)

The most widely known and used procedure is the U.S. EPA EP (extraction procedure) Toxicity Test described in more detail later in this chapter. This method has also been widely condemned for a variety of reasons, but it is still the law in the United States as of this writing. Other test procedures may be used in addition to the EP, and modified tests are also used by the EPA itself for specific scenarios. The major procedures are briefly described later. In comparing leaching procedures, it is useful to use the major variables present in any test as a basis for comparison. These are as follows.

1. Sampling of the waste
2. Preparation of the sample for testing
3. Surface area of the waste
4. Nature of the leaching solution
5. Ratio of waste to leaching solution
6. Time of contact between waste and leaching solution
7. Temperature
8. Agitation method used
9. Test type—single batch, sequential batch, continuous column, etc.
10. Materials of construction
11. Treatment of volatile components
12. Method of liquid–solid separation

These variables are discussed in detail in a working paper for a leaching test workshop [26], in Chapter 3, and in many other references listed in the Bibliography. Mechanisms that can control leachability include chemical reactions between the waste and the leachant, bulk diffusion, and surface layer transfer phenomena. Leaching is generally described as a rate, or sometimes as the cumulative fraction leached. However, in the former case, it is actually expressed as a concentration, that is, milligrams of constituent per liter of leachant after the test is complete. A good theoretical discussion of leaching is given by Cote [27].

Test methods for CFS-treated wastes must address needs at four levels of control (P. L. Cote, private communication, 1988): (1) research, (2) regulatory, (3) treatability study, and (4) operational quality assurance and quality control (QA/QC). This is also true of other test methods, but is especially applicable and more difficult to implement in leachability testing. The difficulty arises primarily at the QA/QC level because of the time necessary to carry out regulatory test procedures. This prevents their use on a real-time basis in process control.

Other ways of categorizing the purpose of leaching tests are as follows.

- System analysis tests designed to approximate disposal conditions as closely as possible, and to generate data for use in mathematical models that attempt to predict long-term environmental effects.
- CFS product optimization tests for use in treatability studies.
- Quality control tests to determine whether the process in use equals the performance of the optimization tests and meets regulatory requirements.
- Accelerated tests that predict long-term behavior in a short-term test; also used in quality control work.

Present test methods are described briefly under the categories used subsequently in this chapter. Other descriptions are presented earlier in this book, in Appendix 2, and in various references in the Bibliography. A comparison of various methods is given in Tables 17-5–17-7. Table 17-8 compares regulatory limits of the various tests for specific hazardous constituents.

U.S. EPA Test Procedures. The EPA has proposed or promulgated several leach testing procedures. The most important is the EP Toxicity Test (EPT).

Extraction Procedure Toxicity Test. The EP test was originally designed to simulate the conditions in a sanitary (RCRA Subtitle D) landfill, where anaerobic decomposition produces organic acids, by leaching the waste product with a solution of acetic acid in a shaker apparatus. The disposal scenario was that of codisposal of hazardous waste in such a landfill, representing a worst-case condition. Nonlisted wastes that pass this test (and also the ignitability, corrosivity, and reactivity requirements) are no longer hazardous by RCRA definition. Listed wastes can be delisted using the EPT (and, sometimes, the MEP and/or OWEP), but with more stringent leaching level requirements (see Chapter 3 and Table 17-8). The codisposal scenario is no longer applicable to most disposal situations, but the test methodology continues, even in the newer TCLP.

The EPT test allows the use of either of two compromises. The waste can be crushed to moderate particle size (9.5 mm) to represent the effects of handling and other environmental conditions. Alternately, a monolithic form can be subjected to an impact procedure (the structural integrity option) and tested in whatever condition results from that procedure. Presumably, any specimen surviving the impact as a monolith would remain that way in the disposal environment, an assumption that is not justified by supporting data. In the EPT, neither preparation method is sufficiently controlled to allow good replication of results, let alone comparison of

TABLE 17-5 Leaching Test Comparisons: Sampling and Sample Preparation

Test Procedure	Sampling Procedure	Preparation of Apparatus	Phase Separation Apparatus
EP Toxicity Test Procedure (EPT) (1980)	Representative sample following ASTM methods or EPA SW-846 procedures. Minimum sample size is 100 g. No preservatives may be used	Not specified	Filter holder: Not specified. Filter media: 0.45-μ membrane filter; alternate is 0.65-μ membrane or glass for difficult to filter materials
Toxicity Characteristic Leaching Procedure (TCLP) (1986)	Sample according to SW-846. Minimum sample size 100 g. No preservatives may be used. Refrigeration allowed unless it causes irreversible change; store at 4°C (39.2°F), extract as soon as possible. Extract within 14 days for volatiles, 40 days for semivolatiles, 28 days for Hg, 180 days for other metals	Not specified	Filter holder: Nucleopore 425900, Micro Filtration Systems 302300, Millepore YT30142HW and XX1004700
Ontario Regulation 309 (1985)	Obtain representative sample equivalent to 100-g solid material	Wash with nonphosphate detergent, rinse twice with reagent water, then 10% HNO_3, then rinse twice. Store bottles filled with 10% HNO_3. For organics, rinse dry bottles with methylene chloride, followed by methanol	Filter holder: stainless-steel 142-mm diam., 1-liter capacity, pressure ability of 5 kg/cm^2. Vacuum filtration may be used. Filter media: 0.45-μ membrane filter of material compatible with leachate. Use Teflon for organic materials, glass prefilter if necessary
Quebec Q-2, R12.1 (1987 draft)	Obtain representative composite sample equivalent to 100 g. Preserve at 4°C (39.2°F). Must arrive 48 hours, be extracted within 30 days, except for phenolics, polycyclic aromatics, within 7 days	Not specified	Filter holder: not specified. Filter media: glass or equivalent membrane filter; 0.45 μ

TABLE 17-5 (*Continued*)

Phase Separation Method	Dry Weight Calculations	Structural Integrity Testing	Particle Size Reduction
Initial pressure 69–103 kPa. Increments of 69 kPa to max. of 517 kPa. Solids extracted above 0.5% by wt. of solid phase	Dry solids content used to determine 0.5% solids limit, not to calculate dry weight of sample into extraction	Yes. Tester has 3.18-cm diam. hammer weighing 0.33 kg with free fall of 15.24 cm. Hammered 15 times after 30 days curing time. No sectioning or dry solids determination	Free-flowing waste ground to 9.5 mm. Includes filter cake if sample was multiphased
Initial pressure 7–69 kPa. Increments of 69 kPa to max. of 345 kPa. Solids extracted above 0.5% by wt. of solid phase. For metals, filters prewashed in HNO$_3$, rinsed in D.I. water	Dry solids content used to determine 0.5% solids limit, not to calculate dry weight of sample into extraction	No	All waste cut, ground, or crushed to pass 9.5-mm sieve
Filter sufficient amount to provide 80-g dry solid. Solids extracted above 0.5% by wt. of solid phase. Use prepurified nitrogen to max. pressure of 490 kPa. Prewash and blank filter media	Calculate dry weight by drying to constant weight at 60°C (140°F). Use on multiphased waste to determine amount to be filtered; on filter cake to determine size of sample for extraction	Same as EPT, except hammered 14 times. Sample may be sectioned or whole sample may be used. Dry sample weight calculated at 60°C (140°F)	Same as EPT
Phase separation to determine % solids if less than 0.5%, no separation above 0.5%. No pressure specified	Dry solids content used to calculate dry weight into extraction, and to calculate 0.5% limit. Dry at 105°C (221°F) to constant weight. Done on filter cake	No. Monolithic wastes treated like other wastes	Free-flowing waste ground to 9.5 mm. Includes monolithic wastes if possible to size reduce. No grinding for volatiles analysis

TABLE 17-6 Leaching Test Comparisons: Extraction

Test Procedure	Vessels	Agitation	Headspace/Volatiles	Quantity of Sample Extracted
EP Toxicity Test Procedure (EPT) (1980)	Not specified	Not specified. Must be sufficient to prevent stratification and continuous contact of all surfaces and liquid	Not defined	Minimum of 100 g
Toxicity Characteristic Leaching Procedure (TCLP) (1986)	Borosilicate glass, PDFE, and Type 316 stainless for all constituents. PE, PP, and PVC for metals only. ZHE (see method for specifications) for volatiles. This applies to extraction and filtration vessels	Any rotary agitation device, with end-over-end rotation of 30 ± 2 rpm	Use ZHE, which allows separation, extraction, and filtration without opening vessel	25 g for ZHE. 100 g minimum otherwise

Ontario Regulation 309 (1985)	Glass or PE bottles with Teflon-lined caps for inorganics. Glass or Teflon only for organics. 1250 ml	Use Solid Waste Rotary Extractor, diagram provided. Rotate end-over-end about a central axis at 10 rpm	Up to 25% free space. No special treatment for volatiles	Sample of solid equivalent to 50 g of dry solids
Quebec Q-2, R12.1 (1987 draft)	Glass for organics, glass or plastic for inorganics. One liter	Rotary extractor turning at 15 ± 5 rpm. Diagram supplied	Use TCLP ZHE equipment and procedures	For solids, use equivalent of 100-g dry solids (50 g for volatiles)

(continued)

TABLE 17-6 Leaching Test Comparisons: Extraction (*Continued*)

Liquid: Solid Ratio	Extraction Fluid	Time/Temperature	pH Adjustment	Separation of Extract
Starting ratio: 16:1. Final ratio: 20:1. Based on weight of undried solid	If pH of sample + 16 parts D.I. water is ≤ 5.0, use D.I. water. If pH > 5.0, add 0.5N acetic acid in D.I. water to reduce pH to 5.0 ± 0.2	24 hours; 28 hours if final pH above 5.2 and all acid not yet added. 20–40°C (68–104°F)	Monitor with Chemtrix-type 45-A controller or adjust at 15, 30, 60 min, moving to next longer interval if pH doesn't need adjustment more than 0.5 units. Adjust for 6 hours, again at 24, 25, 26, 27 hours. Max acid: 4-ml/g solid (2.0 meqH$^+$/g solid)	Same as filtering before extraction
Starting ratio: 20:1. Based on weight of undried solid	Fluid #1: Buffered acetic acid in ASTM Type II water (Type I for volatiles); pH = 4.93 ± 0.05. Fluid #2: Acetic acid in ASTM Type II water; pH = 2.88 ± 0.05 Use #1 for volatiles. Otherwise, use #1 if pH of 5-g sample + 96.5-ml Type II water is <5.0 after 5 min. If pH >5.0, add 3.5-ml 1.0N HCL, cover, heat, and hold at 50°C (122°F) for 10 min. If pH <5.0, use #1 fluid. If >5.0, use #2 fluid	18 hours; 22 ± 3°C (71.6 ± 37.4°F)	No adjustment during extraction. Max. amount of acid added: 2.0 meq H$^+$/g of solid	Metal aliquots acidified with HNO$_3$ to pH <2. Others refrigerated at 4°C (39.2°F). Analysis within 14 days for volatiles, 28 days for Hg, 180 days for other metals

Starting ratio: 16:1. Final ratio: 20:1. Based on weight of dry solid	If pH of sample + 16 parts reagent water is <5.2, use ASTM Type IV water for inorganics, Type I for organics. If pH >5.2, add 0.5N acetic acid to bring pH to 5.0 ± 0.2	24 hours; 20–25°C (68–77°F)	Control during extraction at 15 min, 1,3,6 hours. If pH > 5.2, reduce by adding 0.5N acetic acid. If pH <5.0 ± 0.2, do not adjust until after 6 hours, then adjust volume to 1000 ml with reagent water. Measure and reduce to pH 5.0 ± 0.2 at 22 hours if necessary. Max. amount of acid added is 4-ml/g dry solids (2.0 meq H^+/g dry solids)	Same as filtering before extraction. If analysis is not performed immediately, store at 4°C (39.2°F) after adding appropriate preservatives (guidance provided)
Based on dry solids: above 10% use 10:1	For inorganics, buffered acetic acid in D.I. water; pH 4.5 ± 0.1. Above 10% solids, add D.I. water to 1 liter, and add 100-ml extraction fluid. For organics, add D.I. water to make 1100 ml	24 hours	No adjustment during extraction. Max. acid: 0.82 meq H^+/g dry solids	Decant 30 min. Take unfiltered sample for volatiles analysis. For other constituents, filter through 0.45-μ membrane filter. Prefilters, centrifugin pressure filtration may be used. Appropriate preservation methods specified

TABLE 17-7 Leaching Test Comparisons: Analysis

Test Procedure	Analysis of Total Contaminants	Quality Assurance	Toxicity Testing	Analysis and Interpretation
EP Toxicity Test Procedure (EPT) (1980)	No specific reference	Not specified	Acidic extract could interfere with testing	Add D.I. water to account for differences in liquid:solid ratio. Formula: $V = 20(W) - 16(W) - A$. Where A = ml of acetic acid added during extraction; V = ml D.I. water to be added; W = weight in grams of solid placed in extractor. Separate into component liquid and solid phases. Combine liquid with original liquid separated. Analyze liquid for specified contaminants, following procedures in SW-846
Toxicity Characteristic Leaching Procedure (TCLP) (1986)	If total analysis cannot exceed regulatory thresholds, TCLP need not be run	One blank run after each 10 extractions in a vessel to determine if memory effects occur	Acidic extract could interfere with testing	Nonvolatiles: Filter leachate through new glass fiber filter to separate liquid and solid phases. Acid wash filter if evaluating metals. If waste had no liquid phase, filtered leachate is extract. Follow SW-846 methods in Appendix III of 40 CFR 261. Extracts to be analyzed for metals must be acid-digested.

			Volatiles: In ZHE, separate extract into liquid and solid phases. Filter through glass fiber filter. Follow SW-846 methods of analysis
Ontario Regulation 309 (1985)	No specific reference	Blank run using diluted acetic acid at pH 5.0 ± 0.2 (frequency not specified)	Acidic extract could interfere with testing
			Add enough reagent water to make 1000 ml. Measure pH and record amount of acid added. Separate into liquid and solid phases; discard solids. Use centrifuging if necessary for fine particulates. Analyze liquid for contaminants in Schedule 4, as necessary
Quebec Q-2, R12.1 (1987 draft)	No specific reference	Not specified	Acidic extract could interfere with testing
			Decant 30 min. For nonvolatiles, separate extract into liquid and solid phases through 0.45-μ filter. For volatiles, filter in ZHE as described in TCLP. Analyze for contaminants using appropriate methods of Quebec or Ontario Ministries of Environment, ASTM or USEPA.

TABLE 17-8 Comparison of Regulatory Limits for Various Test Procedures

Contaminant	HWNO	EPT	TCLP	CCWE[a]	1988 1st-3rd Landban TCLP[a]
Acetone			0.59	0.590	
Arsenic	D004	5.000	5.000		0.004
Barium	D005	100.000	100.000		(1)
Benzene	D019		0.070		
Butyl alcohol				5.000	
Cadmium	D006	1.000	1.000		0.066
Carbon disulfide	D021		14.400	4.810	
Carbon tetrachloride	D022		0.070	0.960	
Chlordane	D023		0.030		
Chlorobenzene	D024		1.400	0.050	
Chlorodibenzofurans (all tetra)				0.001	
Chlorodibenzofurans (all hexa)				0.001	
Chlorodibenzofurans (all penta)				0.001	
Chlorodibenzo-p-dioxin (2,3,7,8 tetra)				0.001	
Chlorodibenzo-p-dioxin (all hexa)				0.001	
Chlorodibenzo-p-dioxin (all penta)				0.001	
Chlorodibenzo-p-dioxin (all tetra)				0.001	
Chloroform	D025		0.070		
Chromium (+3)	D007	5.000	5.000		0.094
Chromium (+6)		5.000	5.000		
o-Cresol	D026		10.000	0.750	
m-Cresol	D027		10.000	0.750	
p-Cresol	D028		10.000	0.750	
Cyclohexanone				0.750	
2,4-D	D016	10.000	1.400		
1,2-Dichlorobenzene	D029		4.300	0.125	
1,4-Dichlorobenzene	D030		10.800		
1,2-Dichloroethane	D031		0.400		
1,1-Dichloroethylene	D032		0.100		
2,4-Dinitrotoluene	D033		0.130		
Endrin	D012	0.020	0.003		
Ethyl acetate				0.750	
Ethyl benzene				0.053	
Ethyl ether				0.750	
Heptachlor	D034		0.001		
Hexachlorobenzene	D035		0.130		
Hexachlorobutadiene	D036		0.720		
Hexachloroethane	D037		4.300		
Isobutanol	D038		36.000	5.000	
Lead	D008	5.000	5.000		0.180
Lindane	D013	0.400	0.060		
Mercury	D009	0.200	0.200		0.025
Methanol				0.750	
Methoxychlor	D014	10.000	1.400		
Methylene chloride	D039			0.960	
Methyl ethyl ketone	D040		7.200	0.750	
Methyl isobutyl ketone				0.330	
Nickel					0.048
Nitrobenzene	D041		0.130	0.125	
Pentachlorophenol	D042		3.600	0.010	
Phenol	D043		14.400		
Pyridine	D044		5.000	0.330	
Selenium	D010	1.000	1.000		0.250

TABLE 17-8 *(Continued)*

Contaminant	HWNO	EPT	TCLP	CCWE[a]	1988 1st-3rd Landban TCLP[a]
Silver	D011	5.000	5.000		0.072
Tetrachloroethylene	D047		0.100	0.050	
2,3,4,5-Tetrachlorophenol				0.050	
2,3,4,6-Tetrachlorophenol	D048		1.500		
Toluene	D049		14.400	0.330	
Toxaphene	D015	0.500	0.070		
1,1,1-Trichloroethane	D050			0.410	
Trichloroethylene	D052		0.070	0.091	
2,4,5-Trichlorophenol	D053		5.800	0.050	
2,4,6-Trichlorophenol	D054		0.300	0.050	
1,1,2-Trichloro-1,2,2-trifluoro-ethane			1.050		
Trichlorofluoromethane				0.960	
2,4,5-TP (Silvex)	D017	1.000	0.140		
Vinyl chloride	D055		0.050		
Xylene				0.150	

[a] Non-waste waters, lowest applicable values.

various processes that fracture in different ways. As a result, the sample preparation is very dependent on the technician doing the preparation and the equipment used. For example, the crushing procedure specifies the maximum particle size, but not the minimum size nor the size distribution. Anyone who has actually conducted sample preparation will realize that, within the specifications of the test, the EPT sample can be varied deliberately or accidentally to give widely variable results if the results depend on physical properties. As of this writing, the EPT is scheduled to be replaced, probably for all uses, by the TCLP sometime in 1989.

The methodology of the EPT is given in Appendix 2.

Toxicity Characteristic Leaching Procedure (TCLP). The characteristics of the TCLP and the EPT are compared in Tables 17-5–17-7. The TCLP, a proposed methodology, has never been finally promulgated by the EPA, but has been made official in another way by its use in the 1988 first-third landban regulations. The test is similar to the EPT, but is easier to run and may be more repeatable. Unfortunately, the TCLP procedure does nothing to correct other faults in the methodology, although it does eliminate the variable of the structural integrity option. The variability of test results between laboratories and waste forms, therefore, is not decreased much by the TCLP. The detailed methodology is given in Appendix 2.

Overall, results from the two tests have been much the same, but considerable differences can be experienced with specific wastes and/or CFS processes. Several investigators [28] have found the TCLP to be somewhat more aggressive than the EPT, with leachate concentrations higher by 40 to 80 percent for arsenic, barium, cadmium, and lead. The TCLP has come under serious attack by both industry and environmentalists for being inappropriate in light of present regulations and waste management practices. However, it seems unlikely that the EPA will change the methodology in any substantive way.

The EPA has also used two other test methods for regulation, even though these tests have never been officially promulgated as regulatory requirements on their own. These are the Multiple Extraction Procedure and the Oily Waste Extraction Procedure, both used in "delisting" petitions.

Multiple Extraction Procedure (MEP). The MEP is designed to simulate the long-term leaching effects of acid rain by using sequential extractions of a synthetic acid rain fluid. The EPA estimated that "these extractions simulate approximately 1000 years of rainfall" at sites in Michigan and New Hampshire [22]. The methodology is given in Appendix 2.

Oily Waste Extraction Procedure (OWEP). The OWEP is used for wastes with large amounts of oil. It consists of a preliminary extraction of the organic constituents with a succession of solvents before EPT is run. The concept is that oil and other organics can limit access of the leaching fluid to metal constituents in the waste, thus underestimating their leaching potential. The methodology is given in Appendix 2.

State and Province Test Procedures. The RCRA states that the states may apply any methods in addition to those prescribed by the act. As a result, some states have used other leaching test procedures, and the Canadian provinces have developed some of their own. In the United States, the most well known of these is the California Waste Extraction Test.

California Waste Extraction Test (WET). The WET uses citric acid instead of the acetic acid used in the EPT and TCLP, which creates a much more aggressive leaching environment than do the latter tests—often by one to two orders of magnitude. It also creates another problem. The extraction fluid tends to break up the waste form and create a great deal of colloidal solids that are extremely difficult to filter. The methodology is given in Appendix 2.

Ontario Ministry of Environment Test (LEP) [29]. The LEP is a slightly modified EPT test, designed to give improved precision. It allows for two scenarios: codisposal, as in the EPT and TCLP, and monofill, using deoxygenated distilled water as the leachant.

Other tests. Among others, Pennsylvania, New Jersey, Minnesota, Texas, and Illinois have also developed and used different test procedures at one time or another. The Canadian provinces each have their own test methods, but there is no federal government test as in the United States. The comparative characteristics of several of these tests are given in Tables 17-5–17-7.

ASTM Test Procedures. For more than ten years, ASTM Committee D-34 has been developing and testing various leaching test methods. These range from batch tests similar to the EPT, to sequential batch and column leaching procedures that more closely simulate actual groundwater flow conditions. Some of the developmental ASTM tests are intended to be more applicable to specific disposal situations, approximating the existing ground- or surface water chemistries. One method, ASTM D3987 Shake Extraction Test, has been released as an ASTM standard. The methodology is given in Appendix 2. ASTM also has under development a wide range of other test methods, both chemical and physical. The U.S. EPA and Environment Canada are participating on the ASTM committees doing this work, and it can be expected that a number of new methods will eventually come into the official regulatory framework in the United States and Canada.

Diffusion Modeling Tests. Because this sort of test is still rather uncommon outside of the nuclear waste area, we will discuss it in more detail than the other leaching test methodologies in this chapter. Most of the information presented here originates from several sources both within the nuclear industry [30] and outside it [25,31,32]. The concept of diffusion modeling tests is that, under certain conditions, a proper test methodology can provide the data for a mathematical model of the leaching process that will allow the prediction of long-term effects on the surrounding environment. The equations generally used simplify mass-transport theory by assuming that leaching is diffusion controlled and that the following hold.

1. After leaching commences, the concentration of the species being leached is zero at the surface of the waste form, that is, boundary layer effects are not important
2. Leachant is continuously moving and does not change significantly in its composition during the leaching process
3. Leached waste form is homogeneous and remains essentially unchanged during leaching—the assumption of a semi-infinite medium
4. Waste form surface is smooth and does not deteriorate (e.g., craze, spall, or passivate)
5. Bulk diffusion is the limiting process
6. No time-dependent reaction between leachable constituents and the leachant, matrix, or other constituents

Although these mechanisms do occur to some extent, it is believed (and assumed) that they become rate-determining only at later stages of leaching and that this will be evident from analysis of the data. Other mass transfer equations may be derived for a given waste form (waste and CFS system) from the results of long-term studies. The results of such generic (i.e., waste and CFS system specific) studies provide information on the leaching characteristics of the waste form over the long-term, and thereby elucidate the effective leaching mechanisms. In addition, the studies should determine the long-term stability of the waste form by observing physical and chemical changes in the sample. While the test is normally run on monolithic specimens, particulate wastes can also be tested by compacting the material into a form that is subsequently held together by a mesh basket. The next group of specific tests all attempt to accomplish these objectives.

ANS 16.1, Original and Modified. The ANS 16.1 test consists of a procedure in which the leachant is sampled and replaced at designated intervals. The specimen is prepared in monolithic form, usually as a cylinder. It is suspended in the leachant contained in the leach test vessel so that at least 98 percent of the surface of the specimen is in contact with the leachant. Periodically, the specimen is removed, rinsed, and immediately reimmersed in another fresh leachant batch. The leachants so prepared are analyzed for the constituents of interest. Sufficient leachant is used so that the ratio:

$$\frac{\text{Leachant volume (cm}^3\text{)}}{\text{Specimen surface area (cm}^2\text{)}} = 10 \pm 0.2 \text{ (cm)}$$

is maintained during the leaching test. Leachate replacements are at 2, 7, and 24 hours, and at 24-hour intervals thereafter for the next 4 days. Three additional intervals of 14, 28, and 43 days may be used to extend the test to 90 days total. At the beginning of the test, a short (30 seconds to several minutes) surface "wash-off" period is allowed to remove soluble surface constituents that are normally present in all waste forms and whose inclusion would not meet the requirements of diffusion control.

An effective diffusivity is calculated from the equation

$$D = \pi \left[\frac{a_n/A_0}{(\Delta t)_n}\right]^2 \left[\frac{V}{S}\right]^2 T$$

where

a_n = contaminant loss during leaching period n, mg
A_0 = initial amount of a contaminant present in the specimen, mg
V = volume of the specimen, cm^3
S = surface area of the specimen, cm^2
t_n = nth time interval
T = mean cumulative leaching time of the nth leaching interval
D = effective diffusion coefficient, cm^2/s

Various corrections and conversions are used where the specimen is cylindrical and when more than 20 percent of the constituent of interest has been leached. The effective diffusivity is determined for the ten (or other number of) leaching intervals that are then averaged. The "leachability index" (L or LX), a figure of merit, is then calculated as the negative logarithm of this average. The *leachability index* is defined as a material parameter that characterizes the resistance of the solidified waste to leaching of constituents. It can serve for the comparison of different CFS systems and optimization of the CFS process. To be meaningful, however, L must be related to long-term leaching studies carried out with similar materials (i.e., generic studies) under a range of conditions that determine the actual mass-transport mechanisms.

The test has been modified by Cote and others [33] to change the leachant renewal schedules and to use site water or other leachants to simulate more closely actual field conditions. Cote and Isabel's Static Leaching Test (SLT) [25] uses a leachant renewal schedule appropriate to different waste forms of hazardous wastes as opposed to radioactive wastes. They also generate "pseudo-equilibrium data together with the SLT to ensure that saturation of the leachant does not limit the leaching process."

MCC Tests. The Nuclear Waste Materials Characterization Center (MCC) developed a number of proposed acceptance tests for deposition of wastes in geologic repositories. MCC-1, a static leach test, is somewhat similar to ANS 16.1, but uses higher temperatures and a variety of leachants. MCC-3 tests crushed waste with agitation, in a sequential mode, at above ambient temperatures. Two particle sizes are used: 180 to 425 μ and 74 to 149 μ (-40, $+80$ and -100, $+200$ mesh). Other

MCC tests determine leaching behavior as a function of leachant flow rate. None of these tests has been picked up for use by the nonnuclear stabilization community.

Other Tests. The International Atomic Energy Agency (IAEA) published a suggested standard leaching test in 1971 [34] and, although it was never adopted, it was a basis for later methods, such as ANS 16.1.

Sequential Batch Leaching/Extraction (SBL) Tests. The SBL is a tumbling-type test usually run on specimens ground to −9.5-mm particle size (although monolithic waste forms can also be tested). Therefore, reactions with the leachant do not interfere with the leaching process to as great a degree due to constant removal of any passivating layer from the surfaces. The test can be run with any leachant, including site water, using TCLP-type methodology otherwise in each extraction. Ten to fifteen sequential extractions are usually run on the same sample until the leachate concentration levels for constituents of interest level off, or until reserve alkalinity of the stabilized soil approaches zero. This is a worst-case test method, simulating high gross permeability (or nonmonolithic stabilized soil deposit), rapid groundwater movement through the waste, and particle size degradation. It has been shown [35] that lead and chromium stabilization with cement-based processes can produce low leaching even when the alkalinity of the waste form is reduced to essentially zero and the leachate is acidic. This is believed to be due to the formation of "silicates" where the metal is incorporated into the silica structure. On the other hand, cadmium does not show the latter effect. The SBL can help determine the degree of dependency on waste form alkalinity for the specific waste, CFS process, and constituents of interest.

Solid Waste Leaching Procedure (SWLP). The SWLP was used in at least one EPA treatability study [36] to provide engineering estimates of leachate that might be expected from a Superfund site over the long-term. The test uses deionized water or site water with a number of sequential batch extractions.

Flow-Through Tests. Some of the other tests that have been proposed and used over the years include column leaching tests, field lysimeter tests, and a variety of methods used by vendors in the early CFS years. Several examples of column leaching test methodologies are given in Appendix 2, and a good bibliography of this approach was put together by Jones [37]. The ASTM is making an active effort in developing such a standard test, through Committee D-34. A considerable amount of field lysimeter work has been done, much of it at the EPA's Municipal Environmental Research Laboratory in Cincinnati. This sort of study is expensive and slow, but it is very useful in confirming laboratory tests on a more real-world basis.

European and Japanese Test Procedures. In the early 1970s, Japan's Ministry of Public Welfare developed and used a batch shake test using buffered deionized water as the leachant. This test is described in Appendix 2. Great Britain and other countries in Europe each have test methods of one sort or another, but little has been published here on their procedures.

Special and Experimental Tests. A number of special tests have been used and others are now being developed. One is the equilibrium leach test.

Equilibrium Leach Test. As part of a cooperative study between the U.S. EPA, Environment Canada, and a number of CFS vendors and users, Cote [38] proposed the use of an equilibrium leach test (ELT) to assess concentration of constituents in a leachate at near-equilibrium conditions. The test uses a single batch extraction like the TCLP/EPT, but with distilled water and a specimen ground to -100 mesh ($< 150~\mu$). The leachant/waste ratio is 4:1, and the sample is tumbled for seven days. This procedure is useful to establish the nature of equilibrium conditions for the system being studied, which helps to interpret results of other tests.

While such tests will not likely be utilized in the regulatory mode, they are important both to quality control at the commercial level, and to research in the CFS field. The QA/QC aspects have been discussed in Chapter 3; research tests will be covered briefly in Chapter 18.

REFERENCES

1. U.S. EPA. *Test Methods for Evaluating Solid Waste,* SW-846. Washington, DC: Office of Solid Waste and Emergency Response, 1986.
2. Exner, J. H. A sampling strategy for remedial action at hazardous waste sites. *Hazardous Waste Hazardous Mater.* **2**(4): 503–521 (1985).
3. Brantner, K. A. Priority pollutants sample collection and handling. *Pollut. Eng.* 34–38 (March 1981).
4. ASTM. *Standard Guide for General Planning of Waste Sampling.* Philadelphia, 1985.
5. ASTM. *Standard Practice for Sampling Waste and Soils for Volatile Organics.* Philadelphia, 1987.
6. ASTM. *Standard Practice for Sampling Non-Liquid Waste From Trucks.* Philadelphia, 1987.
7. J. J. Keller & Associates, Inc. *Hazardous Waste Regulatory Guide.* Neenah, WI., 1984.
8. LabelMaster. *LabelMaster 1989 General Catalog.* Chicago, IL, 1989.
9. ASTM. *ASTM Annual Book Of Standards.* Philadelphia, 1989.
10. American Public Health Association. *Standard Methods for the Examination of Water and Wastewater.* Washington, DC, 1980.
11. Moore, J. N., W. H. Ficklin, and C. Johns. Partitioning of arsenic and metals in reducing sulfidic sediments. *Environ. Sci. Technol.* **22**: 432–437 (1988).
12. Bridle, T. R., P. L. Cote, T. W. Constable, and J. L. Fraser. Evaluation of heavy metal leachability from solid wastes. *Water Sci. Technol.* **19**: 1029–1036 (1987).
13. Conner, J. R. Use of computer-based artificial intelligence systems to select detoxification and solidification methods for residues and wastes. In *Proc. 5th International Symposium on Environmental Pollution.* Quebec City, P. Q., Canada: Environmental Pollution Institute, 1985.
14. American Chemical Society. Course on Experimental Design. Atlanta, GA, 1985.
15. U.S. EPA. *Prohibition on the Disposal of Bulk Liquid Hazardous Waste in Landfills—Statutory Interpretive Guidance.* Washington, DC, Mar. 19, 1985.
16. Chemical Waste Management. Verification of chemical reaction during waste stabilization by measurement of strength development with time, Tech. Note 87-124, p. 12, 1987.
17. U.S. EPA. Fourteenth Annual Research Symposium: Land Disposal, Remedial Action, Incineration and Treatment of Hazardous Waste. Cincinnati, OH, 1988.
18. American Public Health Association (APHA). *Standard Methods for the Examination of Water and Wastewater.* 15th Ed. New York, 1980.
19. Tessier, A., P. G. C. Campbell, and M. Bisson. *Anal. Chem.* **51**: 844–851 (1979).

20. EP Toxicity Test Procedure. 40CFR Part 261.24, Appendix II, Federal Register (May 19 1980).
21. Federal Register. **51**(114) (June 13, 1986).
22. Federal Register. **47**(225): 52687 (Nov. 22, 1982).
23. Federal Register. **49**(206): 42591 (Oct. 23, 1984).
24. California Administrative Code, Title 22, 66696: 1800.75-84.3 (1985).
25. Cote, P. L., and D. P. Isabel. Application of a dynamic leaching test to solidified hazardous waste. Presented at the ASTM Symposium Industrial and Hazardous Wastes. (Mar. 7-10, 1983).
26. Environment Canada. Towards the development of a standard leachate extraction procedure, 1988.
27. Cote, P. L. *Contaminant Leaching from Cement-Based Waste Forms Under Acidic Conditions.* Ph.D. thesis, McMaster Univ. Hamilton, Ont., Canada, 1986.
28. WMTC Workshop on Leaching Tests. Oak Ridge, TN, July 21-24, 1987.
29. Ontario Ministry of the Environment. *Interim Guideline for the Interpretation of the Hazardous Waste Definition (Regulation 309).* Toronto, Ont., Canada, 1983.
30. American Nuclear Society. *Measurement of the Leachability of Solidified Low-Level Radioactive Wastes by a Short-Term Procedure,* Working Group ANS 16.1 (Final Draft), Feb. 6, 1986.
31. Cote, P. L., and D. P. Hamilton. Leachability comparison of four hazardous waste fixation processes. In *Proc. 38th Annual Purdue Industrial Waste Conference.* Purdue Univ., West Lafayette, IN, May 10-12, 1957.
32. Cote, P. L., T. R. Bridle, and A. Benedek. An approach for evaluating long term leachability from measurement of intrinsic waste properties. In *Proc. 3rd International Symposium on Industrial and Hazardous Waste.* Cairo, Egypt, June 24-27, 1985.
33. U.S. Army Corps of Engineers. *Guide to the Disposal of Chemically Stabilized and Solidified Wastes.* Cincinnati, OH: U.S. EPA, 1980.
34. Hespe, E. D. (Ed.) Leach testing of immobilized radioactive waste solids, a proposal for a standard method. *At. Energy Rev.* **9**(1): 197-207 (1971).
35. Bishop, P. L. Leaching of inorganic hazardous constituents from stabilized/solidified hazardous wastes. *Hazardous Waste Hazardous Mater.* **5**(2): 129-143 (1988).
36. Barich, J. J., J. Greene, and R. Bond. Soil stabilization treatability study at the Western Processing Superfund site. *Superfund '87:* 198-203 (1987).
37. Jones, B. F. *Bibliography of Column Leaching Test Procedures.* Austin, TX: Radian Corp., 1981.
38. Wastewater Technology Centre and Alberta Environmental Centre. *A Proposed Protocol for the Assessment of Solidified Wastes,* Nov. 1983.

Chapter 18

INFORMATION SOURCES, COMPUTER APPLICATIONS, AND RESEARCH AND DEVELOPMENT

INFORMATION SOURCES

Ten years ago, the available, published, technical information on CFS technology (with the exception of leaching and leachability testing) would have fit easily into one file drawer. Today, the literature in this field has expanded greatly, if not in an orderly fashion. It appears, however, that relatively few scientists and engineers working in CFS technology take full advantage of these resources. While it is not possible in this book to fully describe and list the many different information sources and how they can most efficiently be used, we will discuss several aspects of information acquisition and management as they apply to CFS.

One obvious purpose and use of this book is to provide references and a bibliography for further study. The specific references listed with each chapter are a good place to start if one's interest is in a particular area of CFS, for example, metal fixation in Chapter 4. For more general reading, an extensive bibliography, broken down into several useful categories, is provided at the end of the book. In addition to these sources, the standard information resources available for any technical discipline are applicable in CFS, although many will not have much information in them. These resources are:

Books and journals
Technical society publications
Government publications and project reports
Patent literature
Conferences and seminars
Related technologies
Bibliographies

Because of the relative newness of the field and the way in which it has developed, there is little up-to-date information in the form of text or reference books except for Pojasek [1]. Various environmental, other technical, and trade journals occasionally have papers on CFS, although by far the largest volume has been in the testing area. With regard to testing, the ASTM has published several special technical publications (STPs) in this field. The EPA has published several guides [2] and has a handbook in draft form at the time of this writing. Cote's thesis [3] is also a good discussion of the mechanics of cement-based waste forms as well as leaching test methods.

The best sources of information at present are technical conferences and seminars, the proceedings issuing from some of these conferences, and the patent literature. The EPA holds a yearly conference in Cincinnati to discuss the EPA research program; this usually contains at least several papers on CFS technology, especially case histories of pilot and full-scale projects. As for general hazardous waste meetings, the Hazardous Materials Control Research Institute puts on a yearly conference, there are a number of "HAZMAT" exhibitions/conferences each year, and various trade and technical associations hold meetings that sometimes have useful papers. However, there is no technical or trade association that deals exclusively with CFS, nor any specialized publication of this sort.

One of the best sources of process information is the patent literature. There has been considerable patent activity in the CFS field in recent years. During the 1970s, especially, the Japanese patent publication literature was very heavy with processes for solidification and fixation of hazardous wastes. This source is still an important one, although the information available in English is quite limited, primarily only the *Chemical Abstracts* version. The best single source of information is *Chemical Abstracts*.

COMPUTER APPLICATIONS AND SYSTEMS

Computerized systems are briefly discussed here, since they lend themselves well to CFS technology, due to the empirical nature of CFS test work. In fact, as a laboratory builds its library of test results, a computerized system becomes essential if maximum use is to be made of past experience. With the advent of low-cost microcomputers and their associated software, this has become possible even for small organizations. The computer also allows on-line searching of sources such as *Chemical Abstracts* and the patent data bases, and this information can be downloaded into the user's own data-base system if desired.

WISE™ Data Base

In Chapter 17, we discussed one proprietary data-base system—WISE™—developed by the author [4]. A typical record in this data base was shown in Figure 17-12. When a CFS problem has been defined and information collected, the next step is to analyze this information in light of previous experience. After some years of testing, organizations and individuals have accumulated an enormous amount of data. Experimentation with dozens of different formulations on thousands of distinct waste streams results in a valuable body of knowledge. In the case to be studied here, it included data from the following.

- Actual experimental and field test results and knowledge gathered over nearly 15 years
- More than 100 commercial scale operations
- Current and retrospective worldwide information on CFS technology—patents,

technical literature, unpublished papers and data, private communications—all of which had been critically analyzed for validity by the experts in the field
- Comparative laboratory testing of all CFS system types on more than 1000 distinct waste streams and some 10,000 individual waste samples

It became evident long ago that this mass of data required some sort of automated data processing for optimum utilization. Additionally, it was recognized that if this could be done, it would not only eliminate much repetitive laboratory testing, but could eventually lead to a real understanding of the complex chemistry of CFS systems. WISE™ is now being used to eliminate some routine laboratory work. In the system, each waste stream is represented by a record that consists of a set of waste profile files containing the information about the waste (see the list in the fourth section of Chapter 17). At that stage of development, WISE™ was used in a search/sort fashion. The waste profile data for the subject waste were coded and entered. For numerical data, a range is specified. The data base was then searched for all conforming waste streams, and the conforming test results were printed out. The process could be reiterated, broadening or narrowing the search range as desired. Other programs allowed statistical analysis of selected numerical data, and the data base could be used for nontechnical purposes as well, since it also contained certain customer and business information.

While using the data base in this fashion was helpful in some ways, it was still rather cumbersome because only a few characteristics or properties of a waste could be compared to previous test results at one time. This required a reiteration process that in itself became confusing and time-consuming. In fact, an experienced "expert" could usually look at the data and "guesstimate" a possible solution more efficiently than could the search program.

WISE-SCAN™ Expert System

One development out of the artificial intelligence field of special interest in CFS is the "expert system." Recently, inexpensive, user-friendly expert system software has become available for microcomputers. In 1984, the author used one of these systems to develop and test a system called WISE-SCAN™, which is the first known use of expert systems in CFS technology [4]. Because this system is still unique, and has considerable potential application in the CFS field, it will be discussed here in some detail.

Unlike engineering studies in areas where the process can be analyzed and modeled mathematically, CFS treatability studies are done by empirical laboratory tests that are repeated for each new waste stream. After a time, a good technician develops a sense of what will work, at least within a relatively narrow scope of waste type and treatment system. This expertise is of little help, however, in advancing the state of the science in this field for a number of reasons.

1. The chemical composition of wastes is often extremely complex compared to that of other reaction "raw materials." Therefore, the number of possible chemical interactions among the waste constituents and the treatment chemicals is very large.
2. The physical nature of wastes is usually multiphased, most often with solid particles suspended in a liquid medium such as water. This means that reaction kinet-

ics are affected by physical phenomena, such as solid-state and boundary layer diffusion, as well as by the more definable and often simpler interaction of ions and molecules in solution.
3. Because the waste is often a concentrate produced by treating one or more waste water streams, the hazardous constituents in a given waste are present in much larger concentrations and greater variety than is usually the case for waste water treatment. This makes the body of technical information from the waste water treatment field—much larger, more mature and available in the literature—of limited value here.
4. Commercial enterprises with experience in this field are reluctant to disclose detailed technical information to other investigators.
5. Not only is the chemistry of hazardous waste treatment very complex, but it is doubtful if the interactions can ever be described by the mathematical algorithms used for some of the simpler water treatment systems. Cause and effect must be related by statistical means, but the number of variables makes this difficult, even with today's computers used in the usual fashion.
6. Even "experts" in this field cannot consciously evaluate more than a few waste characteristics at a time in choosing a specific treatment. Therefore, "diagnosis" becomes more intuitive than deductive, not unlike the medical diagnostic process (until just recently, as expert systems are now being applied to medical diagnosis). Consequently, when the expert moves, the organization loses that knowledge; when the expert retires or dies, the knowledge may be lost to society as well.
7. Without reliable cause-and-effect data, it is difficult to develop good experimental plans with critical experiments to pin down the waste characteristics and value ranges that are important. Without being able to identify the causative waste factors from among the many parameters of a typical waste, cause-and-effect relationships can be discovered only by statistical analyses. However, because of the large number of variables, techniques, such as multiple linear regression analysis, are difficult to apply, even with computers.

The expert system is a computerized technique for simulating (in a very limited way) the mental process (conscious and unconscious) of a human expert while combining it with the vast information storage, retrieval, and processing ability of the computer. The expert system has two basic parts: a data or knowledge base, and an "inference engine" or a set of rules that experts apply to the information. The latter part is accomplished by "heuristic" programming, that is, the use of intuitive, informal rules and experiential knowledge rather than the orderly algorithmic progression of most of today's computer programs. There are a number of types and degrees of sophistication in today's expert systems, with the more expensive systems providing greater flexibility and transparency, the ability to use "fuzzy logic" and confidence factors, etc. [5] Two primary differences are whether the system operates on inductive or deductive logic. The system chosen for use in WISE-SCAN™ used inductive logic; that is, it reasons from specific examples to general rules. The reason is obvious: The "rules" for equating waste characteristics to CFS formulations and test results were to be the result of the system, not its inputs. And plenty of specific examples were available.

The system that was chosen [6] uses a series of examples to arrive at a "rule." This rule can then be used by another utility in the program to pick the best CFS process, system, or formulation from those available in its memory, based on the previous examples fed into the program. Rules can also be formulated to integrate factors, such as regulatory requirements and disposal options, into the system. However, this discussion will be limited to choosing a CFS process based on the chemical and physical properties of the waste.

Constructing the Expert System. The expert system requires the following steps to arrive at a rule.

1. Determine the characteristics or attributes of wastes that may have an effect on the choice of CFS process. This will most likely become an iterative process as the unimportant factors are weeded out through progressive development of the final rule. The computer screen has the appearance shown in Figure 18-1 at this point, where each attribute (physical state, viscosity, total solids, etc.) is shown at the top along with the possible values that that attribute can have. Numerical attributes can have any integer value within the range of the program.
2. Enter examples. Each row in Figure 18-2 represents a real waste record (only part of the "example" screens is shown). In theory, a very large number of examples can be used, but practical considerations limit the number to about 200–300 in a microcomputer system.
3. Generate the rule. When all the examples are entered, the program generates a "rule," which is really a logic table or decision tree. Part of a rule is shown in Figure 18-3. Examination of the rule can be quite instructive. For example, it may be seen that some of the attributes have not been used. This means that there is no relationship between these attributes or waste characteristics and the optimum CFS process for the examples given.
4. Set up the query system. This program automatically sets up a "query" system

```
ZDDDDDDDDDDDDDDDDDDDDDDDDDDDDDDDDDDDDDDDDDDDDDDDDDDDDDDDDDDDDDDDDDD?
3 EXPERT-EASE    file: SOLIDIF     13576 bytes left           PHYSSTATE : 0   3
CDDDDDDDDDDDDDDDDDDDDDDDDDDDDDDDDDDDDDDDDDDDDDDDDDDDDDDDDDDDDDDDDDD4
3         logical     integer    integer    logical    logical    integer      3
3         PHYSSTATE   VISCOSITY  TOTSOLIDS  TYPESOLIDS MEDIUM     pHx10        3
3    1    SLDG                              INORGANIC  AQS                     3
3    2    CAKE                              ORGANIC    LTX                     3
3    3    SOLN                              MIXED      OIW                     3
3    4    SUSP                                         OIL                     3
3    5    PWDR                                         SOL                     3
3    6    SOIL                                         MIX                     3
3    7                                                 SLD                     3
3    8                                                 OSW                     3
3    9                                                                         3
3   10                                                                         3
3   11                                                                         3
3   12                                                                         3
3   13                                                                         3
3   14                                                                         3
3   15                                                                         3
CDDDDDDDDDDDDDDDDDDDDDDDDDDDDDDDDDDDDDDDDDDDDDDDDDDDDDDDDDDDDDDDDDD4
3 editing attributes                                                           3
3 , , , ', new, value, delete, change, text ? ('+' for more )          3
3 >                                                                            3
@DDDDDDDDDDDDDDDDDDDDDDDDDDDDDDDDDDDDDDDDDDDDDDDDDDDDDDDDDDDDDDDDDDY
```

Figure 18-1 WISE-SCAN™ expert system: attributes screen.

```
ZDDDDDDDDDDDDDDDDDDDDDDDDDDDDDDDDDDDDDDDDDDDDDDDDDDDDDDDDDDDDDDDDDDDDDDDDDDDDDDDDDDDDDDDD?
3 EXPERT-EASE    file: SOLIDIF      13576 bytes left              50 : TOTSOLIDS     3
CDDDDDDDDDDDDDDDDDDDDDDDDDDDDDDDDDDDDDDDDDDDDDDDDDDDDDDDDDDDDDDDDDDDDDDDDDDDDDDDDDDDDDDDD4
3            logical    integer    integer    logical    logical    integer          3
3            PHYSSTATE  VISCOSITY  TOTSOLIDS  TYPESOLIDS MEDIUM     pHx10            3
3  36        SOLN       1          13         INORGANIC  AQS        34               3
3  37        SLDG       1          15         MIXED      AQS        111              3
3  38        SLDG       1          15         MIXED      AQS        120              3
3  39        SLDG       1          15         MIXED      AQS        122              3
3  40        SLDG       7          15         INORGANIC  AQS        105              3
3  41        SLDG       2          16         MIXED      AQS        102              3
3  42        SLDG       6          18         INORGANIC  AQS        65               3
3  43        SLDG       4          19         MIXED      AQS        69               3
3  44        SLDG       6          19         INORGANIC  AQS        103              3
3  45        SLDG       5          20         MIXED      AQS        38               3
3  46        SLDG       6          20         INORGANIC  AQS        74               3
3  47        SOLN       0          20         MIXED      AQS        88               3
3  48        SLDG       4          21         MIXED      AQS        81               3
3  49        SUSP       4          21         INORGANIC  AQS        68               3
3  50        SOLN       0          23         INORGANIC  AQS        125              3
CDDDDDDDDDDDDDDDDDDDDDDDDDDDDDDDDDDDDDDDDDDDDDDDDDDDDDDDDDDDDDDDDDDDDDDDDDDDDDDDDDDDDDDDD4
3 editing examples                                                                   3
3 , , , !, new, delete, move, change, xpand ? ('+' for more )                  3
3 >                                                                                  3
@DDDDDDDDDDDDDDDDDDDDDDDDDDDDDDDDDDDDDDDDDDDDDDDDDDDDDDDDDDDDDDDDDDDDDDDDDDDDDDDDDDDDDDDDY
```

Figure 18-2 WISE-SCAN™ expert system: examples screen.

that interrogates the user, asking a series of questions, each of which is dependent on the answer to the previous question. This is the user "end" of the expert system. The program used in WISE-SCAN™ also has a text editor that allows the questions to be put in natural language. The initial query screen looks like Figure 18-4, with subsequent screens looking like Figure 18-5. If the program finds a CFS process that fits the waste characteristics entered during the query process, it produces a screen like that shown in Figure 18-6.

```
ZDDDDDDDDDDDDDDDDDDDDDDDDDDDDDDDDDDDDDDDDDDDDDDDDDDDDDDDDDDDDDDDDDDDDDDDDDDDDDDDDDDDDDDDD?
3 EXPERT-EASE    file: SOLIDIF      13576 bytes left              11                 3
CDDDDDDDDDDDDDDDDDDDDDDDDDDDDDDDDDDDDDDDDDDDDDDDDDDDDDDDDDDDDDDDDDDDDDDDDDDDDDDDDDDDDDDDD4
3                                            <3 : S110ADL036                         3
3                                            r3 : S310AAN002                         3
3                                   r7 : PHASELAYER                                  3
3                                      NONE : TOTSOLIDS                              3
3                                            <9 : S110AAL024                         3
3                                            r9 : S110AAL036                         3
3                                      BLAY : VISCOSITY                              3
3                                            <4 : S110ADL024                         3
3                                            r4 : S110AAL024                         3
3                                      MLAY : null                                   3
3                              r17 : SGx100                                          3
3                                  <155 : TOTSOLIDS                                  3
3                                       <22 : TOTSOLIDS                              3
3                                           <19 : S910AAA024                         3
3                                           r19 : TOTSOLIDS                          3
3                                                <20 : S120ABD036                    3
3                                                r20 : S110ADL024                    3
CDDDDDDDDDDDDDDDDDDDDDDDDDDDDDDDDDDDDDDDDDDDDDDDDDDDDDDDDDDDDDDDDDDDDDDDDDDDDDDDDDDDDDDDD4
3 viewing rule                                                                       3
3 , , attributes, files, examples, new, query, print, help ?                     3
3 >                                                                                  3
@DDDDDDDDDDDDDDDDDDDDDDDDDDDDDDDDDDDDDDDDDDDDDDDDDDDDDDDDDDDDDDDDDDDDDDDDDDDDDDDDDDDDDDDDY
```

Figure 18-3 WISE-SCAN™ expert system: rule screen.

```
ZDDDDDDDDDDDDDDDDDDDDDDDDDDDDDDDDDDDDDDDDDDDDDDDDDDDDDDDDDDDDDDDDDDDDDDDDDDDDD?
3 EXPERT-EASE    file: WISE-SCAN    24250 bytes left                           3
CDDDDDDDDDDDDDDDDDDDDDDDDDDDDDDDDDDDDDDDDDDDDDDDDDDDDDDDDDDDDDDDDDDDDDDDDDDDDD4
3                                                                              3
3                WELCOME TO THE WISE-SCAN EXPERT SYSTEM !!!                    3
3                                                                              3
3 WISE WILL GUIDE YOU THROUGH THE PROCESS OF ANALYSING THIS PARTICULAR         3
3 WASTE TREATMENT/DISPOSAL PROBLEM.  PLEASE ANSWER THE QUESTIONS AS THEY       3
3 APPEAR ON THE SCREEN.                                                        3
3                                                                              3
3 TO BEGIN ------ IS THE WASTE HAZARDOUS AND KNOWN TO BE SUBJECT TO            3
3 REGULATION UNDER SUBTITLE "C" OF RCRA?                                       3
3                                                                              3
3                                                                              3
3   1. YES                                                                     3
3                                                                              3
3   2. NO                                                                      3
3                                                                              3
3   3. DON'T KNOW                                                              3
3                                                                              3
CDDDDDDDDDDDDDDDDDDDDDDDDDDDDDDDDDDDDDDDDDDDDDDDDDDDDDDDDDDDDDDDDDDDDDDDDDDDDD4
3 running WISE-SCAN                                                            3
3 more -- press space to continue                                              3
3 >                                                                            3
@DDDDDDDDDDDDDDDDDDDDDDDDDDDDDDDDDDDDDDDDDDDDDDDDDDDDDDDDDDDDDDDDDDDDDDDDDDDDDY
```

Figure 18-4 WISE-SCAN™ expert system: initial query screen.

Development of a usable system may require a number of iterations, weeding out useless attributes and entering new examples. The first stage of WISE-SCAN™ incorporated about 15 percent of the WISE data base and a small number of conscious rules converted into examples (the program cannot take rules as such). This allowed testing of the general design of the system, and refining of the data base. The next stage was to enter the remainder of the data base, along with some more

```
ZDDDDDDDDDDDDDDDDDDDDDDDDDDDDDDDDDDDDDDDDDDDDDDDDDDDDDDDDDDDDDDDDDDDDDDDDDDDDD?
3 EXPERT-EASE    file: WISE-SCAN    24014 bytes left                           3
CDDDDDDDDDDDDDDDDDDDDDDDDDDDDDDDDDDDDDDDDDDDDDDDDDDDDDDDDDDDDDDDDDDDDDDDDDDDDD4
3                                                                              3
3 WHAT IS THE MEDIUM?  (* = most common media)                                 3
3                                                                              3
3   1. AQUEOUS (WATER-BASED)  *                                                3
3                                                                              3
3   2. LATEX                                                                   3
3                                                                              3
3   3. OIL EMULSIFIED IN WATER                                                 3
3                                                                              3
3   4. OIL-BASED  *                                                            3
3                                                                              3
3   5. SOLVENT-BASED  *                                                        3
3                                                                              3
3   6. MISCELLANEOUS MULTI-PHASED MIXTURES                                     3
3                                                                              3
3   7. SOLID  *                                                                3
3                                                                              3
CDDDDDDDDDDDDDDDDDDDDDDDDDDDDDDDDDDDDDDDDDDDDDDDDDDDDDDDDDDDDDDDDDDDDDDDDDDDDD4
3                                                                              3
3 more -- press space to continue                                              3
3 > 1                                                                          3
@DDDDDDDDDDDDDDDDDDDDDDDDDDDDDDDDDDDDDDDDDDDDDDDDDDDDDDDDDDDDDDDDDDDDDDDDDDDDDY
```

Figure 18-5 WISE-SCAN™ expert system: intermediate query screen.

```
ZDDDDDDDDDDDDDDDDDDDDDDDDDDDDDDDDDDDDDDDDDDDDDDDDDDDDDDDDDDDDDDDDDDDDDD?
3 EXPERT-EASE    file: WISE-SCAN    24014 bytes left                  3
CDDDDDDDDDDDDDDDDDDDDDDDDDDDDDDDDDDDDDDDDDDDDDDDDDDDDDDDDDDDDDDDDDDDDDD4
3 AFTER EVALUATING THE INFORMATION WHICH YOU HAVE SUPPLIED, WISE SUGGESTS    3
3 THAT CT'S FORMULA                                                    3
3                                                                      3
3 910AAA BE USED FOR THIS WASTE IN THE ADDITIVE RATIO OF 2 LB. PER GALLON OF  3
3 WASTE.  THIS FORMULA IS PRICED AT $.19 PER LB. IN TRUCKLOAD LOTS.  THE      3
3 ASSOCIATED VOLUME INCREASE IS 8%.                                    3
3                                                                      3
3                                                                      3
3                                                                      3
3                                                                      3
3                                                                      3
3                                                                      3
3                                                                      3
3                                                                      3
3                                                                      3
3                                                                      3
3                                                                      3
CDDDDDDDDDDDDDDDDDDDDDDDDDDDDDDDDDDDDDDDDDDDDDDDDDDDDDDDDDDDDDDDDDDDDDD4
3                                                                      3
3 Enter integer value                                                   3
3 > 100                                                                 3
@DDDDDDDDDDDDDDDDDDDDDDDDDDDDDDDDDDDDDDDDDDDDDDDDDDDDDDDDDDDDDDDDDDDDDDY
```

Figure 18-6 WISE-SCAN™ expert system: final results screen, with formula.

"intuitive" rules that have been codified by careful introspective analysis of our own decision-making processes. These latter rules will be tested statistically against known test results to modify or eliminate those that are not useful. At this point, WISE-SCAN™ was at least 75 percent accurate in predicting, within narrow limits, an optimum formulation for any type of waste and disposal scenario that had previously been tested.

The greatest value of a system such as this—its use in developing the next generation of CFS systems—will only begin to be realized at the end of this second stage. Cause-and-effect relationships between waste characteristics and treatment results—relationships that are not now apparent or are only suspected—can be determined from the large statistical base of the WISE™ data base. Extremely complex interactive chemical relationships, which would require years of empirical testing to determine, could be resolved quickly by using computer-aided experimental design to devise the necessary critical experiments. These experiments, in turn, will allow us to understand the complex chemistry that takes place in real-world CFS systems, and to modify these systems accordingly to achieve the results we want. In economic terms, this means a new generation of sophisticated, highly formulated, CFS processes and products custom-designed for lower cost, better results, easier use, and increased reliability.

RESEARCH AND DEVELOPMENT

As we have seen throughout this book, there are many opportunities for useful and important research in the CFS field. Some of these include a better understanding of the following.

- Speciation of wastes
- Solidification chemistry and mechanisms

- Fixation chemistry and mechanisms
- Speciation of CFS reaction products
- Movement of species through CFS solids
- Fixation/destruction/immobilization of organics
- Effects of curing time on immobilization
- Stability of waste forms toward redox effects
- Biological durability

Process development is important to accomplish a number of improvements in CFS technology, such as the following.

- Lowering of permeability
- Improvement of the physical properties of low solids wastes with minimum volume increase
- More efficient CFS processes for mixed organic–inorganic wastes
- Better, faster test methods
- Achievement of lower leaching levels with less reagent usage

While the use of computer-aided analysis of empirical test data will be very valuable, we need to know more about the speciation of constituents in both raw waste and CFS products. Investigators at Environment Canada's Wastewater Technology Center (WTC) [7] and Louisiana State University (LSU) [8] have approached this in different ways. The WTC workers have used a sequential chemical extraction technique to classify metal compounds into fractions of differing solubility, while the LSU people are working with electron microscopy and microanalytical tools. Some of the methods that seem promising from the results of this and other work are as follows.

Sequential chemical extraction
Scanning electron microscopy
Transmission electron microscopy
Energy-dispersive X-ray analysis
Electron probe microanalysis
Fourier transform infrared spectroscopy
Thermal analysis
Titration analysis

In addition to, and partially resulting from a better understanding of speciation, we need to define the chemistry and mechanisms of reaction in the inorganic CFS systems, especially in cement- and pozzolan-based systems. This is needed both for the solidification reactions and for fixation reactions. At this time, we know very little about the mechanisms of immobilization, and possibly, alteration of organic compounds in CFS systems. And we need more information about how species move from waste forms during the leaching process, both to model field conditions and to have a higher degree of confidence in the long-term stability of CFS products.

One need in the development area is for better, faster leaching test methods. The present regulatory protocols cannot be used for real-time quality control. Another demand is for better permeability control in the CFS process. We require more efficient CFS processes to minimize reagent usage so that valuable landfill space is conserved. And with the present regulatory trends, we need ever better fixation techniques for both metals and organics.

REFERENCES

1. Pojasek, R. B. (Ed.). *Toxic and Hazardous Waste Disposal.* Ann Arbor, MI: Ann Arbor Science Publishers, 1979.
2. U.S. Army Corps of Engineers. *Guide to the Disposal of Chemically Stabilized and Solidified Wastes.* Cincinnati, OH: U.S. EPA, 1980.
3. Cote, P. L. *Contaminant Leaching from Cement-Based Waste Forms Under Acidic Conditions.* Ph.D. thesis, McMaster Univ., Hamilton, Ont., Canada, 1986.
4. Conner, J. R. Use of computer-based artificial intelligence systems to select detoxification and solidification methods for residues and wastes. In *Proc. 5th International Symposium on Environmental Pollution.* Quebec City, P.Q., Canada: Environmental Pollution Institute, 1985.
5. Kinnucan, P. Artificial intelligence: Making computers smarter. *High Technol.* (Nov./Dec. 1982).
6. Michie, D. Expert systems. *Comput. J.* (Nov. 1980).
7. Bridle, T. R., P. L. Cote, T. W. Constable, and J. L. Fraser. Evaluation of heavy metal leachability from solid wastes. *Water Sci. Technol.* **19**: 1029–1036 (1987).
8. Eaton, H. C., M. E. Tittlebaum, and F. K. Cartledge *Techniques For Microscopic Studies of Solidification Technologies.* Baton Rouge, LA: Louisiana State Univ., 1986.

Appendix 1

CFS WASTE DATA BASE (CWDB)*

RCRA CODE: F006   CFS CODE: F006   UCD CODE: 431   SIC CODE: 3471

WASTE DESCRIPTION.   Waste water treatment sludges from electroplating, except for tin, zinc and aluminum on carbon steel, sulfuric acid anodizing, certain cleaning, stripping, etching, and milling

INDUSTRY.   Plating and polishing

PROCESS.   Treatment of waste water

WASTE CONSTITUENTS.   Cadmium, chromium, copper, lead, nickel and zinc; also minor amounts of other metals, including precious metals

APPLICABLE TREATMENT TECHNOLOGY: CFS

RCRA CODE: F007   CFS CODE: F007   UCD CODE: 132   SIC CODE: 3471

WASTE DESCRIPTION.   Spent cyanide plating bath solutions from electroplating

INDUSTRY.   Plating and polishing

PROCESS.   Electroplating from cyanide baths

WASTE CONSTITUENTS.   Cadmium, copper, nickel, and zinc cyanides and other salts

APPLICABLE TREATMENT TECHNOLOGY: Pretreatment/CFS

RCRA CODE: F008   CFS CODE: F008   UCD CODE: 431   SIC CODE: 3471

WASTE DESCRIPTION.   Plating bath sludges from the bottom of plating baths from electroplating where cyanides are used in the process

*Assembled from CONTEC Data Base. (From J. R. Conner, CONTEC Data-Base, Conner Technologies, Atlanta, GA, 1985.)

INDUSTRY.   Plating and polishing

PROCESS.   Electroplating from cyanide baths

WASTE CONSTITUENTS.   Cadmium, copper, nickel, and zinc cyanides and other salts; soluble cyanides

APPLICABLE TREATMENT TECHNOLOGY: Pretreatment/CFS

RCRA CODE: F009   CFS CODE: F009   UCD CODE: 121   SIC CODE: 3471

WASTE DESCRIPTION.   Spent stripping and cleaning bath solutions from electroplating where cyanides are used in the process

INDUSTRY.   Plating and polishing

PROCESS.   Stripping and cleaning of electroplated metal

WASTE CONSTITUENTS.   Metals, soluble cyanides

APPLICABLE TREATMENT TECHNOLOGY: Pretreatment/CFS

RCRA CODE: F010   CFS CODE: F010   UCD CODE: 287   SIC CODE: 3398

WASTE DESCRIPTION.   Quenching bath sludge from oil baths from metal heat treating where cyanides are used in the process

INDUSTRY.   Metal heat treating

PROCESS.   Metal heat treating in oil baths

WASTE CONSTITUENTS.   Metals, soluble cyanides

APPLICABLE TREATMENT TECHNOLOGY: Pretreatment/CFS

RCRA CODE: F011   CFS CODE: F011   UCD CODE: 112   SIC CODE: 3398

WASTE DESCRIPTION.   Spent cyanide solutions from salt bath pot cleaning from metal heat treating

INDUSTRY.   Metal heat treating

PROCESS.   Metal heat treating in salt baths

WASTE CONSTITUENTS.   Metals, soluble cyanides

APPLICABLE TREATMENT TECHNOLOGY: Pretreatment/CFS

RCRA CODE: F012   CFS CODE: F012   UCD CODE: 433   SIC CODE: 3398

WASTE DESCRIPTION.   Quenching waste water treatment sludges from metal heat treating where cyanides are used in the process

INDUSTRY.   Metal heat treating

PROCESS.   Treatment of waste water

WASTE CONSTITUENTS. Metals, complexed cyanides

APPLICABLE TREATMENT TECHNOLOGY: CFS

RCRA CODE: -   CFS CODE: F013   UCD CODE: 143   SIC CODE: 1099

WASTE DESCRIPTION. Flotation tailings from selective floatation from minerals metal recovery

INDUSTRY. Metal ores, NEC

PROCESS. Treatment of waste water

WASTE CONSTITUENTS. Metals

APPLICABLE TREATMENT TECHNOLOGY: CFS

RCRA CODE: -   CFS CODE: F014   UCD CODE: 143   SIC CODE: 1099

WASTE DESCRIPTION. Cyanidation waste water treatment tailing pond sediment from mineral metals recovery

INDUSTRY. Metal ores, NEC

PROCESS. Treatment of waste water

WASTE CONSTITUENTS. Metals, complexed cyanides

APPLICABLE TREATMENT TECHNOLOGY: Pretreatment/CFS

RCRA CODE: -   CFS CODE: F015   UCD CODE: 143   SIC CODE: 1099

WASTE DESCRIPTION. Spent cyanide bath solutions from mineral metals recovery

INDUSTRY. Metal ores, NEC

PROCESS. Mineral metals recovery

WASTE CONSTITUENTS. Metals, soluble cyanides

APPLICABLE TREATMENT TECHNOLOGY: Pretreatment/CFS

RCRA CODE: -   CFS CODE: F016   UCD CODE: 143   SIC CODE: 3312

WASTE DESCRIPTION. Dewatered air pollution control scrubber sludges from coke ovens and blast furnaces

INDUSTRY. Blast furnaces, steel mills

PROCESS. Air pollution control scrubbing

WASTE CONSTITUENTS. Complexed cyanides

APPLICABLE TREATMENT TECHNOLOGY: Pretreatment/CFS

RCRA CODE: -   CFS CODE: F017   UCD CODE: 454   SIC CODE: 3499

WASTE DESCRIPTION.   Paint residues generated from industrial painting

INDUSTRY.   Fabricated metal products, NEC

PROCESS.   Industrial painting

WASTE CONSTITUENTS.   Metal salts, inorganic and organic pigments, resins, solvents, plasticizers

APPLICABLE TREATMENT TECHNOLOGY: CFS

RCRA CODE: -   CFS CODE: F018   UCD CODE: 454   SIC CODE: 3499

WASTE DESCRIPTION.   Waste water treatment sludges from industrial painting

INDUSTRY.   Fabricated metal products, NEC

PROCESS.   Treatment of waste water

WASTE CONSTITUENTS.   Metal salts, inorganic and organic pigments, resins, solvents, plasticizers

APPLICABLE TREATMENT TECHNOLOGY: CFS (low solvent)

RCRA CODE: F019   CFS CODE: F019   UCD CODE: 433   SIC CODE: 3479

WASTE DESCRIPTION.   Waste water treatment sludges from the chemical conversion coating of aluminum

INDUSTRY.   Metal coating and allied services

PROCESS.   Treatment of waste water

WASTE CONSTITUENTS.   Hexavalent chromium, complexed cyanide

APPLICABLE TREATMENT TECHNOLOGY: CFS

RCRA CODE: K001   CFS CODE: K001   UCD CODE: 457   SIC CODE: 2491

WASTE DESCRIPTION.   Botton sediment sludge from treatment of waste waters from wood preserving processes that use creosote and/or pentachlorophenol

INDUSTRY.   Wood preserving

PROCESS.   Treatment of waste waters

WASTE CONSTITUENTS.   Phenol, chlorinated phenols, cresols, creosote, chrysene, naphthalene, anthenes, pyrenes, anthracenes

APPLICABLE TREATMENT TECHNOLOGY: Solidification only

RCRA CODE: K002   CFS CODE: K002   UCD CODE: 431   SIC CODE: 2816

WASTE DESCRIPTION.   Waste water treatment sludge from the production of chrome yellow and orange pigments

INDUSTRY.   Inorganic pigments

PROCESS.   Waste water treatment

WASTE CONSTITUENTS.   Hexavalent chromium, lead

APPLICABLE TREATMENT TECHNOLOGY: CFS

RCRA CODE: K003   CFS CODE: K003   UCD CODE: 431   SIC CODE: 2816

WASTE DESCRIPTION.   Waste water treatment sludge from the production of molybdate orange pigments

INDUSTRY.   Inorganic pigments

PROCESS.   Waste water treatment

WASTE CONSTITUENTS.   Hexavalent chromium, lead, molybdenum

APPLICABLE TREATMENT TECHNOLOGY: CFS

RCRA CODE: K004   CFS CODE: K004   UCD CODE: 431   SIC CODE: 2816

WASTE DESCRIPTION.   Waste water treatment sludge from the production of zinc yellow pigments

INDUSTRY.   Inorganic pigments

PROCESS.   Waste water treatment

WASTE CONSTITUENTS.   Hexavalent chromium, zinc

APPLICABLE TREATMENT TECHNOLOGY: CFS

RCRA CODE: K005   CFS CODE: K005   UCD CODE: 431   SIC CODE: 2816

WASTE DESCRIPTION.   Waste water treatment sludge from the production of chrome green pigments

INDUSTRY.   Inorganic pigments

PROCESS.   Waste water treatment

WASTE CONSTITUENTS.   Hexavalent chromium, lead

APPLICABLE TREATMENT TECHNOLOGY: CFS

RCRA CODE: K006   CFS CODE: K006   UCD CODE: 431   SIC CODE: 2816

WASTE DESCRIPTION.   Waste water treatment sludge from the production of chrome oxide green pigments (anhydrous and hydrated)

INDUSTRY.   Inorganic pigments

PROCESS.   Waste water treatment

WASTE CONSTITUENTS.   Hexavalent chromium, nickel, copper, zinc iron, tin, and aluminum

APPLICABLE TREATMENT TECHNOLOGY: CFS

RCRA CODE: K007   CFS CODE: K007   UCD CODE: 431   SIC CODE: 2816

WASTE DESCRIPTION.   Waste water treatment sludge from the production of iron blue pigments

INDUSTRY.   Inorganic pigments

PROCESS.   Waste water treatment

WASTE CONSTITUENTS.   Hexavalent chromium, complexed cyanide

APPLICABLE TREATMENT TECHNOLOGY: CFS

RCRA CODE: K008   CFS CODE: K008   UCD CODE: 431   SIC CODE: 2816

WASTE DESCRIPTION.   Oven residue from the production of chrome oxide green pigments

INDUSTRY.   Inorganic pigments

PROCESS.   Chrome oxide pigment production

WASTE CONSTITUENTS.   Hexavalent chromium

APPLICABLE TREATMENT TECHNOLOGY: CFS

RCRA CODE: K021   CFS CODE: K021   UCD CODE: 221   SIC CODE: 2869

WASTE DESCRIPTION.   Aqueous spent antimony catalyst waste from fluoromethane production

INDUSTRY.   Industrial organic chemicals, NEC

PROCESS.   Fluoromethane production

WASTE CONSTITUENTS.   Antimony, carbon tetrachloride, chloroform

APPLICABLE TREATMENT TECHNOLOGY: Solidification only

RCRA CODE: K031   CFS CODE: K031   UCD CODE: 241   SIC CODE: 2879

WASTE DESCRIPTION.   By-product salts generated in the production of MSMA and cacodylic acid

INDUSTRY.   Agricultural chemicals, NEC

PROCESS.   Production of MSMA and cacodylic acid

WASTE CONSTITUENTS. Arsenic, organics

APPLICABLE TREATMENT TECHNOLOGY: CFS

RCRA CODE: K032 CFS CODE: K032 UCD CODE: 222 SIC CODE: 2879

WASTE DESCRIPTION. Waste water treatment sludge from the production of chlordane

INDUSTRY. Agricultural chemicals, NEC

PROCESS. Waste water treatment

WASTE CONSTITUENTS. Hexachlorocyclopentadiene

APPLICABLE TREATMENT TECHNOLOGY: Solidification only

RCRA CODE: K033 CFS CODE: K033 UCD CODE: 222 SIC CODE: 2879

WASTE DESCRIPTION. Waste water and scrub water from the chlorination of cyclopentadiene in the production of chlordane

INDUSTRY. Agricultural chemicals, NEC

PROCESS. Waste water treatment

WASTE CONSTITUENTS. Hexachlorocyclopentadiene

APPLICABLE TREATMENT TECHNOLOGY: Solidification only

RCRA CODE: K035 CFS CODE: K035 UCD CODE: 457 SIC CODE: 2879

WASTE DESCRIPTION. Waste water treatment sludges generated in the production of creosote

INDUSTRY. Agricultural chemicals, NEC

PROCESS. Waste water treatment

WASTE CONSTITUENTS. Creosote, chrysene, naphthalene, fluoranthrenes, pyrenes, anthracenes, acenephthalene

APPLICABLE TREATMENT TECHNOLOGY: Solidification only

RCRA CODE: K037 CFS CODE: K037 UCD CODE: 241 SIC CODE: 2879

WASTE DESCRIPTION. Waste water treatment sludges from the production of disulfaton

INDUSTRY. Agricultural chemicals, NEC

PROCESS. Waste water treatment

WASTE CONSTITUENTS. Toluene, phosphorodithioic, and phosphothioic acid esters

APPLICABLE TREATMENT TECHNOLOGY: Solidification only

RCRA CODE: K038   CFS CODE: K038   UCD CODE: 241   SIC CODE: 2879

WASTE DESCRIPTION.   Waste water from the washing and stripping of phorate production

INDUSTRY.   Agricultural chemicals, NEC

PROCESS.   Waste treatment

WASTE CONSTITUENTS.   Phorate, formaldehyde, phosphorodithioic, and phosphorothioic acid esters

APPLICABLE TREATMENT TECHNOLOGY: Solidification only

RCRA CODE: K040   CFS CODE: K040   UCD CODE: 241   SIC CODE: 2879

WASTE DESCRIPTION.   Waste water treatment sludge from the production of phorate

INDUSTRY.   Agricultural chemicals, NEC

PROCESS.   Waste water treatment

WASTE CONSTITUENTS.   Phorate, formaldehyde, phosphorodithioic, and phosphorothioic acid esters

APPLICABLE TREATMENT TECHNOLOGY: Solidification only

RCRA CODE: K041   CFS CODE: K041   UCD CODE: 241   SIC CODE: 2879

WASTE DESCRIPTION.   Waste water treatment sludge from the production of toxaphene

INDUSTRY.   Agricultural chemicals, NEC

PROCESS.   Waste water treatment

WASTE CONSTITUENTS.   Toxaphene

APPLICABLE TREATMENT TECHNOLOGY: Solidification only

RCRA CODE: K044   CFS CODE: K044   UCD CODE: 456   SIC CODE: 2892

WASTE DESCRIPTION.   Waste water treatment sludges from the manufacturing and processing of explosives

INDUSTRY.   Explosives

PROCESS.   Waste water treatment

WASTE CONSTITUENTS.   Ignitable, corrosive, or reactive

APPLICABLE TREATMENT TECHNOLOGY: Solidification only

RCRA CODE: K045   CFS CODE: K045   UCD CODE: 255   SIC CODE: 2892

WASTE DESCRIPTION.   Spent carbon from the treatment of waste water containing explosives

INDUSTRY.   Explosives

PROCESS.   Waste water treatment

WASTE CONSTITUENTS.   Ignitable, corrosive, or reactive

APPLICABLE TREATMENT TECHNOLOGY: Solidification only

RCRA CODE: K046   CFS CODE: K046   UCD CODE: 457   SIC CODE: 2892

WASTE DESCRIPTION.   Waste water treatment sludges from the manufacturing, formulation, and loading of lead-based initiating compounds

INDUSTRY.   Explosives

PROCESS.   Waste water treatment

WASTE CONSTITUENTS.   Lead, organics

APPLICABLE TREATMENT TECHNOLOGY: CFS

RCRA CODE: K047   CFS CODE: K047   UCD CODE: 225   SIC CODE: 2892

WASTE DESCRIPTION.   Pink/red water from TNT operations

INDUSTRY.   Explosives

PROCESS.   TNT operations

WASTE CONSTITUENTS.   Ignitable, corrosive, or reactive

APPLICABLE TREATMENT TECHNOLOGY: Solidification only

RCRA CODE: K048   CFS CODE: K048   UCD CODE: 289   SIC CODE: 2911

WASTE DESCRIPTION.   Dissolved air floatation (DAF) float from the petroleum refining industry

INDUSTRY.   Petroleum refining

PROCESS.   DAF waste treatment

WASTE CONSTITUENTS.   Hexavalent chromium, lead, oil, petroleum products

APPLICABLE TREATMENT TECHNOLOGY: CFS

RCRA CODE: K049   CFS CODE: K049   UCD CODE: 287   SIC CODE: 2911

WASTE DESCRIPTION.   Slop oil emulsion solids from the petroleum refining industry

INDUSTRY.   Petroleum refining

PROCESS.   Petroleum refining

WASTE CONSTITUENTS.   Hexavalent chromium, lead, oil, petroleum products

APPLICABLE TREATMENT TECHNOLOGY: CFS

RCRA CODE: K050   CFS CODE: K050   UCD CODE: 457   SIC CODE: 2911

WASTE DESCRIPTION.   Heat exchanger bundle cleaning sludge from the petroleum refining industry

INDUSTRY.   Petroleum refining

PROCESS.   Cleaning of heat exchangers

WASTE CONSTITUENTS.   Hexavalent chromium

APPLICABLE TREATMENT TECHNOLOGY: CFS

RCRA CODE: K051   CFS CODE: K051   UCD CODE: 286   SIC CODE: 2911

WASTE DESCRIPTION.   API separator sludge from the petroleum refining industry

INDUSTRY.   Petroleum refining

PROCESS.   Waste treatment

WASTE CONSTITUENTS.   Hexavalent chromium, lead, oil, petroleum products

APPLICABLE TREATMENT TECHNOLOGY: CFS

RCRA CODE: K052   CFS CODE: K052   UCD CODE: 447   SIC CODE: 2911

WASTE DESCRIPTION.   Tank bottoms (leaded) from the petroleum refining industry

INDUSTRY.   Petroleum refining

PROCESS.   Cleaning of petroleum tanks

WASTE CONSTITUENTS.   Lead, oil, petroleum products

APPLICABLE TREATMENT TECHNOLOGY: CFS

RCRA CODE: -   CFS CODE: K053   UCD CODE: 251   SIC CODE: 3111

WASTE DESCRIPTION.   Chrome (blue) trimmings generated by the leather tanning and finishing industry: hair pulp and save/chrome tan/retan/wet finish; no beamhouse; through-the-blue; shearing

INDUSTRY.   Leather tanning and finishing

PROCESS.   Trimming hides

WASTE CONSTITUENTS.   Chromium, animal hair, skin, and flesh

APPLICABLE TREATMENT TECHNOLOGY: CFS

RCRA CODE: -   CFS CODE: K054   UCD CODE: 251   SIC CODE: 3111

WASTE DESCRIPTION.   Chrome (blue) shavings generated by the leather tanning and finishing industry: hair pulp and hair save/chrome tan/retan/wet finish; no beamhouse; through-the-blue; shearing

INDUSTRY.   Leather tanning and finishing

PROCESS.   Shaving hides

WASTE CONSTITUENTS.   Chromium, hair, skin, and flesh

APPLICABLE TREATMENT TECHNOLOGY: CFS

RCRA CODE: -   CFS CODE: K055   UCD CODE: 251   SIC CODE: 3111

WASTE DESCRIPTION.   Buffing dust generated by the leather tanning and finishing industry: hair pulp and hair save/chrome tan/retan/wet finish; no beamhouse; through-the-blue

INDUSTRY.   Leather tanning and finishing

PROCESS.   Buffing hides

WASTE CONSTITUENTS.   Chromium, lead, hair, skin, flesh

APPLICABLE TREATMENT TECHNOLOGY: CFS

RCRA CODE: -   CFS CODE: K056   UCD CODE: 251   SIC CODE: 3111

WASTE DESCRIPTION.   Sewer screenings generated by the leather tanning and finishing industry: hair pulp and hair save/chrome tan/retan/wet finish; no beamhouse; through-the-blue; shearing

INDUSTRY.   Leather tanning and finishing

PROCESS.   Waste treatment

WASTE CONSTITUENTS.   Chromium, lead, hair, skin, flesh

APPLICABLE TREATMENT TECHNOLOGY: CFS

RCRA CODE: -   CFS CODE: K057   UCD CODE: 251   SIC CODE: 3111

WASTE DESCRIPTION.   Waste water treatment sludges generated by the leather and finishing industry: hair pulp and hair save/chrome tan/retan/wet finish; no beamhouse; through-the-blue; shearing

INDUSTRY.   Leather tanning and finishing

PROCESS. Waste water treatment

WASTE CONSTITUENTS. Chromium, lead, hair, skin, flesh

APPLICABLE TREATMENT TECHNOLOGY: CFS

RCRA CODE: - CFS CODE: K058 UCD CODE: 251 SIC CODE: 3111

WASTE DESCRIPTION. Waste water treatment sludges generated by the leather tanning and finishing industry: hair pulp and hair save/chrome tan/retan/wet finish; through-the-blue

INDUSTRY. Leather tanning and finishing

PROCESS. Waste water treatment

WASTE CONSTITUENTS. Chromium, lead, hair, skin, flesh

APPLICABLE TREATMENT TECHNOLOGY: CFS

RCRA CODE: - CFS CODE: K059 UCD CODE: 251 SIC CODE: 3111

WASTE DESCRIPTION. Waste water treatment sludges generated by the leather tanning and finishing industry: hair save/nonchrome tan/retan/wet finish

INDUSTRY. Leather tanning and finishing

PROCESS. Waste water treatment

WASTE CONSTITUENTS. Reactive

APPLICABLE TREATMENT TECHNOLOGY: Pretreatment/CFS

RCRA CODE: K060 CFS CODE: K060 UCD CODE: 143 SIC CODE: 3312

WASTE DESCRIPTION. Ammonia still lime sludge from coking operations

INDUSTRY. Blast furnaces, steel mills

PROCESS. Coke making

WASTE CONSTITUENTS. Cyanide, naphthalene, phenolic compounds, arsenic

APPLICABLE TREATMENT TECHNOLOGY: Solidification only

RCRA CODE: K061 CFS CODE: K061 UCD CODE: 431 SIC CODE: 3312

WASTE DESCRIPTION. Emission control dust/sludge from the primary production of steel in electric furnace

INDUSTRY. Blast furnaces, steel mills

PROCESS. Air pollution control

WASTE CONSTITUENTS. Hexavalent chrome, lead, cadmium

APPLICABLE TREATMENT TECHNOLOGY: CFS

RCRA CODE: K062   CFS CODE: K062   UCD CODE: 111   SIC CODE: 3300

WASTE DESCRIPTION. Spent pickle liquor from steel finishing operations

INDUSTRY. Steel mills

PROCESS. Steel pickling

WASTE CONSTITUENTS. Hexavalent chromium, lead, hydrochloric, sulfuric, nitric, and/or hydrofluoric acids

APPLICABLE TREATMENT TECHNOLOGY: CFS

RCRA CODE: -   CFS CODE: K063   UCD CODE: 431   SIC CODE: 3300

WASTE DESCRIPTION. Sludge from lime treatment of spent pickle liquor from steel finishing operations

INDUSTRY. Steel mills

PROCESS. Neutralization

WASTE CONSTITUENTS. Chromium, lead, calcium sulfate, nitrate, chloride, fluoride

APPLICABLE TREATMENT TECHNOLOGY: CFS

RCRA CODE: -   CFS CODE: K064   UCD CODE: 431   SIC CODE: 3339

WASTE DESCRIPTION. Acid plant blowdown/sludge resulting from the thickening of blowdown slurry from primary copper production

INDUSTRY. Metals, nonferrous primary smelting and refining, NEC

PROCESS. Sludge thickening

WASTE CONSTITUENTS. Lead, cadmium, copper

APPLICABLE TREATMENT TECHNOLOGY: CFS

RCRA CODE: -   CFS CODE: K065   UCD CODE: 431   SIC CODE: 3339

WASTE DESCRIPTION. Surface impoundment solids contained in and dredged from surface impoundments at primary lead smelting facilities

INDUSTRY. Metals, nonferrous primary smelting and refining, NEC

PROCESS. Surface impoundment dewatering

WASTE CONSTITUENTS. Lead, cadmium

APPLICABLE TREATMENT TECHNOLOGY: CFS

RCRA CODE: K066   CFS CODE: K066   UCD CODE: 431   SIC CODE: 3339

WASTE DESCRIPTION.   Sludge from treatment of process waste water and/or acid plant blowdown from primary zinc production

INDUSTRY.   Metals, nonferrous primary smelting and refining, NEC

PROCESS.   Waste water treatment

WASTE CONSTITUENTS.   Lead, hexavalent chromium, zinc

APPLICABLE TREATMENT TECHNOLOGY: CFS

RCRA CODE: K067   CFS CODE: K067   UCD CODE: 431   SIC CODE: 3339

WASTE DESCRIPTION.   Electrolytic anode slimes/sludges from primary zinc production

INDUSTRY.   Metals, nonferrous primary smelting and production, NEC

PROCESS.   Electrolytic zinc production

WASTE CONSTITUENTS.   Lead, cadmium, zinc

APPLICABLE TREATMENT TECHNOLOGY: CFS

RCRA CODE: K068   CFS CODE: K068   UCD CODE: 431   SIC CODE: 3339

WASTE DESCRIPTION.   Cadmium plant leachate residue (iron oxide) from primary zinc production

INDUSTRY.   Metals, nonferrous primary smelting and production, NEC

PROCESS.   Zinc production, leaching process

WASTE CONSTITUENTS.   Lead, cadmium, zinc, iron

APPLICABLE TREATMENT TECHNOLOGY: CFS

RCRA CODE: K069   CFS CODE: K069   UCD CODE: 431   SIC CODE: 3341

WASTE DESCRIPTION.   Emission control dust/sludge from secondary lead smelting

INDUSTRY.   Metals, nonferrous secondary

PROCESS.   Air pollution control

WASTE CONSTITUENTS.   Hexavalent chromium, lead, cadmium

APPLICABLE TREATMENT TECHNOLOGY: CFS

RCRA CODE: -   CFS CODE: K070   UCD CODE: SIC CODE: 2299

WASTE DESCRIPTION.   Woven fabric dying and finishing waste water treatment sludges

INDUSTRY.   Textile goods, NEC

PROCESS.   Waste water treatment

WASTE CONSTITUENTS.   Metals, dyes, animal organics

APPLICABLE TREATMENT TECHNOLOGY: CFS

RCRA CODE: K071   CFS CODE: K071   UCD CODE: 431   SIC CODE: 2812

WASTE DESCRIPTION.   Brine purification muds from the mercury cell process in chlorine production, where separately prepurified brine is not used

INDUSTRY.   Alkalies, chlorine

PROCESS.   Brine purification

WASTE CONSTITUENTS.   Mercury, sodium, and potassium chlorides

APPLICABLE TREATMENT TECHNOLOGY: CFS

RCRA CODE: -   CFS CODE: K072   UCD CODE: SIC CODE: 2812

WASTE DESCRIPTION.   Waste water treatment sludge from the diaphragm cell process using graphite anodes in the production of chlorine

INDUSTRY.   Alkalies, chlorine

PROCESS.   Waste water treatment

WASTE CONSTITUENTS:

APPLICABLE TREATMENT TECHNOLOGY: CFS

RCRA CODE: -   CFS CODE: K074   UCD CODE: SIC CODE: 2819

WASTE DESCRIPTION.   Waste water treatment sludges from the production of titanium dioxide pigment using chromium-bearing ores by the chloride process

INDUSTRY.   Industrial inorganic chemicals

PROCESS.   Waste water treatment

WASTE CONSTITUENTS.   Chromium, titanium, chlorine compounds

APPLICABLE TREATMENT TECHNOLOGY: CFS

RCRA CODE: -   CFS CODE: K075   UCD CODE: SIC CODE: 2819

WASTE DESCRIPTION.   Waste water treatment sludges from the production of titanium dioxide pigment using chromium-bearing ores by the sulfate process

INDUSTRY.   Industrial inorganic chemicals

PROCESS.   Waste water treatment

WASTE CONSTITUENTS. Chromium, titanium, sulfate compounds

APPLICABLE TREATMENT TECHNOLOGY: CFS

RCRA CODE: - CFS CODE: K076 UCD CODE: SIC CODE: 2819

WASTE DESCRIPTION. Arsenic-bearing sludges from the purification process in the production of antimony oxide

INDUSTRY. Industrial inorganic chemicals

PROCESS. Purification

WASTE CONSTITUENTS. Arsenic, antimony

APPLICABLE TREATMENT TECHNOLOGY: CFS

RCRA CODE: - CFS CODE: K077 UCD CODE: SIC CODE: 2819

WASTE DESCRIPTION. Antimony-bearing waste water treatment sludge from the production of antimony oxide

INDUSTRY. Industrial inorganic chemicals

PROCESS. Waste water treatment

WASTE CONSTITUENTS. Antimony

APPLICABLE TREATMENT TECHNOLOGY: CFS

RCRA CODE: - CFS CODE: K079 UCD CODE: 454 SIC CODE: 2851

WASTE DESCRIPTION. Water cleaning wastes from paint manufacturing

INDUSTRY. Paints and allied products

PROCESS. Cleanup

WASTE CONSTITUENTS. Metals, inorganic and organic pigments, resins, polymers, solvents, plasticizers

APPLICABLE TREATMENT TECHNOLOGY: Solidification

RCRA CODE: - CFS CODE: K080 UCD CODE: 454 SIC CODE: 2851

WASTE DESCRIPTION. Caustic cleaning wastes from paint manufacturing

INDUSTRY. Paints and allied products

PROCESS. Cleanup

WASTE CONSTITUENTS. Metals, inorganic and organic pigments, resins, polymers, solvents, plasticizers, caustic

APPLICABLE TREATMENT TECHNOLOGY: Solidification

RCRA CODE: -   CFS CODE: K081   UCD CODE: 454   SIC CODE: 2851

WASTE DESCRIPTION.   Waste water treatment sludges from paint manufacturing

INDUSTRY.   Paints and allied products

PROCESS.   Waste water treatment

WASTE CONSTITUENTS.   Metals, inorganic and organic pigments, resins, polymers, solvents, plasticizers

APPLICABLE TREATMENT TECHNOLOGY: Solidification

RCRA CODE: -   CFS CODE: K082   UCD CODE: 454   SIC CODE: 2851

WASTE DESCRIPTION.   Air pollution control sludges from paint manufacturing

INDUSTRY.   Paints and allied products

PROCESS.   Air pollution control

WASTE CONSTITUENTS.   Metals, inorganic and organic pigments, resins, polymers, solvents, plasticizers

APPLICABLE TREATMENT TECHNOLOGY: Solidification

RCRA CODE: K084   CFS CODE: K084   UCD CODE: 243   SIC CODE: 2834

WASTE DESCRIPTION.   Waste water treatment sludges generated during the production of veterinary pharmaceuticals from arsenic or organoarsenic compounds

INDUSTRY.   Pharmaceutical preparations

PROCESS.   Waste water treatment

WASTE CONSTITUENTS.   Arsenic

APPLICABLE TREATMENT TECHNOLOGY: CFS

RCRA CODE: K086   CFS CODE: K086   UCD CODE: 416   SIC CODE: 2893

WASTE DESCRIPTION.   Solvent, caustic and water washes and sludges from cleaning tubs and equipment used in the formulation of ink from pigments, driers, soaps, and stabilizers containing Cr and Pb

INDUSTRY.   Printing ink

PROCESS.   Cleanup

WASTE CONSTITUENTS.   Lead, hexavalent chromium, pigments, driers, soaps, stabilizers

APPLICABLE TREATMENT TECHNOLOGY: CFS

RCRA CODE: -   CFS CODE: K088   UCD CODE: SIC CODE: 3334

WASTE DESCRIPTION.   Spent potliners (cathodes) from primary aluminum production

INDUSTRY.   Aluminum, primary

PROCESS.   Aluminum production

WASTE CONSTITUENTS.   Aluminum, other metals and salts

APPLICABLE TREATMENT TECHNOLOGY: CFS

RCRA CODE: -   CFS CODE: K089   UCD CODE: SIC CODE: 3321

WASTE DESCRIPTION.   Lead-bearing waste water treatment sludges from gray-iron foundries

INDUSTRY.   Foundries, gray iron

PROCESS.   Waste water treatment

WASTE CONSTITUENTS.   Lead, iron, other metals

APPLICABLE TREATMENT TECHNOLOGY: CFS

RCRA CODE: -   CFS CODE: K090   UCD CODE: SIC CODE: 3313

WASTE DESCRIPTION.   Emission control dust/sludge from ferro-chromium-silicon production

INDUSTRY.   Electrometallurgical products

PROCESS.   Air pollution control

WASTE CONSTITUENTS.   Chromium, iron, silicon

APPLICABLE TREATMENT TECHNOLOGY: CFS

RCRA CODE: -   CFS CODE: K091   UCD CODE: SIC CODE: 3313

WASTE DESCRIPTION.   Emission control dust/sludge from ferro-chrome production

INDUSTRY.   Electrometallurgical products

PROCESS.   Air pollution control

WASTE CONSTITUENTS.   Chromium, iron

APPLICABLE TREATMENT TECHNOLOGY: CFS

RCRA CODE: -   CFS CODE: K092   UCD CODE: SIC CODE: 3313

WASTE DESCRIPTION.   Emission control dust/sludge from ferro-manganese production

598   APPENDIX 1

INDUSTRY.   Electrometallurgical products

PROCESS.   Air pollution control

WASTE CONSTITUENTS.   Iron, manganese

APPLICABLE TREATMENT TECHNOLOGY: CFS

RCRA CODE: -   CFS CODE: K098   UCD CODE: SIC CODE: 2879

WASTE DESCRIPTION.   Untreated process waste water from the production of toxaphene

INDUSTRY.   Agricultural chemicals, NEC

PROCESS.   Waste water

WASTE CONSTITUENTS.   Toxaphene

APPLICABLE TREATMENT TECHNOLOGY: Solidification

RCRA CODE: K099   CFS CODE: K099   UCD CODE: 241   SIC CODE: 2879

WASTE DESCRIPTION.   Untreated waste water from the production of 2,4-D

INDUSTRY.   Agricultural chemicals, NEC

PROCESS.   Waste water

WASTE CONSTITUENTS.   2,4-Dichlorophenol; 2,4,6-Trichlorophenol

APPLICABLE TREATMENT TECHNOLOGY: Solidification

RCRA CODE: K100   CFS CODE: K100   UCD CODE: 141   SIC CODE: 3341

WASTE DESCRIPTION.   Waste leaching solution from acid leaching of emission control dust/sludge from secondary lead smelting

INDUSTRY.   Metals, nonferrous secondary

PROCESS.   Lead recovery by leaching

WASTE CONSTITUENTS.   Lead

APPLICABLE TREATMENT TECHNOLOGY: CFS

RCRA CODE: K102   CFS CODE: K102   UCD CODE: 243   SIC CODE: 2834

WASTE DESCRIPTION.   Residue from the use of activated carbon for decolorization in the production of veterinary pharmaceuticals from arsenic or organoarsenic compounds.

INDUSTRY.   Pharmaceutical preparations

PROCESS.   Decolorization

WASTE CONSTITUENTS.   Arsenic, arsenic compounds

APPLICABLE TREATMENT TECHNOLOGY: CFS

RCRA CODE: K104   CFS CODE: K104   UCD CODE: 225   SIC CODE: 2869

WASTE DESCRIPTION.   Combined waste water streams generated from nitrobenzene/aniline production

INDUSTRY.   Industrial organic chemicals

PROCESS.   Waste water

WASTE CONSTITUENTS.   Aniline, benzene, diphenylamine, nitrobenzene, phenylenediamine

APPLICABLE TREATMENT TECHNOLOGY: Solidification

RCRA CODE: K105   CFS CODE: K105   UCD CODE: 225   SIC CODE: 2869

WASTE DESCRIPTION.   Separated aqueous stream from the reactor product washing step in the production of chlorobenzene

INDUSTRY.   Industrial organic chemicals

PROCESS.   Product washing

WASTE CONSTITUENTS.   Monochlorobenzene, benzene, dichlorobenzenes, 2,4,6-Trichlorophenol

APPLICABLE TREATMENT TECHNOLOGY: Solidification

RCRA CODE: K106   CFS CODE: K106   UCD CODE: 431   SIC CODE: 2869

WASTE DESCRIPTION.   Waste water treatment sludge from the mercury cell process in chlorine production

INDUSTRY.   Industrial inorganic chemicals

PROCESS.   Waste water treatment

WASTE CONSTITUENTS.   Mercury, sodium hydroxide, sodium chloride

APPLICABLE TREATMENT TECHNOLOGY: CFS

RCRA CODE: -   CFS CODE: W001   UCD CODE: 214   SIC CODE: 2754

WASTE DESCRIPTION.   Cleanup wastes from rotogravure printing

INDUSTRY.   Commercial printing, gravure

PROCESS.   Cleanup

APPENDIX 1

WASTE CONSTITUENTS. Chlorinated and nonchlorinated solvents, organic binders, organic resins, pigments, metals, sodium hydroxide

APPLICABLE TREATMENT TECHNOLOGY: Solidification

RCRA CODE: -   CFS CODE: W002   UCD CODE: 431   SIC CODE: 2819

WASTE DESCRIPTION. Thorium hydroxide filter cake

INDUSTRY. Industrial inorganic chemicals

PROCESS. Filtration

WASTE CONSTITUENTS. Thorium, lead, chloride, sulfide, sulfate, thorium hydroxide, filter aid

APPLICABLE TREATMENT TECHNOLOGY: CFS

RCRA CODE: -   CFS CODE: W004   UCD CODE: 442   SIC CODE: 2819

WASTE DESCRIPTION. Waste water treatment sludge, calcium phosphate

INDUSTRY. Industrial inorganic chemicals

PROCESS. Waste water treatment

WASTE CONSTITUENTS. Barium, chromium, calcium oxide, sodium hydroxide, phosphate

APPLICABLE TREATMENT TECHNOLOGY: CFS

RCRA CODE: -   CFS CODE: W005   UCD CODE: 445   SIC CODE:

WASTE DESCRIPTION. See F006

INDUSTRY.

PROCESS.

WASTE CONSTITUENTS. Lead, nickel, copper, tin, filter aid

APPLICABLE TREATMENT TECHNOLOGY: CFS

RCRA CODE: -   CFS CODE: W006   UCD CODE: 518   SIC CODE: 3322

WASTE DESCRIPTION. Emission control dust

INDUSTRY. Foundries, other metallurgical sources

PROCESS. Air pollution control

WASTE CONSTITUENTS. Barium, cadmium, chromium, lead, selenium; lime, chlorides, sulfates

APPLICABLE TREATMENT TECHNOLOGY: CFS

RCRA CODE: -   CFS CODE: W006   UCD CODE: 519   SIC CODE: 2911

WASTE DESCRIPTION.   Emission control residues from incineration

INDUSTRY.   Petroleum refining

PROCESS.   Air pollution control

WASTE CONSTITUENTS.   Lead, lead oxide

APPLICABLE TREATMENT TECHNOLOGY: CFS

RCRA CODE: -   CFS CODE: W007   UCD CODE: 431   SIC CODE: 3671

WASTE DESCRIPTION.   Sludge from cathode coating process

INDUSTRY.   Electron tubes, radio and TV receiving types

PROCESS.   Cathode coating

WASTE CONSTITUENTS.   Arsenic, barium, cadmium, chromium, lead, silver; nitrate

APPLICABLE TREATMENT TECHNOLOGY: CFS

RCRA CODE: -   CFS CODE: W008   UCD CODE: 454   SIC CODE: 3499

WASTE DESCRIPTION.   Sludge from industrial painting

INDUSTRY.   Fabricated metal products, NEC

PROCESS.   Industrial painting

WASTE CONSTITUENTS.   Barium, cadmium, chromium, lead, mercury, selenium, silver; solvents, resins and binders, pigments

APPLICABLE TREATMENT TECHNOLOGY: CFS

RCRA CODE: -   CFS CODE: W009   UCD CODE: 271   SIC CODE: 4952

WASTE DESCRIPTION.   Digested sewage sludge

INDUSTRY.   Sewage systems (POTWs)

PROCESS.   Sewage treatment

WASTE CONSTITUENTS.   Chromium, lead, silver, copper, nickel, zinc; nitrogen cpds, phosphates, COD

APPLICABLE TREATMENT TECHNOLOGY: CFS

RCRA CODE: -   CFS CODE: W011   UCD CODE: 245   SIC CODE: 2844

WASTE DESCRIPTION.   Off-spec lotions

INDUSTRY. Toilet preparations

PROCESS. Manufacture of lotions

WASTE CONSTITUENTS. Zinc; fluorides, sulfates, phosphates; fragrances, COD

APPLICABLE TREATMENT TECHNOLOGY: CFS

RCRA CODE: - CFS CODE: W012 UCD CODE: 111 SIC CODE: 2491

WASTE DESCRIPTION. Arsenic wood preservatives

INDUSTRY. Wood preserving

PROCESS. Treatment of wood with preservatives

WASTE CONSTITUENTS. Arsenic, chromium, lead, copper, iron, selenium, antimony; arsenic acid, sulfate

APPLICABLE TREATMENT TECHNOLOGY: CFS

RCRA CODE: - CFS CODE: W014 UCD CODE: 456 SIC CODE: 4953

WASTE DESCRIPTION. Waste water treatment residues from hazardous waste management facilities

INDUSTRY. Refuse systems (hazardous waste treatment systems)

PROCESS. Waste water treatment

WASTE CONSTITUENTS. Arsenic, cadmium, chromium, lead, mercury, selenium, silver, copper, nickel, molybdenum; cyanide, ammonia; phenol, oil

APPLICABLE TREATMENT TECHNOLOGY: CFS

RCRA CODE: - CFS CODE: W015 UCD CODE: 413 SIC CODE: 3369

WASTE DESCRIPTION. Sphincter cone sludge from emission control

INDUSTRY. Foundries, nonferrous, NEC

PROCESS. Air pollution control—scrubbing

WASTE CONSTITUENTS. Barium, cadmium, chromium, lead, copper, zinc, tin; metal oxides; organic carbon

APPLICABLE TREATMENT TECHNOLOGY: CFS

RCRA CODE: - CFS CODE: W016 UCD CODE: 271 SIC CODE: 4952

WASTE DESCRIPTION. Sewage sludge, general

INDUSTRY. Sewage systems (POTWs)

PROCESS. Sewage treatment

WASTE CONSTITUENTS. Chromium, lead, silver, copper, nickel, zinc nitrogen compounds, phosphates, COD

APPLICABLE TREATMENT TECHNOLOGY: CFS

RCRA CODE: - CFS CODE: W017 UCD CODE: 151 SIC CODE: 3699

WASTE DESCRIPTION. Contaminated glass

INDUSTRY. Electrical equipment, machinery and supplies, NEC

PROCESS. Manufacturing

WASTE CONSTITUENTS. Barium, cadmium, chromium, lead, mercury, silver, copper, nickel, zinc; fluoride, sulfate, nitrate, phosphate

APPLICABLE TREATMENT TECHNOLOGY: CFS

RCRA CODE: - CFS CODE: W018 UCD CODE: 431 SIC CODE: 2899

WASTE DESCRIPTION. Waste water treatment residues, metal hydroxide

INDUSTRY. Chemicals and chemical preparations, NEC

PROCESS. Waste water treatment

WASTE CONSTITUENTS. Chromium, copper, nickel, zinc

APPLICABLE TREATMENT TECHNOLOGY: CFS

RCRA CODE: - CFS CODE: W020 UCD CODE: 284 SIC CODE: 2911

WASTE DESCRIPTION. Acid oil sludge

INDUSTRY. Petroleum refining

PROCESS. Acid treatment of petroleum

WASTE CONSTITUENTS. Iron, manganese; sulfide, sulfur; oil, TOC

APPLICABLE TREATMENT TECHNOLOGY: CFS

RCRA CODE: - CFS CODE: W021 UCD CODE: 162 SIC CODE: 3369

WASTE DESCRIPTION. Aluminum or tin dross

INDUSTRY. Foundries, nonferrous, NEC

PROCESS. Nonferrous metal casting

WASTE CONSTITUENTS. Aluminum, tin

APPLICABLE TREATMENT TECHNOLOGY: CFS

RCRA CODE: -   CFS CODE: W022   UCD CODE: 514   SIC CODE: 3568

WASTE DESCRIPTION.   Foundry sand

INDUSTRY.   Mechanical power transmission equipment, NEC

PROCESS.   Metal casting

WASTE CONSTITUENTS.   Lead, copper

APPLICABLE TREATMENT TECHNOLOGY: CFS

RCRA CODE: -   CFS CODE: W023   UCD CODE: 519   SIC CODE: 2911

WASTE DESCRIPTION.   Incinerator ash

INDUSTRY.   Petroleum refining

PROCESS.   Air pollution control, incineration

WASTE CONSTITUENTS.   Arsenic, cadmium, chromium, lead, copper, nickel, zinc; oxides, sulfates, chlorides; COD

APPLICABLE TREATMENT TECHNOLOGY: CFS

RCRA CODE: -   CFS CODE: W024   UCD CODE: 518   SIC CODE: 3499

WASTE DESCRIPTION.   Barrel finishing baghouse dust

INDUSTRY.   Fabricated metal products, NEC

PROCESS.   Air pollution control, dust collection

WASTE CONSTITUENTS.   Chromium, copper, nickel, zinc, iron, calcium; sulfates, oxides

APPLICABLE TREATMENT TECHNOLOGY: CFS

RCRA CODE: -   CFS CODE: W025   UCD CODE: 288   SIC CODE: 3316

WASTE DESCRIPTION.   Rolling mill sludges and mill scale

INDUSTRY.   Cold-rolled sheet steel, strip and bar

PROCESS.   Steel finishing

WASTE CONSTITUENTS.   Iron, chromium, nickel oxides; oil

APPLICABLE TREATMENT TECHNOLOGY: CFS

RCRA CODE: -   CFS CODE: W026   UCD CODE: 111   SIC CODE: 3316

WASTE DESCRIPTION.   Pickle liquor rinsewater

INDUSTRY.   Cold-rolled sheet steel, strip and bar

PROCESS.   Steel pickling

WASTE CONSTITUENTS.   Iron, chromium, nickel, trace metals; sulfates, chlorides, fluorides

APPLICABLE TREATMENT TECHNOLOGY: CFS

RCRA CODE: -   CFS CODE: W027   UCD CODE: 281   SIC CODE: 2999

WASTE DESCRIPTION.   Waste automotive, hydraulic, and cutting oils that may contain solvents and metals

INDUSTRY.   Petroleum and coal products, NEC

PROCESS.   Waste oil collection

WASTE CONSTITUENTS.   Metals, oils, solvents

APPLICABLE TREATMENT TECHNOLOGY: Solidification

RCRA CODE: -   CFS CODE: W031   UCD CODE: 227   SIC CODE: 3631

WASTE DESCRIPTION.   Waste-water-based paint

INDUSTRY.   Household cooking equipment

PROCESS.   Dip painting

WASTE CONSTITUENTS.   Barium, cadmium, lead, mercury, selenium, silver; pigments, resins, binders, plasticizers

APPLICABLE TREATMENT TECHNOLOGY: CFS

RCRA CODE: -   CFS CODE: W032   UCD CODE: 518   SIC CODE: 3325

WASTE DESCRIPTION.   Emission control baghouse dust

INDUSTRY.   Steel foundries, NEC

PROCESS.   Air pollution control

WASTE CONSTITUENTS.   Cadmium, chromium, lead, iron; iron oxide, other oxides

APPLICABLE TREATMENT TECHNOLOGY: CFS

RCRA CODE: -   CFS CODE: W033   UCD CODE: 111   SIC CODE: 3471

WASTE DESCRIPTION.   Chromic rinse from metal finishing

INDUSTRY.   Plating and polishing

PROCESS.   Metal finishing

WASTE CONSTITUENTS.   Chromium, chromic acid

APPLICABLE TREATMENT TECHNOLOGY: CFS

RCRA CODE: -   CFS CODE: W034   UCD CODE: 131   SIC CODE: 3471

WASTE DESCRIPTION.   Waste electroplating baths, acid
INDUSTRY.   Plating and polishing
PROCESS.   Electroplating
WASTE CONSTITUENTS.   Copper, tin, lead, cadmium; ammonia, salts
APPLICABLE TREATMENT TECHNOLOGY: CFS

RCRA CODE: -   CFS CODE: W034   UCD CODE: 132   SIC CODE: 3471

WASTE DESCRIPTION.   Waste electroplating baths, alkaline
INDUSTRY.   Plating and polishing
PROCESS.   Electroplating
WASTE CONSTITUENTS.   Copper, tin, lead, cadmium; salts, strong complexing agents
APPLICABLE TREATMENT TECHNOLOGY: CFS

RCRA CODE: -   CFS CODE: W038   UCD CODE: 241   SIC CODE: 2879

WASTE DESCRIPTION.   Arsenical herbicide wastes
INDUSTRY.   Agricultural chemicals, NEC
PROCESS.   Herbicide manufacture
WASTE CONSTITUENTS.   Arsenic
APPLICABLE TREATMENT TECHNOLOGY: CFS

RCRA CODE: -   CFS CODE: W042   UCD CODE: 255   SIC CODE: 2842

WASTE DESCRIPTION.   Oil/wax emulsion
INDUSTRY.   Polishing, cleaning and sanitation goods
PROCESS.   Oil/wax manufacture
WASTE CONSTITUENTS.   Oils, waxes
APPLICABLE TREATMENT TECHNOLOGY: Solidification

RCRA CODE: -   CFS CODE: W047   UCD CODE: 284   SIC CODE: 2911

WASTE DESCRIPTION.   Waste water treatment residue—rag oil lagoon
INDUSTRY.   Petroleum refining
PROCESS.   Waste water treatment

WASTE CONSTITUENTS.   Acids, oils, metals
APPLICABLE TREATMENT TECHNOLOGY: CFS

RCRA CODE: -   CFS CODE: W048   UCD CODE: 272   SIC CODE: 2911

WASTE DESCRIPTION.   Waste water treatment residues—anaerobic lagoon
INDUSTRY.   Petroleum refining
PROCESS.   Waste water treatment
WASTE CONSTITUENTS.   Oil, metals, organic biological treatment residues
APPLICABLE TREATMENT TECHNOLOGY: CFS

RCRA CODE: -   CFS CODE: W050   UCD CODE: 441   SIC CODE: 2819

WASTE DESCRIPTION.   Lime sludge from hydrofluoric acid neutralization
INDUSTRY.   Industrial inorganic chemicals, NEC
PROCESS.   Acid neutralization
WASTE CONSTITUENTS.   Calcium; calcium fluoride, calcium hydroxide
APPLICABLE TREATMENT TECHNOLOGY: CFS

RCRA CODE: -   CFS CODE: W051   UCD CODE: 518   SIC CODE:

WASTE DESCRIPTION.   Dry electrostatic precipitator dust
INDUSTRY.
PROCESS.   Air pollution control
WASTE CONSTITUENTS.   Calcium oxide, calcium hydroxide
APPLICABLE TREATMENT TECHNOLOGY: CFS

RCRA CODE: -   CFS CODE: W052   UCD CODE: 431   SIC CODE:

WASTE DESCRIPTION.   Sludge from grinding of bearings
INDUSTRY.   Bearing manufacture
PROCESS.   Bearing grinding
WASTE CONSTITUENTS.   Metal particles, metal oxides
APPLICABLE TREATMENT TECHNOLOGY: CFS

RCRA CODE: -   CFS CODE: W053   UCD CODE: 111   SIC CODE: 3479

WASTE DESCRIPTION.   Waste phosphoric acid rinse

608 APPENDIX 1

INDUSTRY. Metal coating and allied services
PROCESS. Acid rinsing
WASTE CONSTITUENTS. Metal salts, phosphoric acid
APPLICABLE TREATMENT TECHNOLOGY: CFS

RCRA CODE: - CFS CODE: W054 UCD CODE: 262 SIC CODE: 2851

WASTE DESCRIPTION. Cleanup wastes from latex paint manufacture
INDUSTRY. Paints and allied products
PROCESS. Cleanup
WASTE CONSTITUENTS. Pigments, resins, binders, plasticizers
APPLICABLE TREATMENT TECHNOLOGY: Solidification

RCRA CODE: - CFS CODE: W055 UCD CODE: 287 SIC CODE: 3316

WASTE DESCRIPTION. Sludge from cold rolling
INDUSTRY. Cold-rolled sheet steel, strip and bar
PROCESS. Cold rolling
WASTE CONSTITUENTS. Metal particles, metal oxides; oils
APPLICABLE TREATMENT TECHNOLOGY: CFS

RCRA CODE: - CFS CODE: W057 UCD CODE: 141 SIC CODE: 4000

WASTE DESCRIPTION. Alodine waste from cleaning and refinishing
INDUSTRY. Tank, truck, and railroad tank cleaning and refinishing
PROCESS. Cleaning and refinishing
WASTE CONSTITUENTS. Aluminum, chromium
APPLICABLE TREATMENT TECHNOLOGY: CFS

RCRA CODE: - CFS CODE: W058 UCD CODE: 535 SIC CODE: 2819

WASTE DESCRIPTION. Waste from BF production
INDUSTRY. Industrial inorganic chemicals, NEC
PROCESS. BF production
WASTE CONSTITUENTS. Boron, fluoride, BF
APPLICABLE TREATMENT TECHNOLOGY: CFS

RCRA CODE: -   CFS CODE: W062   UCD CODE: 131   SIC CODE: 3499

WASTE DESCRIPTION.   Electrochemical machining waste

INDUSTRY.   Fabricated metal products, NEC

PROCESS.   Electrochemical machining

WASTE CONSTITUENTS.   Metal salts, acids

APPLICABLE TREATMENT TECHNOLOGY: CFS

RCRA CODE: -   CFS CODE: W064   UCD CODE: 133   SIC CODE: 3479

WASTE DESCRIPTION.   Waste sulfuric acid pickle liquor

INDUSTRY.   Metal coating and allied services

PROCESS.   Acid pickling

WASTE CONSTITUENTS.   Cadmium, chromium, lead, nickel, copper, zinc, iron; sulfuric acid, metal sulfates

APPLICABLE TREATMENT TECHNOLOGY: CFS

RCRA CODE: -   CFS CODE: W065   UCD CODE: 515   SIC CODE: 1381

WASTE DESCRIPTION.   Drilling mud

INDUSTRY.   Drilling oil and gas wells

PROCESS.   Well drilling

WASTE CONSTITUENTS.   Chromium, barium; barium sulfate, barium carbonate, clay; oil, surfactants

APPLICABLE TREATMENT TECHNOLOGY: CFS

RCRA CODE: -   CFS CODE: W066   UCD CODE: 518   SIC CODE: 3321

WASTE DESCRIPTION.   Baghouse dust from foundry cupola emission control

INDUSTRY.   Gray-iron foundries

PROCESS.   Air pollution control

WASTE CONSTITUENTS.   Cadmium, lead; oxides, sulfates, chlorides

APPLICABLE TREATMENT TECHNOLOGY: CFS

RCRA CODE: -   CFS CODE: W067   UCD CODE: 442   SIC CODE: 2819

WASTE DESCRIPTION.   Acid phosphate filter cake

INDUSTRY.   Industrial inorganic chemicals, NEC

PROCESS. Filtration
WASTE CONSTITUENTS. Phosphoric acid, filter cake
APPLICABLE TREATMENT TECHNOLOGY: CFS

RCRA CODE: - CFS CODE: W073 UCD CODE: 527 SIC CODE: 4400

WASTE DESCRIPTION. Dredging muds and sediments
INDUSTRY. Dredging
PROCESS. Dredging
WASTE CONSTITUENTS. Metals, sediments, inorganics, organics
APPLICABLE TREATMENT TECHNOLOGY: CFS

RCRA CODE: - CFS CODE: W074 UCD CODE: 151 SIC CODE: 3499

WASTE DESCRIPTION. Spent pack hardening compound
INDUSTRY. Fabricated metal products, NEC
PROCESS. Heat treating
WASTE CONSTITUENTS. Barium, chloride
APPLICABLE TREATMENT TECHNOLOGY: CFS

RCRA CODE: - CFS CODE: W075 UCD CODE: 225 SIC CODE: 3293

WASTE DESCRIPTION. Waste water from organic coatings
INDUSTRY. Gaskets, packing and sealing devices
PROCESS. Adhesive coating
WASTE CONSTITUENTS. Lead, mercury, silver, zinc; COD
APPLICABLE TREATMENT TECHNOLOGY: CFS

RCRA CODE: - CFS CODE: W076 UCD CODE: 241 SIC CODE: 2879

WASTE DESCRIPTION. Waste water from pesticide manufacture
INDUSTRY. Pesticides and agricultural chemicals, NEC
PROCESS. Pesticide manufacture
WASTE CONSTITUENTS. Organic gums, atrazine, nortron, surfactants
APPLICABLE TREATMENT TECHNOLOGY: Solidification

RCRA CODE: -   CFS CODE: W077   UCD CODE: 457   SIC CODE: 2869

WASTE DESCRIPTION.   Waste water treatment sludge from organic chemical manufacturing

INDUSTRY.   Industrial organic chemicals, NEC

PROCESS.   Waste water treatment

WASTE CONSTITUENTS.   Aluminum, iron, chromium, lead, zinc, calcium, silicon, copper, magnesium, manganese; phosphate, silica, calcium oxide, magnesium oxide, hydroxides; silicone oils, COD

APPLICABLE TREATMENT TECHNOLOGY: CFS

RCRA CODE: -   CFS CODE: W078   UCD CODE: 456   SIC CODE: 2869

WASTE DESCRIPTION.   Sludge from waste water treatment

INDUSTRY.   Industrial organic chemicals, NEC

PROCESS.   Waste water treatment

WASTE CONSTITUENTS.   Chromium, nickel, iron, calcium; calcium carbonate, calcium chloride, iron oxide; aniline, dinitrotoluene, nitrobenzene nitrophenol, o-toluidine HCl, oil, nitrotoluene, polyols

APPLICABLE TREATMENT TECHNOLOGY: Solidification

RCRA CODE: -   CFS CODE: W078   UCD CODE: 456   SIC CODE: 2899

WASTE DESCRIPTION.   Sludge from waste water treatment

INDUSTRY.   Chemicals and chemical preparations, NEC

PROCESS.   Waste water treatment

WASTE CONSTITUENTS.   Chromium, zinc; chloride, sulfate, nitrite, nitrate, phosphate; methyl-bis-thiocyanate, bromonitrostyrene, TBTO

APPLICABLE TREATMENT TECHNOLOGY: Solidification

RCRA CODE: -   CFS CODE: W078   UCD CODE: 456   SIC CODE: 3411

WASTE DESCRIPTION.   Filter cake from waste water treatment

INDUSTRY.   Metal cans

PROCESS.   Waste water treatment

WASTE CONSTITUENTS.   Arsenic, barium, cadmium, chromium, lead, nickel, copper, zinc, aluminum, calcium, tin, iron; fluoride, sulfate, hydroxide

APPLICABLE TREATMENT TECHNOLOGY: CFS

RCRA CODE: -   CFS CODE: W078   UCD CODE: 456   SIC CODE: 3631

WASTE DESCRIPTION.   Filter cake from waste water treatment

INDUSTRY.   Household cooking equipment, NEC

PROCESS.   Waste water treatment, filtration

WASTE CONSTITUENTS.   Arsenic, barium, chromium, lead, selenium, silver, nickel, copper, zinc, iron, aluminum; phosphate, calcium hydroxide; urea resins

APPLICABLE TREATMENT TECHNOLOGY: CFS

RCRA CODE: -   CFS CODE: W078   UCD CODE: 456   SIC CODE: 4582

WASTE DESCRIPTION.   Sludge from waste water treatment

INDUSTRY.   Airports and flying fields

PROCESS.   Waste water treatment

WASTE CONSTITUENTS.   Cadmium, chromium, lead, nickel, calcium; hydroxides; COD

APPLICABLE TREATMENT TECHNOLOGY: CFS

RCRA CODE: -   CFS CODE: W078   UCD CODE: 456   SIC CODE: 9953

WASTE DESCRIPTION.   Filter cake from waste water treatment

INDUSTRY.

PROCESS.   Waste water treatment, filtration

WASTE CONSTITUENTS.   Barium, cadmium, chromium, lead, nickel, copper, zinc, iron; chloride, fluoride, sulfide, sulfate, nitrate, cyanides; BOD COD, phenol

APPLICABLE TREATMENT TECHNOLOGY: CFS

RCRA CODE: -   CFS CODE: W080   UCD CODE: 441   SIC CODE: 2851

WASTE DESCRIPTION.   Paint stripping waste

INDUSTRY.   Paints and allied products

PROCESS.   Paint stripping with caustics

WASTE CONSTITUENTS.   Barium, cadmium, chromium; sodium hydroxide, pigments; resins, solvents

APPLICABLE TREATMENT TECHNOLOGY: CFS

RCRA CODE: -   CFS CODE: W080   UCD CODE: 441   SIC CODE: 2865

WASTE DESCRIPTION.   Sludge from acid neutralization

INDUSTRY. Cyclic and cyclic intermediates, dyes, organic pigments

PROCESS. Acid neutralization

WASTE CONSTITUENTS. Chromium, mercury, nickel, copper, iron, aluminum; cyanides, chloride, fluoride, sulfide, sulfate, nitrate, phosphate; COD

APPLICABLE TREATMENT TECHNOLOGY: CFS

RCRA CODE: -   CFS CODE: W080   UCD CODE: 441   SIC CODE: 3498

WASTE DESCRIPTION. Neutralized pickle liquor

INDUSTRY. Fabricated pipe and fittings

PROCESS. Waste treatment

WASTE CONSTITUENTS. Arsenic, barium, cadmium, chromium, lead, mercury, selenium, silver, copper, nickel, zinc; cyanides, chloride, fluoride; phenol, COD

APPLICABLE TREATMENT TECHNOLOGY: CFS

RCRA CODE: -   CFS CODE: W080   UCD CODE: 441   SIC CODE: 3631

WASTE DESCRIPTION. Neutralized pickle liquor

INDUSTRY. Household cooking equipment

PROCESS. Waste treatment

WASTE CONSTITUENTS. Arsenic, barium, cadmium, chromium, lead, mercury, selenium, silver, copper, nickel, zinc; cyanides; hydroxides, chloride, fluoride

APPLICABLE TREATMENT TECHNOLOGY: CFS

RCRA CODE: -   CFS CODE: W080   UCD CODE: 441   SIC CODE: 4953

WASTE DESCRIPTION. Sludge from acid neutralization

INDUSTRY. Refuse systems (hazardous waste treatment)

PROCESS. Acid neutralization

WASTE CONSTITUENTS. Lead, copper, nickel, zinc, iron, calcium; chloride, sulfate, calcium sulfate, calcium hydroxide

APPLICABLE TREATMENT TECHNOLOGY: CFS

RCRA CODE: -   CFS CODE: W081   UCD CODE: 144   SIC CODE: 2842

WASTE DESCRIPTION. Cleaning waste

INDUSTRY. Polishing, cleaning and sanitation goods

PROCESS. Cleaning

WASTE CONSTITUENTS.   Chloride, sulfate, nitrate, phosphate

APPLICABLE TREATMENT TECHNOLOGY: Solidification

RCRA CODE: -   CFS CODE: W081   UCD CODE: 144   SIC CODE: 2899

WASTE DESCRIPTION.   Cleaning waste

INDUSTRY.   Coal and petroleum products, NEC

PROCESS.   Chemical manufacturing

WASTE CONSTITUENTS.   Sodium hydroxide; phenol, solvents, pentaerythritol acrylate

APPLICABLE TREATMENT TECHNOLOGY: Solidification

RCRA CODE: -   CFS CODE: W082   UCD CODE: 519   SIC CODE:

WASTE DESCRIPTION.   Emission control dust

INDUSTRY.

PROCESS.   Air pollution control

WASTE CONSTITUENTS.   Lead, zinc, berillium; oxides

APPLICABLE TREATMENT TECHNOLOGY: CFS

RCRA CODE: -   CFS CODE: W084   UCD CODE: 412   SIC CODE: 3691

WASTE DESCRIPTION.   Filter cake from waste water treatment

INDUSTRY.   Storage batteries

PROCESS.   Waste water treatment, filtration

WASTE CONSTITUENTS.   Cadmium, lead, nickel, zinc, calcium; sulfate, hydroxide

APPLICABLE TREATMENT TECHNOLOGY: CFS

RCRA CODE: -   CFS CODE: W085   UCD CODE: 272   SIC CODE: 2865

WASTE DESCRIPTION.   Sludge from biological waste water treatment

INDUSTRY.   Cyclic crudes and intermediates

PROCESS.   Waste water treatment, biological

WASTE CONSTITUENTS.   Lead; calcium carbonate; acetone, acetophenone, acrylamide, benzene, chlorobenzene, dichlorobenzidine, o-chlorophenol, 1,4-dioxane, thiourea, toluene, trichlorophenol, xylene

APPLICABLE TREATMENT TECHNOLOGY: Solidification

RCRA CODE: -   CFS CODE: W085   UCD CODE: 272   SIC CODE: 4952

WASTE DESCRIPTION.   Sewage sludge

INDUSTRY.   Sewage systems

PROCESS.   Waste water treatment

WASTE CONSTITUENTS.   Barium, cadmium, chromium, lead, silver, copper, nickel, zinc, calcium, potassium; phosphate, nitrite, nitrate, ammonia; TOC, pesticides

APPLICABLE TREATMENT TECHNOLOGY: Solidification

RCRA CODE: -   CFS CODE: W086   UCD CODE: 122   SIC CODE: 2819

WASTE DESCRIPTION.   Sodium meta silicate waste

INDUSTRY.   Industrial inorganic chemicals, NEC

PROCESS.   Soluble silicate manufacturing

WASTE CONSTITUENTS.   Sodium metasilicate

APPLICABLE TREATMENT TECHNOLOGY: CFS

RCRA CODE: -   CFS CODE: W087   UCD CODE: 514   SIC CODE: 3471

WASTE DESCRIPTION.   Contaminated soil

INDUSTRY.   Plating and polishing

PROCESS.   Waste disposal

WASTE CONSTITUENTS.   Lead, nickel, copper, zinc, iron; cyanides, fluoride, free cyanide

APPLICABLE TREATMENT TECHNOLOGY: CFS

RCRA CODE: -   CFS CODE: W088   UCD CODE: 122   SIC CODE: 2819

WASTE DESCRIPTION.   Caustic waste

INDUSTRY.   Industrial inorganic chemicals, NEC

PROCESS.   Brine treatment

WASTE CONSTITUENTS.   Calcium carbonate, sodium hydroxide, magnesium hydroxide, phosphate

APPLICABLE TREATMENT TECHNOLOGY: CFS

RCRA CODE: -   CFS CODE: W089   UCD CODE: 144   SIC CODE: 4953

WASTE DESCRIPTION.   Leachate from RCRA TSD landfills

INDUSTRY.   Refuse systems
PROCESS.   Leaching of landfills
WASTE CONSTITUENTS.   Metals; salts; phenol, COD
APPLICABLE TREATMENT TECHNOLOGY: CFS

RCRA CODE: -   CFS CODE: W090   UCD CODE: 165   SIC CODE:

WASTE DESCRIPTION.   Spent catalyst
INDUSTRY.
PROCESS.
WASTE CONSTITUENTS.   Copper, metals
APPLICABLE TREATMENT TECHNOLOGY: CFS

RCRA CODE: -   CFS CODE: W092   UCD CODE: 433   SIC CODE: 3699

WASTE DESCRIPTION.   CaF sludge from acid neutralization
INDUSTRY.   Electrical equipment, machinery and supplies, NEC
PROCESS.   Acid neutralization
WASTE CONSTITUENTS.   Cadmium, chromium, lead, selenium, silver, copper, nickel, zinc, boron, manganese, iron, calcium, silicon, aluminum, tin, sodium, titanium, strontium; silica, calcium fluoride
APPLICABLE TREATMENT TECHNOLOGY: CFS

RCRA CODE: -   CFS CODE: W093   UCD CODE: 614   SIC CODE: 2999

WASTE DESCRIPTION.   Coke dust
INDUSTRY.   Petroleum and coal products, NEC
PROCESS.   Air pollution control
WASTE CONSTITUENTS.   Barium; coke
APPLICABLE TREATMENT TECHNOLOGY: CFS

RCRA CODE: -   CFS CODE: W095   UCD CODE: 262   SIC CODE: 2851

WASTE DESCRIPTION.   Latex paint waste
INDUSTRY.   Paints and allied products
PROCESS.   Cleaning of paint manufacturing equipment
WASTE CONSTITUENTS.   Pigment, resins, binders, solvents, plasticizers
APPLICABLE TREATMENT TECHNOLOGY: CFS

RCRA CODE: - CFS CODE: W096   UCD CODE: 455   SIC CODE:

WASTE DESCRIPTION.   Dye sludge
INDUSTRY.
PROCESS.   Dye manufacture and use
WASTE CONSTITUENTS.   Metals, organics
APPLICABLE TREATMENT TECHNOLOGY: CFS

RCRA CODE: - CFS CODE: W097   UCD CODE: 133   SIC CODE: 3316

WASTE DESCRIPTION.   Spent nitric-hydrofluoric acid pickle liquor
INDUSTRY.   Steel—sheet, strip, and bar
PROCESS.   Stainless steel pickling
WASTE CONSTITUENTS.   Chromium, nickel, iron, molybdenum, vanadium; nitric acid, hydrofluoric acid, nitrates, fluorides
APPLICABLE TREATMENT TECHNOLOGY: CFS

RCRA CODE: - CFS CODE: W100   UCD CODE: 442   SIC CODE: 3479

WASTE DESCRIPTION.   Waste water treatment residue from acid phosphating
INDUSTRY.   Metal coating and allied services
PROCESS.   Phosphate treatment of metals
WASTE CONSTITUENTS.   Zinc, calcium; phosphate
APPLICABLE TREATMENT TECHNOLOGY: CFS

RCRA CODE: - CFS CODE: W101   UCD CODE: 121   SIC CODE: 3471

WASTE DESCRIPTION.   Sludge from alkaline stripping of nickel plate
INDUSTRY.   Plating and polishing
PROCESS.   Alkaline stripping
WASTE CONSTITUENTS.   Nickel, alkalies
APPLICABLE TREATMENT TECHNOLOGY: CFS

RCRA CODE: - CFS CODE: W104   UCD CODE: 221   SIC CODE: 2869

WASTE DESCRIPTION.   Carbon tetrachloride sludge
INDUSTRY.   Industrial organic chemicals, NEC
PROCESS.   Organic chemical manufacture

WASTE CONSTITUENTS. Carbon tetrachloride, other organics

APPLICABLE TREATMENT TECHNOLOGY: Solidification

RCRA CODE: - CFS CODE: W105 UCD CODE: 141 SIC CODE: 3679

WASTE DESCRIPTION. Solder stripping waste

INDUSTRY. Electronic components, NEC

PROCESS. Solder stripping

WASTE CONSTITUENTS. Lead, tin, bismuth; acids, salts; organics

APPLICABLE TREATMENT TECHNOLOGY: CFS

RCRA CODE: - CFS CODE: W109 UCD CODE: 454 SIC CODE: 2851

WASTE DESCRIPTION. Caustic paint kettle cleaning waste

INDUSTRY. Paints and allied products

PROCESS. Cleaning of paint manufacturing equipment

WASTE CONSTITUENTS. Metals; sulfate, sodium hydroxide; resins and binders, vegetable oil, phthlic acid, maleic acid, urethane compounds, pigments, solvents

APPLICABLE TREATMENT TECHNOLOGY: Solidification

RCRA CODE: - CFS CODE: W110 UCD CODE: 525 SIC CODE:

WASTE DESCRIPTION. Aluminum powder

INDUSTRY.

PROCESS.

WASTE CONSTITUENTS. Aluminum metal

APPLICABLE TREATMENT TECHNOLOGY: CFS

RCRA CODE: - CFS CODE: W113 UCD CODE: 225 SIC CODE: 2851

WASTE DESCRIPTION. Waste from caulk manufacture

INDUSTRY. Paints and allied products

PROCESS. Cleaning of caulk manufacture equipment

WASTE CONSTITUENTS. Chromium, lead, calcium, titanium; calcium carbonate, titanium dioxide; bis(2-ethylhexyl) phthalate, acrylic resins

APPLICABLE TREATMENT TECHNOLOGY: Solidification

RCRA CODE: - CFS CODE: W114 UCD CODE: 241 SIC CODE: 2879

WASTE DESCRIPTION. Pesticide cleanup

INDUSTRY. Pesticides and agricultural chemicals, NEC

PROCESS. Cleanup

WASTE CONSTITUENTS. Arsenic, lead, copper, zinc, sodium; clays; toxaphene, surfactants, oil, carbamates, organophosphates, chlorinated hydrocarbons

APPLICABLE TREATMENT TECHNOLOGY: Solidification

RCRA CODE: - CFS CODE: W115 UCD CODE: 241 SIC CODE: 2879

WASTE DESCRIPTION. Emission control dust containing pesticide

INDUSTRY. Pesticides and agricultural chemicals, NEC

PROCESS. Air pollution control

WASTE CONSTITUENTS. Arsenic, barium, cadmium, chromium, lead, mercury, silver, copper, nickel, zinc, boron, iron, manganese, magnesium, cyanide, carbamates, organophosphates, organic sulfur

APPLICABLE TREATMENT TECHNOLOGY: Solidification

RCRA CODE: - CFS CODE: W117 UCD CODE: 431 SIC CODE: 3315

WASTE DESCRIPTION. Neutralized waste pickle liquor

INDUSTRY. Steel wire and related products

PROCESS. Neutralization

WASTE CONSTITUENTS. Chromium, nickel, iron, calcium; fluoride, sulfate, nitrate, calcium hydroxide

APPLICABLE TREATMENT TECHNOLOGY: CFS

RCRA CODE: - CFS CODE: W118 UCD CODE: 113 SIC CODE: 3356

WASTE DESCRIPTION. Waste oxalate solution

INDUSTRY. Nonferrous metal rolling and drawing, NEC

PROCESS. Pickling

WASTE CONSTITUENTS. Arsenic, barium, cadmium, chromium, nickel; fluoride, nitrate, boric acid; oxalic acid

APPLICABLE TREATMENT TECHNOLOGY: CFS

RCRA CODE: - CFS CODE: W119 UCD CODE: 141 SIC CODE: 3356

WASTE DESCRIPTION. Waste permanganate pickling solution

INDUSTRY. Nonferrous metal rolling and drawing, NEC

PROCESS. Pickling

WASTE CONSTITUENTS. Arsenic, barium, cadmium, chromium, lead, selenium, silver, nickel, manganese, potassium; fluoride, nitrate, sodium hydroxide, potassium permanganate

APPLICABLE TREATMENT TECHNOLOGY: CFS

RCRA CODE: - CFS CODE: W120 UCD CODE: 535 SIC CODE: 3499

WASTE DESCRIPTION. Waste weld cleaning solution

INDUSTRY. Fabricated metal products, NEC

PROCESS. Weld cleaning

WASTE CONSTITUENTS. Chromium, iron; sodium hydroxide, nitric acid, ammonium bifluoride, fluoride, nitrate

APPLICABLE TREATMENT TECHNOLOGY: CFS

RCRA CODE: - CFS CODE: W121 UCD CODE: 121 SIC CODE: 3585

WASTE DESCRIPTION. Zinc phosphate solution

INDUSTRY. Refrigeration and heating equipment

PROCESS. Conversion coating of metals

WASTE CONSTITUENTS. Cadmium, nickel, zinc; phosphate, sodium nitrate

APPLICABLE TREATMENT TECHNOLOGY: CFS

RCRA CODE: - CFS CODE: W122 UCD CODE: 254 SIC CODE: 2891

WASTE DESCRIPTION. Adhesive waste

INDUSTRY. Adhesives and sealants

PROCESS. Adhesive manufacture

WASTE CONSTITUENTS. Pigments, organic resins, phenolics, solvents

APPLICABLE TREATMENT TECHNOLOGY: Solidification

RCRA CODE: - CFS CODE: W123 UCD CODE: 262 SIC CODE: 2891

WASTE DESCRIPTION. Waste latex caulking compound

INDUSTRY. Adhesives and sealants

PROCESS. Caulk manufacture

WASTE CONSTITUENTS.   Calcium, titanium; calcium carbonate, titanium dioxide; bis(2-ethylhexyl) phthalate, acrylic resins

APPLICABLE TREATMENT TECHNOLOGY: CFS

RCRA CODE: -   CFS CODE: W125   UCD CODE: 535   SIC CODE: 4953

WASTE DESCRIPTION.   Hazardous organic wastes, NOS

INDUSTRY.   Refuse systems (RCRA TSD facilities)

PROCESS.   Hazardous waste management

WASTE CONSTITUENTS.   Methylene chloride, toluene, 1,1,1-trichloroethane, aromatic hydrocarbons, chlorinated solvents, ethyl benzene

APPLICABLE TREATMENT TECHNOLOGY: Solidification

RCRA CODE: -   CFS CODE: W133   UCD CODE: 281   SIC CODE: 3315

WASTE DESCRIPTION.   Waste drawing lubricant

INDUSTRY.   Steel wire and related products

PROCESS.   Wire drawing

WASTE CONSTITUENTS.   Chromium, nickel, iron; oil, paraffin

APPLICABLE TREATMENT TECHNOLOGY: CFS

RCRA CODE: -   CFS CODE: W135   UCD CODE: 525   SIC CODE: 3630

WASTE DESCRIPTION.   Metal grinding dust

INDUSTRY.   Household equipment

PROCESS.   Metal grinding

WASTE CONSTITUENTS.   Barium, lead, mercury, silver, zinc, iron, manganese, aluminum; oxides

APPLICABLE TREATMENT TECHNOLOGY: CFS

RCRA CODE: -   CFS CODE: W136   UCD CODE: 454   SIC CODE: 3411

WASTE DESCRIPTION.   Waste paint sludge

INDUSTRY.   Metal cans

PROCESS.   Painting

WASTE CONSTITUENTS.   Pigment, resins and binders, solvents, plasticizers

APPLICABLE TREATMENT TECHNOLOGY: CFS

RCRA CODE: -   CFS CODE: W136   UCD CODE: 454   SIC CODE: 3630

WASTE DESCRIPTION.   Waste paint sludge

INDUSTRY.   Household equipment

PROCESS.   Painting

WASTE CONSTITUENTS.   Barium, chromium, lead, selenium, silver, copper, nickel, zinc, manganese, iron, titanium; pigments, lead oxide, zinc chromate; toluene, xylene, resins, chlorinated HCs

APPLICABLE TREATMENT TECHNOLOGY: Solidification

RCRA CODE: -   CFS CODE: W137   UCD CODE: 121   SIC CODE: 3499

WASTE DESCRIPTION.   Waste paint stripper

INDUSTRY.   Fabricated metal products, NEC

PROCESS.   Paint stripping

WASTE CONSTITUENTS.   Chromium, sodium; pigments, sodium phosphate, sodium chromate; 1-butanol, methylene chloride, methyl isobutyl ketone, organic pigments, epoxy resins, ethyl amine

APPLICABLE TREATMENT TECHNOLOGY: Solidification

RCRA CODE: -   CFS CODE: W138   UCD CODE: 131   SIC CODE: 3679

WASTE DESCRIPTION.   Waste printed circuit etching bath

INDUSTRY.   Electronic components, NEC

PROCESS.   Printed circuit etching

WASTE CONSTITUENTS.   Chromium, copper; sulfate, sulfuric acid, chromic acid

APPLICABLE TREATMENT TECHNOLOGY: CFS

RCRA CODE: -   CFS CODE: W139   UCD CODE: 211   SIC CODE: 3679

WASTE DESCRIPTION.   Waste photoresist stripper

INDUSTRY.   Electronic components, NEC

PROCESS.   Caustic stripping of photoresists

WASTE CONSTITUENTS.   Barium, cadmium, chromium, copper, nickel, zinc; pigments, resins and binders, solvents

APPLICABLE TREATMENT TECHNOLOGY: CFS

RCRA CODE: -   CFS CODE: W141   UCD CODE: 151   SIC CODE: 3679

WASTE DESCRIPTION.   Solder flux residues

INDUSTRY.   Electronic components, NEC

PROCESS.   Solder coating

WASTE CONSTITUENTS.   Lead, copper, tin; isopropyl alcohol; salts

APPLICABLE TREATMENT TECHNOLOGY: CFS

RCRA CODE: -   CFS CODE: W142   UCD CODE: 454   SIC CODE: 3411

WASTE DESCRIPTION.   Sludge from spray painting

INDUSTRY.   Metal cans

PROCESS.   Spray painting

WASTE CONSTITUENTS.   Pigments, resins and binders, solvents, plasticizers

APPLICABLE TREATMENT TECHNOLOGY: Solidification

RCRA CODE: -   CFS CODE: W142   UCD CODE: 454   SIC CODE: 3585

WASTE DESCRIPTION.   Sludge from spray painting

INDUSTRY.   Refrigeration and heating equipment, NEC

PROCESS.   Spray painting

WASTE CONSTITUENTS.   Cadmium, nickel, zinc; pigments; aromatic hydrocarbons, alcohols, alkyd and amino resins

APPLICABLE TREATMENT TECHNOLOGY: CFS

RCRA CODE: -   CFS CODE: W145   UCD CODE: SIC CODE: 3315

WASTE DESCRIPTION.   Waste water treatment sludge

INDUSTRY.   Steel wire and related products

PROCESS.   Waste water treatment

WASTE CONSTITUENTS.   Chromium, zinc, iron; calcium carbonate, magnesium carbonate

APPLICABLE TREATMENT TECHNOLOGY: CFS

RCRA CODE: -   CFS CODE: W146   UCD CODE: SIC CODE: 2899

WASTE DESCRIPTION.   Waste liquid

INDUSTRY.   Coal and petroleum products, NEC

PROCESS.   Waste treatment

WASTE CONSTITUENTS.   Chromium; diethylene triamine

APPLICABLE TREATMENT TECHNOLOGY: CFS

RCRA CODE: -   CFS CODE: W147   UCD CODE: 141   SIC CODE: 2899

WASTE DESCRIPTION.   Chromate waste water

INDUSTRY.   Coal and petroleum products, NEC

PROCESS.   Waste treatment

WASTE CONSTITUENTS.   Chromium, zinc; carbonates

APPLICABLE TREATMENT TECHNOLOGY: CFS

RCRA CODE: -   CFS CODE: W149   UCD CODE: 225   SIC CODE: 3221

WASTE DESCRIPTION.   Oily water

INDUSTRY.   Glass containers

PROCESS.   Container manufacturing

WASTE CONSTITUENTS.   Oil

APPLICABLE TREATMENT TECHNOLOGY: Solidification

RCRA CODE: -   CFS CODE: W150   UCD CODE: 227   SIC CODE: 2869

WASTE DESCRIPTION.   Organic-containing waste water

INDUSTRY.   Industrial organic chemicals, NEC

PROCESS.   Washing processes

WASTE CONSTITUENTS.   Sodium; bromide, bisulfides; methanol, mercaptans

APPLICABLE TREATMENT TECHNOLOGY: Solidification

RCRA CODE: -   CFS CODE: W151   UCD CODE: 431   SIC CODE: 3411

WASTE DESCRIPTION.   Waste water treatment residue from can manufacture

INDUSTRY.   Metal cans

PROCESS.   Waste water treatment

WASTE CONSTITUENTS.   Barium, chromium, nickel, zinc, aluminum, calcium, iron; fluoride, sulfate; oil

APPLICABLE TREATMENT TECHNOLOGY: CFS

RCRA CODE: -   CFS CODE: W151   UCD CODE: 433   SIC CODE: 3411

WASTE DESCRIPTION.   Waste water treatment filter cake

INDUSTRY.   Metal cans

PROCESS.   Waste water treatment

WASTE CONSTITUENTS.   Barium, chromium, nickel, zinc, aluminum, calcium, iron; fluoride, sulfate; oil

APPLICABLE TREATMENT TECHNOLOGY: CFS

RCRA CODE: -   CFS CODE: W154   UCD CODE: 272   SIC CODE: 2911

WASTE DESCRIPTION.   Dissolved air floatation sludge

INDUSTRY.   Petroleum refining

PROCESS.   Waste water treatment

WASTE CONSTITUENTS.   Arsenic, cadmium, chromium, lead, copper, nickel, zinc; hydroxides, sulfides, sulfates, chlorides; oil

APPLICABLE TREATMENT TECHNOLOGY: CFS

RCRA CODE: -   CFS CODE: W157   UCD CODE: 454   SIC CODE: 3479

WASTE DESCRIPTION.   Paint sludge from spray painting

INDUSTRY.   Metal coating and allied services

PROCESS.   Waste water treatment

WASTE CONSTITUENTS.   Pigments, resins, binders, solvents, plasticizers

APPLICABLE TREATMENT TECHNOLOGY: CFS

RCRA CODE: -   CFS CODE: W158   UCD CODE: 271   SIC CODE: 4952

WASTE DESCRIPTION.   Primary scum (skimmings) from POTW

INDUSTRY.   Sewage systems

PROCESS.   Waste water treatment

WASTE CONSTITUENTS.   Arsenic, cadmium, chromium, lead, mercury, magnesium, calcium, potassium; chloride, phosphate, nitrogen; BOD

APPLICABLE TREATMENT TECHNOLOGY: CFS

RCRA CODE: -   CFS CODE: W159   UCD CODE: 271   SIC CODE: 4952

WASTE DESCRIPTION.   Primary grit screenings from POTWs and headworks

INDUSTRY.   Sewage systems

PROCESS.   Waste water treatment

WASTE CONSTITUENTS.   Barium, cadmium, chromium, lead, selenium, silver; BOD

APPLICABLE TREATMENT TECHNOLOGY: CFS

RCRA CODE: -   CFS CODE: W161   UCD CODE: 225   SIC CODE: 2399

WASTE DESCRIPTION.   Textile sizing waste

INDUSTRY.   Fabricated textile products

PROCESS.   Textile finishing

WASTE CONSTITUENTS.   Cadmium, chromium, lead, iron, copper, zinc, sodium; chloride, sulfate, nitrogen compounds; surfactants, COD

APPLICABLE TREATMENT TECHNOLOGY: CFS

RCRA CODE: -   CFS CODE: W162   UCD CODE: 454   SIC CODE: 2851

WASTE DESCRIPTION.   Waste water treatment sludge from paint manufacturing

INDUSTRY.   Paints and allied products

PROCESS.   Waste water treatment

WASTE CONSTITUENTS.   Arsenic, barium, cadmium, chromium, lead, mercury, selenium, silver; sodium hydroxide, pigments, resins, solvents

APPLICABLE TREATMENT TECHNOLOGY: CFS

RCRA CODE: -   CFS CODE: W165   UCD CODE: -   SIC CODE: 4952

WASTE DESCRIPTION.   Incinerator ash from sewage sludge incineration

INDUSTRY.   Sewage systems

PROCESS.   Air pollution control, incineration

WASTE CONSTITUENTS.   Chromium, lead, copper, nickel, zinc, calcium, potassium; phosphates, nitrogen compounds, sulfates, oxides; COD

APPLICABLE TREATMENT TECHNOLOGY: CFS

Appendix 2

TEST METHODS

This appendix contains the complete texts of the most active test methods in use in the United States as of this writing. For more detail on proper application of these protocols, as well as specifications for test equipment and data treatment (in the case of ANS 16.1), the reader is referred to the original references and to Chapter 17.

The following test methods are contained herein:

U.S. EPA Extraction Procedure Toxicity Test (EPT)
U.S. EPA Multiple Extraction Procedure (MEP)
U.S. EPA Oily Wastes Extraction Procedure (OWEP)
U.S. EPA Toxicity Characteristic Leaching Procedure (TCLP)
State of California Waste Extraction Test (WET)
American Nuclear Society ANS 16.1 Test

EPA SW-846—METHOD 1310
EXTRACTION PROCEDURE (EP) TOXICITY TEST METHOD AND STRUCTURAL INTEGRITY TEST

1.0 *Scope and application.*

1.1 This method is employed to determine whether a waste exhibits the characteristic of Extraction Procedure Toxicity.

1.2 The procedure may also be used to simulate the leaching which a waste will undergo if disposed of in a sanitary landfill. Method 1310 is applicable to liquid, solid, and multiphase samples.

2.0 *Summary of method.*

2.1. If a representative sample of the waste contains >0.5 percent solids, the solid phase of the sample is ground to pass a 9.5 mm sieve and extracted with deionized water which is maintained at a pH of 5 ± 0.2, with acetic acid. Wastes that contain <0.5 percent solids are not subjected to extraction but are directly analyzed. Monolithic wastes which can be formed into a cylinder 3.3 cm (dia) × 7.1 cm, or from which such a cylinder can be formed which is representative of the waste, may be evaluated using the Structural Integrity Procedure instead of being ground to pass a 9.5-mm sieve.

3.0 *Interferences.*

3.1 Potential interferences that may be encountered during analysis are discussed in the individual analytical methods.

4.0 *Apparatus and materials.*

4.1 Extractor: For purposes of this test, an acceptable extractor is one that will impart sufficient agitation to the mixture to (1) prevent stratification of the sample and extraction fluid and (2) ensure that all sample surfaces are continuously brought into contact with well-mixed extraction fluid. Examples of suitable extractors are available from Associated Designs & Manufacturing Co., Alexandria, Virginia; Glas-Col Apparatus Co., Terre Haute, Indiana; Millipore, Bedford, Massachusetts; and Rexnard, Milwaukee, Wisconsin.

4.2 pH Meter or pH Controller: Accurate to 0.05 pH units with temperature compensation.

4.3 Filter holder: Capable of supporting a 0.45-μm filter membrane and of withstanding the pressure needed to accomplish separation. Suitable filter holders range from simple vacuum units to relatively complex systems that can exert up to 5.3 kg/cm^3 (75 psi) of pressure. The type of filter holder used depends upon the properties of the mixture to be filtered. Filter holders known to the EPA and deemed suitable for use are listed in Table 1.

4.4 Filter membrane: Filter membrane suitable for conducting the required filtration shall be fabricated from a material that (1) is not physically changed by the waste material to be filtered and (2) does not absorb or leach the chemical spe-

TABLE 1. EPA-Approved Filter Holders

Manufacturer	Size	Model Number	Comments
Vacuum filters			
Nalgene	500 mL	44-0045	Disposable plastic unit, including prefilter, filter pads, and reservoir; can be used when solution is to be analyzed for inorganic constituents.
Nuclepore	47 mm	410400	
Millipore	47 mm	XX10 047 00	
Pressure filters			
Nuclepore	142 mm	425900	
Micro Filtration Systems	142 mm	302300	
Millipore	142 mm	YT30 142 HW	

cies for which a waste's EP extract will be analyzed. Table 2 lists filter media known to the agency to be suitable for solid waste testing.

4.4.1 In cases of doubt about physical effects on the filter, contact the filter manufacturer to determine if the membrane or the prefilter is adversely affected by the particular waste. If no information is available, submerge the filter in the waste's liquid phase. A filter that undergoes visible change after 48 hr (i.e., curls, dissolves, shrinks, or swells) is unsuitable for use.

4.4.2. To test for absorbtion or leaching by the filter:

4.4.2.1 Prepare a standard solution of the chemical species of interest.

4.4.2.2. Analyze the standard for its concentration of the chemical species.

4.4.2.3 Filter the standard and reanalyze. If the concentration of the filtrate differs from that of the original standard, then the filter membrane leaches or absorbs one or more of the chemical species and is not usable in this test method.

4.5 Structural Integrity Tester: A device meeting the specifications and having a 3.18-cm (1.25-in.)-diameter hammer weighing 0.33 kg (0.73 lb) with a free fall of 15.24 cm (6 in.) shall be used. This device is available from Associated Design and Manufacturing Company, Alexandria, Virginia 22314, as Part No. 125, or it may be fabricated to meet these specifications.

TABLE 2. EPA-Approved Filtration Media

Supplier	Filter to be Used for Aqueous Systems	Filter to be Used for Organic Systems
Coarse prefilter		
Gelman	61631, 61635	61631, 61635
Nuclepore	210907, 211707	210907, 211707
Millipore	AP25 035 00, AP25 127 50	AP25 035 00, AP25 127 50
Medium prefilters		
Nuclepore	210905, 211705	210905, 211705
Millipore	AP20 035 00, AP20 124 50	AP20 035 00, AP20 124 50
Fine prefilters		
Gelman	64798, 64803	64798, 64803
Nuclepore	210903, 211703	210903, 211703
Millipore	AP15 035 00, AP15 124 50	AP15 035 00, AP15 124 50
Fine filters (0.45 μm)		
Gelman	60173, 60177	60540 or 66149, 60544 or 66151
Pall	NX04750, NX14225	
Nuclepore	142218	142218[a]
Millipore	HAWP 047 00, HAWP 142 50	FHUP 047 00, FHLP 142 50
Selas	83485-02, 83486-02	83485-02, 83486-02

[a] Susceptible to decomposition by certain polar organic solvents.

5.0 *Reagents.*

5.1 Acetic Acid (0.5 N): This acid can be made by diluting concentrated glacial acetic acid (17.5 N) by adding 57 ml glacial acetic acid to 1000 ml of water and diluting to 2 l. The glacial acetic acid should be of high purity and monitored for impurities.

5.2 Analytical Standards should be prepared according to the applicable analytical methods.

6.0 *Sample collection, preservation, and handling.*

6.1 All samples must be collected using a sampling plan that addresses the considerations discussed in this manual.

6.2 Preservatives must not be added to samples.

6.3 Samples can be refrigerated if it is determined that refrigeration will not affect the integrity of the sample.

7.0 *Procedure.*

7.1 If the waste does not contain any free liquid, go to Step 7.9. If the sample is liquid or multiphase, continue as follows. Weigh filter membrane and prefilter to ±0.01 g. Handle membrane and prefilters with blunt curved-tip forceps or vacuum tweezers, or by applying suction with a pipet.

7.2 Assemble filter holder, membranes, and prefilters following the manufacturer's instructions. Place the 0.45-μ membrane on the support screen and add prefilters in ascending order of pore size. Do not prewet filter membrane.

7.3 Weigh out a representative subsample of the waste (100-g minimum).

7.4 Allow slurries to stand, to permit the solid phase to settle. Wastes that settle slowly may be centrifuged prior to filtration.

7.5 Wet the filter with a small portion of the liquid phase from the waste or from the extraction mixture. Transfer the remaining material to the filter holder and apply vacuum or gentle pressure (10–15 psi) until all liquid passes through the filter. Stop filtration when air or pressurizing gas moves through the membrane. If this point is not reached under vacuum or gentle pressure, slowly increase the pressure in 10-psi increments to 75 psi. Halt filtration when liquid flow stops. This liquid will constitute part or all of the extract (refer to Step 7.16). The liquid should be refrigerated until time of analysis.

Note.—Oil samples or samples containing oil are treated in exactly the same way as any other sample. The liquid portion of the sample is filtered and treated as part of the EP extract. If the liquid portion of the sample will not pass through the filter (usually the case with heavy oils or greases), it should be carried through the EP extraction as a solid.

7.6 Remove the solid phase and filter media and, while not allowing them to dry, weigh to ±0.01 g. The wet weight of the residue is determined by calculating the weight difference between the weight of the filters (Step 7.1) and the weight of the solid phase and the filter media.

7.7 The waste will be handled differently from this point on, depending on whether it contains more or less than 0.5 percent solids. If the sample appears to have <0.5 percent solids, determine the percent solids exactly (see Note below) by the following procedure.

    7.7.1 Dry the filter and residue at 80°C until two successive weighings yield the same value.

    7.7.2 Calculate the percent solids, using the following equation:

$$\frac{\text{Weight of filtered solid and filters } - \text{ tared weight of filters}}{\text{Initial weight of waste material}} \times 100 = \% \text{ solids}$$

Note.—This procedure is used only to determine whether the solid must be extracted or whether it can be discarded unextracted. It is not used in calculating the amount of water or acid to use in the extraction step. Do not extract solid material that has been dried at 80°C. A new sample will have to be sued for extraction if a percent solids determination is performed.

7.8 If the solid constitutes <0.5 percent of the waste, discard the solid and proceed immediately to Step 7.17, treating the liquid phase as the extract.

7.9 The solid material obtained from Step 7.5 and all materials that do not contain free liquids should be evaluated for particle size. If the solid material has a surface area per gram of material ≥ 3.1 cm^2 or passes through a 9.5-mm (0.375-in.) standard sieve, the operator should proceed to Step 7.11. If the surface area is smaller or the particle size larger than specified above, the solid material is prepared for extraction by crushing, cutting, or grinding the material so that it passes through a 9.5-mm (0.375-in.) sieve or, if the material is in a single piece, by subjecting the material to the "Structural Integrity Procedure" described in Step 7.10.

7.10 *Structural Integrity Procedure (SIP):*

    7.10.1 Cut a 3.3-cm-diameter by 7.1-cm-long cylinder from the waste material. If the waste has been treated using a fixation process, the waste may be cast in the form of a cylinder and allowed to cure for 30 days prior to testing.

    7.10.2 Place waste into sample holder and assemble the tester. Raise the hammer to its maximum height and drop. Repeat 14 additional times.

    7.10.3 Remove solid material from tester and scrape off any particles adhering to sample holder. Weigh the waste to the nearest 0.01 g and transfer it to the extractor.

7.11 If the sample contains >0.5 percent solids, use the wet weight of the solid phase (obtained in Section 7.6) to calculate the amount of liquid and acid to employ for extraction by using the following equation:

$$W = W_f - W_t$$

where

    W = wet weight in grams of solid to be charged to extractor
    W_f = wet weight in grams of filtered solids and filter media
    W_t = weight in grams of tared filters

If the waste does not contain any free liquids, 100 g of the material will be subjected to the extraction procedure.

7.12 Place the appropriate amount of material (refer to Step 7.11) into the extractor and add 16 times its weight of Type II water.

7.13 After the solid material and Type II water are placed in the extractor, the operator should begin agitation and measure the pH of the solution in the extractor. If the pH is >5.0, the pH of the solution should be decreased to 5.0 ± 0.2 by adding 0.5 N acetic acid. If the pH is ≤ 5.0, no acetic acid should be added. The pH of the solution should be monitored, as described below, during the course of the extraction, and, if the pH rises above 5.2, 0.5 N acetic acid should be added to bring the pH down to 5.0 ± 0.2. However, in no event shall the aggregate amount of acid added to the solution exceed 4 ml of acid per gram of solid. The mixture should be agitated for 24 hr and maintained at 20–40°C (68–104°F) during this time. It is recommended that the operator monitor and adjust the pH during the course of the extraction with a device such as the Type 45-A pH Controller, manufactured by Chemtrix, Inc., Hillsboro, Oregon 97123, or its equivalent, in conjunction with a metering pump and reservoir of 0.5 N acetic acid. If such a system is not available, the following manual procedure shall be employed.

7.13.1 A pH meter should be calibrated in accordance with the manufacturer's specifications.

7.13.2 The pH of the solution should be checked, and, if necessary, 0.5 N acetic acid should be manually added to the extractor until the pH reaches 5.0 ± 0.2. The pH of the solution should be adjusted at 15-, 30-, and 60-min intervals, moving to the next longer interval if the pH does not have to be adjusted >0.5 pH units.

7.13.3 The adjustment procedure should be continued for at least 6 hr.

7.13.4 If, at the end of the 24-hr extraction period, the pH of the solution is not below 5.2 and the maximum amount of acid (4 ml per gram of solids) has not been added, the pH should be adjusted to 5.0 ± 0.2 and the extraction continued for an additional 4 hr, during which the pH should be adjusted at 1-hr intervals.

7.14 At the end of the extraction period, Type II water should be added to the extractor in an amount determined by the following equation:

$$V = (20)(W) - 16(W) - A$$

where

V = milliliters of Type II water to be added
W = weight in grams of solid charged to extractor
A = milliliters of 0.5 N acetic acid added during extraction

7.15 The material in the extractor should be separated into its component liquid and solid phases in the following manner:

7.15.1 Allow slurries to stand to permit the solid phase to settle (wastes that are slow to settle may be centrifuged prior to filtration) and set up the filter apparatus (refer to Steps 4.3 and 4.4).

7.15.2 Wet the filter with a small portion of the liquid phase from the waste or from the extraction mixture. Transfer the remaining material to the filter holder and apply vacuum or gentle pressure (10–15 psi) until all liquid passes through the filter. Stop filtration when air or pressurizing gas moves through the membrane. If this point is not reached under vacuum or gentle pressure, slowly increase the pressure in 10-psi increments to 75 psi. Halt filtration when liquid flow stops.

7.16 The liquids resulting from Steps 7.5 and 7.15 should be combined. This combined liquid (or waste itself, if it has <0.5 percent solids, as noted in Step 7.8) is the extract and should be analyzed for the presence of any of the contaminants specified in 40 CFR Part 261.24 using the analytical procedures as designated in Step 7.17.

7.17 The extract is then prepared and analyzed using the appropriate analytical methods described in this manual.

Note.—If the EP extract includes two phases, concentration of contaminants is determined by using a simple weighted average. For example: An EP extract contains 50 ml of oil and 1000 ml of an aqueous phase. Contaminant concen-

TABLE 3. Precisions of Extraction-Analysis Procedures for Several Elements

Element	Sample Matrix	Analysis Method	Laboratory Replicates
Arsenic	1. Auto fluff	7060	1.8, 1.5 ug/L
	2. Barrel sludge	7060	0.9, 2.6 ug/L
	3. Lumber treatment company sediment	7060	28, 42 mg/L
Barium	1. Lead smelting emission control dust	6010	0.12, 0.12 mg/L
	2. Auto fluff	7081	791, 780 ug/L
	3. Barrel sludge	7081	422, 380 ug/L
Cadmium	1. Lead smelting emission control dust	3010/7130	120, 120 mg/L
	2. Wastewater treatment sludge from electroplating	3010/7130	360, 290 mgL
	3. Auto fluff	7131	470, 610 ug/L
	4. Barrel sludge	7131	1100, 890 ug/L
	5. Oil refinery tertiary pond sludge	7131	3.2, 1.9 ug/L
Chromium	1. Wastewater treatment sludge from electroplating	3010/7190	1.1, 1.2 mg/L
	2. Paint primer	7191	61, 43 ug/L
	3. Paint primer filter	7191	—
	4. Lumber treatment company sediment	7191	0.81, 0.89 mg/L
	5. Oil refinery tertiary pond sludge	7191	—
Mercury	1. Barrel sludge	7470	0.15, 0.09 ug/L
	2. Wastewater treatment sludge from electroplating	7470	1.4, 0.4 ug/L
	3. Lead smelting emission control dust	7470	0.4, 0.4 ug/L
Lead	1. Lead smelting emission control dust	3010/7420	940, 920 mg/L
	2. Auto fluff	7421	1540, 1490 ug/L
	3. Incinerator ash	7421	1000, 974 ug/L
	4. Barrel sludge	7421	2550, 2800 ug/L
	5. Oil refinery tertiary pond sludge	7421	31, 29 ug/L
Nickel	1. Sludge	7521	2260, 1720 ug/L
	2. Wastewater treatment sludge from electroplating	3010/7520	130, 140 mg/L
Chromium(VI)	1. Wastewater treatment sludge from electroplating	7196	18, 19 ug/L

trations are determined for each phase. The final contamination concentration is taken to be:

$$\frac{(50)\ (\text{contaminant concentration in oil}) + (1000)\ (\text{contaminant concentration of aqueous phase})}{1050}$$

7.18 The extract concentrations are compared with the maximum contamination limits listed in 40 CFR Part 261.24. If the extract concentrations are greater than or equal to the respective values, the waste then is considered to exhibit the characteristic of Extraction Procedure Toxicity.

8.0 *Quality control.*

8.1 All quality control data should be maintained and available for easy reference or inspection.

8.2 Employ a minimum of one blank per sample batch to determine if contamination or any memory effects are occurring.

8.3 All quality control measures described in the referenced analytical methods should be followed.

9.0 *Method performance.*

9.1 The data tabulated below were obtained from records of state and contractor laboratories and are intended to show the precision of the entire method (1310 plus analysis method).

10.0 *References.*

1. Gaskill, A., Compilation and Evaluation of RCRA Method Performance Data, Work Assignment No. 2, EPA Contract No. 68-01-7075, September 1986.

EPA SW-846—METHOD 1320
MULTIPLE EXTRACTION PROCEDURE

1.0 *Scope and application.*

The Multiple Extraction Procedure (MEP) described in this method is designed to simulate the leaching that a waste will undergo from repetitive precipitation of acid rain on an improperly designed sanitary landfill. The repetitive extractions reveal the highest concentration of each constituent that is likely to leach in a natural environment. Method 1320 is applicable to liquid, solid, and multiphase samples.

2.0 *Summary of method.*

Waste samples are extracted according to the Extraction Procedure Toxicity Test (Method 1310) and analyzed for the constituents of concern: Maximum Concentration of Contaminants for Characteristic of EP Toxicity, using the 7000 and 8000 series methods. Then the solid portions of the samples that remain after application of Method 1310 are reextracted nine times using synthetic acid rain extraction fluid. If the concentration of any constituent of concern increases from the seventh or eighth extraction to the ninth extraction, the procedure is repeated until these concentrations decrease.

3.0 *Interferences.*
Potential interferences that may be encountered during analysis are discussed in the appropriate analytical methods.

4.0 *Apparatus and materials.*

4.1 Refer to Method 1310.

5.0 *Reagents.*

5.1 Refer to Method 1310.

5.2 Sulfuric Acid:Nitric Acid, 60/40 weight percent mixture: Cautiously mix 60 g of concentrated sulfuric acid with 40 g of concentrated nitric acid.

6.0 *Sample collection, preservation, and handling.*

6.1 Refer to Method 1310.

7.0 *Procedure.*

7.1 Run the Extraction Procedure (EP) test in Method 1310.

7.2 Analyze the extract for the constituents of interest.

7.3 Prepare a synthetic acid rain extraction fluid by adding the 60/40 weight percent sulfuric acid and nitric acid to distilled deionized water until the pH is 3.0 ± 0.2.

7.4 Take the solid phase of the sample remaining after the Separation Procedure of the Extraction Procedure and weigh it. Measure the aliquot of synthetic acid rain extraction fluid equal to 20 times the weight of the solid sample. Do not allow the solid sample to dry before weighing.

7.5 Combine the solid phase sample and acid rain fluid in the same extractor as used in the EP and begin agitation. Record the pH within 5–10 min after agitation has been started.

7.6 Agitate the mixture for 24 hr, maintaining the temperature at 20–40°C (68–104°F). Record the pH at the end of the 24-hr extraction period.

7.7 Repeat the Separation Procedure as described in Method 1310.

7.8 Analyze the extract for the constituents of concern.

7.9 Repeat steps 7.4–7.8 eight additional times.

7.10 If, after completing the ninth synthetic rain extraction, the concentration of any of the constituents of concern is increasing over that found in the seventh and eighth extractions, then continue extracting with synthetic acid rain until the concentration in the extract ceases to increase.

7.11 Report the initial and final pH of each extraction and the concentrations of each listed constituent of concern in each extract.

8.0 *Quality control.*

8.1 All quality control data should be maintained and available for easy reference or inspection.

8.2 Employ a minimum of one blank per sample batch to determine if contamination or any memory effects are occurring.

8.3 All quality control measures suggested in the referenced analytical methods should be followed.

9.0 *Method performance.*

9.1 No data provided.

10.0 *References.*

10.1 None required.

EPA SW-846—METHOD 1330
OILY WASTES EXTRACTION PROCEDURE

1.0 *Scope and application.*

1.1 Method 1330 is used to determine the mobile metal concentration (MMC) in oily wastes.

1.2 Method 1330 is applicable to API separator sludges, rag oils, slop oil emulsions, and other oil wastes derived from petroleum refining.

2.0 *Summary of method.*

2.1 The sample is separated into solid and liquid components by filtration.

2.2 The solid phase is placed in a Soxhlet extractor, charged with tetrahydrofuran, and extracted. The THF is removed, the extractor is then charged with toluene, and the sample is reextracted.

2.3 The EP method (Method 1310) is run on the dry solid residue.

2.4 The original liquid, combined extracts, and EP leachate are analyzed for the EP metals.

3.0 *Interferences.*

3.1 Matrix interferences will be coextracted from the sample. The extent of these interferences will vary considerably from waste to waste, depending on the nature and diversity of the particular refinery waste being analyzed.

4.0 *Apparatus and materials.*

4.1 Soxhlet Extraction Apparatus.

4.2 Vacuum Pump or Other Source of Vacuum.

4.3 Buchner Funnel 12.

4.4 Electric Heating Mantle.

4.5 Paper Extraction Thimble.

4.6 Filter Paper.

4.7 Muslin Cloth Disks.

4.8 Evaporative Flask: 250-ml.

4.9 Analytical Balance: Capable of weighing to ±0.5 mg.

5.0 *Reagents.*

5.1 Tetrahydrofuran: ACS reagent grade.

5.2 Toluene: ACS reagent grade.

6.0 *Sampling.*

6.1 Samples must be collected in glass containers having a total volume of at least 150 ml. No solid material should interfere with sealing the sample container.

6.2 Sampling devices should be wiped clean with paper towels or absorbent cloth, rinsed with a small amount of hexane followed by acetone rinse, and dried between samples. Alternatively, samples can be taken with disposable sampling devices in beakers.

7.0 *Procedure.*

7.1 Separate the sample (minimum 100 g) into its solid and liquid components using the filtration Steps 7.1–7.6 in Method 1310.

7.2 Determine the quantity of liquid (ml) and the concentration of the species of concern in the liquid phase (mg/l) using appropriate analytical methods.

7.3 Place the solid phase into a Soxhlet extractor, charge the concentration flask with 300-ml tetrahydrofuran and extract for 3 hr.

7.4 Remove the flask containing tetrahydrofuran and replace it with one containing 300-ml toluene.

7.5 Extract the solid for a second time, for 3 hr, with the toluene.

7.6 Combine the tetrahydrofuran and toluene extracts.

7.7 Determine the quantity of liquid (ml) and the concentration of the species of concern in the combined extracts (mg/l).

7.8 Take the solid material remaining in the Soxhlet thimble and dry it at 100°C for 30 min.

7.9 Run the EP (Method 1310) on the dried solid.

7.10 Calculate the mobile metal concentration (MMC) in milligrams per liter using the following formula:

$$\text{MMC} = 1000 \times \frac{(Q_1 + Q_2 + Q_3)}{(L_1 + L_2)}$$

where

Q_1 = amount of metal in initial liquid phase of sample (amount of liquid × concentration of metal) (mg)
Q_2 = amount of metal in combined organic extracts of sample (mg)
Q_3 = amount of metal in EP extract of solid (amount of extract × concentration of metal) (mg)
L_1 = amount of initial liquid (ml)
L_2 = amount of liquid in EP (ml) = 20 × [weight of dried solid from Step 9 (mg)]

8.0 *Quality control.*

8.1 Standard quality assurance practices should be used with this method. Laboratory replicates should be analyzed to validate the precision of the analysis. Fortified samples should be carried through all stages of sample preparation and measurement; they should be analyzed to validate the sensitivity and accuracy of the analysis.

9.0 *Method performance.*

9.1 No data provided.

10.0 *References.*

10.1 None required.

TOXICITY CHARACTERISTIC LEACHING PROCEDURE (TCLP)

1.0 *Scope and application.*

1.1 The TCLP is designed to determine the mobility of both organic and inorganic contaminants present in liquid, solid, and multiphasic wastes.

1.2 If a total analysis of the waste demonstrates that individual contaminants are not present in the waste, or that they are present, but at such low concentrations that the appropriate regulatory thresholds could not possibly be exceeded, the TCLP need not be run.

2.0 *Summary of method.*

2.1 For wastes containing less than 0.5 percent solids, the waste, after filtration through a 0.6–0.8-μm glass fiber filter, is defined as the TCLP extract.

2.2 For wastes containing greater than 0.5 percent solids, the liquid phase, if any, is separated from the solid phase and stored for later analysis. The particle size of the solid phase is reduced (if necessary), weighed, and extracted with an amount of extraction fluid equal to 20 times the weight of the solid phase. The extraction fluid employed is a function of the alkalinity of the solid phase of the waste. A special extractor vessel is used when testing for volatiles (See Table 1). Following extraction, the liquid extract is separated from the solid phase by 0.6–0.8-μm glass fiber filter filtration.

2.3 If compatible (e.g., precipitate or multiple phases will not form on combination), the initial liquid phase of the waste is added to the liquid extract and these liquids are analyzed together. If incompatible, the liquids are analyzed separately and the results are mathematically combined to yield volume weighted average concentration.

3.0 *Interferences.*

3.1 Potential interferences that may be encountered during analysis are discussed in the individual analytical methods.

4.0 *Apparatus and materials.*

4.1 Agitation Apparatus: An acceptable agitation apparatus is one which is capable of rotating the extraction vessel in an end-over-end fashion at 30 ± 2 rpm. Suitable devices known to the EPA are identified in Table 2.

TABLE 1 Volatile Contaminants[a]

Compound	CASNO
Acetone	67-64-1
Acrylonitrile	107-13-1
Benzene	71-43-2
n-Butyl alcohol	71-36-6
Carbon disulfide	75-15-0
Carbon tetrachloride	56-23-5
Chlorobenzene	108-90-7
Chloroform	67-66-3
1,2-Dichloroethane	107-06-2
1,1-Dichloroethylene	75-35-4
Ethyl acetate	141-78-6
Ethyl benzene	100-41-4
Ethyl ether	60-29-7
Isobutanol	78-83-1
Methanol	67-56-1
Methylene chloride	75-09-2
Methyl ethyl ketone	78-93-3
Methyl isobutyl ketone	108-10-1
1,1,1,2-Tetrachloroethane	630-20-6
1,1,2,2-Tetrachloroethane	79-34-5
Tetrachloroethylene	127-18-4
Toluene	108-88-3
1,1,1-Trichloroethane	71-55-6
1,1,2-Trichloroethane	79-00-5
Trichloroethylene	79-01-6
Trichlorofluoromethane	75-69-4
1,1,2-Trichloro-1,2,2-trifluoroethane	76-13-1
Vinyl chloride	75-01-4
Xylene	1330-20-7

[a] Includes compounds identified in both the Land Disposal Restrictions Rule and the Toxicity Characteristic.

TABLE 2   Suitable Rotary Agitation Apparatus[a]

Company	Location	Model
Associated Design and Manufacturing Co.	Alexandria, Virginia (703) 549-5999	4-vessel device
		6-vessel device
Lars Lande Manufacturing	Whitmore Lake, Michigan (313) 449-4116	10-vessel device
IRA Machine Shop and Laboratory	Santurce, Puerto Rico (809) 752-4004	16-vessel device
EPRI Extractor		6-vessel device[b]

[a] Any device that rotates the extraction vessel in an end-over-end fashion at 30 ± 2 rpm is acceptable.

[b] Although this device is suitable, it is not commercially made. It may also require retrofitting to accommodate ZHE devices.

4.2 Extraction vessel:

   4.2.1 Zero-Headspace Extraction Vessel (ZHE). When the waste is being tested for mobility of any volatile contaminants (See Table 1), an extraction vessel which allows for liquid/solid separation within the device, and which effectively precludes headspace, is used. This type of vessel allows for initial liquid/solid separation, extraction, and final extract filtration without having to open the vessel (See Section 4.3.1). These vessels shall have an internal volume of 500 to 600 ml and be equipped to accommodate a 90-mm filter. Suitable ZHE devices known to EPA are identified in Table 3. These devices contain viton O-rings which should be replaced frequently.

   4.2.2 When the waste is being evaluated for other than volatile contaminants, an extraction vessel which does not preclude headspace (e.g., 2-l bottle) is used. Suitable extraction vessels include bottles made from various materials, depending on the contaminants to be analyzed and the nature of the waste (See Section 4.3.3). These bottles are available from a number of laboratory suppliers. When this type of extraction vessel is used, the filtration device discussed in Section 4.3.2 is used for initial liquid–solid separation and final extract filtration.

4.3 Filtration devices:

   4.3.1 Zero-Headspace Extractor Vessel: When the waste is being evaluated for volatiles, the zero-headspace extraction vessel is used for filtration. The device shall be capable of supporting and keeping in place the glass fiber

TABLE 3   Suitable Zero-Headspace Extractor Vessels

Company	Location	Model Number
Associated Design and Manufacturing Co.	Alexandria, Virginia (703) 549-5999	3740-ZHB
Millipore Corp.	Bedford, Massachusetts (800) 225-3384	SD1P581C5

filter, and be able to withstand the pressure needed to accomplish separation (50 psi).

Note.—When it is suspected that the glass fiber filter has been ruptured, an in-line glass fiber filter may be used to filter the extract.

4.3.2 Filter Holder. When the waste is being evaluated for other than volatile compounds, a filter holder capable of supporting a glass fiber filter and able to withstand the pressure needed to accomplish separation is used. Suitable filter holders range from simple vacuum units to relatively complex systems capable of exerting pressure up to 50 psi and more. The type of filter holder used depends on the properties of the material to be filtered (See Section 4.3.3). These devices shall have a minimum internal volume of 300 ml and be equipped to accommodate a minimum filter size of 47 mm. Filter holders known to EPA to be suitable for use are shown in Table 4.

4.3.3 Materials of Construction: Extraction vessels and filtration devices shall be made of inert materials which will not leach or absorb waste components. Glass, polytetrafluoroethylene (PTFE), or type 316 stainless steel equipment may be used when evaluating the mobility of both organic and inorganic components. Devices made of high-density polyethylene (HDPE), polypropylene, or polyvinyl chloride may be used when evaluating the mobility of metals.

4.4 Filters: Filters shall be made of borosilicate glass fiber, contain no binder materials, and have an effective pore size of 0.6–0.8 μm, or equivalent. Filters known to EPA to meet these specifications are identified in Table 5. Prefilters must not be used. When evaluating the mobility of metals, filters shall be acid washed prior to use by rinsing with 1.0 N nitric acid followed by three consecutive rinses with deionized distilled water (minimum of 500 ml per rinse). Glass fiber filters are fragile and should be handled with care.

4.5 pH Meters: Any of the commonly available pH meters are acceptable.

4.6 ZHE extract collection devices: TEDLAR® bags or glass, stainless steel, or PTFE gas-tight syringes are used to collect the initial liquid phase and the final extract of the waste when using the ZHE device.

TABLE 4 Suitable Filter Holders[a]

Company	Location	Model	Size (mm)
Nuclepore Corp.	Pleasanton, California (800) 882-7711	425910	142
		410400	47
Micro Filtration Systems.	Dublin, California (415) 828-6010	302400	142
Millipore Corp.	Bedford, Massachusetts (800) 225-3384	YT30142HW	142
		XX1004700	47

[a] Any device capable of separating the liquid from the solid phase of the waste is suitable, providing that it is chemically compatible with the waste and the constituents to be analyzed. Plastic devices (not listed above) may be used when only inorganic contaminants are of concern.

TABLE 5 Suitable Filter Media

Company	Location	Model	Pore Size[a]
Whatman Laboratory Products, Inc.	Clifton, New Jersey (201) 773-5800	GFF	0.7

[a] Nominal pore size.

4.7 ZHE extraction fluid collection devices: Any device capable of transferring the extraction fluid into the ZHE without changing the nature of the extraction fluid is acceptable (e.g., a constant displacement pump, a gas-tight syringe, pressure filtration unit (See Section 4.3.2), or another ZHE device).

4.8 Laboratory Balance: Any laboratory balance accurate to within ±0.01 grams may be used (all weight measures are to be within ±0.1 grams).

5.0 *Reagents.*

5.1 Water: ASTM Type 1 deionized, carbon treated, decarbonized, filtered water (or equivalent water that is treated to remove volatile components) shall be used when evaluating wastes for volatile contaminants. Otherwise, ASTM Type 2 deionized distilled water (or equivalent) is used. These waters should be monitored periodically for impurities.

5.2 1.0 N Hydrochloric acid (HCl) made from ASC Reagent grade.

5.3 1.0 N Nitric acid (HNO$_3$) made from ACS Reagent grade.

5.4 1.0 N Sodium hydroxide (NaOH) made from ACS Reagent grade.

5.5 Glacial acetic acid (HOAc) made from ACS Reagent grade.

5.6 Extraction fluid:

   5.6.1 Extraction fluid #1: This fluid is made by adding 5.7 ml glacial HOAc to 500 ml of the appropriate water (See Section 5.1), adding 64.3 ml of 1.0 N NaOH, and diluting to a volume of 1 liter. When correctly prepared, the pH of this fluid will be 4.93 ± 0.05.

   5.6.2 Extraction fluid #2: This fluid is made by diluting 5.7 ml glacial HOAc with ASTM Type 2 water (See Section 5.1) to a volume of 1 liter. When correctly prepared, the pH of this fluid will be 2.88 ± 0.05.

   Note.—These extraction fluids shall be made up fresh daily. The pH should be checked prior to use to insure that they are made up accurately, and these fluids should be monitored frequently for impurities.

5.7 Analytical standards shall be prepared according to the appropriate analytical method.

6.0 *Sample Collection, preservation, and handling.*

6.1 All samples shall be collected using a sampling plan that addresses the consideration discussed in "Test Methods for Evaluating Solid Wastes" (SW-846).

6.2 Preservatives shall not be added to samples.

6.3 Samples can be refrigerated unless it results in irreversible physical changes to the waste.

6.4 When the waste is to be evaluated for volatile contaminants, care must be taken to insure that these are not lost. Samples shall be taken and stored in a manner which prevents the loss of volatile contaminants. If possible, any necessary particle size reduction should be conducted as the sample is being taken (See Step 8.5). Refer to SW-846 for additional sampling and storage requirements when volatiles are contaminants of concern.

6.5 TCLP extracts should be prepared for analysis and analyzed as soon as possible following extraction. If they need to be stored, even for a short period of time, storage shall be at 4°C and samples for volatiles analysis shall not be allowed to come into contact with the atmosphere (i.e., no headspace).

7.0 *Procedure when volatiles are not involved.*
Although a minimum sample size of 100 g is required, a larger sample size may be necessary, depending on the percent solids of the waste sample. Enough waste sample should be collected such that at least 75 g of the solid phase of the waste (as determined using glass fiber filter filtration) is extracted. This will ensure that there is adequate extract for the required analyses (e.g., semivolatiles, metals, pesticides, and herbicides).

The determination of which extraction fluid to use (See Step 7.12) may also be conducted at the start of this procedure. This determination shall be on the solid phase of the waste (as obtained using glass fiber filter filtration).

7.1 If the waste will obviously yield no free liquid when subjected to pressure filtration, weigh out a representative subsample of the waste (100-g minimum) and proceed to Step 7.11.

7.2 If the sample is liquid or multiphasic, liquid/solid separation is required. This involves the filtration device discussed in Section 4.3.2, and is outlined in Steps 7.3 to 7.9.

7.3 Preweigh the filter and the container which will receive the filtrate.

7.4 Assemble filter holder and filter following the manufacturer's instructions. Place the filter on the support screen and secure. Acid wash the filter if evaluating the mobility of metals (See Section 4.4).

7.5 Weigh out a representative subsample of the waste (100-g minimum) and record weight.

7.6 Allow slurries to stand to permit the solid phase to settle. Wastes that settle slowly may be centrifuged prior to filtration.

7.7 Transfer the waste sample to the filter holder.

Note.—If waste material has obviously adhered to the container used to transfer the sample to the filtration apparatus, determine the weight of this residue and subtract it from the sample weight determined in Step 7.5, to determine the weight of the waste sample which will be filtered. Gradually apply vacuum or gentle pressure of 1–10 psi, until air or pressurizing gas moves through the filter. If this point is not reached under 10 psi, and if additional liquid has passed through the filter in any 2-min interval, slowly increase the pressure in 10-psi increments to a maximum of 50 psi. After each incremental increase of 10 psi, if the pressurizing gas has not moved through the filter, and if no additional liquid has passed through the filter in any 2-min interval, proceed to the next 10-psi increment. When the pressurizing gas begins to move through the filter, or when liquid flow has ceased at 50 psi (i.e., does not result in any additional filtrate within any 2-min period), filtration is stopped.

Note.—Instantaneous application of high pressure can degrade the glass fiber filter, and may cause premature plugging.

7.8 The material in the filter holder is defined as the solid phase of the waste, and the filtrate is defined as the liquid phase.

Note.—Some wastes, such as oily wastes and some paint wastes, will obviously contain some material which appears to be a liquid—but even after applying vacuum or pressure filtration, as outlined in Step 7.7, this material may not filter. If this is the case, the material within the filtration device is defined as a solid, and is carried through the extraction as a solid.

7.9 Determine the weight of the liquid phase by subtracting the weight of the filtrate container (See Step 7.3) from the total weight of the filtrate-filled container. The liquid phase may now be either analyzed (See Step 7.15) or stored at 4°C until time of analysis. The weight of the solid phase of the waste sample is determined by subtracting the weight of the liquid phase from the weight of the total waste sample, as determined in Step 7.5 or 7.7. Record the weight of the liquid and solid phases.

Note.—If the weight of the solid phase of the waste is less than 75 g, review Step 7.0.

7.10 The sample will be handled differently from this point, depending on whether it contains more or less than 0.5 percent solids. If the sample obviously has greater than 0.5 percent solids go to Step 7.11. If it appears that the solid may comprise less than 0.5 percent of the total waste, the percent solids will be determined as follows:

7.10.1 Remove the solid phase and filter from the filtration apparatus.

7.10.2 Dry the filter and solid phase at $100 \pm 20°C$ until two successive weighings yield the same value. Record final weight.

7.10.3 Calculate the percent solids as follows:
Weight of dry waste and filters minus tared weight of filters divided by initial weight of waste (Step 7.5 or 7.7) multiplied by 100 equals percent solids.

7.10.4 If the solid comprises less than 0.5 percent of the waste, the solid is discarded and the liquid phase is defined as the TCLP extract. Proceed to Step 7.14.

7.10.5 If the solid is greater than or equal to 0.5 percent of the waste, return to Step 7.1, and begin the procedure with a new sample of waste. Do not extract the solid that has been dried.

Note.—This step is only used to determine whether the solid must be extracted, or whether it may be discarded unextracted. It is not used in calculating the amount of extraction fluid to use in extracting the waste, nor is the dried solid derived from this step subjected to extraction. A new sample will have to be prepared for extraction.

7.11 If the sample has more than 0.5 percent solids, it is now evaluated for particle size. If the solid material has a surface area per gram of material equal to or greater than 3.1 cm^2, or is capable of passing through a 9.5-mm (0.375-in.) standard sieve, proceed to Step 7.12. If the surface area is smaller or the particle size is larger than that described above, the solid material is prepared for extraction by crushing, cutting, or grinding the solid material to a surface area or particle size as described above. When surface area or particle size has been appropriately altered, proceed to Step 7.12.

7.12 This step describes the determination of the appropriate extracting fluid to use (See Sections 5.0 and 7.0).

   7.12.1 Weigh out a small subsample of the solid phase of the waste, reduce the solid (if necessary) to a particle size of approximately 1 mm in diameter or less, and transfer a 5.0-g portion to a 500-ml beaker or erlenmeyer flask.

   7.12.2 Add 96.5 ml distilled deionized water (ASTM Type 2), cover with watchglass, and stir vigorously for 5 min using a magnetic stirrer. Measure and record the pH. If the pH is ≤ 5.0, extraction fluid #1 is used. Proceed to Step 7.13.

   7.12.3 If the PH from Step 7.12.2 is > 5.0, add 3.5 ml 1.0 N HCl, slurry for 30 s, cover with watchglass, heat to 50°C, and hold for 10 min.

   7.12.4 Let the solution cool to room temperature and record pH. If pH is ≤ 5.0, use extraction fluid #1. If the pH is >5.0, extraction fluid #2 is used.

7.13 Calculate the weight of the remaining solid material by subtracting the weight of the subsample taken for Step 7.12, from the original amount of solid material, as obtained from Step 7.1 or 7.9. Transfer remaining solid material into the extractor vessel, including the filter used to separate the initial liquid from the solid phase.

Note.—If any of the solid phase remains adhered to the walls of the filter holder, or the container used to transfer the waste, its weight shall be determined, subtracted from the weight of the solid phase of the waste, as determined above, and this weight is used in calculating the amount of extraction fluid to add into the extractor bottle.

Slowly add an amount of the appropriate extraction fluid (See Step 7.12) into the extractor bottle equal to 20 times the weight of the solid phase that has been placed into the extractor bottle. Close extractor bottle tightly, secure in rotary extractor device and rotate at 30 ± 2 rpm for 18 h. The temperature shall be maintained at 22 ± 3 °C during the extraction period.

Note.—As agitation continues, pressure may build up within the extractor bottle (Due to the evolution of gasses such as carbon dioxide). To relieve these pressures, the extractor bottle may be periodically opened and vented into a hood.

7.14 Following the 18-h extraction, the material in the extractor vessel is separated into its component liquid and solid phases by filtering through a new glass fiber filter as outlined in Step 7.7. This new filter shall be acid washed (See Section 4.4) if evaluating the mobility of metals.

7.15 The TCLP extract is now prepared as follows:

7.15.1 If the waste contained no initial liquid phase, the filtered liquid material obtained from Step 7.14 is defined as the TCLP extract. Proceed to Step 7.16.

7.15.2 If compatible (e.g., will not form precipitate or multiple phases), the filtered liquid resulting from Step 7.14 is combined with the initial liquid phase of the waste as obtained in Step 7.9. This combined liquid is defined as the TCLP extract. Proceed to Step 7.16.

7.15.3 If the initial liquid phase of the waste, as obtained from Step 7.9, is not or may not be compatible with the filtered liquid resulting from Step 7.14, these liquids are not combined. These liquids are collectively defined as the TCLP extract, are analyzed separately, and the results are combined mathematically. Proceed to Step 7.16.

7.16 The TCLP extract will be prepared and analyzed according to the appropriate SW-846 analytical methods identified in Appendix III of 40 CFR 261. TCLP extracts to be analyzed for metals shall be acid digested. If the individual phases are to be analyzed separately, determine the volume of the individual phases (to 0.1 ml), conduct the appropriate analyses, and combine the results mathematically by using a simple weighted average:

$$\text{Final contaminant concentration} = \frac{(V_1)(C_1) + (V_2)(C_2)}{V_1 + V_2}$$

where

V_1 = The volume of the first phase (1)
C_1 = The concentration of the contaminant of concern in the first phase (mg/l)
V_2 = The volume of the second phase (1)
C_2 = The concentration of the contaminant of concern in the second phase (mg/l)

7.17 The contaminant concentrations in the TCLP extract are compared to the thresholds identified in the appropriate regulations. Refer to Section 9 for quality assurance requirements.

8.0 *Procedure when volatiles are involved.*

The ZHE device has approximately a 500-ml internal capacity. Although a minimum sample size of 100 g was required in the Section 7 procedure, the ZHE can only accommodate a maximum 100 percent solids sample of 25 g, due to the need to add an amount of extraction fluid equal to 20 times the weight

TEST METHODS 647

of the solid phase. Step 8.4 provides the means of which to determine the approximate sample size for the ZHE device.

Although the following procedure allows for particle size reduction during the conduct of the procedure, this could result in the loss of volatile compounds. If possible, any necessary particle size reduction (See Step 8.5) should be conducted on the sample as it is being taken. Particle size reduction should only be conducted during the procedure if there is no other choice.

In carrying out the following steps, do not allow the waste to be exposed to the atmosphere for any more time than is absolutely necessary.

8.1 Preweigh the (evacuated) container which will receive the filtrate (See Section 4.6), and set aside.

8.2 Place the ZHE piston within the body of the ZHE (it may be helpful to first moisten the piston O-rings slightly with extraction fluid). Secure the gas inlet/outlet flange (bottom flange) onto the ZHE body in accordance with the manufacturer's instructions. Secure the glass fiber filter between the support screens and set aside. Set liquid inlet/outlet flange (top flange) aside.

8.3 If the waste will obviously yield no free liquid when subjected to pressure filtration, weigh out a representative subsample of the waste (25-g maximum—See Step 8.0), record weight, and proceed to 8.5.

8.4 This step provides the means by which to determine the approximate sample size for the ZHE device. If the waste is liquid or multiphasic, follow the procedure outlined in Steps 7.2 to 7.9 (using the Section 7 filtration apparatus), and obtain the percent solids by dividing the weight of the solid phase of the waste by the original sample size used. If the waste obviously contains greater than 0.5 percent solids, go to Step 8.4.2. If it appears that the solid may comprise less than 0.5 percent of the waste, go to Step 8.4.1.

   8.4.1 Determine the percent solids by using the procedure outlined in Step 7.10. If the waste contains less than 0.5 percent solids, weigh out a new 100-g minimum representative sample, proceed to Step 8.7, and follow until the liquid phase of the waste is filtered using the ZHE device (Step 8.8). This liquid filtrate is defined as the TCLP extract, and is analyzed directly. If the waste contains greater than or equal to 0.5 percent solids, repeat Step 8.4 using a new 100-g minimum sample, determine the percent solids, and proceed to Step 8.4.2.

   8.4.2 If the sample is ≤25 percent solids, weigh out a new 100-g minimum representative sample, and proceed to Step 8.5. If the sample is > 25 percent solids, the maximum amount of sample the ZHE can accommodate is determined by dividing 25 g by the percent solids obtained from Step 8.4. Weigh out a new representative sample of the determined size.

8.5 After a representative sample of the waste (sample size determined from Step 8.4) has been weighed out and recorded, the sample is now evaluated for particle size (See Step 8.0). If the solid material within the waste obviously has a surface area per gram of material equal to or greater than 3.1 cm^2, or is capable of passing through a 9.5-mm (0.375-in.) standard sieve, proceed immediately to Step 8.6. If the surface area is smaller or the particle size is larger than that described above, the solid material which does not meet the above criteria is separated from the liquid phase by sieving (or equivalent means), and the solid is prepared for extraction by crushing, cutting, or grinding to a surface area or particle size as described above.

Note.—Wastes and appropriate equipment should be refrigerated, if possible, to 4°C prior to particle size reduction. Grinding and milling machinery which generates heat shall not be used for particle size reduction. If reduction to the solid phase of the waste is necessary, exposure of the waste to the atmosphere should be avoided to the extent possible.

When surface area or particle size has been appropriately altered, the solid is recombined with the rest of the waste.

8.6 Waste slurries need not be allowed to stand to permit the solid phase to settle. Wastes that settle slowly shall not be centrifuged prior to filtration.

8.7 Transfer the entire sample (liquid and solid phases) quickly to the ZHE. Secure the filter and support screens into the top flange of the device and secure the top flange to the ZHE body in accordance with the manufacturer's instructions. Tighten all ZHE fittings and place the device in the vertical position (gas inlet/outlet flange on the bottom). Do not attach the extract collection device to the top plate.

Note.—If waste material has obviously adhered to the container used to transfer the sample to the ZHE, determine the weight of this residue and subtract it from the sample weight determined in Step 8.4, to determine the weight of the waste sample which will be filtered.

Attach a gas line to the gas inlet/outlet valve (bottom flange), and with the liquid inlet/outlet valve (top flange) open, begin applying gentle pressure of 1–10 psi (or more if necessary) to slowly force all headspace out of the ZHE device. At the first appearance of liquid from the liquid inlet/outlet valve, quickly close the valve and discontinue pressure.

8.8 Attach evacuated preweighed filtrate collection container to the liquid inlet/outlet valve and open valve. Begin applying gentle pressure of 1–10 psi to force the liquid phase into the filtrate collection container. If no additional liquid has passed through the filter in any 2-min interval, slowly increase the pressure in 10 psi increments to a maximum of 50 psi. After each incremental increase of 10 psi, if no additional liquid has passed through the filter in any 2-min interval, proceed to the next 10 psi increment. When liquid flow has ceased such that continued pressure filtration at 50 psi does not result in any additional filtrate within any 2-min period, filtration is stopped. Close the liquid inlet/outlet valve, discontinue pressure to the piston, and disconnect the filtrate collection container.

Note.—Instantaneous application of high pressure can degrade the glass fiber filter and may cause premature plugging.

8.9 The material in the ZHE is defined as the solid phase of the waste, and the filtrate is defined as the liquid phase.

Note.—Some wastes, such as oily wastes and some paint wastes, will obviously contain some material which appears to be a liquid—but even after applying pressure filtration, this material will not filter. If this is the case, the material within the filtration device is defined as a solid, and is carried through the TCLP extraction as a solid.

If the original waste contained less than 0.5 percent solids (See Step 8.4), this filtrate is defined as the TCLP extract, and is analyzed directly—proceed to Step 8.13.

8.10 Determine the weight of the liquid phase by subtracting the weight of the filtrate container (See Step 8.1) from the total weight of the filtrate-filled container. The liquid phase may now be either analyzed (See Steps 8.13 and 8.14), or stored at 4°C until time of analysis. The weight of the solid phase of the waste sample is determined by subtracting the weight of the liquid phase from the weight of the total waste sample (See Step 8.4). Record the final weight of the liquid and solid phases.

8.11 The following details how to add the appropriate amount of extraction fluid to the solid material within the ZHE and agitation of the ZHE vessel. Extraction fluid #1 is used in all cases (See Section 5.6).

    8.11.1 With the ZHE in the vertical position, attach a line from the extraction fluid reservoir to the liquid inlet/outlet valve. The line used shall contain fresh extraction fluid and should be preflushed with fluid to eliminate any air pockets in the line. Release gas pressure on the ZHE piston (from the gas inlet/outlet valve), open the liquid inlet/outlet valve, and begin transferring extraction fluid (by pumping or similar means) into the ZHE. Continue pumping extraction fluid into the ZHE until the amount of fluid introduced into the device equals 20 times the weight of the solid phase of the waste that is in the ZHE.

    8.11.2 After the extraction fluid has been added, immediately close the liquid inlet/outlet valve, and disconnect the extraction fluid line. Check the ZHE to make sure that all valves are in their closed positions. Pick up the ZHE and physically rotate the device in an end-over-end fashion 2 or 3 times. Reposition the ZHE in the vertical position with the liquid inlet/outlet valve on top. Put 5–10 psi behind the piston (if necessary), and slowly open the liquid inlet/outlet valve to bleed out any headspace (into a hood) that may have been introduced due to the addition of extraction fluid. This bleeding shall be done quickly and shall be stopped at the first appearance of liquid from the valve. Repressurize the ZHE with 5–10 psi and check all ZHE fittings to insure that they are closed.

    8.11.3 Place the ZHE in the rotary extractor apparatus (if it is not already there), and rotate the ZHE at 30 + 2 rpm for 18 h. The temperature shall be maintained at 22 ± + 3°C during agitation.

8.12 Following the 18-hr extraction, check the pressure behind the ZHE piston by quickly opening and closing the gas inlet/outlet valve, and noting the escape of gas. If the pressure has not been maintained (i.e., no gas release observed), the device is leaking. Replace ZHE O-rings or other fittings, as necessary, and redo the extraction with a new sample of waste. If the pressure within the device has been maintained, the material in the extractor vessel is once again separated into its component liquid and solid phases. If the waste contained an initial liquid phase, the liquid may be filtered directly into the same filtrate collection container (i.e., TEDLAR bag, gas-tight syringe) holding the initial liquid phase of the waste, unless doing so would create multiple phases, or unless there is not enough volume left within the filtrate collection container. A separate filtrate collection container must be used in these cases. Filter through the glass fiber filter, using the ZHE device as discussed in Step 8.8. All extract shall be filtered and collected if the extract is multiphasic or if the waste contained an initial liquid phase.

Note.—If the glass fiber filter is not intact following agitation, the filtration device discussed in the Note in Section 4.3.1 may be used to filter the material within the ZHE.

8.13 If the waste contained no initial liquid phase, the filtered liquid material obtained from Step 8.12 is defined as the TCLP extract. If the waste contained an initial liquid phase, the filtered liquid material obtained from Step 8.12, and the initial liquid phase (Step 8.8) are collectively defined as the TCLP extract.

8.14 The TCLP extract will be prepared and analyzed according to the appropriate SW-846 analytical methods, as identified in Appendix III of 40 CFR 261. If the individual phases are to be analyzed separately, determine the volume of the individual phases (to 0.1 ml), conduct the appropriate analyses and combine the results mathematically by using a simple volume weighted average:

$$\text{Final contaminant concentration} = \frac{(V_1)(C_1) + (V_2)(C_2)}{V_1 + V_2}$$

where

V_1 = The volume of the first phase (1)
C_1 = The concentration of the contaminant of concern in the first phase (mg/l)
V_2 = The volume of the second phase (1)
C_2 = The concentration of the contaminant of concern in the second phase (mg/l)

8.15 The contaminant concentrations in the TCLP extract are compared to the thresholds identified in the appropriate regulations. Refer to Section 9 for quality assurance requirements.

9.0 *Quality assurance requirements.*

9.1 All data, including quality assurance data, should be maintained and available for reference or inspection.

9.2 A minimum of one blank for every 10 extractions that have been conducted in an extraction vessel should be employed as a check to determine if any memory effects from the extraction equipment is occurring. One blank shall also be employed for every new batch of leaching fluid that is made up.

9.3 All quality control measures described in the appropriate analytical methods shall be followed.

9.4 The method of standard addition shall be employed for each waste type if: (1) recovery of the compound from spiked splits of the TCLP extract is not between 50 and 150 percent, or (2) if the concentration of the constituent measured in the extract is within 20 percent of the appropriate regulatory threshold. If more than one extraction is being run on samples of the same waste, the method of standard addition need only be applied once and the percent recoveries applied to the remainder of the extractions.

9.5 TCLP extracts shall be analyzed within the following periods after generation: volatiles—14 days, semivolatiles—10 days, mercury—28 days, and other metals—180 days.

STATE OF CALIFORNIA WASTE EXTRACTION TEST (WET)

(a) The WET described in this section shall be used to determine the amount of extractable substance in a waste or other material as set forth in Section 66699(a).

(b) Except as provided in Section 66700(d), the WET shall be carried out if the total concentration in the waste, or other material, of any substance listed in Section 66699 equals or exceeds the STLC value, but does not exceed the TTLC value, given for that substance. The total concentrations of substances listed in Section 66699 shall be determined by analysis of samples of wastes, or other materials, which have been prepared, or meet the conditions, for analysis as set forth in subsections (c) and (d) of this section. Methods used for analysis for total concentrations of substances listed in Section 66699 shall be those given in the following documents or alternate methods that have been approved by the Department pursuant to Section 66310(e):

(1) For metal elements and their compounds, the waste shall be digested according to the indicated methods described in "Test Methods for Evaluating Solid Waste, Physical/Chemical Methods," SW-846, 2nd edition, U.S. Environmental Protection Agency, 1982:

(A) All listed metal elements and their compounds, except hexavalent chromium: Method 3050.

(B) Hexavalent chromium: Method 3060.

(2) For the following substances, the indicated methods as described in "Test Methods for Evaluating Solid Waste, Physical/Chemical Methods," SW-846, 2nd edition, U.S. Environmental Protection Agency, 1982, shall be utilized:

(A) Antimony: Method 7040 or Method 7041.
(B) Arsenic: Method 7060 or Method 7061.
(C) Barium: Method 7080 or Method 7081.
(D) Cadmium: Method 7131.
(E) Total chromium: Method 7190.
(F) Hexavalent chromium: Method 7195, Method 7196, or Method 7197.
(G) Lead: Method 7421.
(H) Mercury: Method 7470 or Method 7471.
(I) Nickel: Method 7520 or Method 7521.
(J) Selenium: Method 7740 or Method 7741.
(K) Silver: Method 7760 or Method 7761.

(L) Trichloroethylene: Method 8010 or Method 8240.

(M) Pentachlorophenol: Method 8040, Method 8250, or Method 8270.

(N) Aldrin, Lindane, Chlordane, DDD, DDE, DDT, Dieldrin, Heptachlor, Toxaphene, and PCBS: Method 8080, Method 8250, or Method 8270.

(O) 2,4-Dichlorophenoxyacetic acid and 2,4,5-trichlorophenoxypropionic acid; Method 8150.

(3) For the following substances, the indicated methods as described in "Methods for Chemical Analysis of Water and Wastes," EPA-600/4-79-020, U.S. Environmental Protection Agency, 1979, shall be utilized:

(A) Beryllium: Method 210.1 or Method 210.2.

(B) Cobalt: Method 219.1 or Method 219.2.

(C) Copper: Method 220.1 or Method 220.2.

(D) Molybdenum: Method 246.1 or Method 246.2.

(E) Thallium: Method 279.1 or Method 279.2.

(F) Vanadium: Method 286.1 or Method 286.2.

(G) Zinc: Method 289.1 or Method 289.2.

(H) Fluoride: Method 340.1, Method 340.2, or Method 340.3.

(4) For the following substances, the indicated methods as described in "Manual of Analytical Methods of the Analysis of Pesticides in Humans and Environmental Samples," EPA-600/8-80-038, U.S. Environmental Protection Agency, 1980, shall be utilized:

(A) Kepone: Section 5,A,(5),(a).

(B) 2,3,7,8-Tetrachlorodibenzo-*p*-dioxin: Section 9,G.

(5) For asbestos, the indicated method as described in the Federal Register, Volume 47, Number 103, Appendix A, pages 23376-23389, May 7, 1982 shall be utilized.

(c) Samples shall be prepared for analysis for total and extractable content of substances listed in Section 66699 (b) and (c) as follows:

(1) Type i: If the waste or other material is a millable solid, the sample shall be passed directly, or shall be milled to pass, through a No. 10 (two millimeter) standard sieve before it is analyzed. If the sample contains nonfriable solid particles which do not pass directly through a No. 10 sieve and which are extraneous and irrelevant as hazardous constituents to the waste or other material, they shall be removed to the extent feasible by mechanical means and discarded. Solids which remain in the waste or other material after removal of the aforesaid extraneous particles shall be milled to pass through a No. 10 sieve and shall then be combined and mixed well with the solids which passed through the sieve without milling. The reconstituted sample shall then be analyzed as prescribed in this section.

(2) Type ii: If the waste or other material is a filterable mixture of liquid and solids in which solids constitute five-tenths (0.5) percent by weight or greater of the sample, the liquid and solids shall be separated by filtration through a 0.45 micron membrane filter. The filtrate so obtained is to be designated as Initial Filtrate. Its volume is determined, and it is retained. The separated solids shall be sieved in a No. 10 sieve and any nonfriable extraneous particles of the kinds described in subsection (c) (1) which do not pass through the sieve shall be removed to the extent feasible by mechanical means and discarded. The solids which remain after removal of the extraneous particles shall be milled to pass through a No. 10 sieve and shall be recombined with solids which passed through the sieve without milling. This recombined solid material shall be extracted following the procedure in subsection

(f). A ratio of 10 milliliters of extraction solution per gram of solid shall be utilized with appropriate modifications for extraction vessel size. After completion of solids extraction, the filtered extractant is combined with Initial Filtrate mixed thoroughly and analyzed as described in subsection (f) (3).

(3) Type iii: If the waste or other material is a nonfilterable and nonmillable sludge, slurry, or oily, tarry, or resinous material, it shall be analyzed as received unless it contains nonfriable extraneous and irrelevant solid particles of the kinds described in paragraph (c) (1) of this section. If it contains such solid particles and they are of such size as not to pass through a No. 10 sieve, they shall be removed to the extent feasible by mechanical means and discarded. The remainder of the sample shall be analyzed as prescribed in this section.

(4) If it is necessary to dry a solid sample or the solids fraction of a sample before sieving, milling, or removal of extraneous solids, or if a sample is dried prior to analysis, all weight losses due to drying shall be determined, and these losses and the conditions of drying shall be reported.

(d) If the waste or other material is a liquid containing less than five-tenths (0.5) percent by weight of undissolved solids, it shall not be subject to the WET procedure, but shall be analyzed directly for the substances listed in Section 66699. The waste shall be classified as a hazardous waste if the total concentration in the waste of any substances listed in Section 66699 exceeds the TTLC value given for that substance. If, however, the total concentration is less than the TTLC but exceeds the STLC when expressed on a milligrams per liter basis, the waste or other material shall be filtered through a 0.45 micron membrane filter, the solids discarded and the filtrate shall be analyzed directly for the substances listed in Section 66699. The waste shall be classified as a hazardous waste if the concentration in the filtrate of any of the substances listed in Section 66699 exceeds the STLC value given for that substance.

(e) The WET extraction solution shall consist of 0.2 M sodium citrate at pH 5.0 ± 0.1, which is prepared by titrating an appropriate amount of analytical grade citric acid in deionized water with 4.0 N NaOH, except that the extraction solution for the determination of chromium (VI) shall consist of deionized water.

(f) The extraction procedure shall be as follows:

(1) Fifty grams of sample, or less if it is a type ii sample prepared pursuant to subsection (c) (2), obtained pursuant to subsection (c) or (d) of this section shall be placed in a clean polyethylene or glass container designated the Treatment, capable of physically withstanding the extraction procedure and which was rinsed previously with, in succession, an aqueous 1:1 ratio by volume nitric acid solution and deionized water. If the extract will be analyzed for any of the organic substances listed in Section 66699(c), a glass container shall be used. Furthermore, a container of the same size, shape and material shall be used for an extraction designated as the Blank, which shall be carried through the same procedure as the Treatment, but without addition of the sample.

(2) Five hundred milliliters of extraction solution, or less if the waste sample is a type ii sample prepared pursuant to subsection (c) (2) shall be added to the Treatment and Blank containers, which shall be then fitted with covered air scrubbers extended well into the extraction solutions and flushed vigorously with nitrogen gas for 15 minutes so as to remove and exclude atmospheric oxygen from the extraction medium. If the sample is to be analyzed for any volatile substance, such as trichloroethylene, the sample shall be added after deaeration with nitrogen to avoid

volatilization loss. After deaeration the containers shall be quickly sealed with tightly fitting caps and agitated, using a table shaker, an overhead stirrer, or a rotary extractor, operated at a speed which shall maintain the sample in a state of vigorously agitated suspension. Required equipment is described in test method 1310 in "Test Methods for Evaluating Solid Waste, Physical/Chemical Methods," SW-846, 2nd edition, U.S. Environmental Protection Agency, 1982. The temperature during the extraction shall be maintained between 20 and 40 degrees centigrade. After 48 hours of extracting, the contents of the Treatment and Blank containers shall be either filtered directly or centrifuged and then filtered. Filtering shall be through a medium porosity prefilter and then through a 0.45 micron membrane filter, using a clean, thick-walled suction flask. For coarser solids, prefiltration shall not be necessary. Pressure filtration shall be an optional alternative to vacuum filtration. If the extracts are first centrifuged, glass or polyethylene bottles shall be used as prescribed for extraction. For very fine solids, centrifuging at as high as 10,000 × G may be necessary. After centrifugation, the liquids shall be decanted, prefiltered if necessary, and then passed through a 0.45 micron membrane filter. All filters shall be of low and identified extractable heavy metals, fluoride, and organic chemicals content.

(3) If the filtered extracts are to be analyzed only for the metal elements listed in Section 66699(b), the filtered extracts from the Treatment and Blank shall be transferred to clean polyethylene bottles and acidified with nitric acid to five percent by volume acid content soon after each extract is filtered. For those wastes or waste materials classified under subsection (c) (2), the Treatment shall be the Initial Filtrate combined with the extract generated by the WET extraction of the initially separated solids. Similarly the Blank in this instance shall be the filtrate generated by the WET Blank accompanying the initially separated solids, to which is subsequently added a volume of deionized water equivalent to that of the Initial Filtrate. These procedures shall be followed prior to acidification of Treatment and Blank solutions with nitric acid to five percent (by volume) acid content. The bottle shall then be stored at room temperature or frozen. If the extracts are also to be analyzed for the organic substances listed in Section 66699(c), or for the organic substances only, the filtered extracts shall be transferred to clean glass bottles. If the extracts are to be analyzed for fluoride, they shall be transferred to clean polyethylene bottles. These extracts, containing organic substances or fluoride, shall not be acidified, but shall be frozen soon after each extract is obtained and held frozen until the day of analysis, unless the extracts are analyzed within 24 hours.

(g) Sample analysis and data treatment shall be as follows:

(1) Each of the filtered extracts from the Treatment and Blank extractions shall have been acidified to five percent by volume nitric acid, and stored at room temperature or frozen in polyethylene bottles or kept frozen without addition of acid in glass bottles until the day of analysis, as prescribed. Each of the extracts shall be thoroughly mixed just prior to being individually analyzed for the substances listed in Section 66699 in order to determine whether the extractable concentration (EC) in the waste or other material exceeds the STLC for any of the substances listed. The extracts shall be analyzed according to the procedures identified in Sections 66700 (b) (2), (b) (3) and (b) (4).

(2) The net EC of a substance in the Treatment sample which is listed in Section 66699 shall be calculated and reported as milligrams per liter of sample (mg/l). This value is derived after subtracting the concentration of the substance in

the appropriate Blank extract from that concentration determined in the Treatment extract.

Note: Authority cited: Sections 208, 25141 and 25150, Health and Safety Code. Reference: Section 25141, Health and Safety Code.

HISTORY:

1. Editorial correction filed 10-5-84; designated effective 10-27-84 (Register 84, No. 41).

AMERICAN NUCLEAR SOCIETY
ANS 16.1 TEST PROCEDURE

Many different testing methods have been used in the past to determine the amount of radionuclide leached from a solidified waste form as a function of time. Various leachant compositions, leachant-renewal frequencies, and test conditions (e.g., temperatures and pressures) have been employed. These factors interact to influence the leach test results. In view of this, the leach test proposed here is purposely uncompromising. It specifies a defined leachant, a set leachant-renewal schedule, a fixed leachant temperature, and other specified test conditions. The procedure can be extended to other leachants and leachant-renewal frequencies (as well as other temperatures, etc.) so that it more nearly represents anticipated conditions under which a solidified waste form may be stored, transported, or disposed. Such more site- or situation-specific long-term, testing efforts may be necessary, but are not part of the Standard.

2.1 SPECIMEN PREPARATION

The method for preparing proper test specimens is specific for each waste type and solidification process. A specific procedure by which specimens, that meet the specifications given below, can be prepared shall be developed for each given type of solid under consideration.

Precautions shall be taken to insure that the specimen is representative of the solidified waste, and that the homogeneity of the test specimen is the same as that of the material in the actual solidified waste form. While a small specimen is desirable to limit the radiation field, it must not be so small as to compromise the specimen homogeneity, require unattainable analytical sensitivities, or provide substantial difficulties for specimen preparation. The test specimen shall be prepared in the same or similar manner as that established for the solidification process. If specimens are obtained by core drilling of waste forms (not recommended), the procedure must be described and the specimen must have surfaces representative of the actual waste form.

In addition to being representative of the solidified waste, the specimen must have a well-defined shape, mass, and volume. The surface condition of the specimen should be representative of the surface condition of the actual full-scale waste product. Where possible, the specimen shall be a monolithic cylinder, parallelepiped, or sphere, the dimensions of which are reported. Cylinders shall have a length-to-diameter ratio in the range of 0.2 to 5. Parallelepipeds shall have a length-to-minimum-thickness ratio in the range of 0.2 to 5. The preferred test specimen geometry is cylindrical. The minimum specimen dimension recommended is 1 cm, unless a need to employ a smaller specimen (e.g., to minimize personnel radiation exposure) is demonstrated.

The representative specimen of the product of a given solidification process shall be prepared, or cast, in such a way that the casting conforms to the sides of the specimen preparation container (to provide a smooth surface), voids within it are eliminated, and homogeneity (as uniform as the character of the material permits) is attained. For glass-like or thermosetting mixtures, this container should be heated to provide a thermal history representative of that which the actual solidified waste form undergoes. The thermal history of the test specimen is to be reported. Immediately after preparation the specimen is to be placed into the specimen container. This may or may not be the same as the specimen preparation container.

The specimen container shall remain sealed during the storage period between preparation and leach testing. The specimen container(s) shall be constructed of material(s) known to be chemically unreactive toward the specimen (e.g., polyethylene, polypropylene, stainless steel, ceramic, glass, etc.). No single container material appears to be superior for all solidified waste products.

2.2 LEACH TEST VESSEL

The vessel in which leaching takes place should be constructed of an "unreactive" material. A material is considered unreactive if:

(a) It does not react chemically with the leachant or the specimen.
(b) It does not sorb chemical species extracted from the specimen or those in the leachant itself. This refers specifically to the species of interest extracted from the specimen during leaching. It also applies, however, to those major species extracted during leaching, which influence the composition of the leachant. Sorption shall be determined by a blank test run.

A leach vessel made of a material which is sorbent toward the extracted species of interest may be used, provided either that the extent of sorption is small ($<5\%$ of the incremental fraction leached) or the sorbed species are removed from the container and analyzed at the same frequency as the leachate is sampled and replaced.
(c) It does not release interfering species which alter the composition of the leachant during the leaching process.
(d) It can withstand the conditions involved in leaching.

The leach test vessel shall be constructed such that excessive evaporation of the leachant ($>2\%$ over 24 hours) is prevented. The dimensions of the vessel shall be such that the entire external geometric surface area ($>98\%$) of the immersed specimen is in contact with the leachant. The dimensions of the leach test vessel shall also be sufficient to hold the leachant while leaving some free volume for convenience in manipulation of both specimen and leachant.

2.3 LEACHANT SOLUTION

In this test the leachate solution is sampled and entirely replaced at designated time intervals. The leachant solution shall be demineralized water with an electrical conductivity of less than 5 µmho/cm at 25°C and a total organic carbon content (TOC) of less than 3 ppm. The leachant solution shall be kept in, and not exceed, the limits of the temperature range 17.5 and 27.5°C during the test.

2.4 LEACH TEST METHOD

After removal from the specimen container and prior to the initiation of the leach test, the test specimen shall be rinsed by immersion in demineralized water for 30 s. The rinse water volume shall be the same as the required leachant volume (described below).

The container used to store the specimen before leaching shall then be rinsed with an amount of water equal to its volume, to recover any radioactivity present in residual liquid or retained on the container walls. This "container rinse" and the specimen rinse water ("wash off") are then to be combined and analyzed to determine the quantity of each radionuclide of interest, i, present. The latter are expressed as fractions $(a/A_o)_i$, defined as the ratio of activity of the radionuclide present in the combined rinse $(a_r)_i$ to the activity of the same radionuclide in the specimen $(A_o)_i$ at the time it is immersed in the first portion of the leachant (i.e., the beginning of the first leaching interval). This fraction is reported for each radionuclide of interest. All radioactivities measured are converted and reported as of a common reference time, for example, the beginning of the first leaching interval.

The specimen shall be supported in the leachant by any convenient device, made from unreactive material (as defined above), that does not interfere with the leachate removal and replacement, does not impede leaching, does not damage the surface of the specimen, and, as mentioned above, does not preclude more than a small fraction of the specimen's external surface ($<2\%$) from exposure to the leachant. Examples of suitable specimen supports include wires for suspension, rigid support stands, or coarse-weave wire-mesh baskets. The specimen shall be located within the leachant so that it is surrounded on all sides by a leachant layer exceeding in thickness 10 cm or the minimum specimen dimension, whichever is smaller.

Sufficient leachant shall be used so that the ratio:

$$\frac{\text{leachant volume (cm}^3\text{)}}{\text{specimen external geometric surface area (cm}^2\text{)}} = 10 \pm 0.2 \text{ (cm)}$$

is maintained during the leaching interval. For example, a 1-cm diameter, 1-cm long right circular cylinder (surface area 4.71 cm^2) would need 47.1 mL of leachant. A ratio of leachant volume to specimen-external-geometry surface area of 10 cm, is usually sufficient to minimize leachant-composition changes during reasonably short leaching intervals, while providing sufficient concentration of extracted species for analysis. The leachant is not stirred during the leaching interval.

At the end of each leaching interval the leachate is removed from the specimen. Quickly removing the specimen from the used leachant and placing it into fresh leachant in a new leachant test vessel is an acceptable and widely employed procedure. After leachate removal, the leach test vessel shall be rinsed with demineralized water to remove residual leachate and contained radioactivity. The specimen may also be momentarily rinsed (<5 s) in demineralized water after leachate removal, but before leachant renewal. The radioactivity in these rinses shall be measured and included with that of the leachate just removed. During leachant renewal the specimen should be exposed to the air for as short a time as reasonably achievable. In no case shall its surface be allowed to dry completely. The specimen shall then be contacted with fresh leachant solution for the specified time of the next leaching interval.

The rate of radioactivity release, and hence the calculated Leachability Index* value, can be a function of the leachant-renewal frequency. A standardized uniform leachant-renewal schedule is thus required. The leachate shall be sampled and the leachate completely replaced after cumulative leach times of 2, 7, and 24 hours from the initiation of the test. Subsequent leachate sampling and leachant replacements shall be made at 24-hour intervals for the next four days. Three additional leach intervals of 14, 28, and 43 days each extend the entire test to 90 (5 + 14 + 28 + 43) days. The leachant change required on Saturday (or Sunday) may be shifted to Friday (or Monday) for the last three renewal periods. The reasoning behind this is that with the protracted leachant renewal periods the boundary conditions of the model are not satisfied and a day under or a day over will not make any significant difference in the results after such extended time periods. The Standard requires that the leachant changeout time be short and insignificant relative to the duration of a leaching interval.

For purposes of quality assessment, a short-time *"abbreviated test"* is defined by the same sequence, etc.; but, without the last three leaching intervals. This "abbreviated test" requires but 5 days of testing a sample.

Note that in this Standard periods of time with the units of hours (or days) require that the words hours (or days) be spelled out since the letters ''h'' or ''d'' are used to represent other quantities.

2.5 LEACHATE ANALYSIS

An aliquot of the leachate shall be taken at the end of each leachate interval to determine, by a suitable method, the amounts $(a_n)_i$ of the species of interest present in the leachate volume. Any generally accepted state-of-the-art analytical procedure may be employed. It is not the intent of this Standard to specify an analytical method. The release of the species of interest will always be determined by measuring the quantity present in the leachate rather than the residual in the specimen.

Leachate aliquots taken for analysis shall be representative of the leachate from which they are taken. The intent is to provide aliquots which will permit a determination of all the radioactivity which has been removed from the test specimen during the leaching interval. This includes any radioactivity associated with particulate solids in the leachate. Stirring of the leachate so as to suspend particulate solids prior to taking the leachate aliquot, or dissolution of such solids by the addition of chemicals to the leachate before sampling, may be necessary. If precipitation occurs in the leachant during the leaching interval, the amount of the extracted species of interest associated with the precipitate shall be determined and added to the amount of the dissolved species. Under no circumstances shall the leachate be filtered and the filter media with the residue be discarded without analysis.

---

*See Chapter 17 for calculations associated with this method.

GLOSSARY

The following terms designated by an asterisk (*) are as defined in the U.S. Environmental Protection Agency (EPA), Rules and Regulations: Hazardous Waste Management System, which appeared in the Federal Register on May 19, 1980; Vol. 45, pages 33073-33076. All other terms are not officially defined by the U.S. EPA or any other organization, but only as they are used in this book.

Accelerated test A test that, based on accepted mechanistic principles, speeds up the testing process when compared to expected field conditions. Generates information as a function of time, on a compressed time scale.

Accuracy Exact conformity to truth related to the actual amount present.

Acid waste Generally an aqueous waste having a pH below 7.

Advection Transference by horizontal current.

Advection transport The movement of water under the influence of a hydraulic gradient. Transference by horizontal current.

Aliquot A known fractional part of a defined quantity.

Alkali waste A caustic, corrosive liquid waste, such as lye, with pH greater than or equal to 12.5.

Aqueous phase Water and any chemical constituent dissolved in water.

Aqueous waste A waste stream that is water-based.

Aquifer* A geologic formation, group of formations, or part of a formation capable of yielding a significant amount of groundwater to wells or springs.

Attenuation capacity (of the surrounding material) The ability of the environment of a waste or at a disposal site to retard movement of contaminants through the adjacent soil or groundwater.

Binder The main component of a solidification formulation, which binds or cements together the components of the waste into a solid mass. Examples are Portland cement, bitumen, and organic polymers.

Bioaccumulative A characteristic of a chemical species when the rate of intake into a living organism is greater than the rate of excretion, or metabolism. This results in an increase in tissue concentration relative to the exposure concentration.

Bioassay An assay method using changes in the biological activity of living organisms such as fish as a qualitative or quantitative measure of a material's effect on other living organisms.

Biochemical oxygen demand **1.** The quantity of oxygen used in the biochemical oxidation of organic matter in a specified time under specified conditions. **2.** A standard test used in assessing waste water.

Biodegradable The ability of a substance to be broken down physically and/or chemically by microorganisms.

Blanks Specimens made of similar material, but with the species of interest absent. A pure, unmingled, unaffected specimen lacking certain characteristics.

Catalyst A substance whose presence increases (or decreases) the rate of chemical reaction without itself being permanently changed.

Cement Generic term describing a material that fastens things together as it hardens.

Channeling The tendency for water to find open, continuous pathways that offer less resistance to flow.

Chemical equilibrium The state of a reversible chemical system in which the products of the forward reactions are consumed by the reverse reactions at the same rate as they are formed.

Chemical stability Quality of a material or component that prevents chemical reactions with the environment.

Classification of wastes Grouping of waste types according to physical and/or chemical characteristics.

Column tests Tests employing vertically oriented cylindrical vessels packed with materials containing leachable species. Columns are operated by passing leachants through them either in a downflow or upflow mode.

Confined aquifer* An aquifer bounded above and below by impermeable beds or by beds of distinctly lower permeability than that of the aquifer itself; an aquifer containing confined groundwater.

Container* Any portable device in which a material is stored, transported, treated, disposed of, or otherwise handled.

Contaminants Typically undesirable (e.g., hazardous or radioactive trace constituents that render another substance impure).

Contamination partition Distribution of a contaminant between the different phases in a waste or leach test system, including sorption on or dispersion in such phases. The driving forces for partitioning may be either physical or chemical.

Contingency plan* A document setting out an organized, planned, and coordinated course of action to be followed in case of a fire, explosion, or release of hazardous waste or hazardous waste constituents that could threaten human health or the environment.

Corrosion Surface dissolution of a material by a chemical process.

Corrosive Wastes that dissolve or wear away materials.

Corrosivity Tending or having the power to wear away by chemical action; usually has a very high or very low pH value. May cause skin reaction on contact. See acid waste and alkali waste.

Cumulative fraction The sum of the fractions leached during previous leaching intervals, plus the fraction leached during the last leaching interval, using the initial amount of the species of interest present in the specimen as unity (100 percent).

Degradation Chemical or biological transformation of a complex compound into a number of simple ones.

Designated facility* A hazardous waste treatment, storage, or diposal facilty that has received an EPA permit (or a facility with interim status) in accordance with the requirements of 40 CFR Parts 122 and 124, or a permit from a state authorized in accordance with Part 123.

Desorption Release of adsorbed substance(s) from a surface. Transfer of species from the solid to the aqueous phase by a variety of chemical and/or physical processes; the opposite of sorption.

Dewatering A physical process that removes sufficient water from sludge so that its physical form is changed from essentially that of a fluid to that of a slurry or damp solid.

Dike An embankment or ridge of either natural or man-made materials used to prevent the movement of liquid, sludges, solids, or other materials.

Dilution Reduction of the concentration of the species of interest in the leachate.

Dirtlike Description of solidified waste used to describe its working characteristics, especially in a landfill. Has properties very much like native clay soils. Will support considerable weight, but can be excavated and moved about with standard earth-moving equipment. Dirtlike materials range from fairly plastic, such as wet clays, to very friable, crumbly material, such as many top soils. They work well in a landfill, are compactable, and can develop considerable loadbearing capacity when compacted.

Discharge* The accidental or intentional spilling, leaking, pumping, pouring, emitting, emptying, or dumping of hazardous waste into or on any land or water.

Disposal* The discharge, deposit, injection, dumping, spilling, leaking, or placing of any solid waste or hazardous waste into or on any land or water so that such solid waste or hazardous waste or any constituent thereof may enter the environment or be emitted into the air or discharged into any waters, including groundwaters.

Disposal facility* A facility or part of a facility at which hazardous waste is intentionally placed into or on any land or water, and at which waste will remain after closure.

Dissolution Transfer of a species from the solid to the aqueous phase through chemical or physical interaction of the dissolvable species with the solvent.

Dissolved solids Theoretically, the anhydrous residues of the dissolved constituents in water. Actually, the term is defined by the method used in determination.

Driving force for leaching The gradient of the chemical potential from that of the mobile species inside the specimen to the solid-liquid interface.

Dump A land site where solid waste is disposed of in a manner that does not protect the environment. Dumping is indiscriminate disposal of solid waste.

Durability The ability of a material, in this case a waste form, to resist physical breakdown and remain intact over time.

Dynamic test A test designed to maintain a driving force for leaching through leachant renewal, thus generating information as a function of time.

Effective porosity The fraction of the total void space that is able to allow passage of water and dissolved species, including colloids. Excludes unconnected pores whose effective diameters are less than that of the moving species.

Effluent A stream of any liquid that is flowing away from the waste source (e.g., the outflow of a sewer). Primarily waste water that can be discharged directly to a sewer or a water pollution control plant without pretreatment.

Elution Removal of sorbed or soluble matter by washing with water or other solvent.

Environmental impact statement A document prepared by the EPA or under EPA guidance (generally by a consultant hired by the applicant and supervised by the EPA) that identifies and analyzes in detail the environmental impacts of a proposed action.

Environmentally persistent waste Any waste that, if exposed to a natural environment, remains hazardous for an extended length of time.

Equilibrium test A test designed to attain or approach chemical equilibrium between the waste form and the leachate. Generates information independent of time.

Existing hazardous waste management facility* A facility that was in operation, or for which construction had commenced, on or before October 21, 1976. Construction had commenced if: (1) the owner or operator has obtained all necessary federal, state, and local preconstruction approvals or permits; and either (2)(a) a continuous physical, on-site construction program has begun, or (b) the owner or operator has entered into contractual obligations—which cannot be canceled or modified without substantial loss—for construction of the facility to be completed within a reasonable time.

Explosive wastes Wastes that are unstable and may readily undergo violent chemical change or explode.

Extract To obtain through the effort of drawing out.

Extraction test procedure A series of laboratory operations and analyses designed to determine whether, under severe conditions, a solid waste, stabilized waste, or landfilled material can yield a hazardous leachate.

Extraction test—single batch A leaching test carried out by exhaustively extracting a single (batch of) sample.

Extraction test—batch A test intended to help identify the presence of a multiple species of interest in a waste form by concentrating the species of interest. The same volume of leachant is exposed to several samples of the waste form in succession to build up the concentration of the species of interest so it can be detected.

Extraction test—chemical A test in which different batches of leachant are used in sequence or the test specimen is exposed sequentially to different leachants.

Facility* All contiguous land, structures, other appurtenances, and improvements on the land, used for treating, storing, or disposing of hazardous waste. A facility may consist of several treatment, storage, or disposal operational units (e.g., one or more landfills, surface impoundments, or combinations of them).

Flammable waste A waste capable of igniting easily and burning rapidly.

Flash point The lowest temperature at which evaporation of a substance produces sufficient vapor to form an ignitable mixture that, when ignited, is capable of the initiation and propagation of flame away from the source of ignition. Propagation of the flame means the spread of the flame from layer to layer independently of the source of ignition.

Fluegas desulfurization The operation of removing sulfur oxides from the exhaust gas streams of a boiler or industrial process.

Fluid A substance that, unlike a solid, can be deformed inelastically (viscously) and adjusts to the shape of the confining space, but unlike a gas does not expand indefinitely. A fluid may be a liquid, a multiphase dispersion, or a flowing powder.

Flyash Particulate matter removed from combustion fluegases. Major sources are pulverized coal-burning boilers.

Food-chain crops* Tobacco, crops grown for human consumption, and crops grown for feed for animals whose products are consumed by humans.

Fraction leached A fraction of the originally present amount of the species of interest that has been removed from the specimen or waste form by the leaching process during a specific leaching time.

Fractional release Amount of contaminants that have leached into the aqueous phase, divided by the total amount originally present in the solid waste.

Free liquids* Liquids that readily separate from the solid portion of a waste under ambient temperature and pressure.

Freeboard* The vertical distance between the top of a tank or surface impoundment dike, and the surface of the waste confined therein.

Grab sample A single sample taken at neither set time, place, nor flow, nor according to a sampling plan.

Gel or gelation In a chemical solidification system, often the first stage in the hardening or curing process. The process wherein the waste transforms from a liquid flowable state to a nonflowable condition.

Generator* Any person, by site, whose act or process produces hazardous waste identified or listed in EPA regulations.

Generic Relating to or characteristic of a whole group or class; for example, materials made from similar-type wastes with the same solidification agent and by the same solidification process.

Groundwater Water below the land surface in a zone of saturation. A body of water generally within the boundaries of a watershed, which exists in the internal passageways of porous geological formations (aquifers) and which flows in response to gravitational forces. Very often the source of water for communities and industries.

Halogen The nonmetals fluorine, chlorine, bromine, and iodine.

Hardening or curing The process in which chemical reactions cause the mixture of waste and solidification chemicals to develop its ultimate properties of hardness, strength, permeability, leaching resistance, etc.

Hazardous waste A waste, or combination of wastes, which because of its quantity, concentration, toxicity, corrosiveness, mutogenicity, or flammability, or physical, chemical, or infectious characteristics may (1) cause or significantly contribute to an increase in mortality or an increase in serious, irreversible, or incapacitating reversible illness; or (2) pose a substantial present or potential hazard to human health or the environment when improperly treated, stored, transported, disposed of, or otherwise managed.

Hazardous waste generation The act or process of producing hazardous waste.

Hazardous waste landfill An excavated or engineered area on which hazardous waste is deposited and covered. Proper protection of the environment from the materials to be deposited in such a landfill requires careful site selection, good design, proper operation, leachate collection and treatment, and thorough final closure.

Hazardous waste leachate The liquid that percolated through or drained from hazardous waste placed in or on the ground.

Hazardous waste management The systematic control of the collection, source separation, storage, transportation, processing, treatment, recovery, and disposal of hazardous waste.

Hydraulic conductivity The reciprocal of resistance to hydraulic flow. The quantity of water that will flow through a unit cube of a given porous substance in a unit of time (in response to a unit hydraulic gradient).

Hydraulic gradient The difference in hydraulic potential between two points, divided by the separation distance. The rate of change of hydraulic pressure as function of distance in a given direction.

Ignitability Tending to be easily ignited or set afire; if liquid, possesses a low flash point temperature; if solid, liable to cause fire by friction or spontaneous chemical changes.

Ignitable Describing a material that can be ignited.

Immersion tests Test carried out by immersion of a specimen into a defined amount of fluid, usually water, under prescribed testing conditions.

Immersion test—flow An immersion test during which the leachant is renewed continuously at a prescribed rate.

Immersion test—multiple batch An immersion test during which the leachant is renewed according to a fixed schedule under conditions prescribed by an approved procedure.

Immersion test—static An immersion test during which the specimen is exposed to the same volume of leachant throughout the time of the test.

In operation* Refers to a facility that is treating, storing, or disposing of hazardous waste.

Incineration An engineered process using controlled flame combustion to thermally degrade waste materials.

Incineration waste The residue from incineration, other than incinerator ash and flyash.

Incinerator* An enclosed device using controlled flame combustion, the primary purpose of which is to thermally break down hazardous waste. Examples of incinerators are rotary kiln, fluidized bed, and liquid injection incinerators.

Incinerator ash The ash residue, other than flyash, resulting from incineration where the waste is reduced to ash.

Incompatible waste* A hazardous waste that is unsuitable for (1) placement in a particular device or facility because it may cause corrosion or decay of containment materials (e.g., container inner liners or tank walls); or (2) commingling with another waste or material under uncontrolled conditions because the commingling might produce heat or pressure, fire or explosion, violent reaction, toxic dusts, mists, fumes, gases, or flammable fumes or gases.

Individual generation site* The contiguous site at or on which one or more hazardous wastes are generated. An individual generation site, such as a large manufacturing plant, may have one or more sources of hazardous waste, but is considered a single or individual generation site if the site or property is contiguous.

Industrial wastes Unwanted materials produced in or eliminated from an industrial operation. They may be categorized under a variety of headings, such as liquid wastes, sludge wastes, or solid wastes.

Infectious Capable of transmitting diseases caused by pathogenic organisms.

Injection well* A well into which fluids are injected. (See also **underground injection**.)

Interface The boundary region between two different phases of a chemical system.

Interrupted flow Liquid phase replenishment that is not continuous.

Land treatment facility* A facility or part of a facility at which hazardous waste is applied onto or incorporated into the soil surface; such facilities are disposal facilities if the waste will remain after closure.

Landfill* A disposal facility or part of a facility where hazardous waste is placed in or on land, and which is not a land treatment facility, a surface impoundment, or an injection well.

Landfill cell* A discrete volume of a hazardous waste landfill that uses a liner to provide isolation of wastes from adjacent cells or wastes. Examples of landfill cells are trenches and pits.

Leach To extract a soluble substance from some material by causing water to filter (down) through the material (by advection, percolation, or diffusion). To dissolve and be washed away.

Leach rate Rate of removal of constituents from the waste form; the amount of a constituent of a specimen or solid waste form that is leached during one time unit (e.g., g/day or μCi/s). Frequently expressed per unit of exposed surface area (e.g., g/cm^2 day).

Leach test specimen/leach specimen The solid body that is immersed into the leachant during the leach test. This body must be representative of the solid that is formed by the combination of waste with a solidification agent by a given process.

Leach test vessel The container in which the leachant and the leach test specimen are placed during the leach test.

Leachability A rate constant (or a combination of several rate constants) that describe(s) the leaching of a constituent from a material under a given set of conditions. The extent to which a soluble constituent can be removed from a solid waste under certain conditions.

Leachability Index An index value (dimensionless) related to the leaching characteristics of solidified waste materials as measured by the leach test defined in the ANSI/ANS-16.1-1986 standard. In this standard the "leachability index" has an exact theoretical meaning only for homogeneous, chemically inert materials, for which bulk diffusion is the predominant rate-determining process during leaching.

Leachant The liquid that contacts the specimen during the course of a leaching test or contacts a solid waste form at a disposal site. Also, the liquid before contact with the waste.

Leachant characteristics Properties of the leachant; for example, pH, E_h, capacity (buffer, complexation, etc.), dielectric constant, ionic strength, and temperature.

Leachant flow rate The amount of leachant that passes through a unit area of a plane at some location inside the leaching specimen (perpendicular to the direction of interest) per unit of time.

Leachate Any liquid, including any suspended components in the liquid, that has percolated through or drained from hazardous waste. Leachant after use (i.e., liquid after contact with the waste).

Leaching Removal of soluble constituents from a solid waste by contact with a liquid; the operation, natural or designed, for producing leachate.

Leaching interval The length of time during which a given volume of leachant is in contact with a specimen of solid waste form.

666 GLOSSARY

Leaching mechanisms Leaching processes (chemical or physical) on whose parameters depend the manner and rate of mobilization and transfer of the species of interest from the specimen into the leachant.

Liner* A continuous layer of natural or man-made materials, beneath or on the sides of a surface impoundment, landfill, or landfill cell, that restricts the downward or lateral escape of hazardous waste, hazardous waste constituents, or leachate.

Low-level waste For the purpose of the ANSI/ANS-16.1-1986 standard, a radioactive waste as defined by 10 CFR 61.

Manifest* The shipping document originated and signed by the generator that contains the information required by EPA regulations.

Manifest system A procedure called for by the EPA in which hazardous quantities are identified as they are produced and followed through further treatment, transportation, and disposal by a series of permanent, linkable, descriptive documents.

Matrix The material constituting the waste form; the basic structure of the solid in which the unreacted waste constituents are suspended, encapsulated, embedded, etc.

Matrix durability The ability of a solidified waste form to resist physical wear and chemical attack while remaining intact for the period of time required for safe disposal.

Mechanistic models Detailed descriptions of leaching processes based on mathematical equations that represent fundamental mechanisms.

Mechanistic studies The part of generic leaching studies aimed at identifying the mechanisms by which the leaching processes occur and then assessing their relative importance in specific situations.

Molecular diffusion The spreading movement of molecules under the influence of a difference in chemical potential. The intermingling of the molecules of two or more substances.

Monitoring **1.** The procedure or operation of locating and measuring radioactive contamination by means of survey instruments that can detect and measure, as dose rate, ionizing radiations. **2.** The measurement, sometimes continuous, of specific properties at a particular point to detect changes that may then be attributed to a nearby operation and controlled at the operation.

Monolithic Massive and solidly uniform.

Monolithic sample A sample consisting of a single massive solid.

Movement* That hazardous waste transported to a facility in an individual vehicle.

New hazardous waste facility* A facility that began operation, or for which management facility construction commenced after October 21, 1976. (See also **Existing hazardous waste management facility.**)

Nonleaching Generally applied to a waste form that will not permit constituents to leach out.

Open dump A site for the disposal of solid waste that is not a sanitary or secure landfill.

Operator* The person responsible for the overall operation of a facility.

Organic Being, containing, or relating to carbon compounds, especially those in which hydrogen is attached to carbon, whether derived from living organisms or not.

Packaging Any material or structure covering the surface of a waste, such as a plastic bag, drum, or concrete cask, but exclusive of a coating or surface treatment.

Partitioning The division (distribution) of a chemical constituent among different phases and interfaces.

Percolation Passing gradually through a porous substance.

pH The measure of the acidity or alkalinity of a chemical solution, from 0 to 14. The reciprocal of the logarithm of the hydrogen ion concentration. A pH of 7 is neutral, less than 7 is acidic, and greater than 7 is alkaline.

Phase changes The transition from one phase to another caused by chemical or physical processes.

Photolytic Decomposition brought about by the action of radiant energy, specifically light.

Pile* Any noncontainerized accumulation of solid, nonflowing hazardous waste that is used for treatment or storage.

Point source* Any discernible, confined, and discrete conveyance, including, but not limited to any pipe, ditch, channel, tunnel, conduit, well, discrete fissure, container, rolling stock, concentrated animal feeding operation, or vessel or other floating craft, from which pollutants are or may be discharged. This term does not include return flows from irrigated agriculture.

Polychlorinated biphenyls (PCBs) A series of hazardous compounds that have been manufactured, for over 40 years, for such common uses as electrical insulation and heating/cooling.

Polymer A substance composed of giant molecules formed by the union of a large number of simple molecules. The simple molecules are known as *monomers*. The process of forming a polymer is called *polymerization*.

Polynuclear aromatics (PNAs) A class of organic materials with a characteristic multiple-ring molecular structure. Often the result of partial thermal degradation and condensation of simpler aromatic compounds. The material is produced in coal-burning boilers and is suspected of causing genetic damage.

Pore structure The makeup and fractal geometry of the connected voids in the sample or specimen.

Porosity Volume fraction of dispersed but connected voids in a solid material. The ratio of the volume of a material's pores to that of its solids content.

Portland cement A specific cement, made by heating a powdered mixture of clay, limestone, and other ingredients to incipient fusion, then grinding to a fine powder. By varying the types and quantities of clay and limestone and other minor additives, different properties may be obtained. Unless otherwise specified, use of the term in this book means Type I (general-purpose) Portland cement.

Precision Related to the reproducibility of a measurement.

Process optimization Development of a process so that it will be able to most effectively achieve the specifications [relative to the optimization parameter(s)] of the product.

Publicly owned treatment works (POTW)* Any device or system that is owned by a state or municipality that is used in the treatment (including recycling and reclamation) of municipal industrial wastes of a liquid nature. This definition includes sewers, pipes, or other conveyances only if they convey waste water to a POTW providing treatment.

Quality assurance (QA) The testing and inspection (of all or a sample) of the documentation of an activity to assure that the pertaining procedures were followed and resulted in a product that meets the specifications set for it.

Quality control (QC) The performance of procedures aimed at verifying the compliance of a material, product, or an activity with the specification.

Radioactive wastes A waste is a radioactive waste if it is not source, special nuclear, or by-product material as defined by the Atomic Energy Act of 1954, as amended, and if a representative sample of the waste has either of the following properties: the average radium-226 concentration exceeds 5 picocuries per gram for solid wastes of 50 picocuries (radium-226 and radium-228 combined) per liter for liquid wastes; or the total radium-226 activity equals or exceeds 10 microcuries for any single discrete source. Most radioactive wastes consist of conventional materials contaminated with radionuclides. Contamination can range in concentration from a few parts per million to as much as 50 percent of the total waste. Depending upon the concentration, wastes are categorized as low- or high-level wastes. Some studies have used "high-level wastes" to refer to those requiring special provisions for dissipation of heat produced by radioactive decay.

Radioactivity The property possessed by certain elements to spontaneously emit radiation by the disintegration of the nuclei of the atoms. The radiation so emitted may consist of alpha particles, electrons, as well as gamma- and X-rays.

Reactive hazardous wastes **1.** Wastes that in themselves are normally unstable and readily under violent chemical change, but do not detonate. Also, wastes that may react violently with water, or that may form potentially explosive mixtures with water, or that generate toxic fumes when mixed with water. **2.** Wastes that in themselves are capable of detonation or explosive reaction, but require a strong initiating source, or that must be heated under confinement before initiation, or that react explosively with water. **3.** Wastes that in themselves are readily capable of detonation, or of explosive decomposition, or reaction at normal temperatures and pressures.

Reactivity Property of a reactive waste.

Reclamation Restoration to a better, more useful state, such as land reclamation by sanitary landfilling, or obtaining useful materials from solid wastes.

Recoverable The capability and likelihood of a material being recovered from waste for a commercial or industrial use.

Recoverable resource Materials that still have useful chemical or physical properties after serving a specific purpose and can therefore be reused or recycled for the same or other purposes.

Release Amount of contaminants that have leached into the aqueous phase (cf. fractional release).

Remaining fraction Unity minus cumulative fraction leached, that is, the fraction still remaining with the specimen or waste form after leaching.

Representative sample* A sample of a universe or whole (e.g., waste pile, lagoon, groundwater) that can be expected to exhibit the average properties of the universe or whole.

Residual liquid Free liquid present in the specimen container at the time the specimen is removed from the container.

Residue Solid or semisolid materials such as, but not limited to, ash, ceramics, glass, metal, and organic substances remaining after incineration or processing.

Rinse Brief immersion of the specimen into water, not to exceed 30 seconds, prior to the beginning of the leaching test, so as to remove loose superficial contamination from the surface of the specimen; the latter being assumed to be no part of the solidified waste material.

Rinsewater Combination of residual liquid, wash-off, and specimen container rinse.

Rocklike or concretelike Most solidification processes can be made to yield very hard, solid materials from the waste. This is accomplished by adding large quantities of the solidification reagents. Unless such physical hardness is required for some particular end use, products of this nature are not cost effective, are usually unnecessary, and from a handling standpoint are often undesirable.

Runoff* Any rainwater, leachate, or other liquid that drains over land from any part of a facility.

Run-on* Any rainwater, leachate, or other liquid that drains over land onto any part of a facility.

Sanitary landfill Constructed to allow for and aid biological and atmospheric degradation of wastes to innocuous dirtlike materials over an extended period of time. Requires protection of groundwater and air from wastes and decomposition products in the form of leachate collection systems and soil covers. Defined by the American Society of Civil Engineers as "a method of disposing of refuse on the land without creating nuisances or hazards to the public health or safety, by utilizing the principles of engineering to confine the refuse to the smallest practical area, to reduce it to the smallest practical volume and to cover it with a layer of earth at the conclusion of each day's operation, or at such more infrequent intervals as may be necessary."

Saturated zone (or "zone of saturation")* That part of the earth's crust in which all voids are filled with water.

Saturation Concentration of the species of interest in the leachant that approaches its solubility limit.

Secure landfill Constructed in cell forms to segregate and isolate hazardous materials from each other and from contact with groundwater or atmosphere. Utilizes low permeability materials, such as clay, plastics, or pozzolanic liners, to form isolated permanent entrapments.

Semi-infinite medium A body of which the outer boundary is considered to be effectively at an infinite distance from the inner region.

Setting agent A chemical agent that reacts with the binder and/or components of the waste itself to produce a solid, and without which the system would not function properly. Examples are sodium silicate in Portland cement systems and peroxides in organic polymer systems.

SIC Stands for "Standard Industrial Classification," which is a recognized two-to-four-digit number that classifies industries by product line.

Simulation of field conditions A testing procedure that attempts to reproduce in detail the environment to which the solidified waste or waste form is exposed at the disposal site.

Sludge* Any solid, semisolid, or liquid waste generated from a municipal, commercial, or industrial waste water treatment plant, water supply treatment plant, or air pollution control facility, exclusive of the treated effluent from a waste water treatment plant.

Solid waste **1.** The fraction of a multiphase waste without free water or liquid. The nonfluid portion of the waste. **2.** Any garbage, refuse, or sludge from a waste treatment (regulatory definition) plant, water supply treatment plant, or air pollution control facility, and other discarded materials, including solid, liquid, semisolid, or contained gaseous material resulting from industrial commercial mining, and agricultural operations, and from community activities, but does not include solid or dissolved material in domestic sewage, or solid or dissolved materials in irrigation return, flows, or industrial discharges that are point sources subject to permits under Section 402 of the Federal Water Pollution Control Act, as amended (86 Stat. 880), or source, special nuclear, or by-product material as defined by the Atomic Energy Act of 1954, as amended (68 Stat. 923).

Solid-waste management The systematic administration of activities that provide for the collection, source separation, storage, transportation, transfer, processing, treatment, and disposal of solid waste.

Solid-waste management facility Any resource recovery system or component thereof, any system, program, or facility for resource conservation, and any facility for the treatment of solid wastes, including hazardous wastes, whether such facility is associated with facilities generating such wastes or otherwise.

Solidification agent Material that is added to the waste during the solidification process to produce a stable solid waste form.

Solidification/solidification process Unit operation that converts a fluid waste into a solid form, typically by the addition of a solidification agent or binder.

Solidified waste A waste liquid, solution, slurry, sludge, or powder that has been converted to a stable solid by a solidification process.

Solvent A substance capable of, or used in, dissolving or dispersing one or more other substances.

Sources and sinks Locations at which mass is either added to or removed from an open system.

Specimen The test specimen.

Spike A known amount of the species of interest deliberately added to a specimen or sample.

Steady-state Constant net flux of a constituent between the solid and aqueous phase. Equilibrium in a closed system. Dynamic equilibrium in an open system having constant and equivalent sources and sinks.

Storage* The holding of hazardous waste for a temporary period, at the end of which the hazardous waste is treated, disposed of, or stored elsewhere.

Surface area of contact The geometric surface area of the specimen that is contacted by the leachant.

Surface impoundment* A facility or part of a facility that is a natural topographic depression, man-made excavation, or diked area formed primarily of earthen materials (although it may be lined with man-made materials) that is designed to hold an accumulation of liquid wastes or wastes containing free liquids, and that is not an injection well. Examples of surface impoundments are holding, storage, settling, and aeration pits, ponds, and lagoons.

Tank* A stationary device designed to contain an accumulation of hazardous waste that is constructed primarily of nonearthen materials (e.g., wood, concrete, steel, plastic) that provide structural support.

Teratogenicity The property of a substance of affecting the genetic characteristics of an organism so as to cause the offspring of the organism to be misshaped or malformed.

Thermal treatment The treatment of hazardous waste in a device that uses elevated temperatures as the primary means to change the chemical, physical, or biological character or composition of the hazardous waste. Examples of thermal treatment processes are incineration, molten salt, pyrolysis, calcination, wet-air oxidation, and microwave discharge.

Tortuosity The circuitous, twisting pathway of connected pores in a solid matrix.

Total solids The sum of dissolved and undissolved constituents in water or waste water, usually stated in miligrams per liter.

Totally enclosed treatment facility* A facility for the treatment of hazardous waste that is directly connected to an industrial production process and that is constructed and operated in a manner that prevents the release of any hazardous waste or any constituent thereof into the environment during treatment. An example is a pipe in which waste acid is neutralized.

Toxic waste A waste that can produce injury upon contact with, or by accumulation in a susceptible site in or on the body of a living organism.

Toxicity The ability of a material to produce injury or disease upon exposure, ingestion, inhalation, or assimilation by a living organism.

Toxicity The property of being poisonous, of causing death or severe temporary or permanent debility of an organism.

Transport The conveyance of chemical constituents in the aqueous phase (of a porous solid).

Transportation* The movement of hazardous waste by air, rail, highway, or water.

Transporter* A person engaged in the offsite transportation of hazardous waste by air, rail, highway, or water.

Treatment* Any method, technique, or process, including neutralization, designed to change the physical, chemical, or biological character or composition of any hazardous waste so as to neutralize such waste, or so as to render such waste nonhazardous or less hazardous; safer to transport, store, or dispose of; or amenable for recovery, amenable for storage, or reduced in volume.

Underground injection* The subsurface emplacement of fluids through a bored, drilled, or driven well; or through a dug well, where the depth of the dug well is greater than the largest surface dimension. (See also **injection well**.)

Unsaturated zone* The zone between the land surface and the water table.

Washoff Liquid containing the mobile surface contamination removed from the specimen by immersing it in demineralized water for (no more than) 30 seconds.

Waste The material (liquid or solid) to be formed into a stable solid by a CFS process.

Waste form/solid waste form The stable solid body formed by the solidification of waste and meeting specifications for emplacement.

Waste management The total process of the collection of waste at its point of generation, transportation, treatment, and final disposal in an acceptable manner.

Water content The amount of moisture present in a waste that is removable by drying.

Water pollution Contamination of any water that will create or is likely to create a nuisance or to render such waters harmful, detrimental, or injurious to public health, safety, or welfare, or to domestic, municipal, commercial, industrial, agricultural, recreational, or other legitimate beneficial uses, or to livestock, wild animals, birds, fish, or other aquatic life, including, but not limited to, such contamination by alteration of the physical, chemical, or biological properties of such waters, or change in temperature, tasks, color, or odor thereof, or the discharge of any liquid, gaseous, radioactive, solid, or other substances into such waters.

Water table The upper surface of the zone of saturation in an unconfined aquifer at which the pressure is equal to atmospheric pressure.

Well* Any shaft or pit dug or bored into the earth, generally of a cylindrical form, and often walled with bricks or tubing to prevent the earth from caving in.

Wet/dry cycle The repeated drying and subsequent rewetting of a waste form and its effects on the eventual state of the waste form.

BIBLIOGRAPHY

GENERAL CFS

Arthur D. Little Inc. *Hazardous Waste Generation, Treatment and Disposal in the Pharmaceutical Industry.* Cincinnati, OH, 1974.

Battelle, Columbus Laboratories. *Assessment of Industrial Hazardous Waste Practices, Electroplating and Metal Finishing Industries.* Columbus, OH, 1975.

Bishop, P. L. Leaching of inorganic hazardous constituents from stabilized/solidified hazardous wastes. *Hazardous Waste Hazardous Mater.* **5**(2): 129–143 (1988).

Bretherick, L. *Handbook of Reactive Chemical Hazards.* Cleveland, OH: CRC Press, 1975.

Calspan Corp. *Assessment of Industrial Hazardous Waste Practices in the Metal Smelting and Refining Industry.* Buffalo, NY, 1975.

Chemical Week. 24–30 (Apr. 17, 1985).

Clendenning, T. G., et al. *Current Technology in the Utilization and Disposal of Coal Ash.* Toronto, Ont., Canada: Ontario Hydro Corp., 1975.

Collins, R. J. DOT/DOE evaluation of kiln dust/fly ash technology. Presented at the 6th Annual International Conference on the Economic and Environmental Utility of Kiln Dust and Kiln Dust/Fly Ash Technology, Feb. 19–21, 1985.

Conner, J. R. Use of computer-based artificial intelligence systems to select detoxification and solidification methods for residues and wastes. Presented at the 5th International Symposium on Environmental Pollution, Quebec City, P.Q., Canada, June 27, 1985.

———. Considerations in selecting chemical fixation and solidification alternatives—The engineered approach. *Chem. Eng.* **93**(21): (Nov. 10, 1986).

———. A critical comparison: ultimate waste disposal methods. *Plant Eng.* (Oct. 1972). (One of the first papers to comparatively describe hazardous waste treatment methods of the time.)

———. Ultimate disposal of liquid wastes by chemical fixation. In *Proc. 29th Annual Purdue Industrial Waste Conference.* Purdue Univ., West Lafayette, IN, May 1974.

———. *The Modern Engineered Approach to Chemical Fixation and Solidification Technology.* Chemfix Inc, Pittsburgh, PA, 1981.

———. Chemical fixation of hazardous material spill residues. In *Proc. 1976 National Conference on Control of Hazardous Material Spills.* New Orleans, Apr. 1976.

———. Stockpiling of metal values in chemical and metal process wastes by the use of chemical solidification and fixation techniques. Presented at the AES SUR/FIN '85 Conference. Detroit, MI, July 16, 1985.

———. Testing of chemically solidified waste for disposal in sanitary landfills or on-site monofills. Presented at the Alberta Environmental Centre Workshop on Environmental Assessment of Waste Solidification. Vegreville, Alta., Canada, November 21, 1983.

———. The present status of sludge problems in the United States. In *Proc. JIPA Conference on Sludge Management.* Tokyo, Nov. 1974.

———. Ultimate liquid waste disposal methods. *Plant Eng.* (Oct. 1972).

———. *Disposal of Concentrated Wastes from the Textile Industry.* Chemfix Inc, Pittsburgh, PA, Dec. 1976.

———. Chemical fixation and solidification of dredged materials. In *Proc. 7th World Dredging Conference.* San Francisco, July 1976.

———. *Ind. Water Eng.* (July/Aug. 1977).

Conner, J. R., and R. J. Polosky. *The Use of Chemically Fixed Solids to Improve Sanitary Landfills.* Pittsburgh, PA: Chemfix, Inc., 1974.

Conner, J. R., and L. P. Gowman. Chemical fixation of activated sludge and end use applications. Presented at the AIChE Conference on WateReuse. Chicago, May 1975.

Conner, J. R., and H. H. Buzbee. Use of chemically fixed sewage sludge as daily, intermediate and final cover in sanitary landfills. Presented at the National Conference on Hazardous Wastes and Environmental Emergencies. Cincinnati, OH, May 15, 1985.

Cote, P. *Contaminant Leaching from Cement-Based Waste Forms Under Acidic Conditions.* Ph.D. Thesis. Hamilton, Ont., Canada: McMaster Univ., 1986.

Cullinane, M. J., R. M. Bricka, and N. R. Francingues, Jr. An assessment of materials that interfere with stabilization/solidification processes. In *Proc. 13th Annual Research Symposium.* Cincinnati, OH, pp. 64–71, 1987.

Cullinane, M. J., and L. W. Jones. *Stabilization/Solidification of Hazardous Waste.* Cincinnati OH: U.S. EPA Hazardous Waste Engineering Research Laboratory, EPA/600/D-86/028, 1986. (EPA's basic handbook on CFS technology and testing.)

Dallavalle, J. M. *Micrometrics.* New York: Pitman Publishing, 1948.

Dean, J. A., Ed. *Lange's Handbook of Chemistry,* 13th ed. New York: McGraw-Hill, 1985.

Design and Control of Concrete Mixtures. Skokie, IL: Portland Cement Association, 1979.

Diamond, S., *et al.* Transformation of clay minerals by calcium hydroxide attack. Presented at the 12th National Conference on Clay and Clay Minerals: 359–379. Lafayette, IN, 1963.

Double, D. D., and A. Hellawell. The solidification of cement. *Sci. Am.* **237**: 82–90 (1977).

Dragun, J. The fate of hazardous materials in soil. *Hazardous Mater. Contr.* 41–65 (May/June 1988).

Falcone, J. S. *Soluble Silicates.* New York: Reinhold Co., 1982.

Falcone, J. S., R. W Spencer, and E. P. Katsanes. *Chemical Interactions of Soluble Silicates in the Management of Hazardous Wastes.* Lafayette Hill, PA: PQ Corp., 1982.

Foster D. Snell Inc. *Assessment of Industrial Hazardous Waste Practices, Rubber and Plastics Industry.* Florham Park, NJ, 1976.

Grim, R. E. *Clay Mineralogy.* New York: McGraw-Hill, 1968.

Grutsch, J. F. *The Chemistry and Chemicals of Coagulation and Flocculation.* American Petroleum Institute Manual on Disposal of Refinery Wastes, Hammond, IN, 1983.

Guide to the Disposal of Chemically Stabilized and Solidified Waste. U.S. EPA SW872, Cincinnati, OH, 1982.

Handbook for Stabilization/Solidification of Hazardous Wastes. Cincinnati, OH: U.S. EPA, 1986.

Haynes, B. W., and G. W. Kramer. *Characterization of U.S. Cement Kiln Dust, Information Circular 8885.* Washington, DC: U.S. Bureau of Mines, 1982.

Iler, R. K. *The Chemistry of Silica.* New York: John Wiley, 1979.

JACA Corp. *Critical Characteristics and Properties of Hazardous Waste Solidification/Stabilization* (Unpublished Draft). Cincinnati, OH: U.S. EPA, 1987.

Jacobs Engineering Co. *Assessment of Industrial Hazardous Waste Practices, Petroleum Refining.* Cincinnati, OH, 1975.

Karol, R. H. *Chemical Grouting.* New York: Dekker, 1983.

Kirk-Othmer. *Encyclopedia of Chemical Technology,* 3rd ed. New York: Wiley, 1979.

Larson, R. J., P. G. Malone, J. H. Shamburger, J. D. Broughton, D. W. Thompson, and L. W. Jones. *Field Investigation of Contaminant Loss from Chemically Stabilized Industrial Sludges.* Springfield, VA: NTIS, 1981.

Lea, F. M. *Chemistry of Cement and Concrete.* New York: St. Martin's Press, 1970.

Leo, P. P., R. B. Fling, and J. Rossoff. Flue gas desulfurization waste disposal study at the Shawnee power station. Presented at the Symposium on Flue Gas Desulfurization. Research Triangle Park, NC, Nov. 11, 1977.

Midwest Research Institute. *Assessment of Industrial Hazardous Waste Practices, Metals Mining Industry.* Cincinnati, OH, 1976.

National Academy of Sciences. *Solidification of High-Level Radioactive Wastes.* Washington, DC: U.S. Nuclear Regulatory Commission, 1979.

Ontario Waste Management Corporation. *Facilities Development Process,* Phase I Report. Ottawa, Ont., Canada, 1982.

Ontario Ministry of the Environment. *An Assessment of a Process for the Solidification and Stabilization of Liquid Industrial Wastes.* Ottawa, Ont., Canada, 1977.

Ottinger, R. S., J. L. Blumenthal, D. F. Dal Porto, G. I. Gruber, M. J. Santy, and C. C. Shih. *Recommended Methods of Reduction, Neutralization, Recovery or Disposal of Hazardous Waste.* Washington, DC: EPA-670/2-73-053-f, 1973.

Pojasek, R. B.. Ed. *Toxic and Hazardous Waste Disposal.* Ann Arbor, MI:Ann Arbor Science Publishers, 1979.

Popovics, S. *Concrete Making Materials.* New York: McGraw-Hill, 1979.

Powers, P. W. *How to Dispose of Toxic Substances and Industrial Wastes.* Park Ridge, NJ: Noyes Data Corp., 1976.

Rosen, M. J. *Surfactants and Interfacial Phenomena.* New York: Wiley-Interscience, 1978.

Rossoff, J., et al. *Disposal of By-Products from Non-regenerable Flue Gas Desulfurization Systems,* Final Report No. EPA-600/7-79-046. Washington, DC: U.S. EPA, 1979.

Sanjour, W. *Cadmium and Environmental Policy.* Washington, DC: U.S. EPA, 1974.

Sax, I. R. *Dangerous Properties of Industrial Materials.* New York: Van Nostrand Reinhold, 1979.

SCS Engineers Inc. *Assessment of Industrial Hazardous Waste Practices—Leather Tanning and Finishing Industry,* SCS Engineers Inc., Reston, VA, 1976.

Sittig, M. *Pollutant Removal Handbook.* Park Ridge, NJ: Noyes Data Corp., 1973.

Skolny, J., and K. E. Daugherty. Everything you always wanted to know about Portland cement. *ChemTech:* 38–45 (Jan. 1972).

Smith, L.M., and H. G. Larew. *User's Manual for Sulfate Waste in Road Construction,* Report N. FHWA-DR-76-11. Springfield, VA: NTIS, 1975.

Sowers, G. B., and G. F. Sowers. *Introductory Soil Mechanics and Foundations.* New York: Macmillan, 1970.

Spencer, R.W., R. H. Reifsnyder, and J. C. Falcone. *Applications of Soluble Silicates and Derivative Materials in the Management of Hazardous Wastes.* Lafayette Hill, PA: The PQ Corp., 1986.

Sturm, W., and J. J. Morgan. *Aquatic Chemistry,* 2nd ed. New York: Wiley, 1973.

Survey of Solidification/Stabilization Technology for Hazardous Wastes. Springfield, VA: NTIS, 1979.

TRW Inc. *Assessment of Industrial Hazardous Waste Practices, Organic Chemicals, Pesticides and Explosives Industries.* Redondo Beach, CA, 1975.

U.S. Army Engineer Waterways Experiment Station. *Guide to the Disposal of Chemically Stabilized and Solidified Waste.* Cincinnati, OH: U.S. EPA, 1982.

Vail, J. G. *Soluble Silicates.* New York: Reinhold Co., 1952.

Versar, Inc. *Assessment of Industrial Hazardous Waste Practices, Storage and Primary Battery Industries.* Springfield, VA, 1974.

———. *Assessment of Industrial Hazardous Waste Practices, Inorganic Chemical Industry.* Springfield, VA, 1974.

———. *Assessment of Industrial Hazardous Waste Practices, Textiles Industry.* Springfield, VA, 1976.

Wapora Inc. *Assessment of Industrial Hazardous Waste Practices, Special Machinery and Manufacturing.* Cincinnati, OH, 1976.

———. *Assessment of Industrial Hazardous Waste Practices, Paint Industry.* Cincinnati, OH, 1975.

TESTING

American Public Health Association (APHA). *Standard Methods for the Examination of Water and Wastewater,* 15th ed. New York: APHA, 1980.

American Nuclear Society. *Measurement of the Leachability of Solidified Low-Level Radioactive Wastes by a Short-Term Procedure* (Final Draft) Working Group ANS 16.1 Feb. 6, 1986.

ASTM. *Standard Practice for Sampling Waste and Soils for Volatile Organics.* Philadelphia, PA, 1987.

———. *Standard Practice for Sampling Non-Liquid Waste From Trucks.* Philadelphia, PA, 1987.

———. *Standard Guide for General Planning of Waste Sampling.* Philadelphia, PA, 1985.

Bishop, P. L. Leaching of inorganic hazardous constituents from stabilized/solidified hazardous wastes. *Hazardous Waste Hazardous Mater.* 5(2): 129–143 (1988).

Brown, T., W. Shively, P. Bishop, and D. Gress. Use of an upflow column leaching test to study the release patterns of heavy metals from stabilized/solidified heavy metal sludges. *Hazardous Industrial Solid Waste Testing and Disposal,* 6th Vol., ASTM STP 933. Philadelphia, PA, p. 79, 1986.

Conner, J. R. Use of computer-based artificial intelligence systems to select detoxification and solidification methods for residues and wastes. Presented at the 5th International Symposium on Environmental Pollution. Quebec City, P.Q., Canada: Environmental Pollution Institute, 1985.

———. Testing of chemically solidified waste for disposal in sanitary landfills or on-site monofills. In *Proc. Workshop Environmental Assessment of Waste Solidification.* Vegreville, Alta., Canada, Nov. 1983.

Cote, P. *Contaminant Leaching from Cement-Based Waste Forms Under Acidic Conditions.* Ph.D. Thesis. Hamilton, Ont., Canada: McMaster Univ., 1986.

Cote, P. L., and D. P. Hamilton. Leachability comparison of four hazardous waste fixation processes. In *Proc. 38th Annual Purdue Industrial Waste Conference.* Purdue Univ., West Lafayette, IN, May 10–12, 1957.

———. Evaluation of pollutant release from solidified aqueous wastes using a dynamic leaching test. In *Proc. Hazardous Wastes and Environmental Emergencies.* pp. 302–308, 1984.

Cote, P. L. and D. Isabel. Application of a static leaching test to solidified hazardous waste. Presented at the ASTM International Symposium on Industrial and Hazardous Solid Wastes. Philadelphia, PA, 1983.

———. In *Proc. 3rd ASTM Symposium on Hazardous and Industrial Waste Management and Testing.* Philadelphia, PA, pp. 48–60, 1984.

Cote, P. L., T. R. Bridle, and D. P. Hamilton. Evaluation of pollutant release from solidified aqueous wastes using a dynamic leaching test. Presented at the Symposium on Hazardous Wastes and Environmental Emergencies. Houston, TX, Mar. 12–14, 1984.

Cote, P. L., T. R. Bridle, and A. Benedek. An approach for evaluating long term leachability from measurement of intrinsic waste properties. Presented at the 3rd International Symposium on Industrial and Hazardous Waste. Alexandria, Egypt, 1985.

Darcel, F. *Recent Studies on Leach Testing—A Review* (Unpublished Draft). Toronto, Ont., Canada: Ontario Ministry of the Environment, June 1984.

Eaton, H. C., M. B. Walsh, M. E. Tittlebaum, F. K. Cartledge, and D. Chalasani. Microscopic characterization of the solidification/stabilization of organic hazardous wastes. Presented at the Energy-Sources and Technology Conference and Exhibition, ASME Paper No. 85-Pet-4. (1985).

EP Toxicity Test Procedure. 40CFR Part 261.24, Appendix II. Washington, DC: Federal Register, May 19, 1980.

Godbee, H., *et al*. Application of mass transport theory to the leaching of radionuclides from solid waste. *Nucl. Chem. Waste Manage.* **1**:29 (1980).

Jones, B. F. *Bibliography of Column Leaching Test Procedures*. Austin, TX: Radian Corp., 1981.

Lowenbach, W. A. *Compilation and Evaluation of Leaching Test Methods,* EPA-600/@-78-095. Cincinnati, OH: U.S. EPA, 1978.

I. U. Conversions Systems. *Simulated Field Leaching Test for Evaluation of Surface Runoff for Land Disposal of Waste Materials.* Horsham, PA, 1977.

Stegemann, J., P. L. Cote, and P. Hannak. Preliminary results of an international government/industry cooperative study of waste stabilization/solidification. *Hazardous Waste: Detection, Control, Treatment.* Amsterdam/New York: Elsevier, 1988.

Swanstrom, C. P. Development of a Stabilized Waste Compression Tester. Presented at the 4th International Hazardous Waste Symposium on Environmental Aspects of Stabilization/Solidification of Hazardous and Radioactive Wastes. Atlanta, GA, May 1987.

U.S. EPA. Federal Register. **49**(206), 42591 (Oct. 23, 1984).

———. *Solid Waste Leaching Procedure Manual,* SW-924. Cincinnati, OH, 1985.

———. Federal Register. **51**(114), 21686 (June 13, 1986).

———. Federal Register. **47**(225), 52687 (Nov. 22, 1982).

———. *Test Methods for Evaluating Solid Waste,* SW-846. Washington, DC: Office of Solid Waste and Emergency Response, 1986.

———. *Stabilization/Solidification of CERCLA and RCRA Wastes: Physical Tests, Chemical Testing, Procedures, Technology Screening, and Field Activities.* Cincinnati, OH: Office of Research and Development, 1989.

Wastewater Technology Centre and Alberta Environmental Centre. *A Proposed Protocol for the Assessment of Solidified Wastes.* Edmonton, Alta., Canada, Nov., 1983.

REGULATORY

California Administrative Code, Title 22, 66696: 1800.75-84.3 (1985).

Chemical Waste Management, Inc. *Characterization Of F006 Electroplating Wastes,* Technical Note 87-116. Riverdale, IL, 1987.

Comprehensive Environmental Response, Compensation, and Liability Act of 1980. PL 96-510, 1980.

Environment Canada. *Towards the Development of a Standard Leachate Extraction Procedure.* Ottawa, Ont., Canada, 1988.

EP Toxicity Test Procedure. 40CFR Part 261.24, Appendix II, Federal Register, May 19, 1980.

Federal Register. **47**(225), 52687 (Nov. 22, 1982).

———. **49**(206), 42591 (Oct. 23, 1984).

———. **51**(102), 19305–19308 (May 28, 1986).

———. **45**(98), 33063–33285 (May 19, 1980).

———. **45**(98), 33119–33133 (May 19, 1980).

———. **51**(114), (June 13, 1986).

———. **45**(98), 33122–33134 May 19, 1980.
———. **53**(121), 23661–23671 (June 23, 1988).
———. **51**(199), 36707–36730 (Oct. 15, 1986).
———. **49**(206), 42591 (Oct. 23, 1984).
———. **51**(114), 21685 (June 13, 1986).
Interim Guideline for the interpretation of the Hazardous Waste Definition (Regulation 309). Toronto, Ont., Canada: Ontario Ministry of the Environment, 1983.
Public Law 93-523, 93rd Congress, S.433, Dec. 16, 1974. Washington, DC: U.S. Government Printing Office, 1974.
Resource Conservation and Recovery Act. PL 94-580, 1976.
Safety of Public Water Systems (Safe Drinking Water Act). PL 93-523. Washington, DC: 93rd Congress, Dec. 16, 1974.
Taylor, J. *Guidance Memorandum,* Georgia Department of Natural Resources, Atlanta, Apr. 20, 1984.
U.S. EPA. Federal Register. **50**(105), 23250 (May 31, 1985).
———. Federal Register. 45CFR: 57332 (Aug. 27, 1980).
———. *Onsite Engineering Report for Waterways Experiment Station for K061* (Draft Report), 1988.
———. *Best Demonstrated Available Technology (BDAT) Background Document for K061,* Vol. I, EPA/530-SW-88-031D. Washington, DC, 1988.
———. Federal Register. **53**(159) (Aug. 17, 1988).
———. *Prohibition on the Disposal of Bulk Liquid Waste in Landfills—Statutory Interpretive Guidance* (Draft). Washington, DC, Mar. 19, 1985.
———. Federal Register. **50**(38), 7896–7900 (Feb. 26, 1985).
———. *Best Demonstrated Available Technology (BDAT) Background Document for F006,* Vol. 13, EPA/530-SW-88-0009-1. Washington, DC, May 1988.

PATENT LISTS

Bahr, W., S. Drobnik, W. Hild, R. Kroebel, A. Meyer, and G. Naumann. U.S. Patent 4,009,116 (Feb. 22, 1977).
Barney, G. S., and L. E. Brownell. U.S. Patent 4,028,265 (June 7, 1977).
Berry, D. U.S. Patent 4,191,584 (Mar. 4, 1980).
Bolsing, F. U.S. Patent 4,018,679 (Apr. 19, 1977).
Bonnel, B., P. Allemand, and P. Versmee. U.S. Patent 3,591,542 (July 6, 1971).
Carlson, J. E. U.S. Patent 4,620,947 (Nov. 4, 1986).
Chappell, C. L. U.S. Patent 4,274,880 (June 23, 1981).
———. U.S. Patent 4,230,568 (Oct. 28, 1980).
———. U.S. Patent 4,116,704 (Sept. 26, 1978).
———. U.S. Patent 4,116,705 (Sept. 26, 1978).
Chen, K. S., and H. W. Majewski. U.S. Patent 4,113,504 (Sept. 12, 1978).
Chestnut, R., J. J. Colussi, D. J. Frost, W. E. Keen, Jr., and M. C. Raduta. U.S. Patent 4,514,307 (Apr. 30, 1985).
Chichibu Cement Co. Japan Kokai 80 165,195 (Dec. 23, 1980).
Chudo, A., T. Sugi, and K. Katsoka. U.S. Patent 4,266,980 (May 12, 1981).
Cocozza, E. P. U.S. Patent 4,049,462 (Sept. 20, 1977).
Columbo, P., R. M. Neilson, and W. W. Becker, U.S. Patent 4,174,293 (Nov. 13, 1979).
Conner, J. R. U.S. Patent 4,600,514 (July 15, 1986).
———. U.S. Patent-Applied-For (1987).
———. U.S. Patent B1 3,837,872 (Feb. 25, 1986).
———. U.S. Patent 4,518,508 (May 21, 1985).
———. U.S. Patent 4,623,469 (Nov. 18, 1986).

———. U.S. Patent 3,837,872 (Sept. 24, 1974).
Conner, J. R., E. A. Zawadzki, and R. J. Polosky. U.S. Patent 4,012,320 (Mar. 15, 1977).
Durham, R. L., and C. R. Henderson. U.S. Patent 4,460,292 (July 17, 1984).
Galer, R. E., and P. C. Webb. U.S. Patent 4,286,991 (Sept. 1, 1981).
Grain, G. E., and R. H. Judice. U.S. Patent 3,213,006 (Oct. 19, 1965).
Hendrickson, T. N., and T. J. Dagon. U.S. Patent 3,594,157 (July 20, 1971).
Iffland, N., H. Isensee, G. Wagner, and H. Witte. U.S. Patent 4,122,028 (Oct. 24, 1978).
Japan JP 62 144,747 (June 27, 1987).
Japan Kokai 80 136,166.
———. 80 132, 633 (Oct. 15, 1980).
Japan Patent 63 35,469 (Feb. 16, 1988).
Japan Kokai 70/40,797 (Oct. 20, 1980).
———. 80 104,687 (Aug. 11, 1980).
Jones, C. T. Canadian Patent 1,034,686 (July 11, 1978).
Joosten, H. U.S. Patent 2,081,541 (1937).
Kaczur, J. J., J. C. Tyler, Jr., and J. J. Simmons. U.S. Patent 4,354,942 (Oct. 19, 1982).
Kawasaki Steel Corp. Japan Kokai 80 136,166 (Oct. 23, 1980).
Kire, K., T. Inoue, and T. Tsutsumi. Japan Kokai 80 48,282.
Kitsugi, K., M. Shimoda, and T. Hita. Japan Kokai 78,117,677 (Oct. 14, 1978).
Kitsugi, K., and M. Shimoda. Japan Kokai 78 12,770 (Feb. 4, 1978).
Kokusai, G. Japan Kokai 81 53,796 (May 13, 1981).
Kotani, K. Japan Kokai 79 147,172 (Nov. 17, 1979).
Krofchak, D. U.S. Patent 4,229,295 (Oct. 21, 1980).
———. Canadian Patent 1,003,777 (Jan. 18, 1977).
Kupiec, A. R., and E. D. Escher. U.S. Patent 4,149,968 (Apr. 17, 1979).
Lancy, L. E. U.S. Patent 3,294,680 (Dec. 27, 1966).
MacMillan, J. B. U.S. Patent 3,502,434 (Mar. 24, 1970).
Marvi, A. Japan Kokai 77 100,744 (Aug. 24, 1977).
Meir, G. U.S. Patent 3,971,732 (July 27, 1976).
Mihara, T., K. Endo, and T. Ando. Japan Kokai 78 64,950 (June 9, 1978).
Minnick, L. J. U.S. Patent 4,397,742 (Aug. 9, 1983).
———. U.S. Patent 3,870,535 (Mar. 11, 1975).
———. U.S. Patent 3,854,968 (Dec. 17, 1974).
———. U.S. Patent 3,753,620 (Aug. 21, 1973).
Mitsubishi Heavy Industries. Japan Kokai 80 109,260 (Aug. 22, 1980).
Miyaharam, S., K. Tayama, and M. Komatsu. Japan Kokai 80 18,209 (Feb. 8, 1980).
Mizumoto, Y., H. Fujita, S. Mitsuoka, and S. Masegama. Japan Kokai 77 42, 469 (Apr. 2, 1977).
Monnel, B. U.S. Patent 3,656,985 (Apr. 18, 1972).
Munster, L. U. S. Patent 4,338,134 (July 6, 1982).
Nakaaki, O., Y. Horie, M. Idohara, and J. Shiraogi. Japan Kokai 75,99,962. (Aug. 8, 1975).
Nakagana, H. Japan Kokai 80 44,309 (Mar. 28, 1980).
Nakanishi, K. Japan Kokai 76 120,974 (Oct. 22, 1976).
———. Japan Kokai 76 120,976 (Oct. 22, 1976).
Nakanishi, M. Japan Kokai 79 107,923 (Aug. 24, 1979).
Neipert, M. P., and C. D. Bon. U.S. Patent 2,885,282 (May 5, 1959).
Nicholson, J. P. U.S. Patent 4,101,332 (July 18, 1978), Reissue 30,943 (May 25, 1982).
Nieuwenhuis, G. J. U.S. Patent 3,493,328 (Feb. 3, 1970).
Nippon Mining Co. Japan Kokai 80 116,497 (Sept. 8, 1980).
Nippon Soda Co. Japan Kokai 80 29,119 (Aug. 1, 1980).
Oda, N., Y. Horie, M. Idohara, and J. Shiraogi. Japan Kokai 75 99,962 (Aug. 8, 1975).
Okada, T. I. Kanai, T. Sato, and M. Mizuno. Japan Kokai 75 87,964 (July 15, 1975).
Onata, H. Japan Kokai 77 62,186 (May 23, 1977).

Ono, M. F. Kitamure and T. Fumio. Japan Kokai 75 131,860 (Oct. 18, 1975).
Onoda Cement Co. U.S. Patent 3,947,284 (Mar. 30, 1976).
Oshikata, T., and K. Ogawa. Japan Kokai 78 15,264 (Feb. 10, 1978).
Patents applied for by Environmental Technology Associates, Lima, Ohio.
Pichat, P. U.S. Patent 4,375,986 (Mar. 8, 1983).
Pool, S. C. U.S. Patent 2,507,175 (May 9, 1950).
Ray, L. F. U.S. Patent 4,353,749 (Oct. 12, 1982).
Richards, R. U.S. Patent 3,371,034 (Feb. 27, 1968).
Saigu, Y., and S. Maida. Japan Kokai 78 14,954 (Feb. 10, 1978).
Sandesara, M. D. U.S. Patent 4,118,243 (1978).
Sano, M., I. Sugano, and T. Okuda. Japan Kokai 75,133,654 (Oct. 23, 1975).
Schroeder, K., R. Gradl, K. P. Ehlers, and W. Scheibitz. Ger. Offen. 2,909,572 (Sept. 25, 1980).
Senda, K. Japan Kokai 75 43,048 (Apr. 18, 1975).
Shimizu, S., and T. Ishii. Japan Kokai 79 62,168 (May 18, 1979).
Shimototki, T., T. Ando, T. Uemura, and T. Ashida. Japan Kokai 78 07,728 (Jan. 24, 1978).
Shinoki, K., T. Katayame, and S. Yamaguchi. Japan Kokai 80 44,355 (Mar. 28, 1980).
Shiraishi, M., T. Moritami, and T. Tasaka. Japan Kokai 77 111,260 (Sept. 17, 1977).
Shockcor, J. H. U.S. Patent 3,692,661 (Sept. 19, 1972).
Simosono, C., Y. Togawa, and Y. Taguchi. Japan Kokai 80,126,557.
Smith, C. L., and W. C. Webster. U.S. Patent Reissue 29,783 (Sept. 26, 1978).
Smith, C. L., and W. C. Webster, U.S. Patent 3,720,609, and Reissue 29,783 (Mar. 13, 1973 and Sept. 26, 1978).
Smith, R. H. U.S. Patent 4,342,732 (Aug. 3, 1982).
Takatomi, H., and S. Tokuda. Japan Kokai 80 01,865 (Jan. 9, 1980).
Takeshita, I. Japan Kokai 79 162,844 (Dec. 24, 1979).
Thompson, S. R. U.S. Patent 3,980,558 (1976).
U.S. Patent 3,592,586.
U.S. Patent 3,947,283 (1976).
U.S. Patent 4,015,997 (Apr. 5, 1977).
U.S. Patent 3,642,503.
U.S. Patent 3,697,421 (1972).
U.S. Patent 4,124,405.
U.S. Patent 3,893,656 (July 8, 1975).
U.S. Patent 3,947,284.
U.S. Patent 4,060,425.
U.S. Patent 3,305,894.
U.S. Patent 4,010,108.
U.S. Patent 4,108,677.
U.S. Patent 4,424,148 (Jan. 3, 1984).
U.S. Patent 3,505,217.
U.S. Patent 3,859,799 (Jan. 14, 1975).
U.S. Patent 4,151,940.
U.S. Patent 3,855,391.
U.S. Patent 3,920,795 (Nov. 18, 1975).
Uchida, Y., and Y. Nagasawa. Japan Kokai 78 102,868 (Sept. 7, 1978).
Uchikawa, H., and M. Shimoda. U.S. Patent 4,132,558 (Jan. 2, 1979).
Unimura, T. Japan Kokai 76 148,667 (Dec. 21, 1976).
Uemura, T., and E. Hirotsu. Japan Kokai 78 130,852 (Nov. 15, 1978).
———. Japan Kokai 78 14,953 (Feb. 10, 1978).
———. Japan Kokai 80 47,251 (Apr. 3, 1980).
Veda, S., and M. Ito. Japan Kokai 77 38,468 (Mar. 25, 1977).
Wakimura, Y. Japan Kokai 80 31,475 (Mar. 5, 1980).

Webster, W. C., and C. L. Smith. U.S. Patent 4,018,619 (Apr. 19, 1977).
Webster, W. C., R. G. Hilton, and R. F. Cotts. U.S. Patent 4,028,130 (June 7, 1977).
Wing, R. E. U.S. Patent 3,294,680 (Sept. 27, 1977).
Yagi, T. Japan Kokai 75,105,541 (Aug. 20, 1975).
Yagi, T., and S. Matsunaga. Japan Kokai 76,123,775 (Oct. 28, 1976).
Yokouchi, H. Japan Kokai 75 44,977 (Apr 22, 1975).
Yoshida, A. Japan Kokai 75 93,264 (July 25, 1975).
Young, D. A. U.S.Patent 4,142,912 (1979).

INDEX

Absorbents. *See* Sorption
Absorption. *See* Sorption
Acetate, 91, 110, 132, 143, 149, 165, 402
Acidity, 203, 265, 266, 531, 550
Acids, 25, 31, 32, 44–46, 53, 55, 56, 63, 71, 72, 76–81, 84, 88, 91, 100, 104–106, 110, 111, 113, 118, 123–125, 129, 133, 139, 143, 149, 174, 177, 178, 181, 192–194, 198, 204, 206, 248, 251, 253, 266, 294, 299, 319, 337, 343, 348–350, 360, 361, 363, 367, 379, 387, 392, 393, 395, 402, 409, 412–415, 424, 446, 451, 453, 458, 460–461, 463, 533, 553, 564
Activated carbon. *See* Carbon absorption
Adsorbents. *See* Sorption
Adsorption. *See* Sorption
Aeration, 94, 106, 177, 256, 300
Alkalinity, 25, 32, 55, 56, 70, 79, 82, 118, 129, 135, 139, 189, 203, 265, 266, 294, 319, 360, 361, 362, 427, 436, 443, 461, 531, 550, 567
Alset process, 494
Alsite, 310
Aluminum, 58, 94, 95, 98, 100, 102, 117, 123, 125, 130, 135, 149, 165, 176, 197, 251, 336, 338, 343, 352, 354, 364, 399–400, 402, 409, 412, 414, 418, 425
  chloride, 180
  hydroxide, 117
  oxide, 291, 293, 339–342, 432–434
  sulfate, 135, 352, 354, 358, 364
AMAX, 495

American Society for Testing and Materials. *See* Leaching tests—ASTM
Ammonium phosphate, 135, 384
Amphoterism, 71, 88, 382
Anefco, 312
Antimony, 11, 58, 61, 70, 81, 88, 103, 104, 159, 164, 165, 251, 252, 254, 318–319, 359, 385, 403, 404, 411, 439
Arsenate, 64, 66, 68, 104, 105, 110, 124, 149, 350
Arsenic, 6, 11, 24, 27–29, 38–40, 43, 44, 58, 59, 61, 70, 71, 78, 81, 83, 102–110, 133, 140, 144, 153, 165, 181, 228, 249–251, 253, 257, 281, 316, 318–319, 321, 344, 359, 399, 405, 411, 420–422, 424, 425, 439, 444, 563
  trioxide, 104, 108
  trisulfide, 105, 181
Arsenite, 64, 66, 68, 105, 107
Asbestos, 184
Asphalt. *See* Organic processes
ASTM. *See* American Society for Testing and Materials
ATCOR, 310, 452
Attapulgite, 94, 96, 367
Attenuation, 60, 94
Australia, 258

Barium, 11, 24, 27–29, 58, 61, 70, 89, 103, 110–113, 124, 129, 228, 253, 281, 316, 318–319, 321, 344, 359, 368, 385, 411, 419, 422, 425, 439, 444, 563
  chloride, 111
  chromate, 129
  compounds, general, 110–112, 129
  sulfide, 110–112

Bases, 1, 95, 105, 125, 149, 256, 280, 299, 323, 350, 409, 536, 541
Battelle Pacific Northwest Laboratories, 476
BDAT. *See* Best Demonstrated Available Technology
Bentonite, 94, 96, 117, 140, 192, 264, 354, 470
Beryllium, 11, 27, 58, 103, 254, 257, 318–319, 439
Best Demonstrated Available Technology (BDAT), 2, 5, 7, 8–13, 60, 110, 141, 144, 150, 154, 156, 280, 281, 317, 419, 438, 442, 445, 450
  constituent list, 8–11
  treatability levels, 12
Bibliography, 673–681
Binders. *See* generic CFS processes
Biocides, 351, 353
Biological oxygen demand, 398
Biological treatment/biodegradation, 2, 5, 14, 21, 33, 55, 100, 101, 176, 179, 189, 191, 194, 206, 217, 220, 255, 256, 258, 266–268, 300, 306, 308, 382, 478, 541, 543, 550
Bitumen. *See* Asphalt
Bondico, Inc., 456
Bonding, 42, 60, 86, 91, 133, 181, 338, 377
Borates, 113, 299, 350
Boron, 65, 103, 331
Boundary layer, 33, 47, 50, 52, 55, 565, 573
Bromide, 132
Browning-Ferris, Inc., 310, 494, 495, 498
Bulk density. *See* Physical properties, density
Burlington Northern, 495

INDEX

Cacodylic acid, 105, 106, 110
Cadmium, 6, 11, 36-38, 40, 58, 70, 85, 95, 103, 112-122, 228, 255, 281, 282, 316, 318-319, 331, 344, 359, 362, 365, 366, 385, 398-400, 405, 411, 419, 421, 422, 425, 440, 444
  carbonate, 40, 113
  compounds, general, 113
  hydroxide, 40, 113, 116
  sulfate, 113
  sulfide, 40, 84, 113
Calcilox, 313, 449-450
Calcium, 65, 107, 398
  aluminate, 130, 220, 341-342, 343, 352, 354
  aluminoferrite, tetra, 341-342
  carbonate, 90, 107, 129, 166, 250, 257, 358, 450, 493
  chloride, 129, 180, 206, 298, 300, 301, 350, 352, 358
  fluoride, 130, 173
  haloaluminate, 352, 364
  hydroxide, 100, 116, 250, 325, 343, 346, 379, 413, 418, 434, 493
  hypochlorite, 153
  oxide, 100, 107, 192, 291, 293, 339-342, 408, 418, 427, 432-434, 436, 460
  silicate, di, 339, 341
  silicate, tri, 339, 341
  sulfate, 34, 111, 112, 220, 248, 250, 257, 300, 306, 336, 338, 339, 341-342, 343, 350, 352, 355, 357, 364, 365, 380, 408, 412, 415, 418, 448, 449, 450
  sulfide, 83, 84, 144, 180
  sulfite, 20, 250, 257, 358
California, 24, 103, 104, 124, 156, 159, 165, 173, 183, 189, 227, 239, 365, 396, 398, 549, 552, 564
California List, 3, 6, 159, 165, 184-189
Canada, 24, 27-29, 46, 103, 141, 184-188, 227, 239, 258, 259, 306, 316, 413, 414, 424, 455, 490, 551, 565, 568, 578
Canadian Waste Technology, Ltd., 497
Carbon absorption, 42, 43, 60, 61, 94, 99, 112, 129, 143, 165, 177, 189, 192, 194, 199, 268, 351, 425
Carbonate, 32, 39, 55, 59, 71, 76, 78, 88-90, 107, 110, 111, 113, 116, 132-135, 139, 149, 165, 250, 257, 415, 450
CCW. See Concentration of Constituent in Waste
CCWE. See Concentration of Constituent in Waste Extract
CECOS, 310
Cellulose, 92, 93, 102, 129
Cement, 18-21, 35, 38, 44, 45, 61-62, 77, 78, 86-92, 100, 106, 107, 108, 112, 114-115, 117, 118, 122, 125, 129, 130, 134, 135, 139, 140, 143, 144, 147, 148, 165, 180, 192, 198, 201, 205, 206, 218-221, 223, 257, 260, 264, 266, 282, 291, 293, 294, 295, 297, 298, 300, 301, 304-306, 309, 313, 314, 316-319, 323, 325, 327-330, 335-375, 376-384, 392, 395-398, 401-403, 408, 412, 414-415, 417, 418, 420, 421, 424-425, 427, 434, 437, 442, 443, 446, 449-453, 456, 460, 463, 464, 470, 479, 481, 486, 493, 501, 505, 567, 571
Cementation, 100, 131, 143, 308, 413, 449, 450
Cement-based systems, 19, 20, 21, 35, 45, 59, 90, 91, 108, 114, 119-120, 126-130, 136-139, 145-146, 148, 151-152, 155, 157-158, 160-164, 190-191, 266, 300, 301, 303, 305, 306, 309, 313, 315, 321-322, 324, 326-329, 335, 343, 346, 348, 352, 353, 355, 357, 359, 363, 365, 366, 368, 383, 403, 408, 437, 446, 451, 452, 464, 469, 567, 571
  accelerators, 351-352, 353
  additives, 335, 336, 349, 351-355
  case studies, 364-373
  cement/clay, 354, 383
  cement/lime, 353-354
  cement types, 337-343
  chemical properties, 337-352, 357-359
  compositions, 130, 337-343
  dispersants, 348
  hydration, 337-339
  inhibition, inhibitors, 348-351
  leaching, 335, 339, 343, 344, 354, 357-364, 366
  phase diagram, 342
  physical properties, 337-352, 355-357
  pilot studies, 359-364
  setting, 335, 337, 343-348, 351, 352, 357, 364, 367
  strength, 356
  volume increase, 351
Cement/flyash processes, 19, 20, 21, 38, 39, 109, 115, 121, 122, 128, 138, 140, 146, 152, 155, 158, 161, 163, 192, 223, 301-303, 305, 313, 319, 321-322, 326-329, 335, 351, 353, 357, 375, 383, 393, 394, 395, 417-426
  additives, 417, 424
  chemical properties, 418-424
  compositions, 418
  hydration, 418, 420
  leaching, 418-421, 424-425
  physical properties, 418, 421-424
  proprietary processes, 424-425
  setting, 417, 420
Cement kiln dust, 220, 282, 325, 427, 428-429, 432, 436, 437, 438, 443, 444-445, 446. See also Kiln Dust and Flyash Based Processes
Cement/soluble silicate processes, 19, 20, 21, 91, 109, 115, 120-121, 126-128, 135-138, 140, 143, 144, 146, 152, 155, 158, 161-165, 190-191, 192, 223, 260, 264, 266, 295, 297, 300, 305, 319, 321-322, 326-330, 353-406, 452, 463, 479
  accelerators, 380
  additives, 397, 403
  case studies, 403-405
  chemical properties, 382
  compositions, 378, 392, 395, 402
  gelation, 377-380
  hardening, 388
  hydration, 377, 379, 380
  inhibition, inhibitors, 380, 380-382, 396, 403
  leaching, 377, 381-383, 386, 387, 392, 393, 397-399, 403, 405
  metal fixation, 380
  physical properties, 383-386, 388-392
  pilot studies, 386-395
  proprietary processes, 395-402
  setting, 378-380, 392, 395, 396, 398, 401, 402
CERCLA. See Comprehensive Environmental Response
Chelation, 40, 74
Chemfix, Inc., 20, 21, 131, 305, 311, 386, 395, 396-400, 406, 507
Chemical fixation. See Fixation
Chemical oxygen demand, 397, 398-399, 404
Chemical potential, 47, 48, 52
Chemical properties, 314-322
Chemical Waste Management, Inc., 310-311, 442-443, 484, 485, 486, 494, 498, 509, 521-524, 536
Chemisorption, 33, 34, 43, 60, 93
Chemline Corp., 496
Chem-Met Services, 311, 409, 496
Chem-Nuclear Systems, Inc., 310, 452
Chem-Technics, 310, 311, 460, 491, 492
Chloride, 89, 92, 94, 106, 110, 111, 113, 132, 134, 140, 142, 144, 149, 153, 156, 172, 179, 193, 206, 255, 257, 266, 298, 300, 301,

684 INDEX

Chloride (cont.)
339, 350, 352, 357, 397, 404, 415, 532
Chlorinated organics. See Organic compounds, chlorinated, halogenated
Chromic acid, 79, 80, 123–125, 458
Chromium, 11, 24, 27–29, 35, 37, 38, 43, 44, 46, 58, 61, 70, 71, 75, 78, 81–83, 85, 88, 91–95, 102, 103, 118, 122–130, 132, 135, 140, 153, 197, 228, 248–253, 255, 257, 261, 281, 282, 299, 316, 318–319, 321, 331, 344, 352, 354, 357, 359, 362, 363, 365, 366, 382, 385, 393, 398–400, 408, 411, 413, 419, 423, 424, 425, 440, 444, 567
Clay, 23, 34, 43, 44, 61, 74, 75, 94–98, 122, 133, 194, 198, 199, 219, 220, 268, 287, 292, 294, 300, 305, 319, 323, 327, 336–338, 343, 349, 351, 353–355, 367, 380, 384, 385, 401, 407, 412–414, 417, 421, 424, 437, 458, 460, 534, 545
Clean Water Act, 4, 246
Cobalt, 28, 58, 103, 159
  chloride, 352
COD. See Chemical oxygen demand
Compaction, 30, 257, 449, 450, 544, 545, 549
Complex/Complexation, 7, 13, 16, 20, 33, 35, 36, 38, 46, 60, 63, 70–75, 85, 89–91, 93, 94, 95–98, 106, 110, 113, 116, 130, 135, 142, 149, 156, 159, 165, 175, 177, 179, 206, 219, 220, 222, 224, 225, 280, 286, 292, 304, 307, 309, 321, 342, 348, 357, 364, 377, 379, 380, 382, 397, 408, 409, 490, 504, 511, 532, 533, 535, 541
Comprehensive Environmental Response, Compensation and Liability Act (CERCLA), 3, 4, 5, 7
Compressive strength. See Physical properties, compressive strength
Computer applications and systems, 571–577
Concentration of Constituent in Waste (CCW), 27–29, 183, 228
Concentration of Constituent in Waste Extract (CCWE), 27–29, 183, 184–188, 189, 228, 562–563
Concrete. See cement-based CFS processes
Contaminated earth, 367
CONTEC Waste Data-Base, 580–626

Convective transport, 33, 48, 55
Conversion Systems. See IU Conversion Systems
Copper, 11, 28, 34, 43, 58, 61, 70, 72, 74, 84–88, 91, 94, 95, 100–105, 112, 113, 124, 144, 147, 153, 154, 156, 159–161, 172–174, 176, 179, 181, 194, 249–255, 282, 298, 300–303, 316–319, 331, 337, 344, 350, 352, 360, 365, 380, 381, 385, 398–400, 411, 419, 423, 424, 425, 440, 444
  hydroxide, 350
  nitrate, 300, 350, 352
  sulfate, 352
  sulfide, 84
Coprecipitation, 143
Corrosion, corrosivity. See Physical properties, corrosivity
Costs, 225–226, 310–312, 322–329
Curing, 14, 90, 99, 189, 198, 206, 216, 222, 282, 286, 287, 290–291, 293–298, 300, 304, 313, 314, 316, 317, 320, 337, 346–348, 351, 363, 379, 403, 414, 434, 437, 442, 453, 454, 513–514, 541, 543–545, 547, 578
Cyan-Cat process, 177–178
Cyanide, 11, 28, 63, 72, 82, 94, 112, 113, 116, 147, 149, 154, 156, 159, 172–180, 248–250, 252, 398–400, 405, 413, 424, 441, 470

Delaware Custom Materials, 311
Delisting, 2, 12, 125, 272, 419, 438, 443, 445, 564
Delivery Systems, 221–223, 310–312, 478–514
Desorption, 41, 42, 177, 178, 443
Dewatering, 46, 86, 216, 217, 249, 251, 282, 309, 378, 412, 448, 463, 464, 478, 488, 541
Diatomaceous earth, 351, 354
Diesel fuel, 190
Diffusion, 30, 33, 47–50, 53, 75, 122, 129, 139, 140, 173, 261, 294, 346, 360, 421, 552, 565, 566, 573
  coefficient, 49, 566
  modeling, 552, 565
Diodochy, 34
Dioxins. See Organic compounds, dioxin
Disposal facility, 223–225
Dissociation constant, 34
Dissolution, 23, 43, 45, 47, 55, 83, 88, 129, 139, 182, 292, 348, 358, 361
Dolomite, 135
Double layer, 51

Dow Chemical Co., 312, 454
Dravo Lime Co., 449
Drinking Water Standards—NIPDWS, 174, 316, 419, 425, 439, 445
Durability, 266, 313, 346, 347, 349, 549, 578

EATA. See Environmental Technology Associates
Ebara-Infilco Co. Ltd., 310
Electrochemical/electrolytic, 79, 100, 101, 129, 177, 178, 179, 267, 541
Elution, 31
Embeddment, 33, 45, 46, 60
Encapsulation, 2, 5, 26, 33, 45, 46, 60, 218, 219, 307, 308, 324, 376, 393, 456, 457
Enreco, 311, 466
Envirite, 311, 498
EnviroGuard, 396, 401, 463
Environmental Technology Associates, 396, 401, 457, 459, 462
Environmental Technology Corp., 86, 310, 365
Envirosafe Services, Inc., 409, 412
Envirostone, 311
Epes Lime Co., 311
Epsom salt, 179
EPT. See Leaching tests, EPT
Equilibrium constant, 33, 42
Equipment, 14, 287
  fixed, 288, 478–490, 491–498, 505, 511
  in-situ, 3, 50, 88, 98, 139, 144, 207, 217, 222, 253, 296, 297, 309, 323, 409, 448, 450, 464–476
  mobile/portable, 289–291, 478, 490, 493, 499–513
ETA Process. See Environmental Technology Associates
Ettringite, 122, 338, 343, 345, 348, 349, 363, 408
European, 552, 567
Evaporation, 177
Expert systems, 572–577
Extraction Procedure Toxicity Test (EPT). See Leaching tests, EPT
Extraction tests. See Leaching tests
Extraction tests, sequential chemical, 70, 532, 550, 578

F-wastes. See Waste types and sources, F-wastes
Ferric chloride, 89, 144, 153, 415
Ferric hydroxide, 107, 166, 192, 414
Ferric oxide, 291, 293, 339–342, 432–434
Ferric nitrate, 358
Ferric sulfate, 107, 144, 358

INDEX

Ferricyanide, 72, 90, 175
Ferrocyanide, 90, 175, 177, 180
Ferrous chloride, 180
Ferrous sulfate, 34, 79-81, 86, 129, 130, 135, 180, 354, 413, 414
Ferrous sulfide, 83, 84
FERSONA, 102, 309
Fixation
  agents, inorganic, 23-57
  agents, organic, 41
  definition of, 2
  in solids and soils, 40
  long-term stability, 75-76
  mechanisms, 32-46, 52, 59, 76-102, 118, 173, 357, 363
  of complex species. See Complex/Complexation
  of metals, 23-57, 58-165
  of nonmetal inorganics, 172-182
  of organics, 40, 183-199
  processes, 23, 125, 129, 144, 173, 334, 379
Flammability. See Physical properties, flammability
Flash point. See Physical properties, flash point
Fluoride, 11, 28, 40, 63, 94, 113, 130, 132, 172, 173, 193, 248, 251, 257
Flyash, 18-21, 38, 61, 70, 77, 94, 95, 117, 118, 122, 135, 140, 192, 205, 219-221, 223, 257, 261, 291, 292, 295, 300, 301, 304-306, 313, 316-319, 323, 325, 327-329, 335, 347, 351, 353, 357, 367, 376, 392, 393, 403, 407-409, 412, 415, 419, 427, 434, 437, 442, 446, 449, 463, 465, 479, 532, 534, 545
  See also Kiln Dust and Flyash Based Processes
Foam separation, 177
Fourier transform infrared spectroscopy, 70
FPL/Qualtec, 311
France, 258, 259, 396, 402, 479, 480-483, 486
Free radical, 195, 196
Free water content, 294, 401
Freeze/Freezing, 14, 30, 177, 217, 314, 337, 355, 385
Fujibeton, 310-311, 398
Fujimasu, 310, 396, 398
Fujisash Industries Ltd., 364, 494
Fulvic acid, 91, 109

Gasoline, 190
Gelation, 266, 376-379, 381, 382, 403, 454
Geo-Con, 469-475
Germany, 258, 259
Glassification, 308, 324, 450, 473
Glossary, 659-672
Groundwater, 47, 50, 52, 53, 55, 122, 134, 140, 204, 330-332, 403, 564, 567
GSX, 494, 498
Gypsum. See Calcium sulfate
Gypsum-Based Processes, 365, 448-449

Halogenated organic compounds (HOCs). See Organic compounds, halogenated
Hardening. See Curing
Harmon Environmental Services, 467-468, 498
Hazardous and Solid Waste Amendments (HWSA), 4, 5
Hazardous waste
  generation, 248, 259
  landfill, 48, 404, 548
  management, 46, 258, 409, 427, 450, 464
Heavy metals. See Individual metals
Hitachi, 396
Hittman Nuclear Corp., 310, 312, 452
Homogenization, 207, 208, 214, 260, 264, 479, 487, 499, 503, 529
HSWA. See Hazardous and Solid Waste Amendments
Humic substances. See Organic compounds, humic substances
Humidity, 297, 298, 314, 543, 544
Hydration, 77, 78, 148, 198, 293, 294, 337-339, 343, 345, 346, 347, 348, 349, 351, 352, 357, 377, 379, 380, 418, 420, 427, 437, 550
Hydraulic conductivity, 314
Hydrazine, 79, 81, 82
Hydrocarbons. See Organic compounds, general
Hydrofluoric acid, 412
Hydrogen bonding, 42
Hydrogen peroxide, 107, 108, 116, 153, 179, 180, 192, 197
Hydrolysis, 40, 191-195, 197, 377
Hydrophobizing, 100, 101, 415
Hydroxide, 35-40, 44, 45, 59, 63, 71, 76-79, 82-85, 87-90, 92, 93, 110-113, 116-118, 122, 124, 125, 129, 132, 134, 139, 140, 149, 178, 181, 204, 248, 250, 298, 300, 301, 304, 343, 346, 350, 352, 361, 363, 379, 381, 382, 413, 414, 418
Hypophorous acid, 143

Ignitability. See Physical properties, ignitability

Illite, 94, 96, 367
Incineration, 41, 74, 75, 116, 177, 178, 189, 225, 252, 256, 281, 397, 479, 493
Industrial Waste Management, 310, 496-497
Information systems/databases, 204, 519-528, 534-541, 570-577
Injection, 246, 250, 376, 464, 467, 470
Inorganic compounds. See individual compounds
In-situ, 3, 50, 88, 98, 139, 144, 207, 217, 222, 253, 296, 297, 309, 323, 409, 448, 450, 464-476
Insoluble starch xanthate, 153
Interface, 49-51, 53, 60, 61, 93, 181
Iodide, 132
Ion exchange, 41, 60, 99-100, 129, 143, 300
Iron, 331, 398, 400, 425
Irradiation, 177
Iso-Clear, 102
Isocyanate, 354
Israel, 258
IU Conversion Systems, Inc., 20, 409, 412, 496

Japan, Japanese, 88, 94, 107-112, 131, 141, 144, 147, 183, 184, 258, 259, 339, 353, 364-366, 396, 398, 405, 417, 424, 425, 456, 490, 512, 552, 567, 571
Japan Organo, Ltd., 396
Jet cement, 125, 129, 130, 352, 358

K-wastes. See Waste types and sources, K-wastes
Kaolin, 94, 95, 122, 413
Kastone process, 177, 179
Kepone. See Organic compounds, pesticides
Kiln dust, 61-62, 118, 135, 190, 192, 220, 282, 300, 305, 306, 313, 317, 323, 325, 327, 368, 407, 427, 434, 437, 442, 443, 446, 449, 465
Kiln dust and flyash based processes, 38, 61, 70, 77, 94, 95, 108-109, 114-115, 117, 118, 119, 120, 122, 126, 127, 136, 137, 140, 145, 151, 155, 157, 158, 160, 162, 164, 180-191, 192, 205, 219-221, 223, 257, 261, 291, 292, 293, 295, 300, 301, 304-306, 313, 315, 316-319, 320, 323, 325, 326, 327-329, 335, 347, 351, 353, 357, 367, 376, 392, 393, 403, 407-409, 412, 415, 417, 418, 420, 421, 424-425, 427-447, 449, 463, 465, 479, 532, 534, 545

Kiln dust and flyash (cont.)
    additives, 442, 443
    case studies, 446–447
    chemical properties, 427, 432–434, 437–442, 444–445
    cure time, 438
    hydration, 427, 437
    leaching, 444–445, 437–441
    physical properties, 427–431, 434–437
    proprietary processes, 442–446
    setting, 437, 442
Kurita, 311, 396

Lagoon, 106, 207, 208, 211, 213–215, 256, 259, 378, 382, 397, 465, 500, 501, 503, 515, 517, 534
Laidlaw Transport, Ltd., 497
Landban, 3, 5–13, 41, 60, 84, 93, 104, 110, 118, 123–125, 141, 150, 154, 156, 183, 184–188, 189, 250, 280, 281, 282, 309, 317, 392, 437, 442, 443, 458, 463, 483, 563
Landfill, 25, 26, 30–32, 48, 52, 60, 61, 75, 85, 94, 173, 181, 206, 217, 218, 222, 223, 225, 226, 246, 248–250, 252, 253, 256, 257, 259, 268, 304, 314, 323, 325, 327, 330, 357, 367, 368, 397, 403–405, 417, 424, 437, 443, 446, 447, 450, 456, 458, 465, 478, 480, 482, 534, 545, 548–550, 553, 578
Leachant, definition of, 23
Leachate, definition of, 23
Leachability, 24–25
    definition of, 23
    factors affecting, 26, 30–33
    index, 50, 110, 566
    measurement of, 24–25
    rate, 23
Leaching, 23–32
    criteria, 25–26
    definition of, 23
    interval, 566
    mechanisms, 33, 46–56, 63, 565
    rate, 23–24, 44, 46–47, 50, 52, 54, 55, 101, 139, 294, 316, 332, 361, 363, 393, 420
Leaching tests, 550–568, 627–658
    accelerated, 553
    agitation, 555–556
    analyses, 32, 560–561
    ANS 16.1, 50, 360, 420, 449, 552, 565–567, 627, 655–658
    ASTM Test Procedures, 1, 58, 110, 121, 146, 152, 163, 167, 301, 337, 338, 422, 515, 530, 531, 543, 545, 548–550, 552, 564, 565, 567, 570

California Waste Extraction Test (WET), 30, 103, 132, 156, 159, 184–188, 218, 246–253, 257, 265, 268, 296, 314, 336, 348, 353, 355, 366, 398, 418, 419, 451, 534, 546, 549, 552, 564, 552, 564, 627, 651–655
column, 393, 398–400, 518, 552, 564, 567
contact time, 31, 541
diffusion modeling tests, 552, 565
dynamic, 321–323
Extraction Procedure Toxicity Text (EPT), 3, 24–32, 48, 75, 93, 101, 103, 107–109, 114–115, 118–121, 126–128, 135–139, 145–146, 151–152, 155, 156–158, 160–164, 184–188, 189, 190–191, 228, 331, 358, 360, 366, 385, 392, 411, 419, 422, 425, 439, 445, 552, 553, 563, 564, 568, 627–634
Equilibrium Leach Test (ELT), 109, 115, 121, 128, 138, 146, 152, 155, 161, 163, 422, 552, 567, 568
European test procedures, 399, 552, 567
flow rate, 567
flow-through tests, 552, 567
Japanese test procedures, 552, 567
long-term, 553, 564–567
MCC Tests, 552, 566, 567
Multiple Extraction Procedure (MEP), 3, 25, 31, 115, 121, 128, 138, 146, 152, 155, 158, 161, 163, 422, 425, 439, 552, 553, 564, 627, 634–636
Oily Waste Extraction Procedures (OWEP), 25, 101, 552, 553, 564, 627, 636–638
Ontario (Regulation 309) Province Test, 552, 554–561, 564
particle size reduction in, 555
pH, 32, 531, 550, 558–559
phase separation, 554–555
quality assurance, 560–561
Quebec, 554–561
separation of phases, 32, 554–555, 558–559
sequential batch extraction tests, 35, 552, 564, 567
shake, 564, 567
single batch, 552, 567
Solid Waste Leaching Procedure (SWLP), 58, 363, 552, 567
special and experimental tests, 567
static, 552, 566
Static Leaching Test (SLT), 110, 552, 566

structural integrity test, 555
temperature effects, 31, 533, 547, 552, 558–559
Toxicity Characteristic Leaching Procedure (TCLP), 3, 24–32, 45, 48, 61, 84, 86, 88, 89, 101, 103, 108–109, 114–115, 118–121, 126–128, 136–139, 144–146, 150–152, 155, 156–158, 160–164, 184–188, 189, 190–191, 228, 229, 281–283, 316–319, 342, 343, 358, 365, 366, 385, 411, 439, 444–445, 549, 552, 553, 562–564, 567, 568, 627, 638–651
vessel, 556–557, 565
water, 358, 411
WET. See California Waste Extraction Test
Lead, 6, 11, 24, 27–29, 35, 37, 39, 43, 44, 58, 61, 70, 71, 74, 75, 81, 84, 85, 87–89, 91, 94, 102, 103, 104, 105, 112, 118, 122–124, 129–140, 154, 159, 165, 176, 193, 228, 248–255, 257, 262, 263, 281, 283, 298–303, 316–319, 331, 337, 344, 350, 351, 354, 357, 360, 362, 363, 365, 366, 380–382, 386, 398–400, 405, 411, 419, 423, 425, 440, 444, 454, 472, 532, 535, 563, 567
    carbonate, 131, 134, 408, 412, 414, 421, 424
    basic, 134
    chromate, 123, 129
    compounds, general, 131–134
    hydroxide, 131, 134
    minerals, 131
    nitrate, 133, 300, 350
    oxide, 131
    phosphate, 134
    sulfate, 131
    sulfide, 84, 132
Ligands, 63, 72, 73, 113, 116
Lime. See Calcium oxide, and Calcium hydroxide
Lime/Flyash, Lime-Based Processes, 23, 45, 55, 108–109, 114–115, 117, 118, 119–121, 126–128, 135–138, 145–146, 151–152, 155, 157–158, 160–164, 192, 220, 300, 301–303, 305, 313, 315, 316, 320–322, 326, 327, 365, 383, 407–416, 421, 646
    additives, 414, 415
    chemical properties, 407–409, 415
    compositions, 407
    inhibition, inhibitors, 408
    leaching, 408, 414–415
    lime/clay, 412–414
    lime/ferric/ferrous ion, 410
    lime/flyash, 409–415
    lime/kiln dust, 407

INDEX 687

lime/sulfide, 410
physical properties, 407, 408–409, 413, 415
proprietary processes, 409–415
setting, 408, 412
Lime Kiln Dust, 220, 305, 325, 327, 427, 430, 433, 436, 437, 438, 446, 460. *See also* Kiln Dust Based Processes
Liner, 208, 223, 518
Lithium, 144
Long term durability, 266, 347
Lopat, 131, 396, 398
LSU. *See* Louisiana State University
Louisiana State University, 198, 312

Magnesium, 58, 78, 84, 87, 95, 156, 179, 248, 298, 299, 339, 350, 352, 354, 376, 381, 393, 402, 414, 418, 437, 450
chloride, 339, 352
oxide, 291, 293, 339–342, 352, 354, 418, 432–434, 436
Manganese, 44, 58, 87, 94, 98, 105, 111, 172, 173, 248, 250, 253, 331, 337, 381, 424
Manifest, 519
MEP. *See* Leaching tests, Multiple Extraction Procedure
Mercury, 6, 11, 24, 27–29, 58, 61, 70, 74, 81, 83–86, 88, 91, 93, 140–144, 147, 148, 181, 228, 248, 250, 251, 253–255, 257, 281, 318–319, 331, 344, 353, 357, 360, 386–398, 405, 411, 419, 423, 425, 440, 444
Mercury compounds, general, 142–143
Mercury sulfide, 84, 147
Mineral absorbents, 217, 354
Mineral By-Products, Inc., 306, 311
Minimum Technology Unit/Requirement (MTU/MTR), 3, 48, 61
Mixing, 25, 45, 74, 79, 100, 125, 189, 205–207, 214, 216, 218, 221, 261, 262, 264–266, 284, 286, 295–297, 355, 364, 369–370, 373, 392, 397, 414, 415, 417, 450, 453–455, 463–465, 467, 468, 469, 479, 480, 483, 487, 493, 541, 543, 544
Molecular diffusion, 33, 47, 48, 50, 360
Molybdenum, 29, 59, 109, 159, 250
Monofill, 3
Monoliths, 15, 26, 381, 454, 553

Morphology, 260–262, 345, 346
MTU/MTR. *See* Minimum Technology Unit/Requirement
Multiple Extraction Procedure. *See* Leaching tests, Multiple Extraction Procedure

Nascent oxygen, 177, 179
National Interim Primary Drinking Water Standards (NIPDWS), 25
National Lime Association, 465
National Pollutant Discharge Elimination System (NPDES), 46, 172, 173
NECO, 312
Nernst potential, 51
Newport News, 312
Nicherecki Chemical Co., 456
Nickel, 6, 11, 36–37, 46, 59, 61, 70, 74, 81, 82, 84, 85, 87–89, 91–94, 102, 103, 110, 113, 132, 148–153, 174, 175, 181, 248–251, 253, 255, 281, 283, 316–319, 331, 344, 361, 365, 381, 382, 387, 398–400, 411, 419, 423, 424, 425, 440, 444
Nickel compounds, general, 148–149
Nickel sulfide, 84, 149
NIPDWS. *See* National Interim Primary Drinking Water Standards (NIPDWS)
Nippon Synthetic Chemical Co., 311, 396
Nitrate, 29, 91, 110, 111, 113, 124, 129, 132, 133, 143, 149, 154, 156, 172, 173, 300, 350, 352, 357, 360
Nitrite, 29, 172
Nitrogen compounds, general, 398
NPDES. *See* National Pollutant Discharge Elimination System
Nuclear. *See* Waste types and sources, radioactive
NuKem, 310
NUS, 310, 494
N-Viro Energy Systems, 311, 446

Oak Ridge National Laboratory, 310, 311, 366, 424
Occupational Safety and Health Administration (OSHA), 4, 132, 531
Off-site, 3, 249, 252, 253, 368
Omni-Fix, 102
On-site, 3, 220, 222, 247–249, 252, 253, 255, 256, 258, 259, 323, 330, 357, 363, 367, 368, 377, 396, 403,

404, 414, 446, 451, 452, 464–476, 486, 501, 502
Onoda Cement Co., Ltd., 310
Ontario Liquid Waste Disposal, 311, 413, 414
Ontario Waste Management Corporation, 46, 239, 497
Optimization, 380, 490, 515, 553, 566
Organoclays, 43, 61
Organic compounds
 acetamides, 194
 acetic acid, 25, 31, 32, 360, 393, 395, 402, 458, 553, 564
 acetone, 533
 acid, 25
 acid chlorides, 299, 349
 acrylamide, 219, 351
 acyl chlorides, 194
 acylating, 194
 adipic acid, 299, 350
 alcohols, 266, 298, 299, 349
 aldehydes, 180, 299, 349
 alkyl chlorides, 193
 alkyl halides, 193, 194
 alkylating, 194
 aluminum stearate, 354
 amides, 194, 298, 299, 402
 amines, 298, 299, 301
 benzene, 190, 191, 192, 197, 299, 350, 458
 bis-ethylhexyl phthalate, 191
 bromophenol, 198, 299, 350
 butadiene, 458
 butyl chloride, 193
 calcium lignosulfonate, 354
 calcium oxalate, 192
 calcium tetraformate, 351
 carbamates, 194
 carbon tetrachloride, 458
 carbonyls, 299, 349
 carboxylates, 349
 chlorinated organics, 192, 194, 197, 200, 218, 219, 248, 249, 253, 299, 304, 307, 349, 351, 450
 chloroform, 193, 458
 chlorohexadecane, 190
 citric acid, 351, 393, 394, 395, 564
 coal, 61, 206, 253, 257, 313, 337, 407, 409, 418, 427, 449, 450, 455, 490
 cresol, 458
 cyanogen chloride, 179
 dichlorobenzene, 458
 diethylether, 458
 dimethylthiocarbamate, 153
 dioxin, 6, 458
 dithiocarbamate, 83
 EDTA, 72, 73, 85, 107, 108, 113, 150, 159, 299, 350, 384
 epoxides, 194, 299
 esters, 194, 402
 ethers, 299
 ethylbenzene, 190, 191
 ethylene glycol, 192, 198, 299, 350, 351, 404
 formaldehyde, 19, 55, 92,

INDEX

Organic compounds (cont.)
  143, 153, 179, 218, 219, 299, 304, 307, 350, 351, 402, 451-453
  fungicides, 105, 125, 142
  gasoline, 458, 461
  general, 27-29, 34, 184-188, 228-229, 460-461
  gluconic acid, 351
  d-gluconolactone, 367
  glycols, 299, 349
  grease, 82, 299, 301-303, 319, 348, 349, 351
  guar gum, 460
  halogenated (HOCs), 6, 8-12
  herbicides, 25
  heterocyclics, 299, 349
  hexachlorobenzene, 298, 299, 301-303, 304, 350
  hexadecylmercaptan, 86, 109, 115, 120-121, 127-128, 137-138, 144, 146, 152, 158, 161, 163, 384
  humic substances, 40, 75, 91, 109, 134
  insoluble starch xanthate, 92, 110
  kepone, 190-191
  ketones, 299, 349
  lactic acid, 351
  lactones, 194
  lignins, 299, 349
  malic acid, 351
  methanol, 299, 350
  methyl bromide, 193
  methyl chloride, 193
  methyl fluoride, 193
  methyl iodide, 193
  B-naphthalene sulfonate, 367
  nitriles, 174, 175, 180, 194
  nitrilotriacetic acid, 192, 201
  NTA, 299, 350
  oil, 4, 82, 100, 101, 183, 203, 204, 247, 249, 250, 252, 257, 260, 264, 268, 298, 299, 301-303, 304, 307, 319, 331, 348, 349, 351, 380, 382, 414, 425, 455, 461, 529, 530, 533, 534, 564
  oxalates, 149
  oxalic acid, 91, 198
  paraffins, 100, 354
PCB. See polychlorinated biphenyls
pesticides, 11, 25
phenol, 82, 192, 197, 200, 201, 250, 251, 299, 301-303, 304, 331, 351, 360, 368, 398-400, 404
phenolic, 252
polychlorinated biphenyls (PCBs), 5, 6, 190, 204, 255, 405
polydithiocarbamate, 143, 146
polyethylene, 91, 218, 307, 352, 456, 458
polyethylene glycol, 352
polyethylene imine, 91, 107

polyoxyethylene, 354
polyvinyl acetal, 354
polyvinyl alcohol, 354
polyvinyl chloride, 458
resins, 79
salicylic acid, 351
semi-volatile, 9-11
sodium lignosulfonate, 354
solvents, 6, 41, 116, 124
starches, 92, 93, 110, 144, 153, 177, 180, 299, 349, 460
sugars, 111, 112, 129, 299, 349
sulfonates, 299, 348, 349
sultones, 194
tannins, 299, 349
tartaric acid, 351
tartrates, 64, 66, 69
thioamide, 83
thiourea, 83, 86, 113, 143, 144, 146, 147, 358
toluene, 190, 191
trichloroethane, 458
trichloroethene, 192
trichloroethylene, 192, 201, 299, 301-303, 350
triethanolamine, 351
triethylene glycol, 352
volatile organics (VOCs), 8-9, 189
xanthates, 83, 86, 92, 93, 102
xylene/xylenes, 189, 191, 299, 350
Organic polymers. See Organic processes
Organic processes, 190-191, 218-219, 266, 307-309, 451, 452
  asphalt, 19, 45, 219, 324, 455, 456
  polymer, 19, 45, 218-219, 312, 324, 448, 449, 452-454, 456, 458
  thermoplastic, 19, 45, 219, 307, 308, 312, 324, 452, 455, 456, 458
  urea-formaldehyde, 19, 55, 218, 451-453
Organization of the book, 15-16
Organo-clays, 192, 199
ORNL. See Oak Ridge National Laboratory
ORP. See Redox
OSHA. See Occupational Safety and Health Administration
OWEP. See Leading tests, Oily Waste Extraction Procedures
Oxalate, 192
Oxidation, 31, 39, 55, 63, 78, 79, 82, 100, 106, 116, 132, 147, 148, 153, 175, 177-179, 181, 189, 191, 192, 195-197, 266, 267, 281, 294, 550
Oxidation/Reduction. See Redox
Oxide, 59, 63, 93, 98, 100, 105, 111, 113, 123, 129, 132, 143, 144, 148, 149, 156, 291, 350, 352, 354, 376, 378, 393, 408, 412, 418, 427, 434
Ozone, 82

Particles/particle size, 26, 30, 33, 41, 44-46, 50, 63, 75, 87, 101, 129, 139, 144, 203, 205, 260-264, 293-295, 319, 337, 359, 361, 381, 393, 428-431, 437, 454, 529, 553, 563, 566, 567
Partitioning, 59, 108
Passivation, 34, 60
PCBs. See Organic compounds, polychlorinated biphenyls
Perlite, 367
Permeability. See Physical properties, permeability
Pesticides. See Organic compounds, pesticides
Petrifix, 311
pH, 25, 31, 32, 35-37, 39-41, 43-46, 50, 55, 59, 61, 63, 71-74, 76-85, 87-89, 91-94, 98, 99, 101, 106, 107, 116, 118, 122, 134, 135, 139, 140, 144, 148-150, 159, 173, 178, 180, 193, 194, 203, 204, 261, 265, 266, 282, 284, 292, 294, 300, 309, 318, 319, 320, 323, 348, 356, 360, 361, 362, 377-382, 388-391, 393, 395, 401, 408, 414, 415, 420, 424, 428-431, 435, 438, 443, 446, 453, 464, 531, 542, 550
Phenolics. See Organic compounds, phenolic
Phosphate, 59, 63, 72, 89-90, 132, 134, 135, 139, 149, 194, 254, 331, 337, 352, 367, 397, 398-399, 417
Phosphonates, 194
Phosphonohalidates, 194
Phosphoric acid, 105, 106, 181, 194, 402, 415, 446
Physical properties/test methods, 313-314, 548-550
  Atterberg limits, 549
  bearing strength, 356, 388-391, 435, 545, 548
  bulk density, 19, 78, 203, 260, 265, 314, 355, 443, 530, 531, 532, 550
  California bearing ratio, 549
  color, 143, 148, 192, 204, 260, 529, 531, 544, 547
  compacted density, 30, 257
  compressive strength, 217, 301, 314, 317, 352, 355, 363, 367, 368, 383, 404, 447, 545, 548, 549
  corrosivity, 46, 105, 113, 123, 124, 130, 132, 228
  dry density, 550

fineness, 130, 346, 418, 428–431
flammability, 203, 530
flash point, 203, 260, 265, 531
freeze/thaw, 14, 30, 217, 218, 314, 337, 355, 385, 549
gel time, 313
grain size, 203, 530, 531, 549
hardness, 450, 545
ignitability, 183, 228, 543, 550, 553
LOI. See loss on ignition
loss on ignition, 418, 432–434, 436, 443
moisture content, 78, 175, 257, 418, 443, 549, 550
moisture-density relationship, 549
odor, 2, 84, 174, 180, 192, 204, 255, 260, 364, 397, 414, 529, 531, 544, 547
penetrometer hardness, 355, 356, 384, 388–391, 435, 543–545, 547, 549
permeability, 45, 46, 48, 52, 53, 55, 87, 101, 173, 219, 220, 298, 306–308, 314, 336, 346, 347, 354, 355, 381–383, 385, 387, 415, 450, 456, 470, 471, 543, 549, 550, 567, 578
porosity, 549
reactivity, 203, 228, 260, 530, 543, 550, 553
shear strength, 51, 205, 263, 264, 297, 454, 533, 548, 549
solids content, 202, 205, 218, 256, 259, 262–264, 282, 283, 294, 295, 327–329, 356, 379, 388–391, 397, 405, 413, 435, 449, 480, 490
specific gravity, 203, 205, 213, 260, 262, 295, 314, 355, 378, 418, 530, 531, 550
strength, 14, 20, 31, 32, 36, 42, 46, 52, 113, 217, 222, 298, 301, 304, 313, 314, 317, 336–338, 343, 346, 349, 351–353, 355, 356, 363, 367, 368, 383, 384, 388–391, 404, 409, 412, 418, 435, 436, 437, 438, 447, 454, 543–545, 546, 548, 549
unconfined compressive strength, 545, 549
viscosity, 203, 205, 206, 260–264, 266, 267, 294–297, 305, 354, 367, 379, 417, 424, 470, 517, 518, 529, 531–534, 542, 544, 574
volume increase, 79, 305, 306, 308, 355, 356, 368, 384, 385, 388–391, 397, 403, 421, 424, 435, 437, 446, 532, 541, 544, 548
water content, 19, 87, 197, 217, 248, 258, 293, 294, 313, 314, 336, 355, 377, 380, 383, 401, 463, 475, 480, 549, 550
wet/dry cycle, 30, 218, 549
Pilot studies, 329–332, 359, 386
PNAs. See Organic compounds, polynuclear aromatics
Polychlorinated biphenyls (PCBs). See Organic compounds, polychlorinated biphenyls
Polymerization, 113, 177, 180, 219, 304, 307, 424, 451–454
Polynuclear aromatics (PNAs). See Organic compounds, polynuclear aromatics
Polynucleolyte, 102
Pore solution, 33, 47, 49, 346, 348
Pore structure, 381, 383
Porosity, 41, 49, 355, 383
Portland cement. See Cement-Based Systems
Potassium, 58, 124, 135, 144, 153, 172, 174–177, 179, 192, 197, 398, 408
Potassium oxide, 432–434
Potassium permanganate, 179, 192, 197
POTW. See Publically Owned Treatment Works
Poz-O-Tec, 409, 412
Pozzolanic reactions, 418
Pozzolans, 77, 260, 293, 335
Precipitation, 30–31, 38, 40, 43, 44, 55, 59, 60, 63, 71, 74, 76, 78, 82–86, 88–91, 94, 98, 106, 107, 109, 111, 116, 118, 129, 134, 135, 144, 147, 153, 156, 159, 181, 192, 217, 300, 343, 346, 352, 376, 379, 381
Pretreatment, 46, 74, 83, 123, 149, 246, 252, 256, 309, 379, 413, 414, 424
Process equipment. See Equipment
ProTek, 396, 401
Pseudoplastic, 205, 263, 264, 297
Publically Owned Treatment Works (POTW), 3, 4, 46

Quality Assurance, 282, 553
Quality Control, 50, 282, 442, 490, 500, 504, 553, 568, 578

Radioactive wastes. See Waste types and sources, radioactive
RCRA. See Resource Conservation and Recovery Act
Reactivity, 203, 228, 260, 449, 530, 543, 550, 553
Reclamation, 254, 404, 405
Redox, 32, 36, 38, 39, 40, 55, 78–82, 106, 122, 140, 181, 198, 265–267, 300, 353
Reduction, 31, 39, 46, 71, 75, 78–82, 90, 91, 98–100, 106, 110, 123, 125, 129–131, 143, 147, 153, 191, 192, 197–198, 227, 247, 255, 261, 267, 294, 336, 348, 363, 366, 378, 387, 393, 409, 414, 449, 454, 490, 550
Reductive resins, 79, 81
Regulatory limits, 562–563
Remedial action, 3, 150, 156, 207, 246, 247, 296, 378, 408, 424, 442, 464, 515
Reprecipitation, 33, 34, 44, 139
Research and development, 577–578
Research Cottrell, 311, 446
Resource Conservation and Recovery Act (RCRA), 1, 3–13, 21, 39, 40, 46, 48, 58, 59, 62, 84, 86, 93, 104, 106, 111, 112, 116, 123, 124, 134, 135, 142, 144, 150, 154, 156, 176, 181, 183, 204, 225, 240, 246, 247, 268, 272, 280, 309, 323, 359, 368, 409, 424, 439, 443, 448, 449, 478, 490, 531, 549, 553, 564
Reverse osmosis, 177
Rheological, 203, 205, 206, 263, 295, 296, 533
Rice hull ash, 354, 401, 460, 463
Rochelle salt, 85

Safe Drinking Water Act, 4, 58, 60
Salt formation, 191, 192, 198
Sampling, 280, 330, 499, 515–519, 552, 554, 555
Sanitary landfill, 3, 20, 256, 268, 367, 397, 404, 404
SARA. See Superfund Amendments and Reauthorizaton Act
SARM, 314–317
Sawdust, 94, 122, 458
Scanning electron microscopy, 45, 70, 532
Scandinavia, 258, 259
Sealosafe, 424, 425
Secure landfill, 3, 61, 217, 223–224, 225, 246, 250, 367, 368, 404, 446, 447
Selenium, 6, 11, 24, 27–29, 43, 59, 61, 70, 92, 102, 103, 153, 154, 156–158, 228, 250, 254, 257, 281, 318–319, 331, 344, 361, 387, 411, 419, 423, 425, 441
Semi-volatile organic compounds. See Organic compounds, semi-volatile

INDEX

Setting, 7, 18-20, 44, 60-62, 99, 206, 216, 220, 222, 260, 266, 267, 290-291, 293-296, 298, 300, 304, 317, 335, 337, 343, 345, 346, 348, 351, 352, 357, 364, 367, 378-380, 392, 395, 396, 398, 401, 402, 408, 412, 417, 420, 437, 442, 453, 479
Setting agent, 378, 402
Sideronatrite, 34
Silica, 35, 43-45, 55, 86-88, 95, 96, 99, 118, 129, 139, 291-293, 295, 297, 301, 309, 331, 336, 338, 342, 346, 352, 354, 363, 376, 378-381, 386, 401, 408, 413, 418, 432-434, 437, 460, 567
Silicate, 59, 71, 77, 86-89, 91, 102, 107, 108-109, 111, 115, 117, 118, 120-121, 127-129, 135, 137-140, 143-146, 148, 152, 155, 158, 161, 163, 165, 192, 197, 205, 206, 219, 220, 223, 260, 264, 266, 295, 297, 298, 300, 301, 305, 306, 313, 319, 323, 325, 327-330, 335-337, 339-342, 343, 345-347, 351-353, 357, 365, 376-384, 386, 387, 392, 393, 395-399, 401, 402, 407, 413, 421, 425, 437, 451-453, 456, 460, 463, 469, 470, 479, 481, 486, 493, 533
Silicate Processes. *See* Cement/Soluble Silicate Processes
Silicate Technology Corp., 102
Silt, 74, 206, 250, 337, 349, 380
Silver, 11, 24, 27-29, 36-37, 59, 61, 70, 81, 85, 87, 103, 113, 154, 156-158, 174-176, 228, 248, 250, 281, 318-319, 344, 361, 381, 387, 411, 419, 423, 425, 441, 444
Slag, 4, 129, 293, 313, 338, 355, 376, 449
Sludge, 173, 249, 250, 252, 256, 267, 268, 365, 415, 437, 534
Sludgemaster, 311, 409
Soda ash, 71, 76, 78
Sodium, 58, 71, 74, 76-78, 80, 83, 86, 88, 92, 102, 104, 105, 107, 112, 113, 122-124, 129, 130, 135, 140, 143, 144, 150, 153, 156, 172, 174-176, 206, 219, 220, 298-301, 304, 306, 309, 323, 325, 331, 337, 342, 346, 349, 350, 352, 354, 355, 357, 363-365, 367, 376, 378, 379, 387, 392, 393, 396, 398, 401, 402, 404, 408, 451, 453, 456, 469, 470, 533
 arsenate, 104, 350
 bicarbonate, 354

 borate, 337, 350, 408
 borohydride, 79, 81, 143, 150, 153
 hydrosulfite, 79, 80
 hydroxide, 71, 76, 78, 92, 140, 178, 298, 300-303, 306, 323, 325, 350, 352
 oxide, 378, 432-434
 metabisulfite/bisulfite, 79, 80
 phosphate, 337
 silicate, 77, 86, 88, 107, 206, 219, 220, 298, 301, 306, 323, 325, 337, 342, 346, 347, 352, 376, 378, 379, 387, 392, 393, 396-398, 401, 402, 456, 469, 470, 533
 sulfate, 34, 300-303, 350, 364, 367, 404, 451, 453
 sulfide, 83, 147, 358, 384
Soft hammer, 6, 7, 229
Soil, 2, 18, 20, 23, 26, 30, 31, 44, 75, 91, 95, 98, 112, 124, 125, 134, 217, 218, 220, 250, 252, 255, 256, 258, 260, 261, 294, 313, 314, 330-332, 355, 363, 364, 366, 376, 402, 403, 413, 414, 424, 467, 469-473, 475, 515, 517, 518, 529, 544, 545, 549, 567
Solid waste, 48, 49, 58, 59, 63, 181, 224, 249, 252, 255, 257, 258, 363, 529, 552, 567
Solidification
 definition, 2
 factors affecting, 298-304
SoliRoc, 399, 407
Solids content. *See* Physical properties, solids content
SolidTek, 396, 401
Solubility
 curves, 36-39
 general, 33, 35, 47
 of metals, 36-39, 63-70
 product, 35, 74
 wet/dry cycle, 30
Soluble silicates. *See* Cement/Soluble Silicate Processes
Soluble Threshold Limit Concentration (STLC), 27-29, 103, 156, 159, 398
Solvent, 177, 183, 202-204, 217, 219, 247, 249, 252, 260, 265, 307, 530, 534
Solvent extraction, 177, 217
Solvents. *See also* Organic compounds, solvents
Sorel cement, 339
Sorption, 5, 13, 34, 60, 61, 93-100, 106, 147, 189, 193, 197, 294, 357, 359, 361, 363, 458, 460-463
South Africa, 258
South Essex Sewage Works, 495
Speciation, 58, 63-71
Specific gravity. *See* Physical properties, specific gravity
Spills, 459

Stabatrol Corp., 496
Stabilization, definition, 2
Stablex, 21, 131, 132, 311, 497
Stern model, 51
STLC. *See* Soluble Threshold Limit Concentration
Strontium, 68-69
Sugar, 367, 417
Sulfate, 79-81, 86, 102, 106, 107, 108, 110-113, 123, 124, 129, 130, 132, 135, 143, 144, 148, 149, 156, 165, 172, 173, 180, 181, 248, 250, 257, 300, 337, 338, 343, 349, 350, 352, 354, 355, 357, 363-365, 367, 397, 398, 404, 407-409, 412-414, 448, 450, 451, 453
Sulfex, 85
Sulfide, 11, 32, 40, 55, 59, 63, 71, 74, 80-86, 88, 91, 105, 107-109, 110-113, 116-118, 125, 129, 130, 132, 135, 139-141, 143, 144, 148, 149, 153, 154, 172, 174, 180-182, 206, 337, 363, 381, 382
Sulfite, 102, 172, 250, 257, 340, 450
Sulfoaluminate cement, 338
Sulfur, 80, 83, 84, 86, 91, 129, 143, 147, 148, 153, 257, 350, 418
Sulfur-based, 190-191, 312
Sulfur dioxide, 79, 80
Sulfuric acid, 20, 56, 102, 110, 177, 409, 412, 414, 458, 533
Superfund. *See* Comprehensive Environmental Response, Compensation and Liability Act
Superfund Amendments and Reauthorization Act (SARA), 4, 5, 7
Surface tension, 260, 264
Surfactant, 42, 180, 265, 354, 380, 398
SWLP. *See* Leaching tests, Solid Waste Leaching Procedure

Takenaka Sludge Treatment System, 366
TCLP. *See* Leaching tests, Toxicity Characteristic Leaching Procedures
Teledyne, 310
Temperature, 31, 32, 35, 41, 52, 73, 75, 87, 110, 150, 153, 174, 178, 189, 193, 197, 220, 263, 297, 298, 300, 306, 313, 336, 337, 348, 380, 381, 397, 413, 443, 455, 533, 547, 552
Test Methods. *See* individual types and groupings
Thallium, 6, 11, 29, 59, 61, 159, 165, 318-319, 441
Thermal analysis, 70
Thermal conductivity, 314

INDEX

Thermoplastic. *See* Organic processes, thermoplastic
Thiocyanate, 64, 66, 69, 154, 156
Thiophosphoric acid, 194
Thiourea, 354
Tin, 58, 132, 298–300, 337, 350, 424
Titration, 70
TJK, 310, 366, 494
Tobermorite, 343, 345
TOC. *See* Total organic carbon
Tortuosity, 49
Total organic carbon, 319, 398
Total Threshold Limit Concentration (TTLC), 27–29, 103, 156, 159
Toxic Substances Control Act (TSCA), 4, 204, 228, 272
Toxco, Inc., 365
Toxsorb, 102, 365
Transmission electron microscopy, 578
Transportation, 183, 222, 223, 225, 226, 258, 322, 323, 325, 327, 368, 417, 446, 451, 452, 478, 500, 501, 547, 548
Treatability levels, 12
Treatability studies/formulation, 541–548
Treatment, Storage and Disposal Unit (TSD), 3, 5, 7, 106, 116, 124, 223–224, 247, 323, 424, 443, 478, 480, 486, 490, 501
Trezek process, 396, 399
Trisodium phosphate, 135, 352
TSCA. *See* Toxic Substances Control Act
TSD. *See* Treatment, Storage and Disposal Unit
TRW, Inc., 312, 456
TTLC. *See* Total Threshold Limit Concentration

Unconfined compressive strength, 301–303, 320
United Kingdom, 258, 259, 424, 425, 486
United Nuclear Industries, 311, 452
Uranium, 103, 156, 159, 254, 368
Urea-formaldehyde. *See* Organic processes, urea-formaldehyde
Urritech, 311
U.S. Waste, 310

Valence, 37–39, 41, 52, 71, 78, 79, 81, 82, 104, 106, 110, 112, 122, 125, 130, 140, 148, 153, 154, 165, 197, 266, 267
Van der Waals forces, 41, 42
Vanadium, 11, 59, 103, 159, 250, 254, 318–319, 441
Velsicol, 311

Vermiculite, 19, 44, 94–96, 131, 143, 147, 217, 354, 355, 451, 458
VHS Model, 3, 13, 50
Viscosity. *See* Physical properties, viscosity
Vitrification, 308, 363, 450, 473, 476
Volume reduction, 247, 463
VOCs. *See* Organic compounds, volatile
Volatile Organic Compounds (VOC). *See* Organic compounds, volatile

Washington State University, 312, 454
Waste Chem, 312
Waste classification systems, 227–246
  CONTEC, 240–246, 267, 580–626
  exempted wastes, 240
  industry groupings (SIC codes, EPA groups), 227, 239, 247–255
  University of California at Davis (UCD), 227, 239, 240
  USEPA, 227–239
Waste homogenization, 208–215
Waste properties, 202–207, 259–267
Waste removal, 208–215
Waste Service Technologies, Inc., 366
Wastestab, 311
Waste transport, 215
Waste types and sources, 207–216, 230–259, 267–280, 580–626
  acetylene, 146, 358
  acid, 248, 251, 253, 266, 425, 496, 536
  acid mine drainage, 412
  air pollution control, 390–391
  aluminum, 312, 425
  amine, 389
  arsenical, 108, 356, 358, 410
  automobile assembly, 267
  battery production, 248, 412
  biological, 119, 126, 145, 151, 155, 157, 160, 162, 388, 389
  cadmium, 120, 358
  calcium fluoride, 356
  calcium phosphate, 356
  caulking compounds, 190, 389
  caustic, 388, 425, 496, 536
  ceramic, 137–138
  characteristic, 3, 4, 14, 75, 96, 129, 148, 180, 181, 183, 228, 262, 272, 454, 518, 530–532, 541, 552, 563
  chlorine production, 146, 248, 358
  chromium, 127–128, 152, 162, 389, 422, 425
  cleaning, 356
  coating, 388–391
  copper, 155, 161
  cosmetic, 388
  cyanide, 410, 425
  dissolved air floatation, 356, 390
  dredging, 146, 258, 267
  D-wastes, 6, 108, 114, 119, 126, 136–145, 151, 155, 157, 160, 162
  electrical equipment, 247
  electric arc furnace dust, 108–109, 115, 119–120, 126–128, 136–137, 145, 151, 155, 158, 160, 162, 164, 358, 384, 410, 438, 484
  electroplating, 108, 114–115, 119–121, 126–127, 136–137, 145, 151–152, 155, 157–158, 160–164, 176, 247, 248, 252, 280, 282, 331, 356, 358, 384, 388–391, 393, 410, 412, 438
  etching, 410, 415, 425
  exempted, 229, 240, 246
  fabricated metals, 247
  flue-gas desulfurization, 20, 257, 493, 496
  flyash, 120–121, 127, 137, 151–152, 161, 163, 384, 410
  food processing industry, 254–255, 267
  form factors, 32–46
  foundry, 250–252, 267
  F-wastes, 3, 12, 108, 109, 114–115, 119–120, 126–127, 136–137, 145, 151, 155, 157–158, 160, 162, 164, 228, 281, 283, 284, 356, 384, 388–391, 410, 438, 445
  glycol, 403–404
  hydrofluoric acid, 412
  incinerator ash, 4, 108, 114, 119, 126–127, 136, 145, 151, 157, 160, 162, 358, 415
  industrial, 239, 240, 246–249, 252, 256, 259, 267
  inorganic chemicals, 248, 435
  iron and steel, 250, 251, 412, 415, 425
  K-wastes, 12, 108, 109, 115, 119–121, 126–128, 136–137, 145–146, 151, 155, 158, 160, 162, 228, 281, 358, 384, 388–391, 410, 438, 445, 484
  landfill leachate, 114, 126, 136, 145, 151, 157, 160, 162, 164, 404, 534
  latex, 143, 388–390, 534
  lead, 137, 358
  leather, 252, 267, 287
  listed, 5, 228, 240, 260, 265, 267, 268, 272, 280
  lubricant, 388
  machinery, 176, 252–253

692 INDEX

Waste types and sources (*cont.*)
 management, 239, 258, 280, 286
 metal finishing, 108, 114, 119, 126, 136, 145, 151, 155, 157, 160, 162, 239, 258, 280, 286
 metal production, 20, 176, 247, 250
 metal refining and smelting, 106, 134, 250, 251, 287
 mining, 19, 176, 240, 246, 253, 254
 municipal sewage sludge. *See* sewage sludge
 nickel, 152, 161, 422
 nuclear. *See* Waste types, radioactive
 oily, 108, 119, 126, 136, 145, 151, 155, 157, 160, 162, 331, 388–391, 415, 496, 536
 organic chemicals, 248, 286, 388, 389, 415, 435, 495, 536
 paint manufacturing and use, 114, 119, 126, 136, 145, 151, 155, 157, 160, 162, 249, 268, 331, 356, 388–390, 425, 435, 496
 pesticide, 390–391, 435
 petroleum refining, 20, 247, 249, 415
 pharmaceutical, 250, 268
 phosphoric acid, 415
 photographic, 177, 389
 pickle liquor, 20, 247, 251, 356, 412, 415, 534
 pigment manufacturing, 137, 358
 plastics and synthetics manufacturing, 252, 263
 power plant, 412
 printed circuit, 356
 pulp and paper, 142
 radioactive (low-level), 19, 240, 254, 264, 450–452
 reactive, 228, 260
 red mud, 107, 129
 rubber, 252, 268
 sewage sludge, 4, 55, 246, 255–257, 259, 388–391, 484, 494, 495
 sodium chromate, 358
 soldering, 389
 solvent, 536
 spills, 75, 249, 258, 458, 459, 463
 synthetics, 252
 tank bottoms, 534
 tanning, 252, 267
 textile, 253, 268, 356
 thorium hydroxide, 356
 weld cleaner, 356
 wood preserving, 109, 127–128, 161, 358, 384
 zinc, 121, 128, 138, 146
Water pollution, 172

Waterways Experiment Station, 305, 317, 322, 403
Wehran Engineering, 311, 443
Werner & Pfleiderer, 312
WET. *See* Leaching tests, California Waste Extraction Test
Wetting, 260, 264–265, 296, 298, 300, 355, 385, 442
Willis & Paul, 311
WISE-SCAN® Expert System, 572–577
WISE® data-base, 202, 571
W.R. Grace Co., 495

Zeolite, 94, 129, 358
Zeta potential, 51, 265, 267
Zinc, 11, 29, 36–37, 43, 44, 59, 61, 70, 71, 74, 75, 84, 85, 87–91, 94, 95, 102, 103, 105, 112, 123, 124, 143, 147, 152, 154, 156, 159, 162–163, 165, 172, 176, 181, 248–255, 283, 300–302, 304, 316–319, 331, 337, 350–352, 357, 361, 365, 366, 380–382, 387, 393, 398–400, 411, 423, 424, 425, 441, 444, 450
 nitrate, 300, 350, 352
 oxide, 350, 352
 phosphate, 391, 422
 sulfide, 84